McGRAW-HILL YEARBOOK OF SCIENCE & TECHNOLOGY

2008

McGRAW-HILL YEARBOOK OF SCIENCE & TECHNOLOGY

2008

Comprehensive coverage of recent events and research as compiled by the staff of the McGraw-Hill Encyclopedia of Science & Technology

McGraw-Hill

New York Chicago San Francisco Lisbon London Madrid Mexico City

Milan New Delhi San Juan Seoul Singapore Sydney Toronto

The McGraw·Hill Companies

On the front cover
The nebula around the massive star Eta Carinae, as seen by the *Hubble Space Telescope*. (*Nathan Smith, University of California, Berkeley; NASA*)

ISBN-13 978-0-07-154834-2
MHID 0-07-154834-3
ISSN 0076-2016

1 2 3 4 5 6 7 8 9 0 DOW/DOW 0 1 3 2 1 0 9 8 7

This book was printed on acid-free paper.

It was set in Garamond Book and Neue Helvetica Black Condensed by Aptara, Falls Church, Virginia. The art was prepared by Aptara. The book was printed and bound by RR Donnelley.

Contents

Editorial Staff

Mark D. Licker, Publisher

Jonathan Weil, Senior Staff Editor

David Blumel, Editor

Stefan Malmoli, Editor

Jessa Netting, Editor

Charles Wagner, Manager, Digital Content

Renee Taylor, Editorial Assistant

Editing, Design, and Production Staff

Roger Kasunic, Vice President—Editing, Design, and Production

Frank Kotowski, Jr., Senior Editing Supervisor

Thomas G. Kowalczyk, Production Director

Consulting Editors

Dr. Milton B. Adesnik. *Department of Cell Biology, New York University School of Medicine, New York.* Cell biology.

Dr. Paul Barrett. *Department of Palaeontology, The Natural History Museum, London, United Kingdom.* Vertebrate paleontology.

Prof. Rahim F. Benekohal. *Department of Civil and Environmental Engineering, University of Illinois, Urbana-Champaign.* Transportation engineering.

Robert D. Briskman. *Technical Executive, Sirius Satellite Radio, New York.* Telecommunications.

Dr. Mark W. Chase. *Molecular Systematics Section, Jodrell Laboratory, Royal Botanic Gardens, Kew, Richmond, Surrey, United Kingdom.* Plant taxonomy.

Prof. Wai-Fah Chen. *Department of Civil and Environmental Engineering, University of Hawaii at Manoa, Honolulu, Hawaii.* Civil engineering.

Prof. J. John Cohen. *Department of Immunology, University of Colorado, School of Medicine, Denver.* Immunology.

Prof. Peter J. Davies. *Department of Plant Biology, Cornell University, Ithaca, New York.* Plant physiology.

Dr. M. E. El-Hawary. *Associate Dean of Engineering, Dalhousie University, Halifax, Nova Scotia, Canada.* Electrical power engineering.

Barry A. J. Fisher. *Director, Scientific Services Bureau, Los Angeles County Sheriff's Department, California.* Forensic science and technology.

Dr. Peter L. Forey. *Department of Palaeontology, The Natural History Museum, London, United Kingdom.* Animal systematics and evolution.

Dr. Richard L. Greenspan. *The Charles Stark Draper Laboratory, Cambridge, Massachusetts.* Navigation.

Dr. Lisa Hammersley. *Department of Geology, California State University, Sacramento.* Petrology.

Prof. John P. Harley. *Department of Biological Sciences, Eastern Kentucky University, Richmond.* Microbiology.

Prof. Terry Harrison. *Department of Anthropology, Paleoanthropology Laboratory, New York University, New York.* Anthropology and archeology.

Dr. Ralph E. Hoffman. *Yale Psychiatric Institute, Yale University School of Medicine, New Haven, Connecticut.* Psychiatry.

Dr. Gary Hogg. *Professor, Department of Industrial Engineering, Arizona State University, Tempe.* Industrial and production engineering.

Prof. A. Gordon L. Holloway. *Chair, Department of Mechanical Engineering, University of New Brunswick, Fredericton, Canada.* Fluid mechanics.

Dr. S. C. Jong. *Senior Staff Scientist and Program Director, Mycology and Protistology Program, American Type Culture Collection, Manassas, Virginia.* Mycology.

Dr. Edwin Kan. *Associate Professor, School of Electrical and Computer Engineering, Cornell University, Ithaca, New York.* Physical electronics.

Prof. Robert E. Knowlton. *Department of Biological Sciences, George Washington University, Washington, DC.* Invertebrate zoology.

Prof. Chao-Jun Li. *Canada Research Chair in Green Chemistry, Department of Chemistry, McGill University, Montreal, Quebec, Canada.* Organic chemistry.

Prof. Donald W. Linzey. *Wytheville Community College, Wytheville, Virginia.* Vertebrate zoology.

Dr. Philip V. Lopresti. *Retired; formerly, Engineering Research Center, AT&T Bell Laboratories, Princeton, New Jersey.* Electronic circuits.

Dr. Dan Luss. *Cullen Professor of Engineering, Department of Chemical and Biomolecular Engineering, University of Houston, Texas.* Chemical engineering.

Prof. Philip L. Marston. *Department of Physics and Astronomy, Washington State University, Pullman.* Acoustics.

Dr. Ramon A. Mata-Toledo. *Professor of Computer Science, James Madison University, Harrisonburg, Virginia.* Computers.

Prof. Krzysztof Matyjaszewski. *J. C. Warner Professor of Natural Sciences, Department of Chemistry, Carnegie Mellon University, Pittsburgh, Pennsylvania.* Polymer science and engineering.

Dr. Orlando J. Miller. *Professor Emeritus, Center for Molecular Medicine and Genetics, Wayne State University School of Medicine, Detroit, Michigan.* Genetics.

Prof. Jay M. Pasachoff. *Director, Hopkins Observatory, Williams College, Williamstown, Massachusetts.* Astronomy.

Dr. William C. Peters. *Professor Emeritus, Mining and Geological Engineering, University of Arizona, Tucson.* Mining engineering.

Dr. Kenneth P. H. Pritzker. *Pathologist-in-Chief and Director, Head, Connective Tissue Research Group, and Professor, Laboratory Medicine and Pathobiology, University of Toronto, Mount Sinai Hospital, Ontario, Canada.* Medicine and pathology.

Prof. Justin Revenaugh. *Department of Geology and Geophysics, University of Minnesota, Minneapolis.* Geophysics.

Dr. Roger M. Rowell. *USDA Forest Service, Forest Products Laboratory, Madison, Wisconsin.* Forestry.

Prof. Ali M. Sadegh. *Director, Center for Advanced Engineering Design and Development, Department of Mechanical Engineering, The City College of the City University of New York.* MECHANICAL ENGINEERING.

Dr. Andrew P. Sage. *Founding Dean Emeritus and First American Bank Professor, University Professor, School of Information Technology and Engineering, George Mason University, Fairfax, Virginia.* CONTROL AND INFORMATION SYSTEMS.

Dr. Alfred S. Schlachter. *Advanced Light Source, Lawrence Berkeley National Laboratory, Berkeley, California.* ATOMIC AND MOLECULAR PHYSICS.

Prof. Ivan Schuller. *Department of Physics, University of California–San Diego, La Jolla, California.* CONDENSED-MATTER PHYSICS.

Jonathan Slutsky. *David Taylor Model Basin, Naval Surface Warfare Center, Carderock Division, West Bethesda, Maryland.* NAVAL ARCHITECTURE AND MARINE ENGINEERING.

Dr. Arthur A. Spector. *Department of Biochemistry, University of Iowa, Iowa City.* BIOCHEMISTRY.

Dr. Anthony P. Stanton. *Tepper School of Business, Carnegie Mellon University, Pittsburgh, Pennsylvania.* GRAPHIC ARTS AND PHOTOGRAPHY.

Dr. Trent Stephens. *Department of Biological Sciences, Idaho State University, Pocatello.* DEVELOPMENTAL BIOLOGY.

Prof. John F. Timoney. *Department of Veterinary Science, University of Kentucky, Lexington.* VETERINARY MEDICINE.

Dr. Daniel A. Vallero. *Adjunct Professor of Engineering Ethics, Pratt School of Engineering, Duke University, Durham, North Carolina.* ENVIRONMENTAL ENGINEERING.

Dr. Sally E. Walker. *Associate Professor of Geology and Marine Science, University of Georgia, Athens.* INVERTEBRATE PALEONTOLOGY.

Prof. Pao K. Wang. *Department of Atmospheric and Oceanic Sciences, University of Wisconsin, Madison.* METEOROLOGY AND CLIMATOLOGY.

Prof. Mary Anne White. *Department of Chemistry, Dalhousie University, Halifax, Nova Scotia, Canada.* MATERIALS SCIENCE AND METALLURGIC ENGINEERING.

Dr. Thomas A. Wikle. *Department of Geography, Oklahoma State University, Stillwater.* PHYSICAL GEOGRAPHY.

Article Titles and Authors

Accoya wood
Edward Pratt
Affinity monolith chromatography
David S. Hage; John E. Schiel
Air-coupled ultrasonic testing
Dale E. Chimenti
AMP-activated protein kinase (AMPK)
D. Grahame Hardie
Ancestral ranges and historical biogeography
Richard H. Ree
Ancient microorganisms in salt
Tim Lowenstein
Anthrax bacillus and the immune response
Marcia M. Pierce
Arc flash hazard
J. C. Das
Asbestos measurement
John Kominsky; Daniel Vallero; Michael Beard
Autonomous microsystems
David Erickson
Auxin and the auxin response
Lynne Armitage; Ottoline Leyser
Avian brain chimeras
Evan Balaban
B chromosomes in plants
James A. Birchler
Barcoding life
Mark W. Chase
Biocontrol of plant diseases with biofungicides
Brian B. McSpadden Gardener
Biodegradable materials for tissue engineering
Minal Patel; John P. Fisher
Biological diversity
R. Geeta; Rodrigo Cogni
Brassinosteroids
Steven D. Clouse
Bronchiolitis obliterans organizing pneumonia (BOOP)
Marcia M. Pierce
Buoyant plumes in crossflows
F. Javier Diez
Capabilities-based planning
Paul Davis
Carbon-based electronics
Jing Guo
Carnivoran evolution
Anjali Goswami
Cellular imaging
Teng-Leong Chew
Chiral Brønsted acid catalysis
Takahiko Akiyama
Climate change and sex determination
Stephanie J. Kamel
Clostridium difficile outbreaks
Marcia M. Pierce
Color-changing inks
Timothy J. Homola
Color management
Abhay Sharma
Complex systems
William B. Rouse
Concentrating solar power
Reuel Shinnar; Francesco Citro
Concept-based user interfaces
George Chin; Deborah Gracio;
Alan Chappell
Cyber forensics
Florian Buchholz
Deep Impact
Hans-Ulrich Käufl
Desorption electrospray ionization mass spectrometry
Justin M. Wiseman
Devonian missing link
Jennifer A. Clack
Dinosaur behavior
David E. Fastovsky
Dispersion modeling in complex urban systems
Daniel Vallero; Alan Huber; Paul Lioy
DNA and animal domestication
Greger Larson
Early spider web
Enrique Peñalver; David A. Grimaldi;
Xavier Delclòs
Earthquake early warning
Charles Scawthorn; Hiroo Kanamori
Effects of carbon dioxide on the upper atmosphere
Stanley C. Solomon
Electronic toll collection
David Levinson
Endocannabinoids
Daniele Piomelli
Equine laminitis
James A. Orsini
Escherichia coli outbreaks
John P. Harley
Ethanol from wood
John I. Zerbe
Ethylene (plant physiology)
Brian M. Kevany; Valeriano Dal Cin;
Harry J. Klee
Financial engineering
David D. Yao; Xun Yu Zhou
Focused sonic booms
François Coulouvrat; Régis Marchiano;
Jean-Louis Thomas
Forensic document analysis
Dan C. Purdy
Fuel cells for automobiles
David A. Masten
Fungal allergies
Reto Crameri; Claudio Rhyner
Gene expression during tooth development
Arthur Veis

Preface

The 2008 *McGraw-Hill Yearbook of Science & Technology* provides a broad overview of important recent developments in science, technology, and engineering as selected by our distinguished board of consulting editors. At the same time, it satisfies the nonspecialist reader's need to stay informed about important trends in research and development that will advance our knowledge in fields ranging from astrophysics to zoology and lead to important new practical applications. Readers of the *McGraw-Hill Encyclopedia of Science & Technology* also will find the *Yearbook* to be a valuable companion publication, complementing the content of that work.

In the 2008 edition, we report on the rapid advances in cell and molecular biology with articles on topics such as barcoding life, cellular imaging, ion channel genetics, organ regeneration, and trans fatty acids. Major developments in the plant sciences are covered, for example, in articles on auxin and the auxin response, ethanol from wood, gibberellin biosynthesis and signal transduction, parasitic plants, and plant metabolomics. Reviews in topical areas of biomedicine such as anthrax bacillus and the immune response, biodegradable materials for tissue engineering, *Clostridium difficile* outbreaks, *Escherichia coli* outbreaks, pathological gambling, preimplantation genetic diagnosis and therapy, and suicidal behavior are presented. In chemistry readers will find overviews on organic reactions in water, shape-memory polymers, affinity monolith chromatography, and asbestos measurement. Advances in computing and communication are documented in articles on cyber forensics, concept-based user interfaces, software-defined radio, spintronics, Web spam identification and blocking, and wireless sensor networks. Noteworthy developments in engineering and technology are reported in reviews on concentrating solar power, electronic toll collection, fuel cells for automobiles, financial engineering, green engineering, hybrid automotive power systems, hydrogen storage materials, and wide-area power systems. In the physical sciences and astronomy we report on high-temperature superconductivity, integrated nanosensors, Sloan Digital Sky Survey, Supernova 2006gy; the Swift gamma-ray burst mission, and Venus Express. And reviews on earthquake early warning, the effects of carbon dioxide on the thermosphere, hydroclimatology, inferring patterns of diversification, new species of Borneo, the optical properties of minerals in the deep earth, and the 2005 Kashmir earthquake are among the articles in the earth and environmental sciences.

Each contribution to the Yearbook is a concise yet authoritative article authored by one or more authorities in the field. We are pleased that noted researchers have been supporting the *Yearbook* since its first edition in 1962 by taking time to share their knowledge with our readers. The topics are selected by our consulting editors, in conjunction with our editorial staff, based on present significance and potential applications. McGraw-Hill strives to make each article as readily understandable as possible for the nonspecialist reader through careful editing and the extensive use of graphics, most of which are prepared specially for the *Yearbook*.

Librarians, students, teachers, the scientific community, journalists and writers, and the general reader continue to find in the *McGraw-Hill Yearbook of Science & Technology* the information they need in order to follow the rapid pace of advances in science and technology and to understand the developments in these fields that will shape the world of the twenty-first century.

Mark D. Licker
PUBLISHER

McGRAW-HILL YEARBOOK OF SCIENCE & TECHNOLOGY

2008

A–Z

Accoya wood

Accoya™ is a new type of wood offering performance characteristics that enable it to replace less sustainable materials and nondurable wood species in exterior applications. Based on more than 80 years of research and development in the field of wood acetylation, combined with leading-edge technology, it is extremely durable, dimensionally stable, and highly reliable. It is also totally nontoxic, recyclable, and made from widely available sustainable wood resources. Accoya wood is now commercially available in Europe and will become available in the United States and other countries in due course.

Finding alternatives to currently available building materials has become a crucial challenge. The world's population is now over 6 billion. It is projected to reach 9.1 billion by 2050 and to have almost doubled, to more than 11 billion, by the end of the century. It is already clear that our choice of materials will need to be more sustainable to support

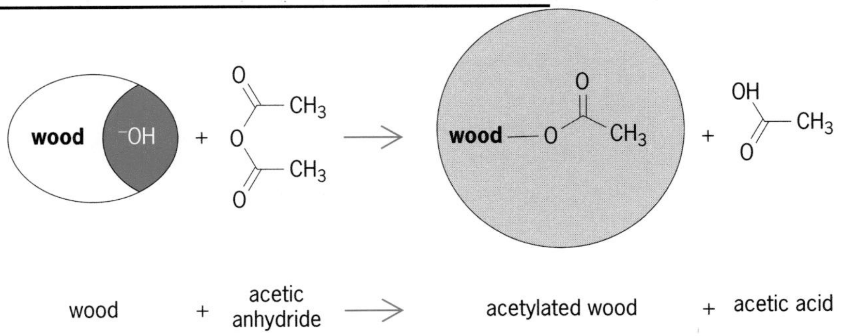

Fig. 1. The chemistry of wood acetylation.

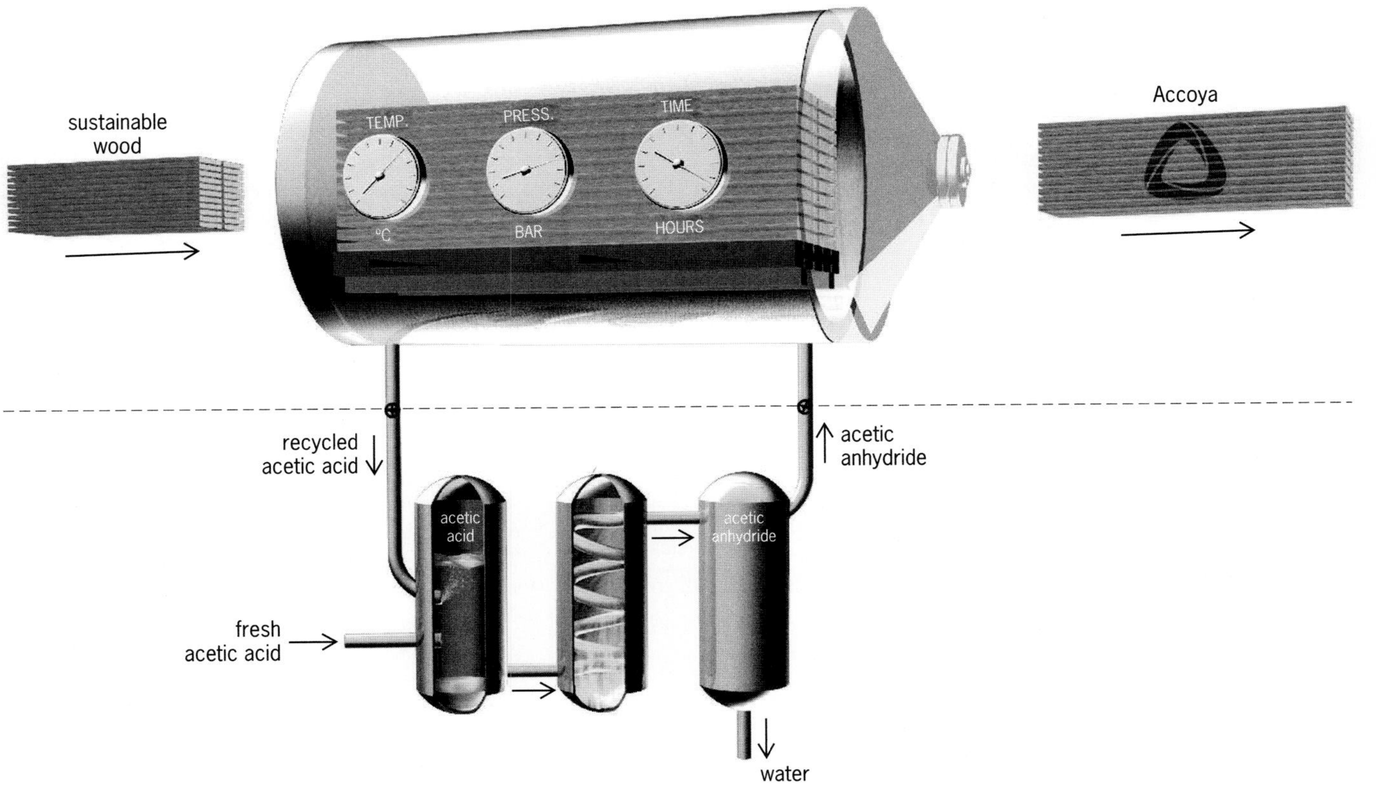

Fig. 2. Diagrammatic representation of the Accoya wood acetylation process.

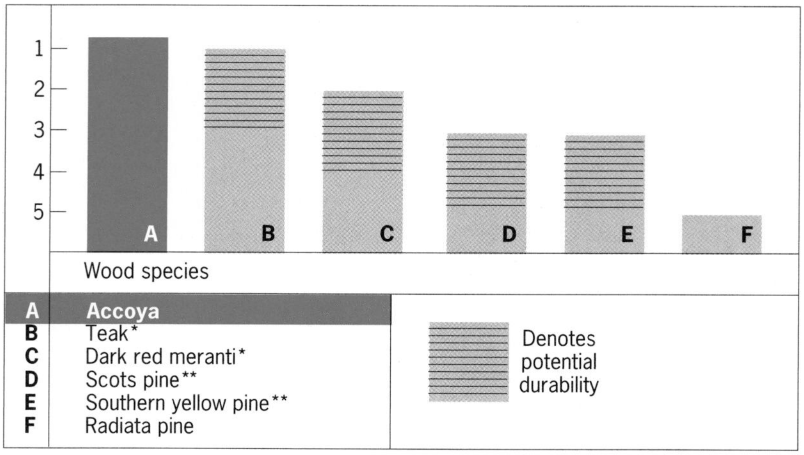

Fig. 3. Durability comparison of Accoya wood with other wood species. *Range caused by variability of species. **Range caused by difference between heartwood and sapwood. EN350-2. Classification tests: EN113, EN252, ENV807.

the demands of so many. Until 2007, the requirement for durable and stable building materials could only be met by slow-growth woods, hardwoods from tropical rain forests, toxic impregnated softwoods, plastics, metals, or concrete. The harmful effects of depleting ancient forests and finite mined resources are now more clearly understood than ever before. Finding a better alternative is a crucial challenge.

Manufacture of Accoya. The physical properties of any material are determined by its chemical structure. Accoya is made through a proprietary chemical process, developed by a multinational team, based on the science of acetylation, which can be described as a single-addition reaction.

Wood contains an abundance of chemical groups called free-hydroxyls, which adsorb and release water according to changes in climatic conditions, making the wood swell and shrink. It is believed that these free hydroxyls are effectively the access points that enable insects and microorganisms to digest the nutrients contained in wood. The Accoya production process takes advantage of wood's natural chemistry, exploiting this "disadvantage" by converting some of the free hydroxyls within the wood cell wall polymers into acetyl groups (**Fig. 1**). Acetyl is an organic hydrocarbon derived from acetic acid. Acetyl groups are naturally present in all wood species; the manufacturing process adds nothing to the wood that does not already naturally occur within it. The replacement of the free hydroxyls makes the wood indigestible to insects, microorganisms, or wood-destroying fungi, giving Accoya its substantially increased durability. It also causes the wood to swell to its "green" size (as it was in the forest before drying), reducing drying stresses and, because the bond between acetyl and wood is irreversible in natural conditions, so the wood will not shrink again.

Owing to complexities in the structure of wood and in the nature of the chemical reaction, which is strongly exothermic (heat-liberating), the reaction between the acetyl and wood takes place under very carefully managed conditions. After certain preparatory steps, wood is impregnated with acetic anhydride (dehydrated acetic acid) in a specially designed reaction chamber (visually similar to a standard impregnation vessel). Under certain temperature and pressure conditions, acetyl molecules form strong covalent bonds with the wood. After the reaction, the by-product acetic acid is recycled (cleansed of any wood extractives and dehydrated again). The Accoya wood is then conditioned and ready for use. This technique can be performed as a continuous, closed-loop process (**Fig. 2**).

The effect of modifying the wood's chemical structure, as opposed to altering its chemical content, is essentially to create a "new species" of wood. By contrast, virtually all other treatments insert chemicals into the cell walls of the wood. Such chemicals generally serve to improve insect resistance, but do not improve dimensional stability and, due to their toxicity, create environmental and disposal problems.

Key properties. Extensive international laboratory and field testing has shown the performance of acetylated wood to be extremely reliable. Accoya wood has been thoroughly tested for dimensional stability, durability, ultraviolet (UV) resistance, paint retention, and in-ground conditions to ensure optimal performance.

Durability. Accoya wood's durability is class 1, matching and even exceeding the performance of the most naturally durable woods, such as teak. The durability of wood is assessed by measuring weight loss over time in exposed conditions (**Fig. 3**).

Dimensional stability. Accoya wood offers dimensional stability in both radial and tangential directions. This resistance to swelling and shrinkage is reduced by 75% or more (**Fig. 4**). From oven-dry to water-saturated conditions, the swelling and shrinkage of acetylated wood is only minimal and actually outperforms the best tropical timbers available.

As a result of this superior stability and improved UV resistance, maintenance intervals are drastically reduced as coatings are not subjected to the same degree of expansion and contraction as with untreated wood, and they therefore last three to four times

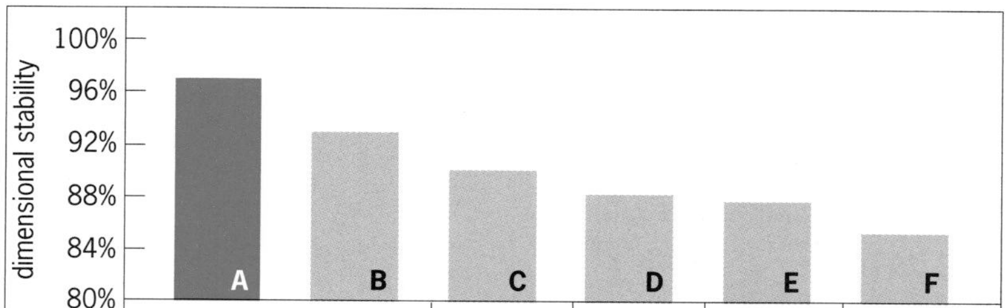

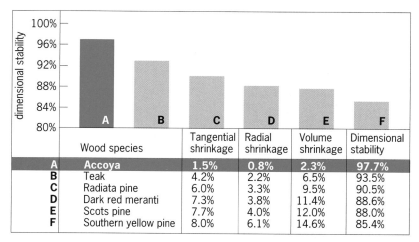

	Wood species	Tangential shrinkage	Radial shrinkage	Volume shrinkage	Dimensional stability
A	Accoya	1.5%	0.8%	2.3%	97.7%
B	Teak	4.2%	2.2%	6.5%	93.5%
C	Radiata pine	6.0%	3.3%	9.5%	90.5%
D	Dark red meranti	7.3%	3.8%	11.4%	88.6%
E	Scots pine	7.7%	4.0%	12.0%	88.0%
F	Southern yellow pine	8.0%	6.1%	14.6%	85.4%

Fig. 4. Dimensional stability comparison between Accoya wood and other wood species. Shown are the maximum swell and shrinkage from fully soaked to oven dry, the most extreme laboratory test.

longer with Accoya. In addition, tests have shown that less paint is required.

Consistent quality. Accoya wood is modified all the way through the section, and its quality is validated in a laboratory after the modification process has taken place to ensure that every batch produced is of consistent quality.

Wood and insect resistance. Accoya wood is indigestible to insects and microorganisms, making it resistant to decay. As well as being the site for water adsorption, the hydroxyl groups in the cell wall are thought to be the site for fungal and bacterial enzyme action. Substitution of these hydroxyl groups with acetyl groups creates a chemical environment that is no longer suitable for hosting these damaging enzymatic reactions. Accoya has improved resistance to brown rot, white rot, and soft rot.

Strength. The Accoya production process does not compromise the wood's strength and its hardness is slightly improved.

Insulating properties. The reduced thermal conductivity of Accoya wood means it is functionally ideal for windows and cladding (a protective, insulated, or aesthetic fixed layer added to the exterior walls of a building), where energy conservation is important. These thermal properties are derived from the wood's chemical stability following treatment.

Natural beauty. The acetylation process does not compromise the wood's natural aesthetics. It does, however, bring about a minor change in color. In addition, Accoya wood has superior resistance to the effects of UV exposure, meaning that the wood's natural appearance is retained for longer.

Nontoxic and recyclable. The addition of acetyl groups in the manufacturing process adds nothing to the wood that does not already naturally occur. Thus, Accoya does not add toxins to the environment and may be safely recycled. The only by-products from the acetylation process and the manufacture of acetic anhydride are a small amount of valuable fertilizer and trace amounts of acetic acid, which can be reused.

Sustainable. By using abundantly available, sustainably grown wood such as radiata pine, Accoya wood avoids depletion of threatened species and rain forests.

Applications. Accoya wood is predicted to last for 60 years aboveground and 25 years in-ground, based upon research by European testing institutes. It can be an ideal material for use in demanding external applications such as window frames, doors, shutters, siding, facades, and decking. As it is nontoxic, Accoya may safely be used for tables, chairs, tree houses, and play frames.

Accoya is already enabling architects, civil engineers, and construction specifiers to reconsider wood as a durable material with a high strength-to-weight ratio that can replace construction materials such as steel. In the Netherlands, for example, Accoya underwent extensive testing before being chosen as the main construction material for two heavy traffic road bridges (**Fig. 5**). For this project, Accoya will be laminated in large sections. Similar projects are being considered in South America.

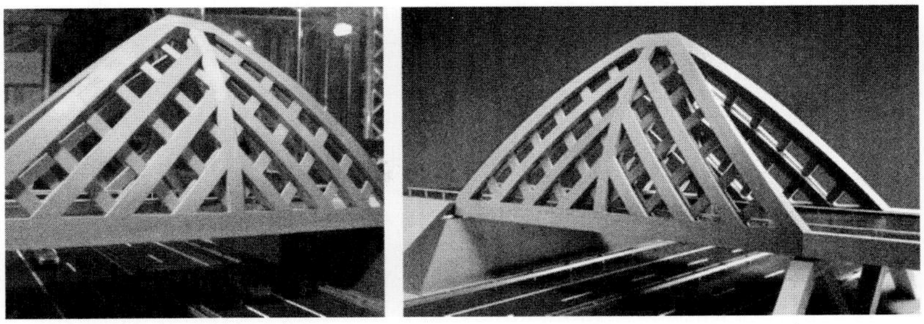

(a) (b)

Fig. 5. Accoya wood has been chosen as the main construction material for two heavy traffic road bridges in the Netherlands: (*a*) scale model of the bridges to be built over the main A7 road in Sneek; (*b*) artist's impression of the Accoya wood road bridges.

(a) (b)

Fig. 6. (*a*) Accoya wood is loaded into one of the reactors at Titan Wood's production plant in the Netherlands. (*b*) The chemical storage tanks at Titan Wood's production plant in the Netherlands.

Also in the Netherlands, Accoya has been used as canal linings, a very demanding application. During trials, Accoya was subjected to 10 years of exposure to water, nutrient-rich soil, and wide temperature variations and emerged with no discernible degradation. Thus, the potential for Accoya is high.

Production. Accoya is produced by Titan Wood, a wholly owned subsidiary of Accsys Technologies PLC, which operates a large-scale technology demonstration plant in the Netherlands (**Fig. 6**) and licenses its technology internationally. Titan Wood's next major initiative is the development of Accoya technology for panel products such as medium-density fiberboard (MDF) and oriented strand board (OSB).

For background information *see* FOREST AND FORESTRY; FOREST TIMBER RESOURCES; LUMBER; MOISTURE-CONTENT MEASUREMENT; WOOD CHEMICALS; WOOD DEGRADATION; WOOD PROCESSING; WOOD PRODUCTS; WOOD PROPERTIES in the McGraw-Hill Encyclopedia of Science & Technology.

Edward Pratt

Bibliography. C. A. S. Hill, *Acetylated Wood—The Science Behind the Material*, University of Wales, Bangor, 2006; C. A. S. Hill, *Wood Modification: Chemical, Thermal, and Other Processes* (Wiley Series in Renewable Resource), Wiley, 2006; R. M. Rowell, Acetylation of wood: Journey from analytical technique to commercial reality, *Forest Prod. J.*, 56(9):4–12, 2006; R. M. Rowell, *Handbook of Wood Chemistry and Wood Composites*, CRC Press, 2005.

Affinity monolith chromatography

Affinity chromatography is a separation method in which a biologically related binding agent is placed in a chromatographic column for use as a stationary phase. If the support material that contains this binding agent is a monolithic media, then the resulting technique is referred to as affinity monolith chromatography (AMC). This combination of a monolithic support with a biologically related stationary phase offers increased performance and ease of structural modification over other forms of affinity chromatography.

General principles. The binding agent that is used in affinity chromatography is known as the affinity ligand. This ligand is generally chosen to be an agent capable of specific and reversible interactions with the desired target. The ligand makes use of the specific binding that is seen in the interactions of an antigen with an antibody or an enzyme with its substrate. In AMC, these interactions are employed by placing the affinity ligand within a monolithic support. The applications for the resulting column will depend on the affinity ligand that is being used and the mobile (liquid) phase conditions that are selected for application and elution of the target. Examples of affinity ligands that have been used in AMC are listed in the **table**.

Ligands used in affinity chromatography and monolith columns

Type of affinity ligand	Retained targets
Biological ligands	
Antibodies	Antigens (drugs, hormones, peptides, proteins, viruses, cell components)
Antigens	Antibodies
Inhibitors, substrates, cofactors, and coenzymes	Enzymes
Lectins	Sugars, glycoproteins, and glycolipids
Protein A and protein G	Immunoglobulins (antibodies)
Nonbiological ligands	
Triazine dyes	Nucleotide-binding proteins and enzymes
Metal-ion chelates	Metal-binding amino acids, peptides, and proteins

In a typical operating scheme for AMC, a sample containing the target is first applied to the affinity column in the presence of an application buffer solution that allows strong binding between the target and affinity ligand, while allowing other sample components to wash from the column. The retained target is later eluted in one of several ways. A strongly retained target is eluted by changing the pH, polarity, or ionic strength of the mobile phase (that is, nonspecific elution) or by switching to a mobile phase that contains a competing agent that is capable of interfering with the analyte/ligand interaction (that is, biospecific elution). These elution methods are commonly used with affinity ligands such as antibodies and enzymes. A target with weak-to-moderate binding to the affinity ligand (that is, as characterized by an association equilibrium constant of 10^6 M^{-1} or less) can sometimes be eluted under isocratic conditions (that is, using a constant composition for the mobile phase) in the presence of the same solvent that is used for sample application. This approach is used in applications such as chiral separations or studies of solute-protein binding.

The support material used in AMC should have low nonspecific binding for sample components, have reactive groups readily available for ligand immobilization, and be stable under the operating conditions necessary for the separation. Traditional affinity separations have been done on large-diameter, nonrigid support materials such as agar-derived agarose beads. A desire for higher performance and faster separations has led to the use of smaller, more rigid materials such as silica particles. More recently, a similar demand for more efficient or faster separations has led to the use of monolithic supports with affinity ligands.

Supports. Monoliths are supports that consist of a single, continuous porous material. This continuous network consists of large, flow-through macropores and smaller mesopores. The resulting structure has good mass-transfer properties and low

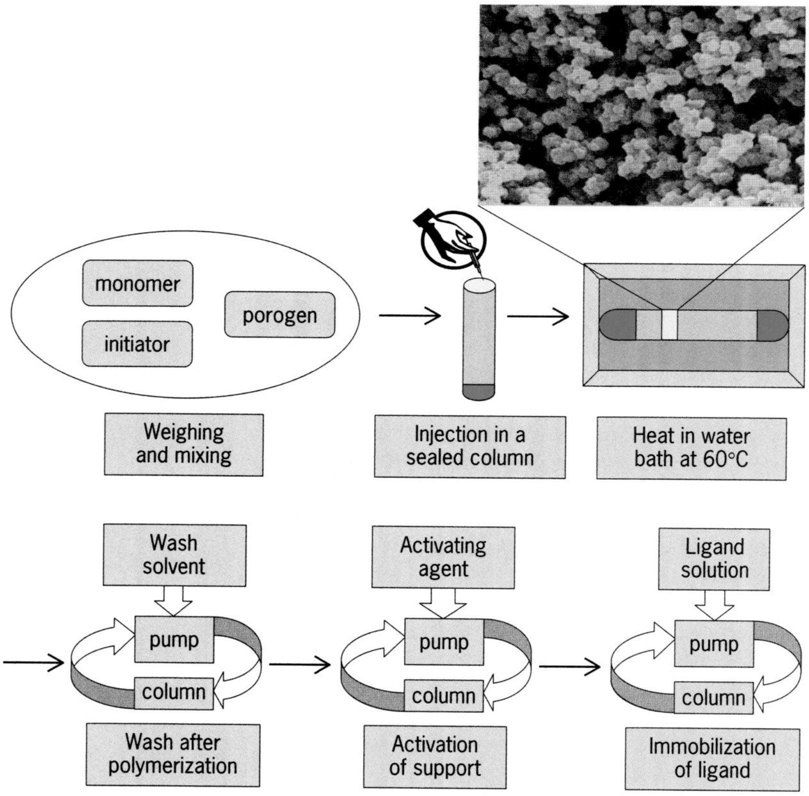

Fig. 1. General format for producing a GMA/EDMA affinity monolith where the reactants are mixed and injected into a sealed column. Heating the reactant mixture within the column results in polymerization and column formation. After heating, the column is washed to remove excess reactants. Finally, the support is activated with necessary chemicals to prepare reactive surface groups for immobilization. Continuous circulation of chemicals for immobilization or of the ligand through the support (with the circulating reagents being changed during each reaction or washing step) completes the production of the affinity monolith. The image in the upper right hand corner is a scanning electron microscope image of a GMA/EDMA monolith.

backpressures. Another useful feature of monoliths in AMC is their ability to be made with a wide range of pore sizes and shapes. In addition, monoliths can be created to contain a variety of functional groups for immobilizing affinity ligands.

A number of materials have been used to prepare monolithic supports for AMC. These materials include copolymers of glycidyl methacrylate and ethylene dimethacrylate (GMA/EDMA), as well as agarose, silica, and cryogels (polymerized at low temperature). GMA/EDMA copolymers have been most frequently used in AMC. This type of monolith is shown in **Fig. 1**. In this case, azoisobutyronitrile (AIBN) is the thermal polymerization initiator and cyclohexanol and dodecanol are used as porogens around which the polymer forms the porous monolith network. These monoliths are relatively easy to prepare, have low nonspecific binding, contain groups that are readily available for modification or immobilization, and can be prepared with a variety of surface areas and pore sizes.

Monolith supports made of agarose have also been used in AMC. This type of monolith is prepared by casting an agarose emulsion, resulting in a monolith with large pores that are 20–200 μm in diameter. The monolith is formed by first heating a mixture of agarose in water, followed by the addition of cyclohexane and Tween® 80 (polysorbate), while shaking. The emulsion is transferred to a plugged column and allowed to cool, forming an agarose gel that takes on the shape of the column housing. These monoliths can be used with a wide variety of ligands, have low nonspecific binding, and are stable over a wide pH range. However, they typically have large pore diameters and low mechanical stability.

The widespread application of silica particles in HPLC and the mechanical stability of these particles have led to the development of silica monoliths for AMC. Silica monoliths are commercially available and can be modified directly for use with affinity ligands by first reacting them with a diol or aminopropyl-containing compound. Silica monoliths can provide good chromatographic efficiencies and have high surface areas that are capable of immobilizing large quantities of ligand. The sol-gel process of silica monolith formation has also been shown to be amendable to ligand entrapment (a noncovalent immobilization technique) that can allow an affinity ligand to be contained in an AMC column in a soluble form.

Another group of supports that have been used in AMC are cryogels. This type of monolith is formed by using monomers that are dissolved in an aqueous phase and polymerized at temperatures below 10°C (50°F). In this method, ice crystals act as the porogen around which the monolith forms. The result is a monolith that is composed of macropores that are typically 10–100 μm in diameter. The large, hydrophobic and interconnected pores of these materials are useful in separations that involve large analytes, such as proteins and matter from microbes or animal cells. However, these large pores also result in small surface areas for ligand attachment.

Various methods can be used to immobilize an affinity ligand in a monolith. Immobilization can be achieved by dipping the monolith in reaction materials or by circulating the reaction materials through the column. Numerous covalent immobilization methods have been adapted for use with monolithic supports, including the Schiff base method, the epoxy method, the glutaraldehyde method, the carbonyldiimidazole method, the disuccinimidyl carbonate method, the hydrazide method, and the cyanogen bromide method. In addition, entrapment and secondary immobilization (for example, the noncovalent adsorption of antibodies to a secondary ligand such as covalently immobilized protein A or protein G) have also been used in AMC methods.

Applications. Affinity-based separations are commonly characterized based on the type of ligand being used. AMC has been used for bioaffinity chromatography, immunoaffinity chromatography, immobilized metal-ion affinity chromatography, dye-ligand and biomimetic chromatography, and affinity-based chiral separations.

Bioaffinity chromatography refers to any separation where the ligand used is some biological compound. For example, protein A and protein G have both been immobilized with GMA/EDMA monoliths. These proteins specifically bind to the F_c region of immunoglobulin G (IgG), making them useful for antibody purification. For example, in one recent report protein A was used along with dispersed cellulose fibers to allow the separation of IgG from human serum albumin in only 30 s based on AMC (**Fig. 2*a***). Such work has application in the rapid analysis of the IgG class of antibodies in clinical samples and in the measurement of antibodies that are used as therapeutic agents.

An important subset of bioaffinity chromatography is immunoaffinity chromatography. This technique refers to the use of an antibody or antibody-related ligand in an affinity column. The resulting support can be used in AMC for the rapid purification or analysis of targets that bind to the immobilized antibodies. An example of this work is shown in Fig. 2*b*, in which a 1 mm × 4.5 mm inner-diameter monolithic disk with anti-fluorescein antibodies was used to quantitatively extract fluorescein in as little as 100 ms. This type of application has recently been explored for the rapid analysis of drugs and hormones in clinical samples.

Chiral separations are another potentially important application of AMC. These separations are used to examine drugs that have another mirror image, or chiral form, where these forms may have different effects within the body. Serum proteins such as human serum albumin (HSA) and alpha-1-acid glycoprotein (AGP) have been used in this work as well as enzymes such as penicillin G acylase. Similar studies have been reported in which AMC columns containing these and other ligands have been used to study drug-protein interactions. Methods that have been used for such work include both zonal elution techniques and frontal analysis (or frontal affinity chromatography). In zonal elution techniques, a small

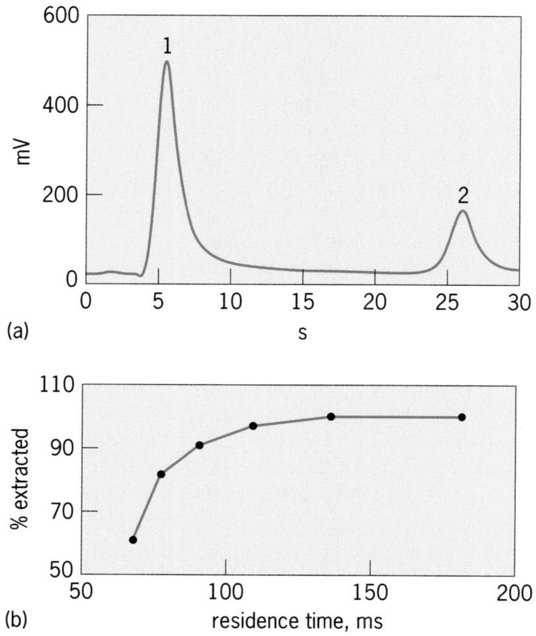

Fig. 2. (*a*) **Representative chromatogram obtained for the fast analysis of human IgG on an affinity monolithic membrane containing immobilized protein A, where peak 1 represents nonretained components in a 10-fold diluted human serum sample and peak 2 is human IgG. (*b*) Extraction of fluorescein by a 1 mm × 4.5 mm inner-diameter monolithic disk containing immobilized anti-fluorescein antibodies, giving an extraction efficiency of 95% in 100 ms. (*Part a is reproduced with permission from D. Zhou et al., Anal. Chem., 71:115, 1999; part b is reproduced with permission from T. Jiang et al., Anal. Chem., 77:2362, 2005*)**

discrete portion of the sample is injected onto the column, while in frontal analysis the sample is continuously applied to the column. In both cases, the time that it takes a drug or other solute to pass through the column is measured and used to provide information on the binding that is taking place between the solute and immobilized ligand. This type of work provides important information that can be used by pharmaceutical chemists to predict the behavior of drugs and other agents within the body.

Outlook. There are several potential advantages that have been reported for AMC when compared to affinity chromatography using more traditional support materials. These advantages include the possibility of fast separations at low pressures, improved mass-transfer properties, and the ability to create monolithic materials with a variety of pore structures, surface areas, chemical structures, and shapes. Continued improvements in techniques for creating monolithic materials with higher surface areas and the adoption of other immobilization methods with these materials should make it possible to further extend the use of AMC with other ligands and for other types of affinity separations.

For background information *see* ANTIBODY; ANTIGEN; ANTIGEN-ANTIBODY REACTION; BUFFERS (CHEMISTRY); CHROMATOGRAPHY; ENZYME; IMMUNOGLOBULIN; LIGAND; LIQUID CHROMATOGRAPHY; POLYMER-SUPPORTED REACTION; PROTEIN; SOL-GEL PROCESS in the McGraw-Hill Encyclopedia of Science & Technology. David S. Hage; John E. Schiel

Bibliography. D. S. Hage, *Handbook of Affinity Chromatography*, 2d ed., CRC Press/Taylor & Francis, 2005; D. Josic and A. Buchacher, Application of monoliths as supports for affinity chromatography and fast enzymatic conversion, *J. Biochem. Biophys. Meth.*, 49:153, 2001; R. Mallik and D. S. Hage, Affinity monolith chromatography, *J. Sep. Sci.*, 29:1686, 2006; G. Platonova and T. Tennikova, Affinity processes realized on high-flow-through methacrylate-based monoliths, *J. Chromatogr. A*, 1065:19, 2005; F. Svec, Monolithic materials: Promises, challenges, achievements, *Anal. Chem.*, 78:2100, 2006; F. Svec, Preparation and HPLC applications of rigid macroporous organic polymer monoliths, *J. Sep. Sci.*, 27:747, 2004.

Air-coupled ultrasonic testing

Ultrasonic testing is the process of introducing high-frequency stress waves, or ultrasound, into an object for determining something about the object's internal structure or makeup, or the presence of material anomalies within the object. This process consists typically of three stages: production of the ultrasonic wave by a device that changes electrical impulses into stress waves and the wave's propagation through some sound-supporting medium to the object under study; interaction of the ultrasound with the object, including the effects of the object's shape, material constituents, and the presence of internal structure; and, finally, reflection, or more generally, scattering of the incident sound wave, either back to the device that originally produced the sound or to another similar device (**Fig. 1**). Sound waves are generated when electrical impulses are transformed into mechanical vibrations by the sending transducer. The resulting sound wave interacts with the object under test and reflects, or scatters, in many directions. Then a portion of the scattered field can be received and detected, either by the same device working in reciprocal mode as a detector or by another similar device. This is the principle of ultrasonic testing.

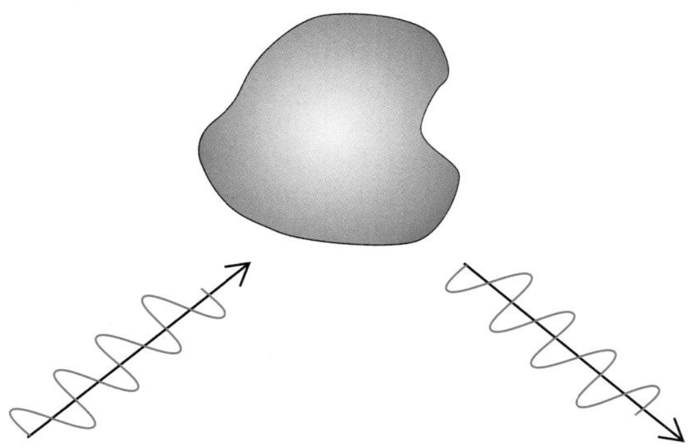

Fig. 1. Schematic drawing showing sound-wave energy propagating from a transducer in a fluid medium (water or air) to an object under test and then scattering from that object and propagating to another transducer.

Conventionally, the sound coupling medium would be water or a type of gel if the transducer were nearly in physical contact with the object. Using distilled, degassed water to couple ultrasound to an object under test works quite well and has many technical advantages, including relatively low signal loss at typical frequencies, almost zero toxicity, and low cost. For many applications the use of water is acceptable and preferred.

There are, however, certain testing applications for which water can be disadvantageous. These situations include materials that are sensitive to contact with water, such as uncured graphite-epoxy composites; extremely large objects whose immersion is impractical; or objects for which rapid scanning is required. In each of these cases, recent technological developments have enabled the replacement of water by a far more ubiquitous sound-coupling medium—air.

Availability of suitable transducers. Ultrasonic testing in air has experienced an upsurge in interest and application recently because of the availability of much more efficient sound-generating devices designed specifically for operation in air. By contrast, in water- or direct-coupled ultrasonic testing, piezoelectric crystal transducers are typically employed to change electrical impulses into sound waves because these materials change their shape slightly when a voltage is applied to them. (The common quartz wristwatch is a major application of piezoelectrics.) Piezoelectric crystals are well suited to the generation of sound in water or in solids because they can push very hard on a liquid or solid over a short distance. In air, however, we need just the opposite. Air is so compliant that a gentle push is quite sufficient to generate sound, but to make high-intensity ultrasound appropriate for ultrasonic testing, we should contrive to push the medium over a much longer distance, that is, a larger displacement.

The new capacitive film transducers do just this. A thin polymer film a few micrometers in thickness comprises one side of the capacitor, and a fixed backplate is the other (**Fig. 2**). Both a constant high-voltage bias and a superimposed radio-frequency ultrasonic testing signal are impressed across the capacitor. The polymer film vibrates with a large displacement from frequencies of about 100 kHz up to nearly 2 MHz, efficiently coupling the electrical impulses to a high-intensity sound wave in the air. The film transducers also operate as reciprocal devices, detecting the waves reflected from or transmitted through the object of the test. Because of the high amplitude of the sound reflected at the interface between air and a solid, however, two separate capacitive film transducers (usually on opposite sides of a solid) are needed to perform air-coupled ultrasonic testing. This phenomenon is explained below. Recently, a new version of the capacitive transducer has been invented that permits native focusing of the sound beam to a minimum circular spot, that is, focusing without the use of lenses, mirrors, or other extraneous means. This new spherically focused capacitive transducer is ideally suited to ultrasonic testing for defect detection and elastic property characterization in laminated materials. *See* MICROMACHINED ULTRASONIC TRANSDUCERS.

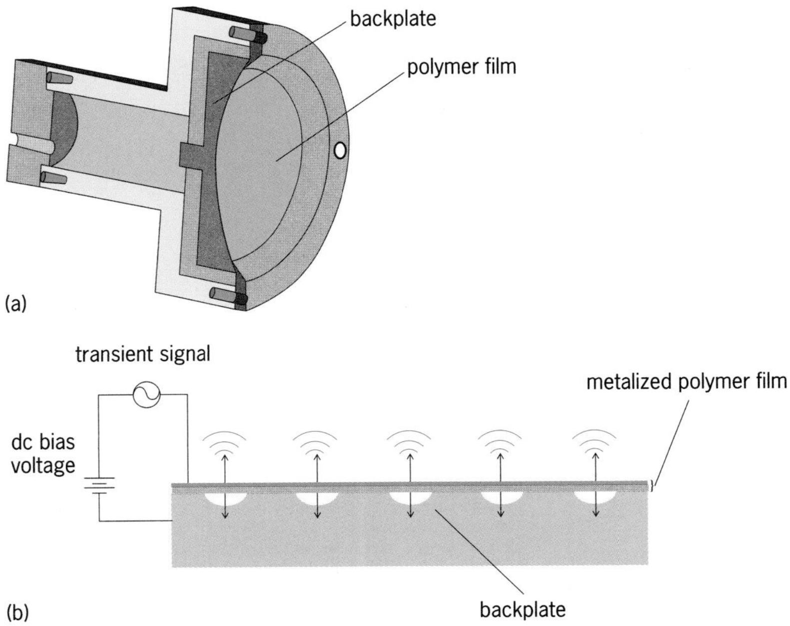

Fig. 2. Natively focused capacitive air-coupled ultrasonic transducer. (*a*) Diagram showing the backplate, the 6-μm polymer film lying over it, and the transducer housing with provisions for electrical connections. (*b*) Detail of capacitive film transducer operation.

Challenges. Air-coupled ultrasonic testing presents subtle challenges to the engineer seeking to employ this technique for defect detection or materials characterization. There is much more to the successful application of this technology than simply emptying the immersion tank of water and repeating the test with different transducers.

Disparity of sound speeds: Extreme refraction. Air is a medium having a sound-wave speed of only 342 m/s (1122 ft/s), much smaller than that of water at 1485 m/s (4872 ft/s), and far smaller than the compressional wave speed of a typical solid, about 2600 m/s (8500 ft/s) for Plexiglas® or 6300 m/s (20,700 ft/s) for aluminum. This large difference in the speeds of sound means that the corresponding sound wavelength in air will be proportionately that much smaller than the wavelength in the liquid or solid. Moreover, the bending or refraction of the sound as it enters a solid will typically be extreme (**Fig. 3**). The wave vector in the air will be much larger than that in the solid because the wave speeds are so different. Snell's law states, however, that the wave-vector projections onto the interface must be the same for both wave vectors, meaning that the critical angle for air-solid transmission will be reached at a small incident angle. An analogous effect is the appearance of bending of a straight pole as it enters a pool of water. The difference in light-wave speeds, however, is no more than 30%, compared to a factor of 10 or more for sound waves in air and solids. What this means for the engineer is that all the interesting sound interaction behavior will be finished beyond an incident angle of about 15° or

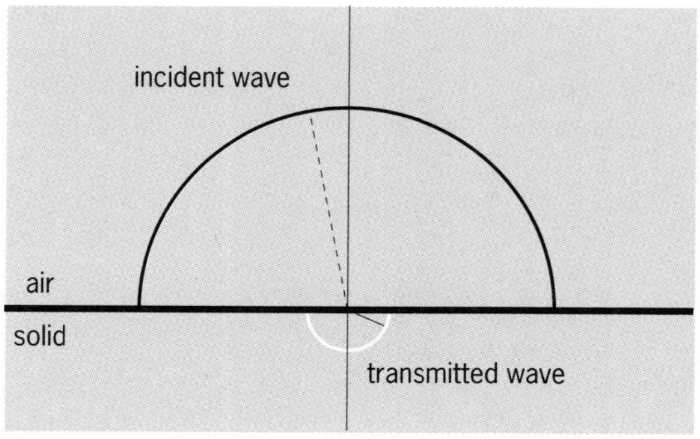

Fig. 3. Wave-vector or slowness plot, showing the interaction at the interface between air and a solid medium. The incident wave in the air is shown as a broken line that could have come from any point on the black circle, corresponding to the slowness in air. The transmitted wave in the solid is limited to points on the white circle below the air-solid boundary. Because the sound speed in the solid is so high compared to air, its slowness is that much smaller. The transmitted wave can no longer propagate once the incident angle in the air has a projection along the interface that exceeds the radius of the white circle. Therefore, only a small range of incident angles from the air can interact with the solid.

20° in most materials. Beyond that critical angle the sound is internally reflected back into the air. This fact implies that focused transducers in air need subtend only relatively small angles to be effective at imaging solids, or in providing a broad angular spectrum of waves. It also means that the incident angles of interest for materials characterization also will be similarly limited.

Strong interface reflections. There is another, more insidious effect of the large disparity between the sound-wave speed in air and solids. Coupling of sound waves from one medium to another occurs when sound in the medium of the incident wave induces wave motion in the medium of the transmitted wave. Ideally, this action will occur best when the wave speeds and mass densities of the two media are about the same. For air and a solid, however, they could hardly be more different. The implication is that, in general, at the air-solid interface, more than 99.99% of the sound amplitude will be reflected back into the air, and only about 0.01% will be transmitted. This unfavorable ratio applies again at the exit of the sound wave from the solid back into the air. For defect detection, both the strong interface reflections and the high refraction must be accounted for, already significant technical challenges. When seeking the material elastic properties of a sample from its sound propagation characteristics, however, there are yet additional considerations.

Materials characterization. To perform materials characterization with air-coupled ultrasonic testing requires a carefully thought-out approach. Because the transducer is only 10 or 20 sound wavelengths in diameter at typical frequencies, energy from the transducer will experience significant diffraction, or the geometric spreading and spatial intensity oscillations of the sound beam caused by interference of portions of the beam with itself. In fact, in many measurements the diffraction effects will seriously complicate the interpretation of the experiment, inducing as much structure in the received signal as is caused by the elastic properties of the material being tested.

One method that has shown considerable success is to negate the diffraction effects of the finite transducers by collecting enough transmitted field data over incident angles and relative transducer positions to suppress the effects of the sound beam characteristics almost completely, leaving the elastic properties of the solid as the only remaining influence on the processed data. This method benefits from focused transducers, where the angle data can be collected in a single series of position measurements, because focusing produces waves at many angles. It remains only to scan the position of one of the two transducers with respect to the other to complete the data acquisition. The result, after processing, is analogous to having made the measurement with much larger transducers, a method known from radar as synthetic aperture imaging. From such a measurement of the transmission coefficient as a function of frequency, the elastic properties of solids can be inferred. An example of a transmission coefficient is shown in **Fig. 4** for a Plexiglas® plate in air at a frequency of 400 kHz. The peaks and valleys show the variation in transmission with incident angle. (There are similar variations with changing frequency.) The solid curve is experimental data, and the broken curve is a theoretical calculation.

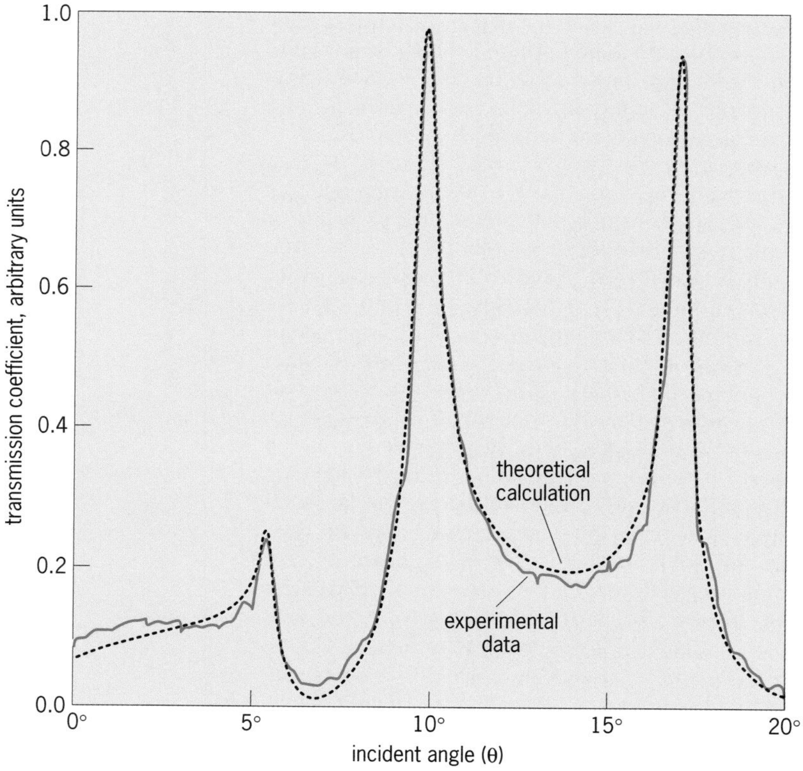

Fig. 4. Transmission coefficient of a Plexiglas® plate at a frequency of 400 kHz. Measurement is solid curve, and calculation is broken curve. Peaks at 10° and 17.5° are incident angles where transverse resonance leads to high transmission. At these conditions, air-coupled ultrasonic testing—whether for defect detection or materials characterization—can be done reasonably efficiently.

Figure 4 demonstrates several of the challenging aspects of air-coupled ultrasonic testing. For example, a very high reflectivity (low transmission) is observed away from the peaks, meaning that air-coupled ultrasonic testing is inherently much less efficient than using water. The implication for engineers is that air-coupled ultrasonic testing works best when performed with two transducers deployed on opposite sides of a platelike structure, to avoid the strong reflection at the front surface, which would swamp receiver circuit electronics. From Fig. 4 it is clear that at some incident angles, namely 10° and 17.5° for the case illustrated here, much higher sound transmission occurs than in general, away from these special angles. At these special angles the plate experiences total transmission, and even air-coupled ultrasonic testing becomes reasonably efficient. The reason for the occurrence of total transmission is slightly technical, but it can be thought of as a transverse resonance of the plate. That is, the sound wave in the plate consists of two components, one traveling down the length of the plate and the other bouncing back and forth between the plate surfaces. At certain frequencies and wavelengths, the bouncing component will form a perfect standing wave, and at these resonances the plate becomes transparent to ultrasound.

With this information, the attractiveness of air-coupled ultrasonic testing for platelike structures becomes clear because only platelike structures have such resonances. Moreover, the utility of a transducer that contains a broad range of incident angles, namely a focused transducer, is now also clear. From Fig. 4, it can be seen that, for a wave at 400 kHz, only sound energy at incident angles of 10° and 17.5° has any real chance of getting through the plate and being detected on the opposite side. At a different frequency, however, these angles will change, and, of course, they will be different for a material other than Plexiglas®. By incorporating focused transducers into the measurement approach, the ultrasonics engineer ensures that nearly any material over almost any frequency range can be examined efficiently using air-coupled ultrasonic testing. Although similar transmission behavior is seen with the plate immersed in water, in that case the ratio of high-to-low transmission is not nearly as large as it is with air.

Air-coupled ultrasonic testing is a technology ready to assume an important role in the nondestructive evaluation of advanced materials. Granted, water is an efficient sound coupling medium, is readily available, and is very inexpensive, but it can also be inconvenient or undesirable in many testing circumstances. For those applications, many relating to advanced plastic-matrix fiber composites, air-coupled ultrasonic testing offers a new and useful alternative, thanks to improved transducer technology, to purpose-built electronics, and to developing methodology designed specifically to exploit this new but challenging nondestructive testing possibility.

For background information *see* ACOUSTIC IMPEDANCE; DIFFRACTION; MICROPHONE; NONDESTRUCTIVE EVALUATIONS; PIEZOELECTRICITY; REFRACTION OF WAVES; RESONANCE (ACOUSTICS AND MECHANICS); SOUND; SYNTHETIC APERTURE RADAR (SAR) in the McGraw-Hill Encyclopedia of Science & Technology. Dale E. Chimenti

Bibliography. D. E. Chimenti and C. M. Fortunko, Characterization of composite prepreg with gas-coupled ultrasonics, *Ultrasonics*, 32:261–264, 1994; T. H. Gan et al., High-resolution air-coupled ultrasonic imaging of thin materials, *IEEE Trans. Ultrason. Ferroelec. Freq. Control*, 50:1516–1524, 2003; M. I. Haller and B. T. Khuri-Yakub, A surface micromachined electrostatic ultrasonic air transducer, *IEEE Trans. Ultrason. Ferroelec. Freq. Control*, 43:1–6, 1996; M. Luukkala and P. Meriläinen, Metal plate testing using airborne ultrasound, *Ultrasonics*, 11:218–221, 1973; D. Reily and G. Hayward, Through-air transmission for ultrasonic nondestructive testing, in *Proceedings of the 1991 IEEE Ultrasonics Symposium*, pp. 763–766, 1991; A. Safaeinili, O. I. Lobkis, and D. E. Chimenti, Air-coupled ultrasonic estimation of viscoelastic stiffnesses in plates, *IEEE Trans. Ultrason. Ferroelec. Freq. Control*, 43:1171–1180, 1996; J. Song, D. E. Chimenti, and S. D. Holland, Spherically focused capacitive-film, air-coupled ultrasonic transducer, *J. Acoust. Soc. Am.*, 119(2):EL1–EL6, 2006.

AMP-activated protein kinase (AMPK)

All living cells carry out many energy-requiring processes such as growth and division, movement, and uptake of nutrients from their environment. To achieve this, they utilize energy stored in the form of a high intracellular ratio of the adenine nucleotides, adenosine triphosphate (ATP) and adenosine diphosphate (ADP). These compounds can be likened to the chemicals in an electrical cell or battery. The oxidation of food molecules such as glucose or fats (catabolism) [and/or photosynthesis in a photosynthetic organism] "charges up" the battery by converting ADP to ATP, whereas energy-requiring processes tend to "flatten" the battery by converting ATP back to ADP again. For living cells to function smoothly, it is essential that these two processes remain in balance. In eukaryotic cells, one of the principal systems carrying out this task is the adenosine monophosphate (AMP)–activated protein kinase, often referred to by its acronym AMPK.

AMPK is a protein kinase, that is, a member of a large family of enzymes that switches the function of other proteins (called target proteins) on or off by covalently attaching phosphate groups to specific sites on their surfaces. Other enzymes called protein phosphatases remove the phosphate groups when the time is appropriate. As its name suggests, AMPK is switched on by a rise in the concentration of a third adenine nucleotide, adenosine monophosphate (AMP), an effect antagonized by high ATP. AMP rises whenever the ratio of ATP to ADP falls (signifying a drop in cellular energy status) due to the reaction 2ADP $\leftrightarrow$ ATP + AMP catalyzed by the enzyme adenylate kinase, which is very active in all mammalian cells. Because of this reaction, small increases in the ADP:ATP ratio are amplified into much larger increases in AMP:ATP. A good example of a physiological situation when AMPK is switched on is in skeletal muscle during exercise, when there is a greatly increased demand for ATP to provide energy for muscle contraction.

Structure and mechanism of activation. In all eukaryotic species, AMPK is a complex of three protein subunits called α, β, and γ. The α subunits catalyze the kinase reaction and contain a domain (region) related to domains on other protein kinases. The β subunits are the scaffold on which the other two subunits assemble and also contain a domain that causes the complex to bind to glycogen, a polymer of glucose that acts as an energy reserve in most eukaryotic cells. This domain may allow the AMPK complex to sense the level of glycogen. The γ subunit contains two tandem domains that bind AMP or ATP in a mutually antagonistic manner. Mutations that interfere with the binding of AMP and ATP to these domains can cause severe heart disease. Another important feature of the regulation of AMPK is that the complex is only active after it has been itself phosphorylated at a specific threonine residue on the α subunit (Thr-172) by upstream protein kinases, one of which is the LKB1 complex. The LKB1 complex appears to continually attach phosphate groups to Thr-172, but these are normally immediately removed by a protein phosphatase. However, binding of AMP to the γ subunit inhibits the removal of the phosphate from Thr-172, thus causing AMPK to switch to the phosphorylated form, which is 100-fold more active than the dephosphorylated form. Binding of AMP also causes a change in shape that enhances the phosphorylated form a further 10-fold (**Fig. 1**). The combined effect of these two mechanisms is to activate the complex by at least 1000-fold. Both mechanisms for AMP activation are also opposed by high concentrations of ATP, which competes for binding at the same sites as AMP, but fails to cause activation.

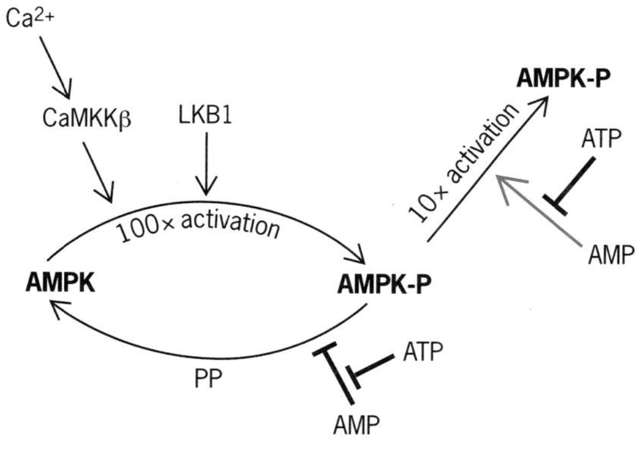

Fig. 1. Mechanism for activation of AMPK by AMP, ATP, and Ca^{2+}. AMPK is converted by upstream kinases (LKB1 and CaMKKβ) from a dephosphorylated form (AMPK, at left) to a form phosphorylated on Thr-172 by upstream kinases (AMPK-P, at center), causing 100-fold activation. AMP binding causes a shape change (AMPK-P, at right) that results in a further 10-fold activation. LKB1 phosphorylates Thr-172 continually, but the phosphate is usually rapidly removed by the protein phosphatase (PP). This dephosphorylation is inhibited by binding of AMP to AMPK-P, thus causing a net switch to the phosphorylated form. Both effects of AMP are opposed by binding of ATP. Arrow from AMP represents activation and lines with bars at the end represent inhibition. In response to a rise in cellular Ca^{2+} ions, CaMKKβ can convert AMPK to the phosphorylated form in the absence of any change in AMP or ATP.

Regulation by calcium ions and cytokines. Although the responses to AMP and ATP described above represent the "classical" regulation of AMPK, it has recently been realized that in certain cell types it can also be switched on in an AMP-independent manner by a rise in cytosol Ca^{2+} ions, due to activation of another upstream kinase, Ca^{2+}/calmodulin-dependent protein kinase kinase-β (CaMKKβ). Since a rise in Ca^{2+} often triggers energy-requiring processes, such as cell movement or secretion of molecules from the cell, this may represent a mechanism to anticipate a subsequent demand for ATP. It has also been reported recently that AMPK activity is modulated by hormones or cytokines (a group of peptides that are released by some cells and affect the behavior of other cells, serving as intercellular signals) that regulate the whole body energy balance, such as leptin, ghrelin, and adiponectin. Leptin is an adipokine (a cytokine secreted by adipose tissue) released by adipocytes (fat cells) and can be regarded as a signal that fat stores are adequate. Leptin has been reported to activate AMPK in skeletal muscle, stimulating fatty acid oxidation and hence energy expenditure, while inhibiting AMPK in the hypothalamus of the brain, promoting a sensation of satiety and hence inhibiting food intake. Conversely, ghrelin is released by the gut and activates AMPK in the hypothalamus, causing a sensation of hunger and promoting feeding. Adiponectin is another adipokine, which appears to be released selectively by adipocytes that have low fat stores. It activates AMPK in muscle and liver, promoting glucose uptake by the former and inhibiting glucose production by the latter.

Downstream targets for AMPK. Once switched on by a rise in the AMP:ATP ratio (signifying a fall in cellular energy status), AMPK phosphorylates many downstream target proteins that cause a metabolic switch toward catabolism (thus generating more ATP) and away from biosynthesis or anabolism (thus conserving ATP). Catabolic processes that are switched on (**Fig. 2**) include the uptake and metabolism of glucose and fatty acids, while anabolic processes that are switched off include the synthesis of carbohydrates (glucose and glycogen), lipids (fatty acids, triglycerides, and cholesterol), and proteins. Activation of AMPK in muscle during exercise is responsible not only for many of the acute effects (for example, increased glucose uptake and fatty acid oxidation), but also for many of the longer-term metabolic adaptations that occur during repeated exercise, especially during training for endurance events. These include increased expression of glucose transporters and increased number or volume of mitochondria. By these means, the uptake and oxidation of glucose and the consequent production of ATP are increased during subsequent exercise. As well as switching off biosynthesis and hence cell growth, AMPK activation can also cause a halt in progress through the cell cycle prior to S phase when DNA replication occurs. These effects may explain why the upstream kinase LKB1 (which is required for AMPK activation in most cells) was originally identified as a tumor suppressor, and why genetic loss of LKB1 in humans (Peutz-Jeghers syndrome) results in the development of benign tumors in the intestine, as well as malignant tumors at other sites.

Obesity and type 2 diabetes. Obesity is a disorder of energy balance characterized by excessive storage of triglyceride (fat). It is usually now defined using the body mass index (BMI), which is the weight of the subject in kilograms divided by the square of his/her height in meters. An individual with a BMI >30 kg/m^2 is considered to be obese and has a 20-fold increased risk of developing type 2 diabetes compared with a lean individual (BMI <23). Diabetes is a disorder characterized by high blood glucose. The type 2 form is not primarily caused by a lack of the hormone insulin (as in type 1); it is caused by organs in the body becoming resistant to the actions of insulin. Insulin resistance can often be initially compensated by increased secretion of insulin from the pancreas, but this may eventually fail, resulting in a permanent increase in blood glucose, when the disorder is classified as type 2 diabetes. In insulin-resistant individuals, muscle no longer takes up as much glucose and stores as much of it in the form of glycogen in response to insulin, while liver no longer shuts down its production of glucose in response to the hormone. Insulin resistance in muscle and liver often occurs when these organs store excessive amounts of triglyceride (fat). As mentioned above, obesity greatly increases the risk of developing type 2 diabetes, perhaps because the capacity of the individual to store fat in adipose tissue is overwhelmed and the individual also starts to store fat in liver and muscle. Because of its many different actions on metabolism, activation of AMPK would be beneficial in individuals with obesity, insulin resistance, and type 2 diabetes. For example, it stimulates uptake and metabolism of glucose by muscle, represses glucose production by the liver, and switches the body from a state where it synthesizes and stores fat to one where it oxidizes ("burns") it instead. Intriguingly, the drug metformin, which has been used to treat type 2 diabetes for about 50 years and is now prescribed to more than 120 million diabetics worldwide, has been recently shown to work by activating the AMPK system. In addition, activation of AMPK in muscle and other organs during exercise may be responsible for many of the health benefits of regular exercise, particularly its ability to protect against the development of insulin resistance and type 2 diabetes.

For background information *see* ADENOSINE TRIPHOSPHATE (ATP); ADIPOSE TISSUE; DIABETES; ENERGY METABOLISM; HOMEOSTASIS; INSULIN; METABOLISM; OBESITY in the McGraw-Hill Encyclopedia of Science & Technology. D. Grahame Hardie

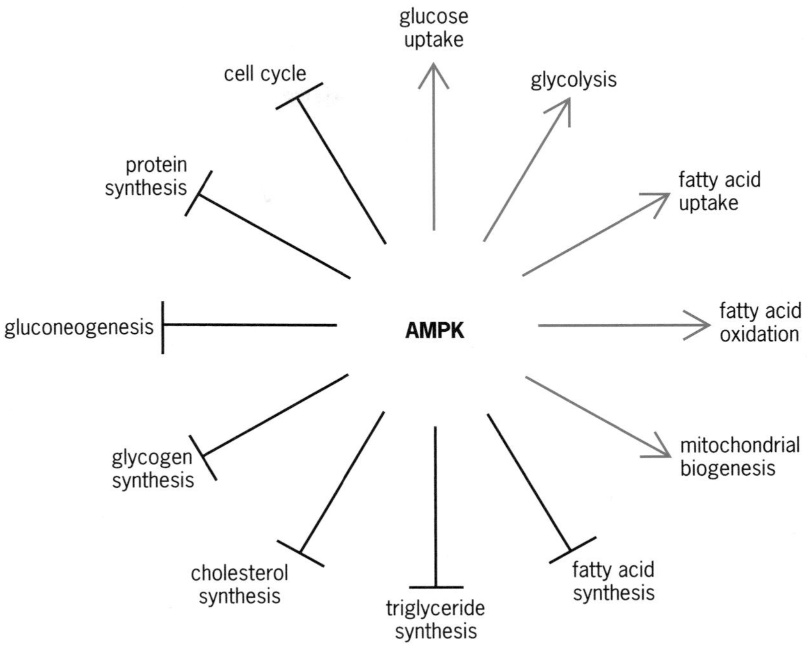

Fig. 2. Metabolic processes that are switched on (arrows) or switched off (lines with bars at the end) by AMPK activation.

Bibliography. D. Carling, The AMP-activated protein kinase cascade—A unifying system for energy control, *Trends Biochem. Sci.*, 29:18–24, 2004; D. G. Hardie, AMP-activated/SNF1 protein kinases: Guardians of cellular energy and beyond, *Nature Rev. Cell Mol. Biol.*, in press, 2007; D. G. Hardie, AMP-activated protein kinase as a drug target, *Annu. Rev. Pharm. Toxicol.*, 47:185–210, 2007; B. B. Kahn, T. Alquier, D. Carling, and D. G. Hardie, AMP-activated protein kinase: Ancient energy gauge provides clues to modern understanding of metabolism, *Cell Metab.*, 1:15–25, 2005.

Ancestral ranges and historical biogeography

The Earth's biodiversity is not distributed evenly across its surface. To understand biogeographic patterns, it is necessary not only to consider current conditions (of climate, ecology, etc.), but also to view them in their historical context, as outcomes of both biotic (for example, speciation or adaptation) and abiotic (for example, geological) evolutionary processes. The field of historical biogeography is grounded in the principle that histories of Earth and life are intimately linked, such that knowledge about one may inform scientific inquiries about the other. One such line of inquiry is focused on unraveling the geographic history of species and their ancestors: where did species occur in the past, and how did biogeographic processes such as dispersal, extinction, and speciation cause their ranges to change over time (that is, evolve), giving rise to what we observe today? Recent developments of new statistical models for how geographic ranges evolve as species descend from common ancestors are allowing this

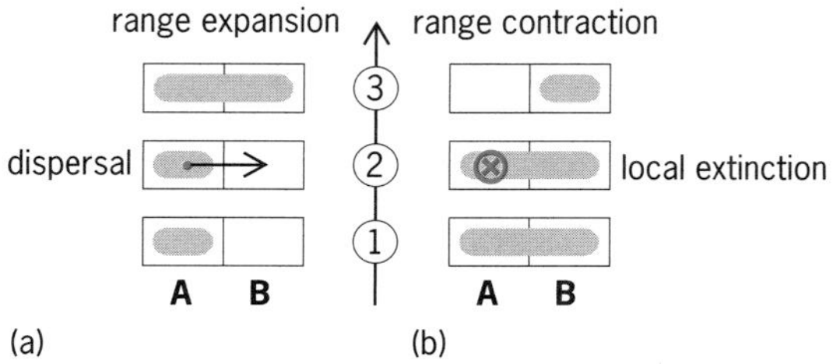

Fig. 1. Events of range evolution for a single species through time: (*a*) dispersal, leading to range expansion of a locally distributed ancestor to a widespread descendant, and (*b*) local extinction, leading to range contraction of a widespread ancestor to a locally distributed descendant. The range of the species is indicated in gray by its presence in areas A and B: (1) before, (2) during, and (3) after each kind of event.

question to be addressed with increasing sophistication.

Research. For many years, historical biogeographers were not in fact much interested in the ancestral ranges of species per se, preferring instead to focus on discovering common geological causes for biogeographic patterns. Of particular influence was the idea of area vicariance, the splitting of one contiguous area into two spatially separate (disjunct) areas. For example, a series of vicariance events characterized the breakup of supercontinents (Pangaea, Gondwana, Laurasia) since the Late Triassic, resulting in the present configuration of continents. Because vicariance can divide entire communities, affecting different kinds of organisms (birds, plants, and so on) in the same way, it came to be viewed as a kind of null model for historical biogeography, in providing a common cause for current biological similarities between disjunct areas. By contrast, other kinds of events, most notably long-distance dispersal (the rare movement of organisms or propagules, such as seeds, across barriers to normal range expansion), were viewed as having relatively little explanatory power in this context. The primary goal of "vicariance" biogeography thus became to reconstruct the sequence of past vicariance events that best explained current biotic distributions—in so doing, focusing on historical area relationships and not the geographic histories of individual species.

Recently, interest has been revived in the historical biogeography of individual species and their close relatives, particularly in the development of new models and statistical inference methods that allow a variety of biotic and abiotic information sources to be brought to bear on the reconstruction of ancestral species ranges and geographic patterns of speciation. For this purpose, the most important biological data come from two sources. One is the fossil record, which preserves direct (if often incomplete) observations of species in time and space. The other is phylogeny, the treelike genealogy that connects all species to their common ancestors. For extant (currently living) species, phylogeny is most commonly estimated by the analysis of DNA sequence data. A phylogenetic tree depicts past speciation events as lineage branching points or nodes. Lengths of branches connecting nodes can be estimated by using DNA sequences as "molecular clocks," calibrated to geological time by placement on the tree fossils of known age.

The phylogenetic tree is a powerful concept for making inferences about evolution. Darwin's revolutionary idea of "descent with modification" compels us to determine how, along the branches of the tree connecting ancestors to their descendants, hypothetical sequences of evolutionary events—in this context, events of biogeographic history, including geographic patterns of past speciation events at nodes—can be arrayed and evaluated with respect to observed data for extant species at branch tips, or "leaves."

Processes of range evolution. Before addressing the question of how ancestral ranges and biogeographic events are inferred on a tree, first principles about process are needed. Three kinds of events can cause the geographic range of a species to evolve (that is, change through time): dispersal, local extinction, and speciation. Dispersal is simply successful establishment of a species in a previously unoccupied area, resulting in range expansion (**Fig. 1*a***). Local extinction in an area is the converse, causing range contraction (Fig. 1*b*). Speciation is the event whereby one ancestral species becomes two daughter species, which are each genetically distinct and geographically cohesive (**Fig. 2**). Among the many causes of speciation, genetic divergence due to geographic isolation between areas is commonly regarded as being particularly important and ubiquitous. For example, divergence of populations separated by a geographic barrier divides the ancestral species range into two distinct daughter ranges, both of which differ from the ancestral range (Fig. 2*b*). Another geographic pattern is peripheral isolate speciation, whereby one daughter species is geographically localized and the other retains the original ancestral range (Fig. 2).

These biogeographic processes—dispersal, local extinction, and speciation—can be used to construct an evolutionary model that allows us to estimate ancestral ranges and geographic patterns of speciation events on a phylogenetic tree of the species of interest. In the model, dispersal and local extinction events occur stochastically through time, according to expectations of a Poisson process (a probabilistic model for a sequence of discrete events, with the defining characteristic being that the time intervals

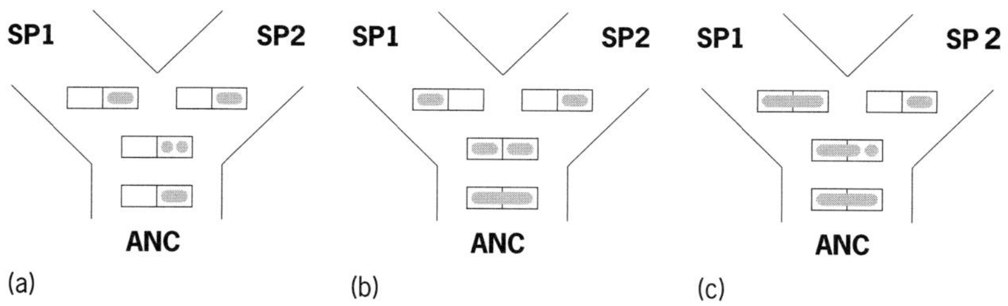

Fig. 2. Geographic patterns of speciation. At cladogenesis (evolutionary branching) events, the ancestral geographic range (ANC) is subdivided and inherited by two daughter species (SP1 and SP2). The subdivision can occur (*a*) within the area of a locally distributed ancestor, (*b*) between areas occupied by a widespread ancestor, or (*c*) within an area of a widespread ancestor.

between successive events are exponentially distributed). Probabilities of geographic ranges changing from ancestors to descendants are calculated as a function of time: the longer the time interval, the more likely it is that dispersal and extinction events will occur. These probabilities are calculated across all branches of the tree, allowing alternative hypotheses about ancestral ranges and geographic scenarios of speciation to be compared statistically. Generally speaking, hypotheses requiring fewer dispersal and extinction events (less evolutionary change) on the tree are more likely than those requiring more, assuming that such events are relatively rare.

Incorporating fossils and abiotic information. Observed ranges of species and their phylogeny represent minimum amounts of information needed to estimate ancestral ranges and geographic patterns of past speciation events; in practice, other sources of relevant information are often available. For example, the spatial arrangement of geographic areas leads us to expect dispersal events between distant areas to be relatively rare compared to areas in close proximity. Geological records may indicate the past rise and fall of temperatures, rainfall, and sea levels, timing of mountain building and erosion, shifts in river catchment (drainage basin), and continental movements. These and other kinds of data tell us about the history of habitats through time, specifically with regard to their suitability for species colonization and persistence, and opportunities for dispersal between them.

To incorporate such information into inferences of biogeographic history, one needs to encode it into the model for range evolution as quantitative constraints. A few examples describing this are as follows. We might define the rate of dispersal to be a function of distance between areas, including terms representing factors such as prevailing directions of winds or ocean currents. Knowledge about the timing of land bridge formation, such as the Isthmus of Panama (connecting Central and South America) or Beringia (the Bering land bridge, connecting the northern points of eastern Asia and western North America), could be used to restrict dispersal between the areas involved to a specific period, with probabilities of range evolution adjusted accordingly for the phylogenetic branches spanning that period. A fossil specimen known to represent a particular branch on the tree means that the range of that ancestor must have included the area of the fossil at the time it was preserved, so the model can thus be adjusted to exclude from consideration other ranges (those that do not contain the fossil's area) at that time.

Use of models to infer the past is not new to evolutionary biology; they are commonly applied to reconstructing ancestral changes in morphological, genetic, and other kinds of biological traits on phylogenetic trees. For historical biogeography, however, they represent a new direction, with much potential for theoretical development and empirical applications. Benefits of using such models include well-understood statistical behavior and specification of uncertainty, important factors in any scientific undertaking. They will allow rigorous tests of biogeographic hypotheses, such as whether direction of dispersal has been significantly asymmetric between areas (for example, "stepping-stone" dispersal from older to younger islands along the Hawaiian archipelago) or to what extent dispersal to new areas has driven speciation and lineage diversification. Continued development of evolutionary models for geographic ranges can thus help move historical biogeography from being a primarily descriptive science to a more quantitative one.

For background information *see* ANIMAL SYSTEMATICS; BIODIVERSITY; BIOGEOGRAPHY; EXTINCTION (BIOLOGY); FOSSIL; ORGANIC EVOLUTION; PHYLOGENY; POPULATION DISPERSAL; SPECIATION; SUPERCONTINENT; ZOOGEOGRAPHY in the McGraw-Hill Encyclopedia of Science & Technology. Richard H. Ree

Bibliography. M. V. Lomolino, B. R. Riddle, and J. H. Brown, *Biogeography*, 3d ed., Sinauer Associates, Sunderland, MA, 2006; R. H. Ree et al., A likelihood framework for inferring the evolution of geographic range on phylogenetic trees, *Evolution*, 59(11):2299–2311, 2005; J. J. Wiens and M. J. Donoghue, Historical biogeography, ecology, and species richness, *Trends Ecol. Evol.*, 19(12):639–644, 2004.

Ancient microorganisms in salt

Natural salt deposits contain microorganisms reported to be up to hundreds of millions of years old. Recent studies have concluded that salt-loving (halophilic) microbes living in ancient surface environments became trapped inside salt crystals and survived in a dormant state within the buried deposits. Protracted survival of microorganisms in ancient salt formations and other materials has redefined views of long-term survival and led to the realization that the biosphere on Earth and perhaps other parts of the solar system extends well into the solid subsurface.

Deposition of evaporites. Salt deposits, called evaporites, form from the evaporation of water at the Earth's surface in marine and inland lake environments with arid climates. Settings with modern evaporites include coastal lagoons, saline lakes such as the Great Salt Lake, Utah, and the Dead Sea, Israel and Jordan, and desiccated "pans" such as Salar de Atacama, Chile. Chemical species dissolved in seawater or nonmarine waters, such as sodium (Na^+), calcium (Ca^{2+}), magnesium (Mg^{2+}), potassium (K^+), sulfate (SO_4^{2-}), bicarbonate (HCO_3^-), and chloride (Cl^-) are concentrated during evaporation until the waters become supersaturated with particular minerals. Typically, the salts formed during evaporative concentration of natural waters precipitate in reverse order of their solubility; that is, relatively insoluble $CaCO_3$ (calcite) and $CaSO_4$ (gypsum) crystallize first, followed by NaCl (halite). The so-called bittern salts composed of magnesium and potassium chlorides and sulfates precipitate last before the brine dries out completely. Ancient evaporites,

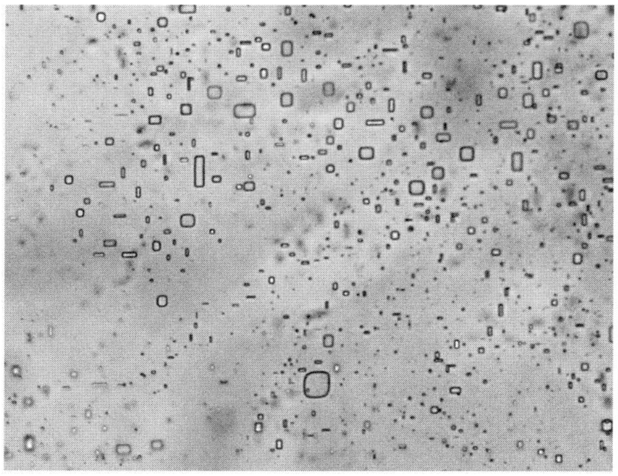

Fig. 1. Cubic and rectangular-prism-shaped brine inclusions in halite crystal. Horizontal field of view is 0.5 mm. (*Courtesy of T. Lowenstein*)

buried underground and accessible in mines and from borehole cores, can underlie large areas with impressive thicknesses. These deposits are exploited today as economic minerals used for production of fertilizer, road salt, and a host of other applications. Buried evaporites may remain in an undisturbed state or may undergo physical and mineralogical changes due to the increased pressures and temperatures of burial.

Fluid inclusions in halite. Fluid inclusions are tiny fluid-filled cavities within crystals. Primary fluid inclusions are so named because they are trapped during crystal growth and may remain in an undamaged state, as a closed system with respect to their surroundings, for millions of years. Secondary fluid inclusions form after crystal growth from healing of tiny fluid-filled cracks within crystals. Fluid inclusions in the mineral halite are particularly abundant and have been studied since the nineteenth century. Primary fluid inclusions in halite are cubic, rectangular prism, or irregular in shape, typically <1 to several hundred micrometers in size, and contain brine incorporated during crystal growth (**Fig. 1**). Such fluid inclusions found in halite crystals formed in ancient oceans, for example, contain evaporated seawater.

Fig. 2. Pink-orange brine color (not shown) of the water is produced by halophilic microorganisms, Saline Valley, California. (*Courtesy of M. Timofeeff*)

Halophilic microorganisms in modern saline environments. Halophilic prokaryotes, notably the halophilic Archaea (haloarchaea) and less abundant halophilic Bacteria, may exist in large numbers (greater than 1 million cells per milliliter) in present-day surface brines, even at salinities greater than 30%, which is ten times more saline than normal seawater (**Fig. 2**). Halite crystals precipitating in these environments trap halophilic microorganisms in the water column, preferentially in brine-filled inclusions (**Fig. 3**). These halite crystals, and the contained microorganisms, may form layered sedimentary deposits that have been buried and preserved for millions of years. Fluid inclusions in halite are an ideal environment for the long-term survival of microorganisms because the dissolved oxygen is low, high salinities decrease the degradation of proteins and nucleic acids including DNA (deoxyribonucleic acid), and with burial microbes are removed from solar radiation.

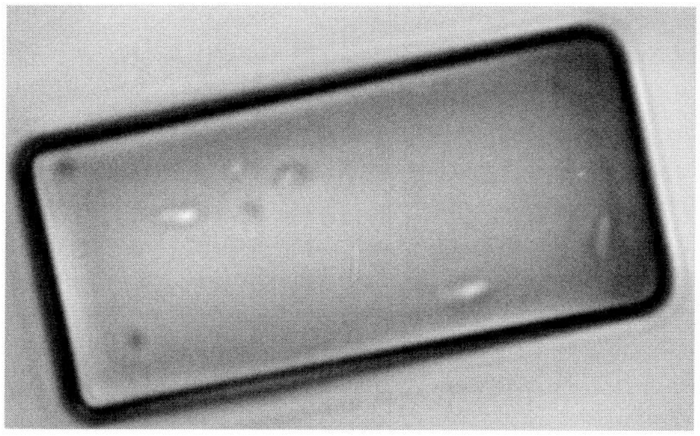

Fig. 3. Brine inclusion with trapped microorganisms in halite crystallized in 2004 from Saline Valley, California. Inclusion is 40×20 μm in size. (*Courtesy of T. Lowenstein*)

Microorganisms in ancient halite. Microorganisms in salt deposits, specifically halite, have been reported since the beginning of the twentieth century. Many recent studies have focused on fluid inclusions in halite. Research falls into two categories. "Reviving" microorganisms trapped inside crystals using culturing methods is by far the most common approach. A second technique involves amplification of DNA fragments that are trapped inside fluid inclusions using the polymerase chain reaction (PCR). In both cases, the largest issue is whether the microorganisms or DNA recovered are as old as the halite in which they are found. Halite crystals must be retrieved from mines or borehole cores and then handled in the laboratory. Therefore, contamination by younger organisms is an important concern. This issue has not been fully resolved, and therefore all reports of ancient microorganisms in halite should be viewed as controversial.

Halite samples analyzed for ancient microorganisms range from 500-g blocks to individual crystals, and in some cases single fluid inclusions within a crystal. In all cases, work is performed in clean laboratory conditions, commonly under a laminar-flow hood, using sterilized equipment. Halite samples must be treated to sterilize the crystal surfaces and destroy surface nucleic acids. Several methods are used. Halite crystals may be soaked in ethanol, which is then burnt off using a flame. Another procedure involves immersing the halite crystals in concentrated sodium hydroxide (NaOH) and then hydrochloric acid (HCl). Once the crystals surfaces are sterilized, they are then dissolved in growth media, if using culturing methods on whole crystals. Growth media vary, but all contain high salt concentrations and a carbon source such as yeast extract, casein-derived amino acids, or glucose, to name a few. If individual fluid inclusions are targeted, then a microdrill is used to breach a fluid inclusion cavity. The brine is then removed with a micropipette and inoculated into growth medium. One study used surface-sterilized halite crystals to sample DNA in fluid inclusions, followed by purification, amplification of 16S ribosomal RNA (ribonucleic acid) genes, and DNA sequencing.

A number of haloarchaea have been cultured from 200–300-Ma halite samples collected from evaporites in England, Germany, and Austria. One of these species, *Halococcus salifodinae*, isolated from geographically separated areas, was interpreted by H. Stan-Lotter and colleagues as a vestige from the marine brines that once covered western Europe. R. H. Vreeland and colleagues reported a spore-forming bacterium, *Virgibacillus* strain 2-9-3, extracted and cultured from a 250-Ma salt crystal found in the Permian Salado Formation of New Mexico. That study has been widely cited because the laboratory work was done using rigorous sterilization techniques. M. R. Mormile and colleagues discovered a halophilic microorganism, identical in its 16S rDNA gene to the Archaeal species *Halobacterium salinarum*, in a fluid inclusion in halite from a 100,000-year-old deposit 85 m below the surface of Death Valley, California. Haloarchaeal and bacterial DNA were recovered by S. A. Fish from halite crystals in evaporite deposits between 11 and 425 million years old from Poland, Thailand, and the United States. These reports of culturing halophilic Archaea and Bacteria and isolating DNA from ancient halites have raised debate on the long-term preservation of microorganisms and DNA. Critics of these studies have argued that recrystallization of halite, movement of brine, and contamination by modern organisms are the most likely explanations for the microbes and DNA recovered in these studies. Others question the results because it is widely believed that DNA molecules should break down to more stable components after millions of years. In addition, the 16S rDNA sequences from the ancient microorganisms are nearly identical to modern organisms, which indicate to some that the ancient organisms may actually be modern laboratory contaminants. For example, the 16S rDNA sequence from the ancient bacterium *Virgibacillus* strain 2-9-3 is nearly identical (differs by only three out of 1559 nucleotides) to a near-modern organism recovered from the Dead Sea named *Virgibacillus marismortui*. The similarities in 16S rDNA of ancient microorganisms and contemporary relatives may, however, also be explained by slow rates of evolution in the stable saline environments that these microorganisms inhabit, and long generation times. Additional reports of ancient microorganisms, particularly with replication of results in separate laboratories, will strengthen the evidence against laboratory contamination. The question of long-term survival of microorganisms has recently taken on extraterrestrial significance because fluid inclusion–bearing halite has been discovered in a meteorite and salts exist on the surface of Mars. That raises the fascinating question of whether it is possible to transport microorganisms encapsulated in fluid inclusions in halite across the solar system.

For background information *see* ARCHAEBACTERIA; BACTERIA; BACTERIOLOGY; HALITE; HALOPHILISM (MICROBIOLOGY); SALINE EVAPORITES; SOLAR SYSTEM in the McGraw-Hill Encyclopedia of Science & Technology.

Tim Lowenstein

Bibliography. S. A. Fish et al., Recovery of 16S ribosomal RNA gene fragments from ancient halite, *Nature*, 417:432–436, 2002; T. J. McGenity, R. T. Gemmell, W. D. Grant, and H. Stan-Lotter, Origins of halophilic microorganisms in ancient salt deposits, *Environ. Microbiol.*, 2(3):243–250, 2000; M. R. Mormile et al., Isolation of *Halobacterium salinarum* retrieved directly from halite brine inclusions, *Environ. Microbiol.*, 5(11):1094–1102, 2003; R. H. Vreeland, W. D. Rosenzweig and D. W. Powers, Isolation of a 250 million-year-old halotolerant bacterium from a primary salt crystal, *Nature*, 407:897–900, 2000.

Anthrax bacillus and the immune response

One week after the attack on the World Trade Center on September 11, 2001, powdered anthrax spores were sent through the public mail system to addresses in the United States, resulting in 22 cases of anthrax with 5 deaths. This act of bioterrorism was perpetrated using letters addressed to public media headquarters, including the major television networks. The first letters were followed a few weeks later by letters sent to the offices of senators Thomas Daschle of South Dakota and Patrick Leahy of Vermont. The infections and deaths sent a wave of fear throughout the American public, and the investigation that followed has yet to lead to conclusive results about the perpetrator(s) of this act of terrorism.

Cause and forms of anthrax. *Bacillus anthracis* is the causative agent of anthrax. This species is a member of the family Bacillaceae, which includes the genera *Bacillus* and *Clostridium*. This family shares the characteristic of forming endospores, that is, structures that allow the bacterium to survive adverse environmental conditions such as radiation, chemical agents, extremely high temperatures, and desiccation. Humans are generally infected through exposure to animals or animal products contaminated with the bacteria or their endospores. This disease has also been known as "woolsorter's disease" because individuals who work with the hides or fleeces of herbivores have a greater risk for the respiratory form of the disease, since they can be exposed to aerosolized spores from the animal products. Before the events of 2001, less than one case of anthrax occurred annually in the United States for more than 20 years. But 20,000–100,000 cases of anthrax are estimated to occur annually worldwide, making this disease of much greater importance than it would seem in the United States.

Three forms of anthrax may occur in those exposed individuals: cutaneous anthrax, which is the most common and least dangerous form of infection with *B. anthracis*; inhalation (also referred to as pulmonary) anthrax, which is very dangerous; and gastrointestinal anthrax, which is very rare in humans but is a common route of infection in herbivores. Cutaneous anthrax starts with the development of a lesion at the site of inoculation that rapidly develops into an ulcer surrounded by vesicles, which proceeds to become necrotic. This necrotic lesion is referred to as an eschar. Mortality in these patients without treatment approaches 20%; however, prompt antibiotic treatment with doxycycline or ciprofloxacin is usually successful, leading to full recovery. Gastrointestinal anthrax, while exceedingly rare in humans, has a mortality rate approaching 100%. Inhalation anthrax can have a prolonged latent period of 2 months or more, while the patient remains symptom-free. The initial symptoms are nonspecific, with fever, cough, shortness of breath, chills, and chest and abdominal pain. Individuals may think they have developed a respiratory infection such as flu or a cold. As a result, medical help is often not sought until the second stage of disease begins, with rapidly worsening fever, edema, and enlargement of the lymph nodes in the chest. After development of the second-stage symptoms, almost all cases lead to death within 3 days. Fortunately, person-to-person transmission does not occur in any form of this disease.

Bacillus anthracis is a large rod-shaped organism (**Fig. 1**) that stains gram-positive and is found either as single or paired rods from clinical specimens or in long, serpentine chains in culture. This species contains a plasmid (a small circular piece of DNA) that codes for the production of a prominent polypeptide capsule made of poly-D-glutamic acid, which is seen in clinical specimens but not when grown in culture. Virulent or dangerous strains of *B. anthracis* carry a second plasmid that codes for the three exotoxins (toxins excreted by a microorganism) that make this species deadly: protective antigen, edema factor, and lethal factor. Protective antigen (so called because its injection into experimental animals results in protective immunity), which is responsible for binding to the host cell, combines with edema factor, which is an adenylate cyclase [an enzyme that synthesizes cyclic adenosine monophosphate (cAMP) from adenosine triphosphate (ATP)], to form edema toxin. This toxin leads to an increase in the amount of interstitial fluid in the patient's lungs. Lethal toxin is formed by the combination of protective antigen with the lethal factor, a zinc metalloprotease. Protective antigen delivers both edema factor and lethal factor to a receptor [the capillary morphogenesis protein-2 (CMP-2) receptor] on the cell membrane of a target host cell (**Fig. 2**). These are then transported into the cell, where they produce toxic effects. The zinc metalloprotease activity of lethal toxin leads to the release of proinflammatory cytokines by the infected cells, including macrophages. These include tumor necrosis factor, which induces cell proliferation, differentiation, apoptosis or programmed cell death, and coagulation. Edema toxin and lethal toxin as well as the polypeptide capsule must be produced by the specific strain of *B. anthracis* for disease to occur.

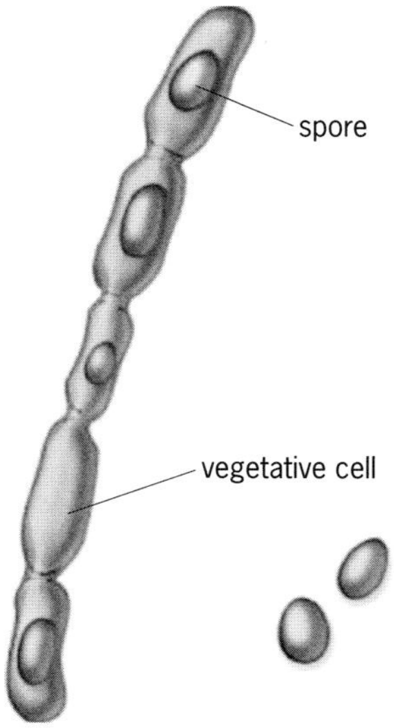

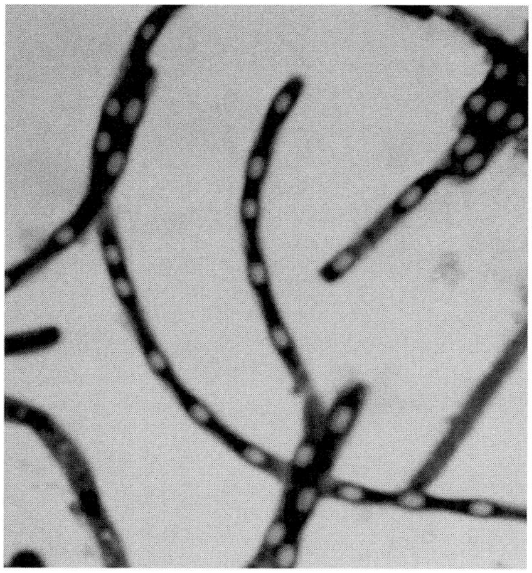

Fig. 1. *Bacillus anthracis* rods arranged in chains. (*Figure 20.18 from Cowan and Talaro, Microbiology: A Systems Approach, McGraw-Hill, New York, 2006*)

Prevention of anthrax. Historically, anthrax has been controlled through the use of vaccines in susceptible herbivore populations, including cattle, sheep, and goats. Robert Koch identified *B. anthracis* as the cause of anthrax in 1877. His discovery was followed by the development of a vaccine to prevent this disease in animals by Louis Pasteur and William Greenfield in 1881. This vaccine consisted of live bacteria that had been weakened or "attenuated" to the point where they could no longer cause disease but were capable of producing an immune response in the cattle. The protection given by the vaccine proved effective, and widespread use of it led to decreased outbreaks of anthrax in cattle herds. Subsequent vaccines developed to protect against anthrax have been based on attenuated cells or anthrax spores, or are cell-free vaccines produced from bacterial supernatants or protein subunits. These vaccines have been administered to thousands of people, mostly members of the armed services.

The current Food and Drug Administration (FDA)–approved anthrax vaccine, AVA Biothrax, is based upon genetically engineered protective antigen (PA) produced by filtering cultures of an avirulent, nonencapsulated strain of *B. anthracis*. PA is released by the bacteria during the growth period; by filtering out the cells, the final product is guaranteed to be free of any *B. anthracis* cells. This renders the vaccine incapable of causing clinical cases of anthrax. It is believed that antibodies produced by the immune system that neutralize PA will give the individual

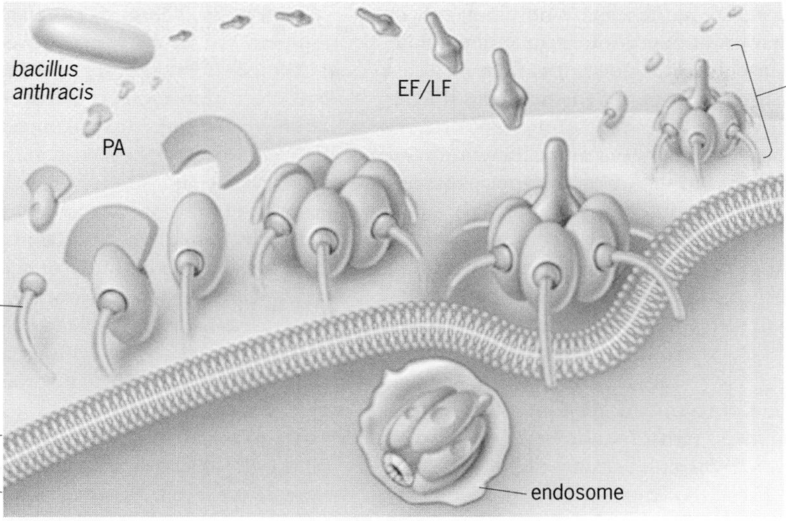

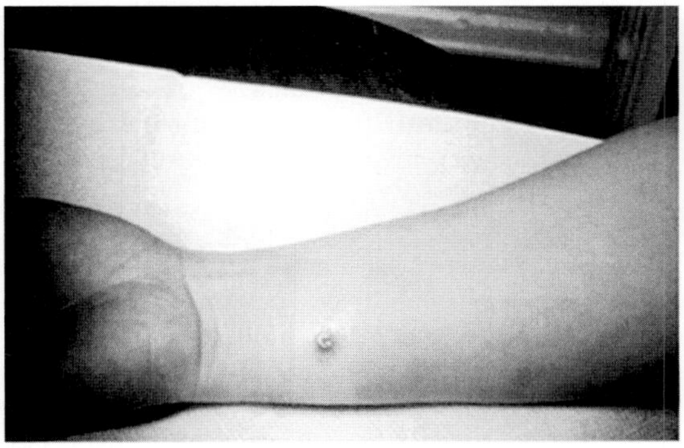

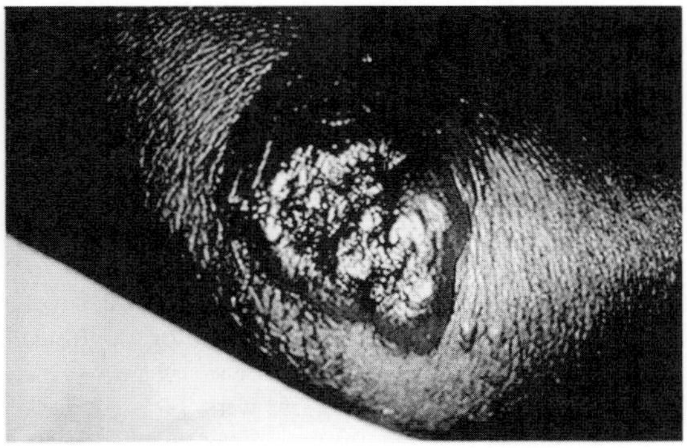

Fig. 2. Detail of a target host cell. (*a*) Protective antigen (PA) delivers edema factor (EF) and lethal factor (LF) to the CMP-2 receptor on the cell membrane. (*b*) A cutaneous anthrax lesion will ulcerate and form (*c*) a necrotic eschar. (*Figure 38.27 from Willey et al., Prescott, Harley, and Klein's Microbiology, McGraw-Hill, New York, 2007*)

protection against the disease. During infection, PA binds to cell surface receptors, where it is cleaved by a membrane-bound protease enzyme. This forms a binding site for the lethal factor and edema factor, which are then carried into the cytoplasm of the cell. The toxic effects are performed by these enzymes, leading to the pathogenesis of the disease.

Treatment for anthrax has relied solely upon the use of antibiotics in those patients who are not vaccinated against it. Researchers are now working to develop a passive form of immunity to treat individuals who have been exposed to or have developed anthrax. These antibodies target the action of PA in an attempt to neutralize its effect. Targets for these antibodies include interference with PA binding or processing, interference with the attachment of edema factor or lethal factor, and internalization of the entire unit (PA with either edema or lethal factor combined) into the host cell. Monoclonal antibodies (MAbs) have been developed in the laboratory using human cells. Monoclonal antibodies are different from the antibodies raised by the natural immune response in animals since they are produced to bind to and neutralize only one aspect of an antigen or foreign particle. Antibodies that develop naturally in the animal are considered polyclonal, binding to many different aspects of the antigen; hence, they have a broader effect but cannot be used to target a single, specific location.

The human MAb produced in the laboratory was chosen based on its ability to effectively neutralize PA outside a host organism. Rabbits were exposed to anthrax spores and then treated with the MAb. The treatment was found to be effective before symptoms started as well as after the onset of disease symptoms. Then, the MAb was tried on monkeys that were deliberately infected with aerosolized anthrax spores. By injecting the MAb, the monkeys were fully protected from developing the disease. During the course of experimentation, a new discovery was also made: the ability of immune cells to take up the neutralized toxin in order to destroy it was greatly reduced when their antibody-binding receptors were blocked. This indicates that the immune cells themselves help in the overall process of neutralizing the PA, leading to prevention of the disease in the presence of appropriate antibodies. These results seem to indicate that when individuals are exposed to anthrax, either in their occupation or as an act of bioterrorism, the use of combined therapy is more likely to prevent them from developing the disease. Individuals who develop symptoms of inhalation anthrax have a high mortality even if treated with antibiotics; in these patients, the use of combined therapy may actually save lives. This research, when applied to the large number of cases of anthrax that occur worldwide each year, may lead to decreased mortality in those who develop the disease.

For background information *see* ANTHRAX; ANTIBIOTIC; BACTERIOLOGY; BIOLOGICALS; CELLULAR IMMUNOLOGY; DISEASE; IMMUNITY; INFECTIOUS DISEASE; VACCINATION; ZOONOSES in the McGraw-Hill Encyclopedia of Science & Technology.

Marcia M. Pierce

Bibliography. K. Chakrabarty et al., Human lung innate immune response to *Bacillus anthracis* spore infection, *Infect. Immun.*, IAI.00046-07v1, 2007; M. K. Cowan and K. P. Talaro, *Microbiology: A Systems Approach*, McGraw-Hill, New York, 2006; E. W. Nester et al., *Microbiology: A Human Perspective*, 5th ed., McGraw-Hill, New York, 2007; L. Vitale et al., Prophylaxis and therapy of inhalational anthrax by a novel monoclonal antibody to protective antigen that mimics vaccine-induced immunity, *Infect. Immun.*, 74(10):5840–5847, 2006; S. Weiss et al., Immunological correlates for protection against intranasal challenge of *Bacillus anthracis* spores conferred by a protective antigen-based vaccine in rabbits, *Infect. Immun.*, 74(1):394–398, 2006; J. M. Willey, L. M. Sherwood, and C. J. Woolverton, *Prescott, Harley, and Klein's Microbiology*, 7th ed., McGraw-Hill, New York, 2007.

Arc flash hazard

The arc flash hazard is a dangerous condition associated with the unexpected release of a tremendous amount of energy caused by an electrical arc within electrical equipment. This release of energy is in the form of intense light, heat, and blast of arc products, which may consist of vaporized components of enclosure material (copper, steel, and aluminum). Intense high sound and pressure waves also emanate from the arc flash, which resembles a confined explosion. Arcing occurs when the insulation between two live conductors breaks down, due to aging, tracking, tree formation, and human error during maintenance. Insulation systems are not perfectly homogeneous. Thermal aging, vibrations, impact damage, fractures, and abraded areas give rise to further development of cavities and partial discharges occur (partial discharge, PD, is a localized electrical discharge that partially bridges the insulation between conductors and which may or may not occur adjacent to conductors.) These discharge patterns, as they grow, resemble a tree and thus the term "treeing." Ultimately the insulation resistance is much lowered and a breakdown occurs. Tracking occurs due to surface contamination. This provides a relatively low impedance path for the electrical arc to strike, say, between the line to ground. An example will be a contaminated insulator under high humidity conditions. Arc temperatures are of the order of 35,000°F (20,000°C), approximately four times the temperature on the surface of the Sun.

The phenomenal progress made by the electrical and electronic industry since Thomas Edison propounded the principle of incandescent lighting in 1879 has sometimes been achieved at the cost of lost human lives and disabilities. Although reference to electrical safety can be found as early as about 1888, it was only in 1982 that Ralph Lee correlated arc flash hazard and body burns with short-circuit currents. Currently, two industry standards, NFPA 70E and IEEE 1584, address arc flash hazard. The hazard is also recognized by the National Electric Code (NEC) and the Occupational Safety and Health Administration (OSHA).

If equipment is maintained in a deenergized condition and remotely operated, worker safety is inherently ensured. However, this procedure is rarely possible. A shutdown of even a small section of a distribution system for maintenance can result in a complete outage of a process plant with colossal loss of production.

Types of electrical hazard. The other electrical hazards are fire, shock, and arc blast.

With regard to fire, the primary objective of the NEC is to avoid sources of ignition from electrical energy.

At an alternating current of about 40 mA, a shock lasting for 1 s can cause ventricular fibrillation and be fatal. An average of about one person is electrocuted every day in the work place in the United States.

While arc flash can be considered a thermal hazard, arc blast is associated with an extreme pressure buildup that may explode some electrical enclosures. Arc-resistant switchgear has been common in Europe and Canada and has recently appeared in the United States. It directs the arc products away from the worker through properly located vents, and has heavy-duty doors and latches that may buckle but will not open up during an arc fault.

Figure 1 shows an arc flash event. Luckily the electrician is not in the direct path of the arc flash and is running for cover. Time-motion studies show that a typical human reaction time is 0.4 s (24 cycles, based upon 60-Hz alternating currents). A person cannot move or react during this period. Most electrical faults are cleared in a lesser time.

Fig. 1. An arc flash event.

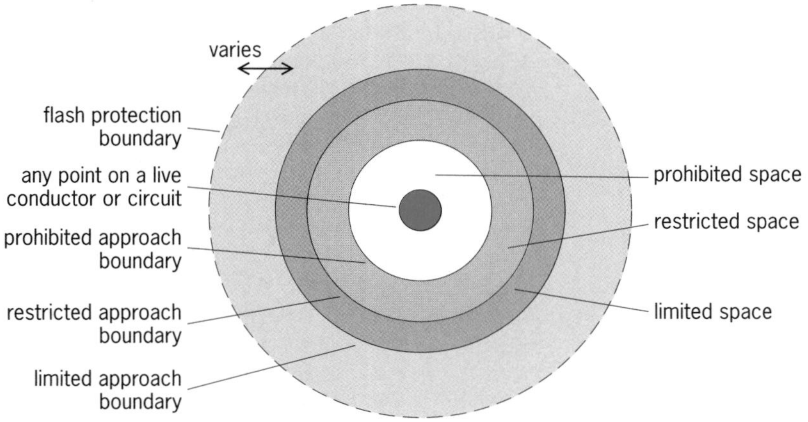

Fig. 2. Boundaries related to arc flash.

Basic concepts. Bolted three-phase fault current is the three-phase symmetrical root-mean-square (rms) fault current at the location where the arc flash hazard is being evaluated, considering zero fault resistance. Knowledge is required of the maximum and minimum bolted three-phase fault currents, whose values depend upon the modes of operation of the plant. For example, a plant generator or a tie circuit may be out of service, which will reduce the fault currents. Thus, to perform an arc flash analysis, a system short-circuit calculation must be first conducted.

Arc fault current. This is the current flowing through the overcurrent protective device (OCPD) for an arcing fault at the instant of the operating time of the device; that is, the decay in the short-circuit currents must be considered. This leads to the second requirement for an arc flash analysis: A protective relaying and coordination study of the distribution system should first be completed. For systems below 1000 V, the arc fault current is of the order of 50–60% of the bolted fault current. For medium-voltage systems, it is somewhat lower than the three-phase bolted current. Below a certain level of bolted fault current the arc fault current can even be higher.

Fault clearance time. Fault clearance time (FCT) is the sum of the protective device trip time plus the opening time of the disconnecting device, that is, a circuit breaker. The faster the fault clearing time, the lower the arc energy release.

System grounding. For the purpose of arc flash analysis, the standard IEEE 1584 brackets ungrounded systems, high-resistance grounded systems, and low-resistance grounded systems together. These give a higher arc flash hazard as compared to solidly grounded systems. Irrespective of system grounding, bolted three-phase fault currents form the basis of arc flash calculations.

Incident energy. This is the energy released at the specified working distance, expressed in J/cm^2 or cal/cm^2. When there are multiple short-circuit sources connected to a bus, for example, a generator and a utility tie, the energy accumulation is based upon the fault clearance time associated with each of the fault current components. The incident energy is calculated at 100% and 85% of the arcing current, and the higher of the two values is used.

Personal protective equipment level. The personal protective equipment (PPE) index specifies the minimum clothing required to protect the worker, depending upon the incident energy release. The NFPA lists five categories (see **table**). Even with the required PPE, a worker may sustain second-degree burns, which are curable. Further, the standards do not provide any guidelines if the energy release is greater than 40 cal/cm^2 (167 J/cm^2). In such cases it is prudent not to carry out any maintenance work when the equipment is in an energized condition. Alternatively, arc flash hazard can be reduced through system and protection modifications.

The standard ASTM F1506 calls for every flame-resistant garment to be labeled with an electrical arc energy rating, ATPV (arc thermal performance exposure value). The rating of the garment is matched with the calculated incident energy level.

Boundaries. **Figure 2** shows the boundaries based upon the standard NFPA 70E. The arc flash protection boundary is the distance at which second-degree burns can occur and the incident energy release is 1.2 cal/cm^2 (5.0 J/cm^2). Inside this boundary the energy levels may be much higher than 1.2 cal/cm^2, as established by calculations. This is the boundary of major interest and should not be crossed without wearing appropriate PPE. This boundary varies with the incident energy release. The PPE outfits are designed to minimize the worker's risk of sustaining an energy greater than 1.2 cal/cm^2; that is, second-degree burns can occur even with the appropriate PPE, and these burns are considered curable. Without appropriate PPE, the burns could be of higher degree and fatal. Unqualified persons, that is, those not specifically trained to carry out the required tasks, are safe when they stay away from the energized parts at a certain distance, which is termed the limited approach boundary. Unqualified persons should not cross the flash protection boundary unless they are wearing the required PPE and are under close supervision of a qualified person. Also, they should not cross the limited approach boundary unless escorted by a qualified person. Under no circumstances should an unqualified person cross the restricted approach boundary, where special shock-prevention techniques and equipment are required. The prohibited approach boundary establishes the space that can be crossed only as if a live contact with the exposed energized conductors or circuits were planned. This boundary is defined solely based

Hazard risk and incident energy release*

Hazard risk level, PPE	Incident energy release	
	cal/cm^2	J/cm^2
0	2.0	8.4
1	4.0	16.7
2	8.0	33.5
3	25.0	104.7
4	40.0	167.5

*From National Fire Protection Association, *Standard for Electrical Safety in the Workplace*, NFPA 70E, 2004.

upon the system voltage. Many establishments have a safety policy that no one should enter locked electrical rooms without wearing the appropriate PPE.

Working distance is defined as the closest distance to a worker's body, excluding arms and hands, that could be exposed to an arc flash hazard. For voltage levels from 15 kV down to 480 V, a working distance of 18 in. (45 cm) is normally considered. The arc flash protection boundary is calculated at the specific working distance.

Labeling requirements. The NFPA and the NEC require that hazard labels be placed on electrical equipment. It is desirable to include the following information on the labels: flash protection boundary, incident energy, working distance, required PPE level, shock hazard voltage (that is, the system voltage), limited approach, restricted approach, prohibited approach, and equipment identification.

Arc flash analysis. It is evident from the above description that the incident energy release is mainly dependent upon (1) the available short-circuit current, (2) how fast the fault is cleared, and (3) the system grounding. None of these parameters can be easily changed in an existing distribution system, though entirely new distribution systems can be designed to reduce incident energy release. The arc flash calculations are fairly complex, and all the power system software vendors in the United States have adopted IEEE and NFPA equations for calculations of arc flash hazard. NFPA calculations are based upon Lee's equations and IEEE calculations are based upon laboratory test results. The NFPA recognizes IEEE calculations; however, the ultimate results obtained by these two methods may differ.

The industry has made many strides in innovations with regard to electrical equipment to reduce arc flash hazard, which are not discussed here. Industrial establishments want a PPE level of preferably no more than 2, and a maximum of 3. A trend in the industry is to reduce the PPE level with equipment innovations and design and engineering of the electrical distribution systems. A level-4 PPE arc flash suit with headgear, eye protection, and supplementary protection for arms and hands looks like a space suit. This level of protection reduces a worker's mobility and his or her capability to handle delicate tasks, for example, terminating small control wires.

The FCT has a marked effect on PPE levels. For example, for a bolted three-phase fault current of 41 kA in a 480-V distribution system, the calculated arc flash current is 20.6 kA. In this system, if the FCT is 0.05 s, the PPE level equals 1; if the FCT is 0.1 s, the PPE level equals 2; if the FCT is 0.3 s, the PPE level equals 3; and if the FCT is 0.4 s, the PPE level equals 4.

A reduction in the system three-phase short-circuit current will, in general, reduce the PPE level. In the last case just considered, if the three-phase bolted fault current is reduced by 50% and the FCT remains the same at 0.4 s, the PPE level is reduced from 4 to 3.

For background information *see* ARC DISCHARGE; BURN; CIRCUIT BREAKER; ELECTRIC PROTECTIVE DEVICES; GROUNDING in the McGraw-Hill Encyclopedia of Science & Technology. J. C. Das

Bibliography. Institute of Electrical and Electronics Engineers (IEEE), *Guide for Arc Flash Hazard Analysis*, IEEE 1584, 2002; IEEE, *IEEE Guide for Testing Medium-Voltage Metal-Enclosed Switchgear for Internal Arcing Faults*, C37.20.7-2001, 2001; R. H. Lee, The other electrical hazard: Electric arc blast burns, *IEEE Trans. Ind. Appl.*, 18:246–251, 1982; National Fire Protection Association, *National Electrical Code*, NFPA 70, 2005; National Fire Protection Association, *Standard for Electrical Safety in the Workplace*, NFPA 70E, 2004; R. Wilkins, M. Allison, and M. Lang, An improved method for arc flash hazard analysis, *Ind. Appl. Mag.*, 11(3):40–48, May–June 2005.

Asbestos measurement

Environmental engineers are generally concerned with two types of air pollutants: gases and particulate matter (PM). Generally, the size categories of PM measured are (1) $\leq 2.5\ \mu m$ in diameter and (2) 2.5–10 μm in diameter. Measurements are taken by instruments called dichotomous samplers, with components that use size exclusion mechanisms to segregate the mass of each size fraction (**Fig. 1**). Particles with diameters $\geq 10\ \mu m$ are of less concern, since their mass is sufficiently large that they rarely travel long distances. Occasionally they are measured if a large particulate emitting source, such as a coal mine, is nearby.

Particles that are highly elongated are usually differentiated as fibers. Such elongation is expressed as a particle's aspect ratio; that is, the ratio of the length to width. Environmentally important fibers include fiberglass, fabrics, and minerals (**Fig. 2**). Exposure to fiberglass and textile fibers is most common in industrial settings. Health problems have been associated with textile workers who were exposed to fibrous matter in high concentration for many years. For example, chronic exposure to cotton fibers has led to the ailment byssinosis, also known to as brown lung disease, which is characterized by narrowing of the lung's airways. When discussing fibers, it is likely that first contaminant to come to mind is asbestos, a group of highly fibrous, naturally occurring minerals with separable, long and thin fibers (Fig. 2).

Fibers are defined by convention, including by regulatory agencies such as the U.S. Environmental Protection Agency and Occupational Safety and Health Administration, to have aspect ratios equal to or greater than 3:1. Methods typically used to collect ambient air samples, such as the system shown in Fig. 1, are not suitable for collection of most fibers for subsequent analysis. Instead, asbestos fibers are typically collected using open-faced cassettes loaded with membrane filters. The Teflon® filter shown in Fig. 1 is not used for asbestos since no solvent is available to extract the fibers. Quartz filters cannot be used since they are made of mineral fibers, which are difficult to differentiate from asbestos fibers. The

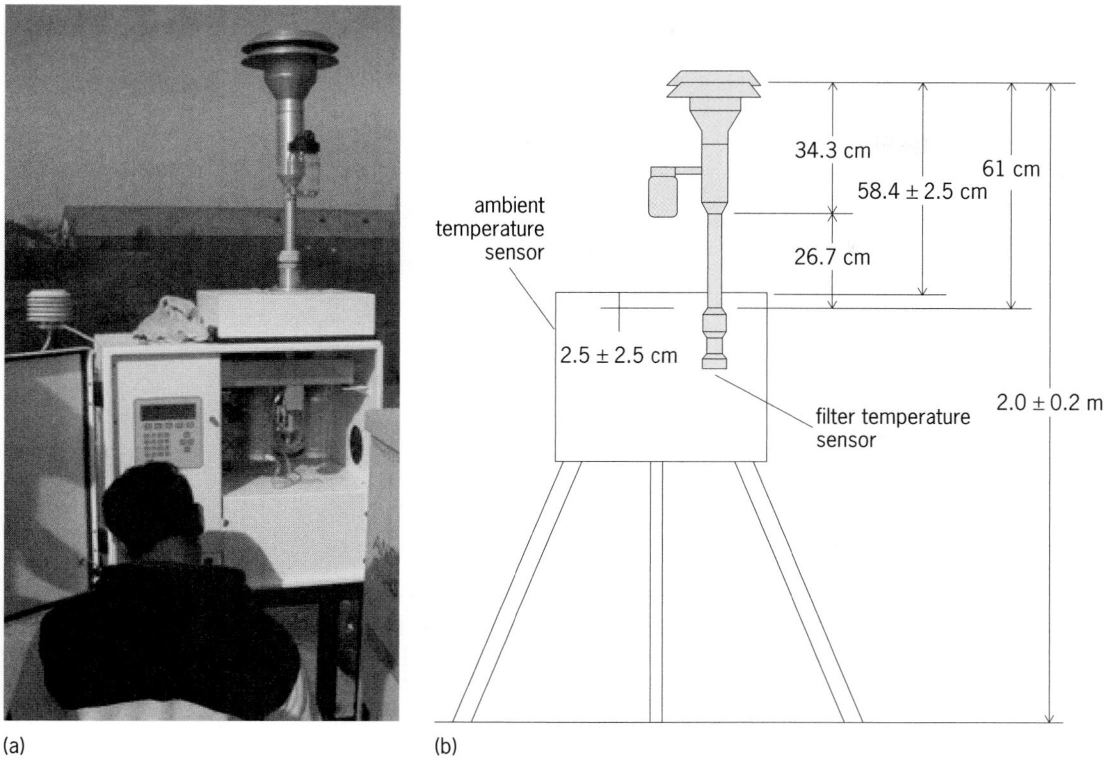

Fig. 1. Sampling device (*a*) photo and (*b*) schematic used to collect particles with aerodynamic diameters ≤2.5 μm. Each sampler has an inlet that extracts a representative ambient air sample and then size segregates the sample into particles below 10 μm (that is, PM_{10}). An impactor downstream of the inlet segregates the PM_{10} size fraction to less than 2.5 μm and this fraction is collected on a Teflon filter. The filter is weighed before and after collection. The Teflon® allows construction for other analyses, such as x-ray fluorescence to determine inorganic composition of the particles. Quartz filters would be used if any subsequent carbon analyses are needed. (*Photo and schematic courtesy of U.S. EPA*)

asbestos fibers are occluded by the larger-diameter quartz fibers as well. This is why mixed cellulose ester and polycarbonate filters in cassettes are used.

Asbestos fibers are strong and flexible enough to be spun and woven. They are resistant to heat and chemicals, making them useful for many industrial purposes. Because of their durability, asbestos fibers that get into lung tissue will remain there for long periods of time.

Fiber type and health effects. There are two general types of asbestos: amphibole and serpentine. Some studies show that amphibole fibers stay in the lungs longer than the serpentine mineral, chrysotile, and this tendency may account for their increased toxicity.

Generally, health regulations classify asbestos into six mineral types: chrysotile (the only asbestos member of serpentine group), which has long and flexible fibers, and five amphiboles, which have brittle crystalline fibers. The amphiboles include actinolite asbestos, tremolite asbestos, anthophyllite asbestos, crocidolite asbestos, and amosite asbestos. Each of the amphiboles includes the term asbestos because, to geologists, amphiboles can also have nonasbestiform habits (that is, the host rock is an amphibole that may contain a mixture of nonfibrous particles, fibers, and cleavage fragments).

The most important risk factors for asbestos-related diseases are length of exposure, air concentration of asbestos during the exposure, and smoking. Cigarette smoking and asbestos exposure are synergistic; that is, the risk of disease is multiplied if a person exposed to asbestos is a smoker (**Table 1**). There is an ongoing scientific debate about the

TABLE 1. Estimated lifetime excess risks due to continuous exposure to asbestos (cases per million population), calculated with a confidence interval = 0.01. Numbers in parentheses are the estimated ranges with a lower limit = 0 and upper limit calculated from a confidence interval = 0.1.

Exposure group	Asbestos dose (fibers m^{-3})				
	8	50	80	500	2000
Male smokers	1 (0–9)	6 (0–55)	9 (0–88)	55 (0–550)	221 (0–2210)
Female smokers	1 (0–5)	2 (0–25)	5 (0–41)	25 (0–250)	101 (0–1010)
Male nonsmokers	1 (0–1)	1 (0–8)	1 (0–11)	8 (0–75)	29 (0–290)
Female nonsmokers	1 (0–1)	1 (0–3)	1 (0–5)	3 (0–28)	11 (0–110)

SOURCE: California Air Resources Board, Staff Report, *Initial Statement of Reasons for Rulemaking*, 1986.

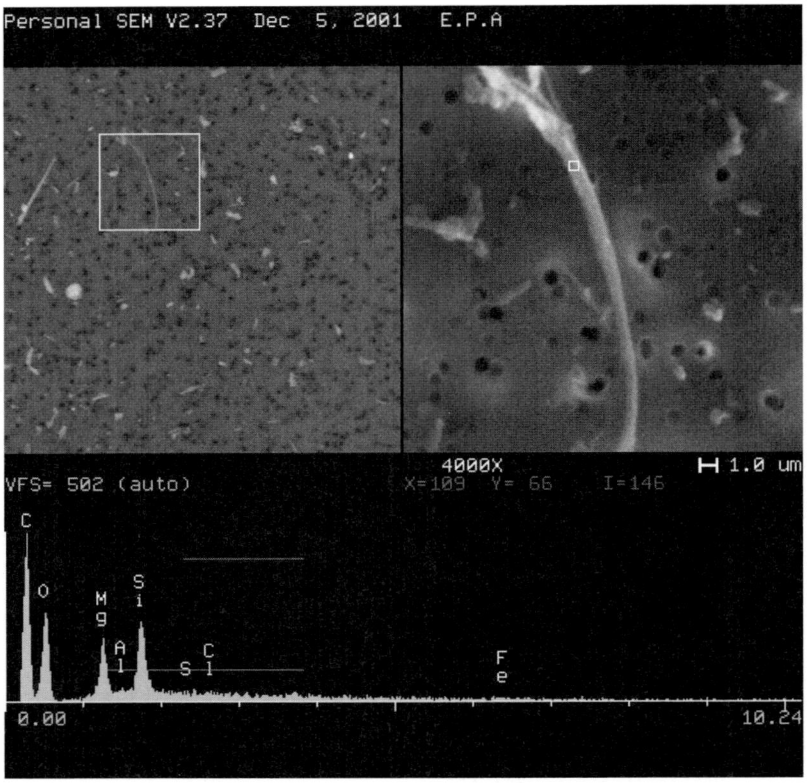

Fig. 2. Scanning electron micrograph of fibers in dust collected near the World Trade Center, New York, in September 2001 (acquired using an Aspex Instruments, Ltd., scanning electron microscope). The bottom of the micrograph represents the elemental composition of the highlighted 15-μm-long fiber by energy dispersive spectroscopy (EDS). The composition (O, Si, Al, and Mg) and the morphology of the fibers indicate they are probably asbestos. The EDS carbon peak results from the dust being scanned on a polycarbonate filter. (*From U.S. Environmental Protection Agency, 2004. Photo courtesy of T. Conner, used with permission*)

differences in the extent of disease caused by different fiber types and sizes. Some differences may be due to the physical and chemical properties of the different fiber types. For example, several studies suggest that amphibole asbestos types (tremolite, amosite, and especially crocidolite) may be more harmful than chrysotile, particularly for mesothelioma. Other data indicate that fiber size (length and diameter) are important factors for cancer-causing potential. Some data indicate that fibers with lengths greater than 5.0 μm are more likely to cause injury than those less than 2.5 μm in length. Additional data indicate that short fibers can contribute to injury. This appears to be true for mesothelioma, lung cancer, and asbestosis. However, fibers thicker than 3.0 μm are of lesser concern, because they have little chance of reaching the lower regions of the lung.

Some groups of people who have been exposed to asbestos fibers in drinking water have higher-than-average death rates from cancer of the esophagus, stomach, and intestines. However, it is very difficult to tell whether this is caused by asbestos or other causes.

Routes of exposure. Ambient air concentrations of asbestos fibers are about 10^{-5}–10^{-4} fibers per milliliter (fibers mL^{-1}), depending on the location. Prolonged human exposure to concentrations much higher than 10^{-4} fibers mL^{-1} is suspected of causing adverse health effects. Asbestos fibers are very persistent and resist chemical degradation (that is, they are inert under most environmental conditions) so their vapor pressures are nearly zero (they do not evaporate), and they do not dissolve in water. However, segments of fibers do enter the air and water when asbestos-containing rocks and minerals are weathered naturally or extracted during mining operations. One of the most serious exposures occurs when manufactured products, such as pipe wrapping and fire-resistant materials, begin to degrade or are improperly handled during renovation or removal activities. Small-diameter asbestos fibers may remain suspended in the air for a long time and be transported advectively by wind or water, before sedimentation. Like spherical particles, heavier asbestos fibers settle more quickly. Asbestos seldom moves substantially via soil. The fibers are generally not broken down to other compounds in the environment and will remain virtually unchanged over long periods.

Although most asbestos is highly persistent, chrysotile, the most commonly encountered form, may break down slowly in acidic environments. Asbestos fibers may break into shorter strands, and therefore into increased number of fibers by mechanical processes (such as grinding and pulverization).

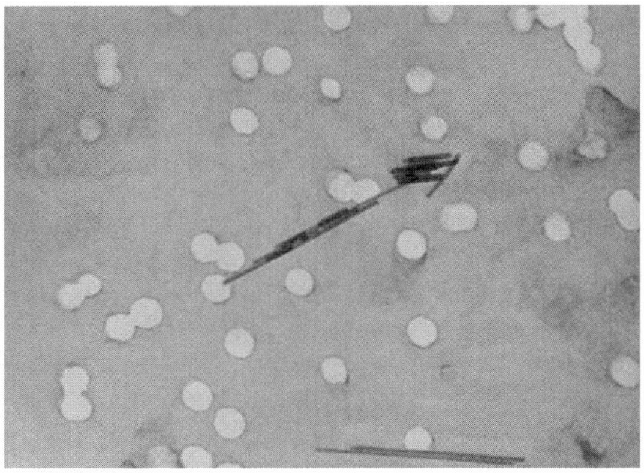

Fig. 3. Transmission electron microscope photograph of a 0.4-μm pore size capillary polycarbonate membrane filter (16,000× magnification) (*From RTI International, S. Doorn, with permission*)

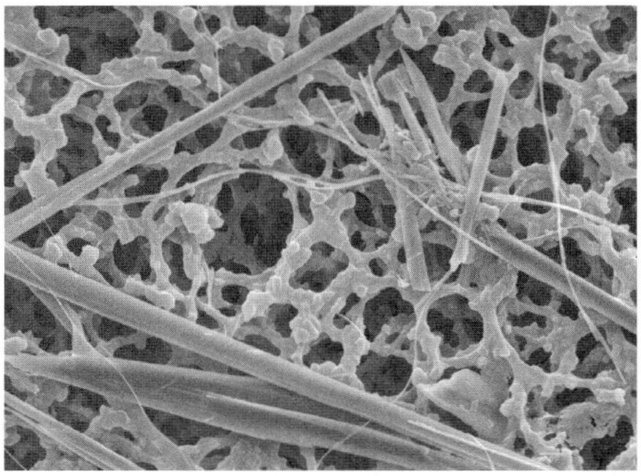

Fig. 4. Scanning electron microscope photograph of a 0.8-μm pore size mixed cellulose ester membrane filter (8000× magnification). (*From MVA Scientific, J. Millette with permission*)

TABLE 2. Methods for analyzing asbestos

Analytical method	Qualitative analysis	Quantitative analysis
Phase contrast microscopy (PCM)	PCM is used for monitoring airborne levels of asbestos in workplace environments where the fiber type is known. PCM can detect fibrous materials but cannot directly distinguish asbestos fibers from other fibrous materials. Air samples are collected on a cellulose ester membrane filter with 0.45–1.2 μm pore size. A portion of the filter is mounted on a microscope slide, cleared using an organic solvent, and fibers are observed in a bright field at a magnification of 100 to 400×. PCM detects fibers as thin as 0.25 μm in thickness.	PCM methods report fibers per cubic centimeter of air sampled (f cm^{-3}). Normally fibers longer than 5 μm in length and having an aspect ratio (length-to-width) of 3:1 or greater are counted. Some protocols use an aspect ratio of 5:1 and count fibers with diameters less than 3 μm. It should be noted that current risk models are based on fibers with a 3:1 aspect ratio, and data from methods employing a 5:1 aspect ratio cannot be used by such models.
Polarized light microscopy (PLM)	PLM is used for determining asbestos and other fibers, and minerals in bulk samples. A portion of the bulk material is examined by stereomicroscopy at a magnification 10 to 60× and subsamples of the various phases in the material are taken and mounted on a microscope slide in various refractive index liquids. Fibers are examined at magnifications 100–500× and identified by their optical properties including morphology (characteristic shape), color, refractive indices, birefringence, extinction angle, and sign of elongation. Characteristic dispersion staining colors are commonly used to identify mineral type.	PLM methods typically report fibers by percent area or percent weight. The percentage may be estimated by visual estimation, comparison with charts showing various area percentages or with gravimetrically prepared standards, or by a point counting procedure. Although point counting is a systematic means of determining the relative amounts of materials in a mixture, it is not a reliable means of determining mass since one point may represent a single thin fiber or a large mineral "boulder." Results using these techniques are usually considered "semiquantitative" since they do not measure quantity directly.
X-ray diffraction spectrometry (XRD)	XRD is used to assist in identifying mineral phases including asbestos and may be used to estimate quantity of the mineral phase. An x-ray beam is directed to the sample and a diffraction pattern characteristic of the mineral phase is produced. The pattern is compared to standard patterns produced by known minerals by means of reference files or through computer analysis. XRD cannot distinguish between asbestiform and nonasbestiform materials of the same mineral phase. XRD results must be confirmed by PLM, SEM, or TEM.	In addition to the qualitative identification, the intensity of the diffraction pattern indicates the amount of material present. A comparison of selected peak heights on the XRD diffractogram to standards of known amounts of a mineral phase may assist the analyst in quantitative analysis of a sample.
Scanning electron microscopy (SEM)	SEM may be used for analysis of fibers in air or in bulk materials (Figs. 2 and 4). Typically, air samples are collected on cellulose ester or polycarbonate membrane filters and prepared for analysis. An image of the surface features on the filter is generated and fibers may be observed and counted. SEM magnifications may range from 2000 to 20,000× and higher. Fibers may be detected that are much thinner than those detected by optical microscopy such as PCM or PLM. An energy dispersive spectrometer EDS) unit, used in conjunction with the SEM, may detect the chemical composition of the fibers.	Airborne fibers measured by SEM are typically reported in fibers per cubic centimeter. Fibers in bulk materials measured by SEM are typically reported as a percentage of the total sample with the percentage determination being made by visual estimation, comparison with standards of known composition, or gravimetry. Counting protocols for fibers by SEM is similar to those by TEM.
Transmission electron microscopy (TEM)	TEM may be used for analysis of fibers in air or in bulk materials (Fig. 3). Air samples collected on membrane filters or bulk samples transferred to membrane filters are carbon coated and a thin carbon layer including the fibers is transferred to a grid for analysis. TEMs provided magnifications from about 5000 to 20,000× or more and fibers with diameters of about 0.01 μm can be detected. An EDS unit can provide the chemical composition. In selected area electron diffraction (SAED), an electron beam is directed to the fiber, producing an electron diffraction pattern representing the crystal structure of the material. TEM is the most definitive technique for detecting and identifying asbestos materials.	

Inhaled fibers may get trapped in the lungs and with chronic exposures build up over time. Some fibers, especially chrysotile, can be removed from or degraded in the lung with time.

Because of its toxicity, it is necessary to have effective monitoring techniques and analytical methods to detect, quantify, and control asbestos in the environment. A number of techniques and methods have been developed to detect and quantify asbestos in bulk samples, air, water, settled dust, and soil.

Analytical methods. Numerous methods are used to measure asbestos: asbestos and mineral fibers in air by scanning electron microscopy (SEM), scanning transmission electron microscopy (STM), asbestos and mineral fibers in air by phase-contrast microscopy, asbestos and mineral fibers in bulk materials, manufactured mineral fibers, asbestos in soils, asbestos in dust, and asbestos in water.

The most widely used techniques are based primarily upon observing the fiber by either optical or electron microscopy. These techniques include phase-contrast microscopy (PCM), polarized light microscopy (PLM), scanning electron microscopy, and transmission electron microscopy. Some methods also use x-ray diffraction spectrometry (XRD) for identifying mineral phases and quantitative analysis. Qualitative and quantitative analysis of fibers by each technique are discussed in **Table 2**.

Over the years, a number of optical and electron microscopy methods have been developed to detect and quantify asbestos in air and other matrices. Each method has its strengths and weaknesses, and must be carefully evaluated to determine how best to detect and quantify asbestos under a given circumstance. Sampling efficiency is essential in exposure assessment.

Typically, mixed cellulose ester (0.45- or 0.8-μm pore size), and to a lesser extent capillary-pore polycarbonate (0.4-μm pore size) membrane filters are used to collect airborne asbestos for count measurement and fiber size analysis. The pore size specification for a membrane filter is an absolute specification only for capillary-pore-type filters, such as polycarbonate (PC). The pore size rating for tortuous-path filters (many entrapments), such as mixed cellulose ester (MCE) filters, is an effective pore size and not a specification that particles exceeding that size be retained by the filter.

The two types of filters differ in their chemical and physical composition. Polycarbonate filters have a smooth filtering surface; the pores are cylindrical, almost uniform in diameter, and essentially perpendicular to the surface (**Fig. 3**). Because of their smooth surfaces, capillary-pore membrane filters are particularly useful for collecting particles that will be observed with a scanning electron microscope. A mixed-cellulose ester filter is a thicker filter with a spongelike appearance and relies on a tangled maze of cellulose ester strands to trap fibers (**Fig. 4**). For microscopic analysis of asbestos deposited on the filter, it is critical that the fibers be in a single plane to ensure they are in focus during analysis. This requirement is simple to achieve for PC filters because of the smooth filtering surface, whereas the MCE filter requires two additional steps in the direct preparation procedure (**Fig. 5**). The MCE filter must be collapsed with an organic solvent, and then the top layer of the collapsed filter material

TABLE 3. Mean asbestos concentration by batch and filter type.

	Mean concentration (structures/mm^2) by filter pore size and nominal loading			
	0.45 μm		0.8 μm	
Batch	Low	High	Low	High
	Fibers $\geq$0.5 μm			
1	321	—	274	—
2	—	958	—	743
3	413	—	388	—
4	—	1512	—	1304
	Fibers >5 μm			
1	84	—	80	—
2	—	373	—	313
3	92	—	100	—
4	—	301	—	333

SOURCE: Data from the U.S. Environmental Protection Agency study of asbestos collection efficiency of mixed cellulose ester filters, 2007 (*J. Kominsky, D. Vallero, and M. Beard, U.S. EPA Report, Comparison of Chrysotile Asbestos Collection Efficiencies on Mixed-Cellulose Ester Filters, 2007*).

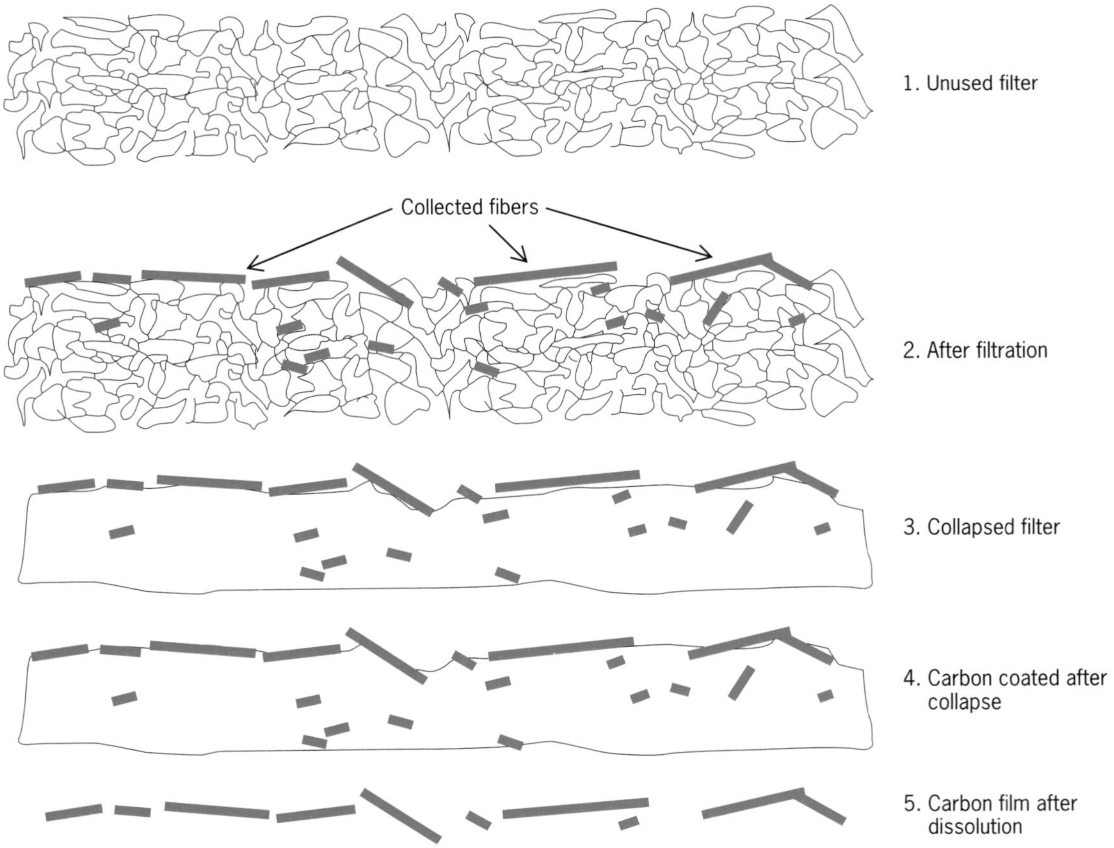

Fig. 5. Steps in preparing a mixed cellulose ester filter for TEM specimen preparation. ***(Adapted from Eric J. Chatfield, "Overview of Measurement Procedures for Determination of Asbestos Fibres in Building Atmospheres," in Asbestos Fibre Measurements in Building Atmospheres, Proceedings of a workshop held at Ontario Research Foundation, Mississauga, Ontario. Sponsored by Environmental Health Directorate of Health and Welfare Canada, March 1985)***

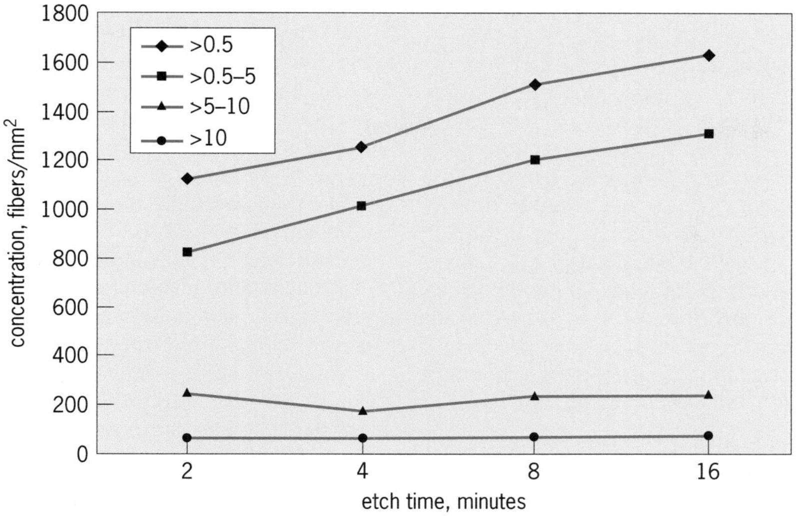

Fig. 6. Asbestos fiber concentrations observed on 0.45-μm pore size MCE filters for variable etching times for four fiber lengths. Data from the U.S. Environmental Protection Agency study of asbestos collection efficiency of mixed cellulose ester filters, 2007. (*From J. Kominsky, D. Vallero, and M. Beard, U.S. EPA report, Comparison of Chrysotile Asbestos Collection Efficiencies on Mixed-Cellulose Ester Filters, 2007*)

must be etched away with a low-temperature plasma asher.

Comparison of air filter types. Collection efficiency of the 0.45-μm pore size MCE filter for aerosols of asbestos fibers ≥ 0.5 μm in length is greater than that for a 0.8-μm pore size MCE filter (**Table 3**). This difference in collection efficiency is not maintained for asbestos structures longer than 5 μm. If the exposure study is focused on fibers <5.0 μm in length, the investigator should use MCE filters with a 0.45-μm pore size. If the exposure study exclusively focuses on structures longer than 5.0 μm, then either filter pore size may be used.

There is a significant difference in the effect of etching times for fibers <5.0 μm and those >5.0 μm in length (**Fig. 6**). The mean concentration of asbestos fibers ≥ 0.5 μm in length increases with etching time (2, 4, 8, and 16 min) of 0.45-μm pore size MCE filters. For fibers >5.0 μm in length, there is no significant difference in numbers of structures counted at the etching times used in these tests. Since most asbestos exposure risk models include fibers >5.0 μm in length, the 8-min etching time specified in ISO 10312:1995 is adequate. However, if an exposure study is focused on fibers <5.0 μm in length, the etching time of 8 min should be reviewed. Further study is needed to determine the etching time beyond which no significant increase in asbestos concentration of fibers <5.0 μm in length is expected.

[*Disclaimer*: The U.S. Environmental Protection Agency through the Office of Research and Development funded and managed some of the research described here. The present article has been subjected to the Agency's administrative review and has been approved for publication.]

For background information *see* AMPHIBOLE; ASBESTOS; CHRYSOTILE; ELECTRON MICROSCOPE; ENVIRONMENTAL ENGINEERING; MUTAGENS AND CARCINOGENS; PHASE-CONTRAST MICROSCOPE; POLARIZED LIGHT MICROSCOPE; RESPIRATORY SYSTEM DISORDERS; SCANNING ELECTRON MICROSCOPE; SERPENTINITE; X-RAY DIFFRACTION in the McGraw-Hill Encyclopedia of Science & Technology.

John Kominsky; Daniel Vallero;
Michael Beard

Bibliography. American Society for Testing and Materials, Standard Test Method for Airborne Asbestos Concentration in Ambient and Indoor Air as Determined by Transmission Electron Microscopy Direct Transfer (TEM), ASTM D 6281, 2006; M. E. Beardand and J. R. Kominsky, Sampling and Analysis of Asbestos Fibers on Filter Media to Support Exposure Assessments: Scoping Effort, report prepared by Environmental Quality Management, Inc., Cincinnati, OH, and RTI International, Research Triangle Park, NC, for U.S. EPA, National Exposure Research Laboratory, task order 0020, contract 68-C-00-186, December 2006; E. J. Chatfield, Measurements of chrysotile fiber retention efficiencies for polycarbonate and mixed cellulose ester filters, in M. E. Beard and H. L. Rook (eds.), *Advances in Environmental Measurement Methods for Asbestos*, American Society for Testing and Materials, *Spec. Tech. Publ.*, 1342, 2000; International Organization for Standardization, Ambient Air: Determination of asbestos fibres—Direct-transfer transmission electron microscopy method, ISO 10312, 1995; International Organization for Standardization, Ambient Air: Determination of asbestos fibres—Indirect-transfer transmission electron microscopy method, ISO 13794, 1999; International Organization for Standardization, Air Quality: Determination of the number concentration of airborne inorganic fibers by phase contrast microscopy—Membrane filter method, ISO 8672, 1993; National Institute for Occupational Safety and Health. Asbestos by TEM, Method 7402, in NIOSH Manual of Methods, issued

August 15, 1994; K. Spurny, On the filtration of fibrous aerosols, *J. Aerosol. Sci.*, 17450, 1986; U.S. Code of Federal Regulations, 40 CFR Part 763, Appendix A to Subpart E, Interim Transmission Electron Microscopy Analytical Methods: Mandatory and Nonmandatory—and the Mandatory Section to Determine Completion of Response Action; U.S. Environmental Protection Agency, Superfund Method for the Determination of Asbestos in Ambient Air, Part 1: Method, EPA 540/2-90-005a, May 1990; G. Yamate, A. C. Agarwal, and R. D. Gibbons, Methodology for the Measurement of Airborne Asbestos by Electron Microscopy, Draft report, EPA contract 68-02-3266 for Environmental Monitoring Systems Laboratory, Office of Research and Development, 2007.

Autonomous microsystems

Fleas, mites, mosquitoes, and other small-scale "living microsystems" have found an ecological niche that is in many ways unparalleled in nature. In the last few years, the convergence of a number of advances in micro-electro-mechanical-systems (MEMS) technology (including power generation, energy storage, communications, sensing, microfluidics, and subcomponent assembly) has opened the door to creating artificial autonomous microsystems that have many of the same capabilities of their living counterparts.

Living microsystems. An autonomous microsystem is an individual functioning of its own accord with the ability to interpret and intelligently interact with its environment, whose fundamental physical dimension is on the order of a millimeter or smaller. To meet this definition, such systems must possess the basic set of capabilities as outlined in the **table**. By comparing this to what we see around us, it becomes apparent that living autonomous microsystems are ubiquitous in nature, particularly among insect species. Insect body size at maturity represents, in the simplest case, a trade-off between overall fitness and functionality (which is generally associated with larger body size) and the amount of time and energy required to reach sexual maturity (the longer the juvenile growth stage, the greater the chance of dying prior to reproduction). In this case, the microsystem paradigm can be thought of as an evolutionary optimal, allowing it to best fulfill its function (procreation) within its ecological niche.

Capabilities defining autonomous microsystems

Capability	Description
Sensing	To interact with the environment the system must obtain useful information about it.
Data processing and decision making	The system must interpret the sensor data and make informed decisions about it based on innate knowledge (hardware) or past experience (memory).
Actuation and/or communication	Interaction with the environment can be direct through mechanical or chemical interaction or indirect through communication.
Energy storage	Driving each of these capabilities requires energetic consumption, which must be extracted from the environment or stored on-board.
Level of functionality	An autonomous microsystem minimizes the level of functionality to just that required to serve its purpose. In living systems, the most basic function is reproduction and continuance of the species.

Artificial autonomous microsystems. The above definition of an autonomous microsystem used the relatively clumsy description of having a size scale "on the order of a millimeter." The main reason for this is that it represents, arguably, the smallest size a system can possess while still being able to physically interact with our macroscale environment at the individual level. (Bacteria and other lower organisms also exhibit a high degree of autonomy. However, their ability to interact with our macroscale environment is somewhat limited.) The other reason, however, is that the fabrication processes used to create living systems in nature (for example, cell division) are largely independent of the final size scale. MEMS technology is fabricated using a set of unique processes that typically involves some form of lithographic feature definition. This niche between nanosystem (molecular) fabrication and macrosystem (mechanical) fabrication allows us to better define an artificial autonomous microsystem as an individual functioning of its own accord with the ability to interpret and interact with its environment where the features that enable its autonomy are fundamentally enabled through microfabrication technology. It is this interaction with the environment that separates a simple system, such as a digital wristwatch, from a truly autonomous microsystem (**Fig. 1**).

The clear commercial trend in the early twenty-first century has been to adapt microsystem technology to portability and omnipresence [ubiquitous wireless fidelity (Wi-Fi), Internet-enabled cell phones] rather than autonomy. Having said that, two emerging applications of autonomous microsystems are autonomous microsensor networks and in vivo and portable diagnostics and therapeutics.

Autonomous microsensor networks. The idea of using small, autonomous sensor nodes that can be dispersed over a large area has long been of interest for applications ranging from military surveillance to ecological monitoring. Such a system could allow information gathering over a wide spatial area and for a long time, without imposing a large footprint. Granting autonomy to the system allows it to operate without external intervention until a detection event has occurred. The use of microfabrication technology and minimizing system functionality to just that required to carry out a basic purpose allows for facile mass production. Given that self-replication is not possible, maintenance of the network does require periodic redispersal of the units, which could be done through a variety of manual or more creative techniques (wind, animal droppings, and so forth). Numerous research groups are working on

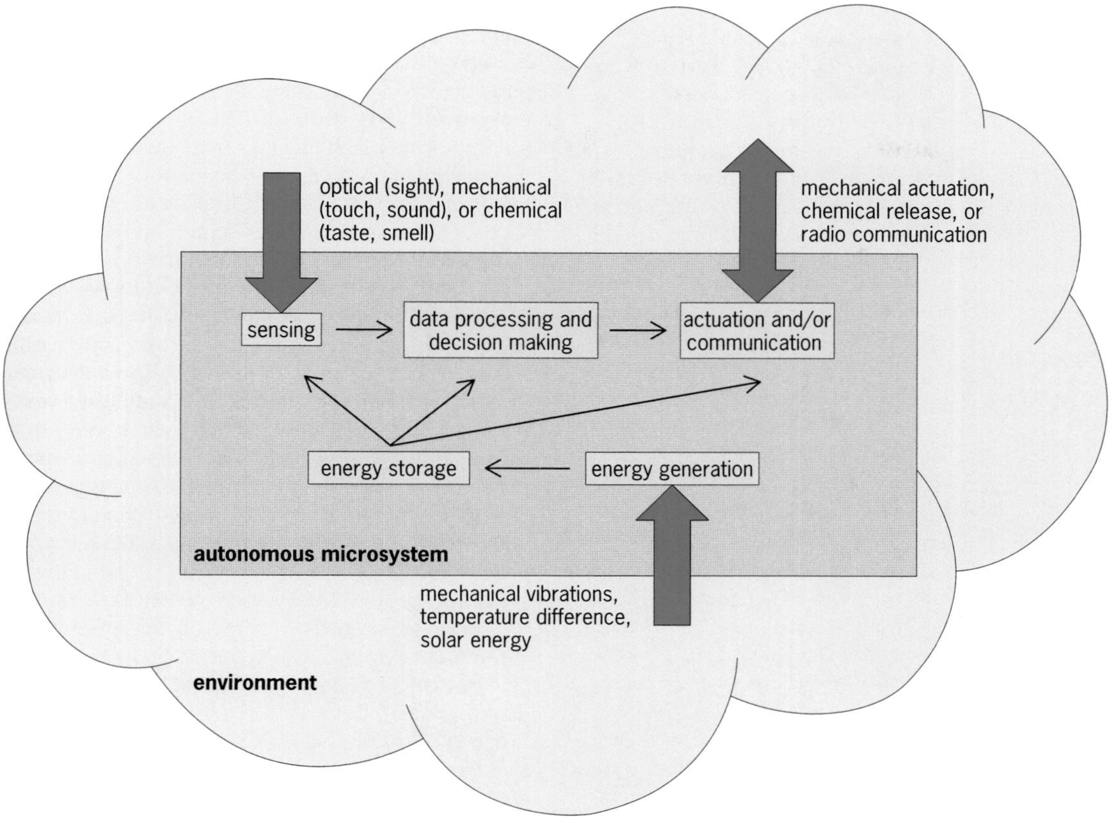

Fig. 1. Operation of an autonomous microsystem. Thick arrows show external interactions with the environment. Thin arrows show internal interaction between subcomponents.

developing such systems. The most well known is likely the "smart dust" program initiated by DARPA (U.S. Defense Advanced Research Projects Agency) in the late 1990s.

In vivo and portable diagnostics and therapeutics. One of the most exciting promises of nanomedicine has been the development of blood-borne nanoparticles functionalized with biomolecular probes specific to various cancer cells. When these nanoparticles are introduced into the bloodstream, they are carried along until they come in contact and bind with the target cells. In contrast to the previous example, this is a case where the emerging opportunity for autonomous microsystem technology is to increase the system size and with it the level of functionality. Exploiting lab-on-chip–based technologies, the ultimate goal of such systems is "fantastic voyage" level functionality, whereby a self-piloted system finds a specific target (through on-board sensing for example), performs a diagnosis, and takes appropriate therapeutic action through electrical, chemical, or mechanical means.

Such systems are still a long way off, but a number of already developed technologies, which are primarily diagnostic in nature, are paving the way. One such example is the PillCam developed by Given Imaging, which is a pill-sized camera that is swallowed by the patient and transmits images taken as the pill passes through the digestive system to a belt worn as the patient goes about a normal routine. A similar system produced by SmartPill uses MEMS technology to collect pH, temperature, and pressure information as it passes through the gastrointestinal tract. As an alternative to these in vivo health-monitoring approaches, there is also significant interest in translating lab-on-chip technology into wearable autonomous microsystems that can periodically take a blood sample and screen it against very low levels of pathogenic (in particular viral) infection. With recent advances in nanobiosensors, it is becoming possible to detect some infections at presymptomatic levels or before the patient becomes contagious. Such systems may prove useful in certain high-risk military and health-care applications. An example of such a system is shown in **Fig. 2**. *See* INTEGRATED NANOSENSORS.

Challenges to developing next-generation systems. The living microsystems discussed above have evolved over millions of years and developed near-optimal motion mechanics, energy-storage, sensing, and communications capabilities. Many of these individual capabilities have proven to be difficult to replicate using existing microsystem technology.

Energy consumption and use. One of the major challenges with autonomous microsystems has been in the generation and storage of energy for use over long periods. There are two basic approaches that can be taken to energy management in such systems. The first is to initially "charge up" the system and then let the energy slowly dissipate over the lifetime of the device. This approach simplifies system architecture but is useful only if the dissipation time is

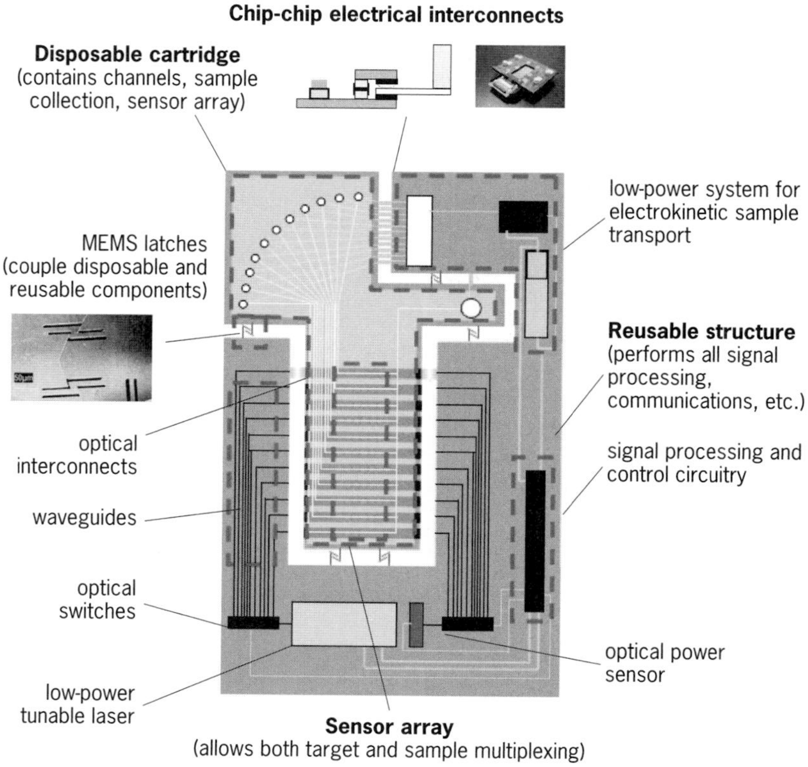

Fig. 2. Envisioned autonomous microsystem for prognostic detection of viral infection.

the same as the desired lifetime of the system. Such an approach has precedence in nature in that some insects spend the majority of their time in early developmental stages eating to store up energy that is subsequently burned off as an adult. The second approach involves the incorporation of an energy-harvesting technology that can extend system life indefinitely. Depending on the system, a number of energy-harvesting mechanisms may be available, including ambient mechanical motion, solar energy, or local temperature variations. Existing devices tend to have relatively low conversion efficiencies or operational limitations (for example, solar cannot be used at night). Recent nanoscopic approaches may serve to improve this.

Information processing, storage, and retrieval. The obvious inherent advantage that artificial systems have over natural ones is the speed with which information can be processed and the density with which it can be stored. The relatively simple tasks that autonomous microsystems are being asked to perform will likely not greatly tax these existing capabilities, and metrics such as energy per operation are likely to be more important. More challenging is being able to retrieve information from an autonomous microsystem. The systems discussed above are able to transmit over short distances (<1 m or 3 ft) or use passive means, whereby conformal changes can be detected by an external optical probe. Long-range radio communication is difficult since the required power increases strongly with the distance the signal must travel. The emergence of omnipresent communication networks may be of significant help by reducing the distance from the system to the nearest reception point.

Integration and large-scale assembly. Microelectronic systems are fabricated using very well defined processes with a very small set of different materials. Autonomous microsystems may use similar fabrication technology but will certainly be made of a greater array of materials. While pick-and-place technologies such as those currently used to assemble computer boards could be used, their serial nature places extreme limitations on the number of units that could be assembled. A promising parallel process is fluidic self-assembly, where components are fabricated separately and the assembly is driven by hydrodynamic forces and geometric interactions. Significant work is required to add functionalities such as error correction to these processes. The extension of these parallel processes to exponential assembly, consistent with what is observed in many living systems with growing populations, remains a difficult technical challenge.

For background information *see* MICRO-ELECTRO-MECHANICAL SYSTEMS (MEMS); MICROFLUIDICS; MICROLITHOGRAPHY; NANOTECHNOLOGY in the McGraw-Hill Encyclopedia of Science & Technology.

David Erickson

Bibliography. M. Gad-el-Hak (ed.), *MEMS: Design and Fabrication*, 2d ed., CRC, 2006; S. E. Lyshevski, *MEMS and NEMS: Systems, Devices, and Structures*, CRC, 2002; M. J. Madou, *Fundamentals of Microfabrication: The Science of Miniaturization*, 2d ed., CRC, 2002.

Auxin and the auxin response

Auxin is a plant hormone involved in regulating growth and development throughout the life cycle of the plant. Auxin plays an essential role in a diverse range of processes, including pattern formation in embryogenesis, organ development, apical dominance (inhibition of lateral bud growth by the apical bud of a shoot), shoot and root growth, lateral root initiation, leaf formation, and tropisms (directional growth in response to environmental stimuli). Auxin itself is a very simple molecule; yet despite over 100 years of research, the mechanisms of auxin biosynthesis, signaling, and action are not fully understood.

Auxin biosynthesis. Auxin was chemically defined in the 1930s as a small organic compound related to the amino acid tryptophan. The most common form of auxin in plants is indole-3-acetic acid (IAA).

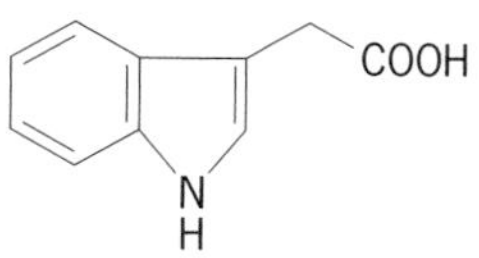

indole-3-acetic acid (IAA)

Other naturally occurring auxins include indole-3-butyric acid (IBA), 4-chloro-indoleacetic acid, and phenylacetic acid (PAA). Synthetic chemicals exhibiting auxin activity in plants such as 2,4-dichlorophenoxyacetic acid (2,4-D) and naphthalene acetic acid (NAA) also exist.

The majority of auxin in the plant is in an inactive form, conjugated with other molecules such as sugars or amino acids, for example, indoleacetyl aspartate (IAA-Asp). Some of these conjugates can be hydrolyzed to produce free auxin. Alternatively, plants can also synthesize auxin de novo. Synthesis occurs at its highest level in actively growing tissues such as the growing tips of shoots and roots. IAA can be synthesized by various methods, either through a tryptophan-dependent pathway from the amino acid tryptophan or through a tryptophan-independent pathway possibly from an indole precursor.

Auxin transport. The transport of auxin is well studied, and indeed directional transport of auxin is important for it to function correctly. Nonpolar transport of auxin along a concentration gradient occurs by diffusion in the phloem (the principal food-conducting tissue in vascular plants). Auxin diffuses from its site of synthesis, the source, to sink tissues where auxin concentrations are low.

In a different mechanism, namely polar auxin transport (PAT), auxin is actively transported around the plant via carrier proteins. In shoots, it travels basipetally (downward) from the shoot tip toward the roots through the xylem (the principal water-conducting tissue and the chief supporting tissue of higher plants). In the roots, it travels both basipetally and acropetally (from the root tip toward the shoots) in separate transport streams. A variety of influx and efflux proteins have been identified that transport auxin into and out of the cell. Auxin in its protonated form (H+ added) enters the cell by diffusion or is actively transported through the influx carrier AUX1 located in the cell membrane. In the higher pH of the cell, auxin is deprotonated/ionized (H+ removed) and can only be transported out of the cell actively. This involves PIN (PINFORMED) proteins and members of a P-glycoprotein (PGP) family, both of which appear capable of transporting auxin out of the cell. The distribution of influx and efflux carriers in the cell can be asymmetric, resulting in a directional flux of auxin. Localization of the influx and efflux carriers can be modulated according to the stage of development and also by environmental conditions such as light and gravity. The distribution of auxin is therefore dynamic and can change rapidly (**Fig. 1**).

Auxin signaling. Auxin acts at the cellular level by altering expression of various genes. Many genes are upregulated in response to auxin. The Aux/IAA family of transcriptional corepressors are central to auxin signaling along with the auxin response factor (ARF) family of DNA-binding transcription factors. Q-rich (glutamine-rich) ARFs promote the expression of auxin-responsive genes by binding to specific auxin-response elements (AREs) in their promoter regions. Dimerization (chemical union of two similar molecules to form a polymer) of ARFs with Aux/IAAs blocks ARF function, resulting in transcriptional repression. In the presence of auxin, the Aux/IAA proteins are degraded, releasing the repression.

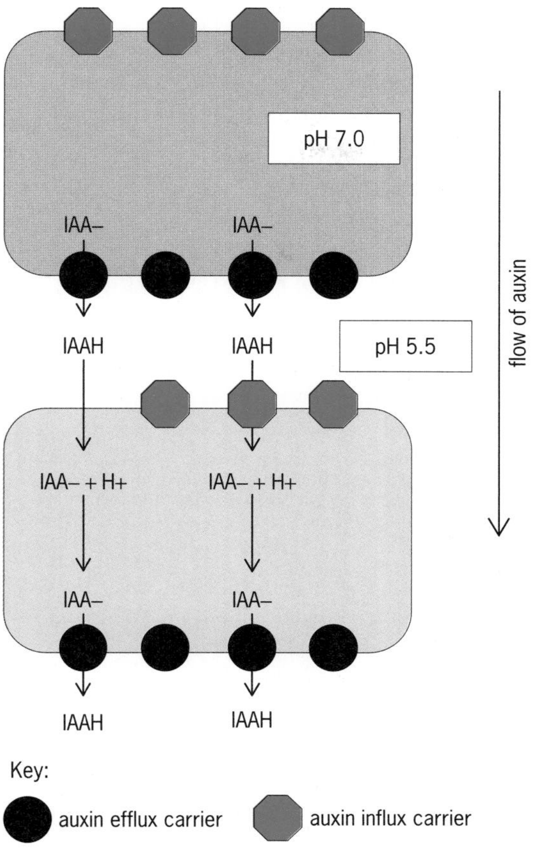

Fig. 1. Polar auxin transport. IAAH (the protonated form of IAA) enters the cell by diffusion or by active uptake via the AUX1 influx carrier. Once in the cell, IAAH is deprotonated (IAA–) and can only be transported out of the cell via efflux carriers such as the PIN proteins. The asymmetric distribution of the efflux and influx carriers directs the flow of auxin through the plant.

After many years of searching, the auxin receptor for this response was finally identified in 2005 as a small family of F-box proteins (which contain a protein motif of ~50 amino acids that functions as a site of protein-protein interaction) typified by TRANSPORT INHIBITOR RESPONSE1 (TIR1) from *Arabidopsis*, a component of the Aux/IAA degradation system.

The addition of a polyubiquitin chain to a protein targets it for degradation by the 26S proteasome. [In ubiquitination, a protein is inactivated by attaching ubiquitin (a small, 76-amino-acid protein) to it. Ubiquitin acts as a tag that signals the protein-transport machinery to ferry the protein to the proteasome for degradation.] This requires three different enzymes: ubiquitin activating enzyme (E1), ubiquitin conjugating enzyme (E2), and ubiquitin protein ligase (E3). It is the E3 that is responsible for specifically selecting the target protein. One class of E3 ubiquitin protein ligase is the SCF complex type (SKP1, cullin, and F-box containing) protein. The F-box component of the SCF confers target specificity. The auxin receptor TIR1 is the F-box protein in an SCF-type ubiquitin protein ligase complex that targets the Aux/IAA proteins for degradation in an auxin-dependent manner. Auxin acts by binding directly to TIR1, promoting the interaction between TIR1 and Aux/IAA proteins. As a result, Aux/IAA proteins are ubiquitinated,

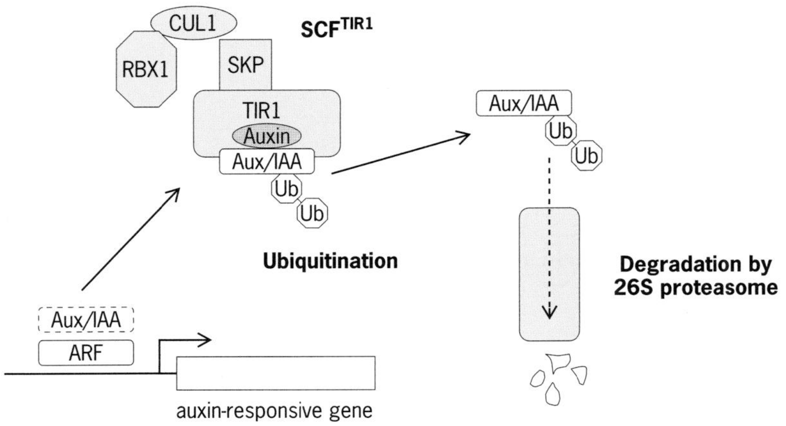

Fig. 2. Auxin signal transduction pathway. Aux/IAAs form dimers with ARFs bound to promoters of auxin-regulated genes, causing transcriptional repression. Repression is released when auxin binds to TIR1, promoting the interaction with Aux/IAAs, which are ubiquitinated and subsequently degraded.

targeting them for degradation. This releases ARFs from repression, allowing them to promote expression of auxin-inducible genes (**Fig. 2**).

Arabidopsis plants lacking the TIR1 protein have only mild defects in growth and development. This suggests that there are other receptors for auxin that can compensate for TIR1. There are over 700 F-box proteins in *Arabidopsis*. At least three other proteins, termed AUXIN SIGNALING F-BOX (AFB), have been identified that are closely related to TIR1 and interact with Aux/IAAs in an auxin-dependent manner. While mutant plants lacking single AFB proteins have mild phenotypes, plants lacking all four of these proteins have severely affected development, leading to embryo lethality. This indicates their overlapping functions in auxin response.

AUXIN BINDING PROTEIN1 (ABP1) is another protein with high affinity for auxin. However, there is no evidence for any involvement of ABP1 in auxin-regulated gene expression, and its role may be limited to membrane-level responses.

Auxin regulation of plant growth and development. Auxin regulates plant growth and development through changes in cell division, differentiation, and elongation. Auxin exerts its effect on the plant early in embryogenesis where PIN proteins direct the flow of auxin from the basal cells, which support the growth of the embryo, to the apical cells, which will go on to form the embryo proper. The polarity of auxin then switches as the PIN proteins relocalize to the basal side of the cell.

Later in development, this polar transport of auxin from the shoot tip down the plant inhibits lateral bud formation in the process called apical dominance. Auxin does not directly enter the bud itself, but acts partly by regulating a second hormone, such as cytokinin, which is a promoter of bud outgrowth, and partly by preventing auxin efflux from the buds, possibly by competing for limited auxin transport capacity in the main stem.

Asymmetric distribution of auxin plays a key role in phototropism, the response of plant shoots and roots to light stimuli. In shoots, auxin promotes cell elongation on the shaded side, causing bending of the shoot toward the light.

In a mechanism similar to phototropism, gravitropism (the response of plants to gravity) is also controlled by an asymmetric distribution of auxin. PIN proteins are redistributed in the cell, directing auxin to the lower side of the root where it inhibits growth relative to the upper side, causing the root to bend downward.

Although this review is by no means an exhaustive list of the role of auxin in the plant, it does show just how vital this simple molecule is in plant development.

For background information *see* APICAL DOMINANCE; AUXIN; CYTOKININS; PHOTOMORPHOGENESIS; PLANT GROWTH; PLANT HORMONES; PLANT MORPHOGENESIS; PLANT MOVEMENTS; PLANT PHYSIOLOGY; PROTEIN in the McGraw-Hill Encyclopedia of Science & Technology.

Lynne Armitage; Ottoline Leyser

Bibliography. O. Leyser, Dynamic interaction of auxin transport and signalling, *Curr. Biol.*, 16:R424–R433, 2006; W. Teale, I. Paponov, and K. Palme, Auxin in action: Signalling, transport and the control of plant growth and development, *Nat. Rev. Mol. Cell Biol.*, 7:847–859, 2006; A. Woodward and B. Bartel, Auxin: Regulation, action and interaction, *Ann. Botany*, 95:707–735, 2005.

Avian brain chimeras

A chimera is an organism made up of cells from two or more genetically distinct sources. While some chimeras (very rarely) arise naturally through the fertilization of an egg by more than one sperm cell, or from the very early fusion of fraternal twin embryos shortly after fertilization, the avian brain chimeras considered here are different because they are produced surgically by the substitution of presumptive nervous system cells between the early embryos of different bird species. The word "presumptive" is used because the transplanted cells have not yet definitively become nervous system cells at the time of transplantation. This is because the cell substitution takes place before major organ systems or blood vessels are formed in the embryo. The surgeries are guided by "fate maps" that identify which groups of embryonic cells give rise to particular parts of the developed brain. Avian brain chimeras are used in basic research examining the mechanisms responsible for the development of, and evolutionary changes in, neural circuits: complex groups of interconnected nerve cells located in many different parts of the brain that regulate particular perceptual, cognitive, and behavioral functions of organisms.

Underpinnings of the technique. This experimental system uses inborn species differences in neural circuit organization to identify and characterize interactions among cells in the developing brain that make important contributions to the development of neural circuit function.

Prenatal brains of different species are built differently. While the ability to predict an individual organism's personal characteristics from its genetic makeup is, and will likely remain, inaccurate, it is much easier to reliably predict particular species-level characteristics of the behavior of organisms. For example, human infants and the young of many vocally imitating birds (species that exhibit the ability for vocal imitation) will spontaneously produce (human) fluent speech or a small number of human speech phrases (vocally imitating birds) when interactively exposed to human speech from an early period in life. Other bird species such as chickens and quails (which do not exhibit vocal imitative abilities) will not; chickens and quails are also not able to learn to produce each other's vocalizations. However, chickens and quails, like dogs, may spontaneously learn (or can be taught) to respond differently to different human vocal commands. It is not that the hearing abilities or sound-producing abilities of chickens and quails are different in character from those of vocally imitating birds. Rather, there is some inborn difference in neural circuits within their brains that makes these species prone to do different things when presented with the same information. The avian brain chimera technique capitalizes on the existence of such circuit differences and uses them to discover the locations of cell populations that make a decisive contribution to circuit development. These are identified by finding groups of cells that, when transplanted from the donor to the host species, transfer donor behavioral characteristics to the host species that are normally not present (**Fig. 1***a*).

Inborn neural circuit developmental decisions. Every cell in an organism's body got there by splitting away from a preexisting cell (starting with the first cell formed by the union of the sperm and the egg) and, through a series of decisions guided by molecules both inside and outside of itself, gradually assuming its developed fate. One of the perennial questions that interests scientists about neural circuit development has been the extent to which these decisions are guided by genetic information (in the form of DNA sequences inside cells inherited from the sperm and egg). That is a question often phrased using the dichotomy of "nature" and "nurture."

Long before people knew about the role of DNA in inheritance, they were already making the distinction between inherited (internal) and environmental (external) sources of developmental information, by raising animals in controlled environments to see which aspects of their natural behavior would appear in spite of the fact that they lacked exposure to either parents or siblings (or to deliberately try to mold their behavior in particular ways). For example, in 1773, the British scientist Daines Barrington published the results of experiments on the developmental origin of bird songs based on these types of studies. Developmental biologists took this same line of experimentation to embryonic cells by the end of the ninteenth century, examining what happens when one of the first two cells that form the embryo is removed: Do you get half an animal or a whole animal? In the 1880s and 1890s Wilhelm Roux and Hans Driesch found that the answer to this question depended on the way the manipulation

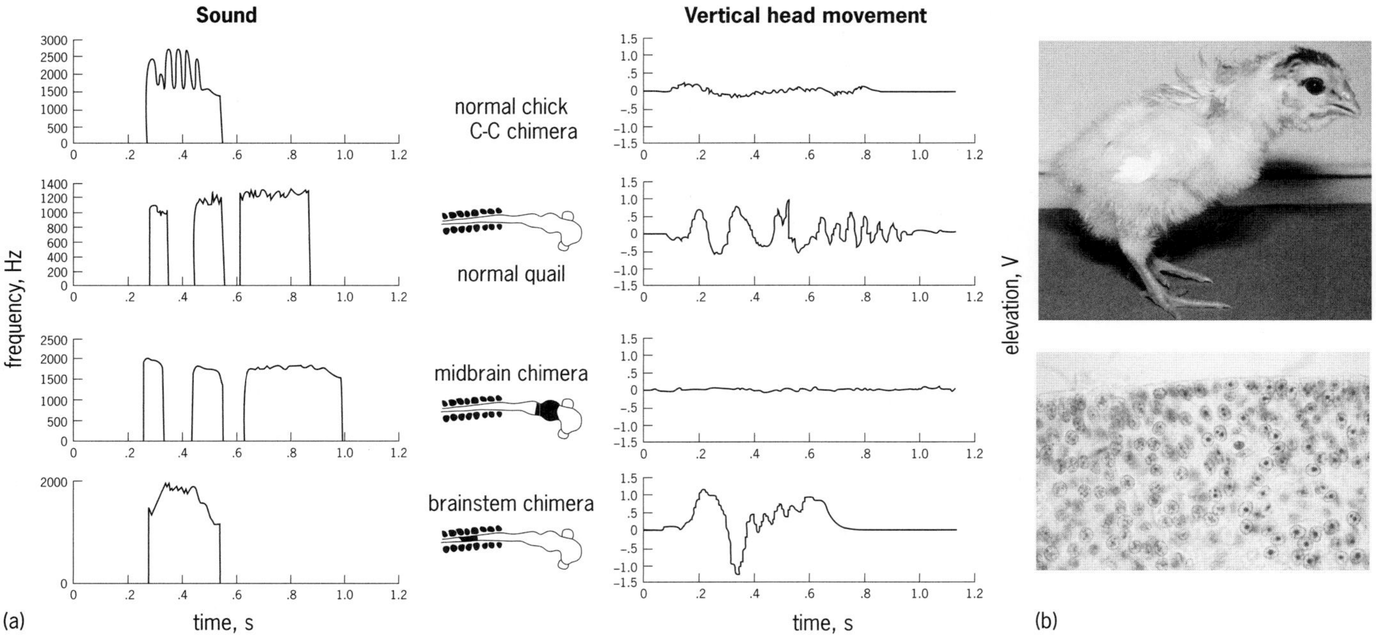

Fig. 1. **(*a*) Quail-chick brain transplants change crowing behavior. (Left) A simplified representation of the form of the crow of juvenile chickens and quails induced by testosterone treatment. (Right) Head movement patterns measured with video object tracking. Midbrain transplants transfer the sound characteristics from quail to chick, while brainstem transplants transfer the head movement characteristics. "C-C chimera" represents animals that received control transplants of these same regions between two individual chicken embryos. (*b*) (Top) Quail-to-chick chimeric animal delivering a crowing vocalization. (Bottom) The quail-chick cell marker: chick cells are on the left part of the image; quail cells have large dark spots and are on the right side of the image.**

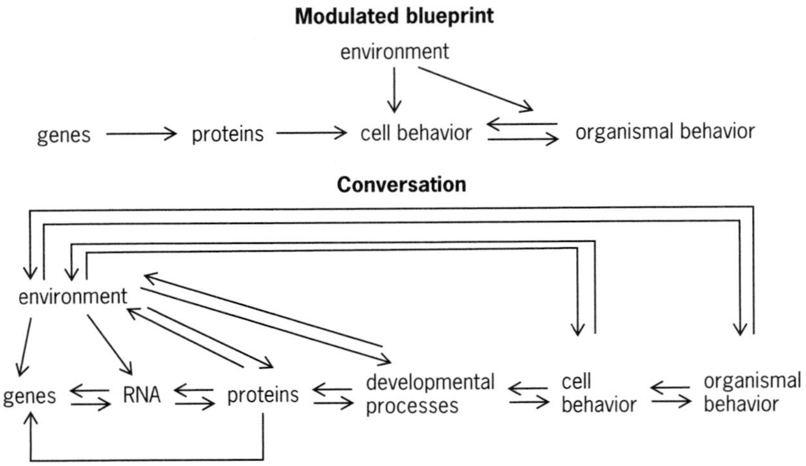

Fig. 2. Two different conceptions (modulated blueprint versus conversation) of the control of cellular decisions during brain development.

was performed. Killing one cell and leaving it attached to the other cell results in half an embryo, while separating the two cells results in a complete embryo of smaller size. The repercussions from this initial realization that embryonic cells make conditional decisions took a while to be realized. Many scientists believed such experiments simply demonstrated limited environmental modulation of mostly autonomous, preordained instructions (a modern version of this idea, the modulated blueprint, is shown in **Fig. 2**). During the past century, scientists have learned enough about the mechanisms of development to understand that the blueprint metaphor and the dichotomy of nature-versus-nurture that underlies it have been deeply misleading. Cell decisions are better visualized as "conversations" rather than blueprints because of the two-way nature of the processes that contribute to them (Fig. 2).

The special challenges to understanding the development of neural circuits were first suggested in the 1920s by Hilde Mangold and Hans Spemann, who examined the emergence of the embryonic brain. They showed that the fate of future brain cells depended on fleeting communication between two preexisting populations of cells during a process called neurulation (differentiation of nerve tissue and formation of a cylindrical embryonic structure called the neural tube). As a result of this earlier communication, one of these cell populations goes on to produce brain cells, while the other population eventually dies and is no longer present at the time of birth. Many cells in the developing brain make fleeting contacts with other cells that they have no contact with later on. We do not know which of these interactions may be important for the function of neural circuits that a particular cell will contribute toward. The chimera technique enables investigators to find embryonic cells that contribute to the function of particular neural circuits; their interactions with other cells can then be followed to study the way in which these cells are involved in building particular neural circuits.

Use of marked cells to distinguish donor and host tissues. Initial work on avian brain chimeras has relied on a naturally existing difference between chicken and quail cells first discovered by the French developmental biologist Nicole Le Douarin, and used for fate-mapping studies in embryos (Fig. 1*b*). For reasons that are still not well understood, the geometry of the arrangement of DNA in the cell nucleus is different in chickens and quails, and any staining method that colors DNA can be used to reveal the species identity of cells. The recent development of nontoxic cell marking substances, and genetic techniques for creating embryos with marked cells, allows chimeras to be made between any species. Birds are preferable for this work because of easy experimental access at a critical time in development (at this same stage, mammalian embryos are establishing connections to the mother's body that manipulations interfere with). Birds also have nervous systems of the same order of complexity as mammals, and exhibit many inborn species differences in both simple and complex behaviors.

Basics of brain switching. To create a brain chimera between two species, eggs are incubated for approximately 2 days to reach the state of embryonic development most conducive to transplants. At this stage, there are no blood vessels in the presumptive brain, and the basic regions of the future brain are laid out, but no cells have yet reached their developed state and are still fairly plastic (if moved they can turn into other kinds of cells than what they would have become if left in place). The eggs are removed from the incubator, turned on their side, and a 1-cm (0.4-in.) hole is cut in the shell to reveal the tiny white embryo on top of the yellow yolk (**Fig. 3**). A solution of sterilized dark food coloring is injected between the embryo and the yolk for visual contrast. An embryo of each species at the same state of development is selected; this can be gauged by the visual appearance of the embryos under a high-magnification surgical microscope. Tiny scalpels made of stainless steel wire are used to cut a small hole in protective membranes overlying the embryo, and the region of the presumptive brain (contained in the neural tube) is mechanically cut out of donor and host embryos using the microscalpels. The dissected fragment from the donor embryo is gently sucked up into a small glass tube filled with sterile fluid, and transferred into the host embryo, substituting for the same fragment removed from the host embryo. Because of the growth happening at this time, the fragment is completely reincorporated into the developing brain within a few hours, with no sign of any surgical intervention. At the end of the transplant, a piece of sterile wound-closure tape is used to close the hole in the egg, and the egg is returned to an incubator to undergo the rest of its development. Well-performed transplants yield a healthy chimeric animal with a morphologically normal brain. Since the brain now contains a population of foreign cells, the immune system of the host animal will reject these cells once cellular immunity is established at about 2 weeks after hatching,

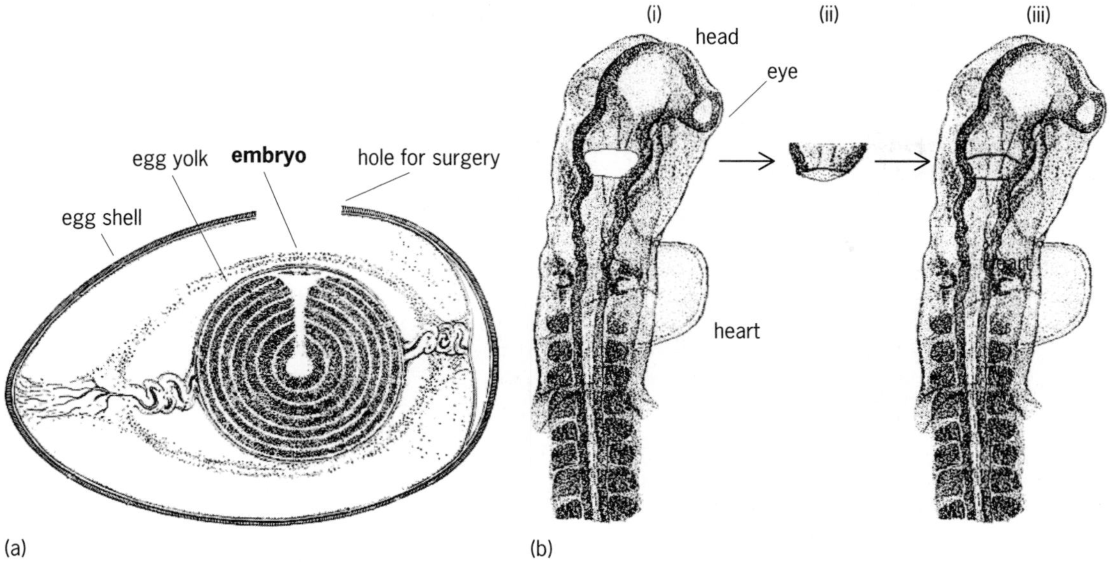

Fig. 3. (*a*) Diagram of a cross section through an incubated egg turned on its side, with a hole cut in the shell to give surgical access to the embryo. (*b*) Magnified view of an embryo at the time of surgery; the locations of the presumptive head, heart, and eye are indicated. (i) The region of the presumptive brain that is going to be transplanted is mechanically excised using a microscalpel from the donor and the host embryo, which are at the same stage of development and thus look very similar. The white area shows the excised region. (ii) The excised fragment from the donor is transferred to the host embryo. (iii) The donor fragment is placed in the incision inside the host embryo, substituting for the host fragment that was previously excised. Within several hours, it has healed seamlessly into place.

unless steps are taken to prevent such rejection from occurring.

These techniques have been used to successfully transplant species differences in a motor behavior (singing) between chickens and quails, and also to study neural circuits controlling inborn auditory perceptual differences between these two species. The technique has also been applied within species to study the etiology of epilepsy using transplants between normal and epileptic strains of chickens, and to study the development of male and female brain differences in quails. Current applications include studying brain differences implicated in vocal imitation behavior using transplants between birds that vocally imitate and those that do not, as well as studying brain mechanisms involved in migratory behavior, and the events that lead to brains first functioning as integrated systems before birth.

For background information *see* AVES; BRAIN; CELL LINEAGE; CHIMERA; DEVELOPMENTAL BIOLOGY; EMBRYONIC INDUCTION; FATE MAPS (EMBRYOLOGY); NERVOUS SYSTEM (VERTEBRATE); NEUROBIOLOGY; NEURULATION; TRANSPLANTATION BIOLOGY; VERTEBRATE BRAIN (EVOLUTION) in the McGraw-Hill Encyclopedia of Science & Technology. Evan Balaban

Bibliography. E. Balaban, Brain switching: Studying evolutionary behavioral changes in the context of individual brain development, *Int. J. Dev. Biol.*, 49:117–124, 2005; E. Balaban, Cognitive developmental biology: History, process and fortune's wheel, *Cognition*, 101:298–332, 2006; N. M. Le Douarin et al., Interspecific chimeras in avian embryos, *Methods Mol. Biol.*, 135:373–386, 2000; K. Long, G. Kennedy, and E. Balaban, Transferring an inborn auditory perceptual preference with interspecies brain transplants, *Proc. Natl. Acad. Sci. USA*, 98:5862–5867, 2001; M. A. Teillet, C. Ziller, and N. M. Le Douarin, Quail-chick chimeras, *Methods Mol. Biol.*, 97:305–318, 1999.

B chromosomes in plants

Each species of multicellular organisms has a set number of chromosomes that carry all genes required for development and living processes. However, some species in both the plant and animal kingdoms have extra chromosomes, called B chromosomes, which are dispensable. They only exist in some individuals of a species and often are variable in number. Variation in number of the normal chromosomes is highly detrimental, but B chromosomes seldom have any detrimental effects unless many copies are present. Thus, they are considered to be mostly inert, being neither required nor detrimental. They exist in populations because they have properties that foster their accumulation; normal chromosomes do not possess these properties. In some plant species, particularly corn, they have been used to make translocations with the normal chromosomes, which have proven to be useful for mutation mapping, chromosomal dosage studies, and recently production of engineered minichromosomes.

Origin and nature of accumulation. The origin of B chromosomes remains a mystery, although it is assumed that they are derived from a normal A chromosome at some point. Because B chromosomes have few if any active genes, inactivation processes must occur to generate these chromosomes. Distinct properties must evolve for the B chromosome to

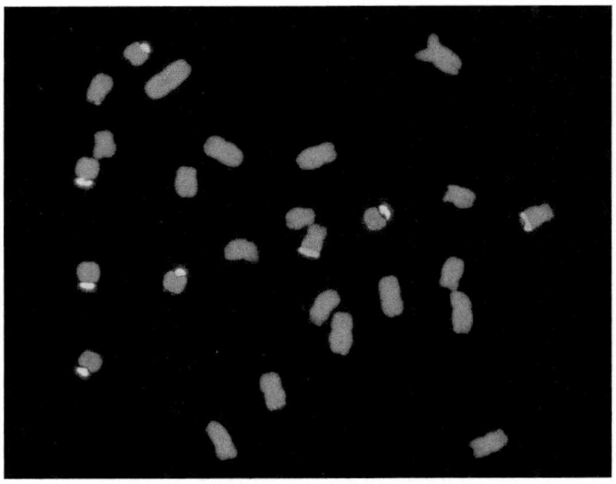

Fig. 1. A somatic chromosome spread of maize with six B chromosomes. The B chromosome-specific DNA repeat is represented in dark gray; the knob heterochromatin is designated by light gray. Chromosomes were stained in blue. (*Photo by Zhi Gao*)

accumulate and be maintained. Thus, it is likely that the mechanisms that initially produce B chromosomes are more common than the observed occurrence of these chromosomes. Nearly 2000 species of plants have been observed to harbor B chromosomes. B chromosomes are almost exclusively reported in diploid rather than polyploid species and in those with an outcrossing mating system as opposed to species with inbreeding.

The nature of accumulation mechanisms of various B chromosomes varies, but this usually involves the process of nondisjunction in which one daughter cell receives two copies of the chromosome and the other daughter cell receives none from a dividing mother cell with one B chromosome. This nondisjunction occurs at a time in the life cycle that will place more copies of the B chromosome into the next generation than were present in the previous generation. Thus, nondisjunction occurs in the developmental lineage leading to formation of gametes.

Corn. The B chromosome of corn is the most thoroughly studied (**Fig. 1**). There are 10 pairs of normal chromosomes in corn. The B chromosome is about two-thirds the length of the smallest normal chromosome and is highly heterochromatic; that is, it stains heavily with chromatin dyes, which is an indication of inactivity. The centromere of the corn B chromosome is near one end, which is different from all of the normal chromosomes. The basis of its accumulation mechanisms is that it nondisjoins at a high frequency during the second pollen mitosis that produces the two corn sperm (**Fig. 2**). The sperm that contains the two B chromosomes then preferentially fertilizes the egg cell rather than the polar nuclei during the process of double fertilization. Because the fertilized egg cell develops into the new generation and the fertilized polar nuclei develop into the dead-end storage tissue called the endosperm, this combination of traits increases the number of B chromosomes. The corn B chromosome divides normally in most other cell divisions in the life cycle. In most genetic backgrounds, 10–15 copies of the B chromosome are tolerated without any effect on the plant. The highest documented number of B chromosomes in corn was 34, which was associated with stunted growth and sterility.

The nature of the nondisjunction property has been studied in corn. The centromeric region is the site of nondisjunction, but other parts of the B chromosome must be present. One of these regions resides at the tip of the other end of the chromosome. When this site is removed, the centromere will behave in a normal manner without nondisjunction at the second pollen mitosis. If this part of the B chromosome is returned to the genotype via genetic crosses, the centromere will again exhibit nondisjunction. In other words, the terminal site is required to be in the same nucleus during the second pollen mitosis, but it need not be present on the same chromosome as the B centromeres that it affects.

The corn B chromosome centromere contains all normal DNA repeats that are typical of a centromere in corn. However, it also has a B-specific repeat that is interspersed throughout the centromeric region and a few other minor sites on the chromosome. There are at least two other repetitive DNA sequences that are found exclusively on the B chromosome. Other sequences on the B chromosome are shared with the normal chromosomes. The normal chromosomes contain many copies of mobile elements. Different families of these mobile elements have had previous periods of major transpositions dating back about 4–5 million years. All of these elements are also represented on the B chromosome, indicating that the B chromosome has been present in the evolutionary history of corn during these periods of transposition.

The B chromosome of corn has been capitalized upon by corn geneticists to develop many research tools. By irradiating corn materials with B chromosomes, translocations have been induced that place a part of a normal chromosome onto the B centromere. This confers the property of nondisjunction to the translocated chromosomal region. Thus, sperm are produced that are missing the chromosomal segment or that have it in duplicate. Upon fertilization, a dosage series for that portion of the chromosome is generated; thus, among the progeny,

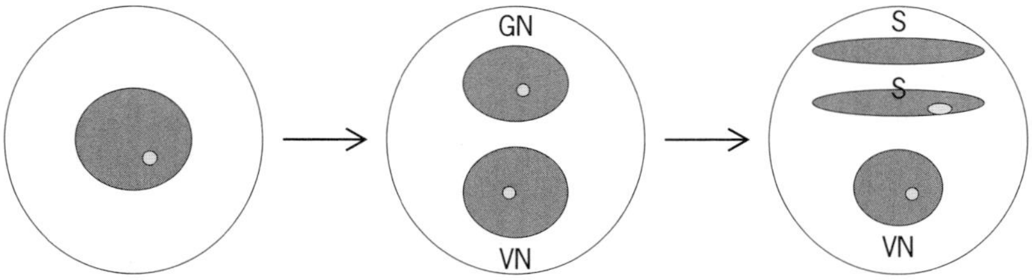

Fig. 2. Process of nondisjunction. At the left is depicted the microspore, the end product of male meiosis in corn. The color dot signifies a B chromosome. The nucleus divides during the first pollen mitosis to produce the generative nucleus (GN) and the vegetative nucleus (VN). During this division, the B chromosome is distributed to both nuclei. During the second pollen mitosis (at right), the two sperm (S) form. The B chromosome nondisjoins so that one sperm has two copies adhered together and the other sperm has no copies of the B chromosome. (*Diagram by Fangpu Han*)

there are individuals with only one copy, some with two copies (in those rare cases in which the B centromere disjoined), and some with three copies of the translocated segment. Recessive mutations in the maternal parent of such crosses are exposed in the progeny that do not inherit the translocated segment. In this way, recessive mutations can be located to the chromosome arm by crossing these materials by a collection of B-A translocations covering the whole genome.

Rye. The B chromosome of rye has also been studied extensively. In this case, the unusual behavior occurs at the first pollen mitosis with a high frequency of directed nondisjunction of the B chromosomes to the generative nucleus, which divides again to produce the two sperm. This behavior generates sperm with a greater number of B chromosomes than the parent plant. There is also directed nondisjunction in the female gametophyte such that the egg cell receives a greater number of B chromosomes. Nondisjunction appears to be mediated by sequences surrounding the centromere, but also requires presence of a heterochromatic region at the tip of the rye B chromosome long arm. This region can act on the centromeric regions of other rye B chromosomes and need not be present on the same chromosome to function in this capacity.

Outlook. B chromosomes in different species often affect the rate of recombination in the normal chromosomes during meiosis. This effect in different species can be positive or negative, but it usually increases in magnitude in relation to copy number of B chromosomes in the cell.

A recent application of B chromosomes is to convert them into artificial chromosome platforms. By introducing the cloned natural ends of chromosomes called telomeres, truncations are produced that remove the terminal end of the chromosome that conditions nondisjunction. At the same time, the truncating sequences are accompanied by recombination sites that will permit further additions to the chromosome. These constructs would permit further additions to the chromosome in a sequential manner so that many new genes could be added to a plant as a newly designed chromosome.

For background information *see* BREEDING (PLANT); CELL CYCLE; CHROMOSOME; CHROMOSOME ABERRATION; CORN; GENE; GENETIC MAPPING; MITOSIS; RYE in the McGraw-Hill Encyclopedia of Science & Technology. James A. Birchler

Bibliography. W. R. Carlson, The B chromosome of corn, *Annu. Rev. Genet.*, 12:5–23, 1978; N. Jones and A. Houben, B chromosomes in plants: Escapees from the A chromosome genome?, *Trends Plant Sci.*, 8:417–423, 2003; N. Jones and H. Rees, *B Chromosomes*, Academic Press, London, 1982; H. Roman, Mitotic nondisjunction in the case of interchanges involving the B-type chromosome in maize, *Genetics*, 32:391–409, 1947; W. Yu et al., Construction and behavior of engineered minichromosomes in maize, *Proc. Natl. Acad. Sci. USA*, 104:8924–8929, 2007.

Barcoding life

In the original science fiction *Star Trek* series on television in the 1960s, the crew of the starship *Enterprise* used a handy little device called a "tricorder" to remotely detect life forms and provide the crew with information about the identity of the organism and if it was dangerous. As is often the case, what was science fiction can several decades later become reality. Although a device such as a tricorder is not yet a reality, it is within reach, and many scientists around the world are working hard to make genetic (DNA) identification via DNA barcodes a reality. Their goal is to produce a handheld device that is the size of a mobile phone that will process DNA and produce the sequence of a small region (or regions, depending on the type of organism; the genetic "barcode") that can be matched to a database (remotely accessed) so that the identity of the organism of interest can be determined. Once a determination has been made and fed back to the device, other information can be downloaded as well, such as images of the organism and any other information that the increasing numerous online databases can provide.

Taxonomy. It took over 2 centuries for scientists to describe 1.7 million species, but this number is grossly less than the number of species that exist. For example, perhaps only 10% of the species of fungi have been named and described by scientists. Not every taxonomist (a scientist who describes and identifies organisms) can identify every organism on the planet; each taxonomist has his or her group of specialization. If you (as a member of the public) find a plant or animal and want to know what species it is, you have to find an appropriate book in which to look it up or find a taxonomist who specializes in that organism. This could be a complicated process if you are unfamiliar with the resources available in the area in which you are located. With a little experience and some work, though, most people can manage to find out at least the general type of organism it is, if not obtain a species name for it. However, if you find only a biological fragment and need to know from what it came, you are almost certainly out of luck. A DNA barcoder could solve all of these problems in a straightforward manner. All you would need to do is to have a small piece of tissue (the size of a pinhead), and the rest would be up to the machine and the online databases. You would not need to know anything about local resources or how to find the person who specializes in that organism to get it identified.

Of course, there are some organisms that you would not feel comfortable about approaching to obtain a small piece of their tissue—a grizzly bear would be one example and a black mamba viper another—but a hair or a piece of shed skin would be adequate. Obviously, most North Americans would know a grizzly bear when they saw one and many Africans would know a black mamba, so identifying such creatures with a DNA barcoder would be unnecessary. Both species are members of the charismatic megafauna (that is, large animals with widespread popular appeal) and are therefore relatively easily identified without recourse to field guides or a taxonomist. However, what about the great majority of plants, beetles, spiders, and nematodes? Most people when faced with these organisms just walk away in frustration—it is simply too much effort to find out what is in your hand. Realistically though, if the organism is an alien introduced pest species, it might be extremely useful to find out in timely manner that you have discovered a pest species in an area in which it had never before been found. This could lead to an invasive species being stopped before it became established. Hence, although most people do not realize it, biological literacy (bioliteracy) could be an immense asset to nearly everyone.

Reference databases. Identification in a timely manner, and linkage of species names to vast online databases, could revolutionize how the average person views the world around them. Species names are the key to unlocking these databases; once you can accurately name an organism, a world of information could be at your fingertips. A DNA barcoder could thus make it possible for everyone to read this "book of life," a revolutionary prospect.

The first thing needed is a reference database of DNA sequences from already identified organisms. This is the current stage of barcoding: collection of the reference sequences against which the DNA of unknowns can be assessed to determine which species they are. This involves sequencing one or a few DNA regions that have been selected because they have high powers of discrimination; by comparing sequences, we find that each has a unique suite of the four component DNA bases (adenine, thymine, guanine, and cytosine) that permits us to know which one we have in front of us. In most groups of animals and marine plants (algae), a portion of the *coxI* (cytochrome oxidase I) gene in the mitochondrial genome (mitochondria are organelles in which respiration takes place—the powerhouses of the cell) has been selected as the part of the genome to act as the single genetic barcode. In land plants and many groups of fungi, the *coxI* gene is not suitable (because of too little variation) and other DNA markers have been selected instead. For example, in land plants, three or four short regions of plastid DNA (DNA in chloroplasts and chromoplasts, the pigmented organelles in plant cells) have been found to serve the purpose. In many groups of organisms, the within-species variability is much less than the between-species variation, which makes it possible to clearly distinguish between species.

Many projects are now under way to produce the reference databases that link genetic barcodes with species names. This linkage of names with databases is essential to the process. Then, once the reference database of known species has been assembled, it enables many other things—for example: (1) making taxonomic products such as descriptions and distributions of species available to the general community by creating standardized and high-tech tools for use in medicine (human and animal parasites and pests), agriculture (invasive species, weeds, insect pests), and commerce (controlling trade in endangered animals, monitoring fisheries, determining biocontaminants in food, etc.); (2) linkage of different life stages of the same species (for example, caterpillars to moths and butterflies, seeds to adult plants); (3) identification of fragments found at crime scenes (forensics); (4) elimination of the enormous burden that carrying out identifications places on taxonomists so that instead they can focus their energies on species discovery and enriching the databases with biological attributes that make identifications more useful; and (5) exploration of species limits. It is this last use that has been among the most controversial for DNA barcoding enthusiasts, but it is clear from early results that extensive use of DNA sequencing of standardized regions leads to identification of new species that had previously been overlooked and assumed to be members of another, highly similar species. In work in Canada on marine brown and red algae, it has been determined that there are roughly 30% more species present than had previously been thought, including totally new entities. What had been thought of as a single morphologically variable species is neatly split by the genetic information into two (and sometimes more) types, which when examined from this perspective are each diagnosable by easily observed features that were previously overlooked. This sort of result has been obtained in group after group and helps to demonstrate that DNA barcoding is both practical and highly useful.

Future research. Once the reference database is well populated with DNA sequences so that identifications of most species are possible, the next stages all involve development of new technologies. The reference database itself will involve a great deal of technological improvement, but it relies first and foremost upon the scaling up of already existing technologies and automation to make feasible the task of producing the reference sequence(s) for the currently known 1.7 million organisms, which is a huge task in and of itself. This phase will require a large investment of time, effort, and funds if it is to be achieved. After this mammoth task has been accomplished, then rapid, small-scale, microchip-based DNA sequencers need to be developed so that handheld devices become practical. Such technology may sound unlikely and difficult to develop, but recent efforts in this direction, under way for reasons other than DNA barcoders, show great promise to deliver this capacity. The marriage of this new DNA sequencing technology in a handheld format to the reference databases of DNA sequences produced with current technology will enable the creation of a bioliteracy tool for use by the general public and taxonomists.

There are many hurdles that remain to be overcome before DNA barcoding becomes a reality; some of these are technological, and others are theoretical. There are still many scientists who doubt that DNA barcoding will work, whereas others fear that efforts to develop DNA barcoding will siphon money and resources away from the already poorly funded researchers and institutes involved in documentation of the Earth's biodiversity. The Canadian government is exceptional in the extent to which they have invested in the development of DNA barcoding, and the early results of these projects have been uniformly exciting. It now seems clear to many people that DNA barcoding has the capacity to help us document the great diversity of organisms on the Earth. This could be the start of one of the greatest periods of biodiversity science.

For background information *see* BIODIVERSITY; CLASSIFICATION, BIOLOGICAL; DEOXYRIBONUCLEIC ACID (DNA); GENETIC CODE; GENETICS; SPECIATION; SPECIES CONCEPT; TAXON; TAXONOMIC CATEGORIES; TAXONOMY; TYPE METHOD; ZOOLOGICAL NOMENCLATURE in the McGraw-Hill Encyclopedia of Science & Technology. Mark W. Chase

Bibliography. P. D. N. Hebert et al., Biological identifications through DNA barcodes, *Proc. Roy. Soc. B*, 270:313–321, 2003; V. Savolainen et al. (eds.), *DNA Barcoding of Life*, Theme Issue, *Phil. Trans. Roy. Soc.*, 360:1803–1980, 2005; V. Savolainen et al., Towards writing the encyclopaedia of life: An introduction to barcoding, *Phil. Trans. Roy. Soc.*, 360:1805–1811, 2005; D. Tautz et al., DNA points the way ahead in taxonomy—in assessing new approaches, it's time for DNA's unique contribution to take a central role, *Nature*, 418:479, 2002.

Biocontrol of plant diseases with biofungicides

Plant diseases significantly limit agricultural production, destroying or spoiling between 10 and 30% of all food and feed produced annually. The majority of plant diseases are caused by several dozen genera of fungi and funguslike oomycetes (filamentous protists such as downy mildews and water molds). Most of these plant pathogens are opportunistic, attacking living plant tissue under conducive conditions but otherwise surviving on plant residues and soil organic matter. Thus, while the relative abundance and distribution of these pathogens vary from place to place, all natural soils contain some plant pathogens, and all plants grown in such soils are infected to some degree. Most plants do not show significant symptoms of disease; however, local epidemics can occur at the field scale, greatly reducing farm productivity and profitability. For this reason, commercial growers apply fungicides (agents that kill or destroy fungi) and other disease control measures to protect their crops.

In the past few years, an increased awareness of the risks and hazards of conventional pesticides and the growing demand for certified organic food have led to increased demand for biopesticide inputs, including biofungicides. Biofungicides include microbes, their metabolites, and other naturally occurring chemicals that can be formulated and applied to suppress a wide variety of plant diseases caused by fungi and oomycetes. Initial development focused on microbial antagonists of fungal diseases, particularly those for which chemical control measures were either unavailable or uneconomical. More recently, microbial fermentations have been viewed as more commercially viable sources of biofungicides that can be readily substituted for more costly and/or hazardous chemicals in conventional disease management programs.

TABLE 1. Types of microbial activities leading to biocontrol of plant diseases caused by fungi and oomycetes

Mechanism	Examples
Hyperparasitism/predation of pathogens	Lytic and some nonlytic mycoviruses *Ampelomyces quisqualis* *Gliocladium virens*
Antibiotic secretion	Polyketides Cyclic lipopeptides
Lytic enzyme secretion	Chitinases Glucanases Proteases
Secretion of metabolic waste products	Ammonia Carbon dioxide Hydrogen cyanide
Physical–chemical interference of pathogen growth	Blockage of soil pores Germination signals consumption Molecular cross-talk confused
Competition with pathogens	Exudate and leachate consumption Siderophore scavenging Physical niche occupation
Improved plant fertility and vigor	Symbiotic nitrogen fixation Associative nitrogen fixation Phosphate solubilization
Induction of plant immunity to disease	Fungal cell wall fragments Pathogen-associated, molecular patterns Phytohormone-mediated induction

Natural plant disease suppression and biofungicides. Diverse soil- and plant-associated microbes can contribute to plant disease suppression through a variety of direct and indirect mechanisms (**Table 1**). Furthermore, individual microbial populations can act to suppress fungal diseases through multiple mechanisms. However, the relative importance of each population or mechanism is difficult to assess in situ because the relative abundance of different soil- and plant-associated microbes can change rapidly over time. Among the biotic factors that are most important for plant disease suppression are the microbial antagonists and their secretions. Potential antagonists that display antibiotic activity in vitro are readily recovered from soils and plant tissues; however, the conditions under which antagonistic activities are expressed in situ are not often apparent. That is one reason why more effort is now being put into identifying novel metabolites that may serve as active ingredients in biochemical mixtures to be marketed as biofungicides. Recent efforts also have been made to develop biofungicide products based on microbial components, such as harpin proteins (a class of proteins produced in nature by certain bacterial plant pathogens), that induce plant host defenses. It is

TABLE 2. Several biofungicides currently marketed in the United States*

Type	Active ingredient	Registrant	Targets
Live microbe	*Aspergillus flavus* AF36	AZ Cotton Research and Protection Council	Aflatoxin-producing strains that infect cotton
	Bacillus licheniformis SB3086	Novozymes Biologicals	Broad spectrum
	B. subtilis MBI00	Becker Underwood	Broad spectrum
	B. pumilus GB34	Bayer Crop Science	Broad spectrum
	Gliocladium catenulatum J1446	Verder Oy	Broad spectrum
	G. virens GL-21	Certis USA	Broad spectrum
	Monopterus albus QST20799	Agraquest	Broad spectrum
	Pseudomonas fluorescens A506	Frost Technology Corporation	Fruit infections promoted by frost damage
	Streptomyces lydicus WYEC 108	Natural Industries	Powdery mildew in greenhouse crops
	Trichoderma harzianum T-22	BioWorks	Broad spectrum
Nonliving	Chitosan hydrolysate	Plant Defense Boosters	Broad spectrum
	Harpin protein	Eden Bioscience	Broad spectrum
	Hydrogen peroxide	Biosafe Systems	Broad spectrum
	Phosphorous acid	Multiple	Downy mildew
	Potassium bicarbonate	Multiple	Broad spectrum
	Saponins from *Chenopodium quinoa*	Heads Up Plant Protectants	Broad spectrum

*A similar number of biopesticides targeting bacterial, nematode, and viral diseases are also registered.

expected that more such compounds will be developed into commercial products as the molecular basis for induced systemic resistance pathways is more fully elucidated.

Living biofungicide agents. Individual microbial populations are considered specific biocontrol agents of disease suppression when they can measurably reduce disease incidence and/or severity when present or added in sufficient densities. Nearly 100 species of microbes with biocontrol activities against plant pathogenic fungi have been described. Among bacteria, specific strains from the genera *Bacillus*, *Lysobacter*, *Pseudomonas*, and *Streptomyces* have been well studied, and some have been developed into products for commercial use. Among fungi, isolates belonging to the genera *Ampelomyces*, *Coniothyrium*, *Gliocladium*, and *Trichoderma* have been most widely studied and commercialized. Additionally, the structure and general activity level of other microbial populations living in and around plants can contribute to general disease suppression, primarily through competition for nutrients and niche space. Interestingly, nonpathogenic strains of common plant pathogenic fungi, for example, *Fusarium oxysporum* and *Aspergillus flavus*, also have been found to be useful for the control of plant pathogenic fungi. Some microbes have been shown to systemically induce one or more pathways of the immune system of plants so as to limit infection by one or more pathogens regardless of the point of infection. These microbes include some of the plant-growth-promoting rhizobacteria (root colonizing bacteria) and some mycorrhizae (which involve mutualistic relationships between fungi and plant roots); however, these types of microbes are not classified as biofungicides for regulatory purposes because their primary benefits to plant health are not presumed to be related to disease suppression. Thus, diverse microbes can contribute significantly to the biological control of plant diseases caused by fungi. These microbes and their metabolites represent a vast reservoir of potential biofungicidal agents, and they are the focus of much of the current bioprospecting for new active ingredients.

Nonliving biofungicides and related compounds. Other nonliving agents that suppress plant diseases have also been classified for regulatory purposes as biofungicides. These include hydrogen peroxide, potassium bicarbonate, and phosphorous acid, all of which have desirable properties as contact fungicides, but they pose no significant environmental contamination issues because they are chemically unstable and break down into innocuous compounds following application. Chitin (a polysaccharide that forms the hard outer integuments of crustaceans, insects, and other invertebrates) and chitosan (a deacylated derivative of chitin) act to promote the growth of microbial antagonists of fungi and insects, as well as to stimulate plant host defenses. Some plant extracts, which can contain a variety of biocides such as the saponins (plant glycosides used as biological detergents), have also been registered as biofungicides. Other regulated compounds, such as gamma butyric acid, several phytohormones (plant hormones), and yeast extract, can affect plant health through a variety of mechanisms, including improved nutrition and/or stimulation of plant immunity to one or more diseases. Other nonregulated materials, such as compost teas, may provide similar benefits to plant health.

Production and commercialization. Currently, biofungicides represent a small but rapidly growing segment of the agricultural input market. In 2007, nearly two dozen active ingredients were registered with the U.S. Environmental Protection Agency for use as biofungicides, although not all of these are currently available to consumers. Those that are available include chemicals with limited toxicity to humans, as well as formulations of living microbes (**Table 2**). Existing products are labeled for use on a variety of crops, especially high-value fruits, vegetables, turf, and nursery plants where current fungicidal applications are both frequent and costly. Several products are also labeled for use as seed treatments for the control of soil-borne pathogens that kill seed and seedlings. Small- and medium-sized companies in over a dozen different countries are responsible for the manufacture of most biofungicides. However, following the recent surge in market demand, several large multinationals are once again taking an interest in the field. Furthermore, the number and variety of biofungicides registered and available for use are expected to increase significantly over the next decade.

Several companies have active research and development programs aimed at expanding product offerings for a variety of markets. These new efforts use modern molecular biology techniques to accelerate and direct selection of candidate microbes prior to the application of more traditional biochemical and biological screening methods that are needed to determine efficacy under field conditions. Particular attention during this screening process is paid to the range of activities of novel agents because broad-spectrum biofungicides are expected to have the greatest value in the marketplace. For living agents, additional attention must be paid to the colonization capacities on the crop(s) to which they are applied, as well as the environmental conditions that promote their activities. The use of mixtures of active ingredients is also of interest to enhance stability and increase efficacy over a range of environmental conditions. In most cases, manufacturers are working toward producing formulations that can be used in both conventional and certified organic production.

For background information *see* AGRICULTURAL SCIENCE (PLANT); AGROECOSYSTEM; FUNGAL BIOTECHNOLOGY; FUNGI; FUNGISTAT AND FUNGICIDE; MICROBIAL ECOLOGY; MYCOLOGY; PESTICIDE; PLANT PATHOLOGY in the McGraw-Hill Encyclopedia of Science & Technology.

Brian B. McSpadden Gardener

Bibliography. S. Gnanamanickam (ed.), *Biological Control of Crop Diseases*, Marcel Dekker, New York, 2002; D. Haas and G. Defago, Biological control of soil-borne pathogens by fluorescent pseudomonads, *Nat. Rev. Microbiol.*, 3:307–319, 2005; G. E. Harman et al., *Trichoderma* species, *Nat. Rev. Microbiol.*, 2:43–56, 2004; B. B. McSpadden Gardener and D. R. Fravel, Biological control of plant pathogens: Research, commercialization, and application in the USA, *Plant Health Progress*, May 2002.

Biodegradable materials for tissue engineering

Tissue engineering involves the use of biomaterials, transplanted cell populations, and molecular signals in the regeneration of diseased or damaged tissues. As a key component in the overall strategy, tissue engineering scaffolds are being extensively researched to repair damaged tissue and promote healing. Scaffolds are typically fabricated in the form of a biological matrix or material and have been used for a variety of biomedical applications, including defect repair, tissue healing, drug delivery, and cell transplantation. Scaffolds can be either nonbiodegradable or biodegradable, and both types have been studied for tissue engineering applications. Because biodegradable scaffolds may be absorbed within the body, they have attracted significant interest since further surgery is not required to remove the scaffold after the initial implantation surgery. The two broad classes of biodegradable scaffolds as defined by the source of the material are naturally derived and synthetically fabricated. Natural biomaterials, such as collagen, chitosan, hyaluronic acid, elastin, and gelatin, have been studied for liver, nerve, bone, and cardiac tissue engineering applications. However, it is difficult to control the physical and chemical properties of natural biomaterials, and this often limits the applications. As a result, many researchers have focused on synthetic biomaterials. The physical and chemical properties of synthetic biomaterials can easily be modified, and may be repetitively produced in similar batches. Commonly studied synthetic, hydrolytically degradable biomaterials are glycolic acid derivatives, lactic acid derivatives, and polyester derivatives. Upon degradation, these biomaterials degrade in the body and release acidic products. Unfortunately, the acidic degradation products have been known to increase inflammation in the surrounding tissue area. Because of this concern, investigators have attempted to create synthetically derived biomaterials with nonacidic degradation products. One group of recently developed synthetic, degradable biomaterials is based upon a cyclic acetal unit. Recent efforts have attempted to develop cyclic-acetal–based degradable biomaterials for tissue engineering and drug delivery applications.

Cyclic acetal. Cyclic acetal biomaterials consist of a six-member ring unit that hydrolytically degrades into carbonyl and diol degradation products. Initial studies have focused on the chemical, physical, and biological properties of the cyclic acetal biomaterial based upon the monomer 5-ethyl-5-(hydroxymethyl)-β,β-dimethyl-1,3-dioxane-2-ethanol diacrylate (EHD; structure **1**). Solid, rigid EH

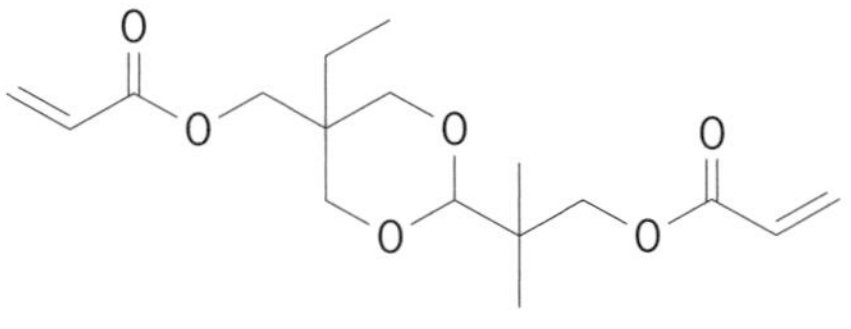

(1)

networks (structure **2**) may be formed by radical polymerization of the diacrylate monomer, using an initiator such as benzoyl peroxide in conjunction with the accelerant *N*,*N*-dimethyl-*p*-toluidine and an acetone diluent. Recent studies have focused on the effects of initiator, accelerant, and diluent content on the physical properties of the polymerized EH networks. Results showed that EH network gelation time varied between 33.3 and 193.9 s, with the gelation time decreasing with increased initiator content. Maximum reaction temperature

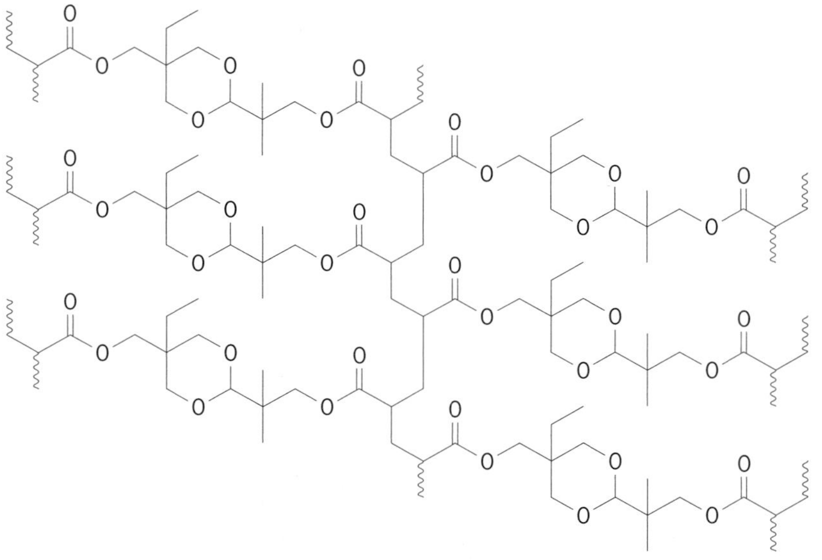

EH (cyclic acetal) network

(2)

also increases from 31.9 to 109.0°C (89.4 to 228.2°F) with an increase in initiator content. Overall, the results indicated that initiator and accelerant had the greatest effect on the rate of the polymerization reaction, as demonstrated by gelation time and maximum reaction temperature. As EH networks are hydrophobic, they do not swell in water. However, swelling in organic solvent can be used to describe network formation. Results showed that EH network swelling varied between 29.9 and 48.3%, while network sol fraction varied from 22 to 45%. Here, the results demonstrated that diluent content had the greatest effect upon swelling degree and sol fraction, and therefore most significantly affecting the extent of the network forming reaction. In addition, recent studies examined whether EH networks could support the adhesion and proliferation of a viable cell population. Results indicated that EH networks could indeed support the adhesion and viability of rat bone marrow stromal cells. There was a significant difference in bone marrow stromal cell viability between all experimental groups and a tissue culture polystyrene control at 4 h. However, viability at 8 h was comparable to the control for EH networks containing high amounts of initiator and diluent. Overall, the results indicated that EH networks may be fabricated with controllable physical properties, and these networks can support a viable bone marrow stromal cell population.

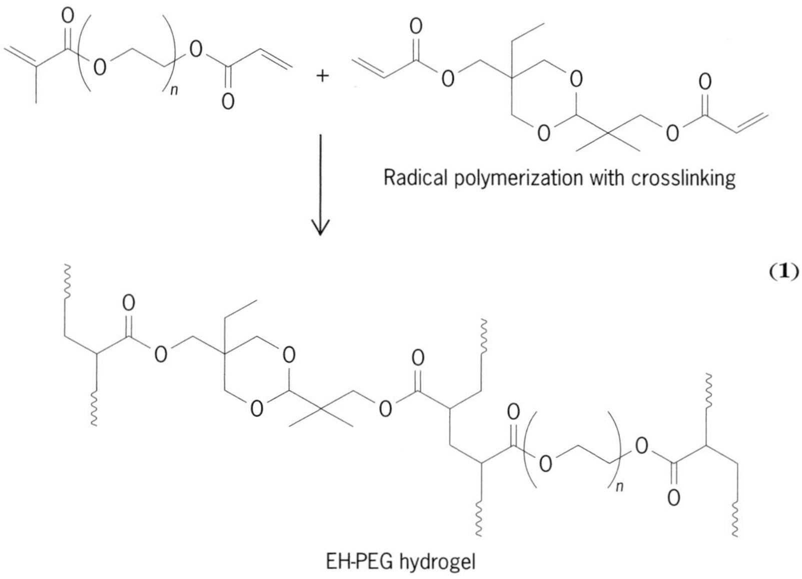

(1)

EH-PEG hydrogels. In contrast to the rigid plastic formed by the radical polymerization of EHD only, cyclic-acetal-based hydrogels may be formed by incorporating poly(ethylene glycol) [PEG] within into the network polymerization reaction. EHD-PEG hydrogels are fabricated by radical polymerization of EHD with poly(ethylene glycol)diacrylate (PEGDA) in aqueous solution [reaction (1)]. Recent studies examined the effect of a hydrophobic monomer, such as the EHD monomer, on the ability to fabricate hydrogels as well as the resulting properties of these hydrogels. The physicochemical properties evaluated included swelling degree, sol fraction, and contact angles. Results showed that the EH-PEG hydrogel swelling degree was notably influenced by the total monomer concentration, with swelling degree increasing as the monomer concentration decreased. Here the results indicated that a solution with low monomer concentration contained a large amount of solvent, allowing the monomers to form a loosely crosslinked network with an increased swelling degree. Further, the results demonstrated that the random order of two monomers, as occurs when their molar contents are similar, allows for the formation of a loosely crosslinked network capable of retaining increased amounts of water. Initiator concentration did not appear to influence the swelling degree of these EH-PEG hydrogels. The results of the sol fraction studies largely reflect the trends observed in the swelling degree studies. One expected observation was confirmed, namely, that a low initiator concentration does not produce a sufficient amount of radicals to propagate thorough crosslinking reactions, leaving many monomers unreacted and a higher sol fraction. A study of water contact angle was done to examine the hydrophilicity of the surface of the EH-PEG hydrogels. The contact angles of the EH-PEG hydrogels were relatively high when the ratio of EHD was high ($71.2 \pm 1.7°$ for EHD/PEGDA = 10.0). However, contact angles decreased as the ratio of EHD decreased. These results followed the expectation that the water contact angle would decrease as the ratio of PEGDA increases, indicating that the addition of PEGDA strongly influenced the hydrophilicity of the material because of its hydrophilic polymer main chain. The range of contact-angle values is within the range 50–75°, where cell adhesion is generally thought to be promoted.

Cell embedding within EH-PEG hydrogels. The biological applications of EH-PEG hydrogels have also been considered. Engineered bone grafts are often fabricated by embedding bone marrow stromal cells within a hydrogel scaffold; however, the components of the embedding system may be cytotoxic. Recent work investigated the efficacy of the EH-PEG hydrogel system for cell embedding. Specifically, this work examined the effect of radical initiators on the viability and metabolic activity of bone marrow stromal cells in monolayer, the ability of the bone marrow stromal cells to osteodifferentiate after initiator exposure, and the viability of bone marrow stromal cells embedded in the EH-PEG hydrogels. EH-PEG hydrogels were fabricated using the water-soluble redox, radical initiation system of ammonium persulfate and N,N,N′,N′-tetramethylethylenediamine. To assess the effect of the initiator system on metabolic activity, bone marrow stromal cells were cultured with the initiators at various concentrations for up to 3 h. At the defined time points, the bone marrow stromal cells were examined using a standard toxicology kit. Results indicate similar levels of activity between the experimental and control groups at early times, and decreased activity for the highly concentrated group. Next, differentiation of bone marrow stromal cells was examined by exposure to the initiator system for 1 min, culture in complete media for 2 days, and then culture in osteogenic media for 4 and 8 days. Bone marrow stromal cells were then isolated and assayed for expression of alkaline phosphatase, an early osteoblastic marker, using an alkaline phosphatase detection kit. Results indicate that exposure to low concentrations of the initiator system does not affect the ability of the cell population to osteodifferentiate. Lastly, bone marrow stromal cells were embedded in EH-PEG hydrogels and cultured in media for 7 days. Each day the cells were analyzed using a fluorescent viability assay. Results qualitatively showed that the majority of the bone marrow stromal cell population was viable at the examined time points. This work indicates that the proposed EH-PEG hydrogel system is a viable approach for bone marrow stromal cell embedding.

Outlook. It is clear from the studies that cyclic acetal biomaterials can be processed, modified, and applied for tissue engineering applications. Advantages of this material include its biodegradation, ease of fabrication, and favorable cell interactions.

For background information *see* BIOCHEMICAL ENGINEERING; BIOMEDICAL CHEMICAL ENGINEERING; BONE; POLY(ETHYLENE GLYCOL); POLYMER; SOL-GEL PROCESS; SURFACE TENSION in the McGraw-Hill Encyclopedia of Science & Technology.

Minal Patel; John P. Fisher

Bibliography. M. W. Betz et al., Cyclic acetal hydrogel system for bone marrow stromal cell encapsulation and osteodifferentiation, *J. Biomed. Mater. Res., Pt. A*, in press; S. Kaihara, S. Matsumara, and J. P. Fisher, Synthesis and characterization of cyclic acetal based degradable hydrogels, *Eur. J. Pharm. Biopharma.*, in press; J. L. Moreau, D. Kesselman, and J. P. Fisher, Synthesis and properties of cyclic acetal biomaterials, *J. Biomed. Mater. Res., Pt. A*, 81:594–602, 2007; L. S. Nair and C. T. Laurencin, Biodegradable polymers as biomaterials, *Prog. Polym. Sci.*, 32(8–9):762–798, 2007; F. H. Silver and D. L. Christiansen, *Biomater. Sci. Biocomp.*, 1999.

Biological diversity

Biological diversity, or biodiversity, can be defined simply as the totality and diversity of life at different levels of biological organization. The tremendous diversity in the manifestation of life is the very basis of biological knowledge and existence. Therefore, it is extraordinarily important to all people, including the scientists who study it. The word "biodiversity," first used in the 1980s, has become ubiquitous in both scientific and common usage (including politics) and is often associated with concerns regarding the natural environment and its conservation.

Global biodiversity. Biodiversity is the inevitable consequence of evolution. The first concrete evidence of life on Earth dates from 3.5 billion years ago. From a bacterium-like ancestor, life on Earth diversified into the variety of organisms we know today. The number of described species of bacteria, plants, animals, fungi, and protists on Earth today is estimated to be 1.5 million, but many species are not yet described; hence, this number is likely to be much higher (estimates vary from 3 to 100 million species). Biodiversity is distributed unevenly across different groups of organisms. For example, in the order Primata (including humans, apes, monkeys, and lemurs) there are fewer than 400 species, whereas in the order Coleoptera (beetles) there are about 350,000 species. Among the most diverse groups of organisms on Earth are bacteria (particularly heterotrophic eubacteria), ascomycete and basidiomycete fungi, orchids, mollusks, insects, arachnids, and crustaceans.

Biodiversity refers both to the number of biological categories and to variation in occurrences of these categories in particular contexts, such as communities and biomes. Species diversity is the variety of species within a habitat or a region. Ecosystem diversity is the variety of ecosystems in a given place. Biodiversity is distributed unevenly across the Earth. It is higher in the tropics and lower in colder areas. The number of species in a particular place depends on factors such as latitude, altitude, area, productivity, and habitat diversity. Tropical rainforests and coral reefs are among the most diverse biomes on Earth. Biodiversity includes measures of species diversity at a local scale (alpha diversity), as well as variation in biological composition among habitats (beta diversity) and landscapes (gamma diversity).

Diversity occurs at different levels of biological organization (for example, genes, individuals, populations, species, communities, and biomes). Genetic diversity is the variety of genes within and across species. Each species consists of individuals that have their own particular genetic composition, so different populations of a species may have different genetic compositions. Different species are expected to differ genetically, so genetic diversity is expected to stretch across all life. To classify this diversity of organisms, scientists use phylogenetic trees, which show the history of descent of a group of species from their common ancestor. The diversity of organisms results from speciation of ancestors. The current uneven distribution of biodiversity across different groups represents the sum total of speciation and extinction events that have occurred throughout the history of life on Earth.

Species interact with each other and with the environment in diverse ways. Almost all organisms on Earth depend on organisms that they interact with for survival and reproduction. Such interactions are an important aspect of the evolutionary forces that lead to diversification and increase in numbers of species on Earth and the wide range of communities that exist today. The uneven distribution of biodiversity on Earth is a function of the diversity of communities and ecosystems that result from varied interactions of organisms with the environment.

Value. The value of biodiversity for humans may be intrinsic (esthetic and religious value; leisure-time activities) or utilitarian. It is the utilitarian value that receives the most attention. For instance, multifarious bioactive chemicals (with biological effects) are produced by living organisms, especially microorganisms, marine invertebrates, and plants. Human beings all around the world have long relied on such organisms as sources of traditional medicines. Today, pharmaceutical companies use extracts of such organisms to manufacture many drugs. The potential for the discovery of new drugs is enormous because only a small fraction of living species has been surveyed for bioactive substances. Biodiversity can be a valuable source of new food to humans as well. At this time, humans consume just a fraction of plant and animal species. Biodiversity is also an incredible source of different kinds of industrial materials, including, for example, fibers, dyes, resins, adhesives, and oils. Plant products used by humans (for example, quinine) usually are specific to particular plants (*Cinchona*) and are the evolutionary result of interactions of that plant in a particular context (South American Andes). As such, the diversity of useful products is a direct result of evolutionary and ecological diversification.

Biodiversity provides several ecological services, such as nutrient recycling and regulation of atmospheric chemistry and water supply. Many scientists consider that the extent of biodiversity determines the extent of stability of ecosystems. The current trend toward homogenization adversely affects the functioning of ecosystems and the ecological services that they provide to humans.

Homogenization of the biosphere, and the consequent threat to biodiversity, is a by-product of more general forces within human society that push it toward increased standardization and uniformity. Technological advances have increased homogeneity in human society at cultural, institutional, and biological levels. Thus, on one hand, uniformity is a desirable quality for exploitation at the industrial scale; on the other hand, preservation of diversity is essential in the long run. This has led to the view of diversity itself as a resource, or "bioresource." It may be this utilitarian view of biodiversity that led to the adoption of the *Convention on Biological Diversity* by world governments.

Threats. The human threat to biodiversity has accelerated steadily. The main threat today is habitat destruction. Habitats are being destroyed around the world for urban development or conversion to pasture, cropland, and orchards. Tropical forests are destroyed at ever-increasing rates; as a result, it is estimated that 10–15% of tropical species will become extinct in the next 30 years. Because most species occur in tropical regions, this extinction will be significant. Humans use powerful technologies, and human population has been growing exponentially. Much of this growth (albeit at a decreased rate) occurs in developing countries, but the highest per-capita impact on the world environment occurs in industrialized countries through the consumption of energy and resources harvested around the world. One consequence of this impact on the environment is global warming, which is causing rapid changes in climate. Most species may not be able to evolve adaptations to these changes. Global warming is likely to cause the extinction of about 18–35% of species on Earth in the next 50 years. Other threats to biodiversity include hunting or overfishing and introduction of exotic species.

Outlook. The intrinsic and utilitarian values of biodiversity are obvious. The negative consequences of biodiversity loss have long been known, and individual governments and groups have made efforts to stop the erosion of biodiversity. However, it was only in 1992, at the United Nations Earth Summit in Rio de Janeiro, Brazil, that a pact was made among the majority of the world's governments to combine conservation of biodiversity with economic development. This pact, the *Convention on Biological Diversity*, has three main goals: conservation of biological diversity, sustainable use of its components, and fair and equitable sharing of the benefits from the use of genetic resources. It recognizes that conservation of biological diversity is "a common concern of humankind" and is an integral part of the development process.

Biodiversity is highest in tropical regions, which also house most developing nations. Historically, this has meant that plants growing in these regions often are valuable—for instance, as sources of drugs—but people living there rarely have seen the benefit of these applications. In part, the *Convention on Biological Diversity* attempts to redress such imbalances by ensuring that benefits are shared fairly in the future. The possibility of making a fortune from gene patents has led to a view of tropical rainforests as "green gold."

Most countries (but not the United States) have ratified the Convention. Signatories to the Convention have adopted a strategic plan since then, the *2010 Biodiversity Target*. This target is a commitment to achieve a significant reduction in the current rate of biodiversity loss at the global, regional, and national levels. Assessment of progress on this front depends upon continuing inventories of biodiversity, but the taxonomic impediment may stand in the way of this.

The *Convention on Biological Diversity* recognizes the major role of scientists for increasing our knowledge on biodiversity. However, a major challenge is the "taxonomic impediment." The taxonomic impediment constitutes both the incomplete knowledge of taxa and the dearth of taxonomists (scientists trained to identify and classify living organisms), especially in developing countries. Mechanisms of the Convention aim to alleviate the acuteness of this problem by (1) promoting the study of taxa in developing countries by taxonomists from developed countries and (2) capacity building in developing countries.

Despite such problems, the future looks promising because this is the first time that national governments have come together to conserve biodiversity.

For background information *see* BIODIVERSITY; BIOME; CONSERVATION OF RESOURCES; ECOLOGICAL COMMUNITIES; ECOLOGY; ECOSYSTEM; ENDANGERED SPECIES; EXTINCTION (BIOLOGY); GLOBAL CLIMATE CHANGE; POPULATION ECOLOGY in the McGraw-Hill Encyclopedia of Science & Technology.

R. Geeta; Rodrigo Cogni

Bibliography. K. J. Gaston and J. I. Spicer, *Biodiversity: An Introduction*, Blackwell, Oxford, UK, 2004; M. J. Novacek (ed.), *The Biodiversity Crisis: Losing What Counts*, New Press, New York, 2001; E. O. Wilson, *The Diversity of Life*, Harvard University Press, Cambridge, MA, 1992.

Brassinosteroids

Plant hormones are naturally occurring organic molecules that regulate multiple aspects of plant growth and development at concentrations much lower than nutrients and other metabolites. At least five classes of plant hormones, including auxins, cytokinins, gibberellins (GAs), ethylene, and abscisic acid (ABA), have been characterized and studied in detail for many decades with respect to their occurrence and biosynthesis in plants, their molecular mechanisms of action, and the resulting alterations in cellular structure and function that lead to defined physiological responses to each hormone. In the past decade, a sixth class of plant hormones, brassinosteroids (BRs), has been added to the list of essential regulators of plant growth and development.

Discovery of BRs. The historical events leading to the discovery of BRs can be traced to early experiments of J. W. Mitchell and colleagues at the U.S. Department of Agriculture (USDA). From 1941 to 1970, they examined organic solvent extracts of pollen from many species in a variety of growth bioassays, and in 1970 reported that an extract of *Brassica napus* pollen, which they referred to as "brassins," resulted in dramatic increases in cell expansion and division in bean stems. Application of brassins to several crop species resulted in pronounced increases in growth, which prompted USDA scientists to isolate and structurally characterize the active component of brassins from *B. napus* pollen. Using single-crystal x-ray analysis, they reported in 1979 the identification of brassinolide, a plant steroid of unique structure that has been shown to be the most biologically active of naturally occurring BRs. Chemically, brassinolide is a polyhydroxylated derivative of 5a-cholestane, namely (22*R*,23*R*,24*S*)-2α,3α,22,23-tetrahydroxy-24-methyl-B-homo-7-oxa-5α-cholestan-6-one (**Fig. 1**). The role of steroids in regulating growth and development and adult homeostasis in animals and insects is well known. The discovery of brassinolide showed for the first time that nanomolar concentrations of a specific plant steroid could regulate essential aspects of plant growth and development.

Natural occurrence in plants. Japanese scientists independently isolated a number of naturally occurring BRs, including castasterone, the immediate biosynthetic precursor of brassinolide. Using analytical techniques such as high-performance liquid chromatography and gas chromatography–mass spectrometry (GC-MS), over 50 structurally distinct BRs have been identified in a wide range of plant species, including monocots, dicots, gymnosperms, and several lower plant species. Thus, BRs are not a novelty of *B. napus*, but appear to be ubiquitously distributed throughout the plant kingdom. Many BRs are biosynthetic precursors or metabolic products of brassinolide, although some, such as castasterone, have independent biological activity in specific plants. Endogenous levels of BRs vary according to species, plant organ type, and tissue age. Pollen and immature seeds contain the highest levels, generally 1–100 ng/g fresh weight, whereas shoots and leaves typically have lower amounts, 0.01–0.1 ng/g.

Biosynthesis. BRs and animal steroid hormones are both products of the isoprenoid [terpene, composed of isopentyl (isoprene) C_5 units] biosynthetic pathway originating with acetyl-coenzyme A and proceeding via mevalonate to squalene-2,3-epoxide. These early biosynthetic steps are conserved between animals and plants, but the conversion of squalene-2,3-epoxide to sterol progenitors differs. In animals, squalene-2,3-epoxide is converted to lanosterol, the precursor of cholesterol and animal steroid hormones (such as testosterone); in contrast, in plants, it is converted to cycloartenol, the parent compound of sterols such as campesterol (Fig. 1). The conversion of campesterol to the 28-carbon BRs, castasterone and brassinolide, involves a series of chemical modifications that have been characterized in detail by feeding labeled biosynthetic intermediates to *Catharanthus roseus* cell suspension cultures, followed by monitoring the conversion of the labeled compounds with sensitive techniques such as GC-MS. The discovery of BR- and sterol-deficient mutants in the model plant *Arabidopsis thaliana* greatly refined the level of understanding of BR biosynthesis. Cloning and sequencing revealed that the mutations were lesions in genes encoding enzymes such as a steroid reductase and several cytochrome P450 steroid hydroxylases required for the stepwise conversion of sterol intermediates to brassinolide. BR-deficient mutants have also been identified in several crop species, including pea, rice, and tomato. The

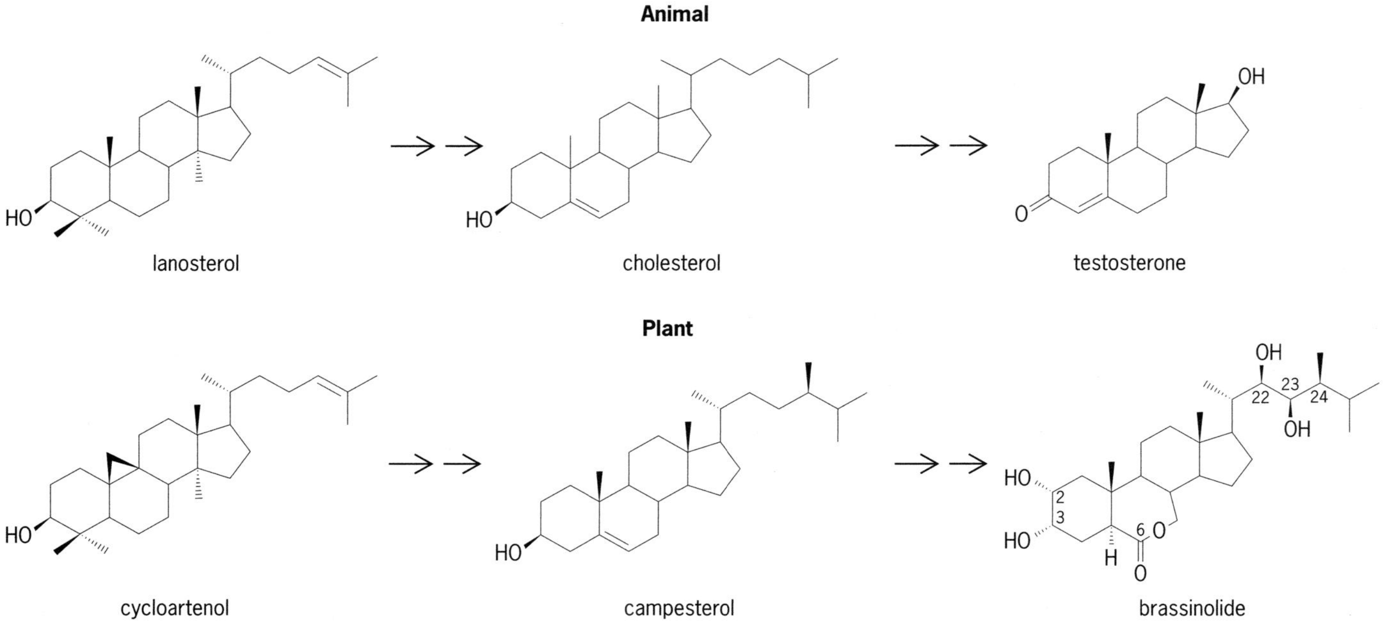

Fig. 1. Comparison of origins and structures of plant and animal steroid hormones. In animals, squalene-2,3-epoxide is converted to lanosterol, which is the precursor of the membrane sterol cholesterol, the progenitor of steroid hormones such as testosterone. In plants, squalene-2,3-epoxide is converted to cycloartenol, the precursor of the membrane sterol campesterol. Brassinolide, the most active BR, is derived from campesterol in a series of steps as described in the text.

phenotype of BR- deficient mutants is quite dramatic and involves multiple developmental defects, including severe dwarfism, altered leaf shape, delayed flowering and senescence, reduced male fertility, and altered vascular structures. All of these phenotypic aberrations can be reversed (rescued) to wild type by exogenous application of brassinolide, but not by other plant hormones such as auxins or gibberellins. These experiments provided genetic proof that BRs are essential for normal plant growth and development, and led to the widespread acceptance of BRs as a sixth class of plant hormones by the scientific community.

Physiological responses. BRs promote stem elongation and leaf expansion. Cells are shorter than normal in BR-deficient mutants and longer than normal in transgenic plants overexpressing genes encoding BR biosynthetic enzymes. Cell growth in plants is restricted by the presence of a rigid, complex cell wall that must transiently yield to allow cell expansion to proceed. One mechanism by which BRs promote cell expansion is to increase the levels and activities of cell wall–modifying enzymes that lead to changes in wall mechanical properties. BRs may also influence cell elongation by affecting transport of water and ions across cell membranes, and by altering the orientation of growth-restraining microtubules (hollow tubelike filaments found in certain cell components). Auxins and cytokinins are required for vascular differentiation in plants, but biochemical and genetic studies suggest that endogenous BRs are also essential for normal development of xylem (the principal water-conducting tissue and chief supporting tissue of higher plants). Similarly, BRs work with auxins and cytokinins to increase rates of cell division in several plant systems. Other physiological effects of BRs include promotion of seed germination, acceleration of senescence, and regulation of reproduction. Reduced fertility or male sterility is commonly observed in BR mutants, and most of these mutants also exhibit an extended life span and delayed senescence when compared to wild-type plants, with the extent of the delayed senescence correlating with reduced levels of fertility. ABA and GAs have antagonistic functions in establishing and breaking dormancy during seed development and germination. BRs can rescue the germination defect in GA mutants, and BR mutants are more susceptible to inhibition of germination by ABA than wild-type forms. Therefore, BR signaling may be required to reverse ABA-induced dormancy and to stimulate germination. Besides promoting growth, BRs also modulate the effects of abiotic and biotic stresses, including responses to extreme cold and heat, salt and drought stress, and pathogen attack.

BR signal transduction. BRs are perceived at the cell surface by BRASSINOSTEROID INSENSITIVE 1 (BRI1), a member of the large family of leucine-rich repeat receptor-like kinases (LRR-RLKs) found in plants. Kinases are important enzymes involved in signal transduction that catalyze phosphorylation, which is the transfer of a phosphate group from adenosine triphosphate (ATP) to serine (Ser), threonine (Thr), or tyrosine (Tyr) amino acid residues. LRR-RLKs have an arrangement of functional domains similar to those found in animal receptor tyrosine kinase and transforming growth factor-β receptors. These animal receptor kinases function by binding a specific ligand in the extracellular domain, which leads to association of identical receptor pairs (homodimerization) or pairing of related, but nonidentical receptors (heterodimerization), which in

Fig. 2. Phenotype of a BRI1 receptor mutant in tomato. The mutant tomato seedling (left) is an extreme dwarf that shows many developmental defects when compared to a wild-type plant of the same age (right). The phenotype is due to a mutation in the cell surface membrane receptor BRI1, which binds the plant hormone brassinolide.

turn promotes phosphorylation and activation of the intracellular kinase domains. Kinase activation results in recognition and phosphorylation of downstream components of the signal transduction pathway, leading ultimately to alterations in gene expression or other nongenomic responses that alter cellular physiology. Recent studies have shown that BRI1 functions in an analogous manner. Brassinolide binds directly to the extracellular domain of BRI1, which promotes heterodimerization with a second LRR-RLK, BRI1-ASSOCIATED RECEPTOR KINASE 1 (BAK1). Both BRI1 and BAK1 are phosphorylated on specific Ser and Thr residues in the kinase domain in a BR-dependent manner, and mutational analysis shows that kinase activity is essential for BRI1 and BAK1 function. A severe *bri1* mutant has the same phenotype as a BR-deficient mutant, but the *bri1* phenotype cannot be rescued by exogenous application of BRs (**Fig. 2**). While mechanistic parallels between BR signaling and animal receptor kinases are apparent, it is interesting to note that plant steroid signal transduction is much different than the prominent steroid signaling pathway in animals where steroids bind directly to an intracellular receptor that functions as a ligand-dependent transcriptional activator. Thus, plant and animal steroid hormones share many structural and functional features, but differ in their primary signaling pathways.

Downstream components of BR signaling have now been identified and studied in some detail. BRI1 KINASE INHIBITOR 1 (BKI1) is a negative regulator of BR signaling that is membrane associated in the absence of BR and binds to BRI1, inactivating its function. BR binding to BRI1 causes dissociation of BKI1 from the membrane, which releases repression of the BR signaling pathway. BR INSENSITIVE 2 (BIN2), a cytoplasmic Ser/Thr kinase, is also a negative regulator of BR signaling that functions by phosphorylating and inactivating BES1 and BZR1, two transcription factors required for BR-regulated gene expression. In the presence of BR, BIN2 activity is apparently inhibited, and the nonphosphorylated forms of BZR1 and BES1 accumulate and bind to the promoters of specific genes involved in the BR response. Classical molecular genetics and global analyses with microarrays have combined to reveal several hundred genes that are regulated by BRs. The current status of BR signal transduction knowledge is summarized in **Fig. 3**.

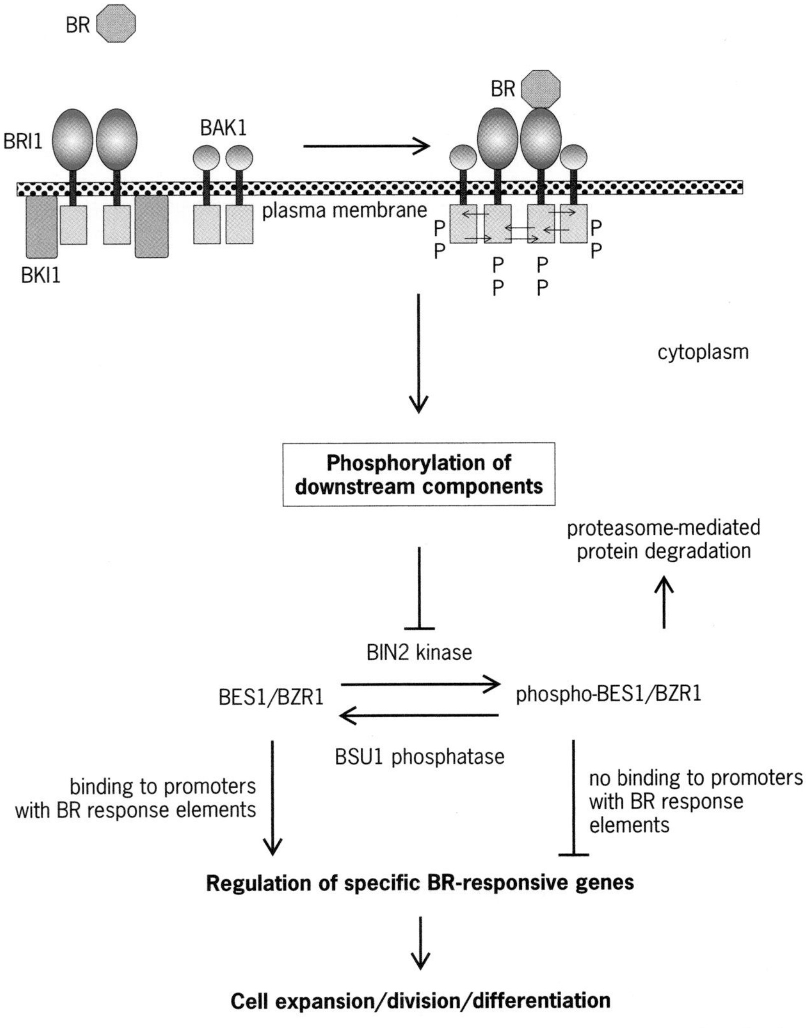

Fig. 3. Brassinosteroid (BR) signal transduction. BR binds to the extracellular domain of the receptor kinase BRI1, which initiates a cascade of phosphorylation events culminating in the transcription and translation of numerous genes involved in regulating cell expansion, division, and differentiation. The cell nucleus is not shown due to some uncertainty in the localization of some downstream components. See text for greater detail and definitions.

For background information *see* AGRICULTURAL SCIENCE (PLANT); LEAF; PLANT GROWTH; PLANT HORMONES; PLANT MORPHOGENESIS; PLANT PHYSIOLOGY; POLLEN; SEED; STEM; STEROID in the McGraw-Hill Encyclopedia of Science & Technology.

Steven D. Clouse

Bibliography. S. D. Clouse and J. M. Sasse, Brassinosteroids: Essential regulators of plant growth and development, *Annu. Rev. Plant Physiol. Plant Mol. Biol.*, 49:427–451, 1998; S. Fujioka and T. Yokota, Biosynthesis and metabolism of brassinosteroids, *Annu. Rev. Plant Biol.*, 54:137–164, 2003; J. Li and H. Jin, Regulation of brassinosteroid signaling, *Trends Plant Sci.*, 12:37–41, 2007; A. Sakurai, T. Yokota, and S. D. Clouse (eds.), *Brassinosteroids: Steroidal Plant Hormones*, Springer, Tokyo, 1999; G. Vert et al., Molecular mechanisms of steroid hormone signaling in plants, *Annu. Rev. Cell Dev. Biol.*, 21:177–201, 2005.

Bronchiolitis obliterans organizing pneumonia (BOOP)

Bronchiolitis obliterans organizing pneumonia was first described in 1985 by Gary Epler as a disease distinct from other disorders of the lungs; it is also known as Epler's pneumonia. It is characterized by the formation of connective tissue masses in the lungs, filling the respiratory bronchioles and making breathing difficult and painful. BOOP presents clinically with initial "flulike" symptoms, followed by progressive dyspnea (shortness of breath), cough, fever, and weight loss. "Crackling" can be heard within the chest using a stethoscope. X-ray imagery shows bilateral patchy infiltrates in the alveoli of the lungs. Computed tomography (CT) shows bilateral areas of lung consolidation. This appears with a "groundglass" haziness that is characteristic for the disease.

BOOP has several known causes, including infection with a number of pathogens known to attack the lungs; use of a number of drugs, both licit and illicit, including amphotericin (an antifungal antibiotic), cephalosporins (a group of antibiotics used to treat infections), and cocaine; association with a number of connective tissue disorders, such as lupus erythematosus and rheumatoid arthritis; occurrence after organ transplantation, such as bone marrow, lung, and kidney transplants; occurrence after radiation therapy; and exposure to a number of environmental hazards, such as textile printing dye inhalation or involvement in a house fire. However, BOOP is usually idiopathic in most patients diagnosed, with no known cause or trigger.

BOOP differs from other lung diseases with chronic inflammatory processes in that it involves both airways and alveoli of the lungs simultaneously, rather than either alone. Chronic obstructive pulmonary disease (COPD) and asthma differ from BOOP in the inflammatory cells, mediators, inflammatory effects, and response to treatment.

Pathogenesis and diagnosis. BOOP differs from other lung diseases such as usual interstitial pneumonia/idiopathic pulmonary fibrosis (UIP/IPF) in that it occurs primarily through an inflammatory process rather than through the formation of fibrous tissue in the lungs. Fibromyxoid (fibrous mucoid) connective tissue is formed in both diseases; however, in BOOP it can be completely reversed by corticosteroid therapy, while in UIP/IPF this tissue causes destruction of the interstitial spaces in the lungs. A possible reason for this difference may be the quantity of capillaries formed within this tissue. In BOOP the fibromyxoid lesions in BOOP appear to have abundant vascular vessels, whereas in UIP/IPF there appears to be minimal vascularization. In addition, programmed cell death, or apoptosis, appears to occur at a higher level in BOOP than in UIP/IPF. This may play a role in the resolution of the connective tissue that is formed in BOOP.

Diagnosis of BOOP is typically made using a video-assisted thoracoscopic procedure to take a lung biopsy. X-rays taken of patients' chests typically show bilateral patchy infiltrates of the alveoli. Infiltrates gradually enlarge from their original site; new infiltrates may also appear. Focal nodular lesions are frequently a sign of BOOP.

Treatment and recurrence. BOOP is usually treated using the corticosteroid prednisone with a gradually decreasing dosage over about 1 year. Permanent recovery is seen in most patients, although it is dependent upon the cause of the disease or whether there are associated systemic disorders (see "Types" below). In one-third of patients treated for less than 1 year, BOOP may recur. However, it can be successfully treated a second or even third time with the same dosage of prednisone. For patients who do not respond to treatment, it is important to determine whether the actual disease is BOOP or whether UIP/IPF is present in the lung. BOOP responds well to corticosteroid therapy, but UIP/IPF generally leads to a progressively deteriorating condition in the patient.

Types. Idiopathic BOOP is the most common type of this disease. Symptoms include a flulike illness, fever, cough, and shortness of breath, with crackles occurring in two-thirds of patients. Prognosis of this form of BOOP is good, with 65–80% of patients treated with prednisone therapy recovering completely.

Postinfection BOOP can occur in relation to a variety of infectious pathogens that target the lungs. These include the bacterial species *Chlamydia pneumoniae*, *Legionella pneumophila*, and *Mycoplasma pneumoniae*, and the viral agents parainfluenza and adenovirus. *Plasmodium*, the genus responsible for causing malaria, and the fungal agents *Cryptococcus neoformans* and *Pneumocystis (carinii) jiroveci* have also been found to cause BOOP. Other pathogens that may cause infections resulting in BOOP include *Coxiella burnetii*, *Pseudomonas aeruginosa*, influenza A virus, measles virus, human immunodeficiency virus (HIV), and parvovirus. Generally, these patients begin to resolve the initial pneumonia upon administration of the appropriate antimicrobial agents, but this improvement ceases and the symptoms and x-ray findings persist. At this point, the pneumonia has become organized into the BOOP lesion, and corticosteroid therapy is necessary to resolve the disease.

Drug-related BOOP has resulted from the use of several different types of medication or drugs. These include cephalosporins, sulfamethoxypyridazine, and amphotericin B (antimicrobial agents); bleomycin sulfate, gold salts, and methotrexate (anti-inflammatory agents); and the illicit use of cocaine.

BOOP has occurred in patients who have received bone marrow, lung, or kidney transplants. BOOP associated with lung transplants has occurred in 10–28% of these patients. It is usually associated with the acute rejection reaction. With successful treatment of the rejection reaction, the process is reversible in these patients. Bone marrow and kidney transplant BOOP is extremely rare, and it is unknown if there is an actual link between the transplantation and the development of BOOP.

Connective tissue disease may lead to an increased risk for BOOP. Rheumatologic or connective tissue BOOP is similar to the idiopathic form and has been seen with all of the connective tissue diseases. BOOP has been seen in patients with lupus erythematosus, rheumatoid arthritis, ankylosing spondylitis (inflammation of the vertebrae), and dermatomyositis (inflammation of skin and muscles).

BOOP now occurs in patients receiving radiotherapy to the breast for breast cancer or small-cell bronchogenic carcinoma. Symptoms occur after completion of radiotherapy, with mild shortness of breath, cough, and fever. X-ray examination of the chest will show the characteristic patchy infiltrates. Interestingly, the infiltrates are often outside the radiation field, having "migrated" to other locations.

A form known as rapidly progressive BOOP has recently been described. This form strikes rapidly, leading to respiratory failure, and is associated with high mortality. In some patients, BOOP was a secondary component of the disease and an underlying fibrotic process was the primary component, leading to death. However, when BOOP is the primary cause of the disease, these patients have a better prognosis despite the rapidity with which the disease progresses.

Patient cases. One patient was a 76-year-old woman who developed rapidly progressive BOOP. She presented to the emergency hospital room with a 3-day history of shortness of breath, cough, and fever, combined with chest pain on the right side of her chest. Crackles were heard in her chest, and her x-rays showed bilateral patchy opacities of the chest. She was diagnosed with pneumonia and treated with antibiotics. Despite treatment, her condition worsened and she was placed on a mechanical ventilator. A lung biopsy revealed that the alveolar spaces had been obliterated with plugs of connective tissue within the terminal bronchioles, alveolar ducts, and spaces. She was then diagnosed as having BOOP. Her clinical features of shortness of breath, cough, fever, and increased white blood cell count were all consistent with BOOP as it usually presents; however, the rapidity with which she declined and the acuteness of her presentation are consistent with the rapidly progressive form of the disease. Her treatment was successful and she was eventually released from the hospital. Follow-up care showed that she was no longer on corticosteroid therapy and was doing well.

Two other cases developed BOOP following radiotherapy for breast cancer. The first patient was a 59-year-old woman who underwent radiotherapy on the left thoracic wall and the axillary lymph nodes. Several months later, she developed shortness of breath and crackles within her lungs. X-rays showed that the left lung had a patchy consolidation zone. She was treated with prednisone and responded well. However, a few months later she was again hospitalized and new patchy infiltrates were found in the right lung. She was then diagnosed with BOOP and treated more aggressively with prednisone. She responded well and improvement was seen within the right lung. However, the left lung continued to have abnormalities visible within the original radiation field.

The second patient was a 92-year-old woman who had developed multiple intradermal metastases from a breast cancer on the right thoracic wall. She was treated with radiotherapy, 3 months after which she developed shortness of breath and fever. Crackles were heard within her right lung, while no abnormalities were apparent on the left side. Treatment with prednisone followed and she was eventually released. Symptoms recurred 2 months later, with the abnormalities spreading to the left lung; at this time, she was diagnosed with BOOP and treated aggressively. Her symptoms resolved within 2 weeks and she was released successfully.

Summary. BOOP in the United States is believed to cause 20–30% of all cases of chronic infiltrative lung diseases. Mortality in patients with BOOP is 10%. Generally, most patients present between the ages of 40 and 70 years, but this disease has been reported in children who are already suffering from an underlying malignancy. BOOP should be recognizable to clinicians when they see a patient with a febrile illness and patchy infiltrates of the lung who has not responded to antibiotic therapy. This disease is a significant cause of illness and mortality in patients; however, BOOP may be treated successfully when recognized promptly.

For background information *see* ANTIBIOTIC; BRONCHUS; COCAINE; LUNG; OPPORTUNISTIC INFECTIONS; PNEUMONIA; RADIATION THERAPY; RESPIRATORY SYSTEM; RESPIRATORY SYSTEM DISORDERS in the McGraw-Hill Encyclopedia of Science & Technology.

Marcia M. Pierce

Bibliography. R. Cornelissen et al., Bronchiolitis obliterans organizing pneumonia (BOOP) after thoracic radiotherapy for breast carcinoma, *Radiat. Oncol.*, 2:2, 2007; G. R. Epler, Bronchiolitis obliterans organizing pneumonia, *Arch. Intern. Med.*, 161(2):158–164, 2001; G. R. Epler et al., Bronchiolitis obliterans organizing pneumonia, *N. Engl. J. Med.*, 312:152–158, 1985; J. L. Myers and M. Selman,

Respiratory epithelium in usual interstitial pneumonia/idiopathic pulmonary fibrosis: Spark or destructive flame?, *Amer. J. Resp. Crit. Care Med.*, 169:3–5, 2004; T. J. Nowak and A. G. Handford, *Pathophysiology: Concepts and Applications for Health Care Professionals*, 3d ed., McGraw-Hill, 2004; F. S. Oymak et al., Bronchiolitis obliterans organizing pneumonia: Clinical and roentgenological features in 26 cases, *Respiration*, 72(3):254–262, 2005; J. P. Utz et al., High short-term mortality following lung biopsy for usual interstitial pneumonia, *Eur. Resp. J.*, 17:175–179, 2001.

Buoyant plumes in crossflows

In environmental flows, the most common way of mixing fluids is through a fluid discharge. Fluid discharges refer in general to releases of liquids, gases, or a combination of both. They are found in many applications, including the following: (1) industrial applications such as furnaces, smoke stacks, or mixing of pharmaceutical drugs; (2) wastewater disposal in rivers, lakes, and oceans; (3) forest fires and building fires; and (4) many common household situations, such as sprinklers, garden hoses, and faucets. Furthermore, based on the geometry of the discharge, they can occur in still environments, coflows, or crossflows, or at an angle. For instance, it is easy to visualize the plume vertically rising from a burning cigarette in a still environment. Similarly, the smoke from a forest fire can be visualized being carried downstream many miles by the wind (that is, in a crossflow). Discharges are also characterized by their amount of momentum or buoyancy. Thus, a jet is a discharge of fluid into a fluid environment of the same density. For instance, a jet ski in a lake will eject water at high speed (that is, momentum) into water to propel itself. On the other hand, if one pours gasoline on the ground and lights it, the smoke will rise due to the difference in density between the air being heated and the cooler surrounding air. Finally, both buoyancy and momentum flux can be present at the source of the injection, and this condition will create a buoyant plume (also called buoyant jet by many researchers). The remainder of this article will focus on the study of buoyant plumes.

Self-preserving properties. The mixing properties of buoyant plumes in crossflowing environments (henceforth described simply as plumes in crossflows) is an important fundamental problem relevant to the dispersion of heat and harmful substances in cases of accidental or intentional release into atmospheric crosswinds. If the release has a very short duration, the resulting discharge is called a thermal rising rather than a plume (for example, a puff produced by a smoker, or the extreme mushroom cloud produced by a nuclear explosion). However, releases are generally of extended duration so that the thermal approaches the self-preserving behavior of a starting plume, and more often than not they occur in the presence of significant crossflows. Unlike other flows, plumes in crossflows have not been measured very often in laboratories due to the complexities of providing well-defined crossflows and the difficulties of measuring the properties of these three-dimensional flows. This article will review the main mixing properties of plumes in crossflows, emphasizing the regions far from the source, where effects of source injection disturbances (nonideal source shape, noise, vibrations, and so forth) are diminished, the flows are largely controlled by their conserved properties (momentum or buoyancy flux), flow properties approximate self-preserving turbulent-flow behavior, and appropriately scaled flow properties became independent of the distance from the source. To say that properties (in particular, velocity or concentration) are self-preserving means that the mean velocity profile or the mean concentration profile at any cross section of the flow can be expressed by the time-averaged velocity or the time-averaged concentration in terms of a maximum value (that is, measured at the plume centerline) and a measure of the width of the plume. Once the self-preserving relations are known from simple laboratory experiments, they can be easily applied to large environmental flows, greatly simplifying research. Thus, the self-preserving region is the main region of interest because the properties of self-preserving turbulent flows provide a compact presentation of measurements that substantially simplifies the interpretation of flow behavior.

Plume structure visualization. A visualization of a typical steady turbulent plume in a crossflow appears in **Fig. 1**. This flow actually involves the injection of dye-containing salt water (the more dense salt water flowing vertically downward) into a freshwater crossflow (flowing from left to right) in a water channel facility. However, the vertical direction has been inverted to show the flow as an upwardly injected turbulent plume with the source density smaller than the ambient density because most people are more familiar with positively buoyant upwardly flowing plumes than with negatively buoyant downwardly flowing plumes. Experiments have shown that the vertical velocities of steady turbulent plumes in crossflows progressively decrease with increasing vertical distance from the source so that the vertical velocity eventually becomes small and the trajectories of steady turbulent plumes in crossflows become nearly horizontal far from the source, where self-preserving behavior is approached. When this condition is reached, the vertical penetration of the flow approximates a two-dimensional horizontal line vortex pair or thermal in a still fluid. Then the vertical motion of the line vortex or thermal, retarded along its sides in the vertical direction by the uniform ambient crossflow, naturally leads to the flow becoming two nearly horizontal counterrotating vortices whose axes are aligned along the axis of the plume as a whole. Evidence for this behavior is provided Fig. 1(*b*), where the darker regions associated with the two vortices are separated by a significantly lighter region dominated by the presence of dye-free ambient fluid that is entrained by the vortex system along its plane of symmetry. Another interesting feature of the visualization appearing in Fig. 1 is that turbulent distortions of the lower surface of the flow (the side facing the source) are smoothed out because this region is stable to buoyant disturbances, whereas the turbulent distortions of the upper surface of the flow are enhanced because this region is unstable to buoyant disturbances. Notably, corresponding visualizations of steady (nonbuoyant) turbulent jets in crossflow do not exhibit this behavior but instead exhibit similar degrees of distortion on the lower and upper surfaces of the flow because mechanisms of buoyant stability and instability are absent in this case.

Most practical releases of turbulent plumes are exposed to crossflow; therefore, there have been a number of attempts to extend the results from turbulent plumes in still fluids to corresponding turbulent plumes in crossflows. Measurements of the mixing structure in these flows in the region near the source have determined their trajectories and the details of how flow undergoes most of its deflection toward the cross-stream direction. Studies of the self-preserving behavior of this flow in the far-field region show that the decay of vertical velocities with increasing distance from the source is relatively rapid so that the flow eventually becomes nearly aligned with the horizontal direction for all source/crossflow velocity ratios, u_0/v_∞; that is, the general appearance of the flow illustrated in Fig. 1 is typical of flows of this type. A parameter that is frequently used to estimate when plumes reach self-preserving behavior is the distance from the origin of the flow (x) normalized by the source diameter (d), that is, x/d, taken to be in the vertical direction (parallel to the source flow). The value of x/d needed for self-preserving behavior depends on the nature of the flow, the properties of the source, and the property for which self-preserving behavior is sought.

Plume mixing visualization. The fine details of the dynamics of the mixing pattern of the source and ambient fluids can be seen from the sequence of planar laser-induced fluorescence (PLIF) images in **Fig. 2**. These instantaneous images show the largely distorted presence of the two counterrotating vortices separated at the plane of symmetry by deeply penetrating ambient fluid. In addition, the presence of ambient fluid being transported deep into the vortex system along its plane of symmetry clearly has an important effect on the flow structure, as mentioned earlier in connection with the discussion of the flow visualization of Fig. 1. Thus, penetration of the flow normal to its axis is two or three times faster for steady turbulent plumes in crossflows than for similar plumes in still fluids for similar vertical distances and initial source conditions. This behavior suggests that there is much more effective mixing between the source and the ambient flows when the source flow is perpendicular to the ambient, which is the case for steady turbulent plumes and jets in crossflows, than when the axis of the source flow is aligned with the direction of penetration, which is the case for similar plumes and jets in still fluids. Thus, plumes in crossflow are the ideal way to dilute and quickly mix wastewaters and smoke stack fumes into the environment.

(a)

(b)

Fig. 1. Visualization of a steady turbulent plume in a crossflow. (*a*) Side view. (*b*) Top view. (*From F. J. Diez, L. P. Bernal, and G. M. Faeth, Round turbulent thermals, puffs, starting plumes and starting jets in uniform crossflow, J. Heat Trans., 125:1046–1057, ASME, 2003*)

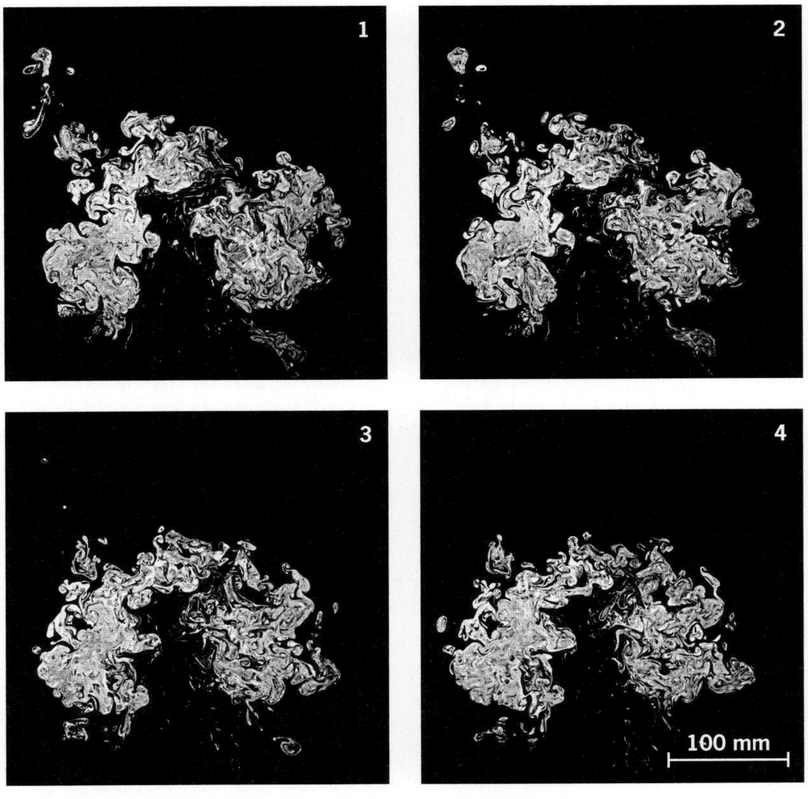

Fig. 2. Instantaneous planar laser-induced fluorescence (PLIF) image series of the cross section of a steady plume in a cross flow.

Recent results. Recent progress in experimental measurements of the mixing properties of plumes in crossflows has provided the following results.

1. Laboratory plumes usually become turbulent at vertical distances of 0–5 source diameters from the source. The onset of self-preserving behavior requires that axes of the counterrotating vortex system be nearly aligned with the crossflow direction, whereas this condition is strongly affected by the source/crossflow velocity ratio, u_0/v_∞. As a result, self-preserving behavior is observed at vertical distances greater than 10–20 source diameters from the source for $u_0/v_\infty = 4$ (the smallest value of u_0/v_∞ considered), increasing to vertical distances greater than 160–170 source diameters from the source for $u_0/v_\infty = 100$ (the largest value of u_0/v_∞ considered).

2. Combining the no-slip convection approximation in the cross-stream direction with self-preserving scaling for a horizontal line thermal in a still fluid in the vertical direction has yielded predictions of both steady penetration properties and the steady structure properties (consisting of the mean and root-mean-square fluctuations of the concentrations of source fluid, respectively) of the flow within the self-preserving region of steady turbulent plumes in crossflows.

3. It has been observed that the self-preserving structure of steady turbulent plumes in crossflows involves a counterrotating vortex system whose axes are nearly aligned with the crossflow and thus is nearly horizontal. The nearly crossflow orientation of the axes of the counterrotating vortex system promotes unusually rapid mixing. The rapid onset of self-preserving behavior for plumes in crossflow at small values of u_0/v_∞, where the axes of the counterrotating vortex system become aligned with the crossflow almost immediately upon leaving the source and the onset of self-preserving behavior is observed at vertical distances of 10–20 source diameters from the source, provides further evidence of unusually rapid mixing due to the crossflow motion of the plume. For example, steady turbulent plumes in still fluids where mixing is limited to the longitudinal direction exhibit self-preserving behavior only at vertical distances greater than 80–100 source diameters from the source.

Unresolved issues. Many issues about these flows, however, still must be resolved: The structure and mixing properties of unsteady thermals and plumes are not well understood due to problems of making measurements of transient flows; the structure and mixing properties of steady plumes in crossflows are not well understood due to problems of making measurements in three-dimensional flows; the behavior of buoyant turbulent flows in stratified environments has received very little attention so that even simple flow penetration measurements for this flow would be helpful; and baseline information about the penetration, structure, and mixing properties of turbulent nonbuoyant puffs and jets should be developed in order to better understand effects of buoyancy in corresponding turbulent buoyant thermals and plumes.

For background information *see* BUOYANCY; ENVIRONMENTAL FLUID MECHANICS; FLUID MECHANICS; JET FLOW; TURBULENT FLOW; VORTEX in the McGraw-Hill Encyclopedia of Science & Technology.

F. Javier Diez

Bibliography. F. J. Diez, L. P. Bernal, and G. M. Faeth, PLIF and PIV measurements of the self-preserving structure of steady round buoyant turbulent plumes in crossflow, *Int. J. Heat Fluid Flow*, 26:873–882, 2005; F. J. Diez, L. P. Bernal, and G. M. Faeth, Round turbulent thermals, puffs, starting plumes and starting jets in uniform crossflow. *J. Heat Trans.*, 125:1046–1057, 2003; F. J. Diez, L. P. Bernal, and G. M. Faeth, Self-preserving mixing properties of round buoyant turbulent plumes in uniform crossflows, *J. Heat Transfer*, 128:1011–1011, 2006; H. B. Fischer et al., *Mixing in Inland and Coastal Waters*. Academic Press, New York, 1979; K. D. Steckler, H. R. Baum, and J. G. Quintiere, Salt water modelling of fire induced flows in multicomponent enclosures. *Proc. Combust. Inst.*, 21:143–149, 1986; J. S. Turner, Buoyant plumes and thermals, *Annu. Rev. Fluid Mech.*, 1:29–44, 1969.

Capabilities-based planning

Capabilities-based planning (CBP) is a general approach to strategic planning, but one associated primarily with the U.S. Department of Defense (DoD). The DoD announced the shift to capabilities-based planning as a major theme in 2001, "to shift the basis of defense planning from the 'threat-based' model that has dominated thinking in the past to a 'capabilities-based' model for the future. This model focuses more on how an adversary might fight rather than specifically who the adversary might be or where a war might occur." A key to planning would be confronting the profound uncertainties that make it impossible to know the identities of future combatant powers or the circumstances of conflict many years in advance. Planning would maintain a broad "portfolio" of military capabilities in functional areas such as power projection and effective use of space and information technology. Another emphasis was to be managing strategic risks. The DoD is in the process of implementing capabilities-based planning. Other federal, state, and national organizations (for example, the Department of Homeland Security) have also adopted this approach, or approaches that borrow from it. What, then, is capabilities-based planning?

Definitions. A capability may be defined as the ability or potential for doing something, as distinct from the way chosen to do so. For example, the capability to move from one place to another might be achieved with a horse, automobile, train, or airplane. A future weapon-system capability might be achieved with, for example, missiles, crewless aircraft, or traditional crewed aircraft. Some military capabilities might be provided by "nonkinetic" means, such as strategic communications or network attacks.

Capabilities-based planning may be defined as planning, under uncertainty, to provide capabilities suitable for a wide range of modern-day challenges and circumstances, while working within an economic framework. It might seem that any planning system, civilian or military, would fit this description. However, organizational planning often pays little attention to uncertainty, and instead focuses on narrow, well-defined visions of the future. Also, planning often assumes that the way to achieve a capability is by upgrading the previous generation of a familiar system, rather than by considering different approaches, especially those involving cooperation with other organizations.

Truly having a capability also means having all the necessary elements of success for whatever mechanism is to be used. Thus, an automobile is useless for transportation without gasoline, or some other energy source, and a driver. An email capability requires a computer, relevant software, an Internet connection, and the skill to use them.

The DoD has attempted to address these matters, insisting that future capabilities be conceived and assessed in terms of joint operations, rather than those of a single service. Further, it now avoids thinking in "platform terms," as in assuming a crewed-aircraft solution when missiles or crewless aircraft might also be considered.

Some subtleties of definition have led to much confusion. One meaning of capability is broad. When previous programs were instituted to create stealth aircraft or precision weapons, "stealth" and "precision" were thought of as broad capabilities that would enable a wide range of operations under different circumstances. In contrast, a definition used by the U.S. Joint Chiefs of Staff is that capability is "the ability to achieve a desired effect under specified standards and conditions through combinations of means and ways to perform a set of tasks." One reason for this second definition is that an "operator" assuring a decision maker that he or she can accomplish something needs to communicate with precision: The operator may be promising a great deal, but not something open-ended or independent of context. Also, military officers taking capabilities-based planning seriously want to ensure that all the critical components of capability will be in place, so that all the supporting tasks can be accomplished. Planning by the DoD therefore addresses doctrine, operations, training, materiel, leadership and education, personnel, and facilities (DOTMLPF), not just high-visibility weapon systems.

Intellectual origins. Capabilities-based planning was the consequence of intellectual developments over the course of the previous decade after the end of the Cold War. A theme of this work was that good planning required confronting deep uncertainty and emphasizing adaptiveness. That is, a "school solution" for dealing with uncertainty is the ability to adapt quickly and effectively. Also recommended was the portfolio-analysis approach in which one recognizes numerous needs, risks, and economic realities, and then seeks to "balance the portfolio of investments" so as to cover the whole in the best way possible: accepting more risks in some areas than others, but revisiting these choices from time to time as developments occur. Although this approach may seem only sensible, it is in stark contrast to "optimizing" for a single objective and for a single set of assumptions about operating context.

Problems and controversies. As expected, the DoD has encountered problems in implementing the concept of capabilities-based planning, including (1) extreme interpretations of the concept, in which offices were admonished not to consider concrete threats and scenarios; (2) omission of the economic dimension (that is, foot-dragging about identifying bill-payers for new initiatives, and making choices within a budget); and (3) organizational and process complexity. In 2007, the DoD was experimenting with improvements, particularly a more streamlined process of capability development with front-end work involving integrated and iterative work by strategists, requirement setters, technologists, systems analysts, and program analysts concerned with budgets.

Achieving adaptiveness. A central concept in capabilities-based planning is planning capabilities that provide flexibility, adaptiveness, and robustness, that is, "FARness." Will future forces be flexible enough to take on missions other than those intended? Will they be adaptive enough to deal with unexpected circumstances? Will they be robust enough to cope well and recover from shocks? These are significantly different attributes. Using an analogy to sports, a star fullback may be robust (able to withstand repeated vicious tackles while punching through the line again and again), but not adaptive enough to deal with a snow-covered field, nor flexible enough to be used as a wide runner and passer. Despite these attributes being distinctive, they are often referred to in shorthand with a single word, either "adaptiveness" or "agility" being the most common.

Whatever the terminology used, an important question is how these attributes are achieved. Across many domains of activity, a key is stressing modularity with the "right" building blocks or components. In the domain of personal computers, we can customize our system by choosing among modular components such as different keyboards, mice, monitors, and storage devices. Within the military, some of the most important building blocks are the units (for example, divisions or air wings). By the mid-1990s, it had become increasingly evident that the military services needed to rethink their building blocks, because they were a legacy from World War II and no longer sufficiently agile. Today, the military services are moving increasingly toward modular approaches to organization and emphasizing smaller building block units, such as newly designed brigades. As another example, the "network centric operations" of the DoD increasingly allow both local and global networks to be assembled and tailored quickly as needed for the operation at hand, rather than being defined only for specific configurations.

Two ingredients in achieving adaptiveness are (1) rapid-tailoring ability and (2) practice. By their very nature, building blocks are standardized and are not always quite appropriate to the need. It then becomes necessary to "tailor," to make the necessary adjustments quickly and effectively. In an information-technology context, this might mean being able to do some problem-specific computer programming that would allow two systems to communicate properly despite having been designed with slightly different protocols in mind. From a planning perspective it is necessary to ensure that systems are developed so as to be modifiable or replaceable, and that people are trained and equipped to do so. Practice is essential.

Capabilities-based planning is evolving and readers should not be troubled if they encounter different definitions, interpretations, and organizational processes that claim the name. They should be wary, however, of anything that undercuts the basic principles.

For background information *see* SYSTEMS ENGINEERING in the McGraw-Hill Encyclopedia of Science & Technology. Paul K. Davis

Bibliography. P. K. Davis, *Analytic Architecture for Capabilities-Based Planning, Mission-System Analysis, and Transformation*, RAND, 2002; P. K. Davis, D. Gompert, and R. Kugler, *Planning for Adaptiveness: a New Framework for Defense Planning*, RAND, 1996; M. Fitzimmons, Whither capabilities-based planning?, *Joint Force Quart.*, 44:101–105, 2004; P. C. Light, *The Four Pillars of High Performance*, McGraw-Hill, 2004; National Research Council, *Naval Analytical Capabilities: Improving Capabilities-Based Planning*, National Academy Press, 2004; Office of the Under Secretary of Defense for Acquisition, Technology, and Logistics, *Joint Defense Capabilities Study, Final Report*, 2003; D. H. Rumsfeld, *Quadrennial Defense Review Report*, U.S. Department of Defense, 2001 and 2006.

Carbon-based electronics

In recent years, significant progress has been made in understanding device physics and in identifying potential applications of carbon-based electronics. In a carbon nanotube (CNT) or a graphene, carriers can travel without scattering at low-bias conditions. The speed at which carriers travel is high, which promises ultrafast device switching and ultrahigh device speed. Deposition of high permittivity gate insulators does not degrade the excellent carrier transport properties. The conduction and valence bands are symmetric, which is advantageous for complementary electronics applications. The conduction-band bottom and the valence-band top are at the same wave vector point, which enables optical emission. A CNT or a graphene nanoribbon (GNR) can be either semiconducting or metallic, depending on its structure, which leads to potential applications in both nanoscale devices and interconnects. Because of these attractive features, carbon-based electronics is receiving much attention for its potential application in future integrated circuits. Compared to mature semiconductor electronics technologies, however, significant challenges, such as synthesis of high-quality materials, variation between devices, and wafer-scale integration, still remain to be addressed for carbon-based electronics.

Band structure. The graphene, which consists of a two-dimensional sheet of carbon atoms in the honeycomb structure shown in **Fig. 1***a*, has been successfully obtained by exfoliation from three-dimensional graphite or thermal decomposition from SiC. The three nearest-neighbor bonds for each carbon atom are made up of the carbon *s* orbital and two of the three *p* orbitals. The strong sp^2 bonding in the plane is what leads to the superior mechanical properties and structure stability. The out-plane p_Z orbital forms the bands near the Fermi level and determines the electrical properties. The conduction and valence bands of the graphene meet at the corners of the Brillouin zone, as shown in Fig. 1*b*, which is also hexagonal. Because the graphene lattice is two-dimensional, its Brillouin zone is also two-dimensional. The graphene is known as a semimetal or a zero-gap semiconductor. The band structure is linear near each Fermi point (the corner of the Brillouin zone), which is similar to the dispersion relation of light, and is different from the parabolic band structure of a classical particle. The linear band structure near the Fermi level, which is similar to that of particles traveling at the speed of light, governed by the theory of relativity, makes it an ideal system to study the relativistic particle properties.

A single-walled CNT (SWNT) can be conceptually viewed as a rolled-up sheet of graphene. In the rolled-up geometry, all bonds of any carbon atom

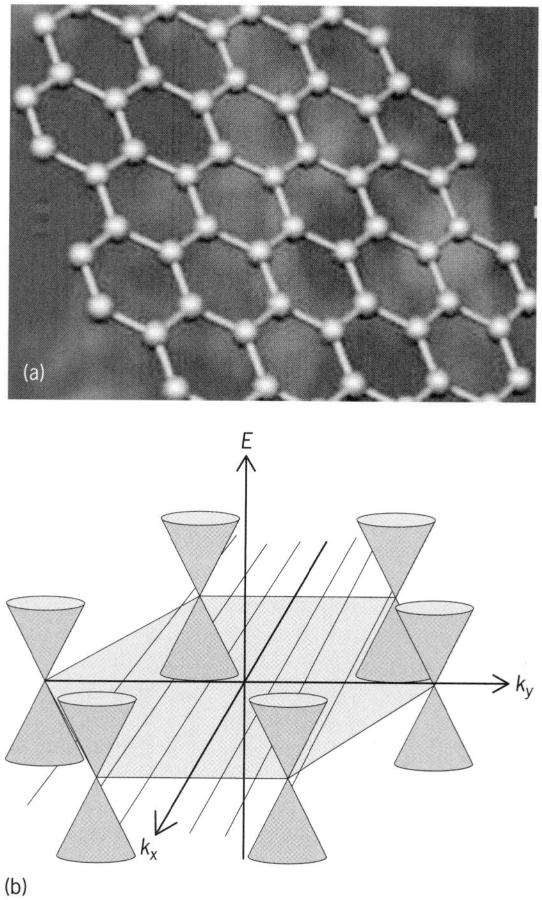

Fig. 1. Graphene: (*a*) lattice and (*b*) the first Brillouin zone.

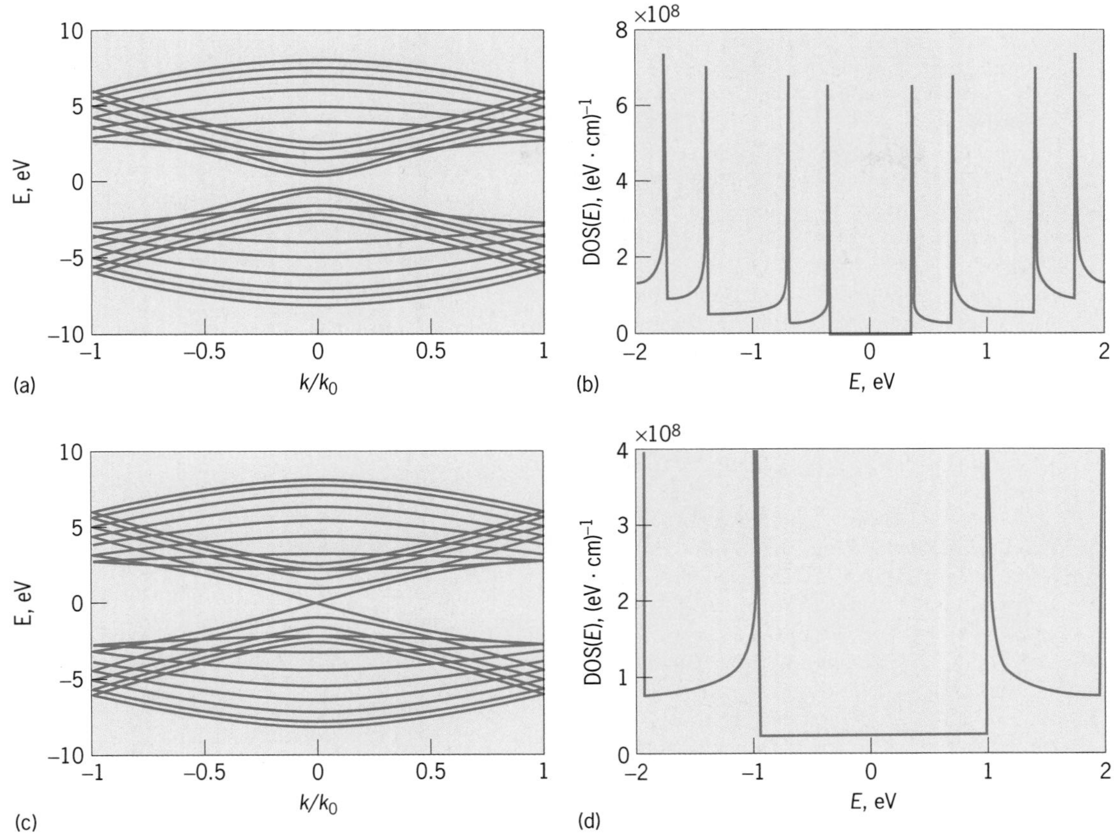

Fig. 2. Electronic structures of CNTs. (*a*) Band structure and (*b*) density of states (DOS) for a semiconducting CNT. (*c*) Band structure and (*d*) DOS for a metallic CNT. In *a* and *c*, the bottom curves are valence bands, which are typically filled by electrons, and the top curves are conduction bands, which are typically empty.

are well satisfied by its nearest-neighboring atoms, which is different from a typical semiconductor-insulator interface, at which the bonds of a semiconductor surface atom are not fully satisfied by its neighbors. The perfect structure of the CNT leads to a smooth surface and excellent transport properties even after application of a high-permittivity gate insulator. The band structure of a SWNT can be extracted by enforcing a boundary condition around the circumference of the SWNT. The solid lines in the 2D *k*-plane in Fig. 1*b* show the allowed *k* states after the application of the periodic boundary condition for a typical CNT. The *E-k* relation of a CNT can be obtained by cutting planes perpendicular to the 2D *k*-plane at the allowed *k*-state lines in the graphene *E-k* space, and each *k*-state line results in a one-dimensional subband. If none of the planes passes the Fermi points, the SWNT is semiconducting. **Figure 2*a*** and *b* shows the band structure and the density of states (DOS) for a typical semiconducting CNT, respectively. A bandgap can be clearly identified near $E = 0$. There are singularities on the DOS plot at the edge of each subband. If one of the slides passes the Fermi point, the SWNT has a zero bandgap and it is metallic, with a typical band structure and DOS as shown in Fig. 2*c* and *d*, respectively. A SWNT, therefore, can be either metallic or semiconducting, depending on its structure, which makes it useful for either metal interconnect or semiconductor device application in an integrated circuit.

CNT field-effect transistors. The first CNT field-effect transistor (FET) was demonstrated in 1998. The device structure, however, was rough. A single-walled CNT channel was inefficiently modulated through a thick bottom gate oxide, and the contact resistance between the CNT channel and the metal source and drain contacts was large. Rapid progress has been made in the past decade in understanding device physics and improving device performance of CNTFETs.

Figure 3*a* shows the cross-section sketch of one of the most mature CNTFETs demonstrated to date, which integrates three advanced fabrication techniques. First, a self-aligned process is used to produce a short channel length (50 nm). Second, a thin hafnium oxide (HfO_2) top gate insulator, which has a permittivity much higher than SiO_2, is integrated, which makes the gate modulation efficient. Third, palladium is used as the source and drain contacts, which eliminates the Schottky barrier between the metal contacts and the CNT channel for holes. These advanced fabrication techniques lead to excellent device performance. The circles in Fig. 3*b* show the experimental current-voltage (*I–V*) characteristics, and the solid lines show the theoretical projection for a ballistic channel (without scattering). The comparison indicates that the CNTFET delivers a near-ballistic DC on-current. The CNTFET is a nanoscale transistor that operates closest to the most recently reported ballistic DC limit.

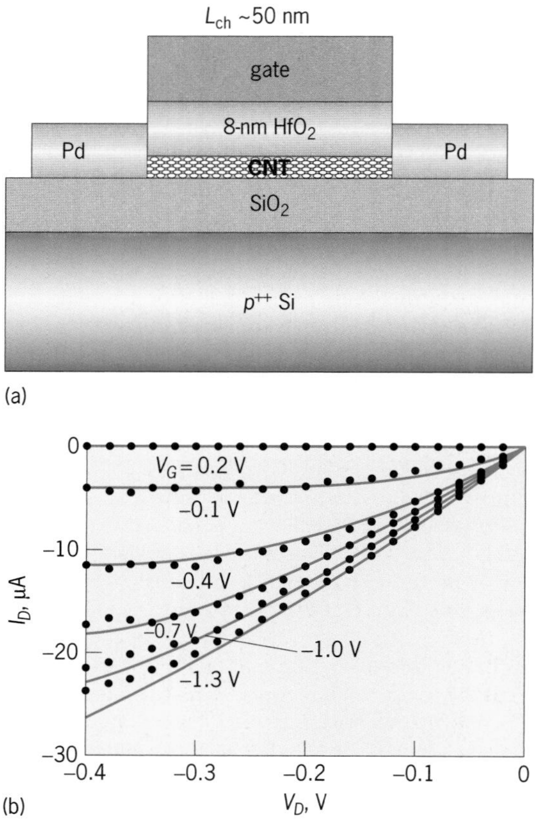

Fig. 3. Self-aligned, top-gated CNTFET with a channel length of 50 nm and palladium (Pd) source and drain contacts. (*a*) Sketch of the device cross section. (*b*) *I–V* characteristics, plotting drain current (I_D) versus drain voltage (V_D) for various values of gate voltage (V_G). The circles are experimental data and the solid lines are theoretical simulation results obtained by a ballistic NEGF simulation. (*Reproduced with permission from A. Javey et al., Self-aligned ballistic molecular transistors and electrically parallel nanotube arrays, Nano Lett., 4:1319–1322, 2004*)

Graphene FETs. The two-dimensional monolayer graphene is a zero-bandgap semiconductor. The on-off current ratio of a monolayer graphene FET is limited to below 10 at room temperature due to its zero bandgap. In order to achieve a higher on-off ratio, engineering the graphene band structure to obtain a bandgap is necessary. Two approaches have been reported. The first approach patterns a two-dimensional monolayer graphene to a nanometer-wide nanostrip, or a GNR. Quantum confinement in the width direction of the GNR creates a bandgap. An FET with a GNR channel width less than 20 nm has been reported. The second approach uses the two-dimensional bilayer graphene as the FET channel. The bilayer graphene itself is a zero-bandgap material, but a bandgap can be opened if a voltage drop is created between the two layers. The voltage drop can be achieved by either an external electric field normal to the bilayer graphene plane or different doping densities between the two layers.

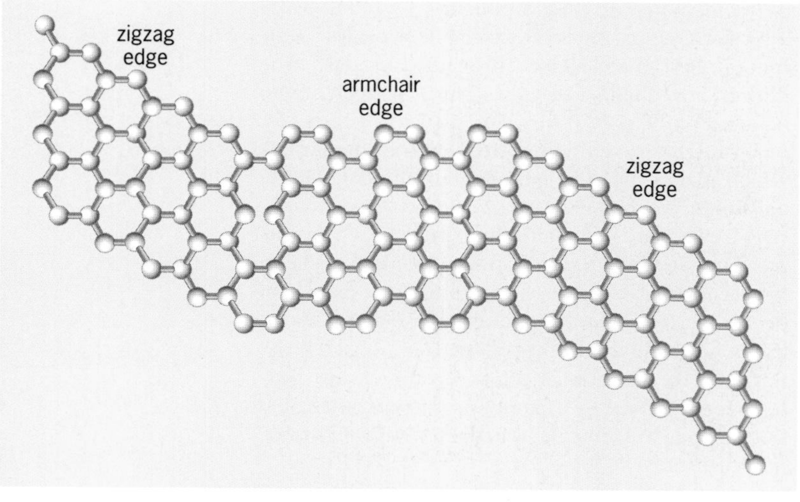

Fig. 4. Part of a graphene nanoribbon, showing types of edges.

Graphene spintronics. A major difference between a GNR and a CNT is that the GNR has edges, which can have an armchair shape, a zigzag shape, or a mixture shape of the two (**Fig. 4**). The zigzag-edge GNR is particularly interesting for potential spintronic device applications due to the existence of ferromagnetic edge states. First-principle simulations indicate that, at equilibrium, both edges of a zigzag-edge GNR are ferromagnetic with opposite polarities. If a transverse electric field is applied, the degeneracy of up-spin and down-spin bands is lifted, and the GNR can become 100% spin-polarized (half-metallic) in an energy range near the Fermi level. The half-metallic material is most ideal for the application as the spin-polarized contacts in spintronic devices. *See* SPINTRONICS.

Device modeling and simulation. Conventional device simulators treat devices using a top-down view. They have evolved to model ever smaller transistors from their origin of describing the physics of micrometer-scale transistors in the 1960s. For CNT and graphene devices, the validity of the top-down approaches is questionable. A graphene or a CNT is a new type of nanomaterial with a different band structure from common semiconductors. The simulation tools for carbon-based devices should incorporate an understanding of the new band structure, quantum effects, ballistic transport, 3D electrostatics, and dissipative scattering. They should also be able to treat open boundaries for devices under biases and nonequilibrium transport conditions. The nonequilibrium Green's function (NEGF) formalism provides an ideal approach for bottom-up device simulations to address the above challenges.

The following simulation capabilities have been developed using the NEGF formalism for CNTFETs or GNRFETs: (1) an atomistic treatment of the transistor channel, (2) a 3D self-consistent solution of the Poisson equation for a complex gating geometry, (3) treatment of ballistic transport or transport with inelastic scattering, (4) modeling either the doped source/drain or the metal source/drain contact, and (5) the capability to include electron-photon coupling for simulating light emission or photocurrent processes. The device simulation has matured to a point that the experiments can be quantitatively described, as shown in Fig. 3*b*.

Pros and cons of carbon electronics. The **table** summarizes the advantages and disadvantages of

Pros and cons of carbon-based electronics, in comparison to conventional electronics

Pros	Cons
Low effective mass (high mobility, carrier velocity)	Low effective mass (increases the transistor scaling limit)
Ballistic transport	Small bandgap (leakage)
Ultrathin body (decreases the transistor scaling limit)	Low semiconductor capacitance (low DOS) variation
Identical conduction and valence bands	
Low power supply operation	
Tunable bandgap and effective mass	
Possibility of 3D integration	

carbon-based electronics, in comparison to conventional electronics based on silicon and III–V compound semiconductor materials, which have dominated electronics for the last 40 years. The low dimensionality, desired crystal structure, and new band structure of a CNT or a graphene offers new opportunities and potentially significant advantages over conventional electronics. Although various challenges, such as control of variation and leakage current, must be addressed before carbon-based electronics strides its way into extensive applications, carbon nanotubes and graphene are among the most promising new materials that will revolutionize electronics in the future.

For background information *see* CARBON; CARBON NANOTUBES; ELECTRONICS; GRAPHITE; NANOCHEMISTRY; NANOSTRUCTURE; NANOTECHNOLOGY in the McGraw-Hill Encyclopedia of Science & Technology.

Jing Guo

Bibliography. P. Avouris et al., Carbon nanotube electronics, *Proc. IEEE*, 91:1772–1784, 2003; S. Das Sarma et al. (eds.), *Special issue*: Exploring graphene—recent research advances, *Solid State Commun.*, vol. 143, no. 1–2, 2007; A. Javey et al., Ballistic carbon nanotube field-effect transistors, *Nature*, 424:654–657, 2003; M. Lundstrom and J. Guo, *Nanoscale Transistors: Device Physics, Modeling, and Simulation*, Springer, 2006; R. Saito et al., *Physical Properties of Carbon Nanotubes*, Imperial College Press, 1998; Y. Son, M. Cohen, and S. Louie, Metallic graphene nanoribbons, *Nature*, 444:347–349, 2006.

Carnivoran evolution

The placental mammal order Carnivora encompasses many charismatic taxa, including dogs and cats, bears, raccoons, weasels, hyenas, civets, mongooses, seals, sea lions, and walruses. With over 260 living species, Carnivora is one of the most species-rich and ecologically diverse groups of mammals. Despite their name, carnivorans range in diet from pure carnivores to species that specialize on fruit, leaves, and insects, as well as the full spectrum of mixed diets. Carnivorans also display a broad range in styles of locomotion, from the fastest land animal, the cheetah, to climbing, burrowing, and aquatic species.

In addition to their living diversity, carnivorans have an excellent fossil record. The earliest carnivorans are known from the early Paleocene [65 to 61 million years ago (mya)] of North America, a time immediately following the Cretaceous-Tertiary (K/T) boundary. It is hypothesized that the K/T extinction event, which saw the demise of two-thirds of the world's species, including dinosaurs, allowed for mammals to become the dominant land animals of the Cenozoic Era (65 mya to the present). The origins of Carnivora are poorly understood, but one possibility is that they evolved from *Cimolestes*, a Late Cretaceous to early Paleocene insect-eating mammal. Although the earliest carnivorans are very different from the forms seen today, most carnivorans, living and extinct, share a characteristic dental modification called carnassials. Carnassials are the bladelike upper fourth premolar and lower first molar, which shear against each other for enhanced meat-slicing ability (**Fig. 1**). While some fruit- or leaf-eating carnivorans have subsequently modified their carnassials, this feature is the diagnostic character of the order.

Recent advances. Because of the living diversity and excellent fossil record, Carnivora has been the focus of many studies in recent years. The greatest advances in the understanding of carnivoran evolution involve resolving the relationships of the living and extinct species, providing a framework for more detailed study of their evolutionary history. The use of molecular phylogenetics, a methodology determining relationships of organisms based on shared gene and protein sequences, has vastly changed the understanding of carnivoran relationships. Among the notable changes is the discovery that pangolins, scaly anteaterlike mammals, are the closest living relatives to Carnivora.

Relationships within Carnivora have also shifted, including the separation of civets into three different families and the separation of weasels and skunks into two families. Molecular data support the close relationship of all of Madagascar's unique

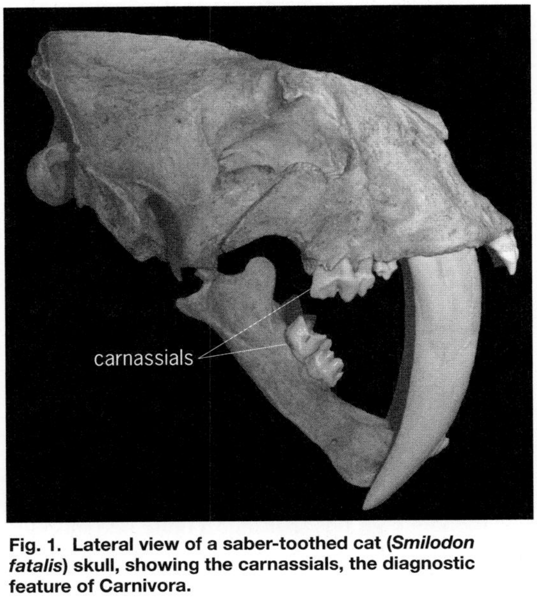

Fig. 1. Lateral view of a saber-toothed cat (*Smilodon fatalis*) skull, showing the carnassials, the diagnostic feature of Carnivora.

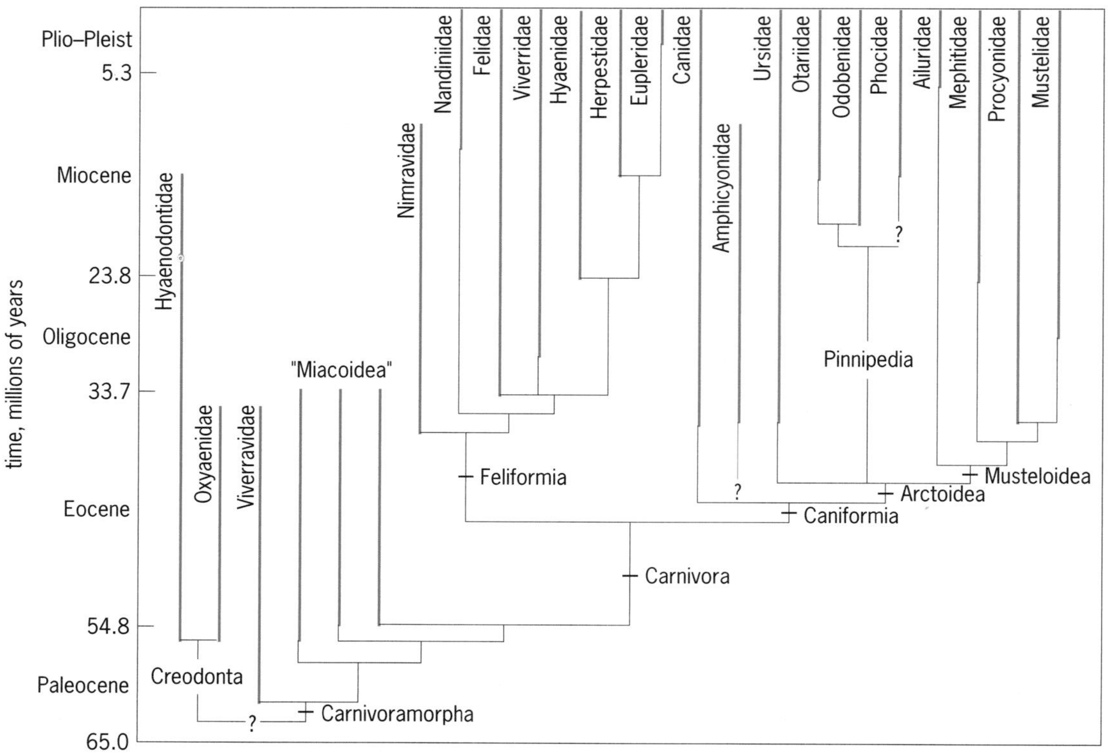

Fig. 2. Carnivoran family tree. Color lines represent known fossil ranges of families. Gray lines represent relationships among taxa.

carnivorans, now unified into a new family, and support a close relationship between Malagasy carnivorans, mongooses, and hyenas, rather than the traditional grouping of hyenas and cats. Molecular data also support a single origin for seals, sea lions, and walruses, which were previously thought to have two separate origins within Carnivora. Fossils have contributed to the changing relationships as well, with new ancient species validating a separate family of red pandas, as opposed to their traditional placement within raccoons. Well-preserved fossils of Paleocene (65–55 mya) carnivorans have also resolved the early history of the group. The new fossils support a single origin of the living carnivoran families from "Miacoidea," which suggests that the living families may have separated almost 15 million years later than previously thought.

The new family tree has been used extensively to reassess the understanding of evolutionary trends in Carnivora. New studies, detailed below, show that most of the modern families started out as small to medium animals, with independent evolution of large body size in several families. New anatomical studies have also shown that the early evolution of carnivorans was fast and that these forms quickly filled in most of the ecological niches that carnivorans still occupy today.

Relationships with other mammals. Recent studies of molecular phylogenetics have found that placental mammals can be divided into four superorders, that is, taxonomic groups that incorporate several traditional orders. Carnivora falls within the superorder Laurasiatheria. Other Laurasiatheria include Perissodactyla (horses, tapirs, and rhinoceroses), Cetartiodactyla (whales and even-toed ungulates), Chiroptera (bats), Soricomorpha (shrews and moles), and Pholidota (pangolins). The other placental mammal superorders are Euarchontoglires (primates, rodents, rabbits, tree shrews, and colugos), Afrotheria (elephants, sea cows, hyraxes, aardvarks, tenrecs, and elephant shrews), and Xenarthra (sloths, armadillos, and anteaters). Together, Laurasiatheria and Euarchontoglires are termed Boreoeutheria, reflecting their hypothesized origin in the Northern Hemisphere.

Family tree. Living carnivorans and their close relatives are divided into two branches (**Fig. 2**): dogs and allies (Caniformia) and cats and allies (Feliformia). Caniformia is composed of the living families of dogs (Canidae), bears (Ursidae), red pandas (Ailuridae), seals (Phocidae), sea lions (Otariidae), walruses (Odobenidae), skunks (Mephitidae), raccoons (Procyonidae), and weasels and allies (Mustelidae). All of the living families except for dogs are grouped in the Arctoidea. Arctoidea includes the three aquatic families (seals, sea lions, and walruses), which are further grouped in Pinnipedia, although the relationship of seals to sea lions and walruses is still debated. The relationships of Musteloidea (red pandas, raccoons, skunks, and weasels and allies) have recently been revised, resulting in the erection of two new families and the separation of skunks from weasels and allies. In addition, one extinct family, bear-dogs (Amphicyonidae), is included within Caniformia, but the exact relationships of bear-dogs to other caniforms are unresolved.

On the other side of the carnivoran family tree, the living feliforms include cats (Felidae), West African palm civets (Nandiniidae), true civets (Viverridae), mongooses (Herpestidae), hyenas and aardwolves (Hyaenidae), and Malagasy carnivorans (Eupleridae). The relationships of feliforms have undergone significant revision in recent years, including the erection of a new family, Eupleridae. Species in this new family include enigmatic animals such as the fossa and falanouc, previously assigned to Viverridae, as well as several genera of Malagasy "mongooses," previously assigned to Herpestidae. Another species previously grouped with Viverridae, the West African palm civet (*Nandinia binotata*), has been identified as the most basal (first branch on the family tree) of the living feliforms. The relative positions of cats and true civets are still unresolved. In addition to the living feliform families, there is an extinct family of false saber-toothed cats, Nimravidae, which independently evolved a catlike anatomy but is not closely related to true cats (Felidae). Nimravidae is probably basal to all living feliform families.

Creodonta. An extinct group with particular relevance to carnivoran evolution is the order Creodonta, composed of two families, Oxyaenidae and Hyaenodontidae. Creodonts were carnivorous mammals that were the dominant predators for much of the early Cenozoic, before going extinct in the late Miocene (~8 mya). Creodonts share carnassials with carnivorans, suggesting common ancestry. However, only the molars of creodonts became carnassials, unlike the premolar and molar carnassial combination observed in Carnivora.

Evolutionary history. The earliest fossil representatives of the living families of Carnivora appeared 37 mya. However, there are earlier fossils with the diagnostic carnassial teeth that do not fall within the living familes (called the "crown" group), but represent the stem leading to the living families. These "stem" carnivorans are part of the more inclusive group, Carnivoramorpha, which includes all species that are more closely related to the living carnivorans than to creodonts. There are two major groups of stem carnivorans: Viverravidae (not to be confused with civets, Viverridae) and Miacoidea. The earliest known carnivorans are viverravids from the early Paleocene (65–61 mya) of North America. Viverravids are probably the most basal group of Carnivoramorpha and were small- to medium-sized terrestrial animals that incorporated insects as a large part of their diet.

Miacoidea is a group of terrestrial and arboreal early carnivoramorphan species that appear to represent a series of intermediate forms between the more basal viverravids and the true carnivorans (Fig. 2). Recent studies have determined that miacoids do not share any diagnostic features to show them all to be closely related, and thus it is likely that Miacoidea is not a valid taxonomic group. It was previously thought that feliforms evolved from viverravids, and caniforms from miacoids. However, new viverravid and miacoid fossils show that all crown carnivorans evolved from miacoids, which pushes the separation of the modern families from the early Paleocene to the middle Eocene, almost 15 million years later than previously thought.

By the late Paleocene (61–55 mya), viverravids and miacoids are known from Asia and North America, spreading to Europe by the early Eocene (55–49 mya). Both Viverravidae and Miacoidea are extinct by the late Eocene (37–34 mya). Also in the late Eocene, the first representatives of several crown group carnivoran families (Canidae, Mustelidae, Ursidae, Amphicyonidae, and Nimravidae) appear on the Northern continents, although modern feliform families do not appear until the Oligocene (34–24 mya). Carnivorans do not invade the Southern continents (Africa and South America) until the Miocene (24–5 mya). While all caniform families have a global distribution, feliforms, except for Nimravidae and Felidae, are largely restricted to the Old World throughout their history.

Evolutionary trends. Because of their lengthy fossil record, carnivorans have been the focus of many studies of evolutionary trends in body size, brain size, diversity, and paleoecology. Body size is of particular interest because it correlates with many life history variables, including gestation length, litter size, home range size, and population density. In addition, living carnivorans have the largest adult body size range of any mammalian order, from the least weasel [30–75 g (1–2.6 oz)] to the southern elephant seal [300–4000 kg (660–8800 lb)]. Families with large modern body sizes are often reconstructed to have similarly large ancestors. However, fossil species are often outside the body size range of their living relatives, particularly as body size began to increase only with the appearance of the modern families in the late Eocene. Using new estimates of fossil body sizes, recent analyses have reconstructed the ancestors of many carnivoran clades, including Caniformia, Arctoidea, and Mustelidae, as small- to medium-sized, with several independent episodes of body size increase, and a few decreases.

Brain size. Casts of the internal surface of fossil carnivoran braincases provide direct evidence of increasing brain size evolution related to enhanced sensory perception, including expanded olfaction and improved depth perception. It has been suggested that carnivorans and herbivores have been engaged in a cognitive "arms race" throughout their evolutionary history. This hypothesis is supported by evidence of alternating increases in predatory and prey brain sizes, although carnivorans usually had larger brain sizes than coexisting herbivores. Extensive studies have demonstrated that modern carnivoran families experienced great increases in brain size from the early Eocene to Miocene (55–5 mya) and a second increase from the Pleistocene to Holocene (2 mya to the present). As the modern carnivoran families must have diverged by the middle to late Eocene, these increases must have occurred during or after the divergence of the modern families. Therefore, most increases in brain size must have occurred independently in the modern families.

Paleoecology. Ecomorphology and competition have been studied extensively in the fossil record of carnivorans. Teeth reflect diet and ecology, and studies of fossil teeth reveal much about paleoecology and its relationship to evolutionary diversity. The early fossil record of carnivoran dentition shows that taxonomic and morphological diversity increased rapidly in the early Cenozoic, and there is no evidence that creodonts suppressed early carnivoran evolution through competition. Interestingly, by the late Eocene/early Oligocene, the early carnivorans had filled most of the same ecological niches that still define the living species. Although different families dominated at different times, entirely new forms and, relatedly, new ecological niches are rare. Instead, catlike forms, wolflike forms, and hyenalike forms evolve multiple times in different families. Even specialized forms such as saber-toothed cats evolved independently in Felidae and Nimravidae. This lack of novelty in the carnivoran record perhaps reflects the stability of prey as a food source, in contrast to the environment-driven shifts affecting herbivore diets.

Locomotor styles also reflect diversity in carnivoran paleoecology, especially when there is significant dietary overlap among coexisting predators. Coexisting carnivorans in modern ecosystems can partition resources by inhabiting different locomotor niches defined by habitat (arboreal or terrestrial) or hunting style (pursuit or ambush). Studies of fossil carnivoran ecomorphology have shown that the locomotor diversity of coexisting carnivorans is similar in fossil and Recent ecosystems. Although the species are different, the forms are similar, demonstrating that extinct taxa partitioned resources similarly to living species.

Large hypercarnivorous forms in particular have evolved several times. Large catlike forms have evolved in at least six different families, from short-faced bear-dogs to leopard-sized mustelids. Bone-cracking forms have evolved at least twice, in hyenas and dogs. Wolflike forms have evolved at least five times, in dogs, bears, red pandas, bear-dogs, and hyenas. However, despite the repeated evolution of hypercarnivorous forms, it has been demonstrated that hypercarnivory is often an evolutionary dead end. Large hypercarnivores diversify quickly, but decline and go extinct after 10–15 million years, often being replaced by another hypercarnivorous group. It has been suggested that this is due to the increasing specialization limiting the group's ability to generalize or expand into other niches, thus increasing their extinction risk. Correspondingly, recent studies have shown that hypercarnivores are always less morphologically diverse than their closest non-hypercarnivorous relatives. Thus, while the saber-toothed cat may be a classic image of the carnivoran radiation, the raccoon may well be the better model for success in carnivoran evolution.

For background information *see* ANIMAL EVOLUTION; ANIMAL SYSTEMATICS; CARNIVORA; CREODONTA; DENTITION; MACROEVOLUTION; MAMMALIA; PALEOECOLOGY; PALEONTOLOGY; PHYLOGENY in the McGraw-Hill Encyclopedia of Science & Technology.

Anjali Goswami

Bibliography. J. J. Flynn and G. D. Wesley-Hunt, Carnivora, in D. Archibald and K. Rose (eds.), *The Rise of Placental Mammals: Origins and Relationships of the Major Extant Clades*, pp. 175–198, Johns Hopkins University Press, Baltimore, 2005; J. L. Gittleman (ed.), *Carnivore Behavior, Ecology, and Evolution*, vols. 1 and 2, Cornell University Press, Ithaca, New York, 1989 and 1996; C. M. Janis, K. M. Scott, and L. L. Jacobs (eds.), *Evolution of Tertiary Mammals of North America*, vol. 1: *Terrestrial Carnivores, Ungulates, and Ungulatelike Mammals*, Cambridge University Press, 1998; K. D. Rose, *The Beginning of the Age of Mammals*, Johns Hopkins University Press, Baltimore, 2006; B. Van Valkenburgh, Major patterns in the history of carnivorous mammals, *Annu. Rev. Earth Planet. Sci.*, 27:463–493, 1999.

Cellular imaging

Proteins, the molecules that perform the majority of biological functions in a cell, are in general colorless. Biologists conventionally use fluorescently tagged antibodies that specifically recognize their proteins of interest in order to indirectly visualize these biomolecules. Antibodies can be introduced into the cells by one of two methods: (1) permeabilization, wherein solvents, detergents, or drugs are used to form gaps in the plasma membrane that allow the passage of the antibodies; (2) microinjection, through which antibodies are introduced with a microneedle. In most cases, permeabilization kills the cells, and immunofluorescence of fixed cells generates only static images, providing a snapshot of otherwise highly dynamic biological structures and processes. On the other hand, microinjection will generate a high local protein concentration of antibodies that may affect the biological functions of the cells. In addition, microinjection is a laborious process that results in relatively few cells suitable for imaging.

The use of green fluorescent protein (GFP) and its color variants, derived from various marine organisms or through point mutations of the GFP, is therefore a revolution in cell biology. Fluorescent proteins (FPs) can be easily conjugated to proteins using recombinant DNA technology, thus making the target protein fluoresce when expressed in live cells. Since FPs are genetically conjugated to their targets, it is easy to express the recombinant proteins in a large number of cells, and even steer the expression to specific tissues in transgenic animals. It is also possible to express more than one FP-tagged protein in the same cell, thus enabling biologists to monitor the dynamic interaction of several proteins in live cells. The techniques of live-cell imaging, likewise, focus on how various aspects of protein dynamics can be studied using FP tags.

Photobleaching techniques. One key property of most fluorophores is their tendency to lose fluorescent intensity after prolonged exposure to excitation light (termed photobleaching). While this property hinders long-term live-cell imaging and complicates fluorescent intensity calculation, it can also be turned into a powerful advantage to study biological processes through ingenious experimental design.

Fluorescence recovery after photobleaching (FRAP). Many proteins exhibit multiple subcellular localization patterns. Some proteins may perform distinct biological functions when targeted to different compartments, such as acting as a receptor on the plasma membrane and affecting gene expression when inside the nucleus. As a regulatory mechanism, proteins can also be sequestered into a different location, away from the site of action. Protein trafficking is thus a critical issue that live-cell imaging is perfectly suited to study. One widely used technique to study protein trafficking is fluorescence recovery after photobleaching (FRAP). Significant advances have been made in photobleaching experiments due to the versatility of laser scanning confocal microscopes, with which one can control the laser intensity within any customized shape in the image.

FRAP operates on a simple principle: In the presence of bidirectional molecular transport between two compartments, if the fluorescent protein in one compartment is optically depleted by photobleaching, the influx of fluorescent protein from the other compartment will cause the fluorescence in the bleached zone to recover. As shown in **Fig. 1*a***, FRAP experiments typically involve bleaching a small area of the fluorescent sample once. The fluorescence recovery within the bleached zone is then analyzed. The rate of fluorescence recovery will be proportional to the turnover rate of FP-tagged protein in the bleached zone.

FRAP is a remarkable technique to reveal and examine protein trafficking. However, there are situations in which the regions of interest are too complex to be targeted by the laser or where the experiment is complicated by the presence of multiple sources from which the FP-tagged protein may be released into the bleached zone. These situations thus underscore the need for other methodologies.

Fluorescence loss in photobleaching (FLIP). FLIP is based on the concept that continuous depletion of fluorescent molecules in the bleached zone will indirectly cause fluorescence loss in the source of protein transport. Technically, FLIP differs from FRAP by the repetitive bleaching of the same region, and is a powerful method to identify other unknown compartments that may be involved in the protein trafficking. As illustrated in Fig. 1*b*, FLIP can be used to identify if the microtubule or the actin network is releasing an FP-tagged protein into the nucleus. The continuous bleaching of the nucleus depletes the fluorescent molecules in the zone. If the microtubule network is mainly involved in this protein trafficking, the microtubule-associated pool will be indirectly and preferentially bleached as well.

Photoswitch techniques. Photobleaching techniques rely on high-intensity lasers that can be detrimental to biological specimens, due to the generation of local concentrations of phototoxic oxidative radicals. To address this problem, a new generation of fluorescent probes has been created to act as photoswitches that can be turned on or off, or change their color upon exposure to a short burst of excitation laser. These new fluorophores thus allow biologists to track a specific pool of protein in living cells without the use of intense laser.

Photoactivation. Photoactivatable GFP (PA-GFP) was generated by mutagenizing GFP into a variant that undergoes an optical enhancement of several orders of magnitude upon activation by 405-nm light (**Fig. 2*a***). This localized and instantaneous increase in fluorescent intensity thus conveniently marks a specific pool of tagged protein whose fate can be easily traced. Technically, it is the exact reverse of FRAP: instead of photobleaching and darkening a specified zone, one activates the protein and brightens the tagged-protein within the zone. However, there is a critical disadvantage of photoactivation: PA-GFP tends to be nearly nonfluorescent before activation, making visualization of the tagged-protein difficult.

Photoconversion. While most fluorophores are single-color dyes, with a single set of excitation and emission spectra, each photoconvertible fluorescent protein has two excitation spectra and two emission spectra. Photoconvertible fluorophores usually

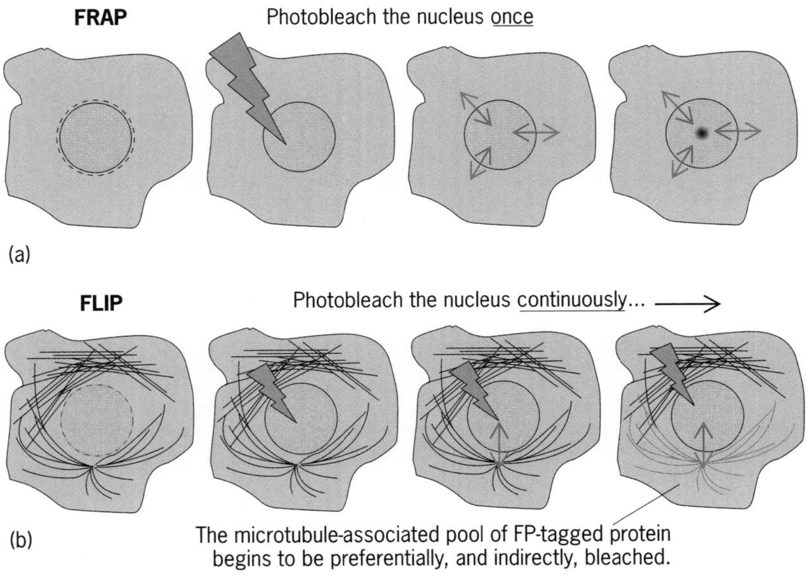

Fig. 1. Photobleaching techniques: FRAP and FLIP. (*a*) Schematic representation of FRAP (fluorescence recovery after photobleaching). To demonstrate that the fluorescently tagged protein of interest (FP) is trafficking between the cytoplasm and the nucleus, a bleached zone is demarcated around the nucleus (dotted line), and the nuclear pool of FP is photobleached once. If FP undergoes bidirectional nucleus-cytoplasm trafficking (two-way arrows), the photobleached pool will gradually be replenished by the fluorescent pool from the cytoplasm and the nucleus will recover its fluorescent intensity, while the cytoplasm will lose part of its intensity due to the exchange of the photobleached pool from the nucleus. (*b*) Schematic of FLIP (fluorescence loss in photobleaching). In this scenario, a nuclear protein is sequestered into two protein networks in the cytoplasm: actin (depicted by the straight lines) and microtubule (curved lines). The nuclear protein is released into the nucleus following some drug treatments, and it is the goal of the experiment to examine if it is the actin-associated or the microtubule-associated pool of the nuclear protein that undergoes trafficking. Since both sources are complex structures not easily demarcated for photobleaching, the "destination" (nucleus) is targeted instead. The nucleus will be repetitively photobleached. If the microtubule pool is involved in the protein transport (two-way arrows), it will be preferentially and indirectly bleached. FLIP thus highlights the source of protein trafficking by repetitively photobleaching the destination.

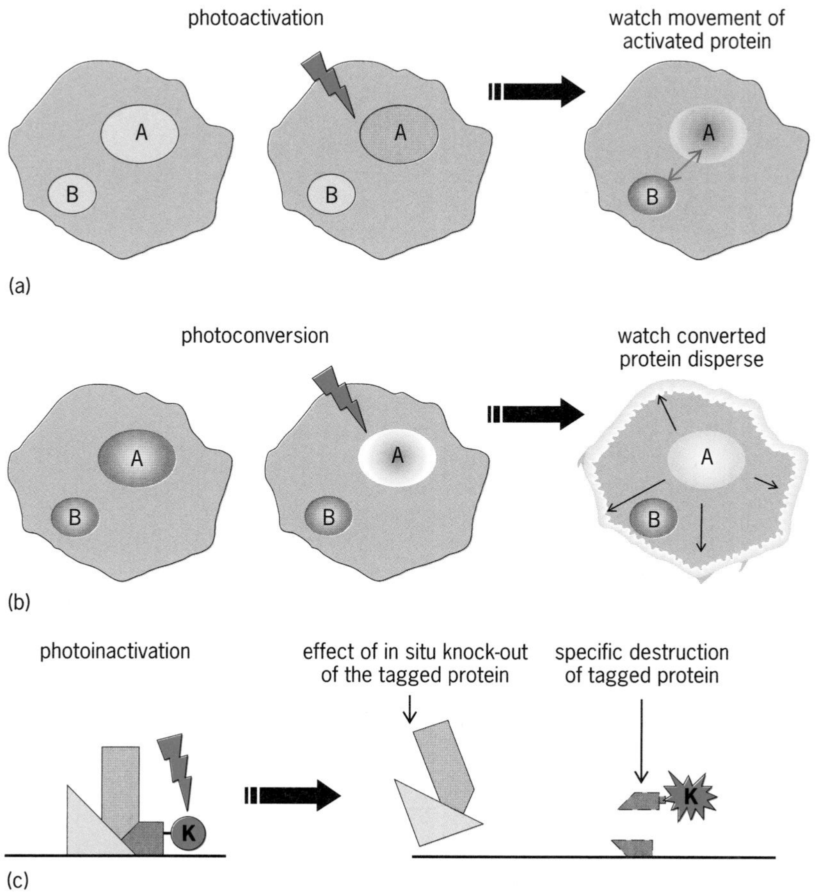

Fig. 2. Photoactivation, photoconversion, and CALI (chromophore-assisted laser inactivation). (*a*) Localized activation of photoactivatable-green fluorescent protein (PA-GFP) tagged to a protein suspected of translocating from subcellular compartment A to compartment B. Immediately following the photoactivation, the fluorescent intensity of PA-GFP increases by two orders of magnitude, allowing biologists to track the path of the specific pool of tagged protein, even though the tagged protein may be ubiquitously localized all over the cell. (*b*) Using the same principle as photoactivation, photoconversion will instantaneously change a green fluorescent protein into a red fluorescent protein. In this schematic example, the converted pool of protein bypasses subcellular compartment B and is redistributed to the plasma membrane, indicating that only compartment A undergoes molecular exchange with the plasma membrane even though both compartments A and B contain the protein of interest. (*c*) Schematic diagram outlining the principle of CALI. The protein of interest is genetically tagged with a photosensitizer "KillerRed" ("K" in a circle). Upon activation, the photosensitizer releases a high local concentration of reactive oxygen species, specifically damaging the tagged protein. This in situ inactivation of a specific protein allows biologists to perform functional knockout to directly examine the role played by the protein without affecting the rest of the cells as would a genetic knockout experiment.

undergo instantaneous and irreversible photoconversion from a green to a red fluorescent form upon activation by 405-nm light (Fig. 2*b*). This property thus makes photoconvertible fluorescent proteins ideal tools for real-time tracking of protein dynamics. Unlike PA-GFP, it does not have invisibility problems prior to conversion. The main drawback is that each photoconvertible fluorescent tag needs the detection of two different colors, thus limiting the choice of additional fluorophore if one also needs to track another protein in the same cell.

Chromophore-assisted laser inactivation (CALI). CALI relies on a group of chromophores (arrangements of atoms forming part of an organic molecule that cause it to be colored) that generate reactive oxygen species (ROS, molecules that contain a very reactive oxygen atom) upon light irradiation. The localized high concentrations of ROS destroy the proteins of interest in situ, while leaving the neighboring molecules intact. CALI thus allows biologists to perform regional knockout experiments to examine the function of a protein. CALI was originally performed using chemical photosensitizing compounds that must be introduced into living systems exogenously, but recently a genetically encoded photosensitizer ("KillerRed") has been derived from a homolog of GFP. The release of ROS by KillerRed can be easily and precisely triggered by excitation light at 540–580 nm. The development of this photosensitizer allows biologists to genetically tag any protein of interest, and to target it for subsequent photoinactivation in situ (Fig. 2*c*).

Fluorescence resonance energy transfer (FRET) techniques. The use of fluorescent proteins offers direct observation of the dynamic properties of specific proteins in live cells by simultaneously providing maximal spatial and temporal resolution. Unfortunately, these fluorophores per se provide no vital information about protein-protein interactions, enzymatic activity, and conformational changes in proteins. It is therefore the ultimate goal of live-cell microscopy to employ biosensors that simultaneously combine the spatiotemporal resolution of fluorescent probes with the functional information offered by biochemical analyses. Significant advances toward this goal have been made through the use of fluorescence resonance energy transfer (FRET). FRET is a distance-dependent interaction between two fluorophores in which excitation energy is transferred from a donor fluorescent molecule to an acceptor molecule without emission of a photon; the energy absorbed indirectly then excites the acceptor to fluoresce with its own emission spectra (**Fig.** 3*a*).

The following two criteria must be considered when choosing a fluorophore pair for FRET: (1) the donor emission spectrum must overlap significantly with the excitation spectrum of the acceptor; (2) the excitation light for the donor must not directly excite the acceptor.

The efficiency of FRET is a function of the inverse sixth power of the intermolecular separation, making it a powerful proximity probe that is sensitive to distances less than 10 nm, which is comparable to the dimensions of biological macromolecules and to large conformational changes within single molecules. The powerful combination of spatiotemporal resolution with the biochemical information provided by FRET has resulted in the innovative design of numerous biosensors (Fig. 3*b*–*d*). Each of these biosensors utilizes one of two detection mechanisms to display the biological event of interest: (1) the gain of FRET as the two fluorophores are brought closer to one another; (2) the loss of FRET as the distance between the two fluorophores increases.

Outlook. With the rapid expansion of new fluorescent probes and the exponential advancement in optical engineering, biologists are at the brink of visualizing many aspects of subcellular architecture and processes with unprecedented spatiotemporal resolution, not only at a single cell level but also in

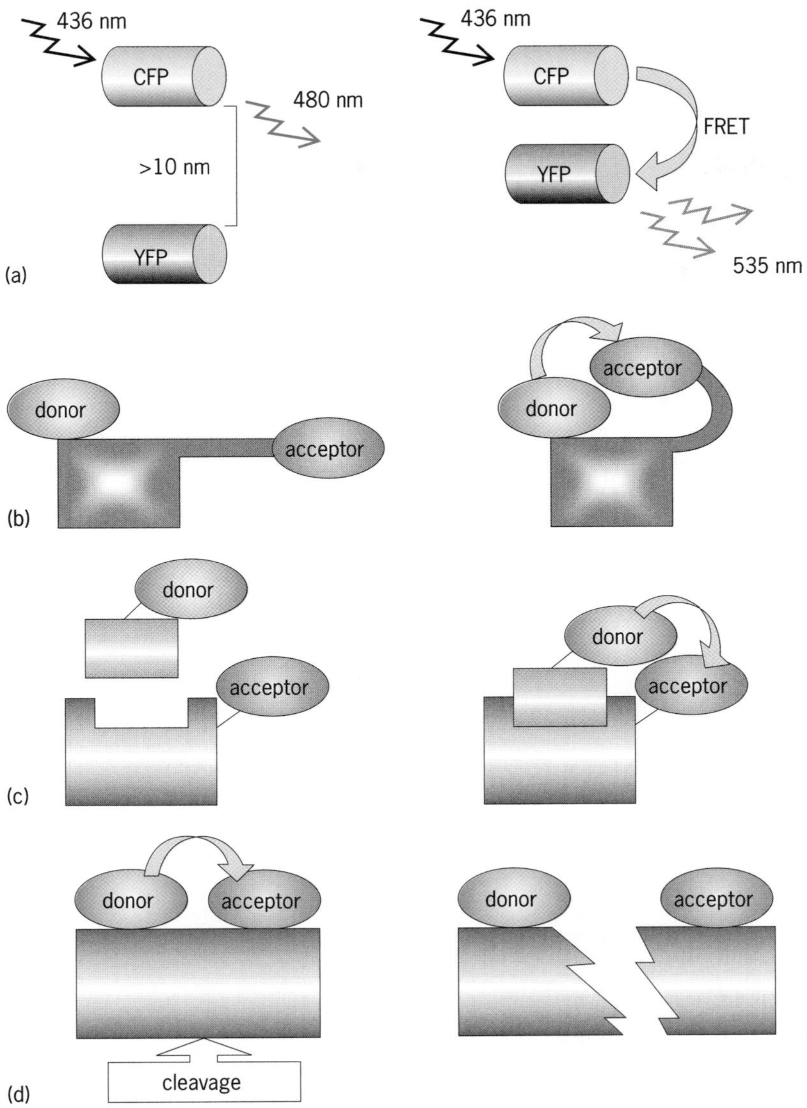

Fig. 3. Principle of FRET and examples of FRET-based biosensors. (*a*) Principle of fluorescence resonance energy transfer (FRET) illustrated by cyan fluorescent protein (CFP, donor) and yellow fluorescent protein (YFP, acceptor), a commonly used FRET fluorophore pair. FRET efficiency is sensitive to the separation between the donor and acceptor. When the distance between the two fluorophores is more than 10 nm, FRET does not occur, and the donor (CFP) emits its own emission spectra upon 436-nm excitation. When the two molecules are within the range of FRET, the highly overlapped spectra of CFP emission and YFP excitation allow energy transfer to YFP (arrow), causing it to fluoresce indirectly at 535 nm. (*b*) One of the most common yet powerful architectural designs of FRET sensors relies on the intramolecular conformational change in the linkers bridging the two fluorophores. The linker can consist of any effector-binding domain that will undergo significant structural change upon binding to its cognate effector, thus directly affecting the efficiency of FRET. (*c*) The interaction of two proteins can also be detected by tagging the two binding partners with FRET fluorophore pairs. The interaction of the two proteins will bring the fluorophores closer to facilitate FRET. (*d*) Protein or polynucleotide processing can be dynamically monitored by flanking the cleavage site with a FRET fluorophore pair. Molecular cleavage will be translated into loss of FRET.

live animals, to help decipher the mysteries of developmental biology and pathogenesis of various diseases.

For background information *see* BIOSENSOR; CELL (BIOLOGY); CONFOCAL MICROSCOPY; FLUORESCENCE; FLUORESCENCE MICROSCOPE; LASER PHOTOBIOLOGY; MOLECULAR BIOLOGY; PROTEIN; VIDEO MICROSCOPY in the McGraw-Hill Encyclopedia of Science & Technology.

Teng-Leong Chew

Bibliography. T. L. Chew and R. L. Chisholm, Monitoring protein dynamics using FRET-based biosensors, in R. D. Goldman and D. L. Spector (eds.), *Live Cell Imaging: A Laboratory Manual*, Cold Spring Harbor Laboratory Press, 2005; D. M. Chudakov, S. Lukyanov, and K. A. Kukyanov, Fluorescent proteins as a toolkit for in vivo imaging, *Trends Biotechnol.*, 23(12):605–613, 2005; B. N. G. Giepmans et al., The fluorescent toolbox for assessing protein location and function, *Science*, 312:217–224, 2006; R. M. Hoffman, The multiple uses of fluorescent proteins to visualize cancer in vivo, *Nature Rev. Cancer*, 5(10):796–806, 2005; O. Shimomura, The discovery of aequorin and green fluorescent protein, *J. Microsc.*, 217(pt. 1):1–15, 2005; R. Y. Tsien, Imagining imaging's future, *Nat. Rev. Mol. Cell Biol.*, 4:SS16–SS21, 2003.

Chiral Brønsted acid catalysis

Preparation of biologically active compounds with high optical purity is very important in the pharmaceutical sciences, as different enantiomers or diastereomers exhibit different biological activity. For example, L-DOPA (3,4-dihydroxy-L-phenylalanine) is used in the management of Parkinson's disease, while the D-isomer does not exhibit such activity. The control of the chirality is also important in the field of material sciences, such as for liquid crystals.

When a compound contains only one enantiomer (stereoisomer that is nonsuperimposable on its mirror image, or chiral) it is said to be optically pure (100% enantiomeric excess). There are a number of methods available for preparing optically pure compounds. One method starts with naturally abundant optically pure compounds, such as amino acids, tartaric acid, and so on. Another uses an achiral compound as the starting material and introduces chirality by means of an external chiral catalyst. Biocatalysts and metal-based catalysts have been used extensively as chiral catalysts. For example, nucleophilic addition to C=O and C=N bonds using chiral Lewis-acid (electrophilic) catalysts provides a useful method for preparing chiral alcohols and amines [reaction (1)]. Lewis acids electrophilically activate

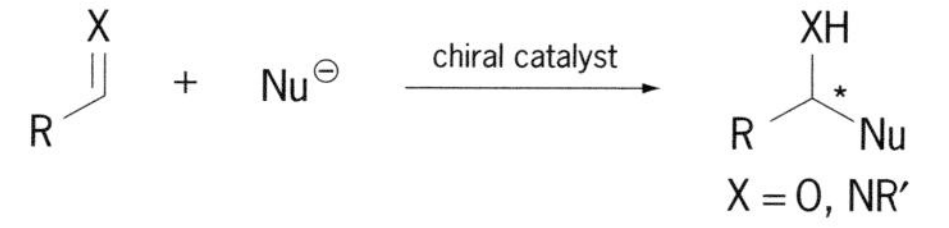

(1)

carbonyl compounds by lowering the lowest unoccupied molecular orbital (LUMO) level.

The reaction of a chiral ligand with the central metal of a Lewis acid results in the formation of a chiral Lewis acid. A number of metal-centered Lewis-acid catalysts have been developed. A range of metal centers, such as B, Sn, Ti, Zr, Mg, Al, Cu, and others, have been investigated as Lewis-acid catalysts.

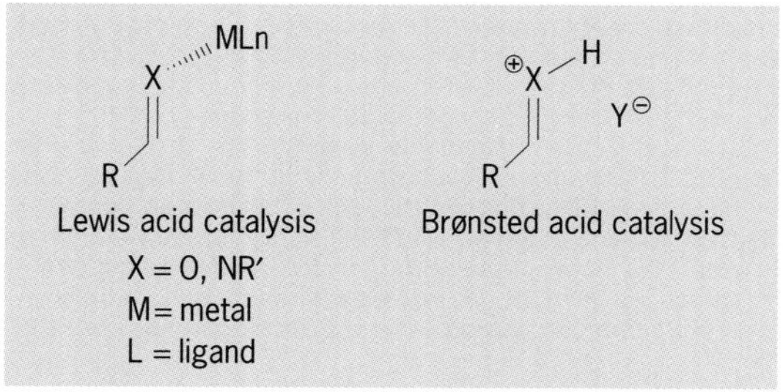

Fig. 1. Lewis acid catalysis versus Brønsted acid catalysis.

The proton of a Brønsted acid (H^+ donor) is the smallest element of the Lewis acid. Brønsted acids activate carbonyl compounds by hydrogen bonding or by forming an oxonium/iminium salt (**Fig. 1**).

In a broad sense, there are three modes of electrophilic activation of carbonyl compounds by Brønsted acid catalysis (**Fig. 2**). Two of them are hydrogen-bonding activation: double hydrogen bonding and single hydrogen bonding. These may be called hydrogen-bond catalysis. The third one is the formation of an oxonium/iminium ion species in the transition state. This is Brønsted acid catalysis, in a narrower sense.

Although Brønsted acids have been used extensively as catalysts for the hydrolysis and/or formation of esters, acetals, and so on, the synthetic utility of Brønsted acids as catalysts for the carbon-carbon bond-forming reactions had not been investigated until quite recently. The last few years have witnessed spectacular advancement in the chiral Brønsted acid catalysis. Thiourea derivatives, $\alpha,\alpha,\alpha',\alpha'$-tetraaryl-1,3-dioxolane-4,5-dimethanol (TADDOL), and phosphoric acids, in particular, have emerged as novel chiral catalysts of choice in a number of synthetic reactions.

Thioureas. In 1998, E. N. Jacobsen of Harvard University found that a thiourea derivative (structure **1**)

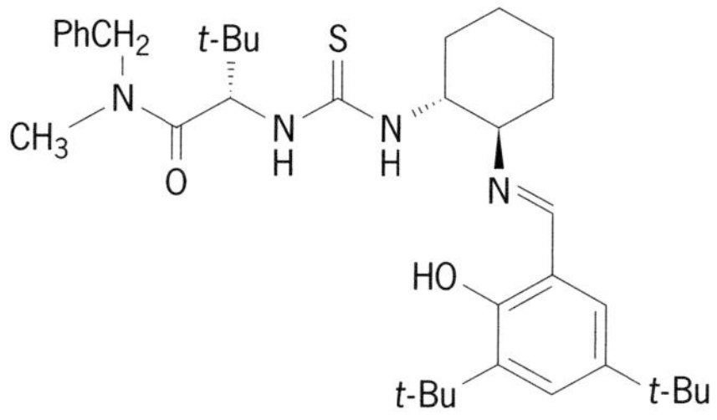

Thiourea catalyst

(1)

was an effective catalyst for the nucleophilic addition to aldimines. HCN, silyl enol ethers, and phosphites have proved to be effective as nucleophiles for the chiral thiourea-catalyzed reactions and the corresponding amines were obtained in high yield with high enantioselectivity. An example is the Mannich-type reaction catalyzed by a thiourea catalyst in reaction (2).

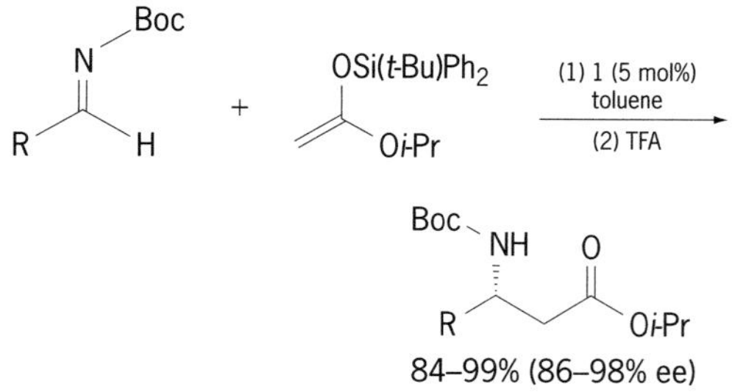

(2)

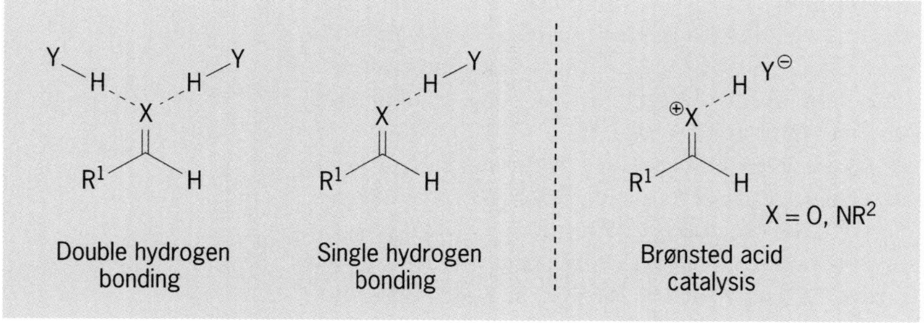

Fig. 2. Modes of activation.

In 2003, Y. Takemoto of Kyoto University developed a bifunctional thiourea derivative (**2**), bearing an amino functionality in addition to a Brønsted acidic site and demonstrated its catalytic activity in

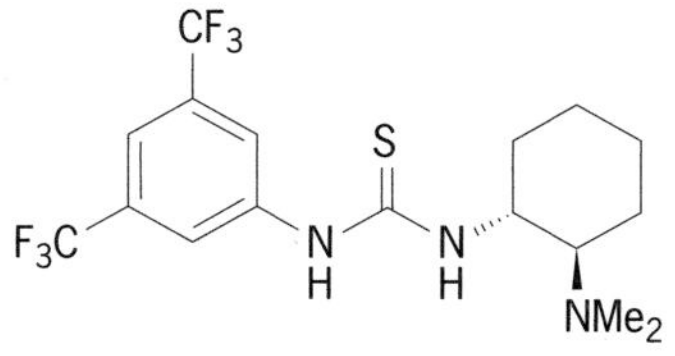

Bifunctional thiourea catalyst

(2)

the Michael addition of malonates to nitroalkenes. The reaction (3) took place smoothly under the in-

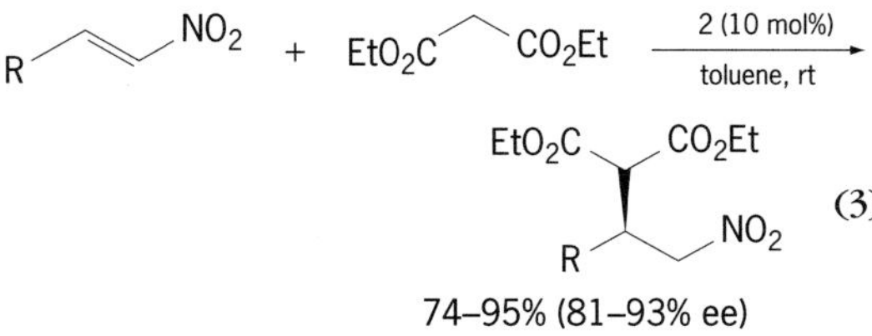

(3)

fluence of Takemoto's catalyst to give the Michael adduct with high enantioselectivity.

TADDOL. In 2002, V. H. Rawal of the University of Chicago found that the hetero Diels-Alder reaction of electron-rich diene with aldehyde was accelerated by an alcohol. In 2003, he reported that the diol, TADDOL (structure **3**), which is readily prepared

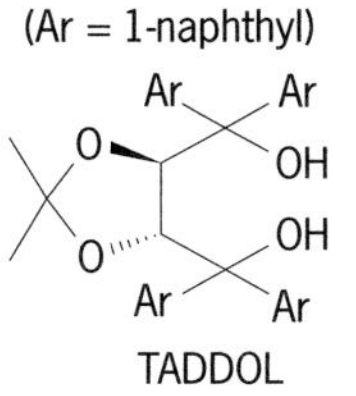

(3)

from tartaric acid, was an effective chiral catalyst for the hetero Diels-Alder reaction (4) of an electron-rich diene with an aldehyde. The corresponding cycloadducts

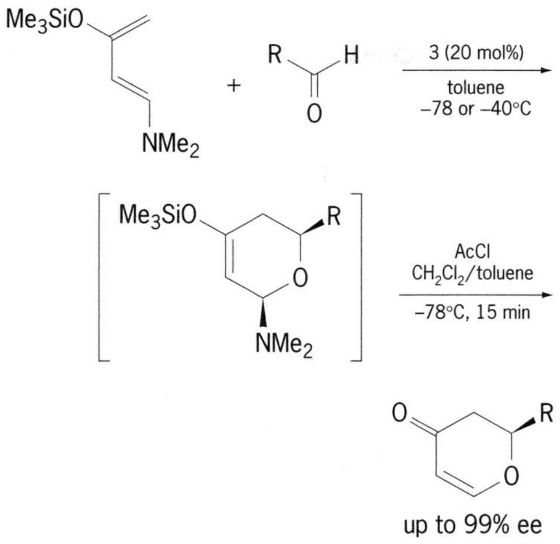

(4)

were obtained with high enantioselectivity. The Diels-Alder reaction of an electron-rich diene with an α,β-unsaturated ketone gave a cyclohexenone derivatives with excellent enantioselectivity. Based on the solid structure of TADDOL, they proposed an internally hydrogen-bonded structure as a transition state (**Fig. 3**).

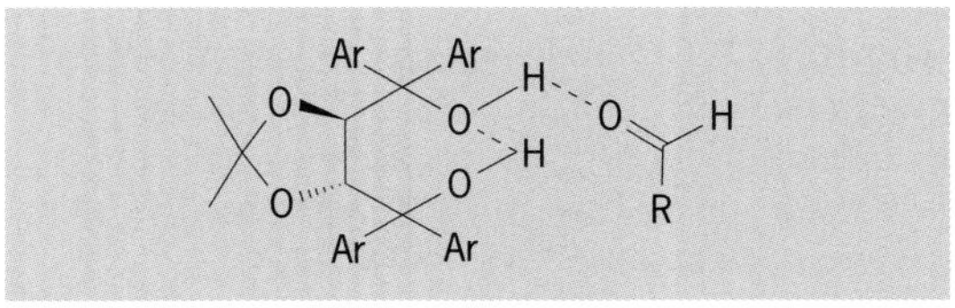

Fig. 3. Mode of activation by TADDOL.

Phosphoric acids. In 2004, T. Akiyama of Gakushuin University, Tokyo, reported that a chiral

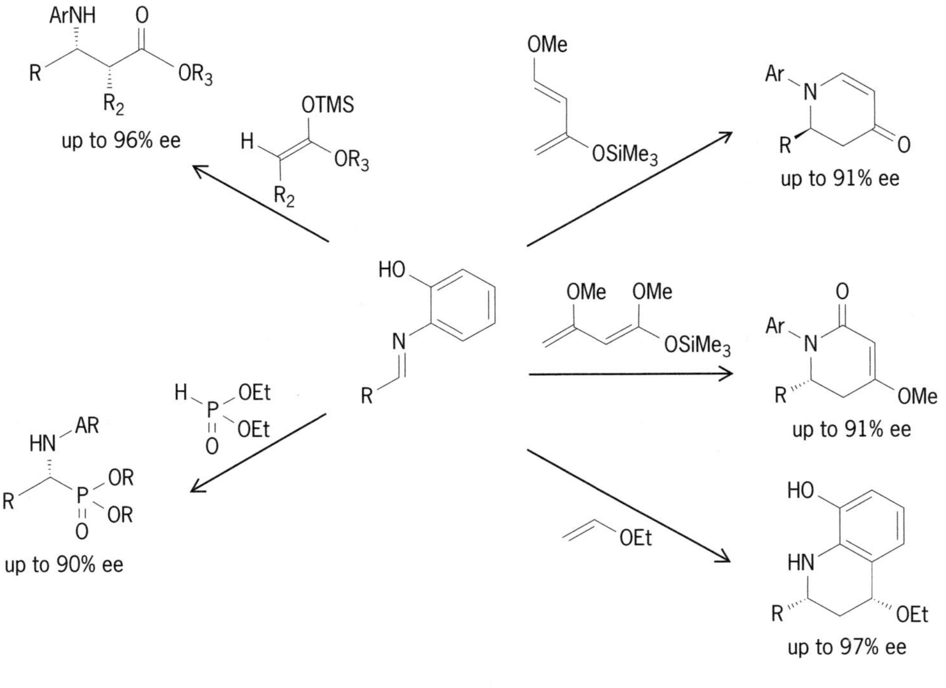

(5)

cyclic phosphoric acid diester (structure 4) was

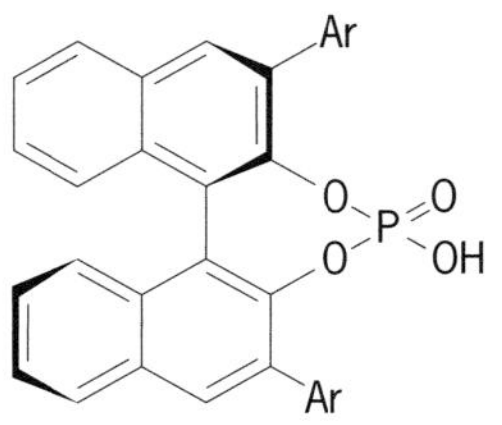

Cyclic phosphoric acid diester catalyst

(4)

an effective catalyst for nucleophilic addition and cyclization reactions (5) with aldimines, generating nitrogen-containing compounds with high enantioselectivity. Subsequently, M. Terada of Tohoku University, Sendai, reported the direct Mannich reaction catalyzed by the chiral phosphoric acid.

After the two reports, a number of phosphoric acid catalyzed enantioselective asymmetric nucleophilic reactions, cycloaddition reactions, and transfer hydrogenations with aldimines have been reported. The proper choice of the 3,3′-substituents is crucial for attaining the high enantioselectivity.

Outlook. So far, chiral Brønsted-acid catalyzed reactions and metal-based chiral Lewis-acid catalyzed reactions work complimentary.

In comparison to metal-based chiral Lewis acids, chiral Brønsted acids are (1) stable to oxygen and moisture, (2) easy to handle, and (3) metal-free and environmentally benign.

The application of chiral Brønsted acid catalysis will be evaluated in industry and more practical asymmetric reactions will be developed.

For background information *see* ACID AND BASE; ASYMMETRIC SYNTHESIS; CATALYSIS; HYDROGEN BOND; LIGAND; MOLECULAR ISOMERISM; OPTICAL ACTIVITY; ORGANIC SYNTHESIS; STEREOCHEMISTRY in the McGraw-Hill Encyclopedia of Science & Technology.

Takahiko Akiyama

Bibliography. T. Akiyama et al., Recent progress in chiral Brønsted acid catalysis, *Adv. Synth. Catal.*, 348:999, 2006; E. N. Jacobsen et al., Asymmetric catalysis by chiral hydrogen-bond donors, *Angew. Chem. Int. Ed.*, 45:1520, 2006; P. R. Schreiner et al., Metal-free organocatalysis through explicit hydrogen bonding interactions, *Chem. Soc. Rev.*, 32:289, 2003.

Climate change and sex determination

Over the next century, global temperatures on Earth are expected to rise from 0.6 to 8°C (1 to 14.6°F). In fact, over the course of the twentieth century alone, the planet has already experienced an increase of approximately 0.6°C (1°F). In response, many plant and animal species have altered the timing of various seasonal activities such as flowering, breeding, nesting, and migration. Species distributions have also been affected, with many temperate zone plants and animals of the Northern Hemisphere moving northward due to increasingly intolerable conditions in the southern parts of their ranges. In the Southern Hemisphere, these species move southward because of the increasingly intolerable conditions found in the northern parts of their ranges. Consequently, scientists are growing ever more concerned with the potential impacts of climate change on natural systems.

An additional cryptic impact of climate change can be seen in organisms whose development is closely linked to temperature. In mammals, birds, amphibians, snakes, most lizards, and some gonochoristic fish (that is, species with separate sexes), specific chromosomes or genes determine the sex of offspring (genetic sex determination). Alternatively, in some reptiles, including all crocodilians studied to date, as well as many turtle and tortoise species and some lizards, the sex of an embryo is determined by its incubation temperature. This process is termed environmental or temperature-dependent sex determination (TSD). Three patterns of TSD have been

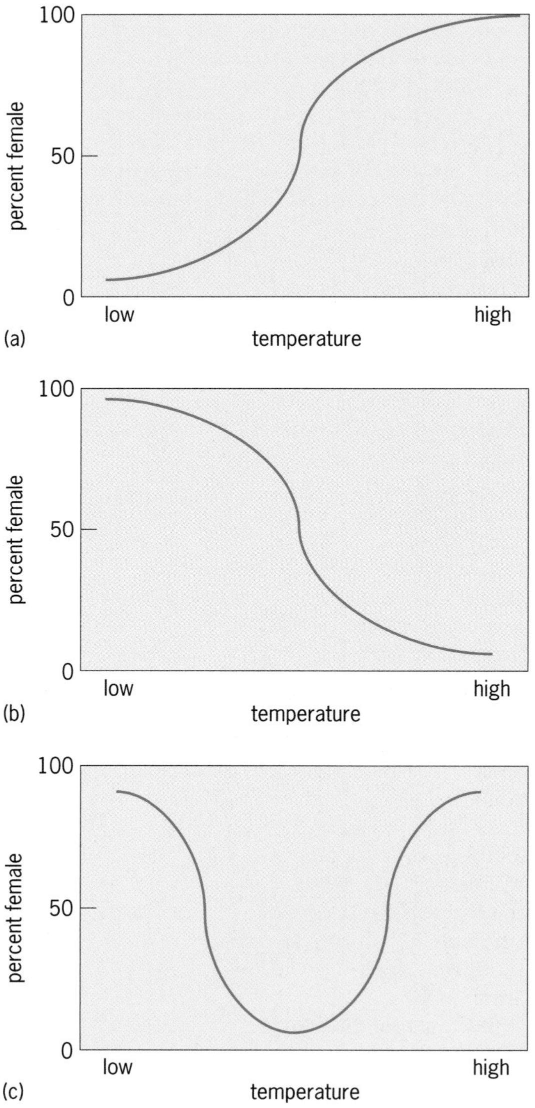

Fig. 1. Patterns of sex ratio in response to incubation temperature: (*a*) type Ia; (*b*) type Ib; (*c*) type II.

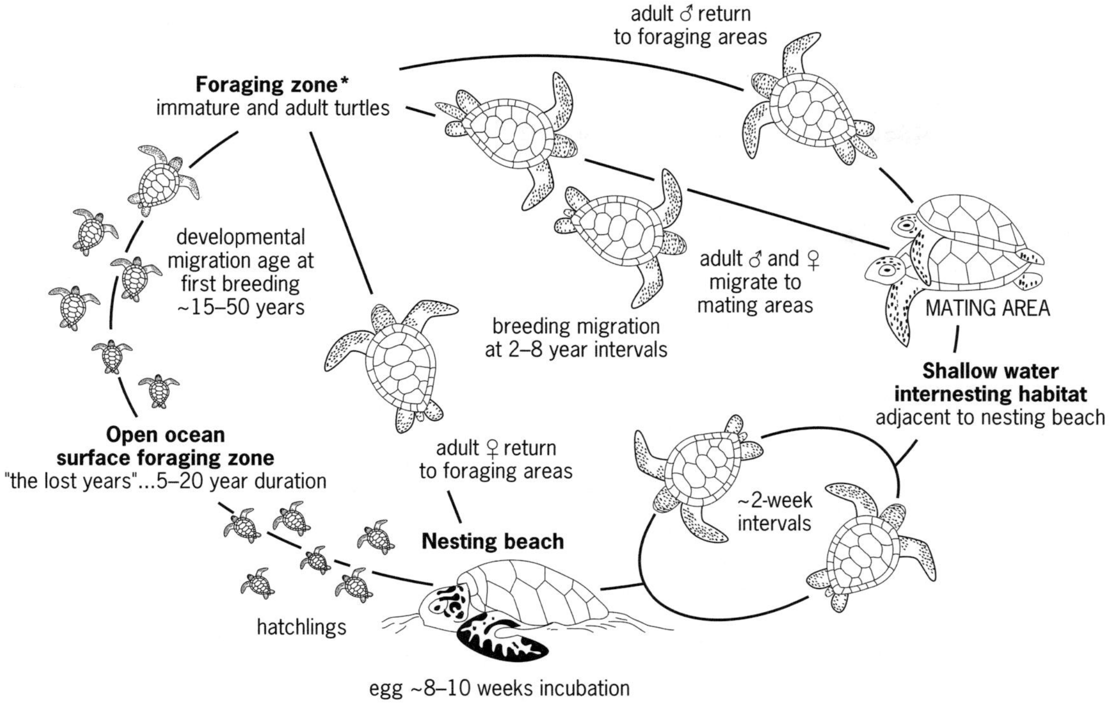

Fig. 2. Generalized sea turtle life cycle; species vary in duration of phases. *Foraging zones, in particular, vary greatly by species.

described: types Ia, Ib, and II (**Fig. 1**). With pattern Ia, males are produced at low temperatures and females at high; pattern Ib is the reverse of pattern Ia. With pattern II, females are produced at both low and high temperatures, with males being produced at intermediate temperatures.

Sea turtles are among the most extensively studied of all animals exhibiting TSD. Of the seven extant species, the leatherback (*Dermochelys coriacea*), hawksbill (*Eretmochelys imbricata*), and Kemp's ridley (*Lepidochelys kempii*) are classified as critically endangered on the World Conservation Union's Red List of Threatened Species, with the olive ridley (*Lepidochelys olivacea*), green (*Chelonia mydas*), and loggerhead (*Caretta caretta*) sea turtles classified as endangered. Flatback (*Natator depressus*) is the remaining sea turtle species and is classified as Threatened. Consequently, sea turtle conservation programs have largely focused on reducing the probability of local population extinctions by limiting adult and egg mortality rates through various practices. However, such interventions are unlikely to effectively address a serious threat to population stability and persistence posed by local climatic changes: the anthropogenic disturbance of primary sex ratios (the number of male and female offspring that are produced) resulting from altered incubation temperatures.

Basic sea turtle biology. All sea turtle species exhibit the same generalized life cycle (**Fig. 2**). They are iteroparous (capable of breeding and reproducing multiple times) and lay large numbers of eggs within a nesting season. Females emerge from the sea and dig their nest chambers on sandy beaches, although the type and composition of beaches can vary greatly. Most females tend to return to or near their natal beaches to nest, but this also varies among species. Sex is determined by the temperature of the nest during the middle third of incubation, with female offspring produced at higher temperatures and males at lower temperatures within a thermal tolerance range of 25–35°C (77–95°F). The pivotal temperature, or the temperature at which equal proportions of each sex are produced when eggs are incubated under constant conditions, is around 29°C (84°F) in these species. A mixture of sexes is produced within the threshold range of temperatures (TRT), which can vary between 1°C (1.8°F) and 4°C (7.2°F). Both the pivotal temperature and the TRT affect primary sex ratio production on sea turtle nesting beaches. However, as a general rule of thumb, nests incubated at sand temperatures less than 29°C (84°F) are mainly male-producing and nests incubated at temperatures higher than 29°C (84°F) are mainly female-producing.

Primary sex ratios. Most sea turtle populations studied to date have highly female-biased primary sex ratios. For example, loggerhead populations in Florida, Brazil, and some parts of the Mediterranean are estimated to be approximately 90% female, with an absence of equally skewed male-producing populations. Unfortunately, numbers of nests or nesting females are often used as indices of population size, and these methods of assessing population stability may in fact be masking a potentially detrimental phenomenon: the decreased production of males.

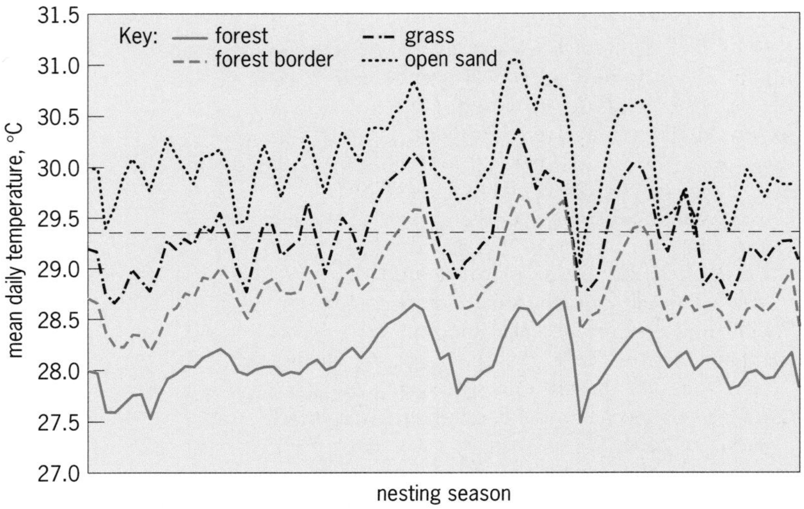

Fig. 3. Seasonal sand temperatures taken at nest depth on a Caribbean hawksbill nesting beach. Temperatures in the lower half are mainly male-producing and temperatures in the upper half are mainly female-producing.

This trend may already be occurring on some nesting beaches. For example, in Antigua, the site of an important hawksbill population, air temperatures (which positively correlate with sand temperatures) have risen by 0.7°C (1.2°F) over the last 35 years. Sex ratio estimates have shown that roughly equal numbers of males and females were produced during the 1990 nesting season, but nests incubated at sand temperatures in 2003 were predominately female-producing. If climate change further biases production of only one sex—for example if no males are produced—then females would have difficulty finding mates and would lay unfertilized eggs, possibly leading to serious population declines. There is evidence that the hatching success of nests laid in Brazil is much lower than on other beaches, a possible reflection of the highly female-biased sex ratios in the region.

However, not all populations are expected to be affected to the same degree, as climate change is expected to be both spatially and temporally heterogeneous. For example, despite global trends, it appears that North Carolina has not experienced pronounced warming over the past three decades and the primary sex ratios of loggerhead turtle populations nesting there are relatively balanced (42% males), especially when compared to other populations. This suggests that the primary sex ratio in North Carolina might not become drastically female-biased in future years, and that male offspring from this region could assume an important role in maintaining viable loggerhead populations in the Western Atlantic.

Responses to climate change. Several scenarios have been proposed with regard to how sea turtles might respond to thermal changes in the environment. First, turtles may alter the placement of their nests along the beach, preferentially using cooler, more male-producing areas. Some hawksbills show a preference for nesting in cooler, vegetated areas such as coastal forests located behind the sandy beach. The sand temperatures in these environments are below the pivotal temperature and seem to be important male-producing areas (**Fig. 3**). Unfortunately, widespread deforestation, especially in coastal areas, may limit the opportunities for individual females to select male-producing nesting areas. Turtles may also place their nests in cooler sand near the water, but this also carries the increased risk of tidal inundation and poor hatchling survival.

A second scenario involves turtles altering the timing of nest placement so as to include cooler periods at the beginning and end of the season to utilize ephemeral male-producing incubation environments. Mathematical models have shown that loggerheads nesting in North Carolina would have to alter the temporal distribution of nesting by only a few days to adapt to 1°C (1.8°F) of warming and by up to a week for 3°C (5.4°F) of warming.

As a final possibility, altered climatic conditions could, in theory, induce an adaptive change in turtle pivotal temperatures. Selection for, and subsequent evolution of, a higher pivotal temperature would then allow males to be produced at warmer temperatures and offset the trend toward female-biased sex ratios. However, the lack of variation in this trait (pivotal temperatures vary little, both within and across species) in conjunction with the relatively long generation times of most sea turtle species indicates that evolutionary modifications would be too slow to respond to the changes in global temperatures projected over the next century.

Other species. The potential impacts of climate change are not limited to sea turtles. For example, an increase of 4°C (7.2°F) in mean air temperature would effectively eliminate the production of male offspring in a population of painted turtles (*Chrysemys picta*) in Illinois. Even more modest increases [less than 2°C (3.6°F)] may drastically alter their sex ratios.

In the Atlantic silverside (*Menidia menidia*), the species that provided the original discovery of TSD in fishes, cooler temperatures like those found at the beginning of the breeding season result in more female offspring being produced. Males are produced later on in the breeding season as temperatures begin to increase. Females benefit more than males from a longer growing season since attaining larger adult size is more important for their reproductive success. Increasing temperatures might thus directly alter the number of females produced or affect the length of the growing season, thereby necessitating range expansions to find cooler waters.

Conservation outlook. While there is still much to be learned about the rate and tempo of climate change, its potential impacts must be considered when formulating species conservation plans. Emphasis should be placed on protecting areas that are important for the production of males, on both regional and local scales. The consequences of any anthropogenic activities that may alter the thermal properties of naturally occurring nest sites should be carefully assessed. In threatened species whose sexual differentiation depends on embryonic temperatures, effective conservation strategies must include the preservation of habitats with appropriate thermal characteristics.

For background information *see* CHELONIA; CLIMATE MODIFICATION; ENDANGERED SPECIES; EXTINCTION (BIOLOGY); GLOBAL CLIMATE CHANGE; MARINE ECOLOGY; POPULATION VIABILITY; REPRODUCTIVE SYSTEM; SEX DETERMINATION in the McGraw-Hill Encyclopedia of Science & Technology.

Stephanie J. Kamel

Bibliography. F. Glen and N. Mrosovsky, Antigua revisited: The impact of climate change on sand and nest temperatures at a hawksbill turtle (*Eretmochelys imbricata*) nesting beach, *Glob. Change Biol.*, 10:2036–2045, 2004; L. A. Hawkes et al., Investigating the potential impacts of climate change on a marine turtle population, *Glob. Change Biol.*, 13:923–932, 2007; F. J. Janzen, Climate change and temperature-dependent sex determination in reptiles, *Proc. Natl. Acad. Sci. USA*, 91:7487–7490, 1994; S. J. Kamel and N. Mrosovsky, Deforestation: Risk of sex ratio distortion in hawksbill sea turtles, *Ecol. Appl.*, 16:923–931, 2006; P. L. Lutz, J. A. Musick, and J. Wyneken (eds.), *The Biology of Sea Turtles*, vol. 2, CRC Press, Boca Raton, FL, 2003.

Clostridium difficile outbreaks

Until the 1970s, the clinical significance of *Clostridium difficile* was underestimated by the medical profession. Although this bacterial species was occasionally isolated from fecal cultures, it was rarely associated with human disease. Today, however, toxin-producing strains of *C. difficile* are recognized as a significant cause of antibiotic-associated gastrointestinal diseases that can range from mild, self-limiting diarrhea to severe, life-threatening pseudomembranous colitis (hemorrhagic necrosis in the intestine). Until fairly recently, individuals who developed symptoms of *C. difficile* infection were most often hospital patients receiving antibiotics for another infection. Recent reports, though, suggest that the rate and severity of *C. difficile*-associated disease are on the increase in the United States and in other, diverse locations, including the United Kingdom and Canada. What is of greatest significance is that this increase is now occurring in populations considered at low risk for the disease.

Antibiotic-associated diarrhea. The *Clostridium* genus includes all anaerobic, gram-positive bacilli capable of forming endospores. Most of the clinically important members of this genus fall within a few species (see **table**). These organisms are ubiquitous, found in soil, water, and sewage. They are also frequently found as part of the normal microbial flora in the gastrointestinal tracts of animals and humans. Clostridia have the following features, which greatly contribute to their ability to cause disease: (1) the ability to survive adverse environmental conditions, such as heat, chemical agents, and desiccation, through spore formation; (2) rapid growth in an oxygen-deprived environment; and (3) production of numerous histolytic toxins, enterotoxins (toxins specifically affecting the cells in the intestinal mucosa, causing vomiting and diarrhea), and neurotoxins.

Clostridium difficile is normally found in the flora of the intestinal tract in a small number of people. When these individuals take antibiotics as a treatment for another, unrelated infection, the normal flora of the intestinal tract are affected by the drug, and many of them are killed. *Clostridium difficile*, however, is relatively resistant to many antibiotics. As a result, it survives treatment and begins to grow in large numbers. This in turn induces production of two toxins by the bacteria: toxin A (an enterotoxin) and toxin B (a cytotoxin). The enterotoxin causes neutrophils (phagocytic white blood cells) to enter the ileum of the intestinal tract. This results in the release of cytokines or cellular proteins, hypersecretion of fluid, and development of pseudomembranous colitis. The cytotoxin causes cellular actin to depolymerize, resulting in destruction of cellular cytoskeleton.

Some *C. difficile* strains have been reported to produce an additional toxin referred to as binary toxin CDT (cytolethal distending toxin). CDT is composed of two unlinked components, both of which are required for the toxin to be active. This toxin is an actin-specific ADP-ribosyltransferase. It belongs to the same group as toxins produced by *C. perfringens* (type E) and the C2 toxin of *C. botulinum*. Binary toxins consist of two independent unlinked protein chains: the binding component and the enzymatic component. The binding component recognizes a cell surface receptor. This in turn triggers the internalization of the enzymatic component into the cell, leading to the ADP-ribosylation of actin in the cell, leading to disorganization of the cytoskeleton. However, since the majority of strains isolated from symptomatic patients produce only the A and B toxins, it may be concluded that CDT is not required for virulence of *C. difficile*. CDT may, however, serve as an additional virulence factor and

Some of the pathogenic *Clostridium* species and their associated human diseases

Species	Human disease	Frequency of disease
C. difficile	Antibiotic-associated diarrhea, pseudomembranous colitis	Common
C. perfringens	Cellulitis, gas gangrene, food poisoning, septicemia	Common
C. septicum	Gas gangrene, septicemia	Uncommon
C. botulinum	Botulism	Uncommon
C. tetani	Tetanus	Uncommon
C. sordelli	Gas gangrene	Rare

may work together with the two toxins to cause disease.

In hospitals and in the community. *Clostridium difficile* is now one of the most commonly detected enteric pathogens in hospitalized patients. A study performed using data from eight health care facilities located in six different states analyzed 187 different isolates of *C. difficile*. These isolates were characterized using pulsed-field gel electrophoresis and PCR (polymerase chain reaction) to detect binary toxin CDT as well as any change in a gene called *tcdC*. Changes in this gene may result in increased production of toxins A and B, leading to greater pathogenicity of the *C. difficile* strain. At least half of the isolates belonged to the same group of *C. difficile*. This group had variations in toxin genes and was more resistant to fluoroquinolones (antimicrobials that kill bacteria or prevent their growth) than historic isolates for *C. difficile*. The states used in this study (Georgia, Illinois, New Jersey, Maine, Oregon, and Pennsylvania) were geographically dispersed, indicating that this new strain has spread widely through health care facilities in the United States. Risk factors for this strain of *C. difficile* included use of gatifloxacin, an antibiotic that is considered more active against anaerobes. This may have led to a greater alteration in bowel flora, making the patient more susceptible to disease caused by *C. difficile*.

While it is most frequently associated with antibiotic use, patients with bowel stasis, those who have had bowel surgery, and those with no known risk factors can also develop *C. difficile*-associated gastrointestinal disease. In fact, the number of cases of *C. difficile*-associated diarrhea occurring in the community has been increasing in recent years.

As to why this is occurring, scientists are proffering two possible reasons: overuse of antibiotics in general, as shown from the six-state study, and the use of powerful new heartburn drugs. A Canadian study examined the link between the use of these drugs (including proton pump inhibitors and H2-receptor antagonists) and diagnosis with *C. difficile*. The researchers found that the use of proton pump inhibitors (which decrease the amount of acid produced by the gastric acid pump in the stomach) increased the risk of being diagnosed with *C. difficile* disease nearly three times, while the use of H2-receptor antagonists (which reversibly reduce output of gastric acid by blocking histamine H2 receptors) made it twice as likely. It was also noted that the number of cases occurring per 100,000 population increased from 1 in 1994 to 22 in 2004. This increase was found to be statistically related to the increase in use of these heartburn drugs. The actual cause of this increase, however, remains suppositional. It may be that by suppressing stomach acid, which is one of the body's main defenses against pathogens, the body becomes more susceptible to disease caused by *C. difficile*.

In December 2005, the U.S. Centers for Disease Control and Prevention (CDC) reported a case involving a 31-year-old woman who was pregnant with twins. The woman experienced intermittent diarrhea over a 3-week period, after which she suffered from cramping and watery stools. At this point, she went to the local emergency hospital department, where they tested her stool for the presence of *C. difficile* toxin. This test was positive and the patient was admitted. Her history indicated that she had received antibiotic treatment some 3 months earlier (trimethoprim-sulfamethoxazole) for a urinary tract infection. Despite aggressive treatment by the hospital, she first lost her fetuses and then lost her life. The autopsy showed evidence of necrotic tissue in the colon, which indicated pseudomembranous colitis had killed this patient. A second case also reported at this time was of a 10-year-old girl who developed *C. difficile* diarrhea despite not having used antibiotics within the previous year. Eventually the girl had to be hospitalized in order to treat her illness, which resolved with the administration of intravenous fluids, electrolytes, and metronidazole (an antibiotic drug used against anaerobic bacteria).

These case studies are part of a study performed by the Pennsylvania Department of Public Health in conjunction with the CDC. Through voluntary reporting, the CDC was able to put together statistics of the number of cases of peripartum *C. difficile*-associated disease (CDAD) as well as of community-associated CDAD (CA-CDAD). Peripartum CDAD was defined as *C. difficile*-associated disease occurring 4 weeks before and after delivery of a child. In this study, 10 peripartum cases occurred in patients from New Hampshire, New Jersey, Ohio, and Pennsylvania. In addition, 23 community-associated cases occurred in the Philadelphia area. Of the 33 cases reported, 8 reported no exposure to antibiotics within 3 months. The strains isolated from two of the patients were characterized and were found to be binary toxin-positive. While these strains were not found to be the "epidemic strain" of recent outbreaks in the United States, the fact that they contain the binary toxin may be relevant. Although its role in human disease is unknown, the binary toxin is now found uniformly in the epidemic strain. Previously, this toxin had been detected in only 6% of isolates; its role may not be defined as yet, but there appears to be a definite correlation between its presence and severity of disease.

The fact that low-risk populations are developing cases of severe CDAD reflects a change in the epidemiology of this disease. Individuals who were not considered to be at risk are now developing disease and, in some cases, dying. The increased resistance of this pathogen to antibiotics, combined with its increased virulence when it causes disease, makes it a concern to the population in general and not just to those hospital patients who were once considered its most likely victims.

For background information *see* ANTIBIOTIC; CLOSTRIDIUM; DIARRHEA; FOOD MICROBIOLOGY; GASTROINTESTINAL TRACT DISORDERS; HOSPITAL

INFECTIONS; MEDICAL BACTERIOLOGY; MICROBIOTA (HUMAN); TOXIN; VIRULENCE in the McGraw-Hill Encyclopedia of Science & Technology.

Marcia M. Pierce

Bibliography. M. K. Cowan and K. P. Talaro, *Microbiology: A Systems Approach*, McGraw-Hill, New York, 2006; S. Dial et al., Use of gastric acid-suppressive agents and the risk of community-acquired *Clostridium difficile*-associated disease, *JAMA*, 294(23):2989–2995, 2005; C. Gonçalves et al., Prevalence and characterization of a binary toxin (actin-specific ADP-ribosyltransferase) from *Clostridium difficile, J. Clin. Microbiol.*, 42(5):1933–1939, 2004; L. C. McDonald et al., An epidemic, toxin gene-variant strain of *Clostridium difficile, N. Engl. J. Med.*, 353(23):2433–2441, 2005; P. R. Murray et al. (eds.), *Medical Microbiology*, 4th ed., Mosby, St. Louis, 2002; E. W. Nester et al., *Microbiology: A Human Perspective*, 5th ed., McGraw-Hill, New York, 2007; J. M. Willey, L. M. Sherwood, and C. J. Woolverton, *Prescott, Harley, and Klein's Microbiology*, 7th ed., McGraw-Hill, New York, 2007.

Color-changing inks

Color-changing inks are most often used to increase value or uniqueness in a variety of applications. The first major commercially successful color-changing ink application was the mood ring of the 1970s, but now even automobiles can change color. Applications and uses continue to expand rapidly. While there are a range of color-changing technologies, as listed in **Table 1**, the most popular and readily printable are thermochromic (changing color with temperature) and photochromic [changing color with ultraviolet (UV) light]. These inks can be reversible or irreversible, as the application dictates. However, irreversible inks are not readily available.

Microencapsulation. Many photochromics and nearly all thermochromics require microencapsulation for protection. As a result, it is beneficial to understand the basics of the process. The most common process for encapsulation is called interfacial polymerization. During the process, the internal phase (material inside the microcapsule), the external phase (wall material of the microcapsule), and water is combined under homogenization (high-speed mixing). The goal here is to make a stable emulsion of the desired particle size, usually 5 micrometers or below. By controlling all the process conditions precisely (temperature, pH, concentrations, mixing speed, etc.) the external phase will surround the internal phase droplet and crosslink. Finished microcapsuled particles under 400× enlargement are shown in **Fig. 1**.

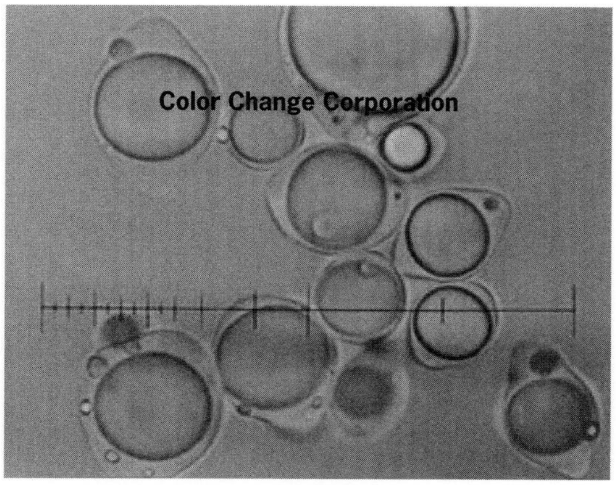

Fig. 1. Microencapsulated particles.

TABLE 2. Thermochromic leuco dye color range.

TLD colors	
Color	Pantone
Red	186C
Rose red	217C
Magenta	675C
Vermillion	1785C
Orange	172C
Yellow	393C
Yellow green	359C
Charm green	373C
Green	3435C
Sky blue	2925C
Turkish blue	320C
Blue	285C
Dark blue	287C
Violet	286C
Black	Black 3C

Thermochromics. There are two predominant reversible thermochromic classes: liquid crystals and leuco dyes.

Thermochromic liquid crystals. Thermochromic liquid crystals (TLCs) are used in the popular mood ring or aquarium thermometer strips and change color from black when cold, to red-orange-yellow-green-blue-violet, and then black upon heating, while the colors reverse upon cooling. They are very sensitive to temperature changes and can be used for medical thermometers, thermal mapping, and toys. TLCs are microencapsulated for protection, with a particle size of 5–50 μm in diameter.

TABLE 1. Color-changing technologies and applications

Technology	Mode of color change	Application example
Thermochromism	Changing color with temperature	Mood ring
Photochromism	Changing color with UV light exposure	Transitions® lenses
Hydrochromism	Changing color with moisture	Water glasses that change from white to clear
Interference Pigments	Changing color by viewing angle	Specialty car paint

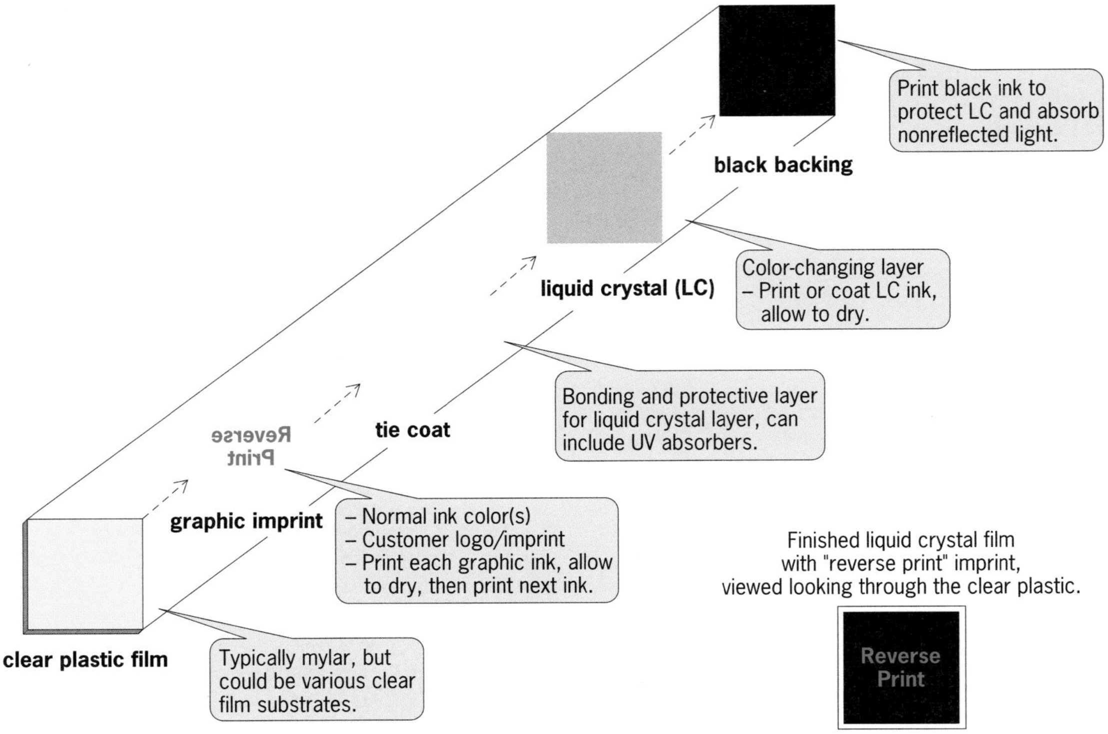
Print black ink to protect LC and absorb nonreflected light.
black backing
Color-changing layer
- Print or coat LC ink, allow to dry.
liquid crystal (LC)
Bonding and protective layer for liquid crystal layer, can include UV absorbers.
tie coat
Reverse Print
graphic imprint
- Normal ink color(s)
- Customer logo/imprint
- Print each graphic ink, allow to dry, then print next ink.
clear plastic film
Typically mylar, but could be various clear film substrates.
Finished liquid crystal film with "reverse print" imprint, viewed looking through the clear plastic.
Reverse Print

Fig. 2. Reverse printing process.

TLCs can be formulated to respond from −30 to 90°C (−22 to 194°F) and require a black background to appear visibly. The full color spectrum can respond over 1°C (1.8°F) or be as wide as 25°C (45°F). This is called the bandwidth. With a narrow bandwidth, the resolution is quite high, maybe 0.2°C (1.36°F) for a 1°C wide mixture (useful for medical thermometers). The resolution drops as bandwidth is increased, so that a 25°C mixture can only resolve about a 5°C (9°F) temperature range (useful for automotive defroster testing). TLC inks are water-based, somewhat difficult to work with, and require such a thick coating that only screen printing is applicable. Special materials and environmental conditions are required to print them. TLCs are most effectively used by reverse printing them upon clear plastic and then protecting them with a black backing, so that they are applied in a strip form (**Fig. 2**). The cost for TLC ink is typically in the $100–200 per kg range in reasonable volume.

Thermochromic leuco dyes. Thermochromic leuco dyes (TLDs) change from one color when in their cool state to translucent when in their warm state. They can be made in most colors, but not white (**Table 2**). They are reversible but usually return to their colored (cool) state a few degrees Celsius below the temperature that made them clear while heating. They are used in a wide range of products, from the hot or cool indicating food and beverage labels, to coffee mugs, toys, and security applications such as check anticopy protection (**Fig. 3**). TLDs are microencapsulated and have an average diameter of 2–5 μm.

TLDs can be formulated to change color from −15 to 60°C (5–140°C). The transition from colored to clear occurs over a 3–10°C (5.4–18°F) range. Consequently, the materials are not suitable for most sensitive thermometer-type applications but work well for general temperature indication such as cold, warm, or hot. TLDs are dyes and not pigments, so they must be printed over a lighter background, and the background color will influence the TLD color if it is any color other than white. For example, a black TLD printed over a red background would appear to change from black to red upon heating (the black TLD revealing the red underneath as it went translucent). A blue TLD printed over a yellow background would appear blue or green depending upon how thickly the blue TLD was printed, and then yellow when warm. TLDs are quite robust and can be used in a wide range of inks (solvent, water-based, UV cured, epoxy, etc.) and printed using most processes

Warm
Cold
Coors
LIGHT
Coors
LIGHT

(a) (b)

Fig. 3. Beverage label printed with a thermochromic leuco dye in their (*a*) colorless and (*b*) colored states.

Fig. 4. Application of a photochromic ink in which customers would expose the container to sunlight to see if the ink changed color, revealing if they were a "winner."

(screen, flexographic, offset, gravure, etc.). Costs for inks range from $75 for large-volume screen-printing inks to $500 per kg for offset inks.

Photochromics. Photochromics (PCs) are relatively new (1990s) compared to thermochromics (1970s), and the underlying technology continues to change rapidly. Most photochromics change from colorless to clear upon exposure to UV light, and then fade back to colorless upon removal from the UV source, as shown in the chemical formula below.

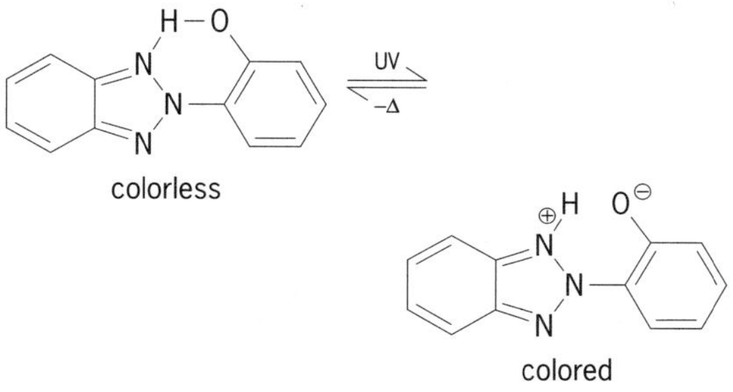

The normal wavelength of excitation is around 360 nanometers. And while sunlight works the best, a fluorescent black light, which emits near-UV (320–400 nm) light, will usually work. There is a full spectrum of photochromic colors available.

The most famous use of PCs is in Transition® photochromic lenses for glasses, but a popular ink application was the Dairy Queen Blizzard® promotion as shown in **Fig. 4**.

Different PC dyes have different kinetics, meaning some will color and fade quickly, while others will color and then fade slowly. The raw PC dyes tend to be quite expensive, ranging from $3 per gram to over $200 per gram. However, they are often used in low concentrations (0.2–1% by weight). Typical inks are expensive, ranging from $100 to over $500 per kg.

Very few PC dyes are water-soluble, so for waterborne applications microencapsulation is required. For many nonaqueous inks, microencapsulation is often preferable because it protects the PC dyes.

The same unique nature that allows PC dyes to change color makes them inherently unstable. Lifetimes for the photochromic dyes can be as short as 1 hour outside without stabilizers. With a stabilization package, lifetimes of about 1 month of outdoor exposure are possible. Because the PCs are dyes, they are most effective on white or very light backgrounds.

For background information *see* DYE; EMULSION POLYMERIZATION; INK; PRINTING; ULTRAVIOLET RADIATION in the McGraw-Hill Encyclopedia of Science & Technology. Timothy J. Homola

Bibliography. P. Bamfield, *Chromic Phenomena: Technological Applications Of Colour Chemistry*, 2001; R. Muthyala, *Chemistry and Applications of Leuco Dyes*, 1997.

Color management

The growth of affordable digital imaging technologies has created a huge demand for digital color reproduction and hence for color management. There is now widespread use of digital cameras, liquid-crystal displays (LCDs), inkjet printers, and photo editing and page layout software. Color imaging devices from different manufacturers use different device components, so the same image may look different on different devices. To achieve accurate and predictable color, it is necessary to have a framework that compensates for device differences.

Color management is the use of hardware, software, and systematic procedures to control and adjust the pixel values in a digital image to ensure accurate color when the same image is displayed on different devices and media.

The most universally used color management framework is that specified by the International Color Consortium (ICC) and implemented via device profiles. ICC color management enables users to achieve accurate color efficiently, thus saving time and money. Today's color management provides new features and opportunities that were very difficult, or even impossible, to obtain in older workflows.

Closed- and open-loop systems. It has always been necessary to modify image pixel data to accommodate individual device differences; what has changed is the way in which this is done. In older closed-loop systems, color was controlled using device-to-device mapping in pair-wise configurations. These proprietary systems worked best with a small number of devices but became unmanageable because modern workflows typically consist of many sources and destinations. Today, color management generally refers to an open-loop system, such as that specified by the ICC. An open-loop color management system uses files, called profiles, to map color information into and out of a central profile connection space (PCS) [**Fig. 1**]. Color management is successful today because this open, common platform has enabled the development of many interconnected software and hardware systems and has been widely implemented in all areas of commercial color imaging from digital photography to prepress and press.

Device-dependent and device-independent color. RGB (red, green, blue) and CMYK (cyan, magenta,

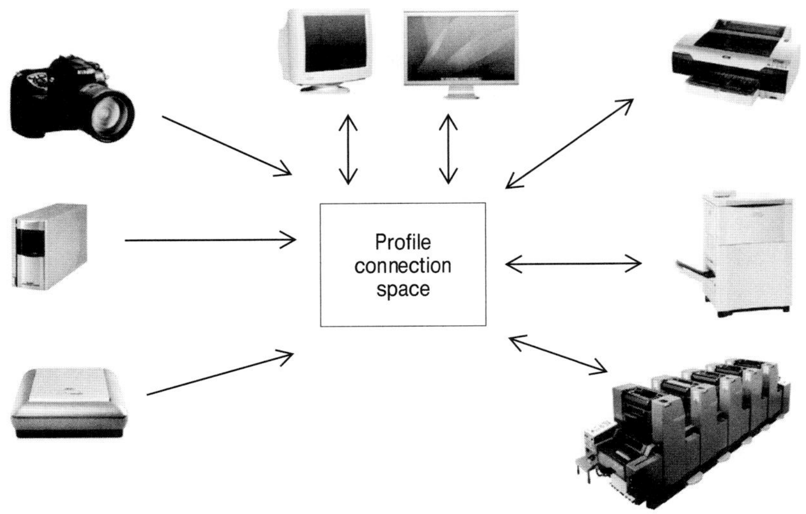

Fig. 1. An open-loop color-managed system uses a central profile connection space to connect many devices. Images arriving from a camera can be sent to a monitor for viewing or a printer for printing. (*Photos left top and left middle courtesy of Nikon, Inc. Melville, New York*)

yellow, black) are examples of device-dependent color specifications. RGB (for a monitor) and CMYK (for a printer) represent pixel values. These values are instructions for a device, and are necessarily in units that a device can understand and use. The color related to each pixel value is not directly specified, as it is dependent on these instructions being interpreted (printed or displayed) on the device being used. Thus the name device-dependent color.

The CIE (Commission Internationale de l'Eclairage or International Commission on Illumination) specifies a number of standardized color measurement systems. One of the most useful CIE systems is CIE 1976 L*, a*, b*, with the official abbreviation of CIELAB. A CIELAB value, such as L*a*b* of 83, −1, 69, for a yellow color, is a device-independent color measurement. A CIELAB value is a measurement of the color itself, irrespective of how the color is produced in practice. Thus, it is not related to any specific device. In a color management system (Fig. 1), the central profile connection space (PCS) is encoded in device-independent units, often CIELAB, while data being sent to or from devices are in device-dependent units such as RGB or CMYK. A profile provides information in the form of a lookup table that is used to convert between device-dependent values (RGB/CMYK) and device-independent values (CIELAB).

ICC profiles. An ICC profile is a data file that represents the color characteristics of an imaging device. ICC profiles can be made for scanners, digital cameras, monitors, or printers. The structure of a profile is standardized by the ICC and strictly regulated so that a wide range of software from many different vendors and at different parts of the workflow (such as image editing, preview, processing, Internet preparation, and printing) can open a profile and act on its contents.

Each device in a color-managed system must have a profile. For example, a camera needs to have a profile, and the profile must accompany that image from that camera. When the image is displayed on a screen, a monitor profile is used. And when the image is printed, a printer profile is used. For a camera-to-printer workflow, image pixel data in the original camera image are converted from camera RGB via the camera profile into CIELAB, and then into appropriate CMYK instructions via the printer profile. Device and media characteristics (differences) are reflected by the content of a profile. Thus color management works by using profiles to correct image data from a camera using a camera profile, to correct for the monitor differences using a monitor profile, and to correct, or account for, printer characteristics using a printer profile.

An image may be processed using a specific sequence of profiles to create different workflows. Soft proofing is used to visualize the printed result on screen. Hardcopy proofing enables a printing press result to be created on an inkjet proofer. Remote soft proofing allows users to use separate monitors to simultaneously collaborate on color-accurate monitor-based evaluation. A camera profile and a printer profile allow a user to create a color-managed print.

Calibration, characterization, and conversion. The practical process of color management is summarized in three stages. Initially, a color-managed system requires some form of calibration to establish system consistency and some form of process control. Next, a device's color characteristics are studied in a process known as characterization. Characterization, commonly known as making a profile, requires printing and measuring some form of test target (**Fig. 2**), followed by profile generation. Finally, during conversion, profiles are applied to an image and image pixel data are converted from one color space to another via the PCS (for example, from RGB to CIELAB to CMYK). The conversion process relies

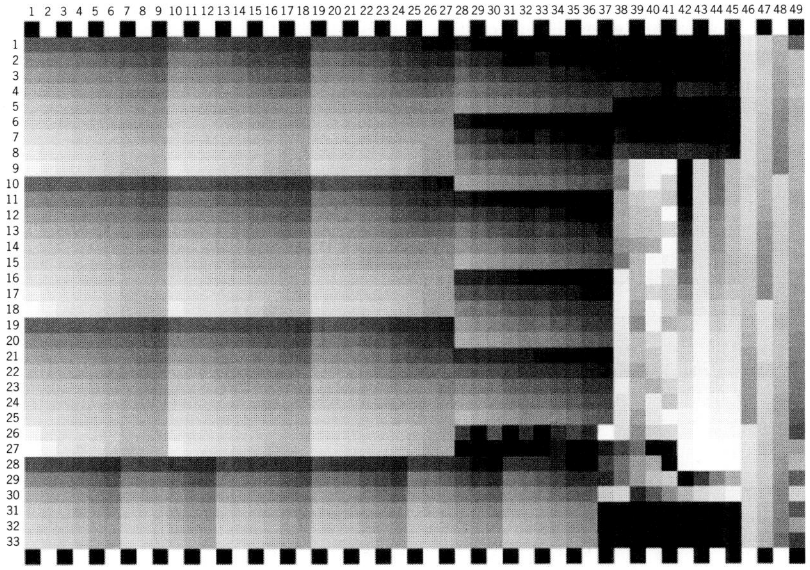

Fig. 2. An example of a printer characterization target (IT8.7/4) that must be printed and measured as part of the process to make a printer profile. The target when shown represents 1617 specially chosen CMYK color patch values.

on application software [such as Photoshop or a RIP (raster image processor)], system-level software (for example, Apple ColorSync), and an underlying color management module (for example, Adobe CMM).

Rendering intent and gamut. One of the main reasons for inaccuracies in color imaging is gamut limitations. As imaging devices work in totally different ways, using additive or subtractive color and with different colorants, they are not able to create an identical range of colors. An accepted generalization is that RGB systems (cameras and monitors) have a larger color gamut and can reproduce more colors than CMYK (printer) systems (**Fig. 3**). This means that often the user will be able to see many more colors in an image on a monitor than can be reproduced in print. In general, all device-dependent color systems have color gamuts that are limited by the color characteristics of the primary colorants for a given device (phosphors for CRT monitors, inks for printing systems, toners for copiers), while device-independent color systems are based on the theoretical responses of a standard human observer and, therefore, encompass all of the colors that can be seen.

As images are passed into and out of the central connection space, they may be sent to a device with a smaller color gamut. If the destination device cannot reproduce a color, a color management system is able to find a suitable replacement. The user can select from different color-replacement schemes called rendering intents. The perceptual rendering intent is useful for general reproduction of pictorial images. The colorimetric intent is used when the user needs to create an exact color match for all in-gamut colors. The saturation intent accentuates the vividness of pure colors and is used for images that contain objects such as charts or diagrams.

Color management cannot alter the gamut of a device, as this is dependent on the fundamental makeup of each imaging device. However, color management can predict the effect that gamut compression

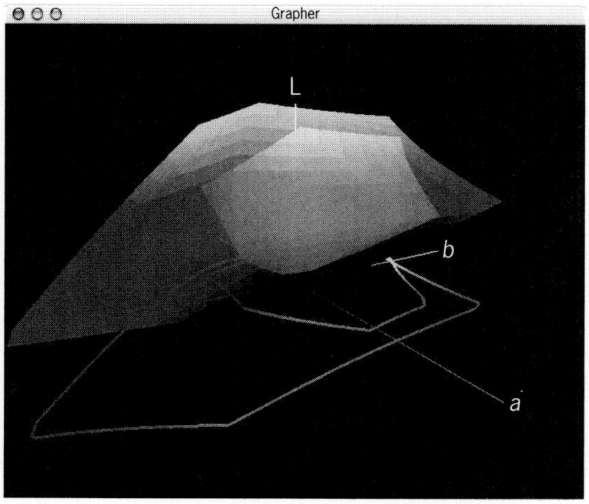

Fig. 3. In color management the gamut of a device is measured during the characterization process. Here the CIELAB diagram shows the larger gamut of an RGB monitor compared to the much smaller gamut of a CMYK print process.

will have on image colors. Gamut information is embodied in an ICC profile, and is thus readily available to software systems that can use this information to warn the user when colors are out of gamut of the destination device, allowing the user to make informed decisions.

International Color Consortium. The International Color Consortium (ICC) coordinates the structure and format of ICC profiles via a technical document called ICC Profile Specification ICC.1:2004-10 and recently released as ISO 15076-1:2005 (Image technology color management—Architecture, profile format, and data structure). This document is one of the reasons for the success of the ICC, as it provides a universally accepted set of rules that all parties adhere to, enabling interoperability of color information between disparate and/or competing imaging systems and companies.

An issue affecting end users is that the specification describes in detail the data structure of an ICC profile but does not describe the contents of a profile in terms of quality or accuracy. Thus, a profile may technically conform to the format of the specification, but there is no guarantee as to the accuracy of the contents of a profile.

Windows color management system. In the early 1990s, Apple's ColorSync technology provided the initial basis and framework for the present-day ICC color management system. Because of this historical connection and because ColorSync continues to be a well-integrated and well-supported part of the Mac OS, the Apple platform is a popular choice among serious users of color management. ICC color management recently has been challenged by a competing architecture from Microsoft called the Microsoft Windows™ color system (WCS), developed in collaboration with Canon.

A number of differences exist between the ICC and WCS frameworks. In most instances ICC profiles have predetermined gamut mapping. Thus, they are hard-wired with a fixed conversion for converting, for example, an image from RGB to CIELAB to CMYK. In contrast to the ICC's fixed model, Microsoft uses a more flexible architecture where the conversion is done at runtime, allowing the system to adapt to the particular source-image RGB gamut (for example, camera) and particular output CMYK gamut (press or proofer). The graphic arts–based ICC has favored predictability and repeatability, while the digital camera-focused WCS is biased toward flexibility that can take advantage of individual camera-printer situations.

The WCS profile is in XML (Extensible Markup Language) format. This means that it is in plain text, which allows users to easily examine and edit a profile. The ability to view and edit plain text greatly enhances the opportunities for third-party development. In some instances we may say that the Windows model has leap-frogged some of the technological issues that hampered development in the ICC.

Outlook. ICC color management is widely used today to provide a flexible way of managing color between multiple source and destination devices. ICC color management is a unifying technology, providing a solution to consumers and professionals alike. However, this universal solution is often problematic as it is often too complex for the home user, and does not provide enough control for the needs of the professional user.

For background information *see* COLOR; COLOR VISION; COMPUTER GRAPHICS; ELECTRONIC DISPLAY; INK; PHOTOGRAPHY; PRINTING in the McGraw-Hill Encyclopedia of Science & Technology.

Abhay Sharma

Bibliography. R. Adams, A. Sharma, and J. Suffoletto, *Color Management Handbook*, 2008; M. D. Fairchild, *Color Appearance Models*, 2005; B. Fraser, C. Murphy, and F. Bunting, *Real World Color Management*, 2004; A. Sharma, Methodology for evaluating the quality of ICC profiles: Scanner, monitor, and printer, *J. Imag. Sci. Tech.*, 50(5):469–480, 2006; A. Sharma, *Understanding Color Management*, 2004; G. Sharma and H. Trussell, Digital color imaging, *IEEE Trans. Image Proc.*, 6(7):901–932, 1997.

Complex systems

The notion of complexity has been of interest for a very long time. However, there has been a relatively recent resurgence of interest in this topic. One reason is people's sense that the world has become more complex, perhaps due to globalization of the economy, increased high-profile terrorism, and even global warming. Everything seems more interconnected, just as we are all becoming more connected via information and communications technologies.

Continually increasing computational power has provided new means to address complexity. Methods and tools for large-scale simulation and data mining have expanded our abilities to model systems as well as model the huge amount of information generated by the organizations associated with these systems. Thus, we are better prepared to understand complex systems.

Nature of complexity. It may seem intuitively obvious whether a system is complex or not. For instance, we might all agree that system A is much more complex than system B. However, if we ask the question, "How complex?" the nature of complexity becomes a bit more elusive.

The National Science Foundation sponsored a recent series of workshops involving 250 experts on complexity. A central conclusion was that the complexity of a system is related to the intentions of the observer. If I use my laptop computer as a paperweight, it is not complex. If I use it as a word processor, it is much more complex. If the laptop fails and someone has to diagnose the source of the failure, the laptop can become very complex.

However, if the person performing the diagnosis is highly skilled, the task may be quite simple. Thus, complexity is related to the intentions of the observer and the expertise with which these intentions are pursued. Complexity concerns the relationship between an observer and a system. When there are multiple observers (or stakeholders) with multiple intentions (objectives) that may be misaligned or conflicting, the notion of complexity becomes even more complicated.

Measures of complexity. The wide variety of definitions and measures of complexity that have been advocated all share a common feature. In order to calculate the complexity of a system, the calculation has to be based on a representation of the system rather than the system itself. The representation chosen might be networks, differential equations, statistical models, computational algorithms, and so forth. System complexity might be defined, for example, as the number of bits (binary digits) needed to define the state of the network model, or perhaps the resources required to execute the algorithm that models the system's operations.

There can be multiple representations of a system, each of which is necessarily an approximation of the real system. Consequently, any measure based on a representation, as all measures must be, provides an underestimate of the complexity of the real system. However one chooses to operationalize complexity, one inevitably disregards some aspects of the system. Of course, to do so may be quite reasonable, given one's intentions.

Impact of complexity. Considerable research has focused on the implications of systems being complex. For example, it has been found that the time to diagnose system failures is highly correlated with an information-theoretic measure of the network model of the system within which the failures occur. As another example, it was found that performance in operating hierarchical communications networks is highly correlated with both complexity measures based on network models and operators' strategies for managing the multiple levels. Another study found that the allocation of resources to improve health care should be based on a network representation of the value streams from medical research to clinical practice. In general, the performance of operators, maintainers, and managers of complex systems tends to degrade as complexity increases, typically measured in terms of characteristics of network models of the systems with which they are involved.

Nature of systems. A system is a group or combination of interrelated, interdependent, or interacting elements that form a collective entity. Elements may include physical, behavioral, or symbolic entities. Elements may interact physically, mathematically, and/or by exchange of information. Systems tend to have purposes, although in some cases the observer ascribes such purposes. Systems are judged to be complex when their perceived complicated behaviors can be attributed to one or more of the following characteristics: large numbers of elements, large numbers of relationships among elements,

Contrasting views of complex systems

Number	View	Approach	Focus
1	Hierarchical mappings	Design decomposition	Engineering solutions
2	State equations	Axiomatic derivation	Control performance
3	Nonlinear mechanisms	Behavior demonstration	Basis of complexity
4	Autonomous agents	Empirical assessment	Emergent behaviors

nonlinear and discontinuous relationships, and uncertain characteristics of elements and relationships.

Views of complex systems. There are multiple perspectives on the "complexity" of systems.

Systems of hierarchical mappings. Complex systems can be viewed in terms of the processes for defining, designing, developing, deploying, and sustaining complex systems. This view tends to be driven by hierarchical decomposition of the design task into component tasks, as well as management of the execution of these tasks and integration of task outcomes. The emphasis is on defining a large number of reasonably straightforward tasks whose outcomes will flow together to create a successful complex system, with appropriate resolution of multiattribute tradeoffs across multiple stakeholders.

Systems of uncertain state equations. Another view of complex systems focuses on the "state" of the system of interest. The mechanisms whereby the state of the system evolves are of central interest as they affect system response and stability. The nature of appropriate feedback mechanisms for controlling system state is a central design issue. The emphasis in this view is on formal depiction and manipulation of mechanisms underlying complex behaviors. This more formalistic approach seldom "scales up" to the types of problems addressed by the hierarchical mapping view of complex systems.

Systems of discontinuous, nonlinear mechanisms. Yet another view focuses on apparently simple underlying phenomena that yield complex behaviors for systems with very few elements, perhaps even just one element with particular interaction terms. The nonlinear or discontinuous nature of the elements of interest lead to behaviors labeled as catastrophes and chaos. Thus, systems that appear simple can produce very complex behaviors relative to our expectations of continuous, linear phenomena. Complexity is understood by exploring underlying mechanisms.

Systems of autonomous agents. A fourth view addresses the emergent properties of complexity. Rather than focusing on decomposition, this view emphasizes composition of large numbers of simple behaviors into overall system behaviors that exhibit characteristics of complex systems. The simple behaviors are created by autonomous "agents" acting independently in pursuit of their individual goals. Reactions of agents to each other's behaviors result in emergent phenomena that could not have been predicted by dissecting the mechanisms within individual agents. Understanding and managing complexity is an experimental rather than axiomatic undertaking.

Contrasting views. The approaches and foci of these four views are quite different (see **table**). However, in some cases, all four views address related phenomena. Consider vehicle behavior and performance in high-density traffic. View number 1 might be employed for designing the vehicle, number 2 to explore the vehicle dynamics, number 3 to model effects of turbulence, and number 4 to understand traffic effects. Of course, the choice of approach would depend on the problem at hand, for example, poor vehicle-handling qualities versus traffic congestion problems.

Complex adaptive systems. Complex adaptive systems learn with experience and change behaviors accordingly. Typically they are composed of independent agents who adapt to each other's behaviors. With differing goals and objectives, agents' behaviors can conflict, leading to unexpected emergent behaviors as agents react to each other. Complex adaptive systems are a class of systems within the autonomous agent's view. However, it is often the case that nonlinear mechanisms are involved and, occasionally, state equation representations are employed for modeling one or more agents' behaviors.

Design of complex systems. The characteristics of complex systems often of most interest are emergent and adaptive behaviors and unintended consequences, as well as characteristics such as robustness, resilience, flexibility, and agility, including tradeoffs among these characteristics. The National Science Foundation initiative noted above resulted in several overarching design questions:

1. What architectures underlie the physical, behavioral, and social phenomena of interest? The goal here is explanations of phenomena of interest in terms of conceptual frameworks, representations, structures, models, and so forth. An example is the architecture of terrorism.

2. How are architectures a means to achieve desired system characteristics? The goal here is methodologies that enable consideration of issues such as robustness versus agility. An architecture for sustainable systems is an example.

3. How can architectures enable resilient, adaptive, agile, and evolvable systems? The goal here is to understand what is fixed and what changes, as well as how to address fundamental tradeoffs. Information system architectures for supporting enterprises with fundamentally changing missions provide a good illustration.

4. How can and should one analytically and empirically evaluate and assess architectures prior to

and subsequent to development and deployment? The goal here is to understand and improve systems prior to and during system operation. An example is a transportation network where the nature of its use cannot be fully projected before it is deployed.

These questions beg a definition of the term "architecture." There are many available. The central constructs, however, in all of these definitions are entities, relationships, behaviors, and performance. To this extent, the concept of architecture is simply an overarching term to capture many constructs that have long been available and employed.

For background information *see* ANALYTIC HIERARCHY; CATASTROPHE THEORY; CELLULAR AUTOMATA; CHAOS; COMPLEXITY THEORY; MODEL THEORY; SYSTEMS ARCHITECTURE; SYSTEMS ENGINEERING in the McGraw-Hill Encyclopedia of Science & Technology.

William B. Rouse

Bibliography. J. L. Casti, *Complexification: Explaining a Paradoxical World Through the Science of Surprise*, Harper Collins, New York, 1994; W. B. Rouse, Complex engineered, organizational, and natural systems, *Systems Engineering*, vol. 10, no. 3, in press, 2007; S. Wolfram, *A New Kind of Science*, Wolfram Media, Champaign, Illinois, 2002.

Concentrating solar power

A main drive for developing reliable sources of renewable energy is the fast-growing worldwide energy demand with consequent depletion of fossil fuels, which supply over 85% of the present energy needs. Among the renewable energy sources used today (such as concentrating solar power, solar cells, wind, geothermal, hydropower, and biomass), concentrating solar power (CSP) is the only technology that can supply several times the world's energy requirements, since it has viable, efficient, and available one-day energy storage methods. Storage systems for other renewable sources, such as photovoltaic and wind, are much less efficient and more expensive. There also is not enough usable land for growing sufficient biomass. CSP plants have to be installed in deserts, because clouds and rain strongly lower their efficiency. Suitable deserts are one or two orders of magnitude larger than needed and, aside from some isolated islands, high-efficiency transmission lines can reach all populated areas. Most potential problems for large-scale installation of CSP are political rather than technical. However, since solar energy is the only large-scale renewable source we have for the future, we have to learn how to overcome such obstacles. North America and many other places do not have such obstacles.

Description of CSP plants. A CSP plant has two distinct, interacting parts (**Fig. 1**). The electricity generation part (dashed line in Fig. 1) is a conventional steam power plant, like those used in coal or nuclear power plants. The novelty resides in the energy generation, or gathering, section (dot-dashed line in Fig. 1). In its most common implementation, CSP uses parabolic-trough collectors (**Fig. 2*a***) to concentrate the Sun's heat on receiver pipes that are located on the focal line of the reflectors. The

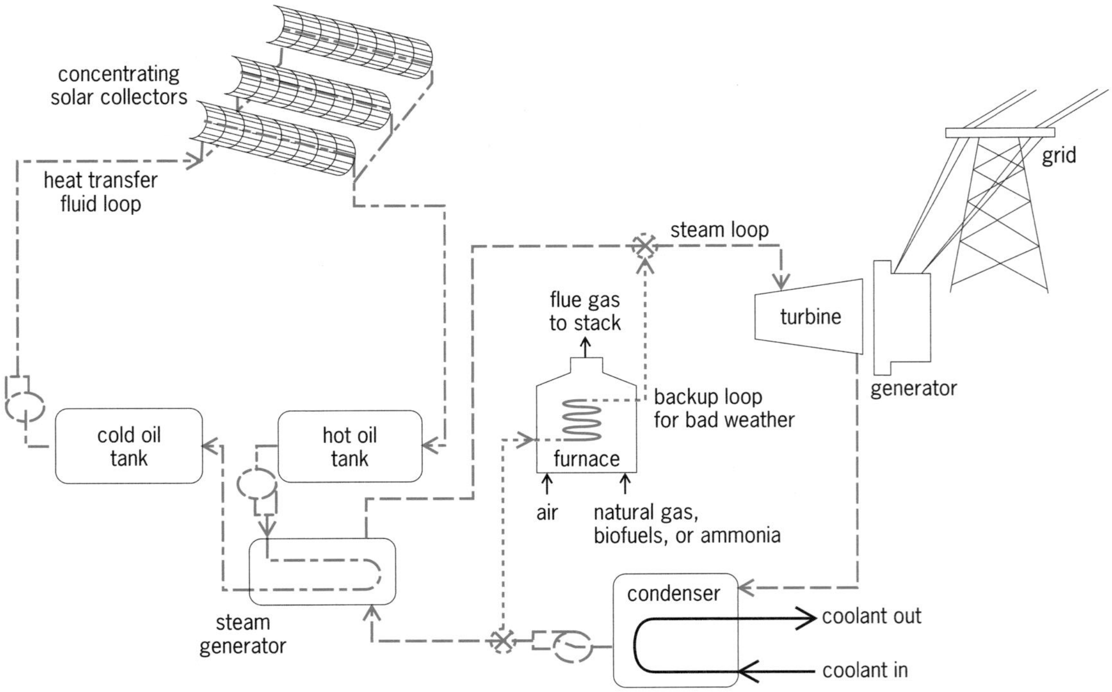

Fig. 1. Concentrating solar power plant with storage. Present installations use conventional steam turbines with water cooling. When many plants are installed, backpressure turbines with air cooling can be used.

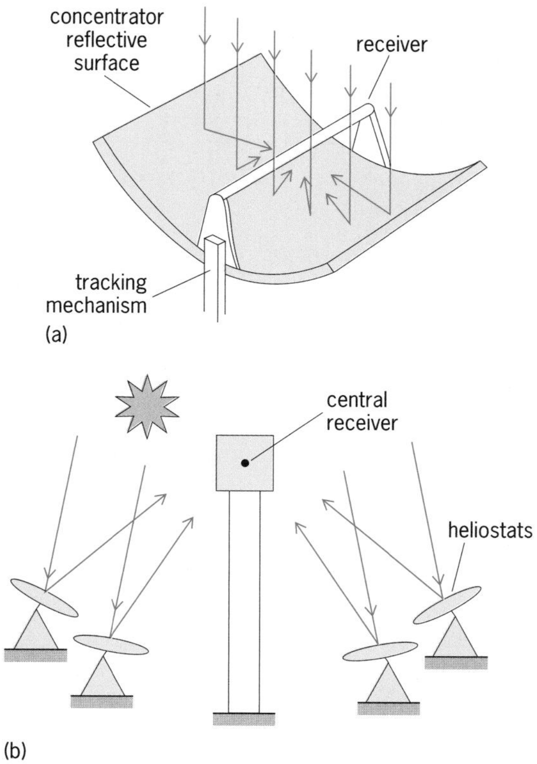

Fig. 2. CSP (*a*) parabolic-trough collector and (*b*) solar tower.

heat is absorbed by the receiver and transferred to a heat-transfer fluid (HTF) flowing inside the receiver. The HTF, which reaches very high temperatures (>800°F; 430°C), is used to raise steam to drive the turbines in the steam power plant. Parabolic-trough collectors are provided with a one-dimensional Sun-tracking mechanism that allows them to follow the Sun throughout the day, so that the mirrors are always perpendicular to the Sun's radiation, which increases the total efficiency. While all present units are water-cooled, in the future when thousands of gigawatts of electrical generating capacity (GWe; gigawatts electrical) are installed, it will be necessary to cool the plants with air, as water sources are scarce in the desert. Cooling with hot air in the desert requires raising the condensation temperature of the turbine using backpressure turbines, which lowers the efficiency. However, if a desert area suitable for large-scale CSP installation is near a water source, water cooling would negatively impact the environment by either releasing large quantities of steam using cooling towers or heating the water source using direct water cooling.

Storage and backup in CSP plants. The most important feature of CSP is that it does not transform the solar energy directly into electricity (like solar cells or wind turbines), but uses a HTF. Therefore, one can heat more HTF than needed during the daytime and store the excess in underground, insulated tanks. If electricity is needed after the sunset, the hot HTF from the storage tanks can be used to raise steam and generate electricity. The cold HTF is stored in separate tanks and reheated the next day. The most common HTFs are flammable oils that operate at a maximum temperature of 800°F (430°C) [for example, Dowtherm A®, a eutectic mixture of biphenyl and diphenyl oxide] and molten salts that operate a temperatures over 1000°F (540°C), allowing higher efficiency. Molten salts can freeze at night or when the plant is shut down for maintenance. To prevent the pipes from clogging with solidified salt there are two solutions. One solution is to electrically heat-trace the whole piping system. The other solution, which is used in present parabolic-trough plants, is to use Dowtherm A in the collectors and heat-exchange it with a molten salt used for storage, as Dowtherm is too expensive for large uses such as storage. New salts with lower freezing points as well as new storage methods are under development. Aside from hydropower, CSP provides the cheapest storage technology compared to currently used or foreseeable methods used for other renewable sources. The importance of a viable storage system is crucial, as this feature makes CSP the preferable renewable energy source of the future.

Having the energy-gathering cycle separated from the electrical-generating cycle allows the use of a backup furnace (dotted line in Fig. 1) to supply heat for those rare cases (<10% of the days) when bad weather persists in the desert longer than the storage provides for. The furnace can be fed with natural gas for the near future and with ammonia when the natural gas resources will have peaked. Ammonia will be produced by the traditional commercial process, but the hydrogen will be obtained by electrolysis using electricity from alternative sources. Although direct use of hydrogen would be more efficient, ammonia (which is practically a hydrogen storage method) is safer and less expensive to store and handle. Also, emissions of nitrogen oxides (NO_x) can be avoided by using the proper catalysts.

Energy storage and the backup furnace allow CSP to supply any desired electricity load, including intermediate (50% of the demand, 8 a.m. to 9 p.m.) using storage in the evening, base (40% of the demand, 24 h/day) providing larger storage, and especially peak (10% of the demand, 1–2 h in the evening) and load following (sudden and unforeseeable electricity requests), the most difficult needs for the electric grid. The ability to load follow makes CSP the potential anchor of a new electric grid totally based on renewable sources, as CSP can stabilize the grid against fluctuating electricity inputs from solar cells and wind, which are inherently variable.

A 354-MWe (megawatts electrical) modular plant using parabolic-trough collectors has operated in the Mojave desert since the late 1980s, while more recent applications have been realized in Nevada and Spain.

Alternative CSP plants design. An alternative to trough collectors are solar towers (Fig. 2*b*). In this implementation a large number of heliostats (flat mirrors, made of glass with silver backing; other materials are being investigated) concentrate the solar radiation on a central receiver within which flows a HTF, usually a molten salt. The functioning is similar to the parabolic-trough collectors. The difference is that the heliostats are each positioned with a different angle, and controlled individually with two-dimensional tracking. Solar towers can reach higher operating temperatures, which translates to higher efficiency.

On the negative side, the present solar towers require about 1.5 times larger areas than parabolic-trough collectors to minimize the shadowing effect between heliostats. There are also efforts, which should be strongly encouraged, to lower the cost and increase the efficiency of CSP plants by improving or designing new storage methods, HTFs, and solar collectors. Any future positive results could contribute, together with the present trough technology (which is already available and proven), to make CSP the major energy source.

Cogeneration of desalinated water. Another crucial benefit that can be gained by using CSP is the cogeneration of desalinated water. Areas next to deserts are in need of potable water, and are sometimes located near seawater or brackish water sources. CSP plants would be especially convenient for such application if they use backpressure turbines, which condense vapor at temperatures higher than conventional steam power plants. Consequently, the waste heat can be used for water desalination. As the major cost in desalination (strongly dependent on production scale) is for the fuel, using the waste heat from CSP plants would represent the cheapest way for water desalination. Producing potable water would compensate for the lower efficiency of the backpressure turbines. Currently this application would be particularly useful for developing countries where drinkable water is scarce. However, as water demand increases, it will benefit the whole world.

Desert area availability. As mentioned, concerns have been raised about desert distributions and availability to supply all the electricity needed. Consider, for example, that a 1-GWe plant requires about 15–20 mi^2 (24–32 km^2) of land (averaged between various loads), including the spacing between collectors. Given the total electricity installed worldwide and considering future incremental electricity demands and the conversion of a consistent part of fossil-fuel uses to electricity, **Table 1** shows that availability of land suitable for CSP installations is not a problem. The only problem would be for Europe, which does not have any significant deserts other than a few small areas in the Mediterranean countries. The CSP plants required by the European countries could be installed in the Sahara desert and the electricity transmitted to them through Gibraltar or Sicily. Modern high-voltage, high-efficiency, direct-current transmission lines would serve for this purpose. Small isolated islands, with no deserts, will need a different source of electricity.

TABLE 1. Total electric power from fossil fuels installed in the world and size of the desert areas available

Continent	Electricity installed from fossil fuels, GWe	Desert areas available, mi^2
North and Central America	841	500,000
South America	57	128,000
Europe	441	—
Asia (includes Russia and Middle East)	1090	1,532,000
Africa	81	3,472,000
Oceania	44	516,000
Total	2554	6,148,000

TABLE 2. Electricity costs for concentrating solar power compared to coal and nuclear

	Base, cents/kWh	Intermediate, cents/kWh
CSP (near term)*	8.0†	8.0
CSP (future)*	6.2†	6.2
Conventional coal power plant (with scrubbers)	4.5‡	8.0
Clean coal	5.6‡	10
Clean coal (with CO_2 sequestration in the plant)	7.1‡	11.5
including ultimate disposal	11.0‡	15.5
Nuclear	6.0‡	10–11

*With proper scale (at least 400 MWe) and mass production (R. Shinnar and F. Citro, 2007).
†Operated 4900 hours/year.
‡Operated 6500 hours/year.
SOURCE: From R. Shinnar and F. Citro (2006).

Cost of CSP electricity. **Table 2** compares the cost of electricity generated with CSP to the cost of electricity from coal and nuclear power. Natural gas is not included in the comparison because it is in short supply and its cost has increased several times over the past 10 years. CSP is already competitive for intermediate loads, and it becomes competitive for base loads when the costs of carbon dioxide (CO_2) separation and its ultimate disposal of waste are taken into account. R. Shinnar and F. Citro have shown that CSP is already the cheapest technology for peak and load following. As large-scale applications and mass production will significantly decrease the cost of CSP, it will become even more competitive.

For background information *see* ELECTRIC POWER GENERATION; ENERGY STORAGE; SOLAR ENERGY; SOLAR HEATING AND COOLING; WATER DESALINATION in the McGraw-Hill Encyclopedia of Science & Technology.

Reuel Shinnar; Francesco Citro

Bibliography. Sargent & Lundy LLC Consulting Group, Assessment of parabolic trough and power tower solar technology cost and performance forecasts, *NREL/SR-550-34440*, National Renewable Energy Laboratory 2003; R. Shinnar and F. Citro, A road map to U.S. decarbonization, *Science*, 313:5791, 1243, 2006; R. Shinnar and F. Citro, Solar thermal energy: The forgotten energy source, *Technol. Soc.*, 29(3):261–270, 2007.

Concept-based user interfaces

Computational researchers have found that scientists often conceptualize their problems and theories in graphical form. From using pen and paper to computer-based drawings, scientists construct diagrams and pictures to convey complex scientific concepts, theories, and processes (**Fig. 1**). Despite differences in symbols, shapes, colors, fonts, and drawing styles, scientists are generally able to comprehend and interpret graphs and diagrams generated by others within the same field or discipline. Graphs and diagrams provide an intuitive, expressive, and universal visual language for scientists. Unfortunately, the graphs and diagrams that scientists develop through typical means are generally static drawings that are not amenable to computation or computational analysis.

Extending from ideas in visual programming and visual modeling, concept-based user interfaces allow users to interactively and intuitively construct visual conceptual models through the use of rich, interactive drawing tools. The resulting conceptual models effectively capture and represent scientific theories and paradigms in a computational form that may be linked to and integrated with scientific data sets and applications. Conceptual models may convey very different kinds of structures and semantics. They may be logical in conveying how individual concepts tie together to form higher theories, analytical in conveying intermediate or final analysis results, or temporal in describing experimental processes in which the execution and order of steps or tasks are defined.

Experimental processes. Research has shown that scientists are naturally capable of drawing graphical depictions of experimental and computational processes to elaborate individual steps and the data or results that are generated and passed within a process. Such experimental process drawings or diagrams are commonly referred to as scientific workflows. A variety of interactive workflow design tools have been developed for specific workflow engines or management systems. Two specific workflow design tools, one developed as part of the Regional Climate Modeling Problem-Solving Environment (RCM-PSE) and the other under development on the Middleware for Data Intensive Computing Initiative (MeDICI) project, have been designed to support multiple levels of abstraction, such that a scientist may depict

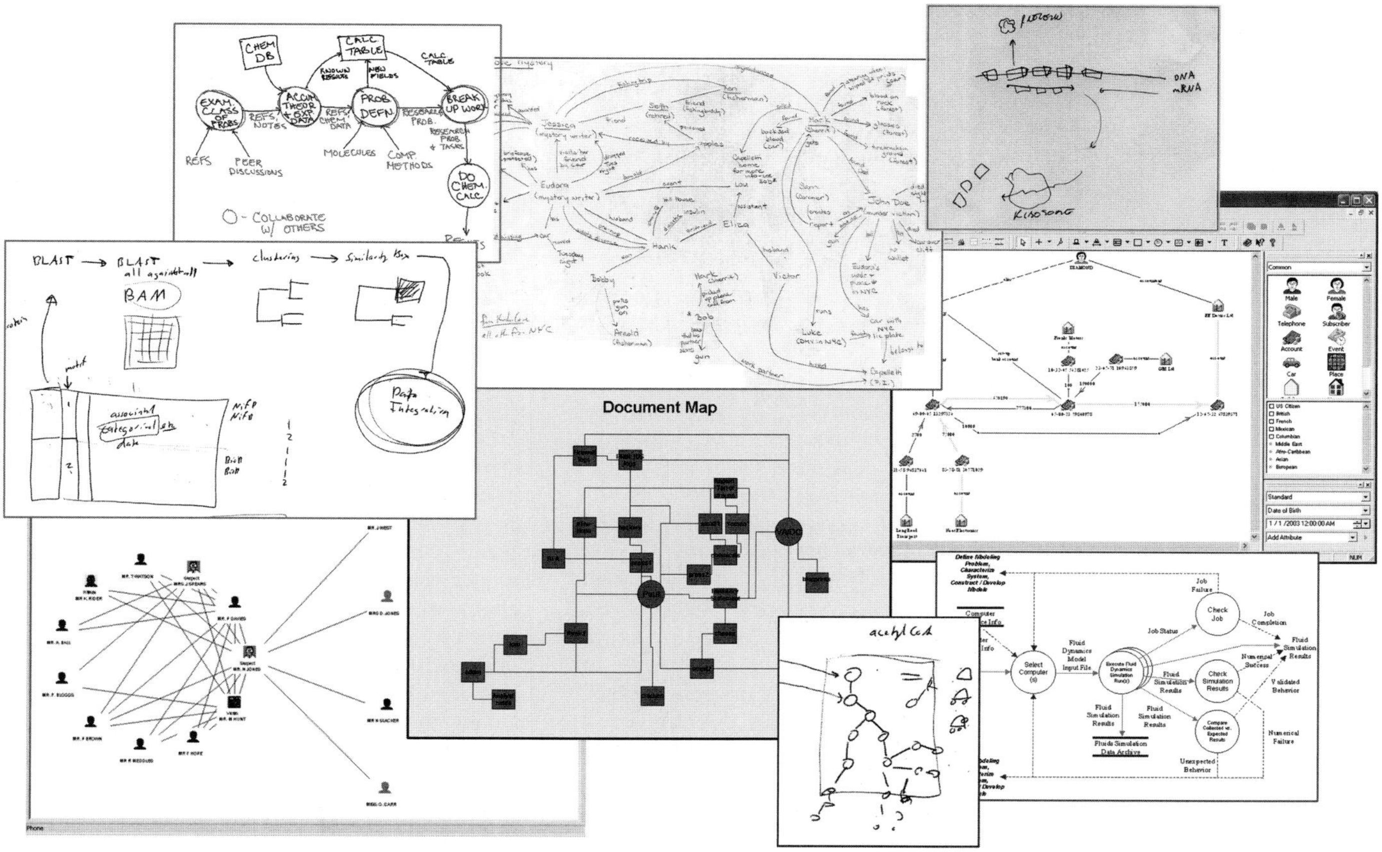

Fig. 1. Examples of scientist-constructed graphs and diagrams conveying various scientific concepts, theories, and processes.

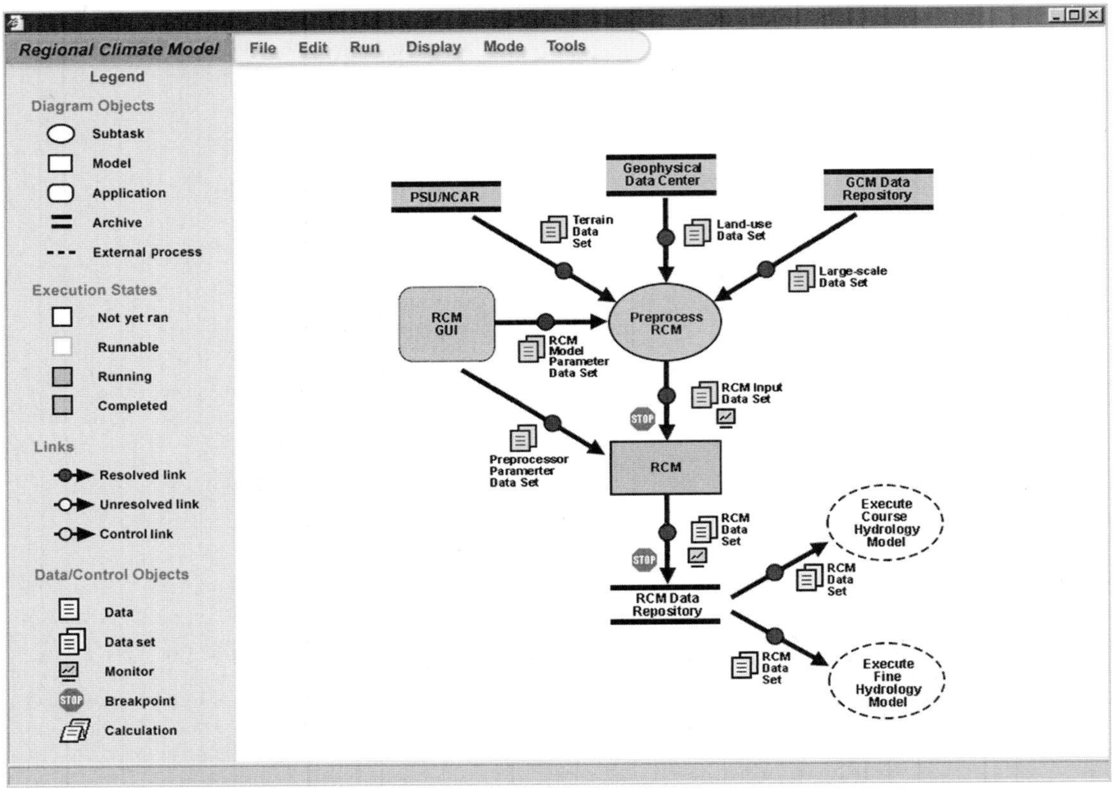

Fig. 2. Computational scientific workflow drawn using the workflow design tool of the Regional Climate Modeling Problem-Solving Environment. The contained workflow shows the steps involved in running a specific regional climate model. Through the workflow design tool, scientists may view, execute, control, automate, and debug the scientific workflow.

both the general experimental process and the underlying detailed data flow and set of computational resources required to implement higher-level experimental tasks.

Once a scientific workflow is expressed in a workflow design tool, it becomes a computational entity that is able to process, execute, and interact as shown in **Fig. 2**. It may be used to semiautomate complex and/or tedious operations. It may be viewed, controlled, and manipulated. It may be debugged by stepping through individual actions, setting breakpoints, and monitoring variable values and intermediate results. These are among the many functions that various workflow design tools provide.

Overall, scientific workflows provide scientists mechanisms for capturing, applying, and exploring procedural knowledge. Scientists are able to pass their procedural knowledge onto others through the saving and sharing of scientific workflow diagrams. Some procedural knowledge may be tacit in the sense that scientists may not be fully aware of the specific tasks, operations, and patterns they follow in their experimental processes. To better reveal this knowledge, the RCM-PSE and MeDICI workflow design tools provide provenance collection mechanisms that are capable of collecting and visualizing the history of workflow execution. Capturing execution history allows scientists to review and examine their own computational and experimental activities, illuminating specific behaviors and tendencies and providing them the opportunity to modify their processes to achieve more efficient operations and/or valid results.

Scientific concepts and theories. Visual tools are also available for scientists to graphically sketch and describe abstract scientific concepts and theories. For example, a computational biology tool known as the Visual Modeling Environment for Biology (VMEB) allows biologists to attach semantics to individual and collections of graphical objects (**Fig. 3**). Using VMEB, a filled box might indicate a protein, whereas a graphical collection of proteins linked together in a particular pattern with an arc might identify a chemical reaction. Semantics are captured from the biologist as a set of rules that infer biological meaning from the graphical attributes and spatial relationships of objects. Once rules have been defined, VMEB may automatically infer biological meaning onto graphical objects as they are constructed by the biologist by traversing and applying the rules in the rules base.

Beyond defining concepts, biologists may want to associate additional information, data, and contexts to concepts. For instance, a biologist may want to link a particular biological concept to publications where that concept originated or is elaborated, or to Web pages of biologists or biology groups who are actively researching that concept. A concept may also be linked to data sets and experimental results related to physical or computational experiments focused on that concept. Additionally, a concept may

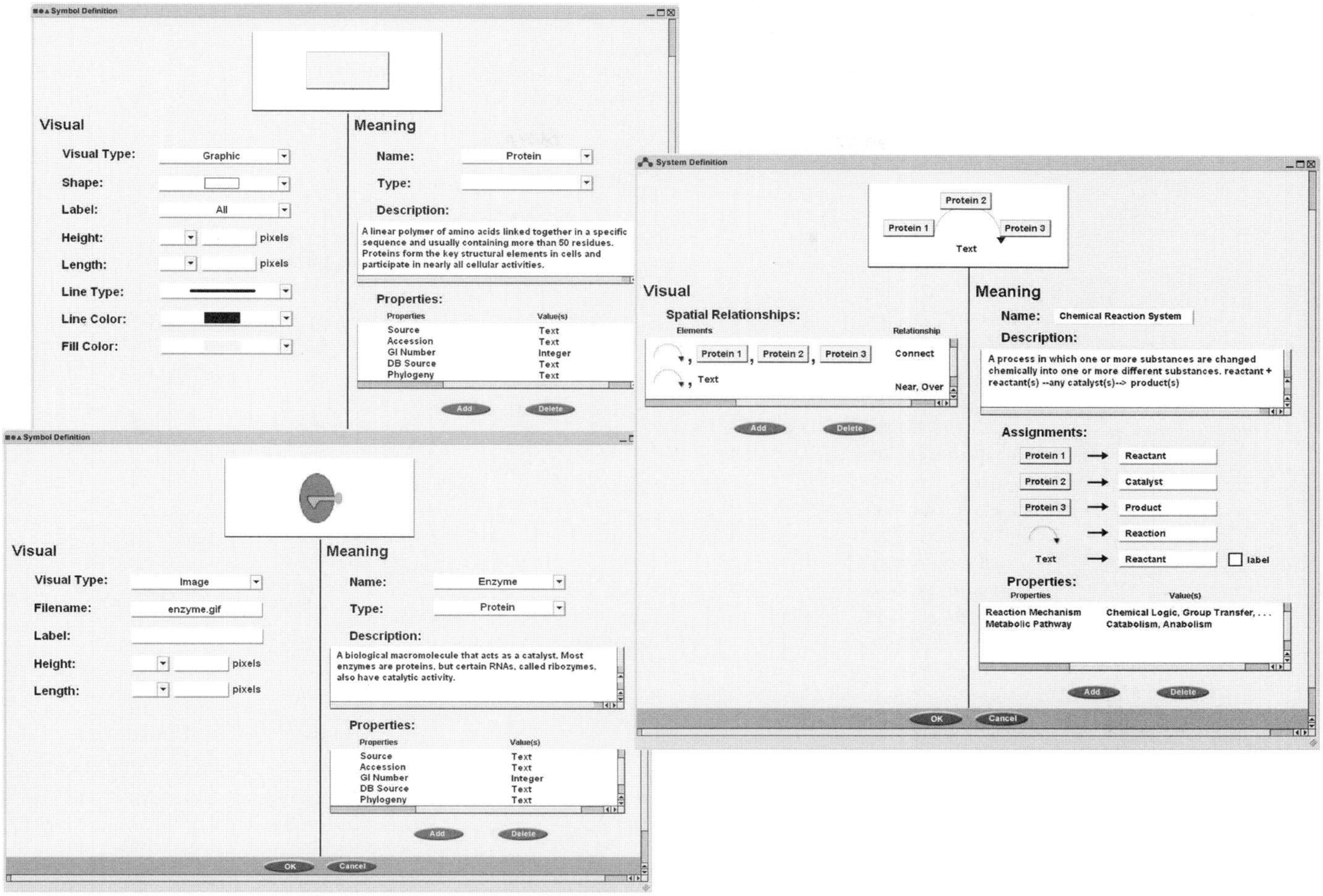

Fig. 3. The Visual Modeling Environment for Biology allows scientists to attach semantics to individual or collections of graphical objects. The visual and semantic properties of a protein, enzyme, and chemical reaction are defined in the figure.

be linked to executable simulations or models that convey dynamic aspects or properties of that concept.

Through its ability to capture and recognize biological concepts, VMEB allows biologists to operate and apply those concepts more directly in their computational research. For instance, biologists may directly query across stored diagrams to identify specific biological concepts. As shown in **Fig. 4**, the biological concept itself (that is, integral member protein) is the search object rather than a keyword or property. In the search, matching concepts are located and the diagrams that contain them are returned via a browser. The biologist may then view the individual diagrams with the matching concepts highlighted.

Overall, VMEB allows biologists to define or infer semantics and semantic associations on graphical depictions of scientific concepts. The goal is to allow biologists to apply and work with computer-based scientific concepts at the same level of abstraction as they naturally think about and operate upon those concepts. This enables a higher-level discourse between biologists and VMEB as well as among biologists sharing information through VMEB, as biologists directly specify, use, and search for scientific concepts such as genes, proteins, cells, and pathways rather than their user-associated images, shapes, colors, sizes, and fonts.

Analysis results. Among the intelligence community, link analysis is a popular form of analysis where analysts graphically draw out and link different kinds of information such as events, peoples, places, locations, facts, weapons, money, and contraband. The Scenario and Knowledge Framework for Analytical Modeling (SKFAM) project is a visual analysis system for capturing and representing link analysis diagrams as graph-based scenarios that may be computationally compared to one another. Graph-based scenarios of past or current intelligence cases may be generated and placed into a reference case library. As a new situation arises, it may be represented as a scenario and compared against the scenario case library to identify other similar or relevant cases. As shown in **Fig. 5**, the graph or subgraphs of the evolving scenario could then be mapped upon each of the matching scenarios from the case library to identify potential missing nodes and links or to predict possible results and outcomes based on prior cases.

With the SKFAM system, a scenario captures a snapshot of an analysis in a graphical form. It may represent the final analysis and conclusions of an

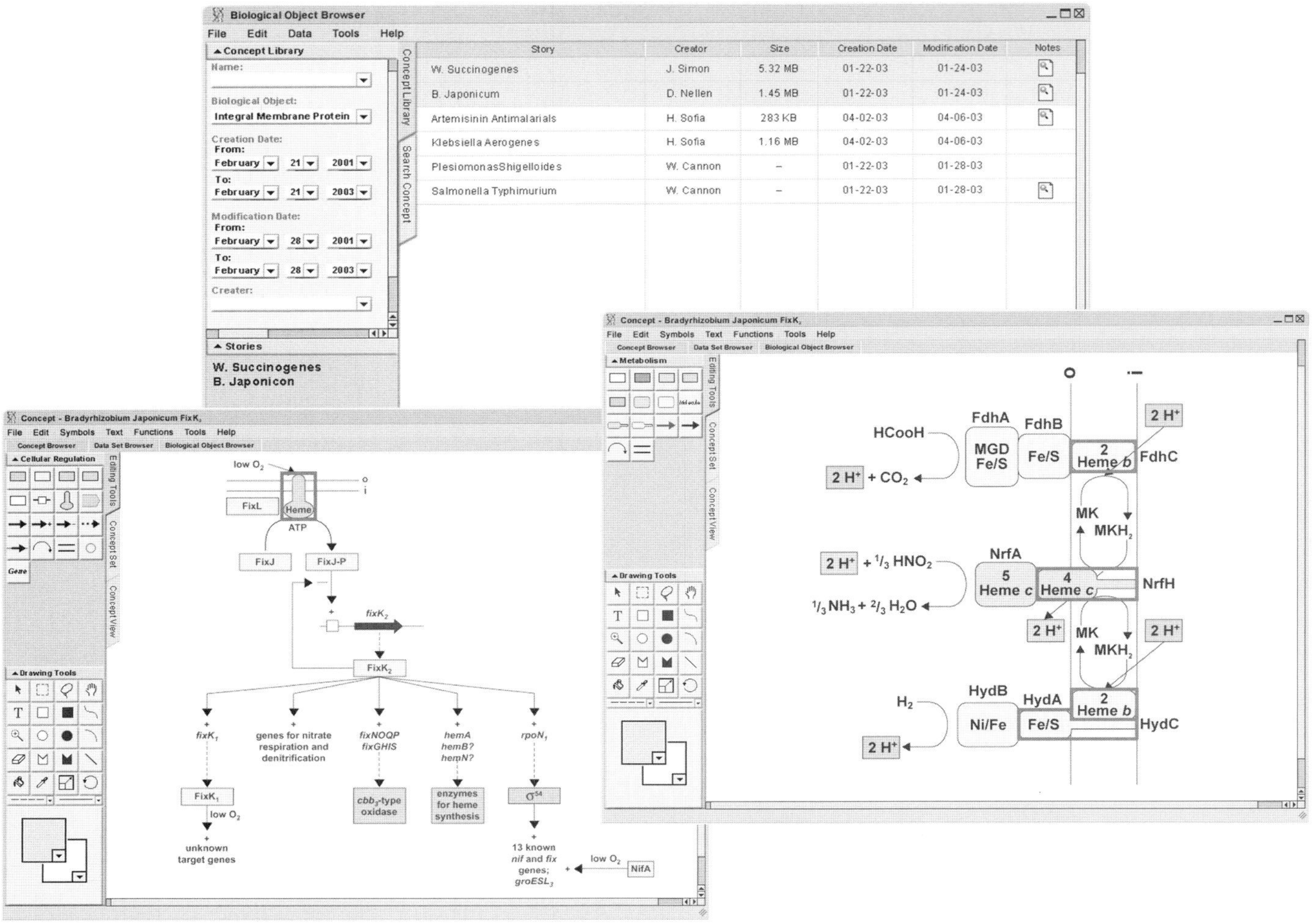

Fig. 4. The Visual Modeling Environment for Biology supports abstract concept searching. The figure shows the searching of integral membrane proteins across stored concept diagrams. Found concepts are highlighted in the diagrams that are returned from the search. (*The diagrams in the search results were reproduced from research directed by biologists D. Nellen-Anthamatten and J. Simon.*)

existing case or a working analysis where certain key nodes and edges are unconfirmed or missing. Additionally, it may represent a hypothesis or proposition where certain nodes and edges are speculated for the sake of projection and comparison. An analyst may develop several competing graphical scenarios based on current intelligence and then monitor which ones are supported or refuted as additional information becomes available.

Outlook. With concept-based user interfaces, scientists tend to devote more time and energy into the upfront development of scientific models and theories. In contrast, the models and theories depicted as diagrams in science journal articles are typically developed after the research has been done, serving mainly as summary and reflective artifacts. Through concept-based interfaces, scientists are required to detail and elaborate their models and theories, which otherwise would remain in partially conceived forms in scientists' heads. Furthermore, as scientists work with their visual models, they are continually reminded of the scientific concepts that enable, support, and constrain their investigations. As a result, research and experimentation becomes more theory-driven.

For the most part, scientific knowledge is generally maintained and managed in a transient form in the scientist's head. An important objective of concept-based user interfaces is to migrate scientific knowledge from its ethereal form to one that is both accessible and computable. With the systems presented here, different kinds of scientific knowledge are captured, represented, and "operationalized." A goal of these visual systems is to represent and convey knowledge at the appropriate level of abstraction such that scientists are not required to encode their theories and processes in a foreign manner, but rather express them in an intuitive and natural way. Scientific knowledge should be maintained and applied in a context and language that is familiar to scientists and consistent with their ongoing research. Through the use of concept-based user interfaces, scientific knowledge becomes concrete, usable, and shareable.

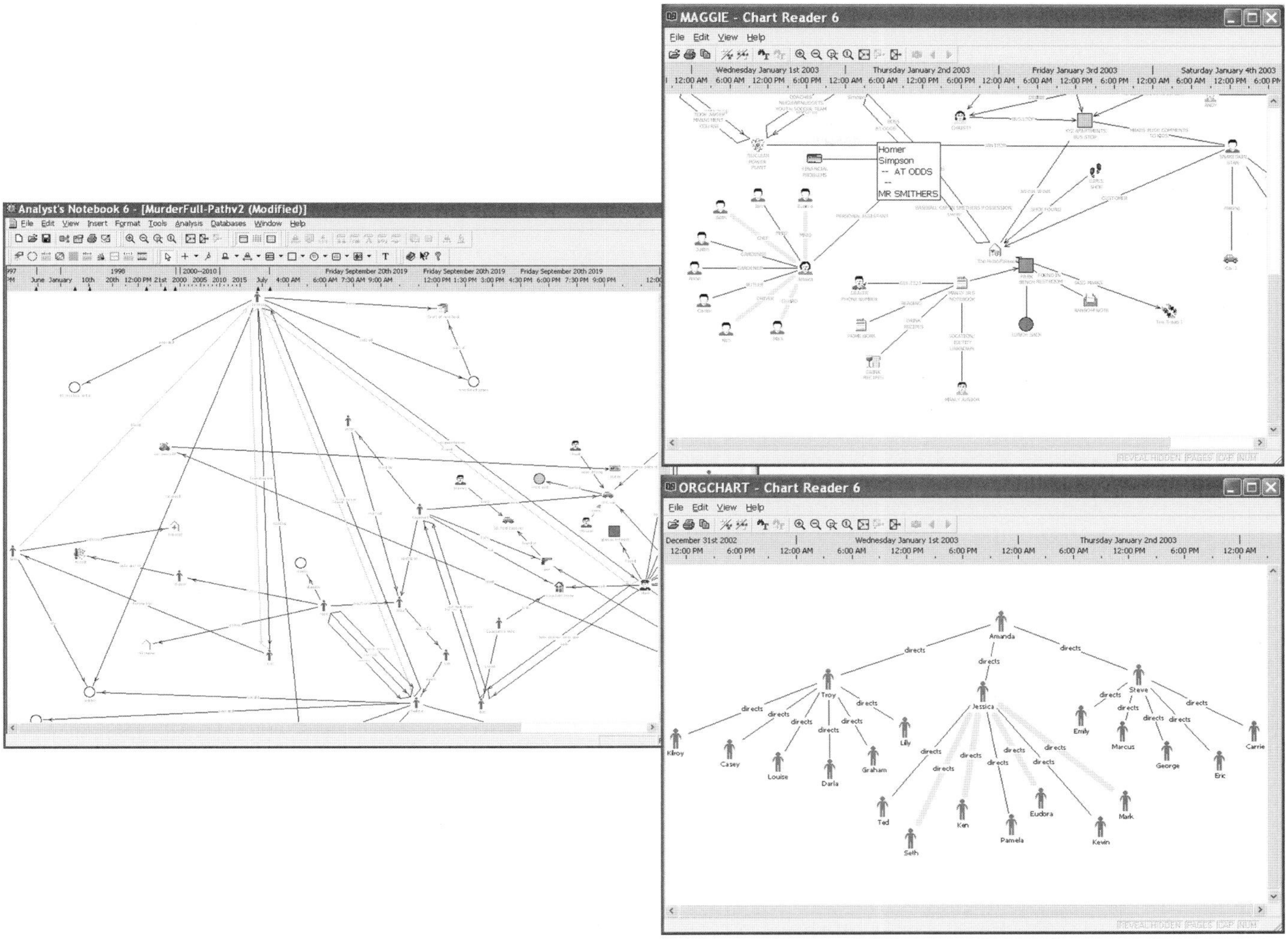

Fig. 5. The Scenario and Knowledge Framework for Analytical Modeling system supports comparisons across graph-based scenarios. In the figure, the query subgraph (highlighted in light gray in left window) matches two scenarios from the case library (matching subgraphs highlighted in dark gray in right windows).

For background information *see* GRAPH THEORY; GRAPHIC METHODS; INFORMATION TECHNOLOGY; MODEL THEORY; SCIENTIFIC METHODS in the McGraw-Hill Encyclopedia of Science & Technology.

George Chin; Deborah Gracio; Alan Chappell

Bibliography. G. Chin Jr. et al., Graph-based comparisons of scenarios in intelligence analysis, in *Proceedings of the 2004 IEEE International Conference on Systems, Man and Cybernetics*, October 10–13, The Hague, The Netherlands, pp. 3175–3180, 2004; G. Chin Jr. et al., New paradigms in collaborative problem solving environments for scientific computing, in *Proceeding of the 2002 International Conference of Intelligent User Interfaces (IUI 2002)*, January 13–16, San Francisco, pp. 39–46, 2002; G. Chin Jr. et al., Participatory workflow analysis: Unveiling scientific research processes with physical scientists, in *Proceedings of Participatory Design Conference (PDC 2000)*, November 28–December 1, New York, pp. 30–39, 2000; G. Chin Jr. et al., Supporting computational visual theories in biology, in *Proceedings of the 2004 IEEE Symposium on Visual Languages and Human-Centric Computing*, September 26–29, Rome, pp. 69–71, 2004; D. A. Thurman et al., SIFT: A component-based integration architecture for enterprise analytics, in *Proceedings of the Sixth International IEEE Conference on Commercial-off-the-Shelf (COTS)-Based Software Systems*, February 26–March 2, Banff, Alberta, Canada, pp. 82–92, 2007.

Cyber forensics

Forensic science is the application of scientific knowledge to legal matters. This means that crime investigations are assisted by processing evidence using scientific methods. Cyber forensics is a subset of forensic science for processing digital evidence. Cyber forensics does not necessarily apply to legal matters. It encompasses a large number of activities, from computer crime investigations to locating specific data on digital devices. Cyber forensics is not limited to crimes or incidents committed with computers. It also includes the processing of electronic data from mobile devices such as cellular telephones, personal digital assistants (PDAs), music players, or microchips with storage capability.

Investigations. Cyber forensic activities can be classified into three categories: computer crime investigation, incident response analysis, and processing digital evidence in support of crime investigation.

Computer crime investigation. In this scenario, a cyber crime is to be investigated, with the ultimate goal of criminal prosecution in court. A cyber crime is an act that involves a computing system as the victim of the crime, or uses a computing device as an instrumentality to commit the crime. Examples of cyber crimes include unauthorized access to computing systems, theft of information, denial of computing services, as well as destruction of digital information.

Incident response analysis. Here no crime may have been committed, so the investigator's objective is to find the series of events that may have led to the incident, or whether an incident actually occurred. This scenario may evolve into a cyber crime investigation, but it may be resolved, for example, by recovery to normal operations, changes in security policies, or ignoring the incident.

Processing digital evidence in support of crime investigation. Here cyber forensics plays an auxiliary role in a crime investigation. The crime itself may have nothing to do with computers, but data stored on computing devices may be important to the investigation, including e-mail data, telephone numbers, Web browsing habits, chat-room activity, and so on, for the purpose of establishing an alibi, motive, or modus operandi of a crime.

Investigative process. The cyber forensics investigative process consists of four major stages: acquisition, recovery, analysis, and presentation. While there exist forensic process models with a more fine-grained differentiation of steps, these four are common to all.

Acquisition. The acquisition phase is concerned with obtaining raw data as digital evidence so that it can be used in a cyber forensic investigation. All investigations should initially be treated as if they pertain to a criminal prosecution, because if digital evidence is not obtained legally, or not handled correctly, it may be ruled as inadmissible if the case is tried in court.

Many factors influence whether digital evidence is legally obtained. The Fourth Amendment to the Constitution protects citizens against unreasonable searches and seizures. The Electronic Communications Privacy Act governs interception of electronic messages between individuals. And other privacy laws exist that may make it illegal to collect certain data without first obtaining a court-issued warrant. In addition, businesses need to have policies and consent forms in place before they may collect or monitor certain information about their employees, such as e-mails or Web-surfing habits.

The next step in the acquisition phase is to generate a forensic duplicate of the original evidence, that is, an exact copy of the original electronic medium at the bit level. The United States Federal Rules of Evidence have established that copies may be used in lieu of the original if it can be authenticated. For this purpose, a cryptographic checksum is used to establish the authenticity of the copy.

The acquisition of digital evidence off "dead" systems (for example, a computer that has been shut down) is fairly straightforward in that data from the permanent storage media of that system is copied. Sometimes, forensic write blockers are used to prevent accidental writing of data to the original evidence. Evidence acquisition becomes more difficult when the data needs to be retrieved from a live or running system. This is due to the nature of volatility of data on any computing system. For example, data changes rapidly on the register level, very quickly on the memory level, and frequently on secondary storage devices. If a system cannot be shut down for forensic acquisition, it is difficult to argue that any forensic duplicate correctly represents the original state of the system because the original evidence was altered by those changes that occurred after the acquisition.

Once the digital evidence has been acquired, its integrity needs to be guaranteed for the duration of the investigation, and all handling of the evidence needs to be accounted for. This is called "chain of custody" and should involve physical safekeeping of the original evidence (evidence locker) as well as an evidence log of anyone accessing the evidence, recorded with a date and time. This ensures that all findings from the subsequent phases of the investigation can be verified with the original evidence and that any forensic duplicates used during that process can be linked back to the originals.

An area of special expertise is the acquisition of evidence from unusual or exotic computing devices, such as mobile telephones, smartcards, or microchips. Often these devices only interface with other computers in a specific way that is not publicly known or that requires special equipment. The acquisition of this type of evidence is therefore reserved to investigators with specialized skills and lab equipment.

Recovery. In the recovery phase, the raw data are converted to information. The bits and bytes of the digital evidence from a computing system are converted into files, metadata, and other information. For this to happen, data structures common to the computing system are applied to the evidence. This includes looking at disk partitions, file systems, data archives, and memory images, as well as extracting information from data that have been deliberately made hard to understand (obfuscated) or encrypted.

The first step is to recover all the regular files from the evidence. Then the files' metadata, such as timestamps, owner, and access rights, need to be retrieved. It is mandatory to always examine the raw data, as opposed to mounting disk partitions directly, because information may be destroyed using the original system during the booting process or in accessing data. Proper recovery techniques often lead to the discovery of deleted and "hidden" files from a file system. Those are files that are no longer available through the file system tree, but whose data are still present on the blocks of the storage device. Furthermore, it may be possible to recover data from slack, or unused, space off a device. This is data that used to belong to old files on the system, which now have been partially overwritten with new data.

Recovery with incomplete information is necessary when damaged media are part of a forensic investigation. When recovering files and information from data structures with incomplete information, advanced file recovery techniques, called data carving, need to be used, which often look for certain characteristics of raw data blocks in the attempt to reassemble them to complete files.

Analysis. In the analysis phase, the investigator needs to interpret the information obtained in the acquisition and recovery phases and apply it to the case at hand. On a computing system, the events that took place often need to be reconstructed. For this purpose, information from logging facilities, system metadata, and application data and logs are combined to form a bigger picture of the events that took place, as well interdependencies between events. Often timestamps found in the logs and metadata are used to create a timeline of events on the system. The investigator can then focus on the times that important events took place, look for unusual activity, or verify if all events that should have happened actually took place.

Given that digital evidence is easy to manipulate, the more independent sources the investigator can find that confirm an event, the higher the confidence that those events actually happened. During the course of the investigation, the investigator will need to form hypotheses pertaining to the case and either verify or disprove them (hypothesis testing). A cyber forensic investigation may be as simple as locating a specific file on a computing device, or it may require the reconstruction of all events that occurred on many systems. The fact that computing systems have differing audit policies, security measures, access, and system clocks set to different times may result in a tedious and time-consuming analysis.

Another important technique that may be used in this phase is program and authorship analysis. In program analysis, unknown computer programs encountered during the investigation are examined to see if they are malicious or have any bearing on the case. This can become a time-consuming process involving disassembling the program and interpreting the machine code. Authorship analysis aims to find the author of a document, e-mail, or computer program. For this, the investigator needs to search for identifying characteristics unique to a specific person or a group of people, or to compare writing styles.

Presentation. When the analysis of the evidence is complete, the investigator needs to report the findings. The extent of such a report ranges from a short executive summary to a court testimony or presentation. It is important that all claims are supported with direct or circumstantial evidence, that the findings of the case are presented in a clear and concise fashion that can be understood by the target audience, and that the report is written so that an independent investigator can repeat the analysis described and come to the same conclusions.

For background information *see* COMPUTER ARCHITECTURE; COMPUTER STORAGE TECHNOLOGY; CRIMINALISTICS; CRYPTOGRAPHY; DATA COMMUNICATIONS; DATA STRUCTURE; DIGITAL COMPUTER; DIGITAL EVIDENCE in the McGraw-Hill Encyclopedia of Science & Technology. Florian Buchholz

Bibliography. E. Casey, *Digital Evidence and Computer Crime*, 2d ed., 2004; D. Farmer and W. Venema, *Forensic Discovery*, 2004; W. Kruse II and J. Heiser, *Computer Forensics: Incident Response Essentials*, 2001; C. Prosise, K. Mandia, and M. Pepe, *Incident Response and Computer Forensics*, 2d ed., 2003.

Deep Impact

The word "comet" is rooted in the ancient Greek word κομήτης (komḗtēs), literally, hairy star. Indeed, this wording is a concise synthesis of naked-eye observations, in line with our modern understanding. A comet consists of a small nucleus of volatile material. Close to the Sun, evaporation forms large tails of gas and dust. In the twentieth century comets were finally recognized as the most pristine remnants of the formation of our solar system, about 5 billion (5×10^9) years ago. Comets are heralds of the physical conditions during the formation of our solar system. Starting from the first orbit determination by Edmund Halley, modern comet research suffered from the paradox that many very specific details were known with high precision, while rather fundamental quantities, such as mass or detailed consistency of the nuclei, had to remain the subject of educated guessing. Fly-by spacecraft imagery, starting with the European Space Agency's *Giotto* passing comet 1P/Halley in 1986 at a distance of 5000 km (3000 mi), is spectacular, but most basic questions can be addressed only by spacecraft directly encountering the nuclei of comets.

The National Aeronautics and Space Administration's (NASA's) *Deep Impact* spacecraft was the first to provide information on the surface and nuclear properties of a comet, 9P/Tempel 1. On July 4, 2005, it separated into a fly-by probe and an impactor; the latter was then obliterated by the much faster-moving comet nucleus leaving an artificial impact crater. Light resulting from the very first hot-plasma-dominated impact phase arrived on Earth at 05h52m03s UTC. Years of coordinated observations by a network of ground-based telescopes and the *Hubble Space Telescope* (*HST*) guided the spacecraft. The impact was witnessed at wavelengths from radio waves to x-rays by the *Deep Impact* fly-by spacecraft itself, the *Hubble Space Telescope*, NASA's *Spitzer Space Telescope*, the European Space Agency's *Rosetta* spacecraft (en route to a soft landing on the nucleus of comet 67P/Churyumov-Gerasimenko in 2014), other

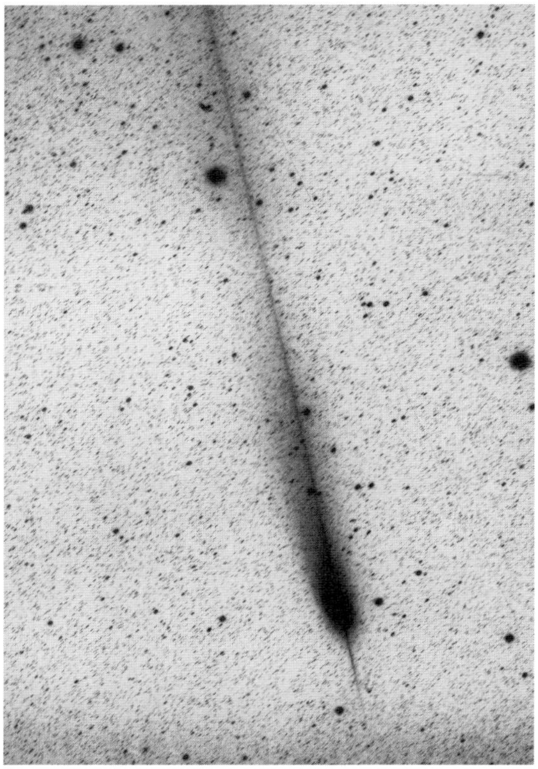

Fig. 1. Wide-field image of comet Hale-Bopp (C/1995 O1) taken in January 1998, exactly when the Earth was passing through the comet's orbital plane. It extends over 5° or some 50 × 10^6 km (31 × 10^6 mi), one-third of the distance from Earth to the Sun. While the dust tail—the quasistraight line in the image—follows Keplerian orbits similar to the comet nucleus, the gas tail—the fuzzy structure—is ionized by solar ultraviolet radiation and interacts with the solar wind to take a completely different direction not constrained by orbital mechanics. The major force on dust particles is radiation pressure. As the mass of the particle scales with volume (r^3) while the force is proportional to the area (r^2) it is obvious that small particles are driven away faster, while larger particles stay close to the orbit of the comet and form an annular cloud giving rise to well-known annual meteor showers (for example, the Leonids or the Perseids). The cometary dust particles range in size from a few micrometers to meters (less than 1/1000th of an inch to ~40 in.). *(Photo courtesy of European Southern Observatory, ESO Press Photo 05a/98)*

satellites, and a network of all major astronomical observatories on Earth.

Properties of comets. Tycho Brahe, in 1566, applying basic trigonometry, deduced that comets were located outside of Earth's atmosphere, and by the nineteenth and early twentieth century statistical data on the orbits of many comets had been obtained. Cometary tails were identified as streams of gas and dust (**Fig. 1**), and the gaseous components have been analyzed over the years with high-resolution spectroscopy and found to consist of ionized fractured simple molecules, for example OH^- (hydroxyl) or CN^- (cyanide) ions. Later, with better instrumentation, the parent molecules of these radicals could be identified in the neutral gas close to the nucleus.

In the early 1950s, Fred Whipple condensed all observational evidence in his famous description of the nucleus as a "dirty snow ball": that is, the nucleus is a conglomerate of material that would be highly volatile at room temperature, mostly water ice mixed with CO and CO_2 plus silicate dust. Astronomers today tend to prefer the term "icy dirt ball" but the model has remained unchallenged.

In May 2006 the fragments of the fractured comet 73P-C/Schwassmann-Wachmann 3 came as close as 11.3 × 10^6 km (7.1 × 10^6 mi) to Earth (30 times the distance to our Moon). Even in this exceptional case the resolution of astronomical telescopes, including *Hubble*, was insufficient to resolve and hence measure the sizes and shapes of cometary nuclei. Estimates, however, can be provided with the standard technique used to measure the sizes of asteroids by comparing the scattered sunlight from the object to its thermal infrared radiation. The comet tails add some confusion, but nuclear sizes are consistently approximately 0.8–30 km (0.5–20 mi). This size range was confirmed by the four spacecraft missions to comets with cameras (including *Deep Impact*). However, the extremely complex shapes of these bodies came as a surprise.

Comets approach the Sun on highly elliptical orbits. Most of the time they are far from the Sun and their average temperature is less than about 60 K (−213°C or −352°F). During passage into the inner solar system, some comets are gravitationally perturbed by Jupiter such that their orbits are then well inside the orbits of Pluto or Neptune ("Jupiter family").

Nagging questions before *Deep Impact*. Among the unanswered questions about comets that the *Deep Space* mission was designed to investigate were the following:

1. What is the density of cometary material? Is it close to 1 g/cm^3, the density of ice, or more like 0.1–0.2 g/cm^2, the density of freshly fallen snow when it is rather cold (−25°C, −13°F)? Or is it more like 0.01 g/cm^2, the density of aerogel, one of the fluffiest structures available?

2. What is the mass of a comet? Size determinations, due to the uncertainty in density, typically resulted in a factor of 100 total uncertainty of the mass. Thus one also had no estimate of the total mass of all comets in our solar system.

3. What is the composition of the crust of a comet? From earlier fly-by imagery it was known that comets had a dark crust, but composition and cohesive strength could at best be guessed from the impact craters on the nucleus of comet 81P/Wild (also known as Wild 2); spontanous splitting of comets (for example, 73P-C/Schwassmann-Wachmann 3 in December 1995); and the spectacular tidal disruption of the Comet Shoemaker-Levy 9 during its close passage of Jupiter, which injected the fragments into an orbit leading to the spectacular impacts in 1994.

4. Why is the outgassing or activity of comets concentrated in a few active regions of the surface? What are the physical processes responsible?

Mission. The idea behind the *Deep Impact* mission was to do the simplest experiment: Create an artificial impact crater on the nucleus of a comet while observing the event from a fly-by spacecraft. Any fly-by spacecraft has only few minutes of time for observations; thus follow-up with ground-based telescopes in the form of a global network and the *Hubble Space Telescope* was planned to have continuous coverage over a few days. The astronomy department of the University of Maryland and NASA collaborated in the form of a "Discovery Mission" (Principal Investigator Michael A'Hearn). Comets close to the Sun move quickly (20–40 km/s or 45,000–90,000 mi/h). In comparison, the acceleration of even the most powerful rockets is rather limited. Thus to freely choose a comet, aim at it, and fire off a rocket is impossible. The major planets in the solar system all move in one plane, the ecliptic, as they formed out of the same protoplanetary disk. Comets, however, are more pristine and many originate from spherical reservoirs; thus they normally do not move in the ecliptic. Leaving the ecliptic requires an extra 10–15 km/s acceleration (22,000–34,000 mi/h). This differential acceleration alone would suffice to reach the Moon. Constraints for an affordable and successful mission, not involving complex and time-consuming swing-by maneuvers using the gravity of planets to gain momentum, are:

1. A well-understood orbit; this rules out newly discovered comets and limits the choice to short-period Jupiter-family-type comets.
2. An orbit that crosses the ecliptic in the vicinity of Earth's orbit, thus defining the encounter point.
3. The ability to be observed from ground-based telescopes (encounter point not in vicinity of the Sun).
4. For ground-based follow-up, the event has to fall into a new-Moon period.

Short-period comets have already experienced many passages close to the Sun; how many, normally, cannot be estimated. Many of the 100–150 possible target comets have undergone such fundamental changes that they are no longer real comets. Comet 9P/Tempel 1 is one of the few fulfilling all requirements. It is a slowly rotating (period of 41 h), medium-size (diameter less than 7.5 km or 4.7 mi), low-albedo (8%), short-period comet, most likely originating in the Kuiper Belt (also known, particularly in Europe, as the Edgeworth-Kuiper Belt). It crosses the ecliptic close to perihelion, which is located approximately at the orbit of Mars, 1.5 astronomical units (au) [140×10^6 mi] from the Sun. The date of the encounter was fine-tuned to have the spacecraft at that moment in contact with two tracking stations for telemetry, and to have the *Hubble Space Telescope* in a good position. All constraints were best met on July 4th, 2005, 5h55 UT.

The 1020-kg (2250-lb) spacecraft was launched on January 12, 2005, from Cape Canaveral on a 5.5-month voyage. Three days before encounter it separated into the fly-by unit and the impactor mass 372 kg (820 lb), mostly copper to minimize the contamination of the ejecta spectra. The impactor hit the comet nucleus at a relative speed of 10.2 km/s (approximately 23,000 mi/h) to release a kinetic energy of 19 gigajoule (5300 kWh), while the fly-by spacecraft passed at a distance of about 500 km (300 mi), observing the event with three on-board remote-sensing experiments: a wide and a narrow-field camera (visible wavelength range) and a near-infrared imaging spectrometer (1–5 μm).

The fly-by spacecraft had to pass the inner coma region configured in safe mode to survive the heavy bombardment by dust particles. Later it could resume operations and play back the data in the on-board mass storage units (approximately 1 gigabyte) via the telemetry link.

Some 10^4 tons of material were released immediately by the impact, most of it accelerated beyond the escape velocity (about 1.5 m/s or 3.4 mi/h) while slower material fell back onto the nucleus. Most likely 5–8 % of all material was water.

During the approach *Deep Impact* took a series of high-quality images and spectra. The combination with ground-based observations and other satellite data make this comet the best-studied such object ever. High-precision albedo, color, and thermal maps of the nucleus exist.

The Coordinated Ground Support Observation Campaign was an important complement to guarantee the success of the mission. All major astronomical observatories participated to observe before, during, and up to several days after the impact. Data were shared in real-time openly through an email exploder and a Web server at the University of Maryland. There was an approximately 72-h video conference with more than 20 sites and a hub at NASA's Infrared Telescope Facility on Mauna Kea, Hawaii.

Results. Really unexpected was finding patches of water ice on the surface, which previously were believed not to be stable. From the gravitational settling of the impact debris, gravity and mass could be determined; the density is approximately 0.3 g/cm^3 (unexpectedly low; similar to freshly fallen snow). The surface crust has different layers and may have a thickness of up to 300 m (1000 ft). In everyday terms, in the impact area, the ratio of the diameter of the nucleus to the thickness of the crust is similar to the ratio of the diameter of an orange to the thickness of its peel. From the surface structure it was concluded that the orbit has changed previously such that the comet had long periods where it did not come close to the Sun to start sublimation.

One of the mysteries for the ground-based observations is that the dust cloud created by the impact vanished with remarkable speed. Its optical properties cannot be described with any model normally used to describe the dust tails of comets. It is unclear if one sees evaporation of particles or if the size distribution was fundamentally different.

Both ground-based observers and the fly-by spacecraft had observed complex patterns of jets in 9P/Tempel 1. To identify the source of any of these jets with surface features (**Fig. 2**) was not possible. The impact had no effect on the gas and dust production after the crater formation had come to an end (in a few hours). Most astronomers had expected the formation of a new active area. Hence what makes part of the surface an active region remains obscure. High-resolution optical and infrared imagery exists for only one part of the nucleus, which may not be representative for the complete comet.

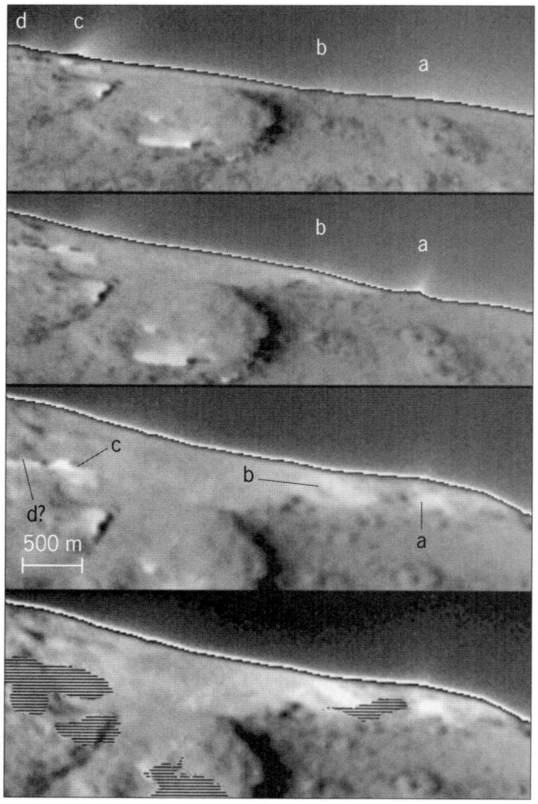

Fig. 2. Samples of the imaging data of the fly-by spacecraft. A small part of the rim of the nucleus is shown for three positions of the spacecraft dashing by the comet (it appears as if the nucleus is rotating). The bottom frame is a repetition of the third with areas of water ice highlighted. In the top, fuzzy structures (a, b, c, d) extend beyond the rim; they are most likely gas and dust streaming off. In the top two panels the bottom points of these jets shift over the horizon. However, there is no obvious surface feature that could shed light on the processes responsible for creating these well-localized jets. (*Photo courtesy of NASA/UM/UAF/Don Hampton and Tony Farnham*)

The conclusions about surface structure, composition, strength, and so forth are based on models that describe the observations, but there is no proof that these interpretations are unique.

Outlook. Two independent proposals for an extended mission (DIXI and EPOCH) were merged into a project for the *Deep Impact* (*DI*) fly-by spacecraft called EPOXI, which has been confirmed by NASA. The plan is to fly past comet 85P/Boethin on December 5, 2008. This comet was detected in 1975 but was not observed during its last apparition in 1996. Astronomers worldwide will soon start another coordinated campaign to recover this object during its next apparition in Summer 2008.

An even more interesting recycled spacecraft is NASA's *Stardust*, renamed *New Exploration of Tempel 1* (NExT). NExT will fly by 9P/Tempel 1 on February 14, 2011, to look for changes in the nucleus after its close approach to the Sun and, with some luck, to image the impact area. It will extend the mapping of 9P/Tempel 1.

The *Rosetta* team now has a better basis for their mission planning. The results of *Deep Impact*, while 10–15 years too late for the preparation of this mission, were most eagerly awaited. It is hoped that many questions *Deep Impact* could not answer will be addressed by this mission in which a lander will be deployed while the mother spacecraft will enter into a stable orbit to follow the comet during its passage through perihelion. This lander, with its onboard chemical laboratory, will answer speculative questions asked in the context of the new field of astrobiology such as: Do comets contain complex organic material, or could they even have brought primitive forms of life to Earth? For results, however, we will have to wait until 2014.

For background information *see* ASTROBIOLOGY; ASTRONOMICAL SPECTROSCOPY; CELESTIAL MECHANICS; COMET; HALLEY'S COMET; HUBBLE SPACE TELESCOPE; KUIPER BELT; SOLAR SYSTEM in the McGraw-Hill Encyclopedia of Science & Technology.

Hans Ulrich Käufl

Bibliography. H. U. Käufl and C. Sterken, *Deep Impact as a World Observatory Event: Synergies in Space, Time and Wavelength*, Springer, 2007 (in press).

Desorption electrospray ionization mass spectrometry

Mass spectrometry (MS) is undergoing significant expansions in its capabilities, in which advances in instrumentation are enabling new applications. In the last 50 years, MS has developed from a method used to measure the masses and abundances of nuclides to a widely used tool for organic structural analysis, environmental monitoring, and metabolite profiling, as well as the principal experimental basis for proteomics. These developments followed directly from advances in mass analyzers, combinations with MS and separation methods, and especially from the invention of new ionization methods. The development of proteomics is based on widely celebrated advances in ionization [electrospray ionization (ESI) and matrix-assisted laser desorption/ionization (MALDI)] techniques and on the less noticed development of tandem mass spectrometry (MS/MS).

More recent developments are in ambient MS ionization methods. Ambient MS is the analysis of samples directly in the free environment, where they are accessible to further chemical or physical manipulation. These methods include atmospheric solids analysis probe (ASAP), direct analysis in real time (DART), electrospray laser/desorption ionization (ELDI), and desorption electrospray ionization (DESI), the principal method described here. As was the case in the development of MALDI, which depended on many years of prior work, the new methods of ambient MS have grown out of the gradual recognition that samples need not be introduced into the vacuum before being analyzed. The requirement for the sample to be introduced into the vacuum system poses potential problems with contamination, speed of analysis, and the ability to provide true in situ measurements. Ambient MS first became apparent in

Typical instrumental parameters used in DESI-MS

Parameter	Value
Electrospray voltage (kV)	4–5
Electrospray flow rate (μL/min)	1–3
Nebulizing gas pressure (psi)*	100–150
Tip-to-surface distance; d_1 (mm)	1–3
Incident angle, α (°)	50–65
Collection angle, β (°)	<20

*The nebulizing gas pressure will vary based on nozzle geometry.

the atmospheric pressure (AP) version of the MALDI technique, in which matrix-treated samples are examined by laser radiation in air. The acceptance of AP-MALDI as an ambient ionization method is limited by the fact that the sample must first be prepared by adding ultraviolet-absorbing compounds to the solution or surface. This step adds complexity to the analysis. The importance of these new methods is that no sample preparation method is done outside the vacuum chamber, while the advantages of mass spectrometry are retained, including its speed, chemical specificity, and low detection limits.

Desorption electrospray ionization. Desorption electrospray ionization, first reported in 2004 by R. G. Cooks and coworkers, uses a pneumatically assisted electrospray ionization source for the analysis of materials present on or near surfaces. **Figure 1** shows the general concept. In the DESI method, charged droplets and ions produced from the electrospray are directed by a high-velocity gas jet to the analyte's surface. The spray impacts the surface and produces ions of the analytes, some of which are adsorbed by primary and secondary droplets. Secondary droplets carrying the analyte are ejected from the surface, collected in the ion transfer tube or atmospheric inlet of a standard commercial mass spectrometer, and mass-analyzed.

There are a number of operational parameters that govern the outcome of the experiment. These factors include geometrical variables [such as the incident angle (α), collection angle (β), and tip-to-surface distance (d_1)], physical variables (such as volumetric flow rate and nebulizing gas flow rate), and solvent/analyte and solvent/surface interactions. The typical operational conditions used in DESI are given in the **table**. Cooks and coworkers examined the effect of sprayer-to-surface distance and volumetric flow rate on signal intensities. Their results showed that the highest signal intensities were obtained when using a sprayer tip-to-surface distance of 1 mm and a solvent flow rate of 3 μL/min. This is consistent with other work by Cooks and coworkers when phase Doppler anemometry, a technique for measuring spray droplet velocity and diameter, was used to characterize the sizes and velocities of the impacting and ejected droplets, while correlating (indirectly) these observations to signal intensities. Their results showed that the droplet velocities are highest closest to the sprayer nozzle. Thus, the droplet impact force is greatest leading to more efficient desorption/ionization at the surface. Further, G. J. Van Berkel and coworkers have described the droplet impact regions that provide the basis for obtaining optimal analysis conditions. **Figure 2** shows a representation of the droplet impact region on a typical surface. Within the impact area is a smaller elliptical region where desorption/ionization is most efficient. A second region is composed of solvent "jets" or secondary droplets that originate from the inner elliptical plume. A third region is composed of larger, secondary slow-moving droplets. Optimal signal levels were obtained when the analyte was sampled in the smaller elliptical region, as opposed to being sampled in the second or third region of the impact plume where desorption/ionization is less efficient.

Applications. A majority of the work reported on DESI have been recorded using commercial quadrupole ion-trap mass spectrometers. However, the common use of ion-trap mass spectrometers is not a consequence of using this method since its implementation simply requires an atmospheric interface. In addition to the use with ion-trap mass spectrometers, DESI has been combined with triple quadrupoles, quadrupole time-of-flight instruments, ion mobility/time-of-flight, and others.

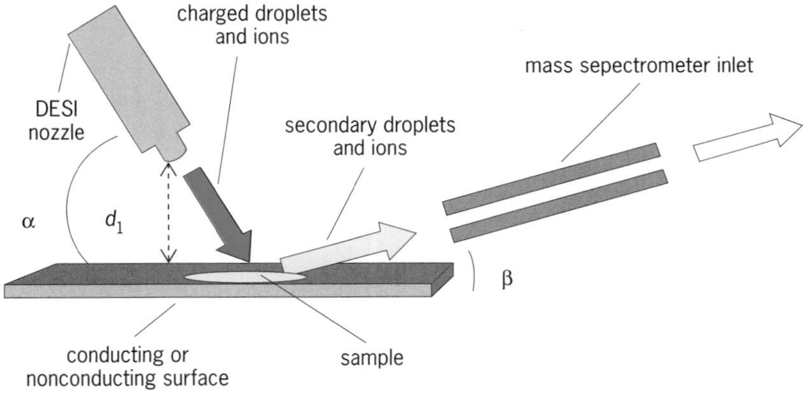

Fig. 1. Schematic representation of the DESI concept, where α is the incident angle, β the collection angle, and d_1 the tip-to-surface distance.

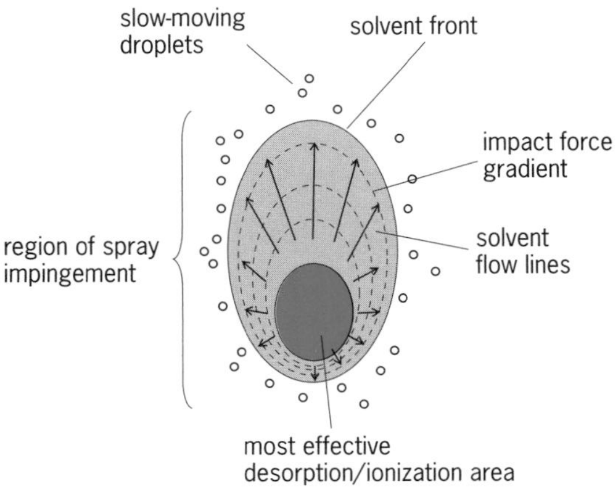

Fig. 2. Idealized representation of the DESI impact plume region. (*Reprinted with permission from Anal. Chem., 79:5956–5962, 2007; copyright 2007 American Chemical Society*)

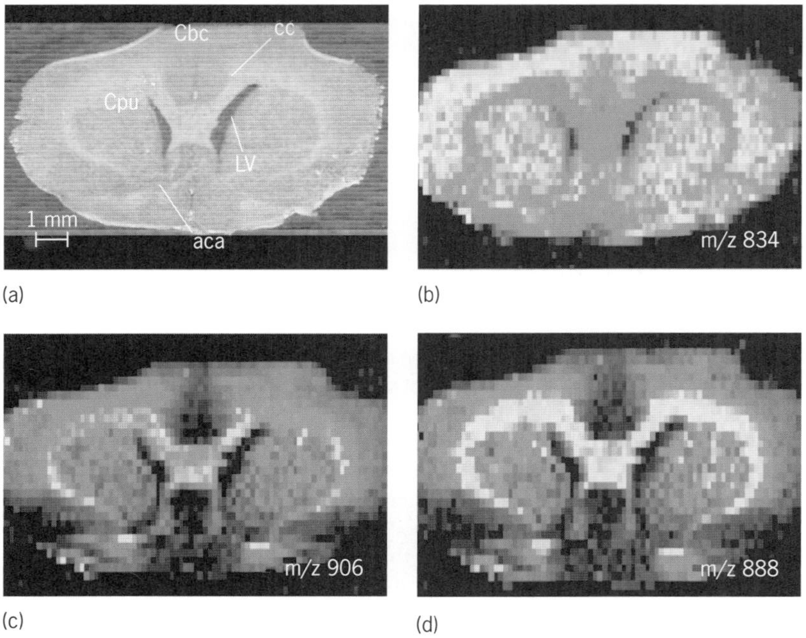

Fig. 3. Optical and selected molecular ion [(M–H)$^-$] images recorded in negative-ion mode of specific lipids from analysis of a 13 × 10 mm^2 area of rat brain tissue section. (*a*) Optical image of the coronal section of the rat brain prior to analysis; cc, corpus callosum; CPu, striatum; Cbc, cerebral cortex; LV, lateral ventricle; aca, anterior part of anterior commissure. (*b–d*) Ion images of phosphatidylserine (40:6), sulfatide (h24:1), and sulfatide (24:1). *m/z* = mass-to-charge ratio. (*Copyright Wiley-VCH Verlag GmbH & Co. phosphatidylserine KGaA; reproduced with permission from Ange. Chem. Int. Edit., 45:7188–7192, 2006*)

So far, DESI applications have focused on high-throughput analysis, chemical-warfare detection, and chemical imaging. DESI has been applied to the rapid analysis of chemical warfare agents and explosives present on many common surfaces (such as paper, plastic, fabric, metal, etc.). Cooks and coworkers determined the limits of detection, reaching levels of picograms to femtograms (10^{-12} to 10^{-15} g) for RDX (trinitrohexahydro-1,3,5-triazine), TNT (1,3,5-trinotrotoluene), and DMMP (dimethylmethylphosphonate). To date, the most widespread use of DESI has been for the rapid analysis of pharmaceutical formulations (such as tablets, ointments, and liquids), without prior chromatographic separation. The direct investigation of pharmaceutical tablets and ointments, as well as abused drugs in tablet form and in plant material, has been successfully demonstrated by a number of groups. One important illustration is the application of DESI to rapid chemical fingerprinting of pharmaceutical tablets for the identification of potentially unsafe or counterfeit tablet formulations. The versatility of the DESI method was explored by F. M. Fernandez and coworkers for the rapid fingerprinting of counterfeit antimalarial (artensuate) tablets. Furthermore, Fernandez and coworkers used a unique aspect of the DESI method that allows chemically reactive species to be introduced into the spray solution to interact with surface molecules, thereby enhancing ionization. In these studies, an alkylamine reagent was used to enhance the ionization efficiency by nearly 170 times, which allowed direct identification of authentic and counterfeit artensuate tablets.

In the past 10 years, imaging mass spectrometry based on MALDI and SIMS (secondary ion mass spectrometry) has become a powerful technique for determining the spatial distributions of molecules on surfaces. MALDI imaging, pioneered by R. M. Caprioli and coworkers, has evolved into a method for imaging sections of tissue samples. The direct analysis of plant and animal tissues has also been demonstrated using DESI. In this application, thin tissue sections prepared for microscopy or native, freshly cut tissue surfaces are used for DESI-MS analysis. Using electronically controlled motion stages and a fine DESI spray, chemical imaging of surfaces is accomplished. Using DESI, the spatial distribution of natural tissue components, such as membrane phospholipids in animal tissues, has recently been demonstrated with a lateral spatial resolution of less than 400 μm by Cooks and coworkers. **Figure 3** shows images of thin coronal rat brain sections using DESI. In this case, distinct anatomical features in the brain are differentiated, as shown by the contrast in the images. Other imaging applications of DESI include the analysis of thin-layer chromatography plates and latent fingerprints, as well as the differentiating of dyes in forged documents.

Portable ambient mass spectrometry. The desire for in situ chemical measurement has driven the development of small MS systems with vacuum interfaces to air. Moving mass spectrometers out of the lab and into the field requires two important advances: (1) removing complex sample preparation steps, and (2) producing mass spectrometers that are small, portable, and inexpensive. Although the latter is accomplished only with parallel advances in vacuum-pump technologies, electronics, and mass analyzers, the former is facilitated in the DESI method. These developments have extended to instruments fitted with ambient ion sources and a portable DESI

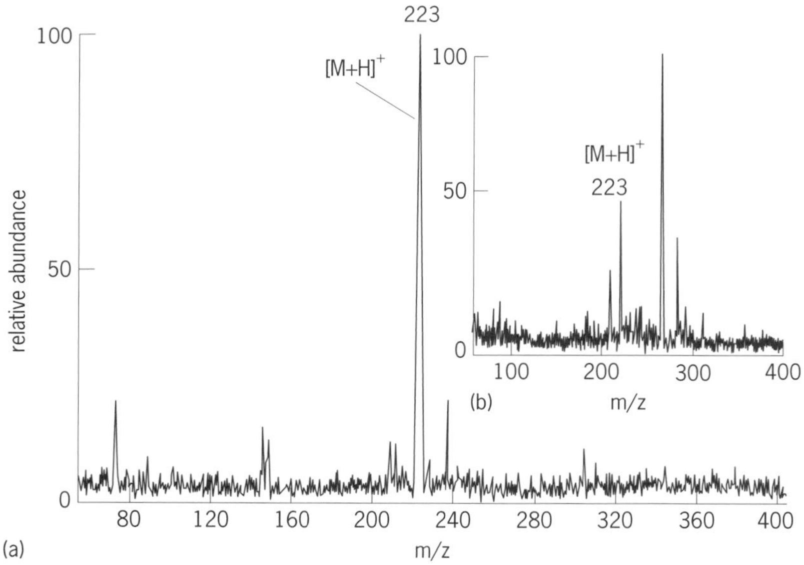

Fig. 4. Positive ion DESI mass spectrum of (*a*) 10 ng of RDX deposited on 1 cm^2 of paper surface, and (*b*) 500 pg of RDX [the limit of detection (LOD) for this study] deposited on 1 cm^2 of paper surface. *m/z* = mass-to-charge ratio (*Chem. Commun., 1709–1711, 2006; reproduced by permission of The Royal Society of Chemistry*)

ion-trap system based on a cylindrical ion-trap analyzer built by Cooks and coworkers. This system is capable of detecting trace levels of explosives and pesticide residues. **Figure 4** shows DESI mass spectra of RDX and DEET (a common pesticide) recorded using a portable MS system. In this example, detection limits were determined to be in the high picogram range for the compounds tested. The combination of the specificity and sensitivity inherent in MS instrumentation and rapid, direct ionization under ambient conditions by DESI presents a very powerful approach to the problems facing homeland security, forensics, pharmaceutical, environmental, and many other fields.

For background information *see* ANALYTICAL CHEMISTRY; MASS SPECTROMETRY; MASS SPECTROSCOPE; SECONDARY ION MASS SPECTROMETRY (SIMS) in the McGraw-Hill Encyclopedia of Science & Technology. Justin M. Wiseman

Bibliography. R. G. Cooks et al., Ambient mass spectrometry, *Science*, 311:1566–1570, 2006; J. B. Fenn et al., Electrospray ionization—Principles and practice, *Mass Spectrom. Rev.*, 9:37–70, 1990; M. L. Pacholski and N. Winograd, Imaging with mass spectrometry, *Chem. Rev.*, 99(10):2977–3006, 1999.

Devonian missing link

A key stage in the evolutionary history of life on Earth is the transition between fish with fins and animals with limbs and digits (such animals are known as tetrapods). This event is thought to have happened between about 370 and 360 million years ago, in the Devonian Period. In 1999, a team of paleontologists mounted their first expedition to Ellesmere Island in the Canadian territory of Nunavut in the Arctic Circle, with the hope of finding fossils that represent this transition. Although this locality had never before been visited by vertebrate paleontologists, the team's visit was an informed guess or prediction that the geological age and type of sediments there were just right for potentially finding relevant material. The sediments were formed in meandering streams and river estuaries in the early part of the Late Devonian about 370 million years ago. Over a series of four expeditions, the collective hunch paid off. After three seasons, the team found a rich seam of vertebrate-bearing rocks that yielded an array of many different kinds of fishes, including some lower jaws and a snout of a previously unknown form but which looked tantalizingly like a transitional form. In the fourth season, they discovered almost complete skeletons and many isolated parts of this creature, which indeed proved to be a spectacular and important new addition to the story of the origin of tetrapods. They called it *Tiktaalik roseae*: *Tiktaalik* is the local Inuktitut name for a large freshwater fish seen in the shallows, and *roseae* honors the benefactor who provided much of the expeditionary funds.

Fossil record. The specimens of *Tiktaalik* were preserved in a reddish siltstone, which had to be removed in the laboratory by a slow process of "mechanical preparation." This involves working under a microscope, drilling, and delicately picking away the siltstone to reveal the bones inside. The process took more than a year, but eventually the fossils were revealed. They proved to be almost three-dimensional, and included three more or less complete individuals that together provide a great deal of detail about the animal's anatomy and add a new dimension to the story (**Fig. 1**).

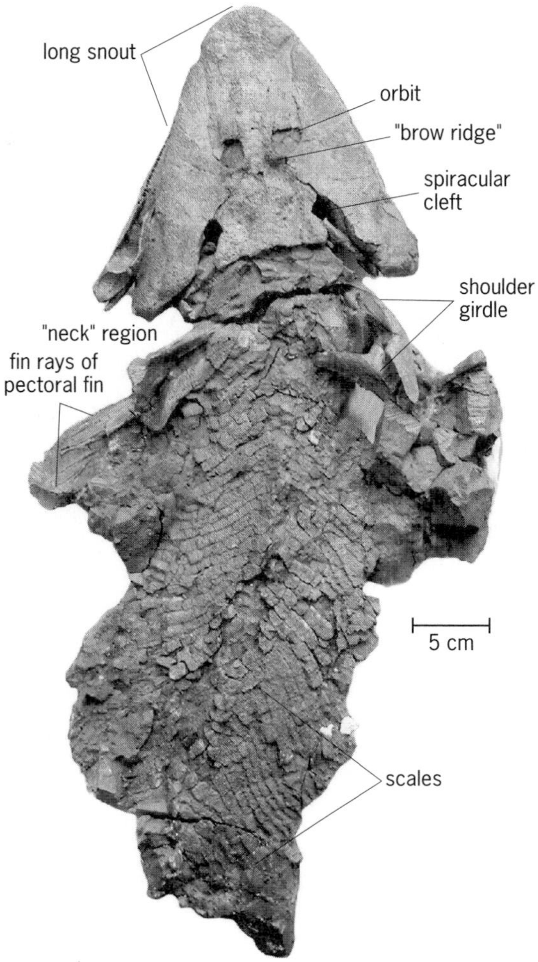

Fig. 1. Photograph of one of the most complete specimens of *Tiktaalik*, seen from above. None of the specimens has been found with the tail and pelvic region intact; this region is missing from the specimen. The specimen shows many of the features mentioned in the text, both fishlike and tetrapodlike.

Tiktaalik fits neatly into a slot between two other creatures that represent the transition. *Panderichthys* is a lobe-finned fish from the Late Devonian of the Baltic States that was previously considered the closest relative of the earliest limbed tetrapods. *Acanthostega*, slightly younger at about 360 million years old, is from the Late Devonian of Greenland and is one of the earliest limbed vertebrates known (**Fig. 2**). Its skeleton is known in considerable detail. In many respects, *Tiktaalik* turns out to be an almost exact intermediate between these two.

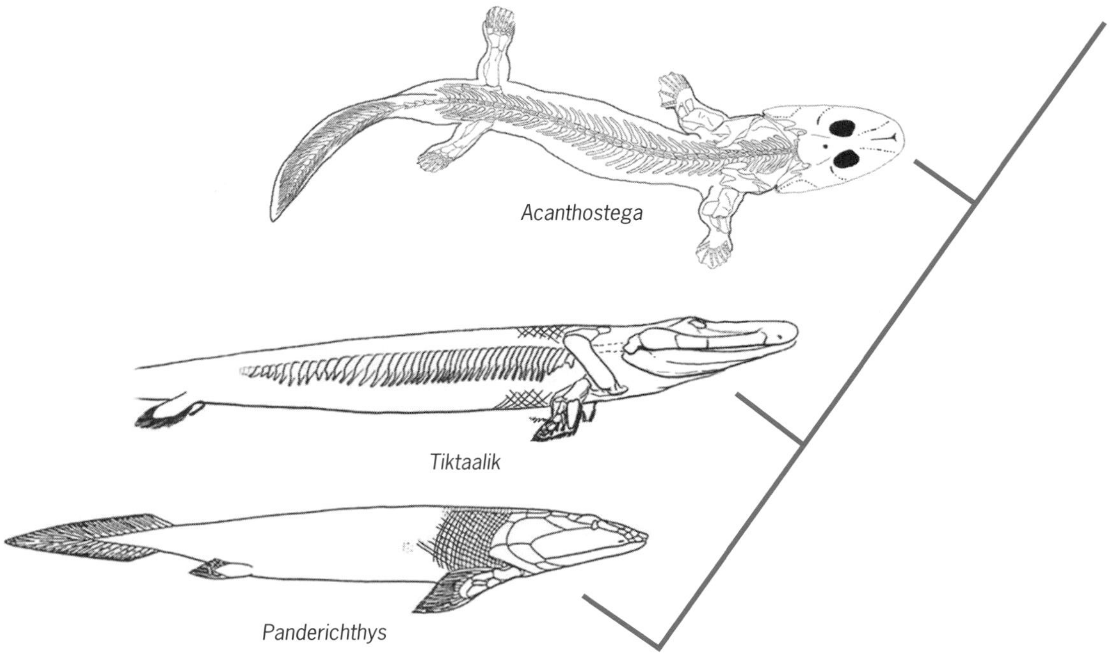

Fig. 2. Part of the "family tree" of early tetrapods and their relatives, showing the body outlines and evolutionary relationships between *Panderichthys, Tiktaalik*, and *Acanthostega*. In *Tiktaalik*, the scale cover has been removed to show the ribs.

The story of the origin of tetrapods, limbs, and terrestrially living vertebrates has become much clearer over the last couple of decades with the discovery of animals like *Acanthostega*, but many questions still remained. These included the order in which certain tetrapod characteristics arose relative to one another, when they did so, and under what circumstances. Many of these questions still remain, of course, but the discovery of *Tiktaalik* has provided new clues and the basis for answers to many of them.

Skeleton and anatomy. The regions of the skeleton of *Tiktaalik* that are fishlike include possession of bony scales covering the whole of its body, fin-rays still present on its fins, the dentition (in which it is *Panderichthys*-like), and some details of the skeletal relationships of the skull to the shoulder region. In the shape of bony components of its fin skeleton, it still shows similarities to earlier members of its group. Those in which it is tetrapodlike include the proportions of the snout, the orbits and the back of the skull, the size of its "spiracular cleft" (see below), the shoulder joint and joint construction in the fin skeleton, and the elongate and overlapping ribs (Figs. 1 and 2).

Tiktaalik has an almost crocodilelike appearance as a result of its relatively elongated snout, with large orbits that lie on top of its head beneath "eyebrow" ridges running along the roofing bones between the eyes. It also has a relatively short region of the skull behind the eyes. These are features seen in early tetrapods. Although their beginnings can be seen in *Panderichthys*, *Tiktaalik* has taken them a step further.

Most members of the group that includes *Tiktaalik* and *Panderichthys* possess a series of bones, called the opercular series, that cover the gill region and help regulate the flow of water through the mouth and over the gills during breathing. Most "fish" have these bones. *Tiktaalik* retains some of these, including those that lie between the lower jaws, called gular plates. Many fishes also have a series of bones that run from the top of the shoulder girdle to contact the back of the skull, and link shoulder and skull into a unit. These bones are called the supracleithral series. *Tiktaalik* retains some of these. Importantly, however, *Tiktaalik* has lost key elements of both these series, including the opercular bones—the topmost and largest elements of the opercular series—and some of the supracleithral series. The result is that apparently *Tiktaalik* has lost the connection between the skull and shoulder girdle and thus has the beginning of a tetrapod feature—the neck. The neck itself was probably not very flexible, and the neck joint was unformed, but the head could probably move more independently than in *Panderichthys* or other fishes. This may be related to another feature: the enlargement of the spiracular cleft. This is an opening that lies between the top of the cheek and the skull roofing bones. In *Panderichthys* it is a narrow slot, but in *Tiktaalik*—and in tetrapods—it is a broad opening with a smoothly rounded apex. Loss of the operculars suggests a lessening reliance on gill breathing, tied up with freedom to raise the head out of the water to gulp air. The spiracular opening may have been involved in its air-breathing mechanism.

The excellent preservation of *Tiktaalik* specimens has allowed the investigation of other key aspects of its anatomy. They include the anatomy of the shoulder girdle, the way that the fin articulated with the shoulder, and the joint movements possible between the bones in its pectoral fin. The shoulder girdle of fishes such as *Panderichthys* is dominated by a large bone called the cleithrum, which is a broad sheetlike bone sheathing the side of the body. Tucked inside toward the base is a small bone called the scapulocoracoid, where the fin bones attach. In *Tiktaalik*, the scapulocoracoid is relatively enlarged, and it bore muscles that strengthened and moved the fin such that the animal could raise the front end relative to the back. That movement could well be connected with the onset of air breathing that involved raising the head out of the water. Although the shapes of the pectoral fin bones are quite similar to those of a fish relative such as *Eusthenopteron* (a genus of extinct lobe-finned fishes from the Late Devonian Period), there are more and smaller elements toward the tip of the fin (**Fig. 3**). The joints between the bones of the fin could be explored because it was possible to extract many of them completely from the matrix and manipulate them one against the other. Thus, the researchers were able to suggest that the fin could bend and rotate in ways that resembled tetrapod elbow and wrist joints. The number and shape of the fin bones were also suggested as comparable to tetrapod digits. The large overlapping ribs probably bore muscles to strengthen the trunk as it was raised out of the water.

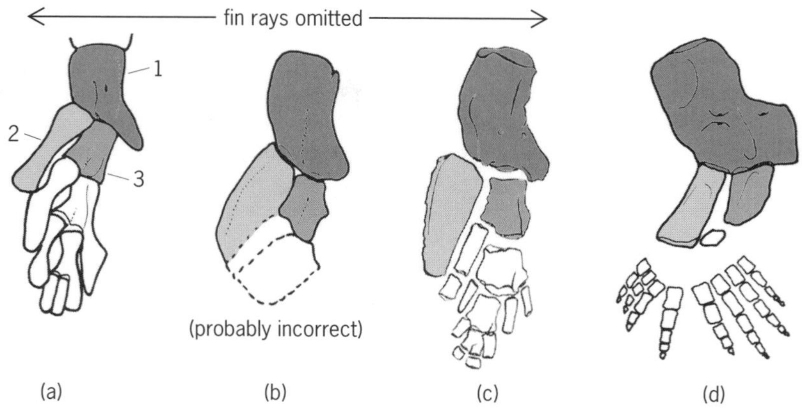

Fig. 3. Pectoral fin skeletons of (*a*) *Eusthenopteron*, (*b*) *Panderichthys*, (*c*) *Tiktaalik*, and (*d*) *Acanthostega*. The reconstruction of the more distal parts of the fin skeleton of *Panderichthys* shown here may be incorrect. Note similarities between each of the four genera in the bones of the (1) humerus, (2) radius, and (3) ulna; *Tiktaalik* has more small elements toward the end of the fin than does *Eusthenopteron*; and *Acanthostega* has true digits but there are eight of them.

Tetrapod hypotheses. *Tiktaalik* has thus stimulated a number of hypotheses in respect to the order in which tetrapod characters arose. It suggests that the opercular series and the bones joining the head to the shoulders were lost before fin-rays disappeared, and before the origin of true digits—in other words, while the animals were still technically "fish." It suggests that many of the adaptations later seen in tetrapods, such as the enlargement of the spiracular cleft and the beginnings of a neck, began as means to aid air breathing, concomitant with a loss of reliance on gill breathing. Even changes to the limbs may have been initiated in response to air breathing rather than directly for locomotion on land. The relatively conservative construction of the fin skeleton in *Tiktaalik*, compared with *Eusthenopteron*, combined with the changes in function that have been suggested, may imply that behavior changed ahead of the morphology that produced true digits (see Fig. 3). The similarities to *Panderichthys* in the jaw and dentition of *Tiktaalik*, however, suggest that the two were still feeding in the same way, which may still have been in water. Finally, the location in which *Tiktaalik* was discovered suggests a hypothesis of the biogeographic origin for tetrapods. *Tiktaalik* was found in deposits along the edge of the "Old Red Sandstone Continent," an area that includes what is now North America, Greenland, and parts of Europe including the Baltic States, which in Devonian times lay across the Equator. This distribution strongly suggests that tetrapods originated from these tetrapodlike fish in this region, in rivers and estuaries not far from the coast, and probably in the early part of the Late Devonian.

For background information *see* AMPHIBIA; ANIMAL EVOLUTION; DEVONIAN; FOSSIL; ICHTHYOSTEGA; PALEONTOLOGY; RESPIRATORY SYSTEM; SKELETAL SYSTEM; SPECIATION; TETRAPODA in the McGraw-Hill Encyclopedia of Science & Technology.

Jennifer A. Clack

Bibliography. J. A. Clack, The emergence of early tetrapods, *Palaeogeog. Palaeoclimatol. Palaeoecol.*, 232:167–189, 2006; J. A. Clack, From fins to limbs, *Nat. Hist. Mag.*, 115:36–41, 2006; J. A. Clack, Getting a leg up on land, *Sci. Amer.*, 293:100–107, 2005; E. B. Daeschler, N. H. Shubin, and F. A. Jenkins, A Devonian tetrapod-like fish and the evolution of the tetrapod body plan, *Nature*, 440:757–763, 2006; N. H. Shubin, E. B. Daeschler, and F. A. Jenkins, The pectoral fin of *Tiktaalik roseae* and the origin of the tetrapod limb, *Nature*, 440:764–771, 2006.

Dinosaur behavior

The last nonavian (nonbird) dinosaur died 65.5 million years ago, and with it any chance of observing nonavian dinosaur (here, "dinosaur") behavior in the flesh. Yet, our understanding of dinosaur behavior has increased with the discovery of new dinosaur species, the number of which has nearly doubled since 1990. A few recent, exciting discoveries are reviewed here.

Feeding. Feeding in many types of herbivorous dinosaurs has been well understood for some years; yet, feeding in large, carnivorous (theropod) dinosaurs has remained very speculative. Although once thought to be solitary predators, a considerably richer picture of large theropod behavior has lately begun to materialize. For example, researchers found distinctive, evenly spaced grooves preserved on the bones of large herbivorous dinosaurs (the sauropod *Rapetosaurus*) from Madagascar. Strikingly, the spacing of these marks exactly matched the spacing of the serrated teeth of the coexisting large theropod, *Majungatholus*. The density of striations on the sauropod skeletons suggests that *Majungatholus* exhibited a marked preference for dining on the axial column (the series of vertebrae forming the axis of the skeleton and protecting the spinal cord). Similarly gnawed bones of *Majungatholus* itself were also found, indicating cannibalistic behavior. Whether *Majungatholus* only scavenged intraspecifically or whether it actively killed (or both) remains unknown.

The large theropod bite has likewise been the subject of recent investigation; here one size surely does not fit all. The application of finite-element analysis to computerized tomography (CT)-derived images allowed scientists to model stress distributions undergone by the skull of the Late Jurassic *Allosaurus*. These revealed an unexpectedly weaker bite than that found in extant crocodiles or previously reconstructed in the Late Cretaceous theropod *Tyrannosaurus*. In *Allosaurus*, and likely many similarly built large theropods, the attack has been reconstructed as a series of aggressive slices (characterized as "slash and tear") with exsanguination (loss of blood) and perhaps subsequent infection (the method used by Komodo dragons). In the slash-and-tear model, killing would not be immediate, and prey would be tracked over a period of time, until loss of blood and perhaps infection sufficiently weakened the animal to minimize the protection conferred by the size disparity between theropods and their sauropod prey.

By contrast, in *Tyrannosaurus rex* for example, a deeper, more robust skull and bulbous teeth suggest a more forceful, bone-crushing bite as the active force in predation. This type of predation would likely involve a more immediate death of the prey.

Tyrannosaurus remains the most celebrated, enigmatic, and debated of all dinosaurs. Its extreme size, tiny but powerful forelimbs, bulbous teeth, reduced running speeds, and an anomalously powerful bite (the last two inferred from robust limb bones and massive skull, respectively) have controversially suggested to some paleontologists scavenging rather than active predation. Brain endocasts (casts of the interior of the brain case) have long revealed large olfactory bulbs, leading to the idea that the animal largely depended upon a well-developed sense of smell, at the expense of visual acuity. Yet, recent work on *Tyrannosaurus* skull morphology shows a narrow snout that permitted a binocular range of 55°, suggesting stereoscopic vision and associated high visual acuity, features generally linked with active predation.

Demonstrating social behavior in large theropods has been elusive; however, the recent discoveries of bone beds of the large Patagonian theropods, *Mapusaurus* and *Giganotosaurus*, hint at social structure and, potentially, pack-hunting. Strikingly, these bone beds reveal individuals at multiple stages of development. The existence of bone beds as well as the size disparity between predators (large theropods) and prey (sauropods) has for some years fueled speculations about large-theropod pack-hunting; these recent discoveries hint further at the existence of some form of social organization in large theropods.

For all this, we may be closer to understanding large theropod feeding in at least its rough outline. Neither scavenging nor predation can be completely ruled out, and the line between the two is blurred if exsanguination and perhaps introduced infection were the means by which killing was effected. Some degree of sociality and pack hunting may have been the norm, and young, superannuated, diseased, and/or wounded animals—including other large theropods that are conspecific (belonging to the same species)—may have been culled from the pack or herd, and preyed upon by hungry, stronger individuals. While individual species of large theropods undoubtedly engaged in modes of feeding that were species-specific, the emerging concept of large theropod feeding behavior is generally one of flexibility within a broad carnivorous diet.

By contrast, ornithomimids, the gracile, long-armed ostrichlike dinosaurs of the Late Cretaceous of Asia and North America, are without teeth. Their large size and powerful grasping claws, paired with toothless skulls, have made their dietary preferences enigmatic. Recent discoveries, however, of beak soft tissue in the Canadian *Ornithomimus* and the Mongolian *Gallimimus* show the beak region to have been highly vascularized, with the upper beak overlapping the lower. Unexpectedly, the upper beak in *Gallimimus* reveals a sievelike structure, suggesting possibly subaqueous, comb-style filter-feeding. However, other research has refuted this, opting for omnivory and herbivory for this group.

Locomotion. Dinosaurs as a group were primitively bipedal running beasts, but the evidence is very clear that their behavioral repertoire far transcended cursoriality (adaptation for running, especially across reasonably large distances). Recently, a trackway from La Virgen del Campo, Spain, was interpreted as a 15-m-long (49-ft-long) record of swimming by a theropod dinosaur. The exposed sandstone bedding plane revealed six pairs of depressions, whose spacing suggested an underwater stride of about 2 m (6.6 ft). Footprint morphology indicated that the track maker was a medium-sized theropod.

Recent reports suggest that some theropods were arboreal (adapted to living in trees) as well. The evidence comes from a juvenile theropod from China, *Epidendrosaurus* (Coelurosauria). Most distinctive is the long digit III on the hand, analogized by the investigators with *Daubentonia*, an extant Madagascar mammal that uses an elongate digit to collect insects from trees. The foot provides further evidence for arboreal affinities: although digit I is not oriented in complete opposition to the other foot digits as it is in modern birds, it sits low, it is long and tipped with a strongly curved claw, and the proportions of the phalanges match those of living arboreal creatures. Living theropods (Aves) have developed a marked suite of arboreal features, and thus it is not unreasonable to predict that highly evolved coelurosaurs might themselves have explored arboreal lifestyles.

Nesting. Knowledge of dinosaur reproduction has vastly increased over the past 25 years as a result of the discovery of nesting and parental care in the duckbilled dinosaur *Maiasaura*, in the Mongolian ceratopsian *Protoceratops*, and, most recently, in the primitive Mongolian ceratopsian *Psittacosaurus*. In this case, an adult *Psittacosaurus* was found, in an area about 0.5 m^2 (5.4 ft^2), surrounded by no less than 34 posthatchling juveniles. Aside from the breathtaking fecundity of the beast, the advanced state of the juveniles indicates parental care and development at the nest.

A newly discovered herbivore has added a new dimension to dinosaur reproductive and nesting behavior. *Oryctodromeus*, a 2-m (6.6-ft), Late Cretaceous, hypsilophodont from Montana, apparently burrowed and nested within a den. The partial remains of an adult and two juveniles were found in a slightly enlarged chamber at the end of a three-dimensional, sinuous, tubular trace fossil (the burrow). The trace fossil cuts across sandy strata and is filled laminated mudstones and sandstones, likely the result of episodic postmortem infilling. The anterior of the snout as well as the shoulder girdle of *Oryctodromeus* appear to be somewhat strengthened relative to closely related genera, and were interpreted as adaptations for digging. The axial skeleton lacks the extensive latticework of ossified ligaments characteristic of ornithopod dinosaurs, likely an adaptation for increasing the flexibility necessary for a fossorial (burrowing) existence. The presence and size of the posthatchling juveniles suggest growth at the nest—in this case, the burrow.

This discovery opens the door to the possibility of dinosaurs and their young seeking underground refugia during times of stress—from prowling predators to the kinds of harsh environmental changes that might have been associated with seasonality in high latitudes, or short-term geological catastrophes.

Resting. Significant to interpretations of dinosaur behavior, as well as relationships, was the recent discovery of *Mei long*, an exquisitely preserved troodontid (theropod) from the Yixian Formation, Early Cretaceous, China. The animal was found curled in an ovoid shape, tail wrapped around the body, and head tucked beneath the forelimb. The pose is decidedly birdlike and atypical of postmortem theropods, in which the head is generally pulled back over the thoracic vertebral column, with the tail and legs extended. The fossil was interpreted as preserving a living pose, sleeping or resting. Heretofore, the tucked position was only known in modern sleeping birds (although some theropod specimens of *Sinornithoides* and *Citipati* hint at this pose) and *Mei long* thus suggests a phylogenetically (if not temporally) earlier appearance than Aves of this "avian" trait.

For background information *see* BEHAVIORAL ECOLOGY; DINOSAUR; ECOLOGICAL COMPETITION; ETHOLOGY; INSTINCTIVE BEHAVIOR; MACROEVOLUTION; PALEONTOLOGY; PREDATOR-PREY INTERACTIONS; REPRODUCTIVE BEHAVIOR; SOCIAL HIERARCHY in the McGraw-Hill Encyclopedia of Science & Technology.

David E. Fastovsky

Bibliography. P. M. Barrett and E. J. Rayfield, Dinosaur feeding: Recent advances and evolutionary implications, *Trends Ecol. Evol.*, 21:217–224, 2006; J. O. Farlow and M. K. Brett-Surman (eds.), *The Complete Dinosaur*, Indiana University Press, Bloomington, Indiana, 1997; J. O. Farlow and T. R. Holtz, The fossil record of predation in dinosaurs, in M. Kowalewski and P. H. Kelley (eds.), *The Fossil Record of Predation*, Paleontological Society Papers, vol. 8, pp. 251–265, 2002; D. E. Fastovsky and D. B. Weishampel, *The Evolution and Extinction of the Dinosaurs*, Cambridge University Press, Cambridge, 2005; J. R. Horner, Behavior, pp. 45–49 in P. J. Currie and K. Padian (eds.), *Encyclopedia of Dinosaurs*, Academic Press, New York, 1997; D. B. Weishampel, P. Dodson, and H. Osmólska (eds.), *The Dinosauria*, University of California Press, Berkeley, 2004.

Dispersion modeling in complex urban systems

Models are used to represent real systems in an understandable way. They take many forms. A conceptual model explains the way a system works. In environmental studies, for example, a conceptual model may delineate all the factors and parameters for determining how a particle moves in the atmosphere after it has been released from a power plant. A conceptual model may also help identify the major influences on where a chemical is emitted and how likely it is to be found in the environment. Such models need to be developed to help target sources of data for assessing environmental problems.

Fig. 1. Example of neutrally buoyant smoke released from the World Trade Center physical model, showing flow from left to right. Natural light is illuminating the smoke and a vertical laser sheet is illuminating the plume near the centerline source. (*From S. Perry and D. Heath, U.S. Environmental Protection Agency, Fluid Modeling Facility, Research Triangle Park, North Carolina, 2006*)

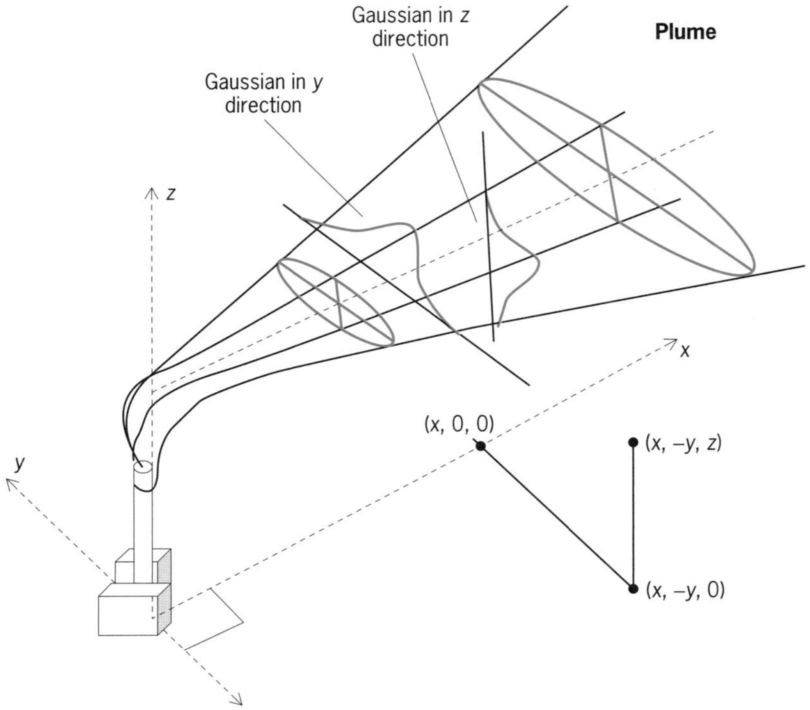

Fig. 2. Atmospheric plume model based upon random (Gaussian) distributions in the horizontal and vertical directions.

In general, developing an air pollution model involves two steps. First, a model of the domain and processes being studied must be defined and mathematical algorithms selected to represent the system. Then, the model boundary conditions are defined to represent the influence of the environment and other factors associated with the study. The quality of the model study is related to the accuracy and representativeness of the actual study.

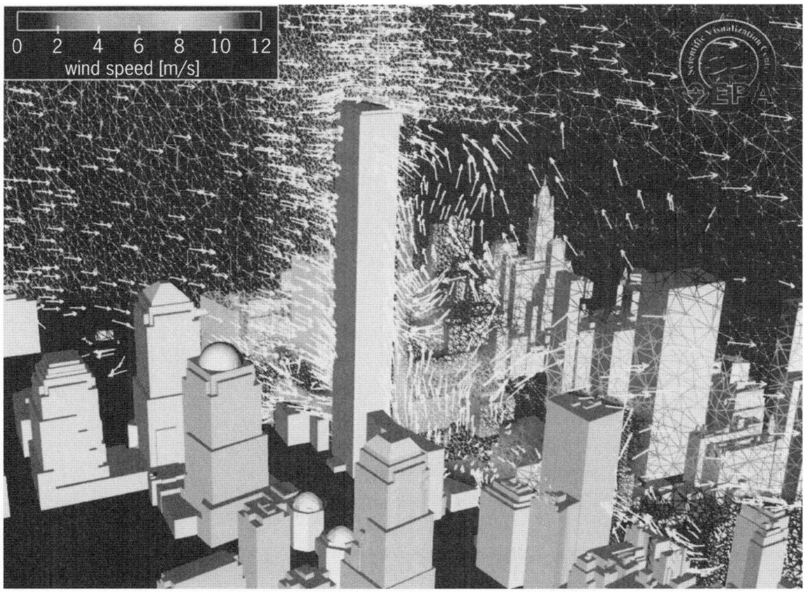

Fig. 3. A vertical slice of the resolved grid and sample of wind vectors in a CFD model of New York City. (*Used with permission courtesy of A. H. Huber, U.S. Environmental Protection Agency, Research Triangle Park, North Carolina*)

Physical models. Research scientists often develop physical, or dynamic, models to estimate the location where a contaminant would be expected to move under specific conditions, only on a much smaller scale. The U.S. Environmental Protection Agency's wind-tunnel facility in Research Triangle Park, North Carolina is sometimes used to study the influences of local buildings and terrain. For example, the wind tunnel housed a scale model of the town of East Liverpool, Ohio, and its surrounding terrain to estimate the movement from an incinerator plume. The plume could be observed under varying conditions, including wind direction and release height. Like all models, the dynamic model's performance is dictated by the degree to which the actual conditions can be simulated and the quality of the information that is used. More recently, the wind tunnel was used to simulate pollutant transport and dispersion of possible plumes of pollutants from the collapse of the World Trade Center (WTC) towers in Lower Manhattan on September 11, 2001. A 1:600 scale model, as it would have appeared approximately 1 week after the collapse, was constructed on a turntable to test different wind directions (**Fig. 1**). Smoke for visualization and tracer gas for measuring dispersion patterns were released from positions throughout the simulated 16-acre site.

Urban street canyons are highly complex systems; thus detailed measurements must be made of flow velocities, turbulence, and concentration distributions. For urban systems, simple dispersion algorithms do not apply since their performance is affected by local turbulence and structures (**Fig. 2**). Urban systems are characterized by vertical venting behind large or tall buildings and channeling down street canyons as well as both horizontal and vertical recirculation associated with individual structures and groups of tall and tightly spaced buildings, such as the Wall Street area in Manhattan.

Numerical simulation models. Numerical models use mathematical expressions to approximate the dynamics of a system. Three basic types of thermodynamics systems are important in air pollution sciences: (1) isolated systems, in which no matter or energy crosses the boundaries of the system (that is, no work can be done on the system); (2) closed systems, in which energy can exchange with surroundings but no matter crosses the boundary; and (3) open systems, in which both matter and energy freely exchange across system boundaries.

Isolated systems are usually only encountered in highly controlled reactors, so they can be ignored for urban dispersion. Most air-pollution systems are open, but with simplifying assumptions some subsystems can be treated as closed. In many cases, one must also be concerned with plume reactivity, which is not easily captured by simple models.

Rapid advances in high-performance computing hardware and software are leading to increasing applications of numerical simulation models that characterize an atmospheric plume as a system. These methods simulate spatially and temporally resolved details of the pathways of air pollutants from source emissions to pollutant concentrations within the local "virtual" microenvironment. This approach to air-pollution modeling applies numerical representation of fluid flow in a fluid continuum. Such an approach is known as computational fluid dynamics (CFD), which determines a numerical solution to the governing equations of fluid flow while advancing the solution through space or time to obtain a numerical description of the complete flow field of interest. CFD models are based on the first principles of physics. In particular, they start with single-phase flow based on Navier-Stokes equations.

High-fidelity fine-scale CFD simulation of pollutant concentrations within microenvironments (such as near roadways or around buildings) is feasible now that high-performance computing has become more accessible. Fine-scale CFD simulations have the added advantage of being able to account rigorously for topographical details such as terrain variations and building structures in urban areas as well as their local aerodynamics and turbulence. Thermal flux values may be added to terrain and building surfaces to simulate the influence of heat exchanges on pollution transport and dispersion at the atmospheric boundary layer. The physics of particle flow and chemistry can be included in CFD simulations. The results of CFD simulations can be used to understand specific case studies as well as to support the development of better, simplified algorithms for use in other modeling systems. For example, CFD simulations with fine-scale physics and chemistry can enhance and complement photochemical modeling. Also, detailed CFD simulation for a complex site study can be used to develop reliable parameterizations to support a simplified and rapid application air pollution model. However, the computational issues are complex, and without good urban tracer studies the CFD are difficult to test for performance.

Currently the main features of CFD application are the inclusion of site-specific geometry and dynamic processes affecting air-pollution transport and dispersion. In the future, refined spatial and temporal details, along with particle flow and chemistry, will be included. A vertical slice of the domain cells and a 10% sample of calculated wind vectors for these cells is shown in **Fig. 3**, demonstrating the tendency for downward airflow on the windward faces of buildings and upward airflow on the leeward building faces. Figure 3 shows two different plume depictions for ground-level point emissions. In one case, the emissions plume is caught in the leeward building updraft leading to significant vertical mixing; the other case shows emissions remaining close to the ground as they move through the street canyons. **Figure 4** shows a vertical slice of concentration for roadway emissions represented as a box along the streets. There are significant differences for the case with a building on only one side of the roadway relative to a street canyon that induces a region of circulation.

Some atmospheric concentrations of pollutants consist of a regional background concentration due to long-range transport and regional-scale mixing as well as specific local microenvironmental concentrations. Concentrations within the local microenvironment often dominate a profile of total human exposure to the pollutant in the local atmosphere.

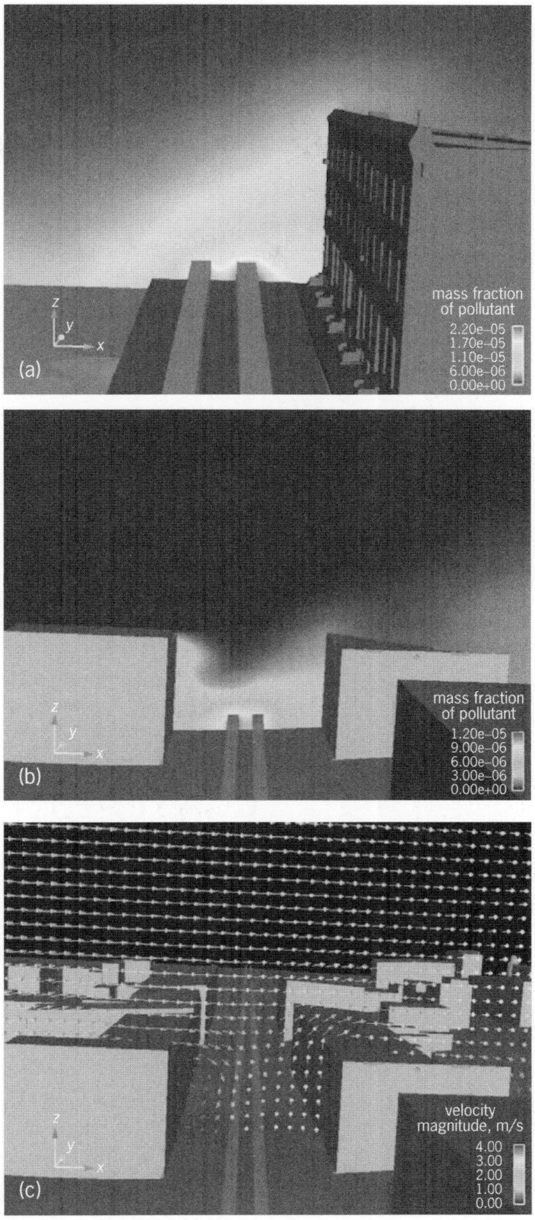

Fig. 4. Vertical slice view of a CFD model of roadway emissions represented as a source box along the roadway for (*a*) concentrations for a street bounded by a building on one side, (*b*) concentrations for a street canyon, and (*c*) a wind vectory for a street canyon. (*Courtesy of Dr. Alan H. Huber, U.S. Environmental Protection Agency, Research Triangle Park, North Carolina*)

Regional air quality applications are normally applied at grid resolutions larger than 10 km. Urban applications are applied at smaller grid scales (4 km) but there is a meaningful limit due to the sub-grid-scale process models. In other words, there is no value in applying the model at fine scales without supporting details on the finer resolution. For example, the present modeling systems support environmental issues where large-sized grid-averaged concentrations are present.

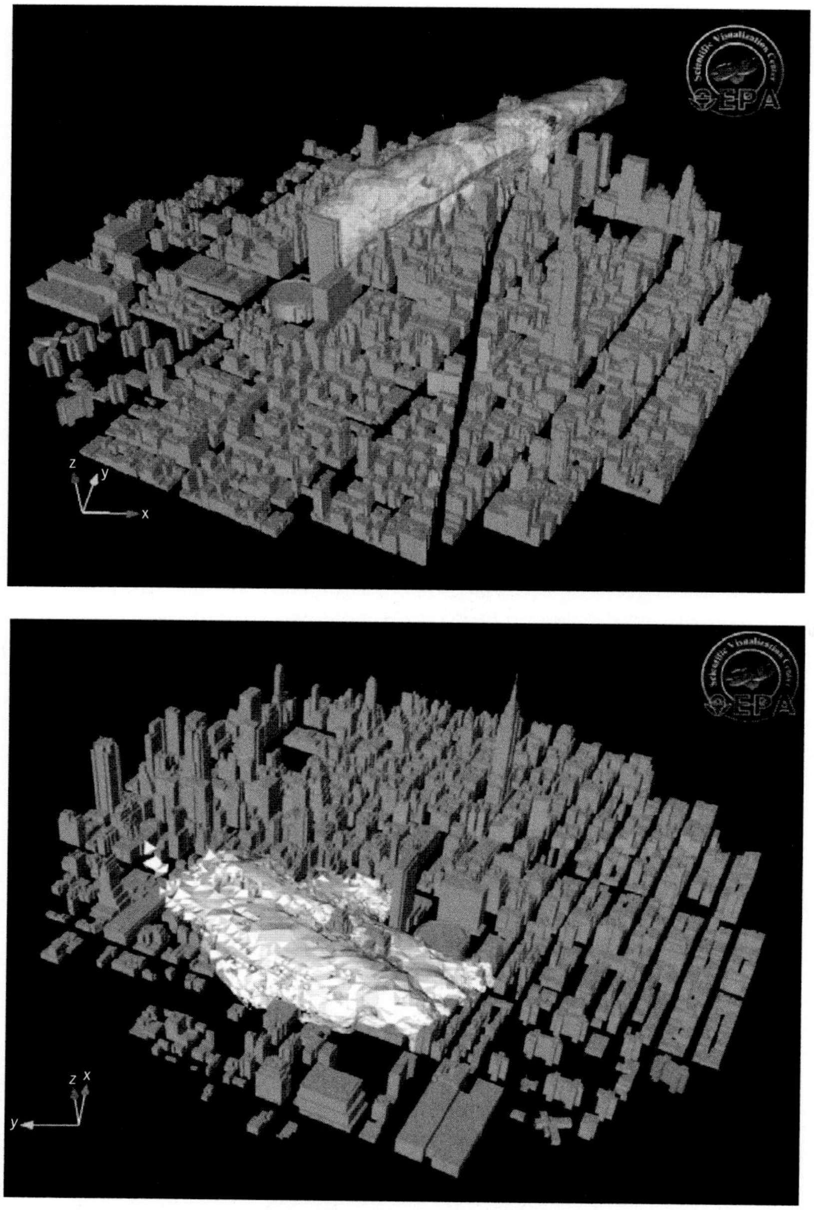

Fig. 5. Plumes from ground-level point source emissions in a CFD model of New York City (midtown area). (*Used with permission courtesy of A. H. Huber, U.S. Environmental Protection Agency, Research Triangle Park, North Carolina*)

The large-grid-size modeling system may be used for estimating profiles of population-based human exposure for pollutants having wider temporal and spatial distributions. Human exposure concentrations on subgrid scales need to apply to a subgrid-scale model, including human exposure factors to estimate expose populations (**Fig. 5**).

Transport and fate models can be statistical and/or "deterministic." In design, statistical models include the pollutant dispersion models, such as the Lagrangian models, which assume Gaussian distributions of pollutants from a point of release (Fig. 2). That is, the pollutant concentrations are normally distributed in both the vertical and horizontal directions from the source. Stochastic models are statistical models that assume that the events affecting the behavior of a chemical in the environment are random, so such models are based upon probabilities.

Deterministic models are used when the physical, chemical, and other processes are sufficiently understood to reflect the movement and fate of chemicals. These are very difficult models to develop because each process must be represented by a set of algorithms. Also, the relationship among systems, such as the kinetics and mass balances, must also be represented. Thus, the modeler must parametrize every important event following a chemical's release to the environment. Often, hybrid models using both statistical and deterministic approaches are used.

Urban dispersion program. The New York City urban dispersion program (NYC UDP) was designed as a four-year research project to study how air flows in an urban environment. The objectives were to (1) improve the permanent network of wind stations in and around New York City to allow better estimates of where contaminants may travel and to enhance the city's emergency response capabilities, (2) conduct field studies to advance knowledge about the movement of contaminants in and around New York City, (3) improve and validate computer models that simulate the atmospheric movement of contaminants in urban areas using data collected from the field studies, and (4) transfer the improved capabilities to New York City emergency agencies.

The program was sponsored by the U.S. Department of Homeland Security, the Defense Threat Reduction Agency of the U.S. Department of Defense, the U.S. Environmental Protection Agency, and the U.S. Department of Energy.

The field studies provided seasonal data for improving and validating atmospheric models to simulate the dispersal of contaminants in and around the city, into building interiors and subways, and into the surrounding region. These atmospheric models are used by emergency management, law enforcement, and intelligence personnel to plan for, train for, and respond to potential accidents or other releases of harmful airborne materials. The New York City field studies built on the knowledge gained from comprehensive field studies in the smaller cities of Salt Lake City in 2000 and Oklahoma City in 2003. For information on the Urban Dispersion Program's New York City field studies, visit http://urbandispersion.pnl.gov/index.stm.

In an urban dispersion field study, numerous portable wind-sampling instruments are placed in and around the study area to measure wind patterns. Additionally, safe-inert tracer gases are released into the air and numerous sensitive sampling instruments are used to measure the movement of the tracer throughout the study area.

The first study within the NYC UDP was done around Madison Square Garden during March 2005. The study examined the local horizontal and vertical dispersion of perfluorocarbon tracers (PFT) released outdoors at four strategic points (**Fig. 6**). PFT releases are an ideal way of studying the nature of potential human contact with released material and identifying activities (in both space and time) that might lead to significant human exposures. To achieve this, a unique prospective personal monitoring study was designed to be completed while the tracer was being dispersed into the atmosphere. The exposure monitoring was accomplished using very sensitive PFT personal air samplers (PAS) from the Brookhaven National Laboratory that were carried in a shirt pocket by the field personnel completing a set of scripted tasks in the area surrounding Madison Square Garden and the adjoining Penn Station. The PAS is a pocket-sized air-sampling unit that is nominally 1 in. wide × 0.75 in. thick × 4 in. high and holds a capillary adsorbent tracer sampler (CATS) tube for passively sampling the tracers. The CATS is a 0.25-in.-diameter (0.6 cm) × 2-in.-long (5.1 cm) glass tube filled with granulated activated carbon to capture the PFTs. With a limit of quantification in the parts per quadrillion (ppqv), PFT samples could be collected for 1 and 10 min in duration and linked with specific activity patterns and the distance from the release point.

In contrast to typical exposure measurement studies of urban exposures to typical air pollutants, during the Madison Square Garden study specific activity patterns (such as walkers evacuating, exiting, or approaching the point of release, emergency workers remaining near the release, etc.) were developed by the investigators. Each route was tailored to the types of activities that could provide both source- and neighborhood-scale information about contact with each PFT. Subsequently, the results could be modeled to simulate contacts with actual chemical, physical, or biological agents. Thus, the study implemented a new approach for simulating human exposure to environmental agents, including highly toxic agents within urban street canyons. The local conditions that affect exposure are clearly driven by the surface winds, the complexity of the urban terrain, and the activities that typically occur in an urban center. An interesting observation was the presence of a "hot spot" for exposure at the northeastern PFT release point B. The nature of the local conditions mentioned above led to perfluoromethylcyclopentane (PMCP) exposures (source or neighborhood scale) that were a factor of 10 greater than observed for the other PFTs. These also led to high levels of exposure to PMCP at release point A (west of the release).

The source-scale exposure data clearly showed that local conditions affected the distribution of each tracer, and consequently the exposure received by each individual. The different 1-min source-scale paths did not show that the semicircle paths or transect path yielded higher exposures. In fact, the results indicated that the higher exposures were not systematic, which reenforced the influence of local surface conditions, including building characteristics and dynamics, traffic, and meteorology on the potential contact with an individual PFT. The range measured for 1-min concentrations/exposures was quite high with values measured at points A and B routinely exceeding 5000 ppqv.

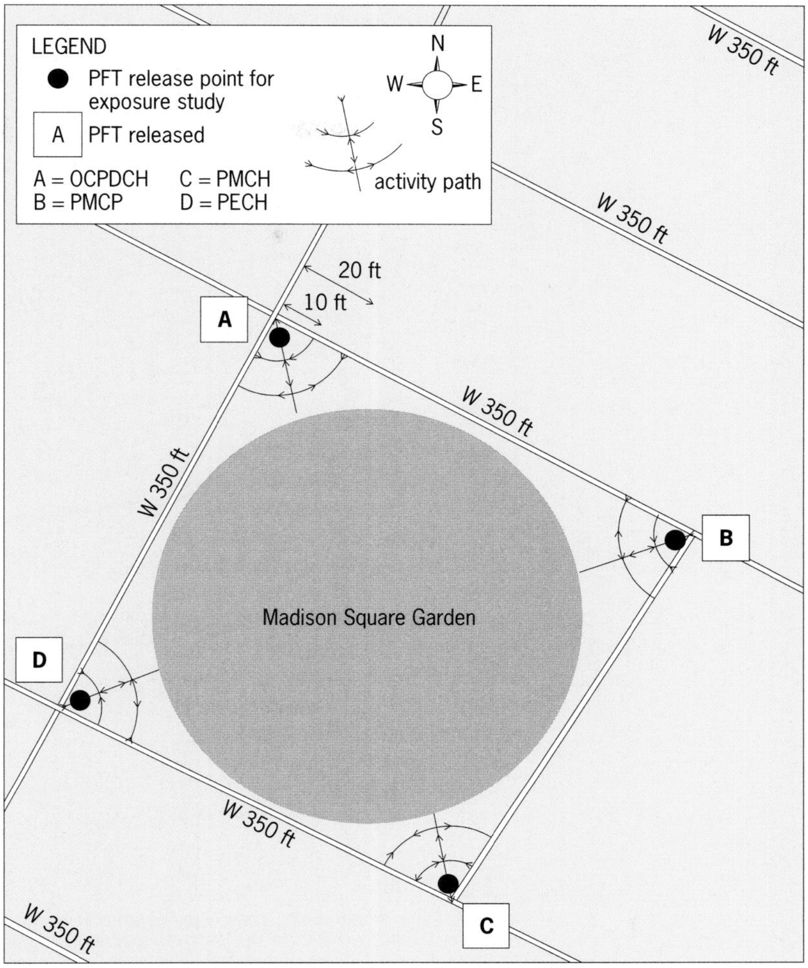

Fig. 6. Location of perfluoride tracer (PFT) releases in March 2005 Urban Dispersion Program study around Madison Square Garden, New York City. The tracers are PMCP (perfluoromethylcyclopentane), oc-PDCH (perfluoro-1,2-dimethylcyclohexane), PMCH (perfluoromethylcyclohexane), and PECH (perfluoroethylcyclohexane). (*From Lioy et al., JESEE, 2007*)

The neighborhood scale (**Fig. 7**) exposures provided a different set of information that was important for quantifying exposures at distances which, based upon the study design, could be as far as six to seven blocks away from the individual release points.

The exposures derived from the neighborhood scripts provided three general observations. (1) The highest concentrations were associated with the 10-min interval samples that were collected as a member of the sampling team returned to a source location. The exposures were not consistently high. In fact, the northeastern corner, which was the location of release point B, always yielded the highest exposures to the PMCP emissions during the source "on" period. In some ways, this was expected since the general west-to-east wind direction and the individuals completing paths BB and CC would return from east to west and at a location near release point B. (2) The local meteorology and street activities influenced the magnitude of the exposures as an individual returned along a scripted path. The levels for the return from path AA to point D, and the return from path CC to point C recorded

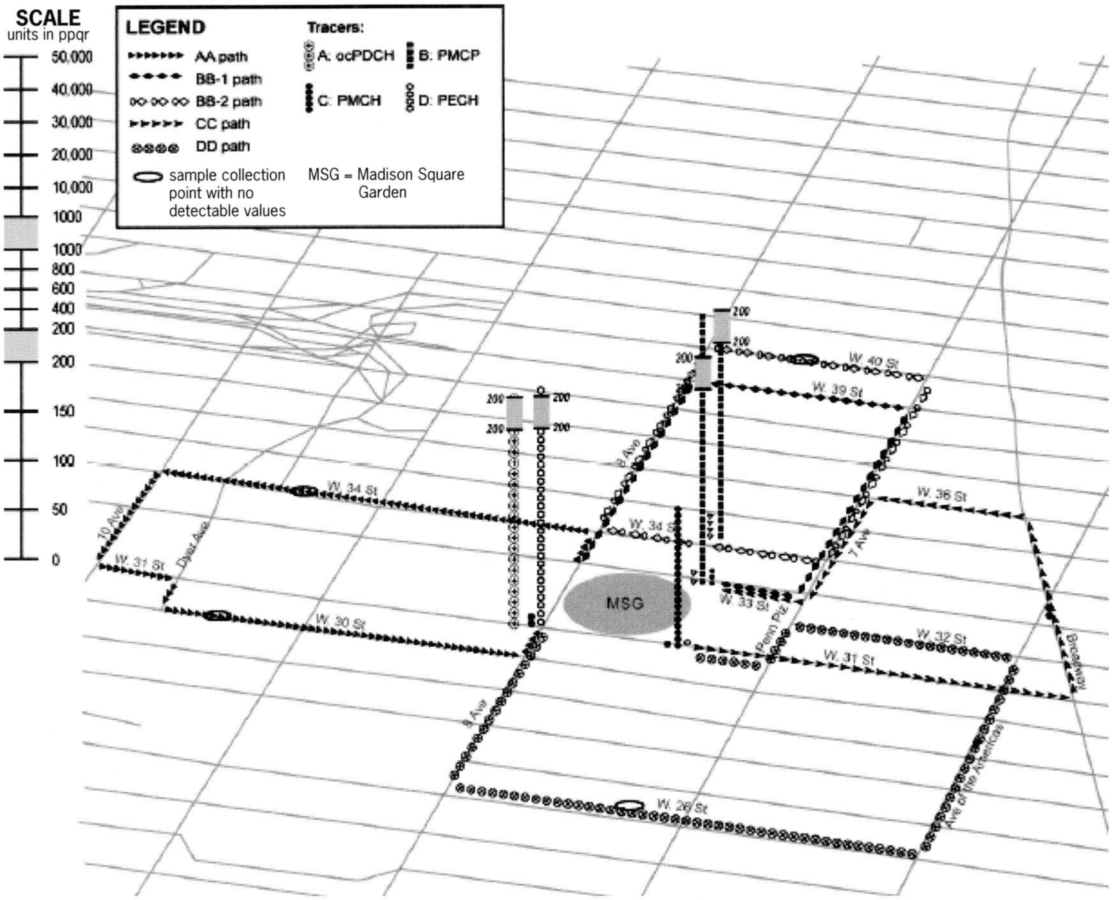

Fig. 7. Neighborhood-scale personal exposure monitoring, 9:00–9:30 a.m., during release of perfluoride tracers (PFTs) on March 10, 2005 releases in the Urban Dispersion Program study around Madison Square Garden, New York City. The tracers are PMCP (perfluoromethylcyclopentane), oc-PDCH (perfluoro-1,2-dimethylcyclohexane), PMCH (perfluoromethylcyclohexane), and PECH (perfluoroethylcyclohexane). (*From Lioy et al., JESEE, 2007*)

detectable exposures during at least one sampling interval. In contrast, the very high exposures were primarily recorded at point B during all sampling intervals. The high exposures recorded at release point B for the neighborhood-scale exposure return paths are supported by the near-source exposure data that showed comparable values for multiple 1-min samples, that is, >5000 ppqv. From the standpoint of all exposure measurements of PMCP, the area between 7th and 8th Avenues on 33rd Street was the most dynamic for future modeling of exposure. In addition to the above observations, the measurement of high levels of PMCP during the first off-period sample (10 a.m.) at point A was due to the transport of PMCP west caused by surface-level turbulence. Since this would be in a direction opposite to the general wind direction (east to west rather than west to east), the local street canyon turbulence influenced the exposure. (3) For the majority of all source- and neighborhood-scale exposure samples, the exposures received by the field team members quickly returned to below the detection limit during both PFT "off" periods. The personal exposures are highly dependent upon the local physical boundary conditions, local turbulent eddy, and the depth of the inversion.

The data collected using near-source and neighborhood personal exposure scripts established a baseline for determining how the release of a gaseous material could affect both the general public and emergency responders. The results indicate that the surface-level structure of the area affects the dynamics of contact with the released material, and that these dynamics can produce significantly high exposures. These results can be used to improve urban-scale CFM and dispersion model performance.

The Madison Square Garden results point to the potential for a hot spot depending upon the boundary conditions and the near-source flux or turbulence. The results also showed clearly that a first responder must be wearing personal protective equipment once inside an area of about one to three blocks radius from an active source.

Urban environments are quite complex. There are a number of models that can be applied to characterize these conditions, but the specific atmospheric conditions unique to urban street canyons must be part of any emergency response actions.

For background information *see* AIR POLLUTION; COMPUTATIONAL FLUID DYNAMICS; ENVIRONMENTAL FLUID MECHANICS; MODEL THEORY; NAVIER-STOKES EQUATION; NUMERICAL ANALYSIS; SIMULATION; WIND

TUNNEL in the McGraw-Hill Encyclopedia of Science & Technology. Daniel Vallero; Alan Huber; Paul Lioy

Bibliography. J. M. Burke, M. J. Zufall, and A. H. Ozkaynak, A Population Exposure Model for Particulate Matter: Case Study Results for PM2.5 in Philadelphia, Pennsylvania, 2007; A. Huber et al., Modeling air pollution from the collapse of the World Trade Center and assessing the potential impacts on human exposures, *Environ. Manage. J. Air Waste Manage. Assoc.*, February, 2004; N. E. Klepeis, Modeling human exposure to air pollution, in *Exposure Analysis*, W. R. Ott, L. A. Wallace, and A. Steinemann (eds.), CRC Press, Boca Raton, 2006; N. E. Klepeis and W. W. Nazaroff, Modeling residential exposure to secondhand tobacco smoke, *Atmo. Environ.*, 40(23):4393–4407, 2006; J. Leete, Groundwater modeling in health risk assessment, chap. 17 in *A Practical Guide to Understanding, Managing and Reviewing Environmental Risk Assessment Reports*, S. Benjamin and D. Belluck (eds.), Lewis Publishers, 2001; C. Lewis, G. Klouda, and W. Ellenson, Cars or Trees: Which Contribute More to Particulate Matter Air Pollution?, U.S. Environmental Protection Agency, Science Forum, Washington, DC, 2003; P. Lioy et al., A personal exposure study employing scripted activities and paths in conjunction with atmospheric releases of perfluorocarbon tracers in Manhattan, New York, *J. Expo. Sci. Env. Epid.*, 17:409–425, 2007; National Bureau of Standards, Oxalic Acid Standard Reference Method SRM 4990B; C. Stallings et al., *CHAD Users Guide: Extracting Human Activity Information from CHAD on the PC*, ManTech Environmental Services, modified by Systems Development Center Science Applications International Corporation, Research Triangle Park, North Carolina, 2002; G. S. Settles, Fluid mechanics and homeland security, *Annu. Rev. Fluid Mech.*, 38:87–110, 2006; U.S. Environmental Protection Agency, *Guideline on Air Quality Models* (also published as Appendix W to Part 51 of 40 CFR), which was originally published in April 1978 to provide consistency and equity in the use of modeling within the U.S. air quality management system; U.S. Environmental Protection Agency, Regulatory Impact Analysis for the Revised Particulate Matter National Ambient Air Quality Standards, 2006; U.S. Government Accountability Office, *Clean Air Act: EPA Should Improve the Management of Its Air Toxics Program*, Report no. GAO-06-669, June 2006.

DNA and animal domestication

Human civilization would not have been possible without the domestication of plants and animals. Domestication is responsible for our clothes, food, and pets, and was also a necessary prerequisite for the development of written language, mathematics, architecture, and every technological breakthrough since stone tools. The study of domestication is therefore the study of the development of humankind over the past 10,000 years. Thus, by understanding how, when, and where animals were domesticated, we gain deeper insights into the foundations of modern society. In addition, domestication is an ideal model for understanding the process of change within evolutionary biology and for understanding the relationship between differences at the DNA level, and the ramifications those differences have on the physical, behavioral, and developmental characteristics of the organism itself.

For most of the past century, the study of domestication has fallen primarily to archeologists, who, by analyzing the bones of animals found at human occupation sites, developed the primary hypotheses on which our fundamental understanding of the timing, geography, and process of domestication rests. The advent of technology that has facilitated the retrieval of DNA from both modern and ancient wild and domestic samples has opened a new perspective on domestication and allowed scientists to test the archeological hypotheses from a different perspective.

Where and how often. Although there are innumerable questions related to the history of domestication, the most fundamental ones rest on where and how many times the process took place. Archeological evidence demonstrates that the Near East, the region where human communities first began to settle down, and the region where wild counterparts of dogs, cows, pigs, sheep, and goats also roamed, was the primary center of early animal domestication in the Old World. The long-standing conventional wisdom holds that domestication was an unusual occurrence, and that the presence of domestic animals outside the Near East was best explained by single domestication "events" followed by an outward spread of domestics, either transported by human populations on the move or traded between adjacent cultures.

Under this model, the genetic variation of domestic animals should be significantly smaller than a reciprocal study of their wild counterparts. Not only that, the genetic signatures of domestics should be more related to each other than any are to different wild lineages, reflecting a single event in which a small subsample of individuals was selected from the pool of wild animals and brought under human control. These animals would then have only been mated with other domestics and never backcrossed with additional wild animals. This sequence of events exactly describes the domestication of the hamster in 1930, when a single wild female and twelve offspring were captured in Syria. These thirteen individuals have since given rise to the entire modern populations of pet and laboratory hamsters across the globe.

Sheep and goats. Recent DNA evidence has challenged this model (and the implied associated intentionality) by demonstrating that the genetic variation within domestic animals is significantly greater than expected. Moreover, in virtually every instance in which a population of domestics is analyzed from a previously unsampled geographic region, the number of identified genetic lineages increases. Domestic goats and sheep, for instance, have each recently been shown to cluster into five distinct genetic lineages found in flocks reaching from Europe to East Asia.

This pattern could be read as evidence for five independent domestications of goats and sheep, but this is likely an oversimplification. The primary unknown factor is the degree of genetic variation within wild goats and sheep 10,000 years ago, and whether there was any geographic pattern to that variation. The idea of five independent events represents the most extreme interpretation, where five unique lineages of wild animals were geographically isolated from each other, and incorporating each lineage into domestic animals required the independent sampling of every available wild lineage in widely separated locales. This is just as unlikely as the opposite extreme in which all five lineages were uniformly admixed across the entire wild population, thus requiring only a single domestication in one place of five animals, each representing a different lineage. The reality most likely lies somewhere in between; however, given the inability of the genetic data to identify which of several potential wild ancestors gave rise to either domestic goats or sheep, the answer remains uncertain.

Pigs. Whereas sheep and goats were initially expected to have been domesticated just once, archeologists and animal scientists have inferred on the basis of marked physical differences between Asian and European pig breeds that pigs were domesticated at least twice. A genetic study of wild and domestic pigs from across the natural range of wild boars (from Portugal to Japan, and from Siberia to Indonesia) demonstrated that, like sheep and goats, wild boars cluster into several highly differentiated genetic lineages. More remarkably given their behavioral plasticity and their ability to travel over long distances, the genetic lineages found in wild boars were exclusively correlated with specific geographic regions. This strong correlation between geography and genetic signature meant that at least four additional domestication centers were identified, including India, Italy, Southeast Asia, and possibly Borneo, when the genetic data from modern domestic pigs were compared with the wild signatures.

Dogs. Wolves have an even greater natural range than wild boars, covering both the Old and New Worlds, thus leaving open the possibility that dogs were independently domesticated numerous times. Genetic sampling of wolves and dogs has revealed four distinct lineages; however, because wolves with distinct signatures are not geographically restricted, it has been difficult to genetically identify unequivocally any one specific region or regions of domestication. The greater variation of domestic signatures in East Asia led one study to conclude that dogs were first domesticated there, and the lack of North American wolf signatures in pre-Columbian dogs led another study to conclude that wolves in the Americas were not independently domesticated. Other genetic studies have shown that wolves and dogs continue to interbreed. Given this propensity to share genetic material, and because DNA from wolves from a variety of locations has been continually incorporated into the domestic pool since domestication first took place, a great deal more data from different regions of the dog and wolf genomes are required before a detailed story of early dog domestication can be revealed.

Horses and cows. Unfortunately, the wild ancestors of both cows and horses are extinct. This has meant that (when dealing with modern samples) geneticists have had to reconstruct the stories of cow and horse domestication by sampling only the domestic variation. The equine story is the least clear of the two. Although there is a great deal of genetic variation, pinning it down to a specific region has been difficult, possibly because horses have historically been used not just for meat (like most domestic animals) but also for transport, thus allowing genetic lineages that may have been geographically isolated to have been spread across continents, blurring any original correlation that may have existed between horse geography and genetics.

A great deal of analyses on the modern and ancient genetics of cows has been brought to bear on the story of cow domestication. Archeological evidence has demonstrated that two different species of cows, Near Eastern taurine cattle (which do not have a hump) and Indian indicine cattle (which do), were independently domesticated. Genetic studies have now shown that two different lineages of indicine cattle were domesticated, one in the north and one in the south of India, and that wild cows (aurochs) were likely also independently domesticated in North Africa. Studies of DNA retrieved from archeological bones of domestic cows and the wild ancestors of cattle have shown that ancient cows brought into Europe certainly hybridized with the indigenous bulls; however, this claim remains contentious as some authors have concluded that European aurochs were independently domesticated.

Studies of other domestic animals such as cats, donkeys, water buffalo, and chickens have revealed that multiple lineages and a high degree of genetic variation within domestic animals are not an exception, but the general rule for domestic animals as a whole. The consistency of this evidence will soon require a significant revision of the assumption that domestication is a rare, remarkable event. More likely, it is the expected outcome of permanent human communities sharing their environment with wild animals.

Other issues. Recent genetic studies have begun using domestic animals to question other elements of domestication. Although numerous studies have applied a molecular clock methodology to their genetic data, conclusions regarding the timing of domestication (answering the question when) have been less than robust. (Molecular clocks work by quantifying the genetic distance between living organisms and calibrating the rate of change using one or more reliable dates from the fossil record for the time of divergence of lineages with living representatives.) Understanding what is responsible for the significant contradictions between dates derived from archeologists and those from geneticists, however, has led to a questioning of assumptions that underlie the molecular clock, which in turn has opened the door to a more nuanced understanding of molecular evolution.

Last, new technology has allowed for the retrieval of DNA from ancient bones that codes for genes underlying important traits associated with domestication, including coat color and muscle development. These methods have already been used to demonstrate that the first European farmers were lactose-intolerant and that the variation in this gene that allows for lactose tolerance quickly spread through European populations, thus illuminating the story of the changing relationship between humans and cows, and the genetic effects that relationship had.

The flood of new data that has already surfaced combined with the results of the continuing spate of publications will certainly require a revision of our understanding of the process of domestication, its triggers, its development, and its ramifications for the genomes of domestic animals and their human coconspirators.

For background information *see* AGRICULTURAL SCIENCE (ANIMAL); BEEF CATTLE PRODUCTION; BREEDING (ANIMAL); DAIRY CATTLE PRODUCTION; DEOXYRIBONUCLEIC ACID (DNA); DOGS; DOMESTICATION (ANTHROPOLOGY); GOAT PRODUCTION; HORSE PRODUCTION; PHYLOGENY; SHEEP; SWINE PRODUCTION; ZOOARCHEOLOGY in the McGraw-Hill Encyclopedia of Science & Technology. Greger Larson

Bibliography. M. W. Bruford, D. G. Bradley, and G. Luikart, DNA markers reveal the complexity of livestock domestication, *Nat. Rev. Genet.*, 4(11):900–910, 2003; K. Dobney and G. Larson, Genetics and animal domestication: New windows on an elusive process, *J. Zool.*, 269(2):261–271, 2006; M. A. Zeder et al. (eds.), *Documenting Domestication: New Genetic and Archaeological Paradigms*, University of California Press, Berkeley, 2006.

Early spider web

Spiders are among the most abundant and important of all insect predators, whose numbers often exceed 100 per square meter, or one million spiders per hectare (10,000 m^2). The ecological success of spiders has been attributed to key innovations in how spiders use silk to capture insects, mainly with different kinds of webs. Spider webs have diverse, intricate architecture resulting from complex adaptations that evolved under various selection pressures. Primitive orb-weavers of the Deinopoidea, for example, produce dry cribellar threads made of thousands of silks (cribellar threads are produced from the cribellum, a specialized, flattened spinning organ), while modern Araneoidea orb-weavers produce viscous adhesive threads with droplets. Both kinds of threads are very elastic and have among the greatest tensile strength of any substance in nature. The stickiness of adhesive thread is more effective, though, which may have contributed among other factors to the dominance in modern environments of araneoids with respect to the deinopoids.

Fossil record. It is not clear if orb webs evolved only once, in a common ancestor of Deinopoidea and Araneoidea, or when it evolved, but it is likely that the two groups of orb-web weavers were already defined in the Early Cretaceous [140 million years ago (Ma)] since the oldest putative araneomorphs are from the Late Triassic (225 Ma). Recent discoveries support a single evolutionary origin since Deinopoidea shares important silk proteins with Araneoidea orb-weavers.

The hallmark features of spiders, the spinnerets (organs that spin fiber from the secretion of silk glands), are known from the Devonian and Carboniferous periods; therefore, fossilized webs should be known from this time. However, they are poorly fossilized, and only amber is known to preserve web silk. In fact, amber preserves spiders and their silks with high fidelity, including microscopic body features such as setae (rigid bristles) and sensilla (sensory nerve receptors), color patterns, and sticky droplets on the silks. In some cases, amber preserves soft internal tissues (cells and organelles). Recently, diverse body fossils of spiders have been discovered in Cretaceous amber from New Jersey (U.S.A.), Manitoba and Alberta (Canada), Álava (Spain), Charente-Maritime (France), the Isle of Wight (England), Jezzine-Hammana (Lebanon), Myanmar (Burma), and Taimyr (Siberia). The oldest records of isolated spider silks are from the Early Cretaceous, namely a single silk strand with droplets in Lebanese amber and an unstudied silk fragment in French amber. Younger Cretaceous records have been found with captured insects, as have more recent ambers from the Cenozoic (65 Ma and younger).

Earliest spider web. The most significant fossil corresponds to the earliest known spider web, which contains an association of trapped arthropods preserved in a transparent yellow mass of amber (**Fig. 1**) from a new outcrop of Spanish amber located in San Just in eastern Spain (Escucha Formation). The amber was produced by an araucariacean conifer and is Early Cretaceous in age (Middle-Lower Albian, about 110 million years old). This is the oldest direct evidence of a spider web and the oldest record of predation on insects. The amber mass is a cylindrical flow or runnel, perhaps formed on a tree branch. It was found in a thin level of lutite (consolidated rock or sediment formed principally of clay or clay-sized particles) that was rich in amber, charcoal, and mainly carbonized woody tissue (fusinite), the last of these formed by an ancient forest fire.

The fossil association in the runnel contains adult winged insects and a nymphal oribatid mite. The insects are a wasp of the genus *Cretevania* (Evaniidae), a wasp of the Mesozoic family Stigmaphronidae, a female *Microphorites* fly (Dolichopodidae), and a beetle (Cucujidae). The piece also contains an insect coprolite (petrified excrement), fungal hyphae (on the *Microphorites* and beetle bodies), unidentified plant trichomes (fine dermal appendages), and a cluster of

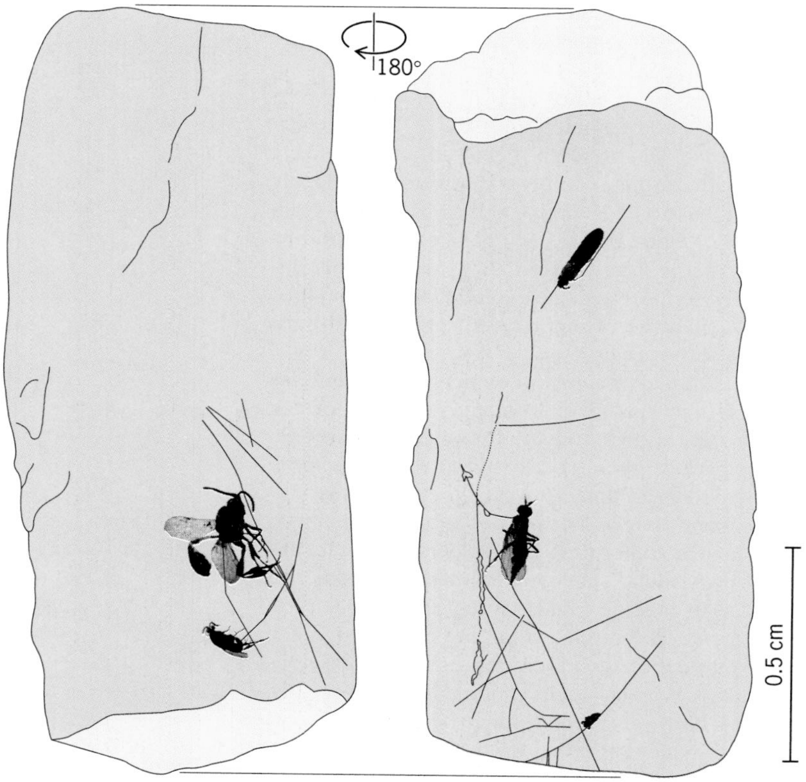

Fig. 1. Map of the silk strands and arthropods in the amber runnel found in the San Just outcrop in Spain. Note the direction of the main strands with respect to the longitudinal axis of the runnel.

strand and its attachment to the wasp indicate that this is a different type of silk. There also are seven silk strands in the same plane, some of them in contact with the bifurcate thread and the wasp.

Droplets on webs can occur from condensation, but the adhering arthropods indicate these were viscid droplets secreted by the spider. Sticky droplets are highly hygroscopic (water-binding) spider secretions, increasing the elasticity of the web. This feature of the fossil web is also very consistent with orb webs. The exceptional record from Spanish amber shows that the silks of the web had great elasticity as well. This characteristic can be inferred from two strands forming hanks, which probably contracted violently due to the release of tension from breaking. In addition, the similar direction of the main strands as the longitudinal axis of the runnel apparently indicates that the web had a vertical orientation. Elasticity and a vertical orientation are very important features for intercepting and capturing flying insects.

The three trapped insects have their abdomens partially destroyed and impregnated with resin, indicating they were consumed by the spider before the resin entombed them. Fungal hyphae on the *Microphorites* thorax and on the beetle's abdomen also indicate extended exposure after death, perhaps

gymnosperm pollen grains. These fossils are in contact with the silk of the spider web grains.

The runnel was prepared into three small sections of amber embedded in epoxy resin to permit optimal study of the specimens (**Fig. 2**), which were near the surface. There are at least 26 silk strands in the three amber sections, which have different orientations and characteristics. Practically all the strands are rectilinear or slightly curved, two of them with glue droplets, and two strands have the shape of hanks (coils). Fungal hyphae are common and readily recognized in amber since they have numerous, short, interlacing fibers. Thus, spider silk and fungal hyphae are easily distinguished.

One of the amber sections, containing the mite and the *Microphorites* specimen, shows 16 silk strands, five of them in the same plane and with similar orientation. These last strands are connected perpendicularly to an incomplete strand, and two consecutive main strands are connected by two very thin strands. The two contiguous thin strands can be interpreted as that of a sticky spiral between two consecutive radii of an orb web. In total, the geometry indicates an orb web. New discoveries in other Spanish amber of the same age (Álava amber), as well as the recent finding of a true orb-weaving spider specimen, support the orb web interpretation.

In another amber portion, there is a thicker, bifurcate thread, composed of numerous thinner strands without droplets, which snagged one leg of the *Cretevania* wasp (**Fig. 3**). The structure of this

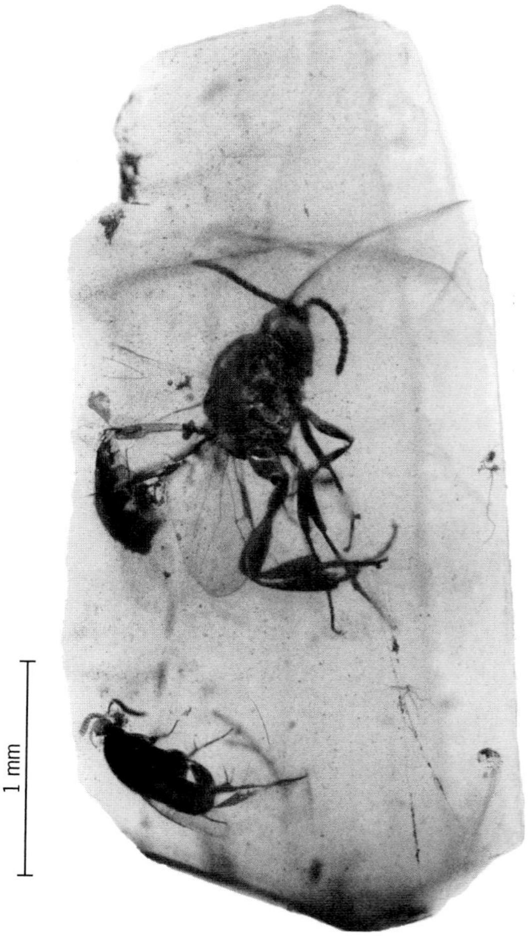

Fig. 2. Amber section with the *Cretevania* wasp embedded in epoxy resin (211 × 390 mm).

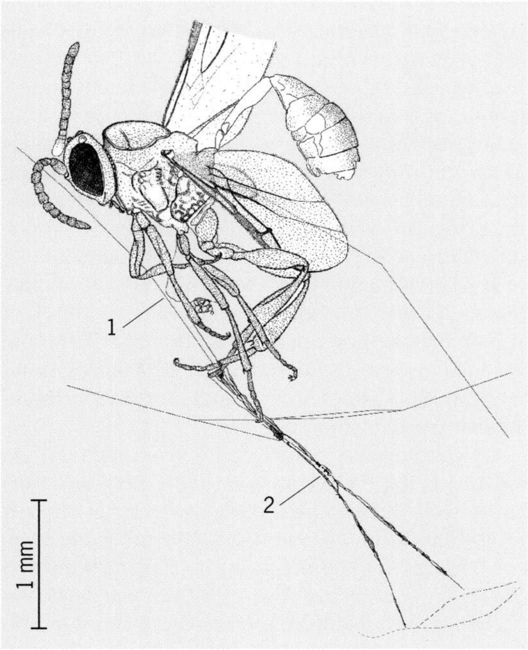

Fig. 3. **Camera lucida drawing of the *Cretevania* wasp (specimen CPT-960, collection of the "Fundación Conjunto Paleontológico de Teruel-Dinópolis") with details of the thicker bifurcate thread that snagged one of the legs (1), comprising numerous thin strands without droplets, and one hank-shaped strand (2).**

because the carcasses remained on an abandoned web.

Evolution. It had been proposed that spider evolution was intimately related to the radiations of insects, mainly winged (pterygote) insects. Pterygotes are at least as old as the Carboniferous, but insects in ancient lineages like odonates (dragonflies) and most polyneopterans (crickets, grasshoppers, stick insects, roaches) are generally too large and powerful to be captured in orb webs. Many paraneopterans (Psocoptera, thrips, hemipterans) could be prey, but the main groups of flying insect prey are those in the Holometabola. The oldest fossil holometabolan insects are from the Permian (Coleoptera), Triassic (Diptera, Hymenoptera), and Jurassic (Lepidoptera), but they have their major radiations approximately during the angiosperm (flowering plant) radiation in the Cretaceous. These radiations involved phytophages (herbivores) and pollinators (both holometabolans and nonholometabolans), including Coccoidea (scales and mealy bugs), Cyclorrhapha flies, termites, and aculeates (ants, bees, other stinging wasps). The small sizes of the insects found in the fossil web are also similar to the sizes of the most abundant type of prey of the orders Diptera, Hymenoptera, and Coleoptera found in modern araneid webs. If a preferred prey is estimated to be 50–80% of the spider's size, this would suggest that the spider that spun the fossil web was ~2.5–4 mm (0.1–0.16 in.) in length. Three araneoid specimens found in other amber pieces of the same Spanish outcrop have a similar size, nearly 3 mm (0.12 in.).

Spider webs thus could have been important selective agents on Cretaceous insects, and it has been proposed that the scaly wings and body of lepidopteran insects primarily evolved as an adaptation against spider predation, allowing these insects to escape from webs. The dating of the Spanish fossil web establishes that the interception of flying insects by spiders has an age coincident with the explosive diversification of the angiosperms and the major pollinating groups of insects.

For background information *see* AMBER; ARACHNIDA; ARANEAE; ARTHROPODA; CRETACEOUS; FOSSIL; INSECTA; NATURAL FIBER; ORGANIC EVOLUTION; PREDATOR-PREY INTERACTIONS in the McGraw-Hill Encyclopedia of Science & Technology.

Enrique Peñalver; David A. Grimaldi; Xavier Delclòs

Bibliography. J. E. Garb et al., Silk genes support the single origin of orb webs, *Science*, 312:1762, 2006; D. A. Grimaldi and M. S. Engel, *Evolution of the Insects*, Cambridge University Press, New York, 2005; E. Peñalver, X. Delclòs, and C. Soriano, A new and rich amber outcrop with palaeobiological inclusions in the Lower Cretaceous of Spain, *Cretaceous Research*, in press, 2007; E. Peñalver, D. A. Grimaldi, and X. Delclòs, Early Cretaceous spider web with its prey, *Science*, 312:1761, 2006; S. Zschokke, Glue droplets in fossil spider webs, pp. 367–374, in D. V. Logunov and D. Penney (eds.), *European Arachnology 2003*, KMK Scientific Press, Moscow, 2004.

Earthquake early warning

Earthquake early warning (EEW) is notification of an earthquake's occurrence in real time, that is, after the earthquake has begun. EEW was first successfully implemented in the UrEDAS (Urgent Earthquake Detection and Alarm System) system for Japan Railways in the 1980s. EEW has safely stopped high-speed trains in Japan and provided warnings in Mexico City. More recently, research has identified characteristics of the seismic P wave that permit warning from a single onsite detector. A fully functional EEW system requires not only seismological detectors and processing, but also decision-making and communications apparatus designed in accordance with social science findings.

Earthquakes are among the most destructive of natural phenomena and are much more fearful than tropical cyclones, floods, or wildfires due to their occurring with virtually no warning. Actually, nature does provide the briefest of warning for an earthquake, in the form of the P (primary or push-pull) body wave, which moves through the Earth faster than the more destructive S (or secondary) body wave and Rayleigh and Love surface waves. Because P-wave velocity is typically several kilometers per second and S waves are only about 60% as fast (and surface waves even slower), the farther away the earthquake is from a given site, the longer the interval between the P–S wave arrival. This fact allows the calculation of the distance from the site back to where the waves initiated (that is, the earthquake epicenter). Over a century ago, recognition of this

P–S arrival interval became one of the cornerstones of seismology, allowing the location of earthquake epicenters. Until relatively recently, however, the P–S interarrival time, being only a few seconds to tens of seconds, could not be used as a means for an earthquake warning. With the advent of modern high-speed computing, it has become possible to exploit the P–S inter-arrival time and other aspects of the seismic wave train for earthquake early warning.

Earthquake warning can refer to several different types of notifications, which may be classified by their temporal order of magnitude. In studies of long-term (centuries to decades) patterns of seismicity, identification of areas of sparse seismicity (or "seismic gaps," or an absence of seismicity in an otherwise clear pattern of seismicity, as along a major fault, implying a buildup of strain and thus an impending earthquake) can lead to a long-term forecast of a large earthquake. Such a forecast's "window," or bounds, on the time of the event's occurrence may be decades. This is the current situation for the Hayward fault in the San Francisco Bay area, the Tokai region of Japan (southwest of Tokyo), and Istanbul, Turkey, where large earthquakes have been forecast as having heightened probabilities of occurrence in the next several decades.

At a smaller time scale, variations of premonitory phenomena such as increased seismicity, topographic deformations, ground-water fluctuations, changes in gravity anomalies, and unusual animal behavior can lead to earthquake prediction, which is generally understood to mean a more specific, smaller window in which an earthquake of specific magnitude range is estimated (with some probability) to occur during a relatively short period, typically measured in days to weeks. Given such a prediction, specialized monitoring equipment may be installed, which might even decrease the prediction window to days, hours, or even minutes. While extremely desirable, such prediction and notification still eludes us. The 1975 M7.3 Haicheng (China) earthquake is regarded by many as the only major earthquake to ever have been successfully predicted, yet recent work has cast even this in doubt. (Earthquake magnitudes are numbers, typically ranging from 2 to about 9, and usually reported to one digit, for example, 7.5. They are preceded by M_w (for example, M_w 7.5)

At an even shorter time scale, it is now possible to provide notification of an earthquake's occurrence after the earthquake has begun. This article discusses earthquake early warning in the sense of real-time or intra-event communication of an earthquake that has begun to occur. We begin first with a brief summary of the mechanics of how an earthquake and its characteristics are identified for warning purposes, then review some important applications of earthquake warning, and conclude with a discussion of earthquake warning policy issues.

Earthquake early warning. The concept of earthquake warning system (EWS) dates back at least to J. D. Cooper, who in 1868 proposed to set up seismic detectors at some distance from San Francisco such that when an earthquake occurred an electric signal would be sent by telegraph to San Francisco, where it would ring a big bell to warn citizens. While Cooper's scheme was never implemented, it contained the essential elements of today's EWS.

A number of major cities, such as Tokyo and Istanbul, lie tens to hundreds of kilometers away from major plate boundaries that historically and someday again will rupture in the form of great earthquakes. If instruments are sited close to the known faults, then once an earthquake rupture begins, the instruments can detect the initial P wave and send a radio or landline signal to a city tens to hundreds kilometers away. Given the relative velocities of the different waves, the warning can arrive at the city ten or more seconds prior to the arrival of the damaging ground motions. This basic concept underlies EWS in Japan, Taiwan, Mexico, and Turkey.

However, simply detecting a P wave can be problematic. Is the P wave for a small earthquake which really will not do much damage far away, or is it for a great earthquake which can do enormous damage hundreds of kilometers away? "Crying wolf" too often may lead to ignoring important warnings. According to H. Kanamori, the answer to this question lies in recognizing that "the P wave carries information and the S wave carries energy." The concept was first used for a warning system for the Shinkansen (bullet train) in Japan, where the frequency content of the initial P wave was used to estimate earthquake magnitude and azimuth and distance of the epicenter. This concept can be used to provide useful warning for a site very close to the source.

Examples/implementation of EWS. The Trans-Alaska Pipeline System has used an earthquake monitoring system (EMS) since the start-up of operations in 1977. Strictly speaking, that system is not an EWS since it does not attempt to provide real-time (that is, intra-event) warnings. Rather, the purpose of the EMS is to detect strong earthquake activity along the pipeline and postprocess the data to determine whether the pipeline should be shut down (which takes some time) and/or inspected.

The best-known example of an EWS is the one initially developed by Japan Railway in the 1960s and significantly improved in the 1980s to slow down or stop high-speed trains before seismic shaking affected them. Y. Nakamura used a single-station approach, where seismic signals are processed locally and an earthquake warning is issued when ground motion exceeds the trigger threshold. This system, called urgent earthquake detection and alarm system (UrEDAS, which phonetically corresponds to "shaking begins" in Japanese), has been widely used by the Japanese railway system.

In Mexico, a Seismic Alert System (SAS) was developed in 1991, with the specific objective of issuing early warnings to the residents and authorities in Mexico City for large earthquakes in the Guerrero seismic gap, approximately 300 km southwest of Mexico City. In preparation for an anticipated major earthquake in Istanbul, an EWS consisting of ten strong-motion stations was installed as close as possible to the fault zone. Continuous online data from these stations via digital radio modem provides early warning for potentially disastrous earthquakes.

Neural networks have been applied to better discriminate earthquake properties. Most recently, not just the event magnitude but rather estimated damage has been proposed as a determinant of whether a warning should be issued.

Socio-political aspects of EWS. A vital aspect of an EWS is planning for when and how the warning is provided. The above discussion has focused solely on the technology of what might be termed the "threat detection." Design of an entire useful EWS involves the technologies used to detect threats; reliability of threat detection technologies; length of time needed to achieve accuracy in forecasts and predictions; reliance on human mediation; types of warning systems and devices used; channels used to issue warnings; familiarity, routinization, and institutionalization of the warning procedure; settings in which systems are used; and system goals and objectives. While the threat detection is the fundamental basis for the warning, warnings can go unheeded if downstream aspects fail.

Mexico established a seismic alert system (SAS) in 1991, following the disastrous 1985 earthquake. As of May 2005, the SAS system had detected more than 1783 earthquakes in the $2.5<M<7.3$ range and generated 57 warning signals with an average of 60 s in advance of earthquake effects, 11 of the warnings being for public use (SAS public warnings have the automatic broadcast support of many commercial AM/FM radio and television stations). Note that none of the 11 warnings corresponded with major damage in Mexico City. While there are concerns that false alarms will create a "cry wolf" effect, the real-time warnings issued in Mexico City provide some evidence that false alerts can function like disaster drills, giving warning recipients an additional opportunity to practice measures that they would have to undertake in the event of a real threat. While Mexico City's is the only public warning system that is currently fully operational, Tokyo is on the verge of implementing a system, and more are sure to follow.

For background information *see* EARTHQUAKE; NEURAL NETWORK; SEISMIC RISK; SEISMOGRAPHIC INSTRUMENTATION; SEISMOLOGY in the McGraw-Hill Encyclopedia of Science & Technology.

Charles Scawthorn; Hiroo Kanamori

Bibliography. R. Allen, *Abstracts, Earthquake Early Warning Workshop*, edited by R. Allen et al., California Institute Of Technology, Pasadena, 2005; R. M. Allen and H. Kanamori, The potential for earthquake early warning in southern California, *Science*, 300:786–789, 2003; M. Erdik, Istanbul earthquake early warning and rapid response system, abstract for R. Allen, *Earthquake Early Warning Workshop*, California Institute Of Technology, Pasadena, 2005; J. M. Espinosa-Aranda et al., Seismic alert systems in Mexico, abstract for R. Allen, Earthquake *Early Warning Workshop*, California Institute Of Technology, Pasadena, 2005; J. M. Espinosa-Aranda et al., Mexico City Seismic Alert System, *Seismol. Res. Lett.*, 66:42–53, 1995; P. Gasparini, G. Manfredi, and J. Zschau, *Earthquake Early Warning Systems*, Springer, 2007; R. J. Geller, Earthquake prediction: A critical review, *Geophys. J. Int.*, 131(3):425–450, 1997; J. D. Goltz, *Introducing Earthquake Early Warning in California: A Summary of Social Science and Public Policy Issues, A Report to OES and the Operational Areas*, Earthquake and Tsunami Program, Disaster Assistance Division, Governor's Office of Emergency Services, Pasadena, 2002; D. G. Honegger et al., Trans-Alaska pipeline system performance in the 2002 Denali Fault, Alaska, earthquake, *Earthq. Spectra*, 20(3):707–738, 2004; I. Iervolino, M. Giorgio, and G. Manfredi, Expected loss-based alarm threshold set for earthquake early warning systems, *Earthquake Eng. Struct. Dynamics*, 36:1151–1168, 2007; H. Kanamori, Real-time seismology and earthquake damage mitigation, *Annu. Rev. Earth Planet. Sci.*, 33:195–214, 2005; J. Kumagai, How to master a seismic disaster, *IEEE Spectrum*, pp. 48–51, June 2007; R. R. Leach and F. U. Dowla, *Real-Time Neural Network Earthquake Profile Predictor*, U.S. Patent no. 5,490,062, 1996; Y. Nakamura, On the urgent earthquake detection and alarm system (UrEDAS), *Ninth World Conf. Earthq. Eng.*, pp.VII-673-678, Japan Association for Earthquake Disaster Prevention, distributed by Maruzen Co., Ltd., Tokyo-Kyoto, 1988; Y. Nakamura and B. E. Tucker, Japan's earthquake warning system: Should it be imported to California?, *Calif. Geol.*, 41(2):33–40, 1988; D. Riopelle, K. Shoaf, and L. Bourque, *Trinet Policy Studies and Planning Activities*, in *Real-Time Earthquake Early Warning: Task 1 Report*, Survey of Potential Early Warning System Users, Center for Public Health and Disasters, University of California at Los Angeles, 2001; K. Tierney, *Implementing a Seismic Computerized Alert Network (Scan) for Southern California: Lessons and Guidance from the Literature on Warning Response and Warning Systems Report, Task 2, Trinet Studies and Planning Activities in Real-Time Earthquake Early Warnings*, Disaster Research Center University of Delaware Wilmington, 2000; K. Wang et al., Predicting the 1975 Haicheng earthquake, *Bull. Seismol. Soc. Am.*, 963:757–795, 2006.

Effects of carbon dioxide on the upper atmosphere

Human activity has caused subtle changes in the Earth's atmosphere over the past several decades due to increased emissions of gases such as carbon dioxide (CO_2) and methane (CH_4). These anthropogenic changes in greenhouse gases have the potential for increasing the temperature in the lower atmosphere due to their ability to absorb infrared radiation. The far-reaching consequences of these changes are now being seen at extremely high altitudes in the outer atmosphere of the Earth, especially in the thermosphere, the high-temperature, near-vacuum region ~90–500 km, which is where many satellites are in orbit.

CO_2 levels. R. G. Roble and R. E. Dickinson predicted in 1989 that a consequence of increasing CO_2 levels would be to decrease the temperature of the

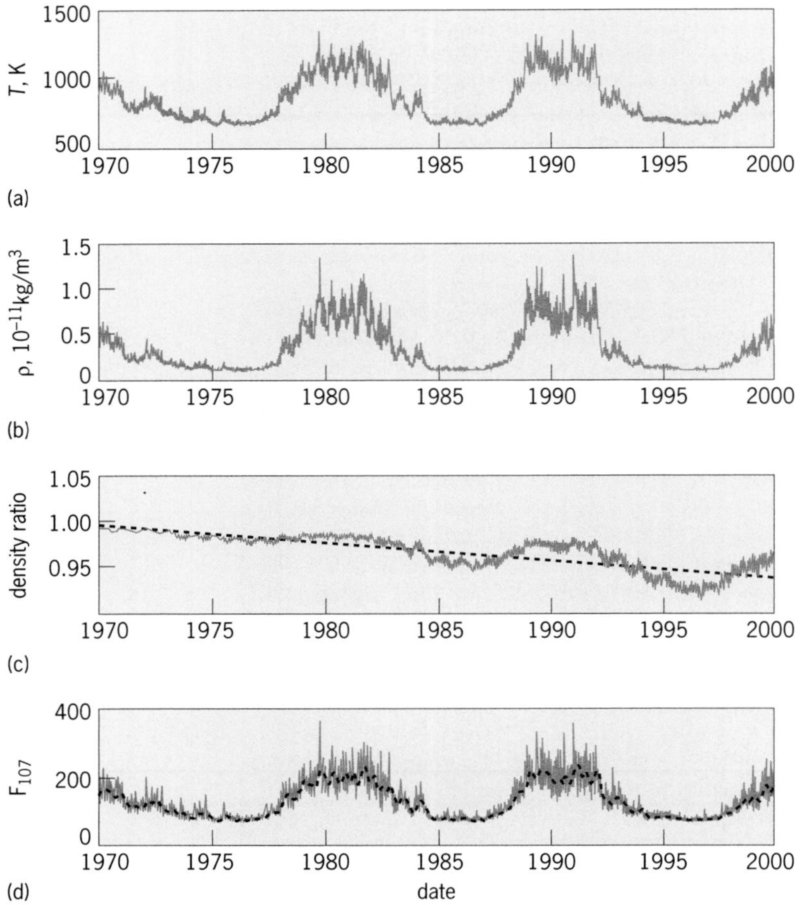

Calculations by L. Qian and coworkers of the thermospheric response to the combined effects of solar activity and anthropogenic change. (*a*) Global mean temperature at 400 km. (*b*) Global mean neutral density at 400 km. (*c*) Gray solid curve = ratio of thermospheric neutral density from two model runs. First run: density calculated with varying CO_2 and solar ultraviolet variation from 1970 to 2000. Second run: density calculated with CO_2 fixed at 1970 levels and with solar ultraviolet variation from 1970 to 2000. The ratio of the two model runs is calculated to remove the influence of solar activity on density variation. Dashed curve = linear regression to the ratio curve. The linear regression shows the average density decrease for the recent three decades is 1.7% per decade. (*d*) Gray solid curve = $F_{10.7}$ index, representing solar activity; dashed curve = 81-day average $F_{10.7}$ index.

upper atmosphere. This is the opposite effect to the response of the lower atmosphere. The reason for this apparent paradox is that CO_2 and other multiatom molecules can emit infrared radiation as well as absorb it. In the lower atmosphere, especially in the troposphere (below ~15 km), CO_2 absorbs radiation coming from the Earth, which excites it to higher vibrational states. Before it can reemit the radiation, it undergoes collisions with other atmospheric gases, transferring the vibrational energy into heat. Any emission that does occur is likely to be reabsorbed by CO_2 at other altitudes. But in the stratosphere (~15–~50 km) and mesosphere (~50–~90 km) the atmosphere becomes increasingly transparent to infrared radiation as densities decrease, and the radiation-excitation-collision process runs in reverse. Instead of absorbing radiation, CO_2 molecules are vibrationally excited by collisions, and then, before additional collisions can remove the vibrational energy, they spontaneously emit in the infrared, which causes radiational cooling of the upper atmosphere.

In the stratosphere, changes in CO_2 have a minor effect on temperature because the energy balance is dominated by processes involving ozone (O_3). In the mesosphere, CO_2 cooling is significant, but long-term changes are difficult to measure. In the thermosphere, one effect of a cooler atmosphere is to cause global density to decrease as the thermosphere cools and contracts. Roble and Dickinson predicted that this could cause a decrease in density of about 40% in the outer atmosphere above ~300 km for a doubling of CO_2 from preindustrial levels, which could happen by 2100. Additional studies by other groups have confirmed these calculations using other models. Changes in density of this magnitude could have significant effects, for example, on the orbits of satellites and space debris, which are affected by drag processes as they pass through the tenuous outer atmosphere. It could also cause changes in the ionosphere, which is the ionized component of the thermosphere, consisting of electrons and positive ions.

Solar cycle. The thermosphere goes through natural, cyclical changes in density driven by the Sun's 11-year activity cycle, heating and expanding at solar maximum, and cooling and contracting at solar minimum. This makes it difficult to detect the smaller and more gradual changes occurring due to increasing anthropogenic emissions. So it may be surprising that during the last several years, three different groups were able to measure thermospheric density changes by observing the effect of atmospheric drag on the trajectories of satellites, some of which have been in orbit since the dawn of the space age. G. M. Keating and coworkers evaluated the long-term orbital decay of five satellites during solar-minimum years and found that density was decreasing at ~5% per decade near 350 km altitude. J. T. Emmert and coworkers derived long-term (secular) trends in upper thermosphere density using 27 near-Earth orbiting objects for all levels of solar activity during the period 1966–2001. Their results indicated that the secular decline ranged 2–5% per decade in the altitude range 200–700 km, with the trend increasing with altitude. F. A. Marcos and coworkers analyzed satellite orbits over the period 1970–2000 and detected an average secular density decrease of 1.7% per decade near 400 km. Since it is expected that temperature change will cause cumulatively larger effects on density at higher altitudes, these observations are in reasonable agreement, except that the initial result from Keating and coworkers indicated somewhat higher trends than either model predictions or other measurements did. To circumvent the problem of the complicating effects of the solar cycle, they confined their study to solar minimum years, which could explain the apparent discrepancy. Emmert and coworkers also observed that the trends were larger at solar minimum than at higher solar activity levels.

Recent work by L. Qian and coworkers employed an extended and updated version of the Roble and

Dickinson model to investigate the interaction of anthropogenic change with the solar cycle. They used daily CO_2 concentration measurements from the observatory at Mauna Loa, Hawaii, and solar ultraviolet variations, to calculate the secular change of thermosphere neutral density for the last three decades. Results show that the average density decrease at 400 km should be 1.7% per decade, in excellent agreement with Marcos and coworkers observations. They also modeled the expected effect under solar minimum and solar maximum conditions, finding larger trends at solar minimum and smaller change during solar maximum, which could explain some of the observational discrepancies. This is because at solar minimum, CO_2 is the primary gas that radiationally cools the thermosphere, but at solar maximum, increases in other minor gases cause CO_2 cooling to be less important. The **illustration** shows results from the 30-year simulations, using the measured CO_2 variations and comparing them to a control run with constant CO_2. Changes in CH_4, O_3, H_2O, and other minor gases may also affect the upper atmosphere, especially by increasing hydrogen and therefore water vapor in the mesosphere, but their effect on thermospheric temperature and density is nearly an order- of-magnitude less than that of CO_2.

Changes in the ionosphere. These should accompany the neutral atmosphere changes, including slight changes in the height and peak density of ionospheric layers, and decreasing ion temperature. Analysis of long-term ionosonde and radar data has proven challenging, but may also reveal trends. These and other possible effects are reviewed in an article in *Science* by J. Lastovicka and coworkers. Mesosphere changes are probably occurring but are difficult to prove conclusively because it is not easy to measure temperature accurately and consistently over many decades. But in the thermosphere, where density is a strong function of temperature, theory and observation are in good agreement with regard to density decreases.

Outlook. Continued monitoring and modeling of upper atmospheric change will be important for understanding the complete picture of global change, and for predicting the trajectories of orbiting objects ranging from the smallest fragments to the *International Space Station*. A beneficial effect could be that it will take less fuel to maintain the orbits of active spacecraft, but defunct satellites, fragments from explosions, and other miscellaneous objects would have even longer orbital lifetimes, which could exacerbate the accumulation of hazardous space debris.

For background information *see* ATMOSPHERE; CLIMATE MODELING; GLOBAL CLIMATE CHANGE; HEAT BALANCE, TERRESTRIAL ATMOSPHERIC; IONOSPHERE; SATELLITE (SPACECRAFT); SOLAR CONSTANT; THERMOSPHERE in the McGraw-Hill Encyclopedia of Science & Technology. Stanley C. Solomon

Bibliography. M. Dikpati, G. De Toma, and P. A. Gilman, Predicting the strength of solar cycle 24 using a flux-transport dynamo-based tool, *Geophys. Res. Lett.*, 33:L05102, 2006; J. T. Emmert et al., Global change in the thermosphere: Compelling evidence of a secular decrease in density, *J. Geophys. Res.*, 109:A02301, 2004; G. M. Keating, R. H. Tolson, and M. S. Bradford, Evidence of long-term global decline in the Earth's thermospheric densities apparently related to anthropogenic effects, *Geophys. Res. Lett.*, 27:1523–1526, 2000; C. D. Keeling and T. P. Whorf, Atmospheric CO_2 records from sites in the SIO air sampling network, in *Trends: A Compendium of Data on Global Change*, Carbon Dioxide Information Analysis Center, Oak Ridge National Laboratory, U.S. Department of Energy, Oak Ridge, Tennessee, 2005; J. Lastovicka et al., Global change in the upper atmosphere, *Science*, 314:1253, 2006; F. A. Marcos et al., Detection of a long-term decrease in thermospheric neutral density, *Geophys. Res. Lett.*, 32:L04103, 2005; L. Qian et al., Calculated and observed climate change in the thermosphere, and a prediction for solar cycle 24, *Geophys. Res. Lett.*, 33:L23705, 2006; R. G. Roble and R. E. Dickinson, How will changes in carbon dioxide and methane modify the mean structure of the mesosphere and thermosphere?, *Geophys. Res. Lett.*, 16:1144–1441, 1989.

Electronic toll collection

The ancient Greeks placed a coin, called Charon's toll, in the mouth or hand of a dead person to pay Charon for ferrying the spirit across the River Styx to the Elysian Fields. In some places, it is still traditional to place pennies on the eyes of a dead person prior to burial. While the first toll road has not been identified, Aristotle's *Oeconomicus* and Pliny's *Natural History* note land tolls in Asia, while the Sanskrit text *The Arthasastra* mentions tolls prior to the fourth century B.C. Strabo in *Geographies*, written at the time of Augustus, reports tolls on the Little Saint Bernard's Pass maintained by the Salassi (a Celtic) tribe. As the Roman Empire declined, the central authority necessary to build and maintain a safe and free (to travelers) road system declined with it, so tolls (not necessarily authorized) became more widespread.

In England, turnpikes were widely deployed in the late seventeenth, eighteenth, and early nineteenth centuries. Raising funds along with travel helped move the British inland transport system from being regarded as extremely backward to among the best in Europe. With the onset of competition from rail and the aversion to stopping to pay tolls, these toll roads eventually reverted to free roads. Today, toll roads operate in many countries, and toll revenues constitute nearly 6% of the highway funds collected in the United States, despite comprising only 5244 mi (8440 km) of the approximately 3.9 million miles (6.3 million kilometers) of public roads.

Electronic toll collection (ETC) describes a set of interrelated technologies used to charge vehicles for using particular roads and bridges. It frequently appears on new toll projects, and has replaced traditional manual toll collection (tollbooths) on many

MnPASS high-occupancy toll lanes on I-394 in the Minneapolis–St. Paul, Minnesota, metropolitan area.

older toll facilities. It is perhaps the most successful and widely adopted example of intelligent transportation systems (ITS).

Technology. Four primary technologies are used in electronic toll collection systems: automated vehicle identification (AVI), automated vehicle classification (AVC), transaction processing, and violation enforcement.

Automated vehicle identification. Automated vehicle identification technologies are used to match vehicles with payment. Several technologies are used for AVI, depending on the type of toll facility.

On most roads using electronic toll collection, vehicles pass under a toll gantry containing an antenna that communicates with a transponder in the vehicle, reading a code that corresponds to its identity, and passing that information to a transaction processing system that debits an appropriate account. The communication takes place using dedicated short-range communications (DSRC) technologies, where there is a radio-frequency identification (RFID) chip on the transponder in the vehicle.

Some facilities, notably Highway 407 in Ontario and the London Congestion Charge, use automated number plate recognition (ANPR) systems. Cameras located along the road are aimed at the license plate, and an optical character recognition (OCR) system digitizes the information. ANPR systems dispense with the need for the vehicle to have an onboard device, and instead rely on the license plate of the vehicle for identification. If the vehicle already has an account, it is noted for use by the transaction processing system; otherwise, payment must be made within some period of time (for example, before the end of the day in the case of London system) or fines accrue. Difficulties arise in automatically reading the numbers on some license plates, especially in poor lighting conditions or bad weather, which may require costly manual review of images to avoid errors.

A third system, used most notably for trucks in Germany, relies upon Global Positioning System (GPS) devices onboard the vehicle [On Board Units (OBUs)] to determine which tolled facilities were used and at what time. The information is communicated electronically to stations placed along the roadway, and accounts are debited. Vehicle without OBUs pay at terminals at rest stops and other locales.

Automated vehicle classification. Automated vehicle classification systems ascertain the vehicle type (for example, whether the vehicle is a two-axle car or five-axle truck), which is important for billing. Ideally, the vehicle type could be stored with the user account. However, the same transponder may be used on different vehicles, or vehicles may change configuration (a tractor may pull different-sized trailers) leading to varying charges. Where vehicle type is a concern, sensors in the road pavement or above ground might be used to determine the vehicle type, for example, by measuring the number of axles using a magnetic induction loop embedded in the pavement.

Transaction processing. Transaction processing is the banking function of an electronic toll collection system. This is where the account debiting takes place. Customers may provide a credit card or bank account number which will be charged directly (or periodically to reduce transaction costs), or may prepay an account with a fixed amount of funds available. In the London Congestion Charge, customers may pay at any one of thousands of retailers offering electronic bill payment systems.

Violation enforcement. Ensuring that travelers do not avoid paying the toll is referred to as violation enforcement. Traditional (nonelectronic) toll collection often used a gate. The word "turnpike" comes from the technology used to enforce payment, whereby a pike (long stick) laid across the road would not be lifted until the traveler paid the toll. With electronic toll collection, using such a gate slows down payers and nonpayers alike, and thus can be seen as a disadvantage, reducing the time savings that electronic toll collection aims to provide.

When a vehicle passes a transponder, a second technology is used to detect its presence. If a detected vehicle does not correspond with payment being made, photographic enforcement or police enforcement are often employed. A photograph can be taken of the vehicle's license plate (much like the automated number plate recognition system described above) and of the driver, and a bill or fine sent to the address corresponding to the license plate. In many locales, traffic offenses are associated with the driver, not the vehicle, so this type of enforcement may be limited. Alternatively, police may be called to issue a ticket.

Economics. Compared with manual toll collection, electronic toll collection increases toll lane capacity, thereby reducing toll processing time and queue lengths at toll plazas. Thus, both delays and the number of toll takers are reduced. The most advanced open-road electronic toll collection can identify vehicles at full speed, reducing vehicle delay due to toll collection to zero, from a time of 10–15 s associated with paying at a tollbooth, and eliminating the concomitant queueing when demand exceeds the available capacity of manual tollbooths. By reducing delays, electronic toll collection increases throughput from 250–350 vehicles per lane through a toll lane with manual collection, up to 2200 vehicles per lane.

Prior to electronic toll collection, toll roads and bridges lost 10–20% of the revenue collected to collection costs, using then-current (labor-intensive) technology. Electronic toll collection can significantly reduce the cost of staffing tollbooths, money handling, and roadway maintenance. The cost of operating an automated lane is about one-tenth the cost of operating a manual toll lane.

Even with the adoption of electronic toll collection on conventional toll roads, when those roads remain embedded in a network of "free" roads, many users will not adopt electronic toll collection, leading to an equilibrium market saturation of about 60%. Toll agencies may need to push discounts and significant time savings to gain wider adoption.

An unintended consequence, at least from the point of view of the user, is that electronic toll collection, by reducing the awareness of paying tolls, allows agencies to increase toll rates beyond what they may have been able to do were the out-of-pocket charges felt directly.

Another advantage of electronic tolling is the ability to impose dynamic or variable tolls, which change by time of day or level of congestion. As of 2006, only 29 facilities worldwide had done so, suggesting the prime motivator to moving toward electronic tolls is the reduction of transaction costs for the producer, and perhaps the ability to raise tolls with fewer complaints rather than improved system management.

A few places have seen congestion charging imposed primarily for traffic or environmental management, rather than as a way to pay for the bonds floated for the construction or to enrich investors. Notably, congestion charges have been imposed in Singapore, London, and Stockholm.

Another use of electronic toll collection has been the conversion of carpool, or high-occupancy vehicle (HOV), lanes to high-occupancy toll (HOT) lanes, allowing noncarpool traffic to buy into the excess capacity on the carpool lanes (see **illustration**). As of 2007, there were five HOT lanes throughout the United States.

For background information *see* HIGHWAY ENGINEERING; TRAFFIC-CONTROL SYSTEMS; TRANSPORTATION ENGINEERING in the McGraw-Hill Encyclopedia of Science & Technology. David Levinson

Bibliography. M. W. Burris, Application of variable tolls on congested toll road, *ASCE J. Transport. Eng.*, 129(4):354–361, July/August, 2003; H. M. Al-Deek, A. A. Mohammed, and A. E. Radwan, Operational benefits of electronic toll collection: Case study, *ASCE J. Transport. Eng.*, 123(6):467–477, November/December, 1997; D. Levinson, *Financing Transportation Networks*, Edward Elgar Publishers, 2002; E. Pawson, *Transport and Economy: The Turnpike Roads of Eighteenth Century Britain*, Academic Press, New York, 1977.

Endocannabinoids

The endocannabinoids are a family of endogenous lipid messengers that engage cannabinoid receptors, that is, the cell surface receptors targeted by the active principle (Δ^9-tetrahydrocannabinol; THC) of marijuana (cannabis). They are produced on demand through cleavage of membrane phospholipid precursors and are utilized in the execution of a variety of short-range signaling processes. In the brain, they activate CB_1-type cannabinoid receptors on neuronal axon terminals to regulate ion channel activity and neurotransmitter release. In peripheral tissues, they activate CB_1 receptors in nerve terminals to reduce pain sensation and increase appetite. They can also engage CB_1 receptors in nonneural tissues to modulate blood pressure (in the vasculature), increase adipogenesis (the formation of fat in white adipose tissue), and stimulate fatty acid accumulation (in liver). In addition, the endocannabinoids bind to CB_2-type receptors expressed in immune cells to influence pain and inflammatory responses. There are two established endocannabinoid transmitters: anandamide and 2-arachidonoyl-*sn*-glycerol (2-AG). The existence of other endogenous ligands has been suggested, but their physiological significance remains unknown.

Cannabinoid receptors. Two cannabinoid receptor subtypes have been molecularly cloned: CB_1 and CB_2. CB_1 is expressed in the neocortex, hippocampus, basal ganglia, cerebellum, and brainstem and accounts for most, if not all, pharmacological actions of THC. It is coupled to $G_{i/o}$ [that is, G_i (the inhibitory G protein) and G_o (a similar G protein of unknown function); so named because they bind to guanine nucleotides, G proteins are intermediaries in intracellular signaling pathways] and initiates signaling events typical of this class of transducing proteins, including closure of Ca^{2+} channels, opening of K^+ channels, inhibition of adenylyl cyclase activity [with its consequent decrease in cytosolic cyclic adenosine monophosphate (cAMP) levels since adenylyl cyclase is the enzyme that synthesizes cAMP from adenosine triphosphate (ATP)], and stimulation of kinases that phosphorylate tyrosine, serine, and threonine residues in proteins. Each mechanism serves distinct functions in translating CB_1-receptor occupation into biological responses.

In the mammalian cortex, CB_1 receptors are found on axon terminals of various interneurons and principal neurons. This expression pattern dominates the neocortex, hippocampus, and amygdala. CB_1 receptors are also densely expressed throughout the basal ganglia and cerebellum. Although in lower numbers, CB_1 receptors are present in many other regions of the brain and spinal cord, in peripheral sensory neurons, and in a number of nonneural tissues including white adipose tissue and liver.

CB_2 receptors are $G_{i/o}$-coupled receptors and share approximately 44% sequence homology with CB_1. They are predominantly found in immune cells (T cells, macrophages, and B cells) and in hematopoietic (blood-forming) cells. Low levels of CB_2 are

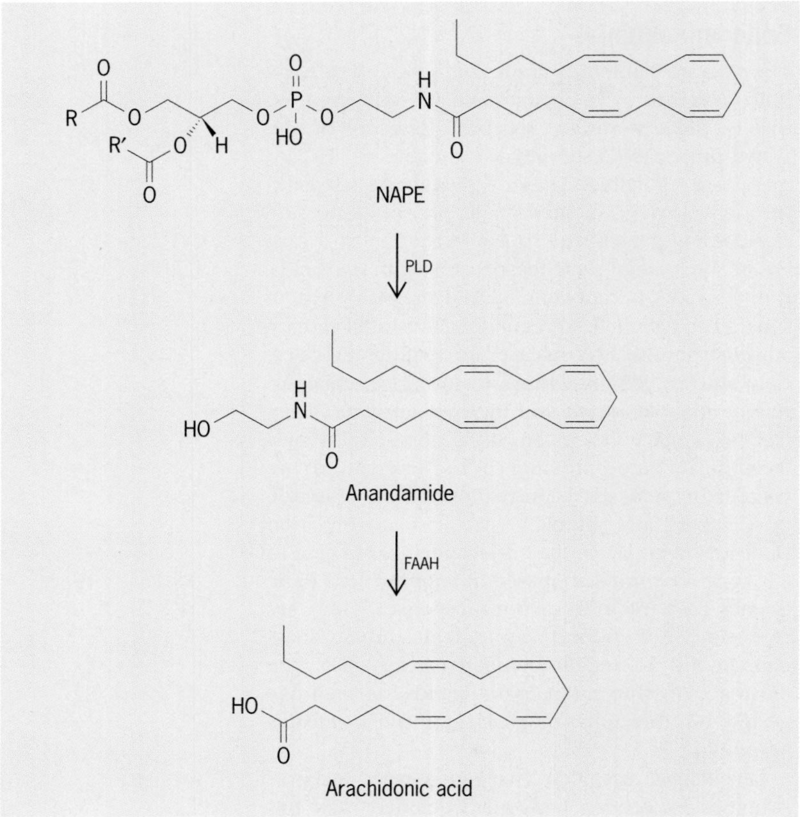

Fig. 1. Mechanism of anandamide formation and degradation in neurons. The cleavage of the anandamide precursor *N*-arachidonoyl-phosphatidylethanolamine (NAPE) by a NAPE-specific phospholipase D (PLD) leads to the formation of anandamide, which is then hydrolyzed by fatty acid amide hydrolase (FAAH) to produce arachidonic acid.

found in the brainstem, where they may mediate the antinausea effects of THC.

Anandamide. Anandamide is the amide of arachidonic acid (a polyunsaturated fatty acid) with ethanolamine (**Fig. 1**) and was the first endocannabinoid ligand to be discovered. Anandamide biosynthesis in neurons requires two steps.

The first step is the stimulus-dependent cleavage of the phospholipid precursor *N*-arachidonoyl-phosphatidylethanolamine (NAPE). This reaction is mediated by a phospholipase D (PLD) and produces anandamide and phosphatidic acid (Fig. 1). A gene encoding a NAPE-specific PLD has been cloned, but it is unclear whether this enzyme is responsible for anandamide formation in vivo. Resting neurons contain tiny quantities of NAPE, but these can be rapidly increased by *N*-acyltransferase (NAT), a Ca^{2+}-stimulated enzyme activity.

In an alternative pathway, anandamide might be generated by phospholipase C-mediated hydrolysis of NAPE to produce phosphoanandamide, which is then further cleaved to yield anandamide. Both Ca^{2+} entry and G-protein–coupled receptor stimulation can trigger anandamide formation in neurons. However, the mechanisms responsible for anandamide formation remain incompletely understood.

2-Arachidonoylglycerol (2-AG). 2-AG (**Fig. 2**), the ester of arachidonic acid with glycerol, is an intermediate in multiple pathways of lipid metabolism. This metabolic role explains its high content in cells (up to 200-fold greater than that of anandamide) and implies that a significant fraction of cellular 2-AG is likely to be engaged in housekeeping rather than signaling functions. Two possible routes of 2-AG biosynthesis have been proposed. The first starts with the phospholipase C (PLC)-mediated formation of 1,2-diacylglycerol (DAG). This intermediate is a substrate for two enzymes: DAG kinase, which attenuates DAG signaling by catalyzing its phosphorylation to phosphatidic acid; and DAG lipase (DGL), which hydrolyzes DAG to produce 2-AG (Fig. 2). Two isoforms of DGL have been molecularly cloned, DGL-α and DGL-β, which are both expressed in the brain and other 2-AG–producing tissues. An alternative pathway of 2-AG biosynthesis starts with the formation, catalyzed by phospholipase A_1, of a 2-arachidonoyl-lysophospholipid, which may be hydrolyzed to 2-AG by a lyso-PLC activity.

Pharmacological and genetic experiments suggest a primary involvement of the PLC/DGL pathway in neuronal 2-AG formation. For example, DGL-α overexpression in mouse neuroblastoma cells increases 2-AG levels and enhances DAG utilization. Conversely, silencing of DGL-α by RNA interference (posttranscriptional gene silencing initiated by the introduction of double-stranded RNA) elicits opposite changes. DGL- mediated 2-AG production in neurons may be initiated by increases of intracellular Ca^{2+}, induced by either membrane depolarization or receptor activation. For example, in acutely dissected hippocampal slices, stimulation of Schaffer collaterals (an excitatory fiber tract) stimulates 2-AG formation. Moreover, in rat brain slices in culture, activation of mGlu5 metabotropic glutamate receptors triggers 2-AG biosynthesis [metabotropic glutamate receptors (mGluRs) respond to glutamate, an important neurotransmitter, by activating proteins inside nerve cells that affect cell metabolism].

Deactivation. Two mechanisms cooperate in attenuating endocannabinoid signaling in the brain: carrier-mediated transport into cells and intracellular hydrolysis.

Endocannabinoid transport. The diffusion of anandamide and 2-AG through lipid membranes is accelerated by a selective carrier system present in neurons and glial cells (the nonneuronal, supporting elements of the nervous system). Unlike other transmitter transport systems, endocannabinoid transport is not driven by transmembrane Na^{+} gradients, suggesting that it might be mediated by a facilitated diffusion mechanism. However, the putative transporter(s) involved remains to be cloned. Endocannabinoid transport is inhibited by various pharmacological agents, which slow down anandamide elimination in vitro and produce analgesic (pain-reducing) and anxiolytic (anxiety-relieving) effects in animal models.

Fatty acid amide hydrolase (FAAH). FAAH is an intracellular membrane-bound enzyme that cleaves anandamide to form arachidonic acid and ethanolamine. It is widely distributed in the brain, where it is

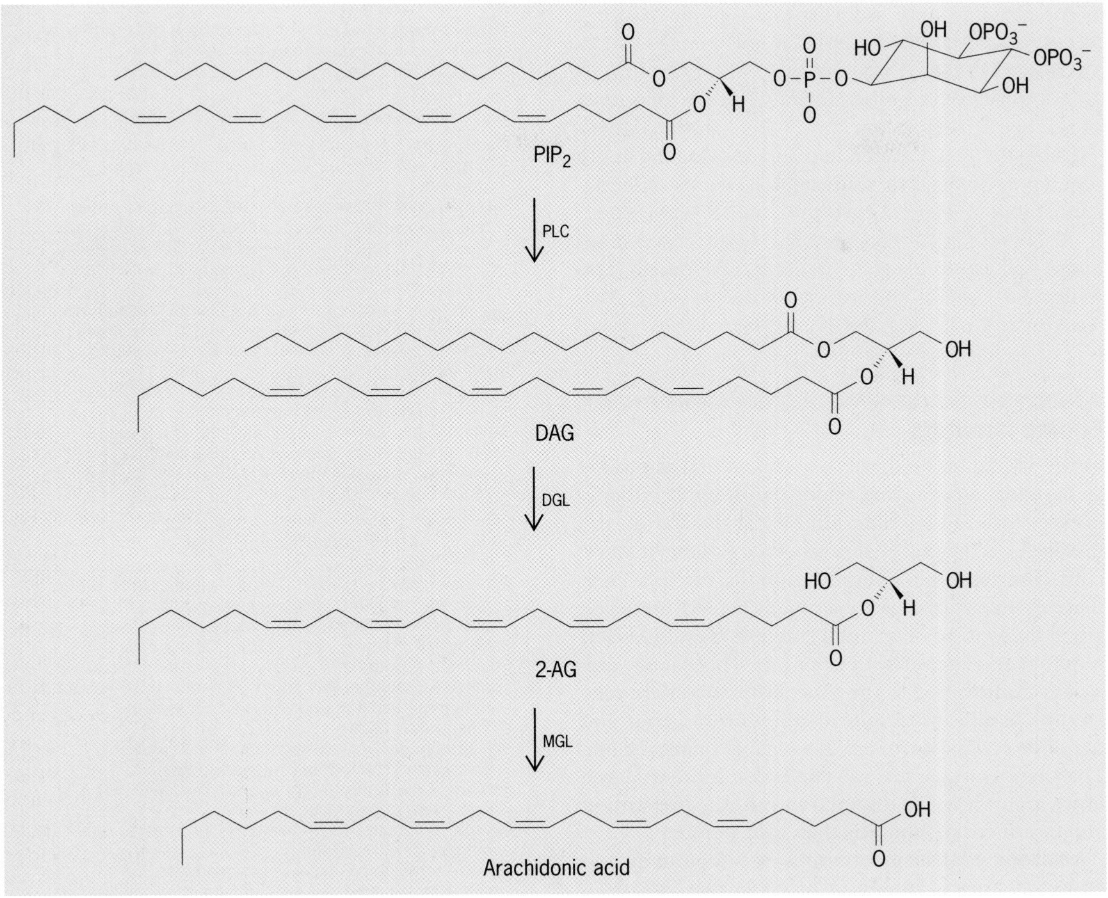

Fig. 2. Mechanism of 2-arachidonoylglycerol (2-AG) formation and degradation in neurons. The sequence of reactions leading to 2-AG formation comprises (1) the cleavage of a phosphoinositide (PIP_2) to yield 1,2-diacylglycerol (DAG), catalyzed by a phospholipase C (PLC), and (2) the subsequent conversion of DAG to 2-AG, catalyzed by diacylglycerol lipase (DGL). 2-AG degradation by monoacylglycerol lipase (MGL) (3) produces arachidonic acid.

expressed at high levels in cell bodies and dendrites of principal neurons. FAAH-positive cell bodies are often juxtaposed to axon terminals that contain CB_1 receptors, suggesting that FAAH participates in the deactivation of anandamide. This idea is supported by experiments using mutant FAAH-deficient mice or selective FAAH inhibitors. For example, the selective FAAH inhibitor URB597 increases anandamide levels in the rodent brain and produces profound analgesic, anxiolytic, and antidepressant-like effects. Interestingly, URB597 was also found to elicit antihypertensive effects, highlighting the broad functions served by anandamide in the body.

Monoacylglycerol lipase (MGL). MGL is a cytosolic enzyme that catalyzes the breakdown of 2-AG and other monoacylglycerols. It has a broad distribution in the brain, which partially overlaps with that of FAAH; however, unlike FAAH, MGL is exclusively associated with presynaptic nerve endings.

Local signaling. The endocannabinoids are produced on demand and act on cells located near their site of synthesis. For example, outside the brain they are generated by circulating leukocytes and platelets and elicit vascular relaxation by interacting with CB_1 receptors located on the membranes of neighboring endothelial and smooth muscle cells.

Similar local actions are thought to take place in the brain, where the endocannabinoids might mediate a retrograde signaling mechanism through which neurons modify the strength of incoming synaptic inputs. A striking example of this mechanism of action was described in the hippocampus. When a neuron in the hippocampus is depolarized, the inhibitory inputs received by that cell are transiently suppressed. This phenomenon is initiated *postsynaptically* by voltage-dependent influx of Ca^{2+} into the neuronal cell body, but is expressed *presynaptically* through inhibition of transmitter release from axon terminals. This suggests that a chemical messenger generated during depolarization of the pyramidal cell travels backward across the synapse to inhibit transmitter release. There is strong evidence that the mediator responsible for this phenomenon is 2-AG. It is envisaged that 2-AG, released from postsynaptic spines, crosses the synaptic cleft, activates CB_1 receptors on the presynaptic nerve ending, and inhibits neurotransmitter release. This process is called "retrograde signaling" because it occurs in an opposite direction (postsynaptic to presynaptic) to that of classical synaptic transmission (presynaptic to postsynaptic).

For background information *see* ANALGESIC; BRAIN; EICOSANOIDS; LIPID; MARIJUANA; NERVOUS SYSTEM (VERTEBRATE); NEUROBIOLOGY; PAIN; SYNAPTIC TRANSMISSION in the McGraw-Hill Encyclopedia of Science & Technology. Daniele Piomelli

Bibliography. B. E. Alger, Retrograde signaling in the regulation of synaptic transmission: Focus on endocannabinoids, *Prog. Neurobiol.*, 68:247–286, 2002; R. A. Nicoll and B. E. Alger, The brain's own marijuana, *Sci. Amer.*, 291:68–75, 2004; D. Piomelli, The molecular logic of endocannabinoid signaling, *Nat. Neurosci.*, 4:873–884, 2003.

Equine laminitis

In the spring of 2006, the world watched in dismay as Barbaro, a promising young Thoroughbred racehorse, shattered his right hind leg at the start of the Preakness Stakes at the Pimlico Race Course in Maryland. After multiple operations on the fractured leg, Barbaro's many fractures eventually healed; however, along the way he developed laminitis, first in his left hindfoot and later in both front feet. This disease ultimately contributed to the loss of the 2006 Kentucky Derby champion because of the chronic pain and the poor prognosis for any reasonable quality of life. The silver lining in this sad tale is that more research funds are now being directed toward the prevention and treatment of laminitis than ever before.

Laminitis explained. Laminitis is a common and potentially devastating foot problem that affects all members of the horse family: horses, ponies, donkeys, mules, and wild equids. The disease process basically involves a breakdown of the cellular bond between the wall of the hoof and the bone within: the distal phalanx (commonly called the coffin bone, pedal bone, third phalanx, or simply P3). The distal phalanx in the horse is equivalent to the bone at the tip of the middle finger. It is completely encased by the hoof, a keratinized bootlike structure that is

Factors implicated in the development of laminitis in horses and other domestic equids

Primary Causes

Carbohydrate overload—excessive dietary intake of soluble sugars and starches, such as grains and high-fructan grasses.
Severe intestinal disease—severe compromise of the intestinal barrier, such as occurs with ischemic bowel disease ("surgical" colic) or enterocolitis; note that carbohydrate overload appears also to involve this mechanism.
Sepsis—any disease that leads to severe bacterial toxemia, for example, pleuropneumonia, septic metritis secondary to retained placenta in postfoaling mares, septic peritonitis.
Exposure to black walnut (*Juglans nigra*) heartwood (usually as wood shavings used for bedding).
Sustained excessive load on one limb as a result of non-weight-bearing lameness in the opposite limb (for example, laminitis in the left hindfoot secondary to injury of the right hindlimb).

Contributing Factors*

Excess circulating glucocorticoid—whether from within (for example, equine Cushing's disease, which is pituitary-dependent hyperadrenocorticism) or outside the body (for example, corticosteroid administration).
Metabolic syndrome—typically associated with obesity; pertinent features in horses include insulin resistance, peripheral cortisol production, and a systemic inflammatory state.
Ingestion of ergot alkaloids, such as are found in endophyte-infested fescue grass or hay.
Inactivity (often coupled with obesity).
Unaccustomed strenuous exercise.
Excessive concussion on the feet from exercise on a hard surface ("road founder"†).
Stress—for example, high-stress occupation or environment, long-distance transport, hospitalization.
Poor hoof conformation or improper trimming or shoeing.
History of laminitis—that is, previous damage to the digital vasculature or the lamellar dermis.

*These factors, while they may not cause laminitis on their own, appear to increase the risk for laminitis; they may cause laminitis when two or more are present concurrently.
†"Founder" is a common horse fancier's term for laminitis.

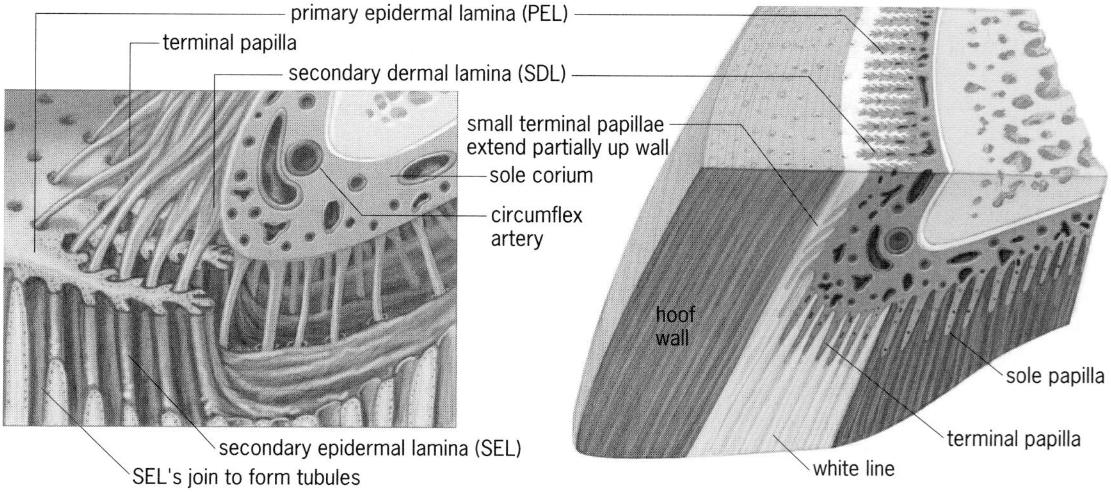

Fig. 1. Representation of the primary epidermal and dermal lamellae that form the strong bond between the bone (third phalanx) and hoof wall. (*From C. Pollitt, Clinical Techniques in Equine Practice, vol. 3, Anatomy and Physiology of the Inner Hoof Wall, p. 11, Elsevier, 2004*)

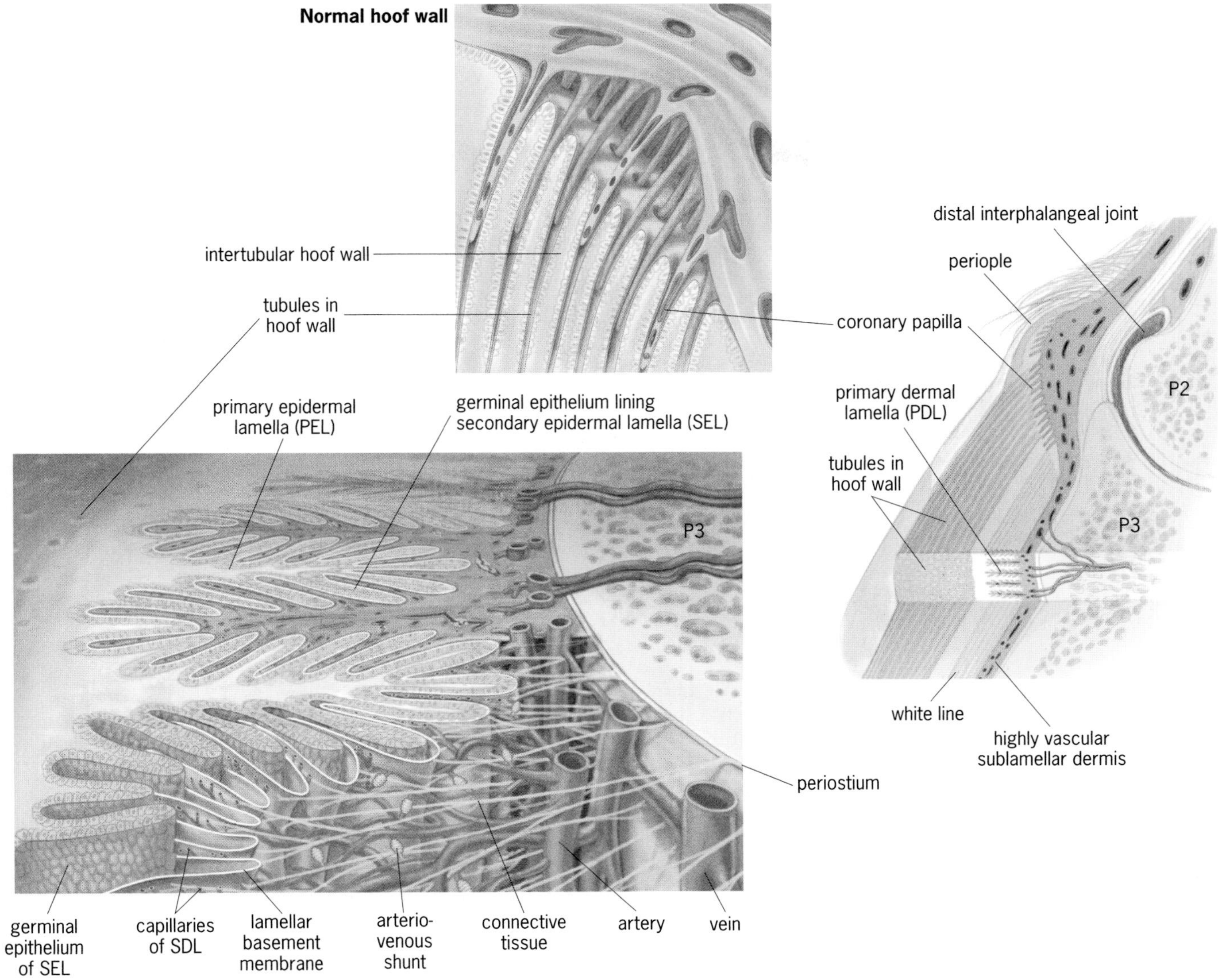

Fig. 2. Portrayal of the infolding of secondary epidermal and dermal lamellae increasing the strength of the hoof wall and bone bond. (*From C. Pollitt, Clinical Techniques in Equine Practice, vol. 3, Anatomy and Physiology of the Inner Hoof Wall, p. 10, Elsevier, 2004*)

essentially a highly modified fingernail. As with the human fingernail, the hoof wall is firmly attached to the distal phalanx by a strong dermal-epidermal bond (**Fig. 1**).

This hoof wall–phalangeal bond is the only thing that prevents the horse's body weight (approximately 500 kg, or 1100 lb, in a horse of Barbaro's size) from driving the distal phalanx through the sole of the hoof. This bond must be strong enough to withstand the forces sustained by the hoof at a gallop pace, yet dynamic enough to allow the hoof wall to grow uninterrupted as it is worn at the ground surface.

In order to maximize the strength of this bond, the dermal-epidermal interface is extensively folded into several hundred lamellae. (*Note:* These folds are often referred to as laminae, although anatomically they are more correctly termed lamellae. The term laminitis is based on the word lamina.) The surface of each lamella is folded into many smaller secondary lamellae, which further increases the surface area and thus the strength of the hoof wall–phalangeal bond (**Fig. 2**).

Just like the skin, the dermal and epidermal cells of the hoof are separate, yet bonded together by a thin layer of extracellular matrix material called the basement membrane. This structure is an essential component of the hoof wall–phalangeal bond, and its destruction and subsequent failure is one of the fundamental events in the development of laminitis.

As it comprises living cells and the material they produce, the hoof wall–phalangeal bond is susceptible to interruption of its nutrient supply and to biochemical degradation. A wide variety of insults can cause or contribute to the development of laminitis in horses (see **table**). Regardless of their other local or systemic effects, each compromises the hoof wall–phalangeal bond through one or more of the

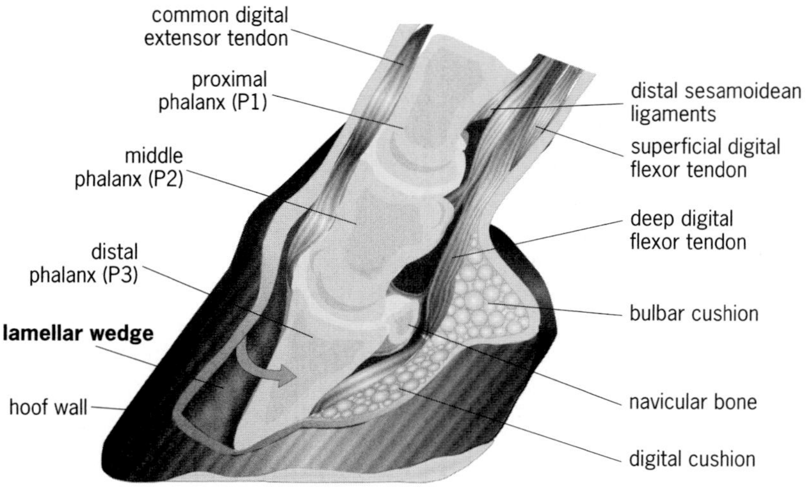

Fig. 3. Rotation of the distal phalanx/coffin bone due to the compromised bond between the epidermal and dermal lamellae tissue along the front of the hoof.

following mechanisms:

- Destruction of the basement membrane by specific protein-degrading enzymes called matrix metalloproteinases 2 and 9 (MMP-2 and MMP-9).
- Interference with glucose utilization by the epidermal cells, which results in detachment of the cells from their basement membrane.
- Interruption of the blood supply (ischemia-reperfusion injury) within the lamellar dermis.

Whatever the mechanism(s), if the hoof wall-phalangeal bond is sufficiently compromised, the hoof wall and distal phalanx separate and displace under the variety of forces including the horse's body weight, the pull of the deep digital flexor tendon where it attaches to the distal phalanx, and stress on the front of the hoof wall from contact with the ground surface at each step. Commonly, the bond is most severely compromised at the front of the foot, so the distal phalanx rotates downward within the hoof capsule, forcing the tip of the bone down onto the dermis of the sole and compromising the vasculature (blood vessels) in that area (**Fig. 3**). Less commonly, but more destructive, the entire distal phalanx separates and drops or "sinks" within the hoof capsule, causing severe pain and extensive vascular injury (**Fig. 4**).

The dermis of the hoof wall and sole has an abundant supply of sensory nerve endings, so severe laminitis is a very painful condition. In fact, it is the extreme and unmanageable pain experienced by horses with severe laminitis that most prompts euthanasia in these cases. Other sequelae of severe laminitis also tip the scales in favor of euthanasia because they substantially worsen the prognosis for a good recovery and include:

- Extensive destruction of the blood supply within the hoof.
- Chronic bacterial infection within the hoof due to low blood flow.
- Bone destruction at the tip of the distal phalanx from abnormal mechanical loading, loss of blood supply, and/or bacterial infection.

Laminitis ranges greatly in severity from mild, where it is virtually unnoticed by the horse owner and only evident by subtle changes in hoof growth or in radiographic (x-ray) findings of the distal phalanx, to death and loss of the entire hoof capsule.

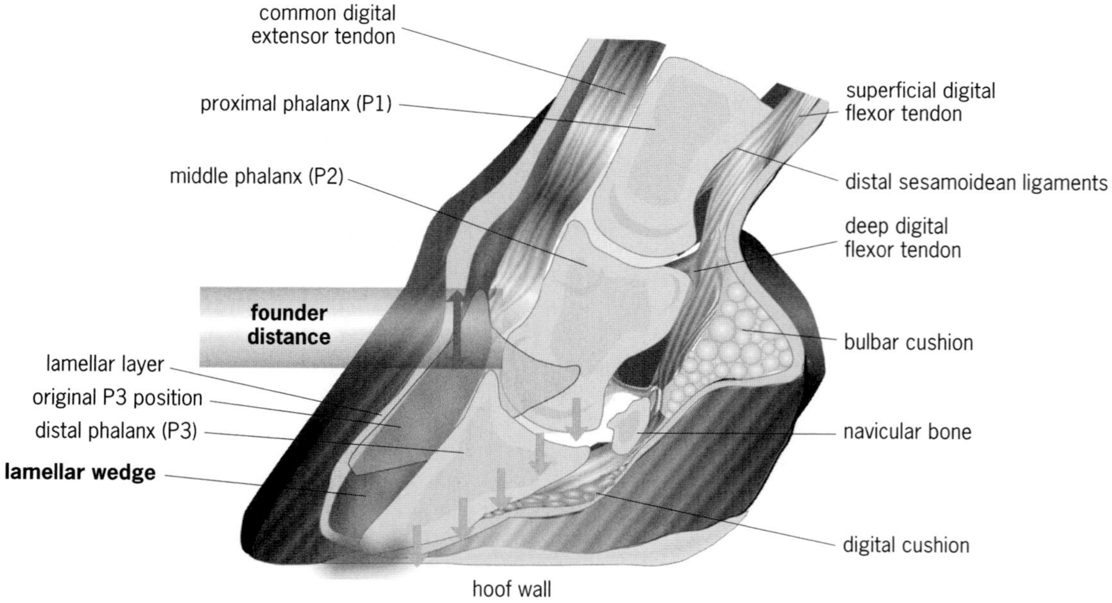

Fig. 4. Sinking of the distal phalanx caused by complete loss of bond between the epidermal and dermal lamellae tissue.

Most cases fall somewhere in between. Prevention is better than treatment with laminitis cases as even mild damage can result in permanent sequela.

Diagnosis and treatment. A clinical diagnosis of laminitis is based on signs and radiographic findings. Laminitis is a painful condition, so the most consistent sign is lameness (gait abnormality). Other signs vary with the severity of the damage and whether the laminitis episode is acute (hours or days old) or chronic (lasting longer than about a week). An increase in pulse pressure in the arteries that supply the affected foot ("bounding digital pulse") is a common finding, although it is not specific for laminitis.

Radiographic changes also vary with the severity and chronicity of the condition. They range from subtle changes within the lamellar dermis (widening of the lamellar zone, change of density indicating fluid accumulation) to marked displacement of the distal phalanx and, later, destruction of the bone (**Fig. 5**). Contrast venography, in which a radiopaque liquid is used to outline the individual blood vessels within the hoof, is a valuable adjunct to regular radiography, as it reveals areas of reduced or absent blood flow (**Fig. 6**).

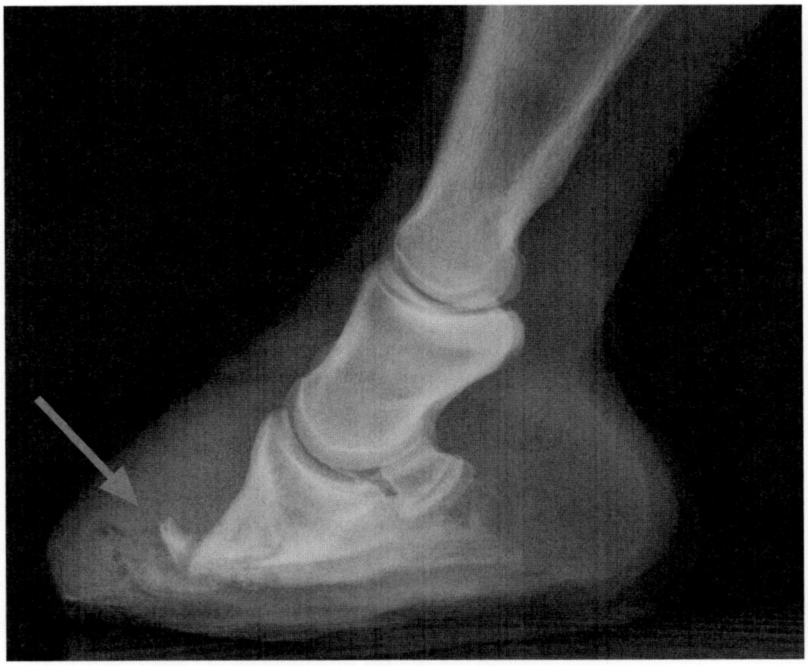

Fig. 5. Digital radiograph from the side of the foot (lateral projection) demonstrating new bone formation due to chronic disease (arrow).

Treatment requires appropriate medical and/or surgical management of the underlying cause, intensive anti-inflammatory and pain therapy, and focused hoof care. In severely affected horses, meticulous nursing care is also essential, as these horses may spend a large part of their time recumbent, a state that is not conducive to good health in any horse. Horses are not designed to stand on three (or fewer) legs nor lie down for long periods. Severe and chronic lameness can have devastating consequences in the horse.

Hoof care primarily involves trimming the hoof and using deep bedding, sole pads, boots, or shoes to restore normal biomechanical alignment of the distal phalanx relative to the rest of the digit and to the ground surface. This strategy helps normalize forces on the remaining hoof wall–distal phalanx bond and blood flow through the hoof, and therefore improves the rate and quality of recovery. A recent study showed that using glue-on shoes improves the outcome in horses with laminitis, presumably because it allows the veterinarian and farrier (a person who shoes horses) to fine-tune the biomechanics of the hoof.

Recent and ongoing research. Laminitis is one of the most studied diseases of horses. Even so, there is still much we do not fully understand. Consequently, prevention and treatment of laminitis remain the challenge.

The period during which we have the greatest potential to influence outcome is the developmental stage—*before* the horse starts showing signs of foot pain. Once the horse shows signs of laminitis, the destructive process is well under way. During this stage of the disease in at-risk horses, cryotherapy (cold therapy) has been shown to be protective to the foot and a means of prevention.

The experimental model used most often in laminitis research is that of carbohydrate overload (CHO): oral administration of a large quantity of starch or oligofructose. With this model, as with naturally occurring CHO-induced laminitis, there is a period of approximately 40 h between ingestion of the carbohydrate and the onset of foot pain. This lag period is called the developmental, prodromal, or preclinical phase of laminitis. With other causes of laminitis, there is also a lag period, although its duration varies with the specific event. For example, it may be as short as 4 h with black walnut-induced laminitis, or as long as several weeks with excessive weight-bearing load (contralateral limb laminitis).

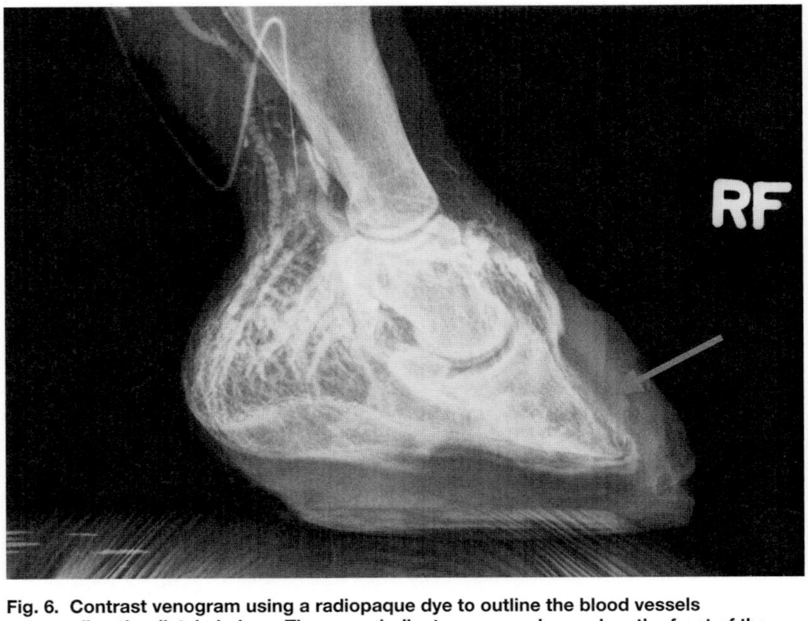

Fig. 6. Contrast venogram using a radiopaque dye to outline the blood vessels surrounding the distal phalanx. The arrow indicates a normal vessel on the front of the bone and the point at which there is frequent disruption of blood flow in the chronic stage of laminitis.

A recent investigation using gene chip technology (DNA microarrays that use DNA sequence information to monitor expression levels of the corresponding genes on a chip) provided a fascinating and unprecedented glimpse into the developmental stage of CHO-induced laminitis. The study showed that over 150 specific genes were upregulated within the lamellar dermis in the first 24–32 h after carbohydrate overload (that is, well before the horses showed signs of foot pain). Genes coding for the production of proinflammatory biochemical or cellular processes and those involved in protein degradation or turnover predominated. Interestingly, several regulatory or anti-inflammatory genes were also upregulated.

Based on this research, it would appear that, at least in the case of CHO-induced laminitis, inflammation is a primary event, rather than a secondary one, because at the time point evaluated the structural changes characteristic of laminitis (including separation of the dermal-epidermal bond) are not yet under way. This research is exciting because it suggests that targeted anti-inflammatory therapy, applied during this preclinical phase in at-risk individuals, may stop the progression of the biochemical and cellular events that ultimately lead to destruction of the hoof wall–phalangeal bond.

Further research is needed, including identification of the specific trigger(s) that can incite the upregulation of these genes. In all likelihood with the CHO model, gut-derived factors are involved. It is well established that carbohydrate overload in horses causes significant changes in the pH and microbial population of the hindgut (cecum and colon), generates vasoactive amines (substances that, if absorbed, could affect blood flow in the feet), and increases the permeability of the bowel lining. Whether the trigger is a microbial fraction or a product that reaches the feet via the bloodstream, or whether it is an aggressive systemic response to the transgression of microbial products across the compromised gut barrier, or both, remains to be determined.

We also need to apply these findings and techniques to other naturally occurring causes of laminitis. If different mechanisms or genes prove to be involved, then different preventive and therapeutic strategies are needed.

For background information *see* BONE; CARBOHYDRATE; DISEASE; EPIDEMIOLOGY; HORSE PRODUCTION; INFLAMMATION; PERISSODACTYLA in the McGraw-Hill Encyclopedia of Science & Technology.

James A. Orsini

Bibliography. C. M. Marr and L. B. Jeffcott (eds.), Laminitis Special Issue, *Equine Vet. J.*, vol. 36, no. 3, April 2004; C. Pollitt, *Laminitis, Clinical Techniques in Equine Practice*, vol. 3, no. 1, Elsevier, 2004; C. Pollitt, New scientific developments in managing laminitis, pp. 52–61, in *Proceedings of the Third International Equine Conference on Laminitis and Diseases of the Foot*, Palm Beach, Florida, 2005; United States Department of Agriculture, *Lameness and Laminitis in U.S. Horses*, USDA:APHIS:VS, CEAH National Animal Health Monitoring System, 2000.

Escherichia coli outbreaks

Escherichia coli O157:H7 is a gram-negative bacterium that causes approximately 73,000 cases of disease, 20,000 hospitalizations, and 60 deaths yearly in the United States. It causes hemorrhagic lesions in the intestines that lead to hemorrhagic colitis with severe abdominal pain and cramps followed by bloody diarrhea. This bacterium is also a frequent cause of hemolytic-uremic syndrome (HUS), consisting of renal insufficiency, anemia, and thrombocytopenia (a low blood-platelet count) that can lead to the permanent need for kidney dialysis or even death. The central nervous system may also be involved.

Children younger than 5 years are most likely to contract HUS, which most commonly occurs around 2 weeks after infection; approximately 8% of children infected subsequently develop HUS. Half of the children diagnosed with HUS ultimately require short- or long-term kidney dialysis, and approximately 4% of those on dialysis die. Limited treatment is with fluid and electrolytes plus the antibiotics doxycycline and trimethoprim-sulfamethoxazole. However, more effective therapies are not available to intervene in the progression of the disease from the gastrointestinal tract to the renal and nervous systems.

O157:H7 is just one of several hundred types of *E. coli* that are found in the intestines of cattle, sheep, and goats. When this bacterium is shed from healthy animals, it can survive for extended periods of time in the environment; survival times measured in days, weeks, and months have been recorded for manure, soil, water, and vegetables such as spinach and lettuce. In the United States, this bacterium is a common cause of food-borne illness. All food-borne diseases are associated with poor hygienic practices.

Bacterium. *Escherichia coli* O157:H7 is an indole-positive, oxidase-negative, lactose-fermenting, gram-negative rod that is placed in the family of enterohemorrhagic *E. coli*. The name includes references to the specific markers (antigens) found on its surface (O157) and flagella (H7), distinguishing it from other strains of *E. coli*. This strain produces toxins that are virulence factors causing disease.

Disease. Infection with this bacterium results in a spectrum of clinical presentations from transient asymptomatic carriage to severe extraintestinal complications and death. The majority of cases start with a predictable series of signs and symptoms. After a 24- to 48-h period of nonbloody diarrhea, abdominal cramping, and a mild fever, infected individuals begin to experience bouts of bloody diarrhea that may last 2–10 days. For most individuals, rehydration therapy results in a complete recovery.

The toxins produced by *E. coli* O157:H7 damage endothelial cells in the kidney and brain, causing renal failure and neurological complications such as seizures. These toxins may also participate in the development of bloody diarrhea by damaging capillaries in the kidneys. However, the precise mechanisms of pathogenesis leading to the diarrhea, renal failure, and central nervous system damage are not fully understood. Effective therapies are not available to intervene in the progression of the disease from the gastrointestinal tract to the renal and nervous systems.

Diagnosis. *Escherichia coli* O157:H7 is diagnosed by a clinical stool specimen taken from the patient. The incubation period of the bacterium ranges from 1 to 8 days following ingestion, but most commonly the symptoms begin 3–4 days after ingestion. Laboratory diagnosis is by selective culture media, rapid identification kits, rapid probe-based identification procedures, serotype-specific probes, and polymerase chain reaction (PCR) techniques. These molecular techniques also enable the detection of a few target cells in large populations. For example, by using the PCR, as few as 10 toxin-producing *E. coli* O157:H7 can be detected in a population of 100,000 bacteria isolated from soft cheese samples.

A major advance in the detection of food-borne pathogens is the use of standardized pathogen DNA patterns or "food-borne fingerprinting." In the United States, the Centers for Disease Control and Prevention (CDC) has established a program called PulseNet, in which pulsed-field gel electrophoresis is used under controlled conditions to determine the distinctive DNA pattern of each pathogen. With this uniform procedure, it is possible to link pathogens, such as *E. coli* O157:H7, with disease outbreaks in different parts of the world from a specific food source. Data from around the world are being used in FoodNet, an active surveillance network, to follow nine major food-borne diseases—enterohemorrhagic *E. coli* O157:H7 is one of them. Using the FoodNet approach, it is possible to trace the course and cause of an outbreak in days and not weeks.

A major problem in maintaining food safety is the need to rapidly detect microorganisms, such as *E. coli* O157:H7, in food products in order to curb outbreaks that can affect large populations. This is especially important because of wide-scale distribution of perishable foods. Standard culture techniques may require days to weeks for positive identification of pathogens. Recent breakthroughs now allow us to identify such pathogens by detecting specific DNA or RNA base sequences with specific probes.

History. *Escherichia coli* O157:H7 is an important food-borne disease microorganism that was discovered in 1982. This pathogen is spread by the fecal-oral route, and an infectious dose is around 500 bacteria. The incubation period ranges from 1 to 8 days following ingestion, but most commonly the symptoms begin 3–4 days after ingestion. *Escherichia coli* O157:H7 has been found in meat products such as hamburger and salami, in dairy products, in unpasteurized fruit drinks, on fruits and vegetables, and in untreated well water contaminated with human or bovine feces. Petting zoos have also been the source of previous outbreaks. Sometimes news sources are filled with reports of million-pound lots of beef being recalled due to *E. coli* O157:H7 contamination. In 1996, more than 9000 school children were infected in Japan from contaminated beef lunches. The outbreak was centered in the western city of Sakai, where 216 children were hospitalized and 7 died. In 1998, apples picked up from manure-laden soil underneath apple trees were pressed into lethal apple juice. Hundreds of visitors to an agricultural fair in Kansas drank the apple juice and became ill. In 1993, over 700 cases of bloody diarrhea were reported in four western states from individuals who ate hamburgers from a fast-food restaurant chain. Approximately 25% were hospitalized and 4 children died. In 2006, the CDC was alerted to a multistate *E. coli* O157:H7 outbreak caused by bagged spinach. Ultimately, 199 cases nationwide were confirmed positive for *E. coli* O157:H7. States reporting the greatest number of confirmed cases were Wisconsin (49), Ohio (25), and Utah (19). Ninety-five percent of the patients reported spinach consumption prior to onset of the illness. Ninety-eight people required hospitalization and 3 died. Thirty cases of HUS occurred nationally among those infected. According to the U.S. Food and Drug Administration (FDA), 18 previous outbreaks have been caused by spinach in the last 10 years. Spinach or lettuce may become contaminated via multiple mechanisms from the farm to consumption (for example, by manure from an infected animal such as a cow; from contaminated irrigation water; from contamination during packaging and processing; or through individual contamination at restaurants or grocery stores).

Prevention. Prevention of food contamination by *E. coli* O157:H7 is essential from the time of production until consumption. Hygiene must be monitored carefully in larger-volume slaughterhouses, where contact of meat with fecal material can occur. Even fruits and vegetables should be handled with care because disease outbreaks have been caused by domestic and imported produce—for example, in the 2006 multistate *E. coli* O157:H7 outbreak caused by bagged spinach. Caution is also essential at the point of use. For example, avoidance of food contaminated by hands and utensils is critical. Utensils used with raw foods should not contact cooked food; proper cleaning of cutting boards and utensils minimizes contamination. Because of the extent and severity of the outbreak that occurred in the western United States in 1993, the Department of Agriculture recommended that hamburgers be cooked to an internal temperature of 155°F (68.3°C).

Closing note. In the most recent outbreak caused by spinach, the FDA traced the source of *E. coli* O157:H7 to a spinach processing company by combining the epidemiological data obtained from a nationwide spinach questionnaire with the laboratory testing performed on the implicated bags of spinach. The outbreak was halted by recalls of the spinach products from the implicated company, and FDA advice to consumers to avoid eating raw spinach until the source of the outbreak could be determined. Future plans include development of an industry-wide plan with steps to prevent contamination, thereby lowering the risk of another outbreak.

For background information *see* BACTERIA; BACTERIAL GROWTH; BACTERIOLOGY; COLIPHAGE; DIARRHEA; ELECTROPHORESIS; ESCHERICHIA; FOOD MANUFACTURING; FOOD POISONING; FOOD SCIENCE;

VIRULENCE in the McGraw-Hill Encyclopedia of Science & Technology. John P. Harley

Bibliography. M. Islam et al., Persistence of enterohemorrhagic *Escherichia coli* O157:H7 in soil and on leaf lettuce and parsley grown in fields treated with contaminated manure composts or irrigation water, *J. Food Prot.*, 67:1365–1370, 2004; J. P. Nataro and J. B. Kaper, Diarrheagenic *Escherichia coli, Clin. Microbiol. Rev.*, 11(1):142–201, 1998; V. L. Tesh, Foodborne disease, in *McGraw-Hill Yearbook of Science & Technology*, pp. 140–142, McGraw-Hill, New York, 1997.

Ethanol from wood

With greatly increased gasoline and diesel prices in the United States, worldwide crude oil prices at record levels, dependence on imports from insecure sources, and environmental concern over accumulation of greenhouse gases from burning fossil fuels, residents and legislators recognize the increasing need for alternative supplies of transportation fuels. Production of ethanol from wood is not yet economically viable, but new legislation supports research to solve problems and provides subsidies that can make manufacturing profitable.

Advantages of wood as a feedstock. In June 2007, the main source for alternatives to gasoline and diesel from petroleum was ethanol from corn grain. However, it is generally agreed that cellulosic ethanol (ethanol from lignocellulosic sources such as wood) has environmental and nonmonetary economic advantages over ethanol made from starch from corn. Cellulosic raw material could come from unused surplus materials such as sawdust and bark residues at manufacturing plants; from tops and limbs of trees left over after harvesting trees for lumber, plywood, and pulp and paper; or from crops such as hybrid poplar trees grown specifically for feedstocks for manufacturing ethanol. Corn ethanol on the other hand uses a raw material that could otherwise be used for food. Before June 2007, because of the high demand for corn to make ethanol, the price of corn doubled in a short time from $2.00 per bushel to $4.00 per bushel. This had a significant increase on food prices. As an example, steaks from grazing animals raised on corn became higher in price. Also, there is much controversy concerning the true benefit of corn in conserving fossil fuel use, using less petroleum specifically, and whether there is overall conservation of energy in comparing energy value of ethanol to energy consumption in making ethanol. It is generally agreed that making ethanol from corn shows a positive balance in end-product energy compared to fossil fuel use in production. However, this ratio would be better for ethanol from cellulosic raw materials.

There has been significant growth in fuel ethanol production since 1980. In 2006, U.S. production was about 6 billion gallons (22.7 billion liters) per year, and in 2007 the estimated production capacity was about 7 billion gallons (26.5 billion liters) per year. In order to boost energy production from alternatives to petroleum and to preserve energy, the U.S. Congress passed the Energy Policy Act of 2005. The Act gives high priority for continued growth of ethanol fuel. Subsidies and mandated targets are provided. For various reasons, the Act emphasizes derivation of ethanol from lignocellulosic sources to reinforce that made from corn grain. There was no ethanol made commercially from lignocellulose in June 2007, but plants were planned.

Making ethanol from corn reduces petroleum use usually at the expense of more natural gas and coal use and, as indicated previously, there is a slight positive energy balance in making ethanol from corn. Usually, there is also a slight saving in fossil fuel usage. This saving is greater when renewable fuel such as wood chips is used for process energy.

Cost differentials. Despite the advantages in making ethanol from cellulosics, there is a major problem in trying to do this economically. Because supplies of wood for making ethanol are dispersed over wide areas and stocks of corn are more concentrated, transportation costs for getting wood feedstock to the plant are higher. Higher concentrations of feedstock around plant sites facilitate construction of larger plants with economies of scale in lower plant costs per unit capacity. Capital costs of wood plants are also higher because of lower yields per ton of feedstock with currently available technology, and the fact that simpler processing of corn results in the need for less expensive equipment and lower costs for manufacturing enzymes and chemicals needed in processing plants in the case of corn. Current fermentation processes produce ethanol of 14–20% concentration with corn, but a concentration of only 4% with cellulosic starting materials. This means that expenditures of costly distillation energy are significantly higher with cellulosic ethanol.

Capital costs can be recovered; however, payoff would be over a significantly longer term for cellulosics. Ethanol yields from cellulosics today are based on yields estimated from technologies proven in use. These are acid and enzyme hydrolysis technologies and glucose and xylose sugar fermentation technologies. There are technologies under research that would gasify cellulosic materials and make ethanol through Fischer-Tropsch synthesis (a catalytic process to synthesize hydrocarbons and their oxygen derivatives by the controlled reaction of hydrogen and carbon monoxide). Fischer-Tropsch and other synthetic procedures take synthesis gas from wood and through the use of catalysts could get yields that are largely ethanol. Although this gasification technology is also being investigated for other fuels such as diesel, gasoline, and mixtures of ethanol and higher alcohols, ethanol yield in itself could be increased to over 100 gal (378.5 L) per dry ton and potentially much more. This would be a higher yield than ethanol from corn grain, which is about 98 gal (371 L) per ton.

Overcoming high costs. Because of the high product costs for making ethanol from wood, commercial plants in the past were more concerned with getting the product at any cost and providing employment than in making a profit. This was the case in the United States during World War I, in Germany during World War II, and in the centrally planned economies of Russia and Bulgaria after World War II. The Energy Policy Act of 2005 (EPACT) provides for supporting research on ethanol, particularly cellulosic ethanol, so that it may become more profitable. Approaches to problem solving for improving economy of production include improving harvesting of woody forest residues to reduce feedstock costs; reducing high capital costs for plants per unit of production through integration of process energy requirements and energy surpluses as through use of lignin for fuel in fermentation ethanol processes, that would otherwise only use cellulose and hemicelluloses effectively, and using outside sources of hydrogen to obtain higher yields in thermochemical processes; gaining higher yields per unit of feedstock through thermochemical processing; and improving enzyme productivity per unit cost in fermentation processing; reducing consumption of energy for distilling ethanol from water through thermochemical production; or developing membrane filter technology to separate water from ethanol.

In February 2007, to follow through on provisions in EPACT, the U.S. Department of Energy announced investment of up to $385 million for six biorefinery projects over the next four years. When fully operational, the biorefineries are expected to produce more than 130 million gallons (492 million liters) of cellulosic ethanol per year. This production will help further the goal of making cellulosic ethanol cost-competitive with gasoline by 2012 and, along with increased automobile fuel efficiency, reduce America's gasoline consumption by 20% in 10 years. Combined with the industry cost share, more than $41.2 billion will be invested in these six biorefineries.

One of the plants in LaBelle, Florida, with a grant of up to $33 million, will produce 13.9 million gallons (52.6 million liters) of ethanol and 6255 kilowatt-hours of electric power per year, as well as 8.8 tons of hydrogen and 50 tons of ammonia per day. For feedstock, the plant will use 770 tons per day of yard, wood, and vegetative wastes, and eventually energy cane (sugarcane of high fiber content).

In Irvine, California, a plant is to receive up to $40 million to use 700 tons per day of sorted green waste and wood to produce about 19 million gallons (72 million liters) of ethanol per year.

In Shelley, Idaho, a Canadian company that has been working on ethanol from biomass for more than 20 years has an allocation of up to $80 million to produce 18 million gallons (68 million liters) of ethanol per year. Wood will not be a feedstock initially, but could become one later.

Near Soperton, Georgia, a plant will use up to $76 million to produce about 40 million gallons (151 million liters) of ethanol per year and 9 million gallons (34 million liters) of methanol per year from wood chips and unmerchantable Georgia pine trees and forest residues. This company will not use a fermentation technology that limits conversion of only cellulose and hemicellulose in wood, but a thermochemical process that converts all of the constituents in wood including lignin and bark. In a two-step process, wood is first gasified and synthesis gas is then converted to products.

However, fermentation technology development also gained a significant boost when, in March 2007, the U.S. Department of Energy announced an additional follow-on grant award package to provide $23 million in federal funding for five projects focused on developing highly efficient fermentative organisms to convert biomass material to ethanol.

More money to further the ethanol from wood cause became available in June 2007 when the Department of Energy announced that it would allocate $125 million to each of three new research centers (in Oak Ridge, Tennessee; Madison, Wisconsin; and near Berkeley, California) for investigation into new ways of turning wood and plants such as switchgrass into fuel. Research projects at these new centers will bring together scientists from 18 universities, seven Department of Energy national laboratories, one nonprofit organization, and several private companies. The collaborations are aimed at improving cellulose to fuel processes and making these processes more cost-effective.

With investment in projects such as these, there is hope for research breakthroughs that could make ethanol from wood economically competitive with ethanol from corn grain. The U.S. Department of Energy has a goal for reducing the cost of cellulosic ethanol to $1.07 per gallon by 2012, which probably would be less than the cost of making it from corn grain. At long last, the possibilities for using excess wood to manufacture transportation fuel profitably appear promising.

For background information *see* ALCOHOL FUEL; ALTERNATIVE FUELS FOR VEHICLES; BIOMASS; CELLULOSE; CORN; ETHYL ALCOHOL; FISCHER-TROPSCH PROCESS; LIGNIN; RENEWABLE RESOURCES; WOOD ANATOMY; WOOD CHEMICALS in the McGraw-Hill Encyclopedia of Science & Technology. John I. Zerbe

Bibliography. A. E. Farrell et al., Ethanol can contribute to energy and environmental goals, *Science*, 311:506–508, 2006; R. D. Perlack et al., Biomass as feedstock for a bioenergy and bioproducts industry: The technical feasibility of a billion-ton annual supply, Oak Ridge National Laboratory, U.S. Department of Energy contract DOE/GO-102005-2135, April 2005; N. Schmitz and meó Consulting Team, Renewable raw materials, in *Bioethanol in Deutschland*, Series No. 21, Federal Ministry for Consumer Protection, Nutrition, and Agriculture (BMVEL), BMVEL no. 22009200, 2003; J. I. Zerbe, Thermal energy, electricity, and transportation fuels from wood, *Forest Prod. J.*, 56:4–12, 2006.

Ethylene (plant physiology)

The phytohormone (plant hormone) ethylene is an important signaling molecule that is involved in many plant processes including, but not limited to, abscission (the physiological process promoted by abscisic acid whereby plants shed a part), leaf and flower senescence, germination, sex determination, and fruit ripening. Ethylene also functions in both biotic and abiotic stress responses. Exposure to environmental stresses—such as flooding, wounding, herbivory (the consumption of plants without killing them), chilling, or pathogen attack—can enhance ethylene production. This ethylene then slows growth until the stress is removed. Interest in ethylene's importance as a plant hormone has resulted in thousands of peer-reviewed publications in the last 100 years and has laid the foundation for a real understanding of ethylene's involvement in plant growth and development.

Description. Ethylene is a small, gaseous, two-carbon molecule ($H_2C{=}CH_2$) with the ability to diffuse through hydrophilic and hydrophobic environments. This property allows it to pass into any compartment in the plant cell. Responses to ethylene can occur at very low levels of the hormone, with 1 ppm (parts per million) having a significant effect on most ethylene-regulated processes.

Early application experiments suggested that the amino acid methionine is a precursor of ethylene, but it was subsequently determined to be 1-aminocyclopropane-1-carboxylic acid (ACC), which is derived from methionine, is the immediate precursor of endogenous ethylene. ACC is produced from S-adenosyl methionine (SAM) by a pyridoxal phosphate (vitamin B_6)-requiring enzyme termed ACC synthase (ACS). The conversion of ACC to ethylene is subsequently performed by the oxygen-requiring enzyme, ACC oxidase (ACO).

Most tissues synthesize low levels of ethylene. Synthesis can be stimulated by a number of means, including wounding, submergence, chilling, and pathogen attack. Synthesis of ACC is considered to be the rate-limiting step in ethylene production. Thus, increased ethylene production requires modulation of ACS expression and/or activity.

Involvement with fruit ripening. The ability to alter plant development has been known for centuries, with farmers from many cultures using smoke and wounding to induce flowering and ripening. While ethylene is often characterized as the ripening hormone, not all fruit require ethylene to complete the ripening process. Species are often characterized by the presence or absence of a large increase in ethylene production concomitant with increased respiration at the onset of ripening. Species whose fruit exhibit these increases are termed climacteric; those that do not are nonclimacteric. Climacteric species include apple, avocado, banana, peach, and tomato, while nonclimacteric species include strawberry, grape, cherry, and citrus varieties. The increase in ethylene production associated with climacteric ripening is essential for complete ripening. Blockage of either ethylene biosynthesis or perception results in the inability of the fruit to complete its ripening program.

Ethylene's involvement in ripening has been implicated in fruit softening, volatile production, and accumulation of lycopene (a red, crystalline hydrocarbon that is the coloring matter of certain fruits, such as tomatoes). Ethylene biosynthesis during fruit development generally goes through three distinct stages. There is a slight burst of synthesis after successful pollination, which then falls to low levels until onset of climacteric ethylene production at the onset of ripening. Ethylene production during immature fruit development has been termed system I and is characterized as low-level production, which can be inhibited by treatment with exogenous ethylene. Ethylene biosynthesis in mature fruit, referred to as system II, is autocatalytic, meaning it can induce its own synthesis. In tomato, there is a large increase in expression of the ethylene receptor family at the onset of ripening (see receptor description below). In addition, autocatalytic ethylene production from one ripening fruit can induce the ripening of mature fruit within close proximity. The induction of ethylene synthesis at the onset of ripening is believed to be due to developmental induction of an ethylene-inducible ACS.

The advent of controlled atmosphere systems has allowed for the precise control of fruit ripening in a number of species, including apple and pear. Low O_2 levels and low temperature, as well as higher CO_2 levels, are used to delay ripening in these species. These conditions are used because they are known to suppress ethylene production, thus exploiting the necessity of ethylene for the ripening of these species.

Role in flower senescence. While ethylene's involvement in fruit ripening often garners the most attention, its involvement in flower senescence has great implications, both biologically and commercially. Flowers are often characterized using the same terminology as that used with fruits: climacteric and nonclimacteric species. Climacteric species experience an increase in ethylene production after pollination that often leads to flower senescence. In nonclimacteric species, there is no increase in ethylene production, but ethylene exposure at this stage can cause abscission or affect flower opening, ovary development, or pedicel elongation. While these responses have specific biological purposes, they are often not desirable for commercial floriculture. In an attempt to alleviate some of these issues, floriculture research has focused on two types of control, inhibiting ethylene synthesis or action. 1-Aminoethoxyvinylglycine (AVG) and aminooxyacetic acid (AOA) inhibit ACS activity, thus preventing ethylene synthesis. The chemicals, silver thiosulfate (STS) and 1-methylcyclopropene, prevent ethylene binding at the receptor, thus preventing the plant from recognizing that ethylene is present (**Fig. 1**). Some flower auctions require pretreatment with an ethylene-inhibiting compound, owing to the importance of regulating ethylene-related responses in flowers.

Involvement during seedling emergence. In addition to ethylene's role in fruit development, it also plays an important part in seedling emergence. During germination, seedlings must be able to force their way through any soil between them and a light source. When a seedling encounters a barrier in the soil, the restriction can induce ethylene production. Dark-grown seedlings, such as those found underground, are often tall and spindly in the presence of air alone.

Fig. 1. Different varieties of carnation treated overnight with 0.2 mM STS solution. Photograph was taken after 10 days of vase life. Note that cultivar Chinera (darker shaded), with reduced sensitivity to ethylene, benefits less from the STS pretreatment. (*Adapted from Serek et al., 2006*)

Upon exposure to ethylene, the seedling growth habit changes and exhibits growth that is referred to as the "triple response." This response manifests as a shortening of both the hypocotyl and root, radial thickening of the hypocotyls, and an exaggeration of the apical hook. These changes allow the seedling to push through any barriers without damaging the meristem. While this mechanism has evolutionary importance, the ability to exploit this response has revolutionized the ethylene biology field by allowing researchers to screen for mutants in ethylene biosynthesis and signaling.

Signal transduction. Besides control at the level of synthesis, ethylene response is also finely controlled at the level of perception. Much of the initial ethylene perception and signal transduction research was done in *Arabidopsis thaliana*, and the *Arabidopsis* system has been exploited to identify orthologous genes in agronomically important crops. The *Arabidopsis* ethylene receptor *ETR1* was the first phytohormone receptor cloned in plants, and was isolated from a mutagenized population that was screened for plants deficient in the triple response. Ethylene-insensitive mutants grow tall and spindly even in the presence of ethylene, whereas constitutive ethylene response mutants will show a triple response in the absence of ethylene (**Fig. 2**). An ethylene-insensitive mutant, *etr1-1*, was isolated in one of these screens. It was subsequently cloned and shown to encode an ethylene receptor with homology to bacterial two-component sensors (many bacterial signal transduction pathways involve a two-component design). In subsequent work, a total of five receptors were cloned from *Arabidopsis* (*ETR1*, *ETR2*, *ERS1*, *ERS2*, and *EIN4*).

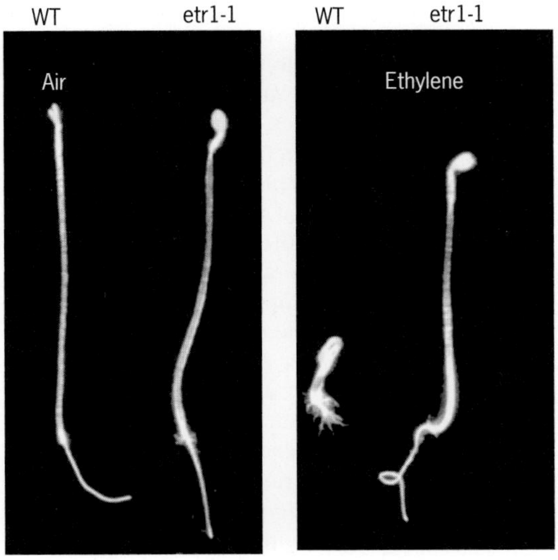

Fig. 2. Response of dark-grown wild-type (WT) and *etr1-1 Arabidopsis thaliana* seedlings to ethylene exposure. (*Adapted from Schaller and Kieber, 2002*)

Based on all of the genetic data available, the receptors appear to function as negative regulators of the ethylene response. In the absence of ethylene, the receptors actively block ethylene response; however, upon ethylene binding, this suppression is removed and the response is able to occur. A subset of these receptors, *ERS1*, *ETR2*, and *EIN4*, are inducible upon exposure to ethylene. While this may seem counterintuitive, since an increase in receptor levels would actually lead to a reduction in ethylene sensitivity, this appears to be a common theme in the regulation of plant hormone responses aimed at maintaining homeostasis.

The ethylene signal transduction pathway in *Arabidopsis* is believed to be relatively linear, but it is not known if all of the elements have been identified (**Fig. 3**). Epistatic (gene network) analysis has allowed researchers to putatively order the components starting with the receptors. The next component is a serine/threonine protein kinase, CTR1 (constitutive triple response 1), which has been shown to physically interact with the receptors. CTR1 has significant homology to MAPKKKs (mitogen-activated protein kinase kinase kinases, which are enzymes involved in signal transduction); further, although no MAPKKs (mitogen-activated protein kinase kinases) or MAPKs (mitogen-activated protein kinases) have been found to be involved in ethylene signal transduction, their involvement in this pathway cannot be ruled out. EIN2, a protein showing homology to Nramp metal transporters, is the next member of the pathway. The role and activity of this protein in the pathway is unknown, but it is absolutely necessary since knockouts [organisms with a gene (or genes) disabled via genetic engineering] show complete ethylene insensitivity in every assay tested. The end of the ethylene signal transduction pathway is composed of the transcription factors EIN3, EIL1-3, and ERF1. EIN3 loss-of-function (LOF) mutants show partial ethylene insensitivity, which is probably due to redundancy within the gene family containing at least three members. In the absence of ethylene, the EIN3 protein is targeted for degradation in the 26S proteasome by a pair of F-box proteins, EIN3-binding factors EBF1 and EBF2. Upon ethylene binding, this repression is released and EIN3 binds to the promoter of ERF1, activating its transcription. ERF1 is a member of the large AP2 gene family of transcription factors and is involved in regulating the transcription of ethylene-responsive genes. ERF1 overexpressers show a slight constitutive

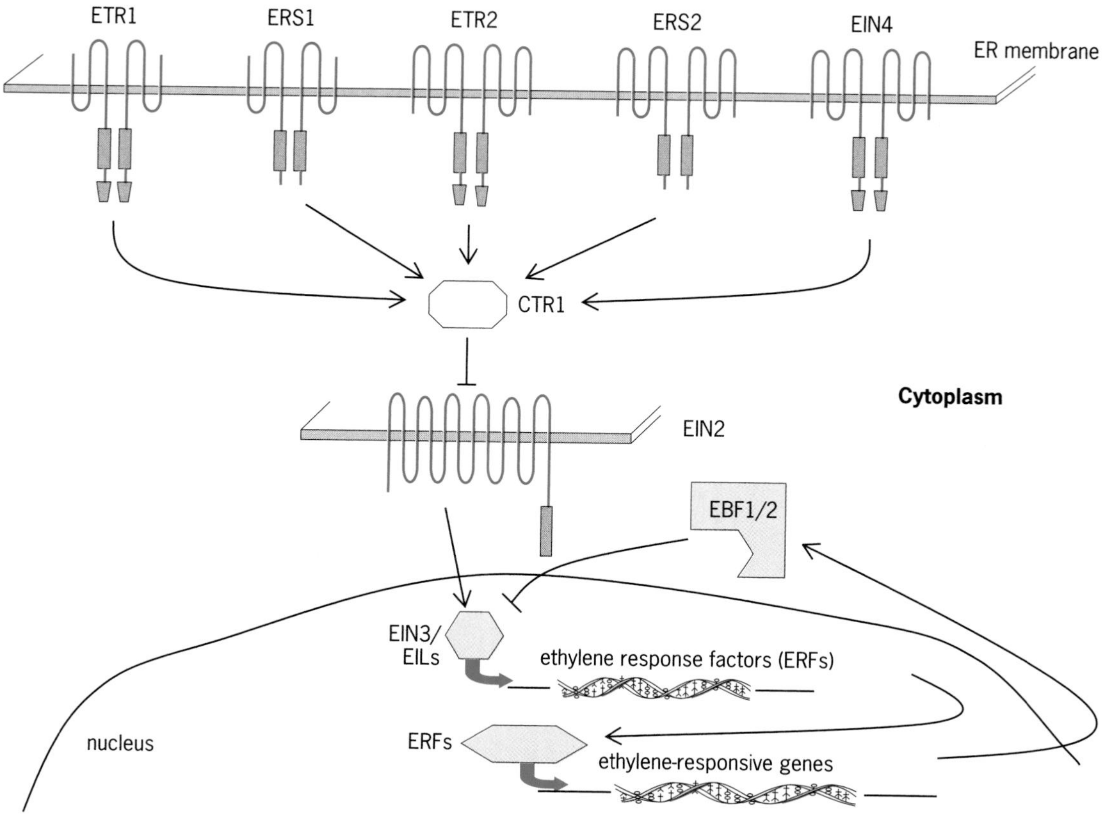

Fig. 3. The ethylene signal transduction pathway in *Arabidopsis thaliana*. The ethylene receptor proteins are embedded in the endoplasmic reticulum (ER) membrane. CTR1 protein negatively regulates EIN2 protein activity. EIN2 protein is embedded in a membrane of unknown location, but putatively signals to EIN3/EIL transcriptional regulators in the nucleus, inducing expression of ERFs. This ultimately increases expression of ethylene-responsive genes. EIN3/EIL protein levels are negatively regulated by EBF1/2 proteins.

ethylene response, suggesting that there are other important players in the transcriptional control of ethylene-responsive genes—potentially other members of the AP2 gene family.

The initial identification of each member of the signaling pathway in *Arabidopsis* has allowed for the isolation of homologues in other species. Tomato represents the best-characterized system outside of *Arabidopsis*. In addition to its genetic plasticity, tomato also offers the ability to study ethylene's involvement in fleshy fruit development. A homologue of *ETR1*, the *Never-ripe* (*NR*) gene, was isolated from a tomato mutant that showed a greatly reduced response to ethylene. The *Nr* mutation causes semidominant ethylene insensitivity, which prevents floral wilting and abscission, alters leaf senescence, and prevents fruit ripening. Subsequent work has identified six receptor family members in tomato (*LeETR1-6*, with *LeETR3* corresponding to the *NR* gene). Two of these family members, *LeETR4* and *LeETR6*, have been shown to be intricately involved in the timing of ripening, with reduced expression lines showing significantly reduced time from anthesis (flowering) to the initiation of ripening. Subsequent work has shown that ethylene binding results in a rapid turnover of these receptor proteins, and that treatment of immature fruit with ethylene reduces the time to ripening by decreasing the levels of these negative regulators of the ethylene signaling pathway. This type of analysis would be impossible in *Arabidopsis* and thus points to the importance of elaborating the differences between species in order to understand the universal themes.

For background information *see* ABSCISSION; AGRICULTURAL SCIENCE (PLANT); ETHYLENE; FRUIT; PLANT GROWTH; PLANT HORMONES; PLANT MORPHOGENESIS; PLANT PHYSIOLOGY; TOMATO in the McGraw-Hill Encyclopedia of Science & Technology.

Brian M. Kevany; Valeriano Dal Cin; Harry J. Klee

Bibliography. F. B. Abeles, P. W. Morgan, and M. E. Saltveit, *Ethylene in Plant Biology*, 2d ed., Academic Press, San Diego, 1992; A. B. Bleecker and H. Kende, Ethylene: A gaseous signal molecule in plants, *Annu. Rev. Cell Dev. Biol.*, 16:1-18, 2000; J. J. Giovannoni, Molecular biology and fruit maturation and ripening, *Annu. Rev. Plant Physiol.*, 52:725-751, 2001; H. J. Klee, Ethylene signal transduction: Moving beyond *Arabidopsis, Plant Physiol.*, 135:660-667, 2004; G. E. Schaller and J. J. Kieber, *The Arabidopsis Book*, American Society of Plant Physiologists, Rockville, Maryland, 2002; M. Serek et al., Controlling ethylene responses in flowers at the receptor level, *Biotech. Advances*, 24:368-381, 2006; S. F. Yang, *Plant Senescence: Its Biochemistry and Physiology*, pp. 156-165, American Society of Plant Physiologists, Rockville, Maryland, 1987.

Financial engineering

Engineering is about making a product (such as a laptop computer), designing a process (such as turning data into digital packets and transferring them on the Internet), or providing a solution (such as how to manage and optimize a firm's supply chain). Financial engineering (FE) is also concerned with products (such as options, futures, and other derivatives), processes (such as optimizing a portfolio over time, managing a pension or a hedge fund), and solutions (such as how to hedge against risks in exchange rates and in defaults). Engineering is built upon scientific laws and principles and makes extensive use of mathematical and computational tools. Financial engineering is no different, only with a slight variation; its guiding principles are drawn mainly from mathematical sciences and economics as opposed to physical sciences. Engineering design often starts with empirical data, continues with analysis and synthesis, and ends with some systemwide considerations regarding implementation and execution. Likewise, a sound FE model is usually well calibrated by data. Financial data are among the most widely available and most extensively studied data forms in a modern society. In implementation and execution as well as in design, FE uses many engineering ideas and practices. No engineering design is complete without carefully weighing the tradeoff between cost and benefit. In the same spirit, any FE product, process, or solution is ultimately a managed (or hedged) balance between risk and return.

In recent years, some leading engineering schools in the United States have started offering degree programs in FE. For instance, Columbia University's Fu Foundation School of Engineering and Applied Science offers a Master of Science degree in FE. The curriculum includes courses in stochastic processes, optimization, Monte Carlo simulation, data analysis, and computation. Students also study portfolio theory, derivatives pricing, term structure, financial risk analysis, and other advanced topics and tools. In addition to courses within the engineering school, students can also take electives from various schools within the university, in particular, Columbia Business School. Graduates from the program have taken up positions in the securities, banking, and financial management and consulting industries, as well as in corporate treasury and finance departments of manufacturing and service firms.

Derivative pricing. An important subject in FE is the pricing of derivative products in the financial market, such as futures, options, swaps, and CDOs (collateralized debt obligations). These are called derivatives as they are built upon other primary financial assets such as stocks and bonds.

Consider option pricing as an example. A call (or put) option is a right (but not obligation) to buy (or sell) a financial asset (the underlying), for example, a stock at a future date (maturity) at a specified price (strike or exercise price). More precisely, this is a European option. The European option can be exercised at maturity only, while an American option allows the holder of the option to exercise at any time up to the maturity.

To price such an option, let's start with a simple case. At time zero (present), the price of the underlying stock is \$80. Suppose at maturity, the stock price will either go up to \$120 or go down to \$60, with equal probability (0.5). We want to price a call option on this stock with a strike price of \$90.

It is tempting to think that since the payoff of the option is either \$30 (=120-90 in the up scenario) or zero (in the down scenario). So with equal probability, the option should be priced at \$15 [=30(0.5) + 0(0.5)], which is the option's expected (that is, average) payoff.

A more sophisticated approach is to reason as follows. Suppose we put together at time zero a portfolio that consists of two assets: a long position of (that is, the portfolio owns) half a share of the stock and a short position of \$30 in cash (that is, borrowed money). Then, this portfolio will match exactly the payoff of the option in both scenarios at maturity: in the up scenario, the portfolio's value is $\frac{1}{2}(120) - 30 = \$30$, whereas in the down scenario, its value is $\frac{1}{2}(60) - 30 = \$0$. Since the value of the portfolio at time zero is \$10 (half a share of the stock, \$40, minus \$30 of borrowed cash), and since it can deterministically replicate the payoff of the option regardless of the random up and down scenarios, we claim the option should also be priced at \$10.

To support this claim, suppose the option is priced higher than \$10, say \$15 as given earlier. Then at time zero, we can short the option (that is, borrow and sell it) and long the portfolio, pocketing the difference of \$5 (=15 − 10) as profit. At maturity, we can use the portfolio to cancel out the short position on the option and still keep the \$5 as profit. Next, suppose the option is priced lower than \$10, say \$8. Then we can short the portfolio and long the option at time zero, and make a profit of \$2 (=10 − 8). At maturity, the long position on the option and the short position on the portfolio again cancel out each other, leaving our \$2 profit intact. In both cases, we can make a sure profit without undertaking any risk (that is, the uncertainty of a loss). To the extent that this violates a basic premise of an efficient financial market, the principle of no arbitrage (or "no free lunch"), we have completed a proof by contradiction.

It is worth noting that the probabilities (0.5, 0.5) that characterize the randomness of the stock price play no role in the above reasoning. Indeed, if we must take average (expectation) on the option payoffs to come up with its price, then the probabilities should be modified to $(\frac{1}{3}, \frac{2}{3})$, since $10 = 30(\frac{1}{3}) + 0(\frac{2}{3})$.

Notwithstanding its simplistic features, the above example captures some of the key ingredients in option pricing, such as a replicating portfolio (or hedging strategy) that can track the value of the option (via dynamically adjusting the stock and cash positions in the portfolio), the no-arbitrage argument, and the modification of the probability law (or measure) that governs the original stock-price dynamics to what's called a risk-neutral (or martingale) measure.

The most basic and best-known model for option pricing is the Black-Scholes-Merton (BSM) model. Its starting point is to assume that the stock price (at time t), denoted S_t, follows the so-called geometric Brownian motion. This price dynamics can be expressed in the form of a differential equation (1),

$$\frac{dS_t}{S_t} = \mu dt + \sigma dB_t \tag{1}$$

where B_t denotes the Brownian motion, and μ and σ are two parameters representing, respectively, the growth rate and the volatility of the stock price. Brownian motion is named after a botanist, Robert Brown, who was studying the movement of pollen particles floating on the surface of water. Since then, Brownian motion has been used as a basic mathematical model to study many physical processes by scientists including Albert Einstein and Norbert Wiener. (Brownian motion is equally well known as the Wiener process.) Differential Eq. (1) is a form of the so-called diffusion equation, widely used in engineering to capture the dynamics of diffusive objects and materials. The equation can be explicitly solved to yield the solution $S_t = S_0 e^{(\mu - \sigma^2/2)t + \sigma B_t}$.

The BSM model then prices the call option with maturity T and strike price K as in Eq. (2).

$$c = e^{-sT}\mathbf{E}_r[\max\{S_T - K, 0\}] \tag{2}$$

Here $\max\{S_T - K, 0\}$ is the payoff of the call at maturity. If the stock price is above the strike price, the difference is the payoff; otherwise, the payoff is zero and the option will not be exercised. Hence, the price of the option (at time zero) is the expected payoff discounted to its present value, with r being the risk-free interest rate. It is important to note that the expectation here, $\mathbf{E}_r$, is in terms of a modified probability measure, with the stock's intrinsic growth rate μ replaced by r, in the same spirit as (0.5, 0.5) must be modified to ($^1/_3$, $^2/_3$) in the earlier example. This expectation can be worked out to yield the following pricing formula (3), where Φ denotes the distribution function of the standard normal variate, whose values are widely available in table form (or spreadsheet), and

$$c = S_0\Phi(d_1) - Ke^{-rT}\Phi(d_2) \tag{3}$$

$$d_1 := d_2 + \sigma\sqrt{T} \qquad d_2 := \frac{\ln(S_0/K) + (r - \sigma^2/2)T}{\sigma\sqrt{T}}$$

Portfolio optimization. Another important area in FE is asset management. It is primarily concerned with portfolio choice, namely, to distribute a given amount of wealth among several securities (or assets) to achieve a certain financial goal. The return of a security (for example, a stock) typically involves uncertainty, as discussed above. This calls for a proper balance between return (as characterized by its mean) and risk (or volatility, as represented by the variance of the return). In addition, two securities in the same portfolio may be related to each other in that their prices may go up and down in the same direction or in opposite directions. For example, a bank stock and a property stock may exhibit similar price movements due to the mortgage products that they both relate to. On the other hand, the stock of an upstream oil firm that specializes in crude oil production and the stock of a downstream firm that processes crude oil into petrochemical products may move in opposite directions, as the price increase in crude will benefit the former while hurting the latter. This kind of correlations can be measured by the covariance of the returns of the two security prices. This leads to the practice of diversification, as the loss of one security may be compensated by gains of other securities in the same portfolio.

These concepts are formalized by H. Markowitz in his celebrated mean-variance model. Suppose there are n assets in a portfolio, indexed by $i = 1, \ldots, n$. Each dollar invested in asset i becomes R_i dollars at the end of the period (our investment horizon), where R_i is a random variable. Suppose at time zero we have a total of \$1 million to invest in the portfolio, and we want to decide the amount to invest in each asset i, denoted w_i. Once the w_i values are determined, the value of the portfolio at the end of the period is $\Sigma_{i=1}^{n} w_i R_i$. We want to minimize the variance of this value (the terminal wealth of our portfolio) and at the same time ensure that its mean reaches a certain target level.

A standard derivation leads to the following, the mean of the terminal wealth, $\Sigma_{i=1}^{n} w_i r_i$, where $r_i := \mathbf{E}(R_i)$ is the mean return of asset i; and the variance of the terminal wealth $\sigma^2 = \Sigma_{i,j=1}^{n} w_i w_j \sigma_{ij}$, where σ_{ij} denotes the covariance between two assets i and j. Our objective is to minimize the variance, while requiring the mean to achieve a prespecified value, denoted as r [Eq. (4)].

$$\begin{aligned} &\min \sigma^2 := \sum_{i,j=1}^{n} \sigma_{ij} w_i w_j \\ &\text{s.t.} \quad \sum_{j=1}^{n} r_i w_i = r, \quad \sum_{j=1}^{n} w_i = 1 \end{aligned} \tag{4}$$

Therefore, the mean-variance model aims at striking the right balance between risk and return: for any desired rate of return (r), we want to find a minimal level of risk represented by the variance of the portfolio (σ^2) via asset allocation (that is, splitting \$1 million among n assets, with w_i invested in asset i). The model in Eq. (4) is known as a quadratic program, one of the most readily solvable optimization problems. As the value of r (in the constraint) changes, so does σ, the objective value. This way, (σ, r) traces out a curve, the so-called efficient frontier. A remarkable implication of the model is that any efficient portfolio can be constructed from asset allocation between one risk-free asset (for example, a money-market fund or a bond) and a set of risky securities (for example, a mutual fund in stocks), with the latter following a certain mixture among the securities, via solving the optimization problem in Eq. (4). Furthermore, this mixture is independent of the desired risk-return profile, which will only affect the allocation between the risk-free asset and the set of risky securities. This is known as the mutual fund theorem, or separation theorem, since any efficient portfolio can be separated as a combination of a mutual fund (a set of risky securities with a certain mix that achieves efficiency) and a risk-free asset. This has significant consequences in a macroeconomy, since the homogeneous demand for stocks results in market equilibrium and in turn fair pricing of the securities, the so-called capital asset pricing model (CAPM).

The mean-variance model involves a single investment period; that is, the investor makes a portfolio choice at the beginning and then liquidates the portfolio at the end of the period, without any change of asset allocation in between. Thus, it is a static model. Yet, in practice portfolios are usually updated from time to time, sometimes rather frequently in response to the changing market conditions. This calls for dynamic models, including those with multiple periods and continuous-time models. Interestingly, extensions to the dynamic setting (continuous-time in particular) in the asset allocation literature have taken a considerable departure from the mean-variance model. Instead of the mean-variance tradeoff, the expected utility maximization has become the new investment criterion, with a general utility function representing the investor's preference. In particular, the utility theory assumes that investors are rational, hence they are risk averse, leading to a concave utility function. The risk preference, or degree of risk averse, is captured (albeit implicitly) by a choice of the utility functions (such as a power, a logarithmic or an exponential function). In the continuous-time setting, the model is a stochastic control problem, the centerpiece of which is a partial differential equation called the Hamilton-Jacobi-Bellman (HJB) equation, which characterizes the optimal solution. Robert Merton showed via the HJB equation that when the investment opportunity set—the set of return and volatility rates of the available securities—is deterministic (though possibly time varying), an optimal investment strategy is to continuously rebalance the portfolio to keep a prespecified mix among risky securities all the time, whereas this mix is completely determined by the opportunity set and not by the individual utility function. In other words, the mutual fund theorem in the single-period setting still holds. And this forms the foundation for intertemporal CAPM.

The stochastic control problem in the continuous-time setting usually requires a sophisticated numerical method to solve. D. D. Yao, X. Y. Zhou, and S. Zhang have developed an approach using semidefinite programming (SDP) and applied it to a specific portfolio optimization problem of tracking a market index or a prespecified growth rate using a relatively small number of securities. This is a useful tool that a fund manager would want to have. For example, the Janus Twenty Fund (JAVLX) is a mutual fund that consists of a select set of 20–30 large cap stocks. The performance of the fund is usually benchmarked against the S&P 500 index, however; the fund manager may also have some target growth rate in mind.

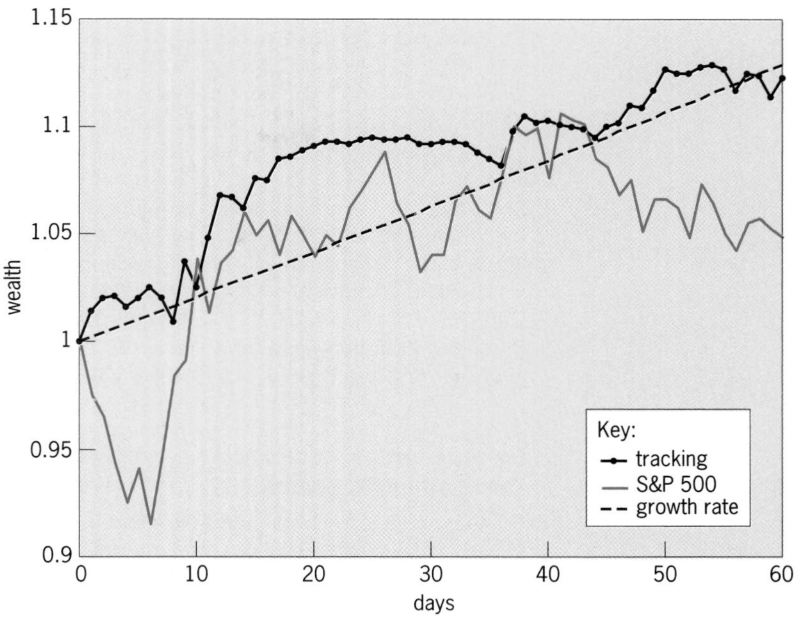

Tracking a 50% (annualized) growth rate using five randomly selected S&P 500 stocks (JPM, EK, KG, HOT, JLB) from October 1, 2002 for 60 days. (Target rate = 50%; risk-free rate = 4%; portfolio updated daily.)

As an illustrative example, consider a group of five stocks randomly generated from the constituents of the S&P 500 index: J.P. Morgan Chase & Co. (JPM), Eastman Kodak (EK), King Pharmaceuticals (KG), Starwood Hotels & Resorts (HOT), and Jabil Circuit (JBL). The tracking period starts from October 1, 2002 and lasts for 60 trading days. Suppose we want this small portfolio to yield an annualized 50% growth rate. The tracking performance of the stochastic control/SDP approach applied to this portfolio, using real market data, is shown in the **illustration**, along with the performance of the S&P 500 index. Evidently, the portfolio achieves the desired growth rate (at the end of the tracking period) and significantly outperforms the S&P 500 index. What the tracking technique does is to do optimal allocation (that is, rebalancing) among the five stocks from time to time (say, every day or every week), including shorting the nonperformers. The idea is very similar to guiding a rocket in space, overcoming the random disturbances from the atmosphere, so that it will follow a desired trajectory to reach a designated target. Here, the portfolio is the rocket, and the stochasticity comes from the financial market, whereas the control is executed via asset allocation.

For background information *see* BROWNIAN MOVEMENT; INDUSTRIAL ENGINEERING; MODEL THEORY; OPTIMIZATION; STOCHASTIC CONTROL THEORY in the McGraw-Hill Encyclopedia of Science & Technology.

David D. Yao; Xun Yu Zhou

Bibliography. F. Black and M. Scholes, The pricing of options and corporate liabilities, *J. Polit. Econ.*, 81:637–654, 1973; J. C. Hull, *Options, Futures, and*

Other Derivatives, 6th ed., Prentice Hall, 2005; H. Markowitz, Portfolio selection, *J. Finance*, 7:77–91, 1952; R. Merton, Lifetime portfolio selection under uncertainty: The continuous time case, *Rev. Econ. Stati.*, 51:247–257, 1969; R. Merton, The theory of rational option pricing, *Bell J. Econ. Manage. Sci.*, 4:141–183, 1973; W. F. Sharpe, Capital asset prices: A theory of market equilibrium under conditions of risks, *J. Finance*, 19:425–442, 1964; D. D. Yao, S. Zhang, and X. Zhou, Tracking a financial benchmark with a few assets, *Oper. Res.*, 54:232–246, 2006; X. Zhou and D. Li, Continuous-time mean-variance portfolio selection: A stochastic LQ framework, *Appl. Math. Optim.*, 42:19–33, 2000.

Focused sonic booms

A sonic boom is the loud impulsive sound produced by any object moving faster than the local sound speed. Human-made sonic booms have existed for a long time, due to cracking whips or artillery shells, but they were not recognized as such before World War I, when booms interfered with Ernest Esclangon's attempts to localize German guns by acoustical goniometry (direction finding). Then sonic booms quickly returned to obscurity until the birth of the "supersonic age" in 1947 when Charles Yeager broke the sound barrier on the aircraft Bell X-1. This event was followed by the design of supersonic fighters and the launch in the 1960s of civil supersonic projects (the B2707 in the United States, the Soviet TU144, and the British-French Concorde), at a time of cold war competition when the future of air travel was confidently forecast as supersonic. But high costs, increasing environmental concerns, surging oil prices, and the political turmoil of the 1970s contradicted optimistic anticipations. The Concorde alone achieved the dream of supersonic travel, but only for wealthy customers flying transatlantic. Indeed, the Federal Aviation Administration banned civil supersonic flight over U.S. territory in 1973 as a protection against undesirable booms (at that time, the Concorde's). This ban was also adopted by other countries.

Recent studies confirm that low boom levels remain unachievable for commercial jets. However, in 2003 the U.S. "QSP" (Quiet Supersonic Platform) project provided the first in-flight demonstration of the mitigation (by about one third) of the boom produced by an existing aircraft (the F5-E). This reduction was achieved by a modification of the front fuselage according to shaping rules. This event was the starting point for several aircraft manufacturers to launch low-boom design programs for supersonic business jets, for which boom abatement appears easier than for Concorde-sized carriers.

Boom formation by cruising aircraft. A sonic boom is nothing but the long-range evolution of the aerodynamic flow around the aircraft that determines its performance. Pressure disturbances propagating at the sound speed lag behind the supersonic aircraft. They are thus confined within the Mach cone, whose apex angle varies inversely with the Mach number (**Fig. 1**). The Mach cone determines the initial wavefront that is then propagated through the atmosphere down to the ground. Propagation satisfies the laws of refraction: The sound path, or acoustical ray, minimizes the travel time. Because temperature, and hence sound speed, varies with altitude, sound rays are curved, but atmospheric winds also strongly deviate rays. A real aircraft produces several Mach cones, each associated with the aircraft's main geometrical elements such as the fuselage nose and tail, the wing leading and trailing edges, the engine air intakes, and so forth. These cones correspond to shock waves of different amplitudes, propagating at speeds slightly different from the sound speed. These small differences make the different shocks progressively merge with one another, so that there usually remains at the ground level only a front shock and a rear one. This leads to the typical "N" wave shape of the booms recorded from many supersonic objects as different as the Concorde, the Soviet SR-71 (the Black Bird), or the Space Shuttle.

Boom formation by accelerating aircraft. The above process is used to predict boom levels at cruising speeds. However, before reaching cruise, an aircraft accelerates because it also has to take off. During unsteady acceleration, the above picture becomes more complex (**Fig. 2**): The Mach cone closes up, and acoustical rays launched perpendicular to the Mach cone are no longer parallel, as they would be in cruise flight. They converge onto a surface called a caustic, which forms an envelope of rays and separates them from a "silent" zone. Because of the conservation of sound intensity, the pressure amplitude is related to the distance separating two adjacent rays. At the caustics, rays converge, the separation distance vanishes, and the pressure amplitude increases infinitely. Caustics are thus areas of sound focusing.

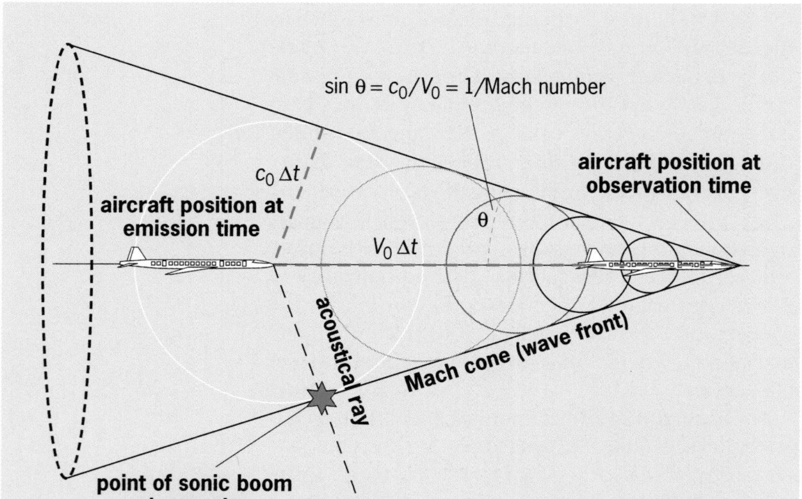

Fig. 1. The Mach cone. Here c_0 is the sound speed, V_0 is the aircraft speed, and Δt is the elapsed time from the emission time to the observation time.

Caustics are well known in optics, where they are easily observable as bright local surfaces. Depending on their topology, they are classified according to the theory of catastrophes, initiated by René Thom. An acceleration boom is the simplest "fold" caustic. That denomination comes from the local cusped shape of the wavefront near the caustic, displaying two sheets orthogonal to the two families of rays just before or after they tangent the caustic. This shape is universal, independent of the physical nature of the wave and the way focusing is obtained. In optics, a historically important fold caustic is the rainbow, caused by the refraction of light within waterdroplets. Refracted optical rays are deviated with an angle that has a maximal value, with this maximum corresponding to an envelope of refracted rays. In 1838, studying the rainbow, Sir George Airy discovered the diffraction behavior that (1) prevents light from being unbounded in intensity on the caustics and (2) matches continuously the shadow zone to the illuminated zone with optical rays.

Distinctive features of focused booms. Compared to the rainbow, the focused sonic boom has two distinctive features. First, atmospheric sound propagation is not dispersive and all sound frequencies focus at the same geometrical location. On the contrary, the dispersion of the optical refractive index shifts the caustic position with color, thus making the rainbow a colorful spectacle instead of a uniformly bright white circle. The second difference is that a sonic boom is a shock wave. Applying Airy theory to it would result in a singularity because the high-frequency part of the wave spectrum contained in the shocks would be overamplified. Beyond Airy diffraction, an additional phenomenon is required: the nonlinear dependence of sound speed on wave amplitude, which also explains the "N" shape of usual booms. Taking this phenomenon into account, Jean-Pierre Guiraud established in 1965 that an equation called the nonlinear Tricomi equation governs the focused boom. Though apparently simple, it nevertheless took more than 35 years to solve. In particular, it is a mixed hyperbolic-elliptic equation, reflecting the change across the caustic in the nature of wave propagation. This property makes it similar to the equations of transonic aerodynamics, for which the first numerical algorithms were developed only in the 1970s.

Solution of the governing equation. The solution was finally found in 2000–2003 by combining techniques inspired from both aeronautics and acoustics. The numerical solution provides the pressure field as a function of observation time and distance from the caustic. Folded wavefronts associated with the front and rear shocks of an incoming N wave are clearly visible (**Fig. 3**). Maximum amplification (here 383 pascals or 3.83 mbar) is observed near the geometrical caustic but not directly on it. Values are typical for the Concorde focused boom. A key feature is the transformation from an "N" to a "U" wave after the boom has gone through the caustics, with sharp amplification at the focus by a ratio of about 3 (Fig. 3).

Validation. Experimental observations were then used to validate the model. As quantitative comparisons with test flight recordings are extremely difficult, a laboratory-scale experiment was indispensable. The best candidate for optimal scaling from real-size flights turned out to be shock waves in water, which obey the same physics as in air. Working with typical conditions for current medical ultrasonic therapy [1-MHz shock waves of 5-MPa (50-bar) amplitude], an almost ideal scaling of 1:100,000 can be reached, resulting in a "virtual wind tunnel," with 1 m (3 ft) of water replacing 20 km (12 mi) of atmosphere. The aircraft was mimicked by an array of piezoelectric transducers, and pilot acceleration by a network of high-performance electronic channels able to reproduce the caustic geometry using sophisticated field-control techniques. An essential point was the low absorption of sound in water compared

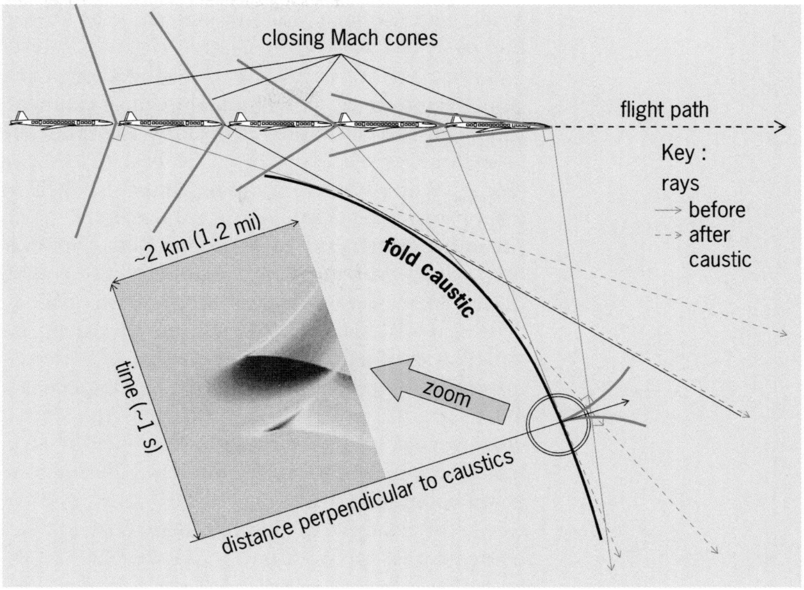

Fig. 2. Principle of focused sonic boom. Black-and-white-level inset displays local pressure field with pressure peaks in white and troughs in black.

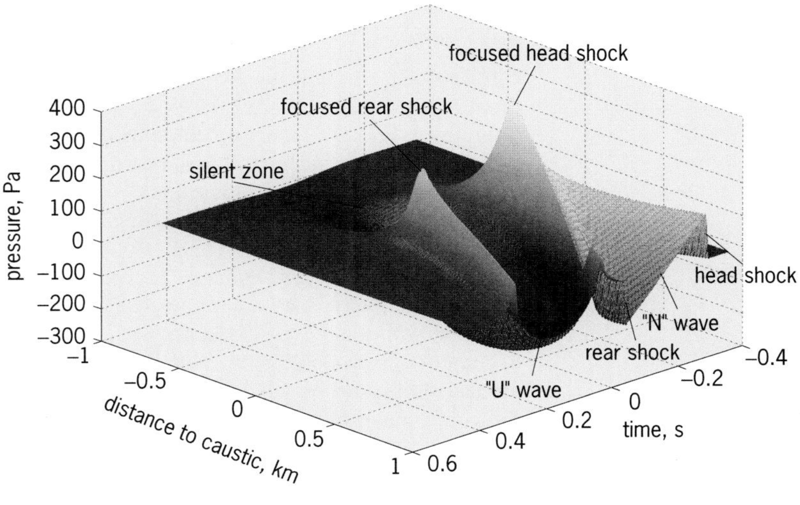

Fig. 3. Three-dimensional view of numerical simulation of nonlinear Tricomi equation, showing transformation of an "N" wave into a "U" wave. 1 km = 0.62 mi. 100 Pa = 1 mbar = 2.1 lb/ft^2.

to air, with the same order of magnitude as for real booms.

Predictions. Once validated, the model could be qualified for use in predictions through an extensive campaign of simulations to quantify the influence of meteorological variability. One year (1993 as a "mean" year considering the height of the thermal tropopause) of meteorological data over the North Atlantic Ocean between Paris and New York was used, along with Concorde transonic acceleration procedures over Normandy and the English Channel. Statistical analysis showed a surprisingly large variability of the focused boom compared to the cruising one. Although the mean value was as expected—410 Pa (4.1 mbar or 8.5 lb/ft^2), four times the Mach-2 cruising value of about 100 Pa (1 mbar or 2 lb/ft^2) typical for a Concorde-sized aircraft—a significant number of booms displayed much higher values. The surface of the ground area receiving at least one focused boom per year was estimated at 7750 km^2 (2992 mi^2). For the considered route, that area was almost entirely overseas, but with no safety margin.

Prospects. Focused sonic booms due to acceleration remain a major obstacle for aircraft designers when targeting unrestricted overland supersonic flight. The impact of aircraft low- boom design on focused booms remains unexplored today. It is likely to reduce boom levels, but will this be sufficient for overland acceleration? Acceleration over oceans or over sparsely populated regions with carefully defined protection procedures will probably still be required. Innovative focused low-boom design is conceivable but will have to be traded off with constraints of other design points at take-off, landing, and subsonic and supersonic cruise, including other environmental requirements governing community noise and emissions.

For background information *see* ABERRATION (OPTICS); CATASTROPHE THEORY; DIFFERENTIAL EQUATION; DIFFRACTION; NONLINEAR ACOUSTICS; RAINBOW; SHOCK WAVE; SONIC BOOM; SUPERSONIC FLIGHT; TRANSONIC FLIGHT in the McGraw-Hill Encyclopedia of Science & Technology.

François Coulouvrat; Régis Marchiano; Jean-Louis Thomas

Bibliography. M. V. Berry, Waves and Thom's theorem, *Adv. Phys.*, 25:1–26, 1976; R. Blumrich, F. Coulouvrat, and D. Heimann, Variability of focused sonic booms from accelerating supersonic aircraft in consideration of meteorological effects, *J. Acoust. Soc. Am.*, 118:696–706, 2005; D. J. Maglieri and K. J. Plotkin, Sonic boom, in H. H. Hubbard (ed.), *Aeroacoustics of Flight Vehicles, Theory and Practice*, vol. 1: *Noise Sources*, Acoustical Society of America, Woodbury, New York, pp. 519–561, 1995; R. Marchiano, F. Coulouvrat, and R. Grenon, Numerical simulation of shock waves focusing at fold caustics, with application to sonic boom, *J. Acoust. Soc. Am.*, 114:1758–1771, 2003; R. Marchiano, J.-L. Thomas, and F. Coulouvrat, Experimental simulation of supersonic superboom in a water tank: Nonlinear focusing of weak shock waves at a fold caustic, *Phys. Rev. Lett.*, 91(18):184301, 1–4, 2003.

Forensic document analysis

Forensic document analysis, or forensic document examination as it is often called, is one of the oldest fields of forensic science. The examination of contested documents in North America gained notoriety during the early 1900s largely through the efforts of Albert S. Osborn and other founding members of the American Society of Questioned Document Examiners.

Forensic document examiners (FDEs) draw on their skill, knowledge, and experience to determine if a questioned document is legitimate or fraudulent. Proven scientific methods are used to determine if a specific pen, typewriter, computer printer, photocopier, or other instrument produced the document. A FDE might also be consulted to determine if a document was fabricated in its entirety or whether it was altered by adding or removing information. Contested documents are not necessarily prepared on or about their purported dates. For example, it can be very important to determine when key documents related to a tax fraud investigation or a medical malpractice case were produced. The FDE can identify the presence of dating anachronisms after thoroughly examining the evidence.

Some specialists limit the scope of their analyses to one or two specific areas such as ink analysis or paper identification, while other FDEs examine a broad range of document problems, including photocopy comparisons, the identification of computer-printed and facsimile-transmitted documents, the comparison of rubber stamp impressions, the restoration of indented impressions, writing sequence determinations, charred document decipherment, and a host of other examination types. Falsified documents are often exposed by evidence uncovered by more than one of the above analyses.

Handwriting identification. The term *handwriting* refers to cursive writing, initials, signatures, numerals, and hand printing that can become the subject of investigation when dealing with matters that involve fraud. These entries are often written on paper using common writing instruments such as ballpoint, roller ball, fountain, gel and felt tip pens, as well as pencils and felt markers. Lipstick, chalk, or spray paint are occasionally used to write graffiti on walls, vehicles, or mirrors. On rare occasions, questioned documents may contain secret messages written with onion or lemon juice, milk diluted with water, saliva, urine, or other substances that need to be treated or developed before they can be read.

Handwriting comparisons are usually done in several stages. The first step entails an independent assessment of the questioned writing to determine if it contains features indicative of genuineness or spuriousness. The second step involves an inspection of the specimen material to identify any entries written by someone other than the known writer. The specimens are also assessed to determine if they are suitable for comparing with the questioned entries. The questioned and specimen groups of writing are then compared and all significant writing features

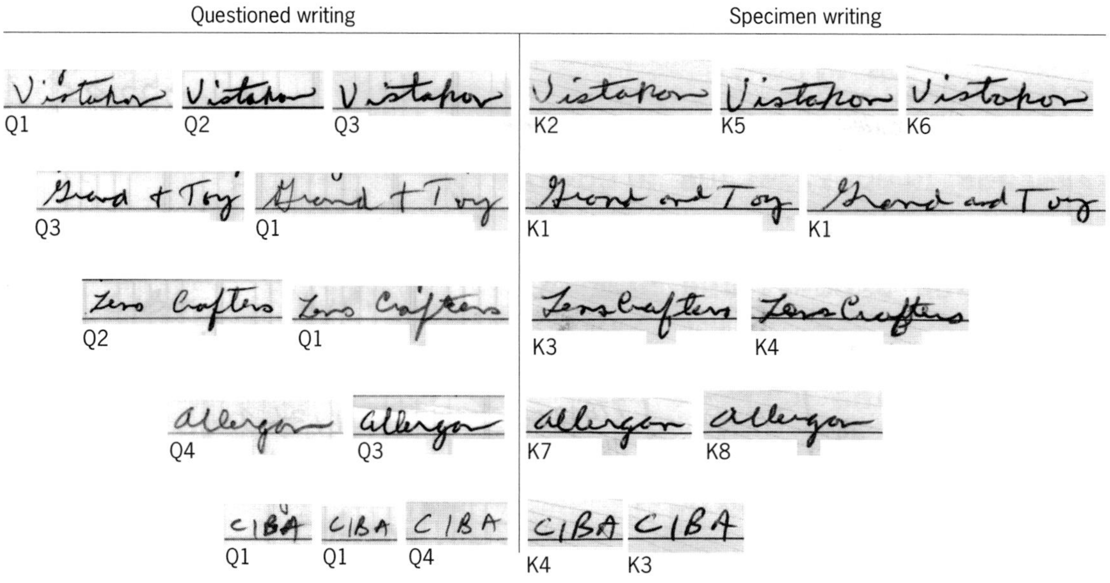

Fig. 1. An illustrative chart showing similar features that appear in both the questioned and specimen handwriting.

are noted. In the final step, the document examiner carefully reviews his notes and expresses an opinion or conclusion that is supported by the evidence.

Illustrative charts are very useful for demonstrating the basis of a handwriting comparison. **Figure 1** shows an illustrative chart containing several words from the questioned and specimen handwriting. The notation beneath each word indicates the document in which it is located. This chart does not show all the writing features examined but it does contains many of the features that support the opinion of identification that was expressed in this instance.

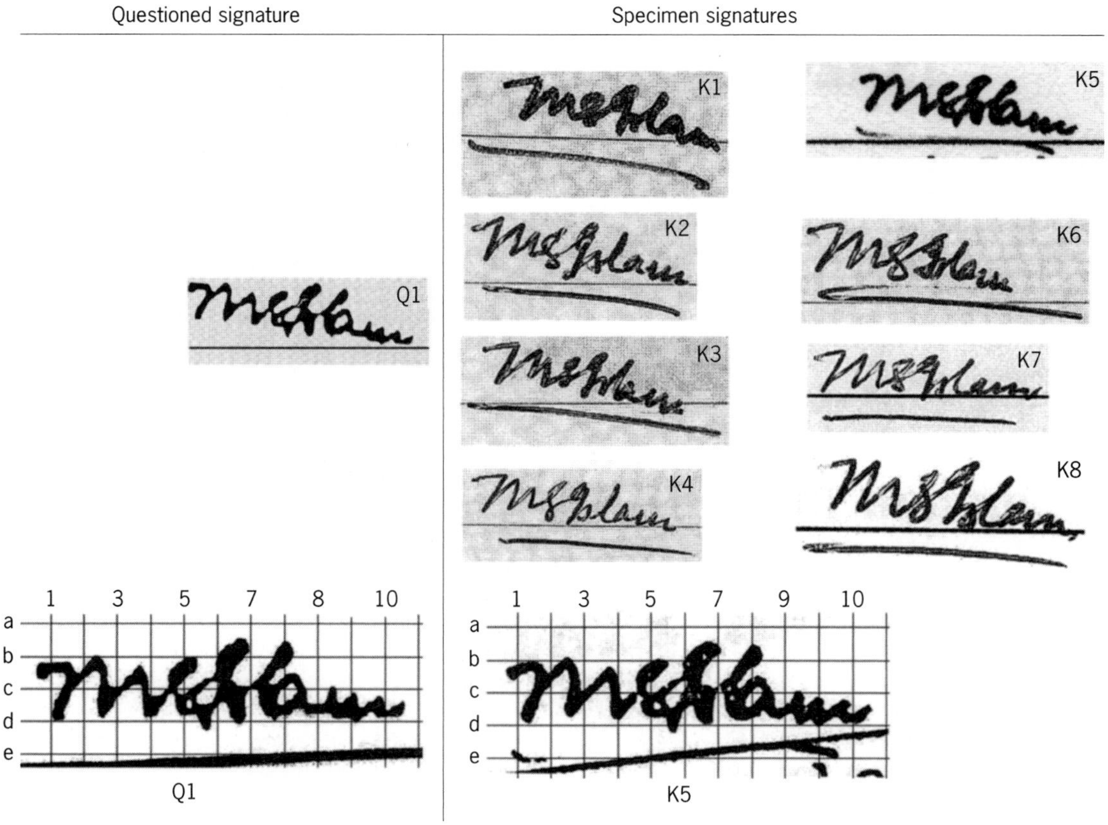

Fig. 2. The signature on a photocopied contract that was inserted by cutting and pasting a genuine signature from item K5 into position on the contested document. The overlaid ruled graphs confirm that the signatures on items Q1 and K5 are duplicates of one another with the exception of the underscore.

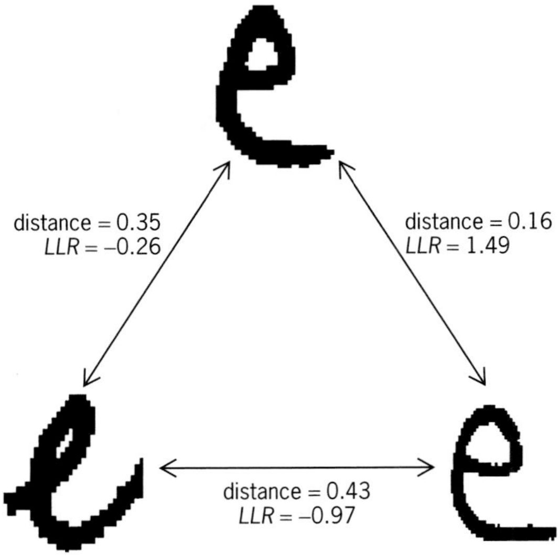

Fig. 3. The *D* and *LLR* values among three examples of the letter "e" written by two writers.

The plaintiff in another case produced a multigeneration photocopy of a contract bearing a questioned signature and stated that the original document was either misplaced or accidentally destroyed. Several documents bearing samples of the victim's signature were presented for comparison purposes. The contested signature in **Fig. 2** contains many features that also appear in the adjacent specimen signatures. This evidence was sufficient to prove that the questioned document is a montage constructed by cutting and pasting the genuine signature from document K5 into position on the photocopied contract. The fraudster failed to include the underscore that appears below the K5 signature and each of the specimens.

Forensic document examiners are not the only specialists who have an interest in handwritten documents. CedarTech, a company that specializes in pattern recognition, has developed a software program that compares handwriting and verifies the authorship of documents in an efficient and objective manner. Their software allows document examiners to search large numbers of exemplars for samples that closely match a questioned handwritten document or to compare writing on two or more documents to determine if they were (or were not) written by the same person.

The degree to which two samples match (or do not match) is expressed objectively as two variables: distance (*D*) and log likelihood ratio (*LLR*). **Figure 3** shows information extracted from three examples of the letter "e" that were executed by two writers. A high *D* value and negative *LLR* value suggest two letters were likely produced by different writers, while a low *D* value and positive *LLR* value suggest the pair were in all likelihood written by the same individual. **Figure 4** shows the histograms and parametric probability density functions of the same and different writers for the letter "e."

Admissibility of handwriting evidence. Document examiners have presented expert testimony with respect to handwriting identification in hundreds of thousands of judicial proceedings during the past century. An example from the early 1900s is *United States v. Ortiz*, where an expert witness acquired special qualifications from familiarity gained over the course of time by dealing with official records and the signatures they bear.

Forensic document examiners generally agree that handwriting identification is based on the following fundamental propositions: (1) no two people write in exactly the same manner, (2) although everyone's writing varies, the same series of recurring characteristics or habitual features appear in the handwriting of each individual, and (3) characteristics that serve to identify each person's handwriting vary to a lesser extent than the same characteristics in samples authored by other individuals.

Although these propositions are widely accepted, the admissibility of opinion evidence with respect to handwriting identification has recently been challenged in some United States courts. The challengers have raised a number of issues, but most claim that handwriting identification has not been established as a reliable expertise and that it lacks a methodology that is both objective and reliable.

The most detrimental arguments were presented in the case *United States v. Starzecpyzel*, where it was argued that no strong statistical evidence supported the aforementioned principles. In 1999, a district court ruled that the trial judge is responsible for determining if an expert's scientific testimony is relevant and reliable. The court further noted that Federal Rules of Evidence 702 and 703 gave the expert witness a degree of latitude not afforded to other witnesses because his opinion "will have a reliable basis in the knowledge and experience of his discipline."

Despite these and other favorable rulings, forensic document examiners have taken steps to ensure their profession, especially as it relates to the identification of handwriting, is viewed as a credible, scientific endeavor. In one study, Professor Moshe Kam of Drexel University compared the performance of professional forensic document examiners with nonprofessionals in their ability to match examples of signatures. The results clearly demonstrated that laypersons made far more errors than professional examiners. A later test done by Dr. Kam yielded similar results with respect to hand- and nonhand-printed material.

Ink analysis. The date when certain handwritten entries were made on a suspect document can be very significant in civil and criminal litigation. For example, the document examiner may be asked to determine if the evidence proves that a contract was signed after its alleged date or whether entries on a page from a diary or appointment book were all produced in one sitting or whether information was added some time later.

Two approaches can be used to answer these questions, one static and the other dynamic. In the static approach, a sample of ink extracted from the questioned entry is analyzed to determine its chemical composition. A database of writing inks can then be searched to identify the manufacturer and when the ink first appeared on the market. Some ink formulations are not unique to one particular manufacturer and can be produced by several companies. This and the near-impossible task of collecting samples of each ink manufactured limit the ink chemist's ability to draw definitive conclusions regarding the source of a particular writing ink.

Some companies have adopted the practice of adding trace amounts of unique chemicals to their writing inks. The composition of these tags is usually changed on an annual basis. Tags detected in a suspect entry can indicate if the ink was available when the document was written or whether it was written later than its date suggests. **Figure 5** shows three inks analyzed by thin layer chromatography. Ink 1 contains a fluorescent tag added to black ballpoint pen inks manufactured in 1981 and 1991 while no dating tags were detected in inks 2 and 3.

The dynamic approach to ink dating relies on the principle that certain ballpoint pen inks age or change with time as they dry on a document. The resin components of many ballpoint pen inks also undergo a transformation as they harden and volatile solvents, such as benzyl alcohol or phenoxyethanol, evaporate or are absorbed by the paper as the ink dries. The rate of evaporation is much faster during the initial stages of aging. A quantitative analysis of an extracted ballpoint ink sample by gas chromatography–mass spectrometry (GC–MS) will discriminate fresh ink from dry ink and indicate when the ink was applied to the paper.

Printers and copiers. Document examiners are frequently asked to determine what equipment was used to produce a document or whether a document was printed on a specific printer or photocopier. Few areas have experienced as much growth and development as the office products industry. New photocopiers, fax machines, and printers tend to operate faster, cost less, and produce better quality output than their predecessors.

Color printers and photocopiers operate by applying multicolor and black toner or ink to paper or another supporting medium. Many color machines generate a repeated sequence of inconspicuous yellow dots, each less than a millimeter in diameter, over the entire surface of the document. Barely visible to the naked eye, these dots are easier to see in nonprinted areas when the document is illuminated with ultraviolet light (**Fig. 6**). The location and relationship of each dot to its neighbors indicate the serial number of the machine as well as the time and date when the copy was produced. Industry representatives claim that these codes represent a significant deterrent to counterfeiting.

Certification update. Most qualified document examiners receive their training from federal, state, municipal, or other government forensic laboratories. These training programs are highly structured and typically take 2–3 years of full-time study to complete. Not only do they cover the theoretical aspects of document examination, they also place great emphasis on applying learned principles to practical case problems. Qualified instructors administer these programs and monitor the progress of trainees on a frequent basis.

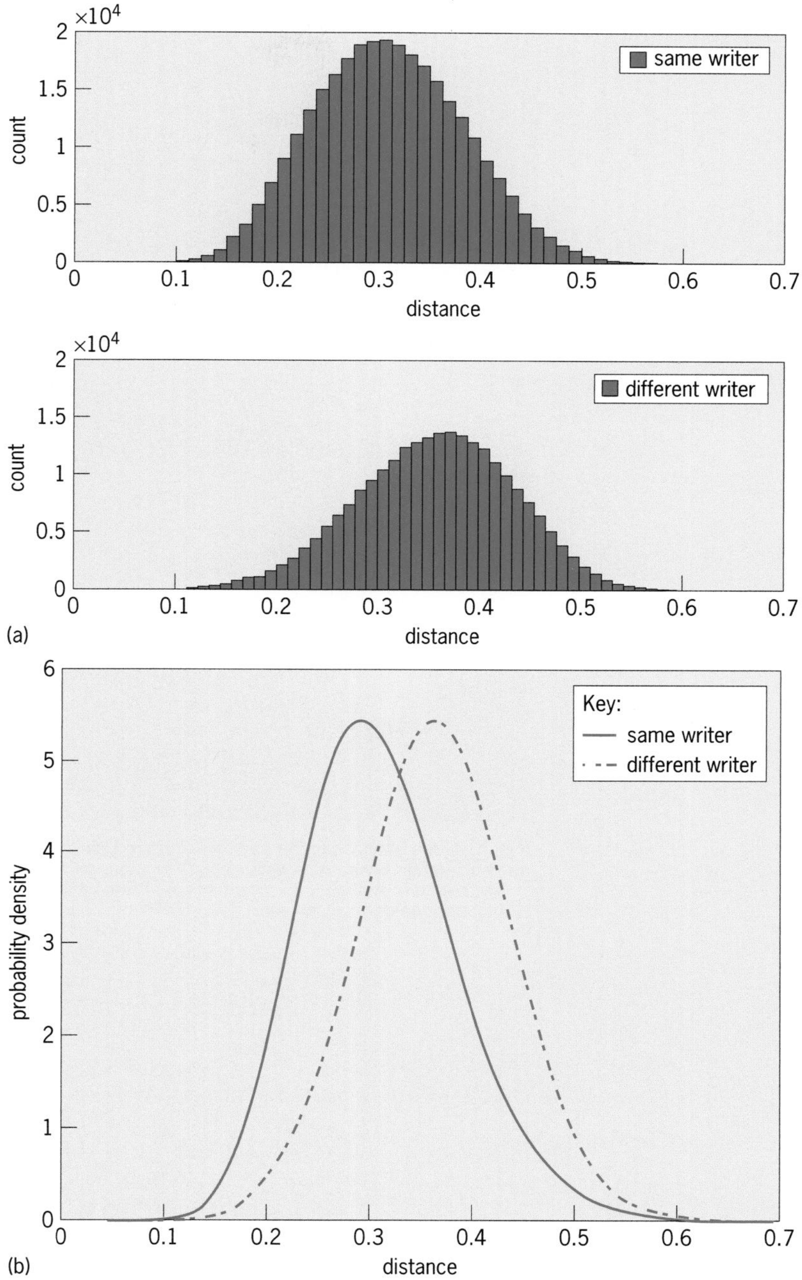

Fig. 4. Histograms and parametric probability density functions: (*a*) same and different histograms for the letter "e" and (*b*) same and different probability density functions for letter "e."

No degree of personal supervision and no amount of sophisticated forensic equipment will compensate for a poorly trained expert. That is, document examiner trainees must receive their groundwork from effective training programs that provide the necessary knowledge, skills, and perceptive reasoning they need to perform their job.

Training acquired by some so-called experts is not as rigorous. Many individuals received their instruction in a relatively short time through either online courses or correspondence courses. Self-directed

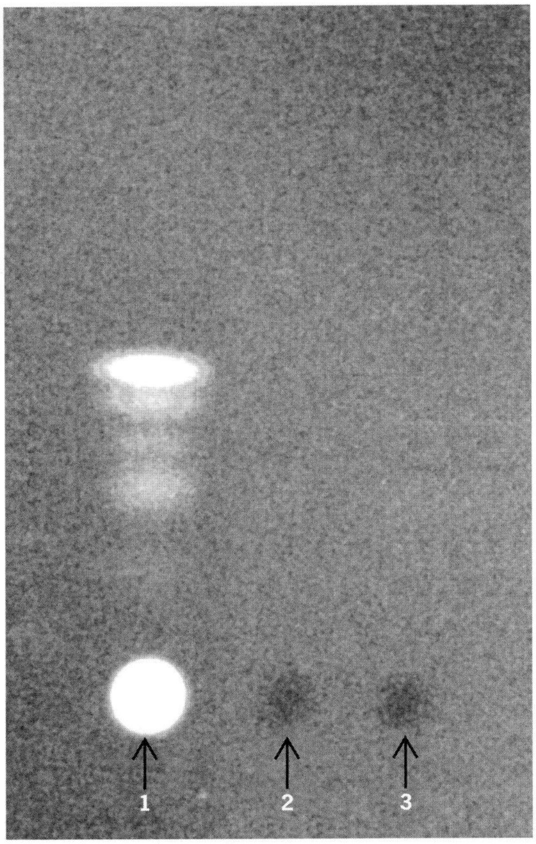

Fig. 5. Three black ballpoint pen inks analyzed for dating tags using thin layer chromatography. The tag found in ink 1 was added to writing inks manufactured during 1981 and 1991, while no dating tags were detected in inks 2 and 3.

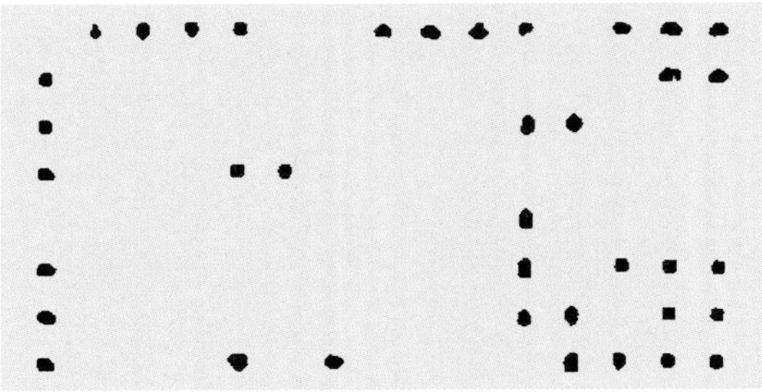

Fig. 6. Dots appearing in the background of documents produced by Xerox 5750 digital color copiers and printers.

distance learning is no substitute for more rigorous long-term training programs. A lack of day-to-day student-instructor interaction has also been found to limit performance and compromise training objectives. The more traditional pedagogical approach adopted by most government forensic laboratories clearly has its advantages.

In 1977, the American Board of Forensic Document Examiners (ABFDE) was formed to identify document examiners who meet or exceed its stringent certification requirements. In addition to this service, the ABFDE provides professional development opportunities to qualified examiners through workshops and seminars.

Candidates seeking certification must be engaged in the full-time pursuit of forensic document examination and pass a series of written, practical, and oral examinations. Certified examiners must subject their credentials to a thorough review every 5 years to maintain their certification.

In February 2007, the ABFDE's certification program was approved by the Forensic Specialties Accreditation Board, an independent agency supported and funded by the American Academy of Forensic Sciences, the National Forensic Science Technology Center, and the National Institute of Justice.

For background information *see* CHROMATOGRAPHY; COMPUTER PERIPHERAL DEVICES; CRIMINALISTICS; FORENSIC CHEMISTRY; GAS CHROMATOGRAPHY; INK; INKJET PRINTING; MASS SPECTROMETRY; PHOTOCOPYING PROCESSES; PRINTING in the McGraw-Hill Encyclopedia of Science & Technology.

Dan C. Purdy

Bibliography. M. Kam and E. Lin, Writing identification using hand-printed and non-hand-printed questioned documents, *J. Forensic Sci.*, 48(6):1391–1395, 2003; M. Kam et al., Signature authentication by forensic document examiners, *J. Forensic Sci.*, 46(4):884–888, 2001; *Kumho Tire Co., Ltd., v. Carmichael*, 526 U.S. 137, 119 S. Ct. 1167, 1999; *United States v. Ortiz*, 176 U.S. 422, 429, 1900; *United States v. Starzecpyzel*, Cite as 880 F. Supp. 1027 (S.D.N.Y. 1995).

Fuel cells for automobiles

With increasing global emphasis on reducing greenhouse gas and pollutant emissions and concern over potentially unstable petroleum supplies, automotive manufacturers have been actively researching and developing powertrains to address these concerns. Many believe that fuel cells will be the dominant powertrain, as they promise higher efficiency, zero emissions, and fuel feedstock flexibility. Small fleets have been demonstrated and are in use worldwide. Predictions for full commercialization range from 2010 to 2020 and beyond. Today's fuel-cell vehicles now are approaching the performance and packaging of production vehicles with approximately twice the fuel economy. Remaining challenges include cost, durability, fuel storage, and fuel infrastructure.

Fuel cells convert a fuel and an oxidant directly into electricity by an electrochemical process. Those adopted for automotive engines universally use hydrogen and oxygen (from air) to produce electricity, with water as the exhaust product. Since fuel cells are not heat engines, they are not subject to the Carnot efficiency limitations and can operate at much lower temperatures than internal combustion engines (ICEs), with efficiencies typically 2–2.5 times greater. Compared to battery-powered electric vehicles, they can have greater range due to the higher energy density of hydrogen and can be refueled quickly. Other desirable consumer features include quiet operation, near-instant torque response,

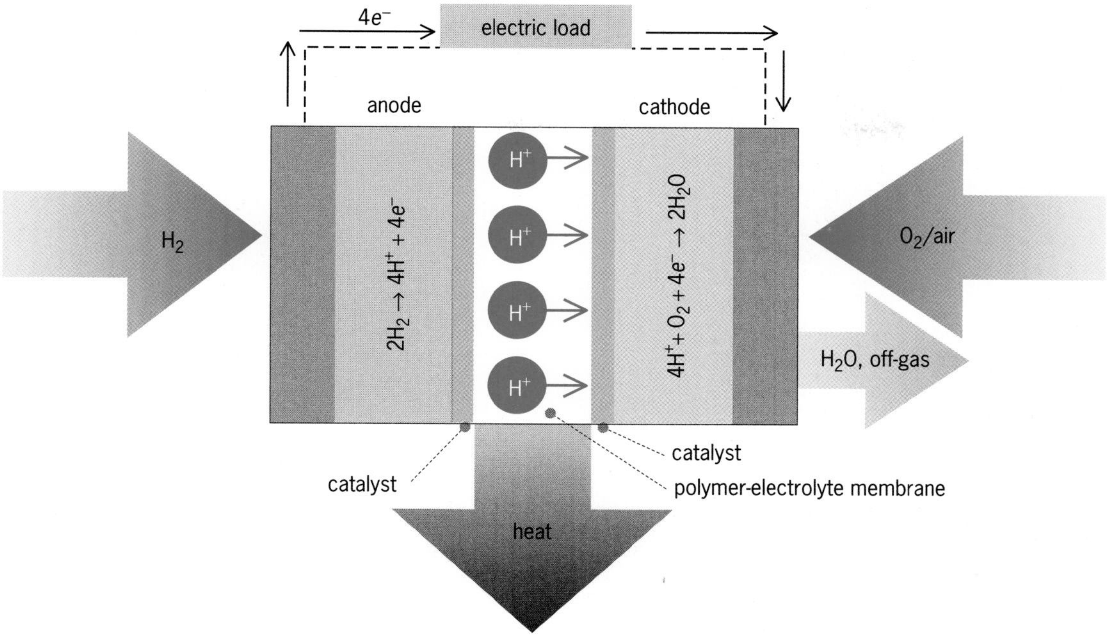

Fig. 1. Principle of operation of a fuel cell.

and the potential for auxiliary electric generator capability.

Principle of operation. Similar to batteries, fuel cells electrochemically produce current at a voltage. Whereas batteries convert the electrodes, requiring recharging or replacement, fuel cells consume fuel and can run as long as it is supplied.

While there are numerous fuel-cell types, all automotive fuel cells are of the PEM (proton exchange membrane, or polymer electrolyte membrane) type. Positive attributes include compactness, no moving parts, high power density, efficient and capable of starting and operating in cold conditions. Due in part to their low-temperature operation, hydrogen is the only practical fuel with sufficiently high kinetic activity. It must be nearly contaminant-free to avoid degrading electrode performance. Platinum (Pt) is the only known catalyst material that achieves the desired activity and power density at these operating temperatures.

The core of a fuel cell consists of a membrane electrode assembly (MEA), which is placed between two flow-field plates. The MEA consists of two electrodes (anode and cathode), each coated with a catalyst layer and separated by a proton exchange membrane. The flow-field plates direct hydrogen to the anode and oxygen (from air) to the cathode.

Figure 1 shows the operation of a fuel cell. Hydrogen gas is fed to the anode flow field through channels in a bipolar plate made of an electrically conductive material, typically a graphite-carbon composite or thin metal plate. The hydrogen is evenly distributed through a conductive porous gas diffusion layer (carbon paper) to the anode's electrode catalyst layer, which typically consists of finely dispersed platinum on high-surface-area carbon. The Pt catalyst dissociates the hydrogen molecule into two protons and two electrons [reaction (1)]. The solid polymer electrolyte, a thin (18–50-micrometer) proton-conducting plastic membrane (typically a perfluorosulfonic acid/polytetrafluoroethylene copolymer, similar to DuPont's NAFION™) allows proton conduction to the cathode electrode. Air is fed through a similar flow-field plate and diffusion layer to the cathode electrode. Oxygen is dissociated and recombined with the protons and electron current passed through an external circuit (the load) on a catalyst surface, typically of higher platinum loading, to produce water [reaction (2)]. A positive voltage and current is created through the electrochemical conversion of the hydrogen/oxygen heat of reaction. The overall reaction (3) is the same as in a hydro-

Anode reaction: $$2H_2 \rightarrow 4H^+ + 4e^- \quad (1)$$

Cathode reaction: $$O_2 + 4H^+ + 4e^- \rightarrow 2H_2O \quad (2)$$

Overall reaction: $$2H_2 + O_2 \rightarrow 2H_2O \quad (3)$$

gen/air combustion engine, but at lower temperature, higher efficiency, and with no pollutants from incomplete combustion.

The theoretical maximum cell voltage, determined by the reaction Gibbs energy, is approximated 1.23 V at ambient conditions; however, open-circuit voltage is typically near 1 V due to unavoidable kinetic losses. Operating single-cell voltages range from open-circuit 1V to design maximum power voltages of about 0.6–0.7 V. Losses at high current are driven by cathode kinetics, mass transport limitations, and both electronic and protonic (membrane) resistance. In automotive applications, cell efficiency may be defined as the cell voltage divided by the theoretical voltage derived from the lower heating value of the hydrogen fuel (1.254 V at standard conditions).

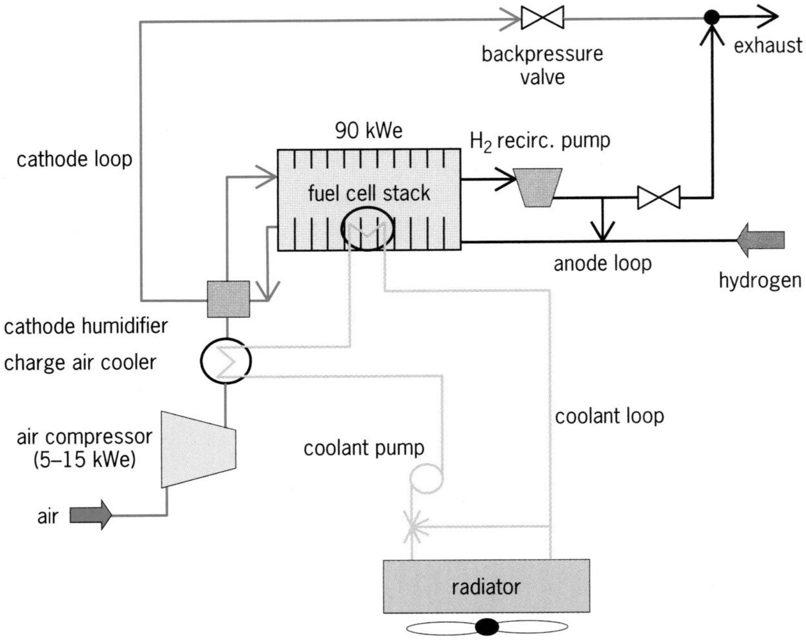

Fig. 2. Example of a fuel-cell system.

Thus, a cell at a typical drive cycle midpower operating point producing 0.75 V would be 60% efficient. Cells (200–400) are combined in series in a fuel-cell stack, or substacks, to yield the practical voltages required for efficient, compact power electronics and motors. Typical maximum stack power for today's automotive fuel cells range from about 60 to 100 kW (80 to 134 hp) to meet continuous top-speed and grade-climb requirements.

PEM membranes conduct protons in an aqueous phase and require the membrane to be at relative humidity approaching 100% for reduced protonic resistance loss and increased efficiency. This, as well as membrane chemical and mechanical properties, dictates nominal operating temperatures near 60–80°C. A primary area of membrane research is higher temperature and dryer operation, which is desired to enable system simplification.

Fuel-cell and vehicle configuration. Key subsystems in an automotive fuel-cell system (**Fig. 2**) include the fuel-cell stack, the anode subsystem to supply the reactant hydrogen, the cathode subsystem to provide the air, a cooling subsystem to remove waste heat and maintain proper fuel-cell temperature, and the electrical subsystem to deliver and supply the various electric power loads. Humidifiers may be required on the cathode and/or anode supply gases. The actual system design, components, and operating conditions vary by manufacturer, each optimizing for particular system requirements.

Key components common to most systems include an electrically driven air compressor, or blower, to supply high airflow to the cathode at operating pressure. Pressures range from near ambient to as high as 0.3 MPa absolute, depending on stack and system. In a high-pressure system, this can consume as much as 10–20% of the fuel cell's power output, but does yield balancing gains in stack performance. While more costly, high pressure may allow for reduced humidification requirements and stack and system size. Special humidifiers are being developed to capture moisture from the exhaust air and feed it back to the cathode inlet air, typically via a membrane water-vapor transport device. Hydrogen is supplied in excess of the amount required for the electrochemical reaction to improve gas distribution and water management. Being too valuable to waste, the exhaust hydrogen is redirected back to the inlet, typically with an electric hydrogen anode recirculation pump or an ejector (a passive pump), or combination thereof. Stack temperature is controlled using a cooling loop with an electric pump, radiator, and thermostat-type valve similar to the cooling circuit

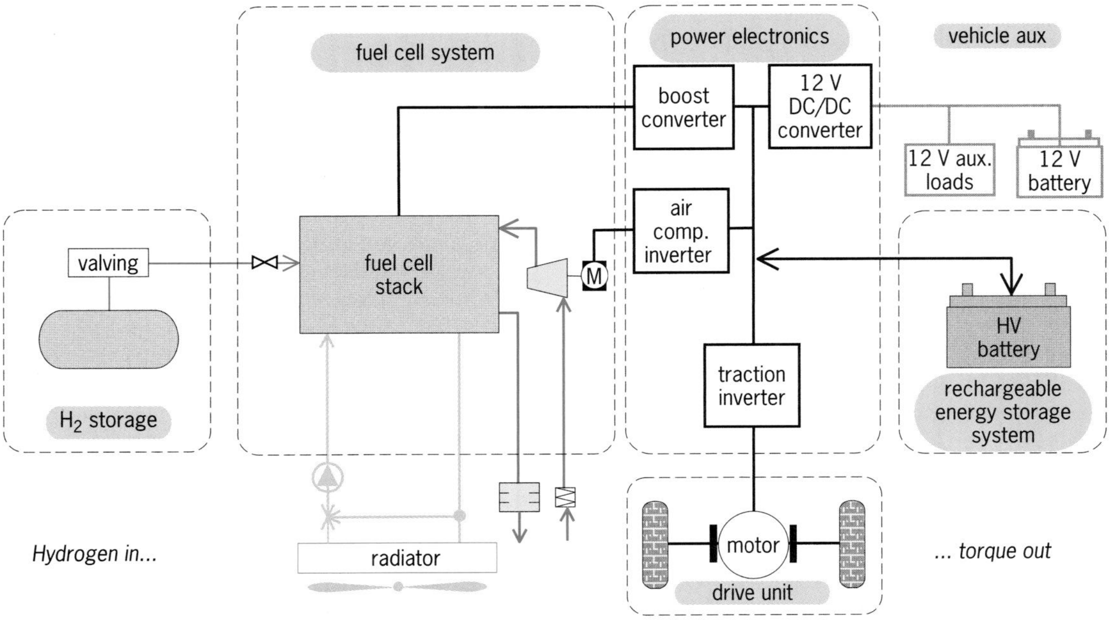

Fig. 3. Example of a fuel-cell propulsion system.

Targets and status of key subsystems. Derived from U.S. Department of Energy fuel cell technology roadmaps. Status is based on current technology at high-volume. For H_2 storage status, the first number is for compressed, the second in parenthesis for liquid.

	Units	Status	2010	2015
Fuel cell system (80 kW)				
Cost, excluding H_2 storage	\$/kW	110	35	25
Full-power efficiency	%	50%	50%	50%
Quarter-power efficiency	%	58%	60%	60%
Durability	h	1000–2000	5000	5000
Stack				
Power density	kW/L	≥2	2	2
Precious metal loading	g/kW	0.4	0.3	0.15
H_2 storage				
Cost	\$/kWh	18 (6)	4	2
Energy density	kWh/L	0.8 (1.2)	1.5	2.7
Specific Energy	kWh/kg	1.6 (1.7)	2	3
Battery (25 kW)				
Mass	kg	25	40	40
Volume	L	20	32	32
Cost	\$	750–900	500	500
Life	# cycles	300,000+	300,000	300,000
Traction motor + inverter				
Cost	\$/kW	35	19	12
Specific power	kW/kg	0.9	1.06	1.2
Power density	kW/L	1.8	2.6	3.5
Efficiency	%	90	90	93
Distributed H_2 production				
Cost	\$/gge*	<\$3		\$2–3

*gge = gallon of gasoline equivalent (on a per mile basis).

in an ICE. However, because of the relatively low temperature of the system (80°C compared to ICE max temperatures near 120°C) oversized radiators and coolant flows are required to reject the waste heat. The radiator size is package-limited and requires very high airflows for adequate stack cooling in periods of extended high-power operation. This requires larger, more costly high-power electric fans (more than 1 kW compared to ICE's 0.2–0.5 kW).

This fuel-cell system integrates with the fuel-cell propulsion system, which will include some or all of the parts shown in **Fig. 3**. In most systems, some form of rechargeable electrical energy storage subsystem is used to capture regenerative braking energy, supply peaking power for acceleration, and balance power demands on the fuel-cell stack for enhanced efficiency and durability. The U.S. Department of Energy (DOE) has standardized on a 25-kW peak power output, though manufacturer systems vary from about 20 to over 50 kW, depending on acceleration and regenerative energy capture targets. Similar to ICE hybrids, most use nickel-metal hydride batteries, with transition to lithium-ion expected as the technology matures. Alternately, systems may use ultracapacitors, which offer high peak loads per unit mass but over shorter periods. General Motor's HydroGen3 was designed to operate solely on fuel-cell power to demonstrate its capability to meet real-world power demands, with the potential for reduced cost (fewer intermediary power electronic components). Ballard/DaimlerChrysler buses have similarly demonstrated batteryless operation. However, given advances in automotive batteries, and the desire for greater fuel economy, most systems will likely use lithium (Li)-ion batteries.

A high-power drive motor (ac asynchronous, permanent magnet, or similar) delivers traction power to the wheel, typically from 60 kW to more than 120 kW (80 to +160 hp) depending on vehicle application. Peak drive motor efficiencies approach 90% with compact dimensions and specific power approaching 1 kW/kg. Present capability is nearly sufficient for automotive commercialization. However, advanced development continues to increase efficiency and reduce cost, mass, and volume.

Power electronics are used to control and distribute the high-voltage power to the drive motor, battery/ultracapacitor, and auxiliary component motors (stack air compressor, hydrogen recirculation pump, coolant pump, radiator fan, air conditioner compressor). The technology is well developed, but is subject to continued development to reduce cost.

The hydrogen storage unit allows for the refueling, storage, and delivery of the hydrogen to the fuel cell stack. As fuel-cell vehicles have demonstrated EPA-cycle fuel economies of 45–70 mi/kg-H_2, target storage ranges about 4–8 kg to achieve their desired 300+ mi range. The present status of most vehicles is 3–5 kg, consuming more packaging space, weight, and system cost than desired. Since fuel cell vehicles will be 2–2.5 times as efficient as gasoline vehicles, about half of ICE gasoline energy storage is required for similar range. Note that 1 kg of hydrogen contains almost the same amount of energy as 1 gal of gasoline. Thus, a FC vehicle storing about 6 kg of hydrogen may equal the range of an ICE vehicle with 12-gal tank.

Targets for hydrogen storage, as well as other key subsystems derived from DOE fuel-cell technology roadmaps, are shown in the **table**. Today's systems fall far short of long-term automotive storage and cost targets. Liquid hydrogen (at cryogenic temperature of about −253°C) and compressed hydrogen at

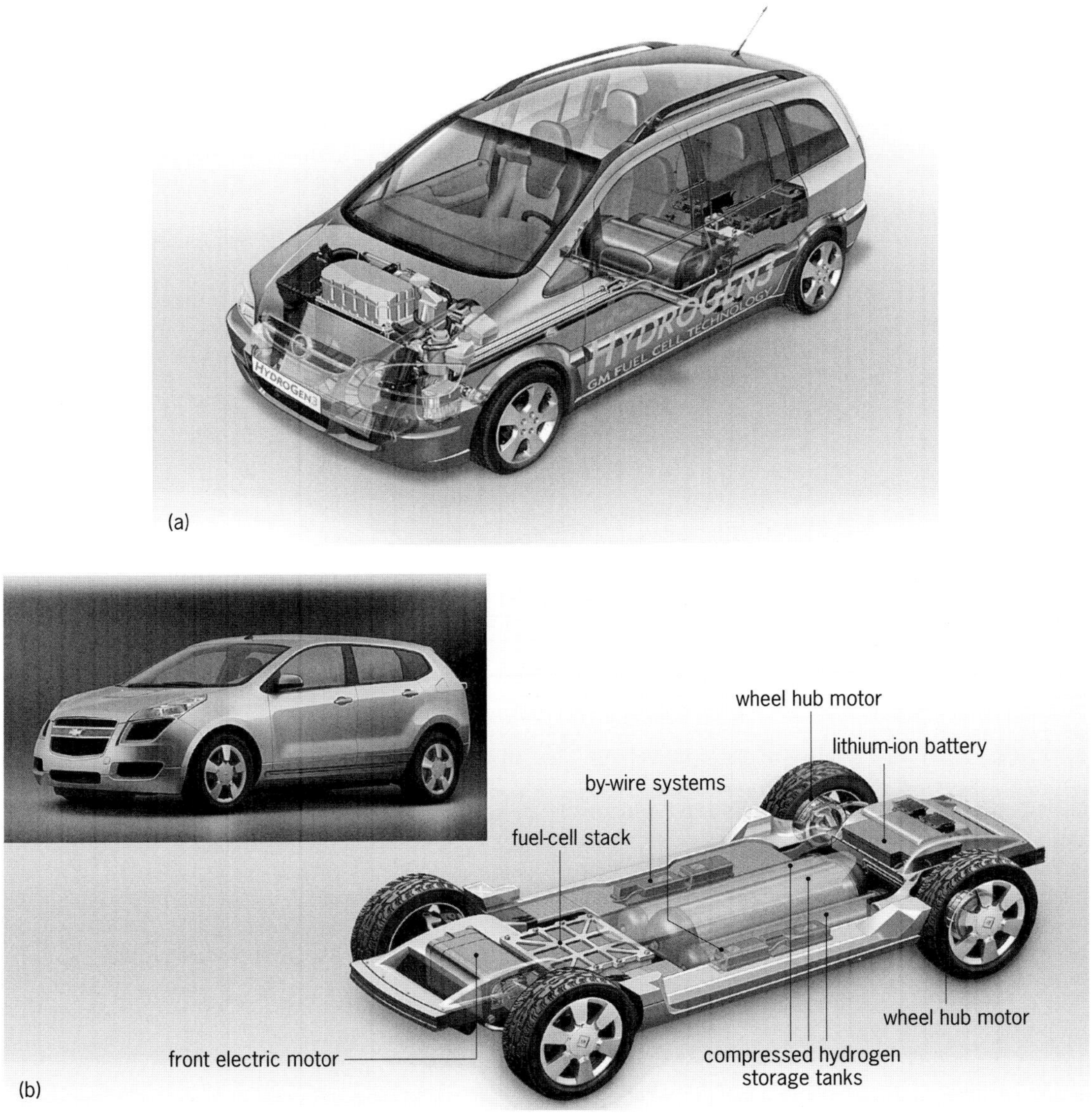

Fig. 4. Fuel-cell vehicle packaging showing alternative fuel cell placed (*a*) underhood and (*b*) underfloor.

pressures of either 5000 or 10,000 psi, are currently the only viable storage technologies. Storage energy density of hydrogen in either technology is much lower than that of gasoline (0.8–1.2 kWh/L versus approximately 6 kWh/L), but far superior to that of batteries (0.07–0.15 kWh/L). Liquid hydrogen offers superior volumetric storage compared to compressed. For example, General Motors HydroGen3 vehicles are capable of storing either 3.1 kg of 10,000 psi of compressed hydrogen or 4.6 kg of cryogenic liquid hydrogen. Issues with eventual liquid boil-off and loss in upstream fuel efficiency (liquefaction consumes energy equal to approximately 30% of that stored) has led to near-universal adoption of high-pressure hydrogen as the storage method.

Research on alternative materials promises increased storage capacity, including metal hydrides, carbon-based materials, cryoadsorbents, and chemical hydrides. In addition to high storage density, the systems must allow for fast refill (3 min target) and ability to supply hydrogen below −25°C. Adsorption-based materials, requiring considerable heat for hydrogen release, should operate at less than the fuel-cell temperature to allow for use of stack low-grade waste heat. Otherwise, excess hydrogen would need to be combusted or converted into electricity to supply the desorption endotherm, resulting in undesirable reduction in system efficiency and range. In addition, enthalpies of adsorption must be low to avoid impracticably large heat-transfer systems for heat removal or supply on fast refuel or full-power operation.

Although current storage density is less than half that of long-term targets, today's fuel-cell vehicles store sufficient hydrogen (typically 4–5 kg) for early commercialization.

Hydrogen production, infrastructure, and safety. Presently, 50 million tons of hydrogen are produced worldwide annually, with 95% produced from natural gas reforming. While it is consumed primarily for petroleum refining and fertilizer production, in theory this hydrogen could supply the needs of 200 million fuel-cell vehicles. Used as the feedstock and production pathway, analyses indicate that this could result in perhaps a 50% reduction in fossil-energy consumption, greenhouse gas reduction, and near-zero petroleum use compared to today's ICE technology. The DOE forecasts that hydrogen from this pathway will be competitive with present gasoline costs on a dollar-per-mile basis.

Challenges remain in distribution due in part to the low energy density of hydrogen as well as hydrogen embrittlement concerns and hydrogen purity requirements. Distribution systems for hydrogen already exist in some parts of the United States in the form of a limited pipeline network (about 700 mi), attesting to technological feasibility. The primary challenge is one of infrastructure capital cost and commitment.

Like gasoline, hydrogen is a fuel, and all fuels need careful handling to be safe. Hydrogen has been produced for many decades and is handled safely every day. Hydrogen is not toxic, poisonous, or corrosive. Hydrogen refueling connections are air tight, so there are no fumes escaping at the fuel station and no harm to the environment. In the event of a leak, hydrogen would disperse up and away in the air almost immediately due to its low density. Its release is not known to contribute to atmospheric or water pollution. Next-generation hydrogen fuel cell vehicles are being designed to be demonstrated on roads around the world to meet applicable safety standards, including the rigorous standards for safety of pressure vessels to store the fuel.

Status. The technology is at a state that allows for vehicles that match the performance of typical four-cylinder engines, allowing for real-world automotive performance (100 miles per hour, 0–60 mph times of about 12 s) with 60–100 kW engines usually augmented with battery power. EPA-certified fuel economy of 57 mpg has been demonstrated, with significantly higher mpg anticipated as increased fuel-cell system efficiencies and lighter-weight systems and vehicles are introduced. Stack power densities have increased by more than an order of magnitude since 1990, resulting in packageable underhood systems. Passenger and cargo space is similar to commercial vehicles with small allowances for hydrogen storage requirements (**Fig. 4**). Crash tests have been successful. General Motors recently demonstrated 300-mi range with two Sequel concept vehicles, with 8 kg of on-board hydrogen storage. More typical vehicle platforms adapted to fuel-cell powertrains have been limited to 3–5 kg of hydrogen, resulting in a real-world range of about 150–250 mi. A few vehicles have used alternative metal or chemical hydride storage, though none are considered near commercialization.

DOE cost estimates extrapolated to full-scale manufacture are shown in the table, however, there is considerable range in such estimates due to differing assumptions and uncertainties in rapidly developing technologies. Current vehicle costs are much higher due to small-scale production and, to a lesser extent, the use of more conservative catalyst platinum loading.

Key challenges. The primary challenges are reduced fuel-cell material cost, increased durability, and high-capacity and low-cost hydrogen storage. The key off-vehicle challenge to commercialization is the development of a hydrogen production, distribution, and supply infrastructure. The latter is more a cost and commitment issue, as there are no fundamental technology hurdles.

Primary fuel-cell stack challenges are reduced platinum-group metal loading (cost) and increased membrane and electrode durability under severe, dynamic automotive conditions. These are fundamental material challenges. Progress has been made with novel Pt-alloy catalysts, demonstrating twice the activity of pure platinum. Research in inexpensive nonprecious-metal catalysts has resulted in markedly improved kinetic activity, but is still very far from automotive viability.

Membrane materials for higher-temperature operation (midterm target 95°C, long-term target 120°C) and lower humidification are required for full auto-competitive performance, while continuing to meet low-temperature start requirements.

Fuel-cell balance of plant requires cost reduction, particularly for the air compressor. This may be achieved through continuous engineering improvements and economies of scale production.

Most automotive fuel-cell systems incorporate supplemental battery assist. Continued advancements in battery cost, tolerance-abuse (Li-ion), and cold-start performance are required for true competitiveness with current automotive system targets. Since the battery requirements are nearly identical to those for ICE hybrids, advancements in technology and scale-up will be fully leveraged. Hydrogen storage capacity and cost remain key challenges, although they are sufficient for initial commercialization.

Outlook. While nearly all manufacturers agree that the fuel-cell vehicle will ultimately replace the ICE, there is considerable debate as to when. Estimates for commercialization range from commencing in the 2010–2015 time frame, to 2020 and beyond. Deployment and hydrogen infrastructure development plans remain a key issue. We may see the development of dedicated government and/or private-sector fleets, or a more automotive industry and consumer-driven growth of the market as seen with hybrids. However, it is likely that national commitment will be needed to develop the infrastructure, including hydrogen production and fueling, for full consumer acceptance. Such challenges are not insurmountable nor have any fundamental showstoppers been identified. Thus the hydrogen fuel-cell is anticipated by most OEMs to be the automotive powertrain of the future.

For background information *see* ALTERNATIVE FUELS FOR VEHICLES; AUTOMOTIVE ENGINE; BATTERY; ELECTROCHEMISTRY; FREE ENERGY; FUEL CELL; HYDROGEN; INTERNAL COMBUSTION ENGINE in the McGraw-Hill Encyclopedia of Science & Technology.

David A. Masten

Bibliography. F. Barbir, *PEM Fuel Cells: Theory and Practice*, Academic Press, 2005; J. Larminie and A. Dicks, *Fuel Cell Systems Explained*, 2d ed., Wiley,

Chichester, United Kingdom, 2003; M. Mathias et al., Two fuel cells in every garage?, *Interface*, vol. 14, no. 3, 2005; U.S. Department of Energy, Office of Fossil Energy, National Energy Technology Laboratory, *Fuel Cell Handbook*, 7th ed., 2004.

Fungal allergies

Over 20% of the population in industrialized countries suffers from allergic disease, and a wide number of allergenic sources associated with different forms of allergy have been reported. Among these, fungal allergens play a predominant role because they cover more than 17% of all allergen structures known to induce immunoglobulin E (IgE)–mediated sensitization. [IgE is one of the body's five classes (isotypes) of immunoglobulins (any of the glycoproteins in the blood serum that are induced in response to invasion by foreign antigens and that protect the host by eradicating pathogens; antibodies belong to this group of proteins.] In contrast to pollen allergens, which show a typical seasonal variation, fungal allergens are present in our environment all year long and often in far greater concentrations than pollen grains during the seasonal peak.

Since only some individuals develop allergy even though the whole population is exposed to perennial allergens, a genetic predisposition along with environmental exposure is necessary for the occurrence of the atopic (allergic) state. A variety of pathological conditions, including impaired immune functions or disrupted skin barriers, are believed to cause host susceptibility to fungal allergens. The allergic response can involve different organs, and symptoms can include nasal congestion, shortness of breath, or skin rash, resulting in rhinoconjunctivitis, asthma, urticaria (hives), eczema development, or even anaphylaxis and death.

Epidemiology, prevalence, and clinical relevance. As strictly heterotrophic organisms, fungi occupy almost all ecological niches, resulting in a ubiquitous presence. Three groups of fungi include most of the species relevant for allergic disease: Zygomycota, Ascomycota, and Basidiomycota. Spores deriving from these three groups are the major contributors to the fungal spore load responsible for the aforementioned allergic complications. The major allergic manifestations related to fungal exposure include asthma, rhinitis, allergic sinusitis, allergic bronchopulmonary mycoses, hypersensitivity pneumonitis, and atopic eczema. Additional disorders that have been attributed to fungi include sick building syndrome, farmer lung disease (an allergic disease usually caused by breathing in the dust from moldy hay), and pulmonary hemorrhage/hemosiderosis [a lung disorder in which bleeding (hemorrhage) into the lungs leads to abnormal accumulation of iron]. A special case of fungal allergy is allergic bronchopulmonary aspergillosis [ABPA, an allergic lung reaction to a type of fungus (*Aspergillus fumigatus*) that causes cough, wheezing, sometimes fever, and (if untreated) chronic lung damage], which, with exception of invasive mycoses, represents the most severe disease due to fungal exposure.

TABLE 1. Estimated prevalence of fungal sensitization in patients suffering from respiratory complications

Fungal species	%	% monosensitized*	% polysensitized
Alternaria	10–14	70	30
Aspergillus	2–3	1	99
Candida	6–9	59	41
Cladosporium	2–3	1	99
Penicillium	1–2	1	99
Saccharomyces	1–2	1	99
Trichophyton	1–2	5	95

*Percent of individuals monosensitized to the fungal species reported in column 1 calculated as percent of column 2.

The exact prevalence of fungal sensitization has not yet been reliably established (**Table 1**). This may be due to the sizable number of fungal species, the lack of standardized extracts, and the difficulty in deciding which are clinically relevant and important. Most cases of fungal allergies are associated with severe asthma or, in the case of sensitization to yeasts, atopic eczema. Even in the case of ABPA, which probably represents the best-studied fungal disease affecting almost exclusively patients suffering from asthma or cystic fibrosis, our best understanding of its prevalence is quite contradictory. ABPA could be present in 0.25 to 0.8% of all asthma patients. Criteria for diagnosing of ABPA in cystic fibrosis are perhaps even more hazardous in view of preexisting bronchopulmonary pathology and infection. Extrapolations from published studies suggest an ABPA rate of no more than 11% in cystic fibrosis; however, there is a huge discrepancy in the prevalence reported by different studies.

One of the difficulties in assessing the true prevalence of fungal sensitization is that prevalence is age-dependent, peaking at around 8 years, followed by a decline with increasing age. Therefore, a reasonable estimate of the prevalence of fungal allergy should involve a large cohort of patients suffering from respiratory diseases covering all classes of age. The results of one such study involving a cohort of 4962 individuals suffering from respiratory complications between 3 and 80 years of age indicate a prevalence of sensitization of fungal allergy to be about 20%. Our best understanding of fungal sensitization under the existing limitations suggests rates of around 3% for the general population, whereas the sensitization rate might increase up to 30% in selected patients with asthma.

There is clear evidence that fungal allergy is of clinical relevance in contributing to the severity of many diseases, and the extent of the problem might be underestimated. Specific IgE and positive skin test reactions to *Malassezia sympodialis* (a yeast fungus) are almost specific for atopic eczema patients. A clear correlation between exposure and exacerbation of eczematous reactions has been shown in atopy patch tests. Moreover, many case reports show that treatment of the affected skin with antimycotics results in fast improvement of the disease conditions.

Interestingly, the yeast quantitatively releases allergens at a pH around 6.5, corresponding to the pH of the skin in areas with a disrupted skin barrier. In contrast, at a pH of 4.5, corresponding to that of healthy skin, *M. sympodialis* is unable to release allergens into the environment, indicating a clear role of host–pathogen interaction in a subset of patients suffering from atopic eczema. It is well documented that ABPA occurs only in patients sensitized to *A. fumigatus* and that molds of the genera *Alternaria* and *Cladosporium* are related to severe asthma. Therefore, it would be important to exclude fungal sensitization in all asthmatic patients.

Diagnosis and recombinant fungal allergens. The diagnosis of fungal allergy is, as in the diagnosis of all type I allergic reactions, always associated with the direct or indirect detection of allergen-specific IgE. Standard skin prick tests and in-vitro determination of mold-specific IgE against fungal extracts are the most common diagnostic procedures used daily in clinical practice. The major breakdown in the diagnosis of allergy in general and of fungal allergy in particular derives from the fact that the experimental results directly depend on the quality of the extract used. An additional complication in diagnosing fungal allergy is the limited number of molds that can be tested. The number of fungi able to induce allergic responses exceeds by far the panel of extracts that can reasonably be tested in routine assessments. Moreover, in contrast to pollens and mites, fungal extracts recognized by the World Health Organization (WHO) as diagnostic standards are not available. This might partly explain the huge differences in fungal sensitization reported in different studies. However, aerobiological surveys, skin test studies, and allergen characterization suggest that at least *Alternaria alternata*, *Aspergillus fumigatus*, *Cladosporium herbarum*, *Epicoccus nigrum*, *Fusarium roseum*, *Penicillium chrysogenum*, and the yeasts *Candida albicans* and *Malassezia sympodialis* should be considered as a minimal testing set when fungal allergy is suspected.

Some efforts have been undertaken to characterize the repertoires of mold allergens at the molecular level. A vast variety of fungal allergens have been identified, cloned, sequenced, and produced as highly pure recombinant proteins in different expression systems. The official allergen list of the International Union of Immunological Societies (IUIS) reports to date 107 characterized fungal allergens shown to be relevant in clinical tests. The best-characterized fungal allergen repertoires are those of *A. fumigatus* and *M. sympodialis*, which include 12 and 13 different allergen structures, respectively. Diagnostic applications of these complex allergen panels have been shown to be superior to fungal extracts for diagnosing allergy to *A. fumigatus* and *M. sympodialis* due to the lack of false-positive and false-negative test outcomes. Therefore, the use of recombinant allergens eliminates problems related to contaminations of extracts with natural IgE-binding components, altered allergenicity due to extraction procedures, and variability of allergen contents deriving from the preparation of extracts from natural sources. Although powerful, the widespread use of recombinant allergens for the diagnosis of allergic conditions is presently limited by the almost complete lack of commercially available products and by legal restrictions, as in many countries their application for the in-vivo diagnosis of allergy in standard skin tests is not yet legalized.

TABLE 2. Selection of cross-reactive pan-allergens showing IgE-mediated reactivity to the homologous human self-antigens

Protein class	Allergen	GeneBank no.*	Protein Data Bank no.†
Cyclophilins	Asp f 11	Q9Y7F6	2C3B
	Asp f 27	Q4WWX5	—
	Mala s 6	O93970	2CFE
	Human	AAO64722	1BCK
MnSOD	Asp f 6	Q92450	1KKC
	Mala s 11	Q873M4	—
	Sac c	P00447	—
	Human	P04179	1AP5
Ribosomal P_2 protein	Alt a 5	P42037	—
	Asp f 8	Q9UUZ6	—
	Cla h 5	P42038	—
	Fus c 1	Q8TFM9	—
	Human	P05387	—
Thioredoxin	Asp f 28	Q1RQJ1	—
	Asp f 29	Q4WV97	—
	Cop c 2	Q9UW02	—
	Cur l	Q1EHHO	—
	Fus c 2	Q8TFM8	—
	Mala s 13	Q1RQI9	2J23
	Human	P10599	1ERU

*Sequences can be retrieved from the ExPASy (Expert Protein Analysis System) proteomics server (www.expasy.org).
†Structures are accessible through the Protein Data Bank (www.pdb.org).

Cross-reactivity. An important aspect of fungal allergens is their extended cross-reactivity. It has been shown that many structures deriving from phylogenetically highly conserved proteins are present as IgE-binding components in different fungal extracts. Classical examples are manganese-dependent superoxide dismutases, cyclophilins, thioredoxins, and heat shock proteins. These structures have been demonstrated to be fully cross-reactive in terms of IgE binding and might contribute to an essential reduction of the number of allergens needed for a clear-cut diagnosis of fungal allergy. The present view is that complex fungal extracts contain a vast variety of cross-reactive structures (**Table 2**) [besides a few species-specific allergens such as Asp f 1, Alt a 1, or Mala s 1, which seem to be strictly limited to the genera *Aspergillus*, *Alternaria*, and *Malassezia*]. Progress in the development of recombinant allergen-based microarrays for the diagnosis of allergic diseases might allow the development of fast and reliable tests for fungal sensitization in the near future.

Treatment. In spite of the fast progress achieved in characterizing the allergenic components of fungal extracts, new therapeutic concepts for the treatment of fungal allergies are not yet emerging. Avoidance of exposure, whenever possible, should be considered for individuals at risk (for example, asthmatic patients). Fungal sensitization might be treated with corticosteroids to avoid deterioration of the lung function, especially in asthmatics with concomitant sensitization to *A. fumigatus*, which represents a risk for the development of a life-threatening ABPA.

The results of allergen-specific immunotherapy, the only treatment able to cure allergic disease, are not very encouraging. Clinical efficacy of specific immunotherapy with fungal extracts has been shown in 79 actively treated patients in four controlled trials, with only two fungal species, namely *Alternaria alternata* and *Cladosporium herbarum*. The lack of success in many other trials might be related to the quality of the extracts used, to an insufficient diagnosis of the fungal species involved, or to the treatment of polysensitized individuals known to be more refractory to allergen-specific immunotherapy than monosensitized ones.

In conclusion, the use of recombinant fungal allergens might create new prospects in diagnosis and specific immunotherapy for fungal allergy.

For background information *see* AIR POLLUTION, INDOOR; ALLERGY; ASTHMA; FUNGI; HYPERSENSITIVITY; IMMUNOLOGY; MEDICAL MYCOLOGY; MYCOLOGY in the McGraw-Hill Encyclopedia of Science & Technology. Reto Crameri; Claudio Rhyner

Bibliography. M. Breitenbach, R. Crameri, and S. B. Lehrer (eds.), Fungal allergy and pathogenicity, *Chem. Immunol.*, 81:1–310, 2002; R. Crameri et al., Fungal allergies: A yet unsolved problem, *Chem. Immunol. Allergy*, 91:121–133, 2006; A. Mari et al., Sensitization to fungi: Epidemiology, comparative skin tests, and IgE reactivity of fungal extracts, *Clin. Exp. Allergy*, 22:1429–1438, 2003; L. Romani, Immunity to fungal infections, *Nat. Rev. Immunol.*, 4:1–23, 2004.

Gene expression during tooth development

Teeth are very complex organs. To meet their mechanical requirements, vertebrate teeth are formed from two major distinct layers: an outer, hard, highly mineralized brittle layer of enamel, and a mineralized, but softer, less brittle layer of dentin. In a particular species, the pattern of dentition is fixed, but each tooth is formed independently of all others with a unique size, shape, and function. The relative location of each tooth is specified early in embryonic development by a distinct set of gene expressions, followed by other gene-regulated signals that specify the timing of tooth growth, and still other gene-regulated processes that determine the ultimate shape of the tooth. The cells in each layer are programmed to produce uniquely different proteinaceous structures that subsequently mineralize relatively late in tooth formation. Initial assembly of the tooth germ requires constant communication between the ectodermal cells of the oral epithelium, which differentiate to become ameloblasts and produce the enamel, and the underlying ectomesenchymal cells (also called mesenchymal cells), which become the dentin-producing odontoblasts. Ectomesenchymal cells are derived from the cells of the neural crest. Neural crest cell migrations that establish the oral cavity are rapid but highly controlled, and their cells have undergone several localized gene-regulated morphogenetic differentiation and commitment steps in reaching their positions. In the human fetus, the oral cavity is formed by week 4, with the cells of the oral ectoderm covering the connective tissue formed by the neural ectomesenchymal cells.

At that point, the outlines of the maxilla (upper jaw) and mandible (lower jaw) have become defined. Although the cells are still embryonic at this early stage, they have differentiated far from the initial stem cell level and their fates have been mostly determined. By week 5, at ~37 days, the epithelia of the maxilla and mandible begin to thicken in horseshoe-shaped bands marking the positions of the dental arches and forming the dental lamina. It is at this point that the story of specific tooth development begins, by directed thickenings of the oral epithelium within the dental lamina.

Cytodifferentiation of the tooth germ: bud, cap, and bell stages. Mammalian tooth formation follows a highly regulated pathway through distinct stages of development: determination of tooth location; positional thickening of the dental lamina and ingrowth into the mesenchyme; determination of tooth type (molar, incisor, etc.); determination of specific tooth shape (number of cusps and roots); terminal differentiation of enamel and dentin layers; maturation of the mature functional tooth crown and root; and eruption of the crown into the oral cavity. Mice, rather than humans, have been used in most gene expression and morphogenic analyses because of their short 21-day gestation yielding postnatal day 1 erupted teeth, and the opportunity to make genetic manipulations related to gene knockout, mutation, or overexpression. **Figure 1** presents a schematic view of the developmental stages of the mouse molar and indicates the timing and location of the expression of the various gene products, correlated with the corresponding structures. The bottom row depicts the developmental morphological changes in the generalized molar structure, shown as a cross section in the coronal-apical, labio-lingual orientation; the aligned upper set of boxes contain lists of the principal transcription factors operative at that stage in either the oral ectoderm (clear boxes) or in the subjacent odontogenic mesenchyme (gray boxes). Alongside each box, the main signaling molecules are shown, with the arrows depicting the targets of their activity. Note that the same main signaling factors appear at each stage, but their effects are context and concentration dependent, so it is the field or balance of their activities that determines behavior at any point along the developmental pathway.

The lists of factors in Fig. 1 are an oversimplification. Each factor has many family members; for example, there are more than 15 bone morphogenetic proteins (BMPs), of which BMPs 2, 4, and 7 are prominently involved in tooth development. Similarly, as shown in **Fig. 2**, the products of six distinct *Wnt* gene expressions are seen in both dental ectoderm and the mesenchyme at all stages from dental lamina thickening to the bell stage. [Note that tooth development

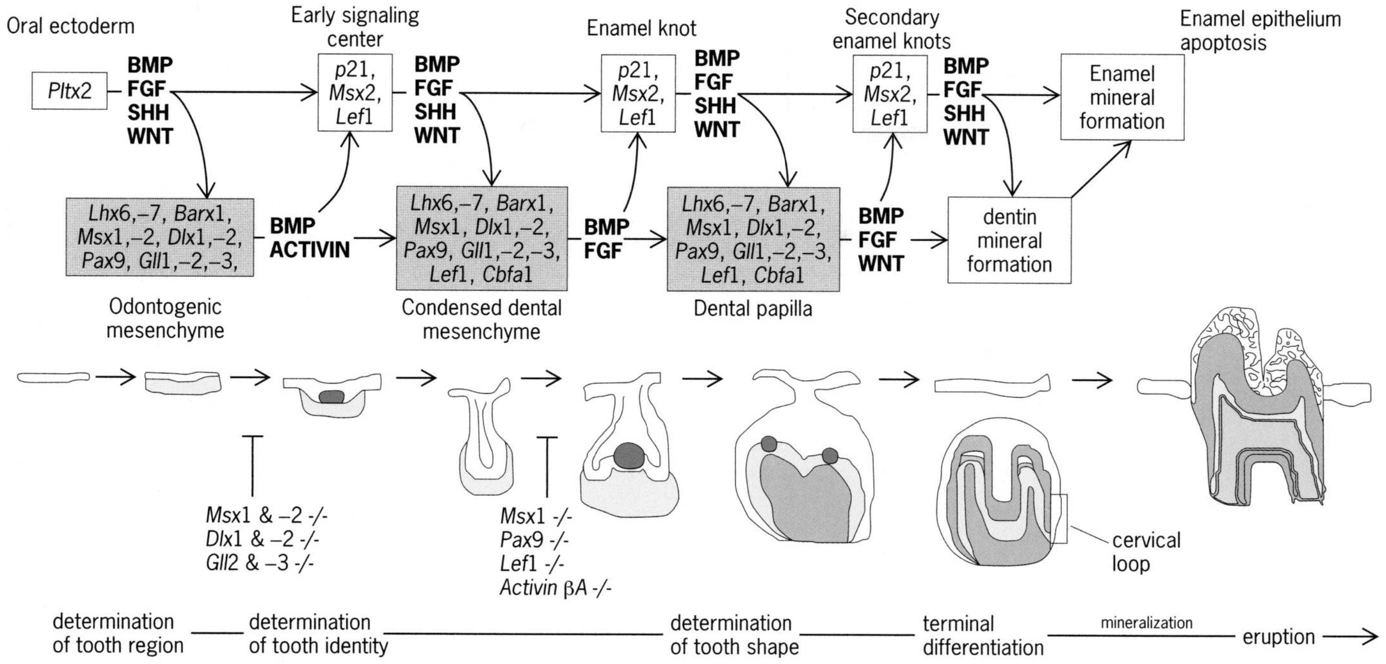

Fig. 1. Schematic correlation of morphological changes in tooth development and the expression and use of transcription and growth factors. The upper tier boxes list the most important transcription factors and, to their right, the most prominent growth factors in the oral epithelium; the lower boxes provide the same information for the odontogenic mesenchyme. Horizontal arrows show the direct effects of the factors within the same layer of cells; curved arrows show the direction of epigenetic signaling between layers. The corresponding tooth structures are shown below, with information on the effects of gene knockout studies that block development at the specified points. The region noted in the tooth terminal differentiation step as the cervical loop area is shown in expanded form in Fig. 3. (*Adapted from Jernvall and Thesleff, Mech. Develop., 92:19–29, 2000*)

is divided into three stages, called bud, cap, and bell, based on the overall morphology of the enamel organ at each stage (Fig. 2).] Similarly, at least 6 *Dlx* genes are expressed in these same stages. Further complicating the picture, each of these transcription factors and signaling proteins has specific receptors and antagonists that are differentially expressed and thus modulate the activities and net effects of the various factors.

The enamel knot. The specific thickening of the dental lamina at the site where a tooth is to be formed takes place by cell replication in the direction perpendicular to the dental lamina in response to epithelial factors. *Wnt* gene regulation of segmental polarity is clearly involved. *Wnt-3* and *Wnt-7b* are expressed in the oral epithelium, but are absent from the sites of dental lamina where thickening takes place. In contrast, *Wnt-10b* is expressed specifically in the mesial cells of the thickenings, but not elsewhere in the oral epithelium (Fig. 2). These invaginating cells from the initial tooth bud compress and ultimately enclose the underlying dental mesenchyme. The epithelial cells directly at the tip of the invaginating bud form a signaling center, expressing BMP, FGF, Hh, and Wnt family members, in response to signaling from the cells of the condensed odontogenic mesenchyme, prominently including BMP4. The epithelial cells become the aggregate known as the enamel knot (EK, the colored dots in Fig. 1), and the signals from it control the folding of the epithelium that leads to the cap stage. The tooth crown shape is determined by expansion of the epithelium to enclose more of the mesenchyme, essentially forming two distinct cell aggregates: the enclosed dental papilla and the surrounding dental follicle. The epithelial cells facing the dental papilla (the inner enamel epithelium, IEE) differentiate into enamel-producing ameloblasts, while the opposing mesenchymal cells mature into odontoblasts and produce the dentin. The cells of the EK are transient and do not proliferate, but undergo apoptosis (cell death). Although the EK expresses FGFs (fibroblast growth factors), which are important for cell division and growth of the IEE, the EK cells do not produce FGF receptors. The FGF family members are an excellent example of the reciprocal signaling between epithelium and mesenchyme. *Fgf10* is expressed only in the mesenchyme of the dental papilla during the cap and early bell stages, *Fgf3* is expressed in both EK and the mesenchyme of the papilla and follicle, while *Fgf4* and *Fgf9* are expressed only in the EK. FGF4 and FGF9 diffuse and stimulate cell proliferation in both the epithelium and mesenchyme, whereas FGF10 stimulates mitogenic activity in the dental epithelium but not in the odontogenic mesenchyme. In teeth destined to have multiple cusps, multiple new secondary EKs develop and regulate the mitogenesis related to those shapes, with FGFs playing a major role and with *Fgf4* expression restricted to the EKs.

Terminal differentiation and tooth eruption. The epithelia related to the enamel organ include the outer enamel epithelium (OEE), stellate reticulum, and stratum intermedium as well as the IEE. The

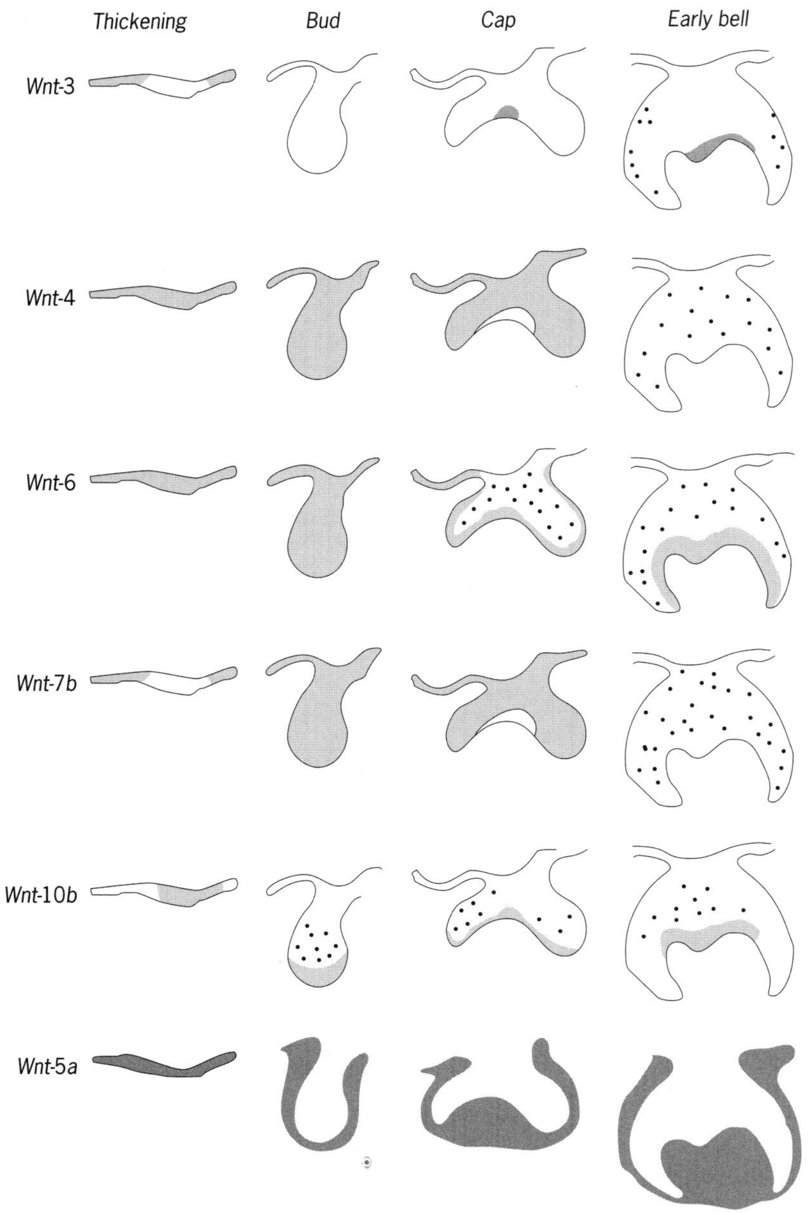

Fig. 2. A diagrammatic representation of the specificity of expression of Wnt pathway genes at different stages in tooth development. Key: light grey, expression within the epithelium; dark gray, expression in the mesenchyme. (*Modified from Sarkar and Sharpe, Mech. Develop., 85:197–200, 1999*)

cells connecting the OEE and IEE form the cervical loop. When mitogenic expansion of the IEE in the apical direction is completed, the mature size of the tooth crown is achieved. Little is known about this aspect of size control. The odontoblasts of the dental papilla continue to increase in number and extend in the apical direction to form the root, interfacing with the dental follicle cells and the dental mesenchyme. Signals from the odontoblasts trigger the formation of another mineralized part of the system, the cementum, from the dental mesenchyme, anchoring the attachment of the tooth to the bone of the maxilla and mandible. Family members from the same basic set of BMPs and FGFs, as well as others, are undoubtedly involved. However, at this point in tooth maturation, a new group of signaling events is brought into play to regulate the rates of formation of the mineralized enamel and dentin. The lack of detail on the right-hand side of Fig. 1 at the terminal differentiation step reflects the current state of knowledge. **Figure 3** (outlined area in Fig. 1) illustrates the structures of the preameloblasts (preAm) and preodontoblasts (preOb) as they mature to secretory stage. Proteinases released from the preOb degrade the intervening basement membrane and allow the passage of epithelial signaling molecules from the preAm to initiate odontoblast maturation, cell polarization, and organization for directed secretion of the dentin collagen and other extracellular matrix macromolecules. These organize into the predentin and, at the same time, communicate with the preAM to delay ameloblast maturation. When the predentin finally begins to mineralize, these epithelial-mesenchymal communications are blocked, and the preAM then proceed to mature to their secretory state, deposit their amelogenin-rich matrix, and form the mineralized enamel. The consequence of this delay has the physiological value of ensuring that the enamel mineralized layer, which is essentially a permanent, nonremodeling, nonliving structure in the mature tooth, does not form until the proper softer, living, and responsive dentin base is present. The disorder dentinogenesis imperfecta (an inherited dental disorder that causes defective dentin formation) illustrates the imperative of a proper dentin base for enamel since the enamel that is formed on a nonmineralized dentin matrix is readily fractured and removed. The signaling proteins passed at this crucial stage are under intense investigation, but appear to be molecules related either to specific protein products of spliceforms of the amelogenin gene expressed transiently in the odontoblasts or to members of the SIBLING (small integrin-binding ligand N-linked glycoprotein) protein family that may be expressed in either epithelium or mesenchyme. Dentin matrix protein 1 (DMP1), dentin matrix protein 2 (DMP2, DPP), and dentin sialoprotein (DSP) are major SIBLINGs of the tooth, but are not restricted to the tooth. DMP1 has been found to be of importance in bone, and recent reports have linked DPP to early activity in kidney ureteric bud and lung alveolar bud development.

The early gene-regulated development of the teeth is similar to that of all segmented systems in that the same set of gene expressions and growth factors common to all are used repetitively at all stages, but in different combinations at each point. As the tissues mature in both cyto- and morphodifferentiation, it is the local permutations of reiterated expression of these genes that guide development. At the later stages of maturation, tissue-specific types of signaling macromolecules are expressed, and these may also be involved in important regulatory events by their involvement with the integrin-related signaling pathways.

For background information *see* DENTITION; DEVELOPMENTAL GENETICS; GENE; GENE ACTION;

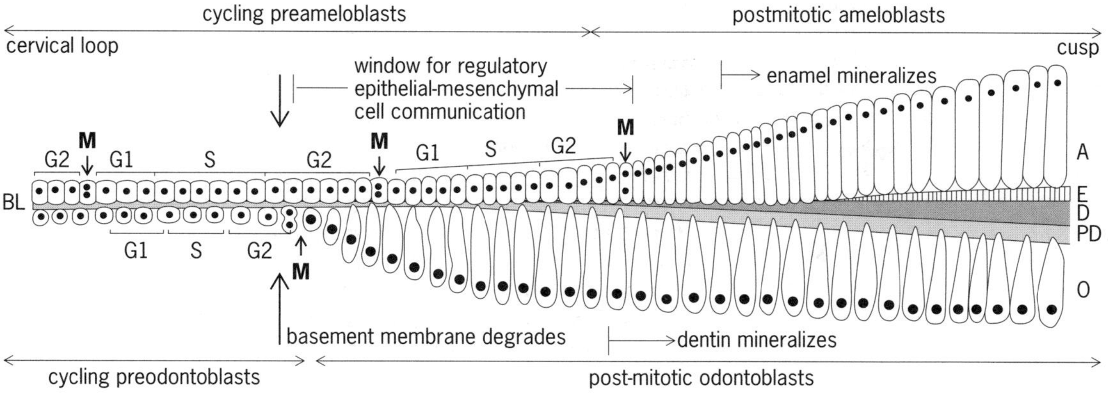

Fig. 3. The timing in the maturation of preameloblasts and preodontoblasts relative to the completion of their mitotic cycles and subsequent progression to the secretory stage. Mitotic expansion ceases in the cells of the IEE at the cervical loop, whereas the odontoblasts continue to form the tooth root dentin. The delay between the last mitosis in the ameloblasts and odontoblasts, between the time when the basal lamina is degraded and the development of the dentin mineral, is the crucial final period for epithelial-mesenchymal signaling using tissue-specific ameloblast and odontoblast proteins. Over the period of opening of this regulatory window, the diffusing epithelial signals (large downward arrow) stimulate the maturation of the odontoblasts; the return signals from the maturing odontoblasts (large upward arrow) inhibit preameloblast maturation until shut off by the mineral of the dentin. (*Adapted from A. Veis, Cell. Mol. Life Sci., 60:38–55, 2003*)

GENETICS; GERM LAYERS; MORPHOGENESIS; NEURAL CREST; TOOTH; TOOTH DISORDERS in the McGraw-Hill Encyclopedia of Science & Technology. Arthur Veis

Bibliography. K. Alvares, Y. S. Kanwar, and A. Veis, Expression and potential role of dentin phosphophoryn (DPP) in mouse embryonic tissues involved in epithelial-mesenchymal interactions and branching morphogenesis, *Dev. Dynam.*, 235:2980–2990, 2006; J. Huang et al., Dentin matrix protein 4, a novel secretory calcium-binding protein that modulates odontoblast differentiation, *J. Biol. Chem.*, 282:15357–15365, 2007; J. Jernvall and I. Thesleff, Reiterative signaling and patterning during mammalian tooth morphogenesis, *Mech. Develop.*, 92:19–29, 2000; J. V. Ruch, Tooth crown morphogenesis and cytodifferentiations: Candid questions and critical comments, *Connect. Tissue Res.*, 32:1–8, 1995; L. Sarkar and P. T. Sharpe, Expression of Wnt signalling pathway genes during tooth development, *Mech. Develop.*, 85:197–200, 1999; K. Tompkins, Molecular mechanisms of cytodifferentiation in mammalian tooth development, *Connect. Tissue Res.*, 47:111–118, 2006.

Giant panda

Ever since Manhattan socialite Ruth Harkness set foot in San Francisco in 1936 holding a baby giant panda—the first living panda to enter the United States—the world has been enamored with this enigmatic black and white bear. Yet, despite the giant panda's immense global popularity, this animal is listed as endangered under the World Conservation Union's Red List of Threatened Species (International Union for Conservation of Nature and Natural Resources, IUCN)—the de facto catalog of the world's most endangered species. Thus, how is it that the world's oldest surviving bear species—with an ancestry stretching back 7–8 million years—is considered endangered?

Current status and threats to survival. At one time, giant pandas (**Fig. 1**) were found across much of south-central China and even northern Vietnam. However, as the forests in these regions were replaced with fields of rice, towns, and cities, the panda's range steadily shrunk to its present-day refuge—China's temperate and tropical forests covering the steep mountains that rise up to form the Tibetan plateau (**Fig. 2**). There are approximately 24 separate panda populations spread across six mountain ranges. The total remaining global habitat is less than 14,000 square miles (36,260 square km)—about the size of the state of Maryland (Fig. 2). According to the Third National Giant Panda Survey (conducted by China's State Forestry Administration and the World Wildlife Fund), completed in 2004, there are as few as 1590 pandas left in the wild.

The giant panda's scientific name—*Ailuropoda melanoleuca*—means "panda foot, black-and-white," referring to its distinctive coloration as well as its unique "pseudothumb." Bamboo comprises

Fig. 1. Giant panda (*Ailuropoda melanoleuca*). (*Photo by Gary M. Stolz/U.S. Fish and Wildlife Service*)

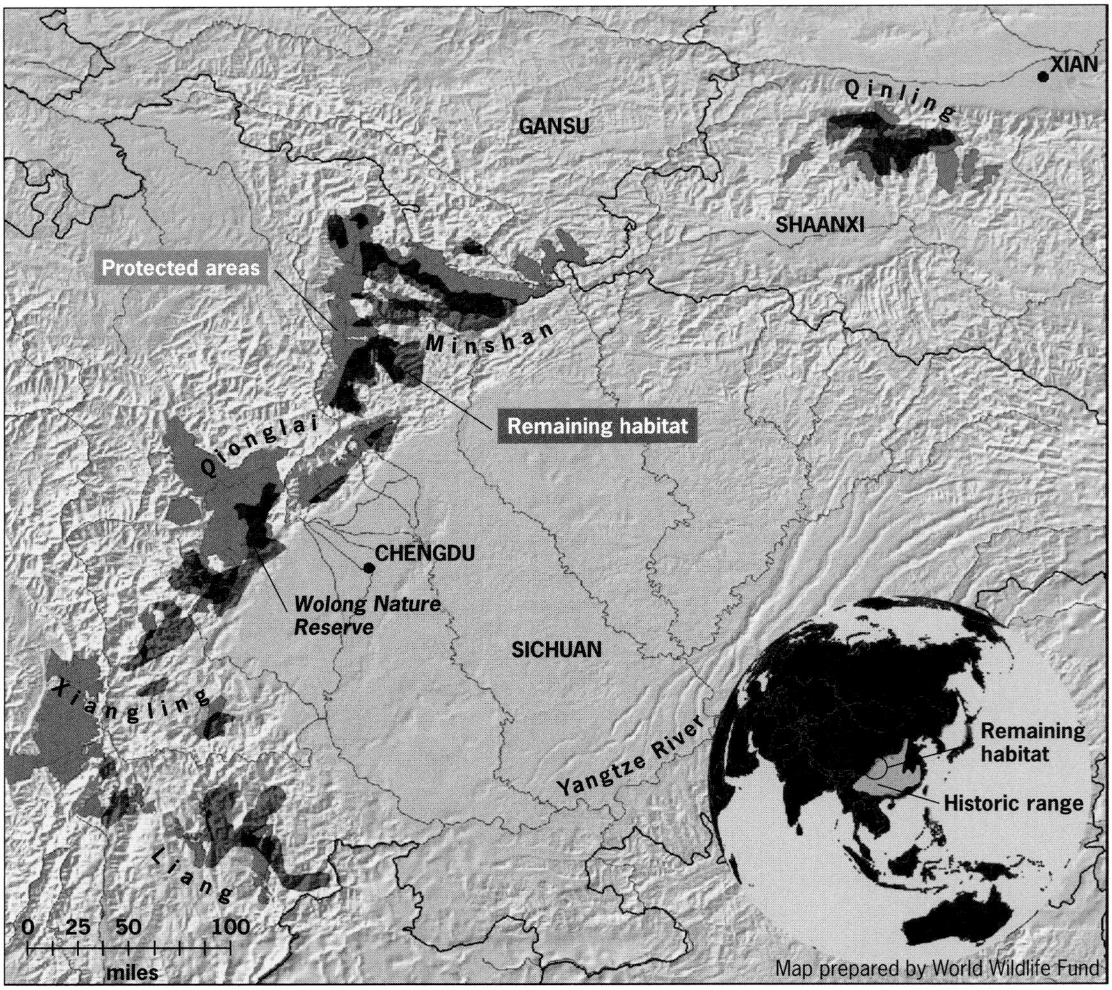

Fig. 2. Present distribution of the giant panda. Protected areas that overlap with giant panda distribution are shown in dark gray. The inset globe shows the historic distribution of the giant panda, with remaining habitat given for reference. (*World Wildlife Fund*)

99% of the panda's diet; in order to allow the panda to grab and hold bamboo stalks, a sixth digit—actually a wrist bone—evolved into a pseudothumb. The panda's most pressing need for survival is continued access to its bamboo habitat. Bamboo grows best under the protective cover of forests, linking the survival of pandas in the wild to these bamboo forests. Occasionally, wide swaths of bamboo die simultaneously. This natural event occurs when a bamboo species flowers. Bamboo flowering and subsequent die-off happens periodically every 5–40 years. When bamboo dies in one area, pandas need to move to new bamboo forests or face starvation.

Until recently, most of the panda's remaining bamboo forest habitat was steadily being cut down and replaced with agricultural crops, such as cabbage, or planted forest plantations, which are ill suited for bamboo survival. Furthermore, forests were being cut down by adjacent communities so they could use the wood to heat their homes, cook their food, or cook the food they feed farm animals such as pigs. Taken together, this steady erosion of panda habitat has had considerable implications for the future of pandas, as they are being forced into the steepest, most isolated forest areas, typically support have less bamboo.

The fragmentation of remaining habitat into ever smaller pieces compounds the existing pressure. Whereas habitat can be fragmented by natural obstacles (for example, large rivers), it also has been increasingly fragmented by human development such as roads, railroads, and dams. China is steadily implementing development programs to open up western provinces for greater economic development. In 2006, China completed the world's highest operating railroad connecting Beijing to Tibet's ancient capital, Lhasa. Road and dam development that accompanied this initiative adversely affects giant panda habitat and threatens to impose barriers to panda movement and reduce population size.

Poaching and unregulated tourism pose additional threats to the giant panda. While death penalties or long prison terms have largely deterred most panda poachers and smugglers, the pandas still inadvertently get caught in wire snares set to capture deer or other animals. Wolong Nature Reserve—China's giant panda flagship panda reserve—receives more than 100,000 tourists every year. Many will stay overnight and require food and lodging. The natural resources used to care for tourists are often obtained from the local forests—the very forests that the pandas need for their survival.

Conservation action and recent policy changes. The panda's long-term conservation needs include the permanent protection of its remaining habitat and restoration of fragmented agricultural areas back to thriving contiguous forests. These are formidable hurdles, but promising developments in recent years provide guarded optimism. China has enacted several forward-looking conservation programs involving an investment of roughly $1 billion annually for over a decade.

In 1998, massive flooding, the worst experienced in more than 40 years in China, affected large parts of China. More than 81,000 square miles (209,780 square km) of land were flooded, affecting more than 240 million people and causing more than $20 billion worth of damage. The Chinese central government, responding to the extensive damage done by floods, declared a national logging ban of all natural forests. While this policy's intended purpose was to prevent future large-scale flooding, it also inadvertently secured a future for pandas by giving temporary protection to all remaining panda habitat. Since this ban was enacted, timber production has fallen 97% within China.

In the wake of the 1998 floods, the Chinese government also instituted a "Grain-to-Green" policy. The Grain-to-Green policy is attempting to restore forests on hillsides that had been cleared of timber and converted to agriculture fields. The impetus behind this policy was to limit downstream flooding by planting trees to uptake water and limit soil erosion. In areas where this policy is being implemented adjacent to nature reserves or intact forests, the long-term benefit to pandas may be increased habitat or corridors of habitat between protected areas. However, the net result of this policy's effects on pandas is not likely to be known for several more years, as the planted forests need time to grow.

A third Chinese policy initiative—the Wildlife Conservation Protection Program—establishes nature reserves specifically to protect pandas. China established its first panda reserves in 1963. By the 1980s, there were fewer than 20 reserves, despite a clear need to conserve more forests. Under the Wildlife Conservation Protection Program, the number of reserves doubled from 30 in 2002 to 60 in 2006. The Wildlife Conservation Protection Program is attempting to increase the protection of wild pandas in nature reserves from 75% today to 90% in the coming years. Past research has demonstrated that populations will rebound and grow if wild pandas are given proper protection, thus avoiding the breeding difficulties common to zoo populations.

The work of the Chinese government is being carried out in collaboration with a number of Chinese and international nongovernmental organizations (NGOs) and international zoos dedicated to ensuring that pandas survive in the wild. The organizations often complement the work done by the government by providing training to the reserve staff in reserve management and monitoring or work on developing innovative methods to conserve remaining habitat outside of existing reserves, such as providing financial incentives for local communities to protect the panda.

Captive breeding and reintroduction. Views on the significance of captive panda populations to the conservation effort are varied. From a research standpoint, captive populations provide scientists access to pandas that would be otherwise impossible with wild pandas. Research in zoos and captive breeding centers has allowed scientists to examine and understand valuable information on panda life history, disease, communication, and reproduction.

The advances made in understanding panda biology and reproduction, combined with the use of artificial insemination, have sparked a recent "baby boom" in captive-bred pandas. For example, in 2006, China's collective giant panda captive breeding facilities produced more than 30 cubs—a record. More than 200 captive-bred pandas were living in zoos at the end of 2006, most of those in China. In the United States, four captive pandas have been born in three different zoos from 1999 to 2005.

At 120 captive pandas, the Wolong breeding center has the largest captive panda population in the world. The goal of the breeding center is to eventually introduce captive-bred pandas into the wild to bolster wild populations. However, efforts to reintroduce captive pandas into the wild have been unsuccessful, underscoring the daunting challenge of reintroducing captive-bred individuals into wild populations.

Outlook. Conservation of the giant panda has made significant strides in the 70 years since Ruth Harkness introduced giant pandas to the Western world. The panda's future—and that of numerous other rare or endangered species that share its habitat—in the wild is better secured today than a mere decade ago. This is largely due to the numerous successful collaborations among NGOs, governments, and zoos to infuse cutting-edge science into policy decisions that have yielded measurable progress in understanding and addressing the biological and ecological needs of giant pandas.

For background information *see* CARNIVORA; ECOLOGICAL MODELING; ECOLOGY; ENDANGERED SPECIES; MAMMALIA; PANDA; POPULATION ECOLOGY; POPULATION VIABILITY in the McGraw-Hill Encyclopedia of Science & Technology. Colby J. Loucks

Bibliography. D. Lindburg and K. Baragona, *Giant Pandas: Biology and Conservation*, University of California Press, Berkeley, 2004; S. Lumpkin and J. Seidensticker, *Smithsonian Book of Giant Pandas*, Smithsonian Institution Press, Washington, D.C., 2002; G. Schaller, *The Last Panda*, University of Chicago Press, Chicago, 1993; L. Zhi, *Giant Pandas in the Wild: Saving an Endangered Species*, Aperture Foundation, Hong Kong, 2002.

Gibberellin biosynthesis and signal transduction

The gibberellins make up a large class of carboxylic acids, certain members of which function as endogenous growth regulators in plants. They are diterpenoids, derived from four C_5 isoprene units, containing 19 or 20 carbon atoms arranged in a four-ring structure (**Fig. 1**). As well as occurring in higher plants, gibberellins are produced by some fungal and bacterial species and were originally discovered as secondary metabolites of the fungus *Gibberella fujikuroi*. There are currently 136 fully characterized gibberellins that have been isolated from all sources. These are named gibberellin $A_{1...136}$ in order of discovery, commonly abbreviated GA_1, GA_2, etc., with GA used as a general abbreviation for the whole class.

The predominant biologically active GAs in higher plants are GA_1 and GA_4, with GA_1 the more abundant form in most species. However, GA_4 is the major active form in some species, such as members of the Cucurbitaceae and in *Arabidopsis thaliana*, in shoots of which it is present at 10 times the concentration of GA_1. *Gibberella fujikuroi* produces mainly GA_3, also known as gibberellic acid, and is used for the commercial production of this compound, which has a number of applications in agriculture, including the production of malt and seedless grapes. However, although GA_3 is endogenous to some higher plant species, its concentration in plants is low compared with GA_1. The major activity of GAs in higher plants is to induce organ elongation or expansion, including that of roots, hypocotyls, leaves, stems, floral organs, and fruit. In addition, GAs promote seed germination and induce some developmental switches, such as between the juvenile and adult growth phases and between vegetative and reproductive development. Gibberellin signaling responds to environmental cues, including changes in light conditions, temperature, or stress, allowing these extrinsic signals to be translated into developmental changes.

Biosynthesis. As diterpenoids, GAs are synthesized from *trans*-geranylgeranyl diphosphate (GGPP) produced in plastids, predominantly via the methylerythritol phosphate (MEP) pathway (an important pathway for biosynthesis of isopentenyl diphosphate, the central precursor of isoprenoids). There is also some contribution from the cytosolic mevalonic acid pathway (another important cellular metabolic pathway), presumably due to movement of isopentenyl diphosphate across the plastid membrane. The conversion of GGPP to the final biologically active GAs, such as GA_1, requires the activity of terpene cyclases, cytochrome P450 monooxygenases (a family of enzymes mediating a wide range of oxidative reactions involved in the biosynthesis of plant secondary metabolites), and 2-oxoglutarate-dependent dioxygenases (Fig. 1). GGPP is converted in the plastid to the tetracyclic hydrocarbon *ent*-kaurene by the sequential action of two terpene cyclases: *ent*-copalyl diphosphate synthase (CPS), which forms rings A and B via a proton-initiated cyclization, and *ent*-kaurene synthase (KS), which forms rings C and D following cleavage of the diphosphate group. The *ent*-gibberellane structure is produced from *ent*-kaurene by the sequential activity of two cytochrome P450 monooxygenases. The first of these, *ent*-kaurene oxidase (KO), produces *ent*-kaurenoic acid by oxidation of C-19 from a methyl to a carboxyl moiety. The second enzyme, *ent*-kaurenoic acid oxidase (KAO), converts *ent*-kaurenoic acid to GA_{12} in three steps through oxidation at C-7, C-6, and again at C-7, with the intermediate step resulting in the contraction of ring B from six to five C atoms. These enzymes are associated with membranes, with KO thought to be located in the outer plastid envelope, while KAO is present in the endoplasmic reticulum.

Gibberellins A_4 and A_1 are produced in parallel pathways, the former from GA_{12} and the latter from its 13-hydroxylated analogue, GA_{53}. There is evidence for direct conversion of GA_{12} to GA_{53}, although 13-hydroxylation may also occur earlier in the pathway. Both P450-type and dioxygenase-type 13-hydroxylases may be present in plants. The conversion of GA_{12} and GA_{53} to the bioactive end products is catalyzed by the soluble dioxygenases, GA 20-oxidase (GA20ox) and GA 3-oxidase (GA3ox), the former producing the C_{19}-GA intermediates, GA_9 and GA_{20}, via C-20 alcohol and aldehyde intermediates and loss of C-20, and the latter enzyme converting these precursors to GA_4 and GA_1 by 3β-hydroxylation. In plants, GA_3, when present, is a result of side reactions of GA 3-oxidases, in which GA_{20} is desaturated to GA_5 and then hydroxylated. In *G. fujikuroi*, GA_3 is produced by a different pathway involving mainly cytochrome P450s. The pathways in fungi and higher plants would appear to have evolved independently.

Several mechanisms for inactivation of bioactive GAs have been identified. The best characterized is 2β-hydroxylation, which is catalyzed by another group of dioxygenases, the GA 2-oxidases (GA2ox). Inactivation can also result from oxidation of the 16,17-double bond, catalyzed by class CYP714 cytochrome P450s. In rice, an enzyme of this type, encoded by the *EUI* gene, produces 16,17-epoxides that are then hydrated to 16,17-dihydro-16,17-dihydroxy derivatives (dihydrodiols). In *Arabidopsis*, two genes encoding GA methyl transferases (GAMT), which produce inactive methyl esters, are expressed in developing seeds and may contribute to GA inactivation in these organs. Conjugation of GAs with glucose is potentially another inactivation mechanism that may serve to sequester GAs in the vacuole.

The enzymes catalyzing early steps in the pathway are generally encoded by single or small numbers of genes. For example, *Arabidopsis* contains single copies of *CPS*, *KS*, and *KO*, and two *KAO* genes. Loss-of-function mutations in these single-copy genes result in a severe GA-deficient phenotype. The dioxygenases (GA20ox, GA3ox, and GA2ox) are encoded by small gene families, with members showing distinct temporal and spatial expression patterns. Loss of one

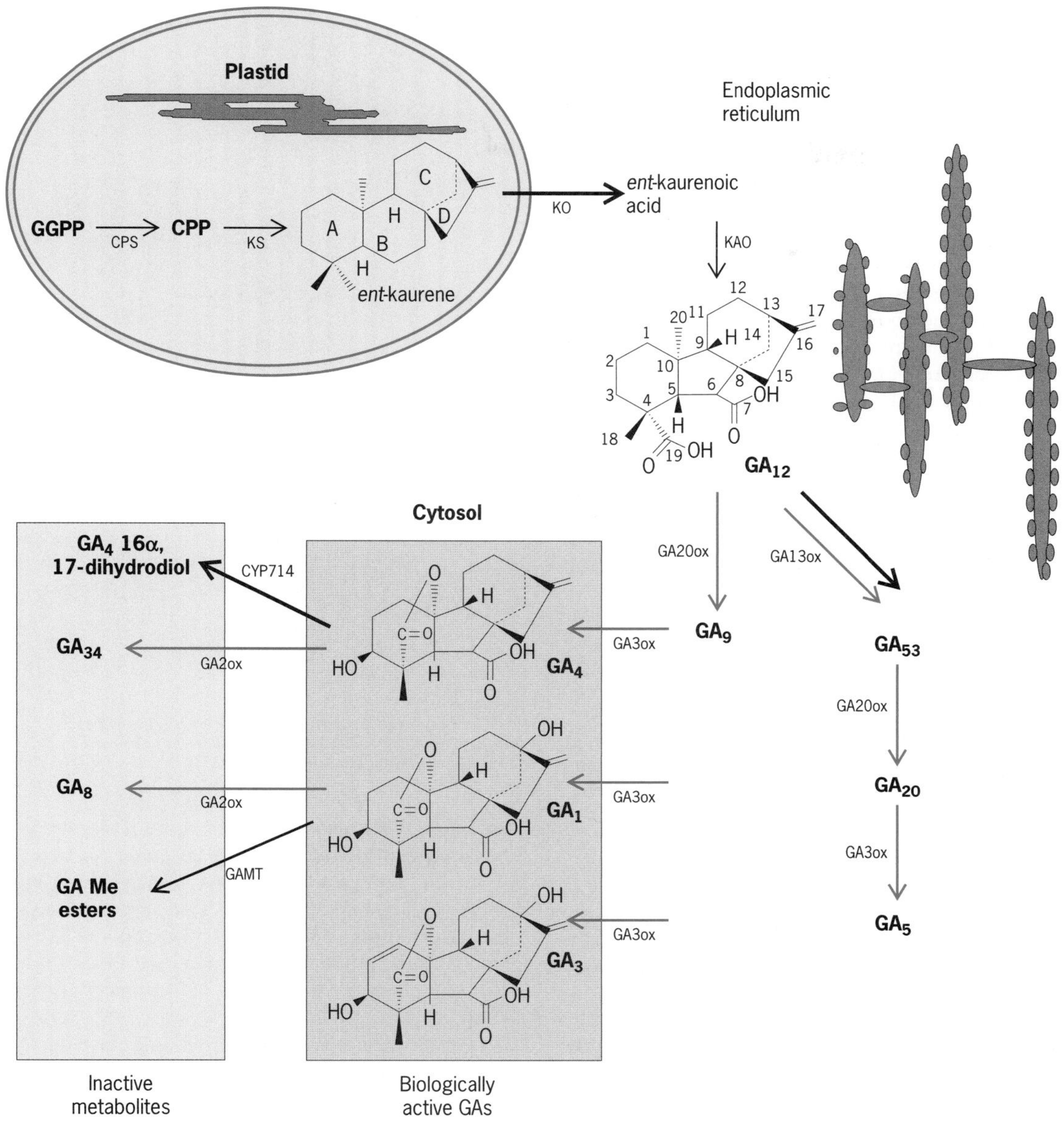

Fig. 1. Simplified overview of the GA-biosynthetic pathway from GGPP, showing subcellular compartmentalization. The biologically active GAs (GA_1, GA_3, and GA_4), are boxed in the center under "cytosol" and inactive metabolites are boxed to the left. Reactions catalyzed by terpene cyclases and gibberellin methyl transferase (GAMT) are shown with black arrows, cytochrome P450-catalyzed reactions with heavy black arrows, and reactions catalyzed by 2-oxoglutarate-dependent dioxygenases with colored arrows. The ring designations are shown for *ent*-kaurene and the C-atom numbering for GA_{12}.

of these enzymes has minor effects, often restricted to certain organs and developmental processes, indicating some redundancy between the gene family members as a result of overlapping expression or movement of GAs between tissues. The *GA2ox* genes form the largest group, which divides into three clades, two of which encode enzymes that act on C_{19}-GAs, while the third is specific for C_{20}-GAs and so acts earlier in the pathway.

The dioxygenase genes are major sites of regulation for GA biosynthesis, their expression being influenced by numerous developmental and environmental factors. For example, GA homeostasis is achieved by suppression of the expression of certain *GA20ox* and *GA3ox* genes and enhancement of *GA2ox* expression by GA activity. Stimulation of germination of photoblastic (light-sensitive) seeds by red light or by stratification includes induction of *GA3ox* and suppression of *GA2ox* expression, whereas de-etiolation (the greening response of plants grown in the dark) of dark-grown seedlings involves rapid light-induced induction of *GA2ox* expression followed by suppression of *GA3ox* expression. The dioxygenase genes are also the targets for auxin stimulation of GA production in fruit and stems. Gibberellin levels in the shoot apical meristem are regulated by KNOX-type homeodomain transcription factors (proteins involved in the regulation of gene expression that bind to the promoter elements upstream of genes and either facilitate or inhibit transcription) through up- and downregulation of *GA20ox* and *GA2ox* expression, respectively.

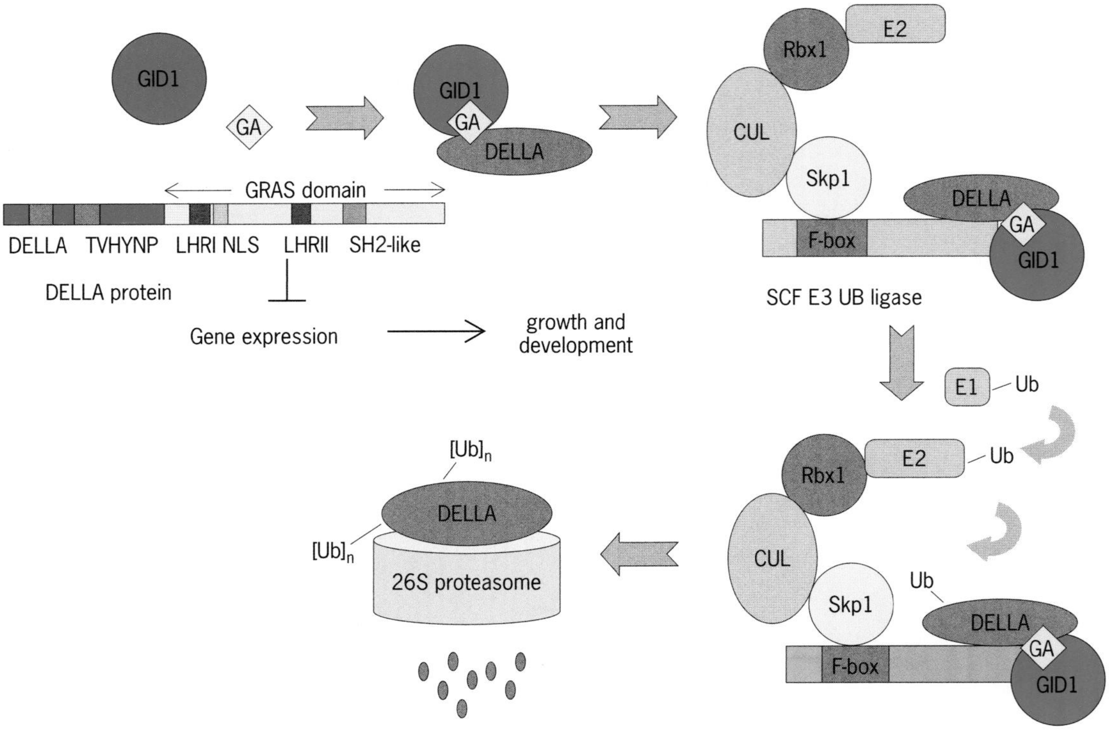

Fig. 2. DELLA protein showing conserved domains, and a model for DELLA degradation via the SCF E3 ubiquitin ligase, which consists of a DELLA-specific F-box, Skp1, Cullin1 (CUL), and Rbx1 subunits. Binding of the GID1-GA complex to DELLAs promotes association with the F-box, allowing polyubiquitination and degradation of DELLAs by the 26S proteasome.

Signal transduction. Gibberellins act through modifying the expression of genes in their target tissues. The change in expression of certain genes, such as the GA-biosynthetic genes *GA20ox* and *GA3ox*, occurs within 15–30 min after GA treatment, while induction of genes encoding the starch-degrading enzyme α-amylase in the cells of the aleurone, a layer surrounding the endosperm in cereal grains, takes several hours and is considerably downstream of the primary GA response.

Key components of the GA signal transduction pathway are the DELLA proteins, which are members of the GRAS family of transcriptional regulators, but uniquely contain a highly conserved N-terminal sequence, known as the DELLA domain, that is absent from other GRAS proteins. Cereals possess a single DELLA protein, while *Arabidopsis* contains five DELLAs which differ somewhat in their distribution, although there is considerable redundancy between them.

DELLAs are not typical DNA-binding proteins; instead, they are thought to control transcription as part of protein complexes, with their different partners imparting specificity of function. DELLAs act as growth repressors, although the molecular basis for this activity is still unknown. Gibberellins function by promoting the degradation of DELLA proteins in a fairly direct manner (**Fig. 2**). The hormones promote binding of DELLAs to a soluble nuclear protein (GID1), thereby facilitating the association of DELLAs with the F-box component (a protein motif of approximately 50 amino acids that functions as a site of protein-protein interaction) of an SCF E3 ubiquitin ligase. The E3 ubiquitin ligase covalently attaches ubiquitin (a small, 76-amino-acid, highly conserved protein present in the cytoplasm and nucleus of all eukaryotes) to a lysine residue on a target protein, a process important in polyubiquitination (that is, the binding of many ubiquitin molecules to the same target protein). The resulting polyubiquitination marks DELLAs for degradation by the 26S proteasome (a large proteolytic particle that is the site for degradation of most intracellular proteins).

The GID1 proteins, which function as GA receptors, are related to the hormone-specific lipases

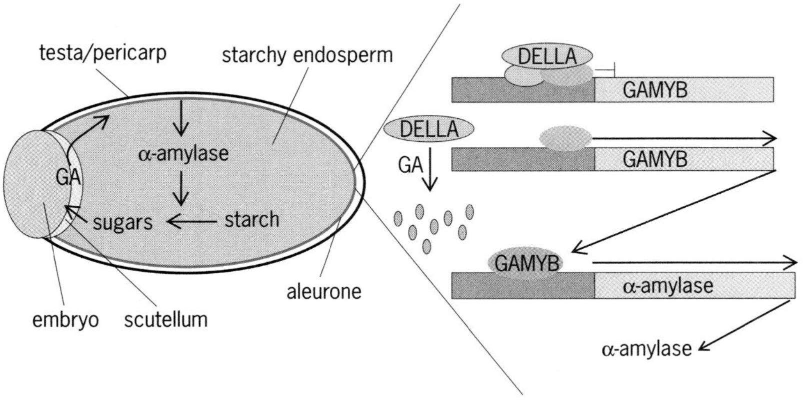

Fig. 3. Model for GA-induced production of α-amylase in the aleurone of germinating cereal grains. GAs produced in the scutellum diffuse to the aleurone cells, where they stimulate transcription, processing, and secretion of α-amylase and other hydrolytic enzymes. GAs induce expression of a GAMYB transcription factor that activates α-amylase transcription. DELLA proteins suppress *GAMYB* expression, although the mechanism is not yet fully understood.

(HSL), but lack lipase activity due to an amino acid substitution at the catalytic site. The GID1-GA complex binds to DELLA proteins via their DELLA and TVHYNP domains, mutation of which results in loss of GA-induced degradation, but otherwise does not affect DELLA function, which resides in the C-terminal GRAS domain. Such gain-of-function mutations give rise to GA-insensitive dwarfs, which have been identified in several species, most notably wheat, in which the introduction of *Rht* alleles to produce semidwarf high-yielding varieties was an important factor in the Green Revolution, that is, the dramatic increases in cereal-grain yields in many developing countries beginning in the late 1960s, due largely to use of genetically improved varieties.

The gene targets of the GA signaling pathway and how their activity translates into physiological responses are still being elucidated. Stem extension, a classical GA-stimulated process, is the result of cell elongation and/or cell division, and GA has been shown to induce upregulation of genes that mediate both processes. These include genes encoding xyloglucan endotransglycosylase/hydrolase (XET) and expansins, which enable cell-wall loosening, as well as cyclin-dependent kinases that promote cell division. In cereal grains, GA acts as a paracrine signal between its site of synthesis in the scutellum (the flattened cotyledon of a monocotyledonous plant embryo, such as a grass) and the aleurone cells of the endosperm, where it induces production and secretion of hydrolytic enzymes, including α-amylase, that break down the macromolecular reserves in the endosperm. Expression of α-amylase genes is activated by an MYB transcription factor (GAMYB), whose transcription, and possibly also posttranslational activation, is induced by GA through release of DELLA suppression (**Fig. 3**).

Gibberellin action also causes a rapid increase in cytosolic Ca^{2+} concentration and in the abundance of calmodulin, a Ca^{2+}-binding protein, as well as cGMP. The role of these factors in GA signaling is not clear, but Ca^{2+}/calmodulin may be involved in the synthesis and secretion of α-amylase, which is a metalloenzyme (a protein enzyme containing a metallic atom as an inherent portion of its molecule that catalyzes important cellular reactions).

For background information *see* AGRICULTURAL SCIENCE (PLANT); AUXIN; GIBBERELLIN; PLANT GROWTH; PLANT HORMONES; PLANT MORPHOGENESIS; PLANT PHYSIOLOGY; TERPENE in the McGraw-Hill Encyclopedia of Science & Technology.

Peter Hedden

Bibliography. P. Hedden and A. L. Phillips, Gibberellin metabolism: New insights revealed by the genes, *Trends Plant Sci.*, 5:523–530, 2000; N. Olszewski, T. P. Sun, and F. Gubler, Gibberellin signaling: Biosynthesis, catabolism, and response pathways, *Plant Cell*, 14:S61–S80, 2002; T. P. Sun and F. Gubler, Molecular mechanism of gibberellin signaling in plants, *Annu. Rev. Plant Biol.*, 55:197–223, 2004; S. G. Thomas and P. Hedden, Gibberellin metabolism and signal transduction, in P. Hedden and S. G. Thomas (eds.), *Plant Hormone Signaling*, pp. 147–184, Blackwell, Oxford, UK, 2006; M. Ueguchi-Tanaka et al., GIBBERELLIN INSENSITIVE DWARF1 encodes a soluble receptor for gibberellin, *Nature*, 437:693–698, 2005.

Green engineering

Green engineering is a much-needed approach to transform existing engineering disciplines and practices to those that promote sustainability. The concept of sustainability is to develop and implement technologically and economically viable products, processes, and systems that meet the needs of humanity, while protecting the environment. Green engineering is governed by the following principles:

1. Use the least amount of energy to achieve any given task.
2. Generate as much energy as possible using renewable resources.
3. Generate the least amount of pollutants and by-products during energy generation.
4. Use renewable and biodegradable materials to a maximum extent for building structures and fabricating products.
5. Reduce waste during construction and fabrication.
6. Design structures and products to maximize their life spans and minimize maintenance.
7. Design for easy deconstruction and facilitate the reuse of components and materials from obsolete structures and products in new construction and fabrication.
8. Make the least impact on the environment.

The obvious question is: Why are these principles not followed? The answer is: Because of economics, convenience, ignorance, and affluence, with economics playing the major role. For example, thermal power plants are still a more economical source for electrical energy compared to solar energy. However, the depletion of raw materials and the cost of controlling pollution and by-products are resulting in a steady increase in the cost of electricity produced by thermal power plants. This, in combination with the improved efficiency of solar cells, is making solar cells a viable alternative. Similar arguments can be made for all the other principles. Another example is the use of steel versus timber for structures. Industrialization has created a demand for large factory and office buildings that cannot be built with timber and bricks. Even though steel consumes much more energy to produce and fabricate, it has to be used to construct tall buildings and cover long spans. In this case, the economics of doing business and crossing large rivers takes priority over the consumption of energy.

In the area of energy production, the fraction of energy produced by renewable resources is still very small. The popular sources are coal, natural gas, oil, and nuclear. Hydroelectric power plants are a long-standing renewable source. The growing sources of energy production are solar and wind. Fossil fuels generate carbon dioxide and large amounts of residues, such as fly ash and bottom ash. Nuclear power plants generate less waste, but, it is very expensive to dispose of nuclear wastes. Philosophically, most of the energy we are using came from the Sun. For example, coal, oil, oil shale, and tar sand were produced over millions of years from forest growth. If we could harness solar radiation, then most of the world's energy needs could be met. To achieve this, the efficiency of solar cells has to be increased considerably. It is estimated that the Sun provides about 120 quadrillion watts of energy daily, while worldwide consumption is about 13.5 trillion watts per year.

In the area of energy consumption, buildings and other infrastructures consume the major share. If the consumption for maintenance is included, the energy consumed by the building industry is estimated at 40%. The other major consumer is transportation. The transport of people and goods around the world is steadily increasing, and economic growth in Asia could accelerate this trend.

Buildings and infrastructures also consume the lion's share of materials. The most common construction materials are concrete, asphalt, steel, and clay. Timber is used widely in North America and in some parts of Australia. It is estimated that 12.5 billion tons of concrete are used annually and that it will be the dominant structural construction material for the near future. Portland cement, the most expensive ingredient in concrete, is produced worldwide, with the production cost ranging from (U.S.) $0.07 to $0.15 per kilogram. Almost every country in the world has the expertise for using this material, including design, fabrication, and maintenance. However, cement production results in the emission of carbon dioxide, which is a major drawback. Asphalt is a by-product of oil conversion to various fuels, and its use parallels that of oil. Timber is the only renewable structural construction material, but it is available in large quantities only in a few countries such as the United States and Canada. Steel consumes the most energy to produce but has mechanical properties that cannot be matched by any other structural material. Therefore, its use will continue at the current or slightly accelerated rate over the next few decades, especially in developing countries like China and India.

Built environment. The built environment uses the majority of the world's nonrenewable and much of its renewable material resource as well as contributes significantly to pollution, habitat destruction, and other forms of environmental damage. As the developing world's economies continue to grow, resources will be depleted even faster and exhausted sooner.

The National Science Foundation (NSF) of the United States supports research for developing building systems that are compact, minimally intrusive of the natural environment, resilient, economical, and easily recyclable, as well as systems that minimize the use of nonrenewable materials and nonrenewable energy. Specific research projects focus on developing pervious pavement that meets strength and durability requirements and is easily recyclable. Pervious concrete is an alternative paving material that is being considered as a best management practice for stormwater runoff and nonpoint source pollution control. NSF sponsored a project by the University of South Carolina for developing and implementing pervious concrete on several roads throughout South Carolina. The demonstration sites allowed researchers to gather data on the performance of this highway material.

Another renewable material being evaluated is recycled tires used in manufacturing sidewalk tiles. These sidewalks provide a surface that is more durable than traditional concrete sidewalks, since cracks caused by root growth do not occur because the material can flex. Compared to concrete, rubberized sidewalks provide a softer surface for running and walking.

On the larger scale of civil-built infrastructure, there is continued research on sustainable systems. For example, researchers are working on better systems for integrated water use for individual structures such as the reuse of so-called grey water in irrigation as well as rain collection and use in heating and cooling. Additional research includes developing local electrical generating systems and the means of connecting them to the larger grid.

Environmentally friendly materials. Research to develop novel bio-based products for use in various applications lies in the larger scope of renewable materials. Bio-based materials include industrial products, but not food or feed, made from renewable agricultural and forestry feed stocks, including wood, wood wastes and residues, grasses, crops, and crop by-products. Such eco-friendly biomaterials are desired for their recyclability and triggered biodegradability. Applications include natural fiber–reinforced composites, which could replace artificial fibers such as carbon and glass in various applications in the automotive industry. Researchers at Daimler-Chrysler have developed a plant-fiber-based material that has reduced the weight and cost of insulation in car door panels.

Energy devices. Another area of green engineering is the development of materials and processes that are biomimetic in nature. Researchers from the Massachusetts Institute of Technology, the U.S. Naval Research Laboratory, and the University of Tennessee have mixed biology and electronics in solar cells that use photosynthetic spinach leaf molecules or photosynthetic bacteria to convert light to electricity. Potentially, the method could be used to fabricate low-cost solar devices on plastic and other thin, flexible surfaces using inexpensive spray-on techniques.

Environmentally friendly manufacturing processes. Manufacturing and assembly processes are active areas of green engineering. The chemical process industry is developing methods for creating products necessary for daily life that do not cause collateral damage to the environment or society as a whole. Another area of interest is in disassembly of electronic components for reuse. By embedding a desired disassembly process in the products, disassembly is achieved simply.

Fundamental research needs. Research needs essentially follow the principles of green engineering. We need to generate more energy from solar and wind power, design and build structures and products to last longer, use renewable and biodegradable materials, and use less energy for maintenance of structures and transportation needs. In the case of solar power, the research needs are to increase the efficiency of energy consumption and develop new materials other than silicon for energy conversion. Developments in nanotechnology are creating new opportunities for introducing new photovoltaic materials.

In the area of structures, recent developments in self-cleaning and depolluting material combinations are opening up new research areas. By incorporating nano-sized anatase titanium dioxide (TiO_2) in concrete and surface coatings, researchers have shown that self-cleaning and depolluting surfaces can be created. This opens up avenues to use buildings to clean the air around them. Energy and resources are saved by reducing maintenance. A short review of the current state of the art and the research needs is presented in the following section.

Self-cleaning and depolluting construction systems. Photocatalysis, which is a reaction mechanism used between a semiconductor catalyst such as titanium dioxide and natural or artificial light, is the scientific basis for both self-cleaning and depollution mechanisms. The focus of most of the current work is on the use of titanium dioxide. Titanium dioxide is available in rutile and anatase forms; the anatase form is preferred for its photocatalytic activity.

Both indoor and outside air purification systems have been attempted. Most studies in Japan focus on indoor pollution. Building sizes with an airflow range of 100 to 1,500,000 m^3/h have been used in experimental studies. European studies are focusing on the outdoor environment. They are evaluating plaster, mortar, concrete, and coatings containing TiO_2. One of their recently reported activities was the use of 7000 m^2 of road surface in Milan, Italy, for evaluating photocatalytic cement, resulting in up to 60% reduction in the concentration of nitrogen oxides.

In the United States, researchers at Rutgers University are evaluating an inorganic polymer coating containing photocatalytic materials. The advantage of this system is that it can be applied to both new and old structures. The composition is compatible with common construction materials such as concrete, steel, timber, and clay bricks. **Figure 1** shows the self-cleaning properties of a coating made using nanoparticles. It is seen that the coated surface is free of mold, compared to the adjacent uncoated surface.

Sustainable construction. Realizing the importance of green engineering, the Directorate for Engineering of NSF is requesting proposals for research in the area of resilient and sustainable infrastructures (RESIN). This area was chosen as one of the two emerging frontiers of research innovation for 2007 by the Directorate. One of the focal areas addresses green engineering directly. The solicitation states, "Sustainable technologies for the design, analysis, and construction of resilient physical infrastructure networks that can be deconstructed and reconfigured or recycled, without generating waste and which require orders of magnitude less nonrenewable resources for their function." Although the focus is on infrastructure, the principles are applicable for manufactured products as well.

One of the important aspects of sustainable construction is the use of renewable structural material, and timber is the primary and truly renewable structural material. The major concerns with using timber are related to fire resistance, low stiffness and strength, creep, and durability. Recent developments in high-strength composite materials can be used to overcome most of these deficiencies. For example, research at Rutgers University indicates that carbon fibers and inorganic polymer can be effectively used to improve mechanical properties and fire resistance. Using normal and high-modulus fibers, they improved stiffness up to 20 times and the strength up to 10 times. The fiber has a modulus of 900 GPa, which is three times higher than steel. They also showed that 3-mm-thick structural insulation is sufficient to pass the Federal Aviation Administration's requirements for high-temperature exposure. **Figure 2** shows balsa wood with the coating after fire exposure. Experimental and analytical investigations covered the topics of beams and columns and various types of wood.

Another promising area is the use of nanotechnology and ceramics to create alumina-silicate coatings for concrete structures. It is well known that ceramic surfaces can last for thousands of years. Terracotta (baked clay) soldiers (figures) have been discovered in China that are more than 2000 years old. We now have the technology and materials to create complex shapes with terracotta properties with no need for high-temperature processing.

Fig. 1. Self-cleaning coating (on left) made of nanoparticles.

(a)

(b)

Fig. 2. Balsa wood with an inorganic polymer coating (*a*) before and (*b*) after high-temperature exposure.

Photovoltaic cells. The use of solar cells for energy production is increasing steadily. The most recent advances include the development of non-silicon-based solar cells, including flexible sheets and energy-producing paints. In the area of efficiency, researchers at the University of Delaware achieved 42.8% efficiency in 2007. They used a high-performance crystalline silicon solar-cell platform to achieve this efficiency at standard terrestrial conditions. The goal is to achieve 50% efficiency. These developments could lead to a net efficiency that is 30% greater than existing module efficiency and twice the efficiency of standard silicon solar-cell modules.

[The authors would like to acknowledge the contribution of Matt Carnavos, who compiled the research projects being supported by NSF.]

For background information *see* BIODEGRADATION; CERAMICS; CONCRETE; CONSERVATION OF RESOURCES; CONSTRUCTION ENGINEERING; CONSTRUCTION METHODS; ELECTRIC POWER GENERATION; ENERGY SOURCES; GREEN CHEMISTRY; NANOPARTICLES; RENEWABLE RESOURCES; SOLAR ENERGY; STRUCTURAL MATERIALS in the McGraw-Hill Encyclopedia of Science & Technology.

P. N. Balaguru; Ken Chong

Bibliography. M. Anpo, Utilization of TiO_2 photocatalysts in green chemistry, *Pure Appl. Chem.*, 72(7):1265–1270, 2000; M. Ehses et al., *Workshop on Utilization of Photocatalytic Coatings for Clean Surfaces*, Saarbruecken, Germany, 2001; A. Fujishima, K. Hashimoto, and T. Watanabe, *TiO_2 Photocatalysis: Fundamentals and Its Applications*, BKC, Japan, 1999; J. Giancaspro, P. Balaguru, and R. Lyon, Use of inorganic polymer to improve the fire response of balsa sandwich structures, *J. Mater. Eng.*, ASCE, 18(3):390–398, May–June 2006.

Gypsum megacrystals

Gypsum ($CaSO_4 \cdot 2H_2O$) is a common mineral in sedimentary environments. The crystal structure of this calcium sulfate dihydrate can be defined as alternating double-sheet layers of sulfates bound covalently by calcium ions and single-sheet layers of water molecules linked by weak hydrogen bonds. The mineral gypsum has several varieties that differ by the shape of the crystals and their textural arrangements. The variety called gypsum alabaster is made of fine-grained crystals and is used for craftworks. The satin spar variety is made of tiny fibrous crystals and has an attractive silky luster. The most famous variety of gypsum is called selenite, which is characterized by colorless and transparent crystals. Large selenite crystals were very valuable in Roman times because they were used for covering windows, in therma (baths) and palaces. Roman stonemasons took advantage of a well-known physical property of transparent selenite crystals, namely cleavage, the ease with which some crystals split along definite planes where the atoms are weakly bonded to the adjacent layers of atoms, thus creating smooth surfaces. In the case of gypsum, that plane is pinacoid (010), which corresponds to the plane parallel to the layers of molecules in the crystal structure. According to Pliny the Elder, the largest, high-quality selenite crystals were found in Segobriga, central Spain. This was the main source of crystals for window coverings until the introduction of the flat glass technology in the Roman Empire at the end of first century A.D. Amazing as these crystals were, they cannot be compared in size or quality with the crystal wonderland recently discovered in Naica, a mining town located 112 km (70 mi) SE of Chihuahua City in Northern Mexico.

Naica crystals. The limestone of the Naica Mountains contains one of the richest lead and silver mines in the world, which has been exploited since the nineteenth century. The ore minerals (silver-lead-zinc sulfides and sulphosalts) formed when the carbonate rocks were intruded by hot, acidic magma about 26 millions years ago. In addition to metal sulfides, the hydrothermal fluid circulation formed the mineral anhydrite ($CaSO_4$), which occurs massively in Naica at depths below −240 m (−790 ft). The mining district is still under a thermal anomaly that keeps the rocks and ground waters close to 55°C (131°F). The phreatic level (below the water table) is located at a depth of −120 m (−394 ft), so that the main engineering problem for mining and exploration activities in Naica is draining the water from the aquifer. This is achieved by pumping water at average rate of 55 m^3/min (14,500 gal/min). Noticeably, the water is slightly supersaturated in calcium sulfate

Fig. 1. Cave of the Swords, Naica mine, Mexico. Level = −120 m.

with respect to gypsum and slightly undersaturated with respect to anhydrite.

The underground mine of Naica can be described as a set of exploration and exploitation galleries arising from a central road helicoidally descending toward the mine front, which is currently at −800 m (−2600 ft). During more that 150 years of industrial mining, cavities containing large gypsum crystals were accidentally discovered. The largest cavities are always close to faults that formed when ore mineralization took place, guiding the main fluid flow in the rocks. The most famous of these cavities, called the Cave of the Swords, was discovered in 1910 at −120 m (−394 ft). It is a 75-m (246-ft) corridor along a fracture covered by flattened crystals of gypsum of a few centimeters long (crystal blades). In addition, the floor is spotted with bunches of meter-size elongate crystals called crystal swords (**Fig. 1**). In 2000, exploration at −290 m (950 ft) uncovered a few new cavities. The largest of them was called the Cave of Crystals by the miners. The number of crystals per unit volume in this cavity was much smaller than in Cave of the Swords. The floor was covered with stubby crystals of up to 2 m (6 ft) in size but the walls and the ceiling were just spotty with crystals, allowing the walls of limestone to exhibit a red coating containing celestite (strontium sulfate) and iron oxide, among other minerals. Even more amazing, giant elongate gypsum crystals up to 11 m (36 ft) in length, called beams, crossed the cavity from side to side (**Fig. 2**).

Formation of giant crystals of gypsum. The analysis of the fluids trapped inside has shown that the crystals grew from low-salinity solutions at a temperature of ~54°C (129°F), slightly below the temperature (58°C; 136°F) at which the solubility of anhydrite equals that of gypsum. Sulfur and oxygen isotopic compositions indicate that Naica's gypsum crystals grew from solutions formed by dissolving the anhydrite found in the mine. These and other analyses suggest that the megacrystals formed when the Naica district cooled below 58°C, thus triggering the dissolution of anhydrite and the precipitation of gypsum. The small number of crystals in the caves can only be explained by formation at very low supersaturation. Therefore, the problem is to explain how the required low supersaturation can be sustained for a long time without large fluctuations that would provoke substantial nucleation. **Figure 3** shows the variation of the solubility of anhydrite and gypsum as a function of temperature. Considering that the nucleation flow increases exponentially with supersaturation (or undercooling), for the Cave of the Crystals it is clear that undercooling should never have been very large, keeping the temperature always very close to 58°C. Larger undercooling would have yielded a large number of relatively smaller crystals, as is the case of the shallower (and therefore cooler) Cave of the Swords. A large undercooling would provoke the formation of millions of microcrystals of gypsum. Gypsum nucleation kinetics calculations based on laboratory data show induction times longer than 1 m.y. for the temperature of fluid inclusion (54°C). Therefore, this mechanism can account for the formation of these giant crystals, yet only when operating within the very narrow range of temperature identified by the fluid inclusions. These conditions are very difficult to accomplish, particularly the smooth cooling to

Fig. 2. Cave of the Crystals, Naica mine, Mexico. Level = −290 m. (*Photo by Javier Trueba*)

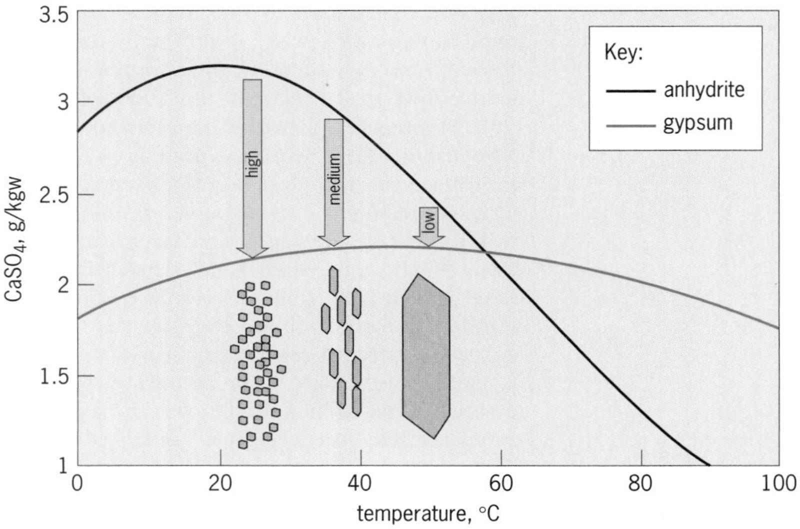

Fig. 3. Variation of solubility of anhydrite and gypsum as a function of temperature. Small differences in undercooling will provoke large changes in nucleation density.

Fig. 4. Giant gypsum crystals in El Teniente mine, Chile. (*Photo by Javier Trueba*)

Fig. 5. Crystal of the giant gypsum geode of Pilar de Jaravia, Spain. (*Photo by Javier Trueba*)

keep nucleation flow very low. When all of these conditions take place simultaneously, an extraordinary crystallization phenomenon occurs that creates a mineral wonderland.

Other giant selenite crystals. In addition to Segobriga and Naica, giant gypsum crystals have been found in two other places, both located in mining districts. One of them is the copper mine of El Teniente, close to Rancagua in the Andes of central Chile. El Teniente is the world's largest underground copper mine. El Teniente is a porphyry Cu-Mo deposit related to calc-alkaline felsic to intermediate intrusions of Miocene and Pliocene age. Most of the ore formed at the late stages of the magmatic episode (5.9–4.9 Ma). The center of this deposit is the so-called Braden Pipe, a massive granitic breccia pipe intruded during a late hydrothermal stage (4.8–4.4 Ma). The Braden breccia presents a system of vertical faults more intense at the boundaries of the intrusion. Several cavities developed in the fault zones containing transparent gypsum crystals of up to 7 m (23 ft) in length and 0.5 m (1.5 ft) thick. The largest cavity is located at level 6 and is about 300 m^3 (10,600 ft^3) in volume. The fracture seems to extend to other levels with more than 170 m (560 ft) of vertical continuity. The cavity has walls that are covered by centimeter-size euhedral crystals of pyrite and barite and contains elongate gypsum crystals up to 5 m (16 ft) in length (**Fig. 4**). The origin of these crystals is currently under study, but the presence of a large amount of primary igneous anhydrite and the evidence of a cooling thermal history suggest that the mechanism explaining the formation of Naica crystals could also apply for this locality.

The second case is an 11-m^3 (388-ft^3) ovoid geode hosted by carbonates that was discovered by mineral collectors in 1999 at Mina Rica in the extinct iron-lead mining district of Pilar de Jaravia, Almeria, SE Spain. The mineralization occurred when the dolomites of the Alpujarride complex were intruded by Neogene volcanism of Cabo de Gata, and a complex Fe-Pb-Sb-Ag-Ba-Hg mineralization formed within the carbonates due to the hydrothermal fluid circulation. The walls of the geode are covered by stubby gypsum crystals with an average length of 0.5 m (1.6 ft). In the center of the geode there are a couple of elongate crystals up to 2 m (6.6 ft) in length that are similar to the crystal beams found in Naica and El Teniente, which certainly grew later on (**Fig. 5**). The genesis of these large gypsum crystals in the geode is not yet clear. However, the presence of celestite needles at the base of the gypsum crystals, such as recorded in Naica, suggests that dissolution of preexisting anhydrite, slight supersaturation in gypsum of the solution, and subsequent formation of the crystals could also have take place.

For the background information *see* ANHYDRITE; BRECCIA; CRYSTAL STRUCTURE; CRYSTALLIZATION; GEODE; GYPSUM; NUCLEATION in the McGraw-Hill Encyclopedia of Science & Technology.

Juan Manuel García-Ruiz; Angels Canals; Carlos Ayora

Bibliography. J. Cannell et al., Geology, mineralization, alteration, and structural evolution of the El Teniente porphyry Cu-Mo deposit, *Econ. Geol.*, 100:979–1003, 2005; W. Foshag, The selenite caves of Naica, Mexico, *Am. Mineral.*, 12:252–256, 1927; J. García-Guinea et al., Formation of gigantic gypsum crystals, *J. Geol. Soc.*, 159:347–350, 2002; J. M. García-Ruiz et al., Formation of natural gypsum megacrystals in Naica, Mexico, *Geology*, 35:327–330, 2007; C. Klein, *Manual of Mineral Science*, 22d ed., Wiley, 2002; F. Palero, F. Gómez, and J. M. Cuesta, Pilar de Jaravía: La Geoda Gigante de la Mina Rica, *Bocamina, Rev. Miner. Yacimientos España* 6:54–67, 2000; Pliny the Elder, *Natural History, X: Books 36-37* (D. E. Eichholz, transl.), Harvard University Press, 3d ed., 1992; P. C. Rickwood, The largest crystals, *Am. Mineral.*, 66:885–907, 1981.

High-temperature superconductivity

In 1911, G. Holst and H. Kamerlingh Onnes discovered superconductivity—that is, electrical conductivity without resistance—in mercury at a temperature, called the critical temperature (T_c), just below the boiling point of helium. Hundreds of superconducting materials are now known, including many

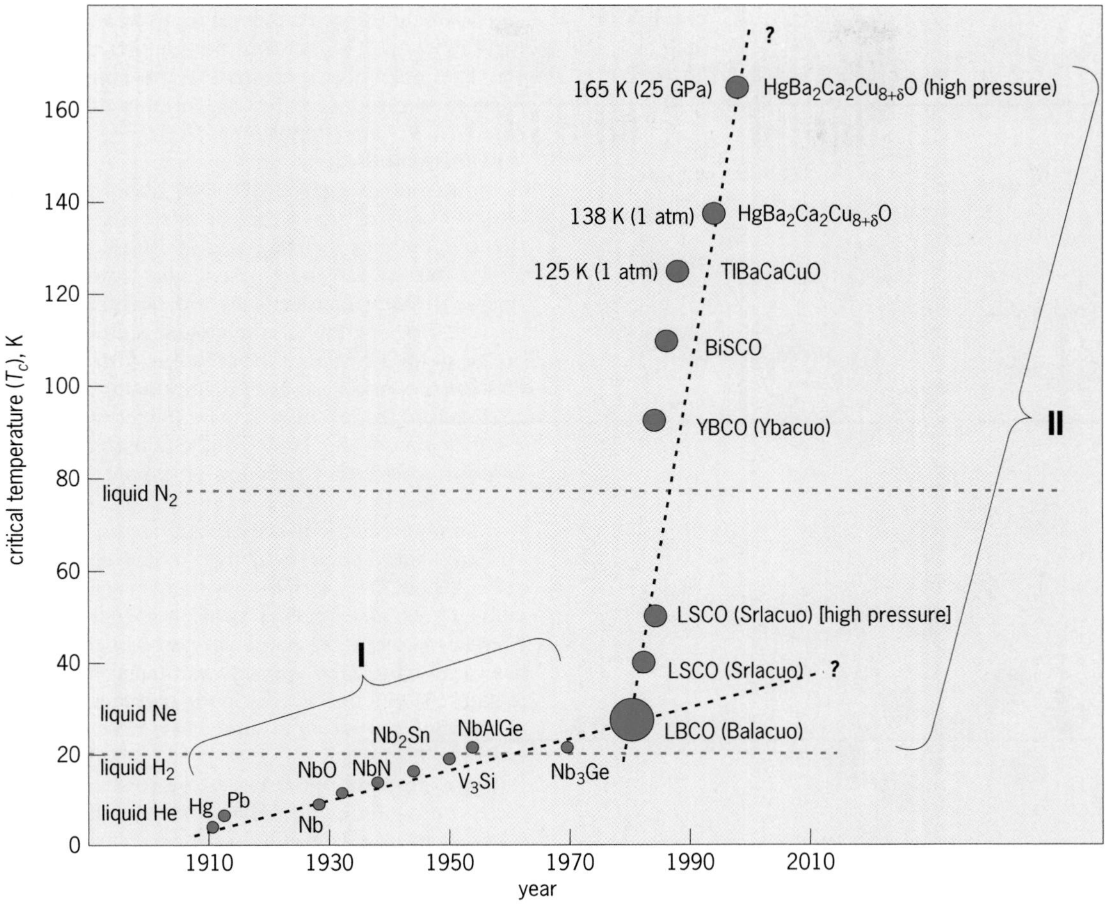

Fig. 1. Evolution of critical temperatures of superconducting materials with time, showing the dramatic change with the discovery of high-temperature superconductivity. Evolution approximately followed line I before this discovery and line II after it. $T(^\circ C) = T(K) - 273.15$; $T(^\circ F) = 1.8 \times T(K) - 459.67$.

elements, such as niobium, lead, and tin; intermetallic compounds, such as Nb_3Ge; and nonmolecular solids such as the spinel $LiTi_2O_4$ and the perovskite $Ba(Pb_{1-x}Bi_x)O_3$. In parallel with that progress, a number of applications of superconducting materials have been envisaged, and some of them currently find extensive use, such as magnetic resonance imaging (MRI).

However, although many superconducting compounds were found prior to the 1980s, they all had lower critical temperatures than Nb_3Ge, the material with the highest critical temperature, 22.3 K (−250.8°C or −419.5°F), of the so-called classical superconductors. Between 1940 and 1970, the critical temperature increased at an average rate of only 3 K per decade, so that reaching the simpler, and cheaper, cryogenics of liquid nitrogen remained a prospect for the distant future (**Fig. 1**).

Discovery. Matters changed substantially in 1986, when J. G. Bednorz and K. A. Müller found that LBCO or Balacuo (acronyms for $La_{2-x}Ba_xCuO_4$) and its cognates with strontium and calcium appeared to have T_c values around 30–40 K. Curiously, these materials were already rather well known at that time, except for their superconducting properties.

However, the door had been opened to the new field that is now called high-temperature superconductivity (HTSC). In fact, in the work of Bednorz and Müller, and even more in the work that followed, several aspects of materials science that are important in the search for new materials were already clear: structure and composition, but also the synthesis conditions, which affect not only structure and composition but also the microstructure. This was the path to follow to reach ever higher critical temperatures.

It was, in fact, the pressure that led to the really high-temperature materials: C.-W. Chu and his colleagues observed that the critical temperature of $La_{2-x}Sr_xCuO_4$ (with $x \sim 0.15$) clearly increased with mechanical pressure: from $T_c \sim 35$ K at room pressure to $T_c \sim 50$ K at about 40 GPa (400 kilobars).

The idea was then to apply chemical pressure by replacing the lanthanum by a smaller ion in a similar oxidation state, and yttrium was chosen. This led Chu and his colleagues to a mixture of phases, among which a minority one was superconducting at the then absolutely unexpected critical temperature of ~90 K (−183°C or −298°F). This was more than double the maximum critical temperature established for any previously known superconducting material and 13 K above the boiling point of nitrogen at room pressure, or ~77 K (−196°C or −321°F). The chemical formula for the superconducting

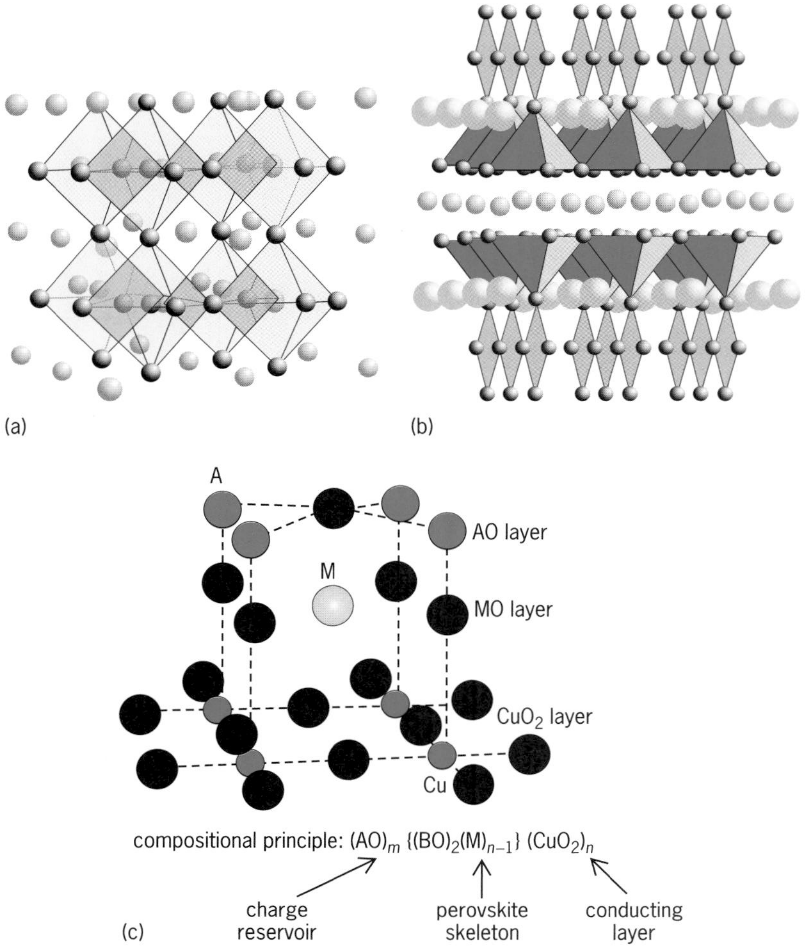

Fig. 2. Structures of superconducting compounds. (*a*) Perovskite structure. (*b*) YBaCuO structure. (*c*) Common features of high-temperature cuprate superconductors.

material, $Ba_2YCu_3O_7$ (nicknamed YBCO, YBaCuO, or even "123"), was established by a number of scientists who assigned the material an orthorhombic cell. The structure was fully established simultaneously by a number of groups. As with all high-temperature superconductors, it is a derivative of the well-known perovskite-type structure (**Fig. 2*a*** and *b*).

Compositional and structural principles. All HTSC materials are characterized by a group of common features whose consideration makes it possible to understand the situation, and also to suggest modifications of existing materials in attempting to realize and eventually improve superconducting properties. In particular, there exist both a compositional and a structural principle that are evidently interrelated (Fig. 2*c*).

The common structural features consist of a conducting plane (CP), $\{CuO_2\}_n$; a charge reservoir (CR) layer, $\{AO\}_m$; and a perovskite-based structural support or skeleton, $\{M_{n-1}(BO)_2\}$. Putting all this together, one has the formula

$$\{AO\}_m\{M_{n-1}(BO)_2\}\{CuO_2\}_n \qquad (1)$$

where A is one of a wide number of single or multiple ions that range from copper itself to mercury, carbonate, and many others; B is usually an alkali-earth ion; and M is a trivalent ion. In this mode, the way to predict novel materials of the cuprate type consists of introducing into this formula different elements in various proportions.

In the simplest case, in which $m = 0$ and $n = 1$, formula (1) becomes $(BO)_2CuO_2$, and if B = La, one has $La_2^{3+}Cu^{2+}O_4$. Although this material is not superconducting, it can be made so, as Bednorz and Müller did, by replacing part of the lanthanum (a typical trivalent ion, La^{3+}) by a divalent ion, such as Ca^{2+}, Sr^{2+}, or Ba^{2+}. In chemical terms, this decrease of the valence of the B cation, if the oxygen content is constant, implies that as many Cu^{2+} ions are oxidized to Cu^{3+}. This can be reflected in a general formula $La^{3+}_{2-x}Sr^{2+}_x Cu^{2+}_{1-x}Cu^{3+}_x O_4$. In the electron-hole formalism, this oxidation of copper is said to create holes in the CP layer, and this is associated with superconductivity in these cuprates. It was later observed that superconductivity could also be induced in LBCO by introducing oxygen interstitials, either at a high oxygen pressure or electrochemically. It is generally observed that the number of holes has to be within the approximate range 0.05–0.25 to realize superconductivity in the cuprates, but this proposition has been questioned in at least one case, as will be discussed.

There are three perovskite-based structures with the formula AB_2O_4 and different ion coordinations. All of them have in common the presence, already mentioned, of the Cu-O conducting planes. However, these planes can also be part of either a square pyramid or an octahedron, as in LBCO. In the other two cases, represented by formulas such as $RE_{1-x}M_xCuO_4$ or even $RE_{2-x-y}MCuO_{4-z}$, where M is a tetravalent metal such as thorium or cerium and RE is a rare-earth ion, the superconducting carriers are electrons instead of holes. For this reason these very interesting materials are often called electronic superconductors. However, so far the maximum T_c value attained in this family is below around 30 K.

YBCO. The famous YBCO ($Ba_2YCu_3O_7$) corresponds to the family term with $m = 1$ and $n = 2$, that is, $\{CuO\}\{(BaO)_2Y\}\{CuO_2\}_2$. This is best formulated as $CuBa_2YCu_2O_{7-\delta}$, since that structural formula indicates that copper atoms occupy two different crystallographic sites: one in the conducting layer and the other, square planar, in the charge reservoir (Fig. 2*b*). This is one of the most studied nonmolecular chemical compounds. Besides its preeminence as the first superconducting material above 77 K, it shows many interesting physical-chemical characteristics. Its oxygen content can be varied as a function of the thermodynamic variables, in particular temperature and oxygen chemical potential, and it is therefore also a very good example of a nonstoichiometric compound. Moreover, as is always the case with these types of compounds, compositional variations modify its properties: In this case, T_c changes drastically with δ: For $\delta \sim 0.5$, $T_c \sim 60$ K, and for $\delta > 0.75$, the material is no longer superconducting. However, the superconducting properties can be restored by reoxidation.

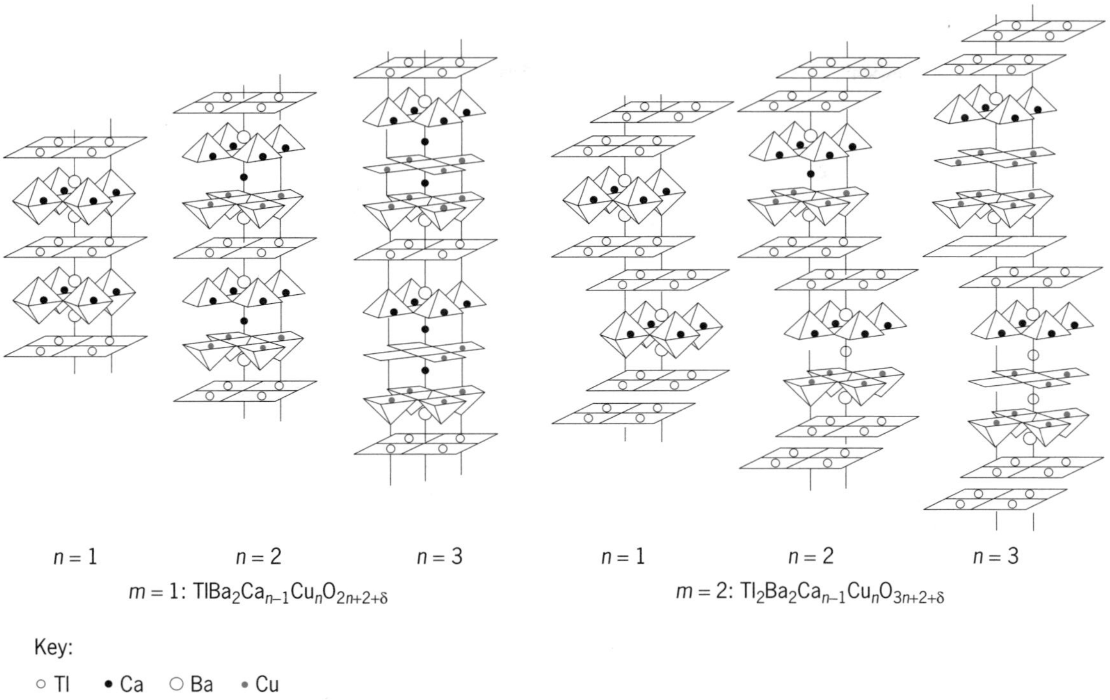

Fig. 3. Schematic representation of the two families of thallium superconductors: $Tl_mBa_2Ca_{n-1}Cu_nO_{xn+2}$.

Bismuth cuprates. The next step in this "high T_c race" was the discovery of the bismuth cuprates. Several aspects made this family interesting: The charge reservoir layer is often a double Bi-O layer, the structure is a rather elaborate incommensurate modulated one, and one of the members of the Bi-Ca-Sr-Cu-O system, discovered by H. Maeda, showed for the first time a T_c over 100 K (−173°C or −280°F). The acronym adopted for these materials is usually BiSCO. Also, the layerlike structure is more marked here than in other cuprates due to the disposition of the so-called inert electron pair which is characteristic of bismuth. This property is of more than academic interest since it helps in the processing of these materials, which is important in view of applications.

Thallium superconductors. Another very interesting case is that of the so-called thallium superconductors, which, although rather toxic and consequently less adequate for applications, are useful because they demonstrate how T_c can be modified by using formula (1).

Indeed, as shown by Z. Z. Sheng and A. M. Hermann, and others, the thallium superconductors comprise two families of compounds (**Fig. 3**), with the general formula

$$Tl_mBa_2Ca_{n-1}Cu_nO_{xn+2} \quad (2)$$

When $m = 1$ and $n = 1$, 2, or 3, one has the family given by the formula

$$TlBa_2Ca_{n-1}Cu_nO_{2n+2+\delta} \quad (3)$$

while for $m = 2$ and $n = 1$, 2, or 3, one has the corresponding formula

$$Tl_2Ba_2Ca_{n-1}Cu_nO_{3n+2+\delta} \quad (4)$$

and T_c increases with n in both families. This is an important result since increasing n, the number of conducting planes in the structure, increases the critical temperature, up to 110 K in $TlBa_2Ca2Cu_3O_9$. But increasing m, the number of charge reservoir layers, also increases T_c. Consequently, increasing simultaneously both n and m markedly increases T_c, which is 125 K (−148°C or −235°F) for $Tl_2Ba_2Ca_2Cu_3O_{10}$, a material that, with two charge reservoir layers and three conducting planes, had the highest known T_c until the discovery of the mercury cuprates.

At this point it was evident that, even if the ultimate goal—a room-temperature superconductor—was not yet in sight, a substantial increase in the pace of discovering superconducting materials had been achieved (Fig. 1). One could then hope that, by simply increasing the two coefficients m and n, the road was clear to achieving an ever-increasing T_c. However, it was soon realized that there was a limit to these trends in both m and n.

Mercury cuprates. Yet additional resources were still available for the main goal of increasing T_c, and this increase was achieved by again playing with the chemical nature of the charge reservoir layer, as demonstrated by E. V. Antipov, S. N. Putilin, and colleagues. In particular, by introducing mercury as the metallic element in the charge reservoir layer, they discovered the first few members of the family of cuprates that, up until now, show the highest values of T_c: $HgBa_2Ca_{n-1}Cu_nO_{2n+2+\delta}$. The structures of these materials correspond to that of the single-layer thallium compounds; yet there are important differences with respect to composition since, in the thallium case, the thallium ions Tl^{+} are located in the center of an octahedron and there is little variation in the oxygen content. On the contrary, in the mercury family of cuprates, the mercury ions, Hg^{2+}, adopt a dumbbell coordination, typical of these ions, and so there is much less oxygen in the charge reservoir layer. Even more, in modifying the oxygen content of the charge reservoir layer by chemical means, that is, by a red-ox process, one observes a dome-like behavior of the critical temperature with respect to the oxygen excess for all members of the family.

The maximum critical temperatures so far known in the high-temperature superconductors are observed in this family. At optimum doping, that is, for the apex of the dome, one has, for $n = 1$-3, $T_c = 97$, 122, and 135 K (−138°C or −217°F), respectively, the last value being an absolute maximum at room pressure that becomes 165 K (−108°C or −163°F) at 30 GPa (300 kilobars). Once again, after $n = 3$, T_c decreases, and for $n = 4$-6, $T_c = 127$, 110, and 107 K, respectively.

As just indicated, mechanical pressure has a very marked influence on T_c, and both types of behavior—the influence of chemical and mechanical pressure—appear to be related to the particular nature of the mercury coordination. This geometry allows a better matching between the different parts of the structure and is at its best for $n = 3$, where the conducting planes are practically flat and allow a very good orbital overlapping between oxygen and copper in two dimensions.

Possibility of CP oxygen vacancies. While the presence of electron holes in the conducting plane is considered the paradigm of HTSC, it is also common wisdom that one cannot have oxygen vacancies in those planes. However, there is an interesting example that is at present the object of serious debate: The T_c of $Sr_2CuO_{4-\delta}$ can be markedly increased by annealing processes that order the oxygen vacancies implied by the oxygen deficiency δ. The debate is about whether these oxygen vacancies are in the charge reservoir layer or in the conducting planes. If, as it seems probable, the vacancies are in the conducting planes, this example will contradict all models for superconductivity in the cuprates that are based on the stability and completeness of the Cu-O_2 planes; and this for a hole concentration of around 0.8, very far from the "magic" one of around 0.15. This is a promising result, and it may well be that it has a more general application and can be used to enhance T_c in other cuprates. However, more work is required before the paradigm is modified.

Prospects. Since the discovery of the mercury cuprates almost 15 years ago, there has not been any real improvement in the value of T_c. However, a number of superconducting materials have been discovered, and many of them are quite interesting for different and varied reasons, even if they are not high-temperature superconductors: MgB_2, $RuSr_2LnCu_2O_8$, lithium-inserted nitrohalides, boron-doped diamond, lithium metal at room pressure, carbon nanotubes, and more exotic materials such as $PuCoGa_5$, with a T_c of 18.5 K.

The search for superconductors has been very fertile so far, and there is no reason why it cannot be so in the future. However, it is perhaps worth recalling that all high-temperature superconductors known so far are cuprates. That is, in both structural and compositional terms, we have, using a musical analogy, "variations on an original theme." Yet, it is interesting to speculate on the possibilities if a new theme could be found.

For background information *see* CARBON NANOTUBES; NONSTOICHIOMETRIC COMPOUNDS; PEROVSKITE; SUPERCONDUCTING DEVICES; SUPERCONDUCTIVITY in the McGraw-Hill Encyclopedia of Science & Technology. Miguel Á. Alario-Franco

Bibliography. E. Antipov, A. M. Abumakov, and S. N. Putilin, Chemistry and structure of Hg-based superconducting Cu mixed oxides, *Supercond. Sci. Techn.*, 15:R31–R49, 2002; J. G. Bednorz and K. A. Müller, Possible high T_c superconductivity in the Ba-La-Cu-O system, *Z. Phys. B—Cond. Matter*, 64:189–193, 1986; K. A. Müller, On the superconductivity in hole doped cuprates, *J. Phys. Cond. Matter*, 19:251002, 2007; C. P. Poole, Jr. (ed.), *Handbook of Superconductivity*, Academic Press, San Diego, 2000; I. K. Schuller and J. D. Jorgensen, Structure of high T_c oxide superconductors, *MRS Bull.*, 14(1):27–32, January 1989; H. Shaked et al., *Crystal Structures of the High-T_c Superconducting Copper Oxides*, Elsevier B. V., Amsterdam, 1994; S. Tanaka, High-temperature superconductivity, *Jpn. J. Appl. Phys.*, 45:9011–9024, 2006.

Honeybee genome

The Western honeybee, *Apis mellifera*, is a key model for social behavior and is essential to agriculture and global ecology because of its pollination activity. There are striking differences in the size of gene families in the honeybee genome relative to other sequenced genomes for several important protein and domain families. These differences, which involve both larger and smaller gene numbers, as well as other novel features of the honeybee genome, have been suggestively related to the social lifestyle of the honeybee.

Compared with other sequenced insect genomes, the *A. mellifera* genome has high DNA contents of A+T (adenosine plus thymidine); lacks major transposon families (a transposon is a genetic element that comprises large discrete segments of DNA capable of moving from one chromosome site to a new location); has evolved more slowly; and is more similar to vertebrates with regard to circadian rhythm, RNA interference, and DNA methylation genes, among others. Furthermore, *A. mellifera* has fewer genes for innate immunity, detoxification enzymes, cuticle-forming proteins; and gustatory receptors; more genes for odorant receptors; and novel genes for nectar and pollen utilization—consistent with its ecology and social organization. Compared with *Drosophila*, genes in early developmental pathways differ in *Apis*, whereas similarities exist for functions that differ markedly, such as sex determination, brain function, and behavior. Population genetics suggests an African origin for the species *A. mellifera* and provides insights into whether Africanized bees (also known as "killer bees") spread throughout the New World via hybridization or displacement.

Social behavior. Honeybees are known as paragons of sociality, living in societies that rival human societies in complexity and cohesion. Honeybees are eusocial, which means their colony's population is differentiated into queens that produce offspring and nonreproductive altruistic workers that communally gather and process food, care for young, build nests, and defend their hive.

Social evolution endowed honeybees with impressive traits. Queens and workers come from the same genome, but queens—usually one per colony—have 10 times the life span of workers (typically queens live for several years), lay up to 2000 eggs per day, and store sperm for years without losing viability. Workers, numbering tens of thousands per colony, display sophisticated cognitive abilities, despite a brain containing only 1 million neurons, which is five orders of magnitude less than the human brain. Worker bees can learn to associate a flower's color, shape, scent, or location with a food reward, thereby increasing the efficiency with which they gather nectar and pollen. Worker bees can even learn abstract concepts such as "same" and "different," which presumably also increases their ability to home in on the most profitable flower patches. Honeybees that find a good source of food return to the hive and communicate their discovery with a "dance language," the only known nonprimate symbolic language, in which information regarding the location of a food source is transmitted from a "finder" bee to others within the hive.

Benefits to humans. Honeybees benefit humankind in exceptionally broad ways. In agriculture, honeybees are the most important pollinators of food and fiber crops, with a value of about $15 billion dollars annually in the United States alone. Of course, they also produce honey. In biology and biomedicine, honeybees are popular model research organisms in diverse areas including allergic disease, development, gerontology, neuroscience, social behavior, and venom toxicology. However, honeybees are also threatened by human activity, perishing due to insecticides that indiscriminately may kill both pests and beneficial insects, and exotic parasitic mites vectored around the world by human commerce. The sequencing of the genome of the Western honeybee, *A. mellifera* (see **illustration**), was carried out in order to advance basic biology and applied apiculture.

Genome sequencing project. With primary financial support from the National Institutes of Health–National Human Genome Research Institute and a contribution from the U.S. Department of Agriculture, the honeybee genome sequencing project began in December 2002. The results of this sequencing effort, led by the Baylor College of Medicine's Human Genome Sequencing Center, but with contributions from over 100 research laboratories from 16 countries, were published in over 50 papers in many prominent scientific journals.

Honeybee on a DNA fragment analysis map. (*Photo by Peggy Greb/USDA*)

The genome of the honeybee contains a total of approximately 250 million bases of DNA. Approximately 10,000 genes have been identified to date, primarily on the basis of computer programs for gene prediction. This is lower than the 13,000 genes identified from the genome of the fruit fly (*Drosophila melanogaster*), which is one of the most intensively studied genomes in all of biology. It is expected that the number of genes identified in the honeybee genome will increase in the future.

The honeybee genome contains a greater proportion of the adenine (A) and thymine (T) nucleotides than does the *Drosophila genome*. Moreover, genes are not distributed evenly throughout the bee genome, but show a tendency to appear in (A+T)-rich regions. This is the exact opposite of the situation in the human genome, in which the preponderance of genes are located in regions of the genome that contain greater proportions of the guanine (G) and cytosine (C) nucleotides. Understanding the basis of these fundamental differences promises to illuminate our understanding of genome organization and illustrate the benefits of comparative genomics.

It often takes some time for the results of a genome sequencing project to lead to major scientific discoveries. However, already some tantalizing findings have emerged from the honeybee genome that might reflect the bee's intense social life. Five are listed below.

Pace of evolution. It appears that the honeybee genome evolved more slowly than the genomes of two other insects that have had their genomes sequenced, the fruit fly (*D. melanogaster*) and the malaria mosquito (*Anopheles gambiae*). One consequence of that slower evolutionary pace is that the bee genome contains versions of some important genes found in mammals that have been lost from these other two insect genomes. Is the honeybee more slowly evolving than most organisms, or have the fly and mosquito (both members of the order Diptera) evolved faster? Moreover, if the honeybee is evolving more slowly, is that because of the bee's social lifestyle? These questions can only be answered by future analyses of genome sequences for more species.

Killer bee invasion biology. Population genetic analyses based on the honeybee genome have generated exciting new insights into the longstanding controversy of whether Africanized bees ("killer bees," *Apis scutellata*) spread throughout the New World via hybridization or displacement. The answer is *both*!

Africanized honeybees were introduced to Brazil from Africa in 1956 in order to breed a strain more suited for the tropical climate. The plan was to breed out the aggressiveness of *A. scutellata* before releasing the bees, but they were accidentally released before this could happen. Africanized bees spread throughout the New World, reaching the United States in 1990. This spread of Africanized honeybees has been one of the most spectacular examples of a biological invasion.

Analyses of the honeybee genome reveal that this biological invasion has involved extensive hybridization with European subspecies, but the genomes of some subspecies appear to be more resistant to domination than others. Genes from *A. scutellata* have largely replaced many genes from one previously dominant subspecies of honeybee, *A. mellifera ligustica* (the "Italian bee"), while *A. m. mellifera* (the "German black bee") genes have been essentially unchanged. It will be fascinating to learn why these two subspecies show different "susceptibilities" to Africanization, and what this might mean for the genetics of aggressive behavior.

Genome deficiencies and risk. Relative to the fruit fly and mosquito, honeybees show a remarkable reduction in the size of gene families associated with the detoxification of harmful chemicals encountered in the environment. Honeybees also show a similar reduction in the size of gene families that encode components of the immune system. Why this is so is a mystery, especially since life in a densely populated beehive would seem to put bees at special risk for environmental toxins, pathogens, and parasites. Perhaps this is why honeybees are extremely vulnerable to many types of insecticides and have suffered major population losses in some agricultural regions of the world, including the recent reports of devastating losses due to Colony Collapse Disorder. On the other hand, it appears that bee social evolution has also led to novel behavioral mechanisms of protection, such as the ability of some "nurse bees" to detect and remove diseased larvae from the hive, and the collection by foragers of plant-produced resins with antimicrobial activity that are used to coat the walls of the beehive.

Sensible sensory genes. Honeybees have a smaller number of genes encoding taste receptors, but a larger number of genes encoding smell receptors. The limited gustatory receptor repertoire perhaps reflects their lessened need to avoid toxic chemicals in their food—unlike herbivorous insects, bees and plants enjoy a strongly mutualistic relationship. Plants need bees for pollination and produce nectar as a reward to entice them to visit. There are very few poisonous nectars in the world. By contrast, honeybees rely heavily on olfaction to find flowers outside and to exchange and interpret chemical messages in the hive. Chemical communication is especially well developed in the bee society, with over a dozen pheromones identified and many others remaining to be discovered.

Effect of methylation. The honeybee is the first insect known to possess all the genes necessary to encode a complete, vertebratelike system for "epigenetic" control of gene activity by methylation (in general, DNA methylation has been found to have a repressive effect on gene activity). Epigenetics (the study of those processes by which genetic information ultimately results in distinctive physical and behavioral characteristics) is increasingly recognized as a potent force in the regulation of genome activity in humans and other organisms, implicated in disease and normal processes such as nutrition, brain development, and social behavior. It is not yet clear why the honeybee possesses this form of genome regulation whereas some other insects do not, or what exactly it does in the honeybee, but these issues are fertile topics for future investigation. Further, the ability to use an insect to study methylation is expected to result in improved understanding of the process in general.

Outlook. As these results illustrate, the sequencing of the honeybee genome is expected to usher in a bright era of bee research for the benefit of agriculture, biological investigations, and human health.

For background information *see* BEE; BEEKEEPING; GENETIC CODE; GENETIC MAPPING; GENOMICS; INVASION ECOLOGY; POLLINATION; POPULATION GENETICS; SOCIAL INSECTS in the McGraw-Hill Encyclopedia of Science & Technology. Gene Robinson

Bibliography. J. D. Evans et al., Immune pathways and defence mechanisms in honey bees *Apis mellifera, Insect Mol. Biol.*, 15:645–656, 2006; Honeybee Genome Sequencing Consortium, Insights into social insects from the genome of the honeybee *Apis mellifera*, *Nature*, 443:931–949, 2006; H. M. Robertson and K. W. Wanner, The chemoreceptor superfamily in the honey bee *Apis mellifera*: Expansion of the odorant, but not gustatory, receptor family, *Genome Res.*, 16:1395–1403, 2006; Y. Wang et al., Functional CpG methylation system in a social insect, *Science*, 314:645–647, 2006; C. W. Whitfield et al., Thrice out of Africa: Ancient and recent expansions of the honey bee, *Apis mellifera, Science*, 314:642–645, 2006.

Hybrid automotive power systems

Hybrid vehicles use two or more sources of energy for propulsion. Typically, one source is a liquid or gaseous fuel and the other is energy stored in an energy storage system such as batteries or ultracapacitors. Other options for energy storage include compressed gas, pressurized liquids (hydraulics), and mechanical energy stored in a flywheel. The most common type of hybrid vehicle is the hybrid-electric vehicle (HEV), which typically uses a liquid fuel (gasoline, diesel, or a gasoline/ethanol blend) to power an internal combustion engine that works with batteries, or ultracapacitors, to power an electric motor.

Why use hybrid vehicles? Technology in vehicles has progressed significantly since their introduction around the beginning of the twentieth century. Recent technological improvements have not been used to reduce fuel consumption, but rather to increase performance to meet consumers' demands for larger and heavier vehicles. This is seen by the widespread use of minivans and sport utility vehicles over the past three decades (**Fig. 1**). From 1981 to 2003, vehicle weight increased by 24% and accelerated from 0 to 96 km/h (0 to 60 mi/h) 29% faster, but increased fuel economy by only 1%. Meanwhile, oil demand is increasing significantly in developing countries while oil production is declining in a number of countries, leaving most of the world's significant reserves of conventional oil in only a few places. Consequently, it is useful to consider alternative means of personal and commercial transportation that use less oil. HEVs are one such alternative technology that can use less fuel while maintaining the performance levels of current vehicles. Of the alternative powertrain options, HEVs appear to be some of the best choices for providing near-term reductions in oil use, as evidenced by their increasing commercial availability from many manufacturers.

How do HEVs save oil? To understand why HEVs save energy, one must consider how to minimize the energy losses in conventional vehicles.

Efficiently satisfy the vehicle driving forces. In physics, energy is defined by the force exerted over a distance. For vehicle energy use, four forces dominate. Newton's Second Law of Motion is shown in the equation $\mathbf{F} = m\mathbf{a}$, where $\mathbf{F}$ is the force required for a specified mass m to be accelerated at a rate of $\mathbf{a}$. As suggested here, the force (and hence the fueling rate) required to move a vehicle increases for heavier vehicles and for frequent rapid vehicle speed changes (which create large accelerations). Much of the energy used to accelerate a vehicle then is stored as kinetic or motion energy, which is equal to ${}^1/_2 mv^2$, where v is the velocity of the vehicle. Conventional vehicles waste nearly all of that kinetic energy by dissipating it as thermal energy with friction brakes. HEVs, on the other hand, recover that kinetic energy through regenerative braking. That is, the wheels turn the shaft of an electric motor to generate electricity, which is then stored in the batteries (or other electrical energy storage

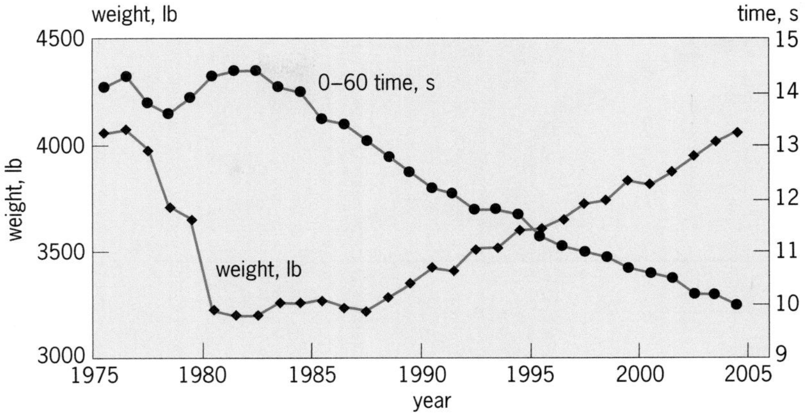

Fig. 1. Change in vehicle weight and acceleration performance from 1975 to 2004 (*U.S. EPA Light-Duty Automotive and Fuel Economy Trends: 1975–2004, April 2004*).

device). However, the power generated by braking can be much higher than the power rating of the motor and batteries. Consider that the kinetic energy of a midsized vehicle weighing 1500 kg (3360 lb) traveling at 120 km/h (75 mi/h) is about 833 kJ. A braking event of 8 s to stop the vehicle would produce more than 100 kW of power, which is higher than the recharge rate that many systems could accept.

The additional force a vehicle must overcome when it climbs a hill is $F = mg \sin(\theta)$, where g is the gravitational constant and θ is the angle between the horizonal and the road surface. Although force is required to climb a hill, the vehicle's potential energy increases with increasing elevation. When a conventional vehicle descends a steep hill, the mechanical brakes again convert the potential energy to waste heat energy to maintain the vehicle at a safe speed. However, an HEV descending the same hill benefits from the ability to convert the potential energy and store it as electricity through the regenerative braking process.

Vehicles also face an opposing force to push through the surrounding air. This is known as aerodynamic drag and is equal to ${}^1/_2 \rho C_d A_f v^2$, where ρ is the density of air, C_d is the coefficient of drag for that vehicle's design (including the effects of the vehicle shape, wheel wells, underbody, etc.), and A_f is the frontal area of the vehicle. This equation again illustrates that vehicle design and use (for example, high-speed driving) affect the force and hence the fuel required to operate the vehicle. Design improvements to minimize aerodynamic drag can be applied to any vehicle platform, but occur with greater frequency in HEVs, because they emphasize efficiency improvements. Typical drag coefficients range 0.4–0.45 for light trucks and 0.3–0.35 for cars; the drag coefficient for the Toyota Prius HEV is at 0.26 and the Honda Insight HEV is at 0.25. In comparison, an F-16 Fighting Falcon (jet) has a C_d of 0.0175.

The last significant force is overcoming the rolling resistance of the vehicle, which is equal to $C_{rr}mg$, where C_{rr} is the coefficient of rolling resistance for the tires, m is the vehicle mass, and g is the

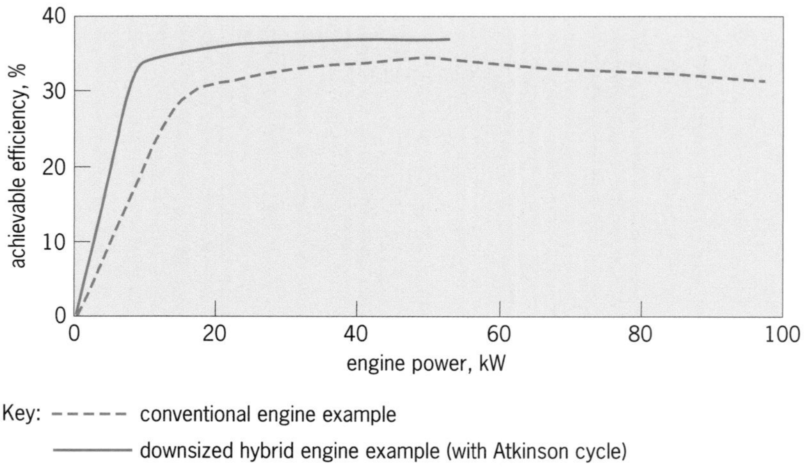

Fig. 2. Engine efficiency curve showing peak efficiency at high power levels and low efficiency at low levels. A smaller engine lowers the inefficient operation power region. An Atkinson cycle engine can also achieve higher peak efficiency.

gravitational constant. As with drag reductions, efforts to reduce tire rolling resistance are not unique to HEVs, but may be included in those vehicles as part of the overall fuel-saving package. Rolling resistance for car tires on a hard paved surface ranges 0.007–0.013. Train wheels on steel have rolling resistances of less than 0.0025.

Efficiently convert fuel to motion energy. A conventional vehicle ultimately derives all of the energy to satisfy the vehicle driving requirements from its liquid fuel. The same is true of a standard HEV, except that the electric components can help the engine use the fuel more efficiently. The engine size in a conventional vehicle is typically dictated by the largest expected power requirement, such as that to satisfy a 0–97 km/h (0–60 mi/h) acceleration in 8 s. During other driving situations, the vehicle power is often quite a bit lower than this peak value (power is calculated by multiplying vehicle speed by the sum of the force equations). Engines are most efficient when operated at substantial power levels, and efficiency drops off significantly at low power levels (see the conventional engine efficiency curve in **Fig. 2**). This creates the counterintuitive phenomenon of vehicles often achieving better fuel economy while cruising at 65–75 km/h (40–47 mi/h) than at 25–35 km/h (16–22 mi/h). Even though the road load power demands increase with increasing speed, the increase in engine efficiency more than compensates for it by moving up the steep portion of the efficiency curve (Fig. 2).

In a comparable HEV, the engine can be smaller because the electric motor assists the engine during the momentary peak power events. Compared with a larger engine, the lower efficiency region of operation for a downsized engine shifts to lower power levels (Fig. 2). This means that the smaller engine will use less fuel to satisfy the significant amount of vehicle operation that requires relatively low engine power output. Also, for very low power levels, an HEV engine may turn off altogether and allow the electric motor to propel the vehicle. At slightly higher but still fairly inefficient power levels, the engine may be used to power the vehicle's wheels and to turn the shaft of the electric motor, generating electricity to store in the batteries. The added load of the motor on the engine requires additional fuel, but helps the engine operate at a higher efficiency and stores electrical energy that the motor can use later to help propel the vehicle. This process is known as "load leveling" the engine. Finally, the added control capability provided by the electric drive components can enable an HEV to use engine technologies with higher peak efficiencies but a narrower operating range than those typically employed in conventional vehicles. This is the case with the Atkinson cycle engines used in the Toyota Prius and Ford Escape Hybrid.

Efficiently power the vehicle's ancillary loads. A conventional vehicle engine continues to burn fuel even when the vehicle is idling and requires no power for propulsion. This keeps the engine operating so that it can provide power as soon as the driver accelerates, and keeps powering the vehicle's ancillary loads, which run off belts attached to the engine crankshaft. Ancillary power is required for low-voltage electronics such as the lighting and entertainment systems, for power steering and power braking assistance, and for climate control (up to 6 kW during cabin cool-down). In an HEV, these ancillary loads can be electrified, which alleviates their drag on the engine and allows them to operate more efficiently (rather than being tied to the speed of the engine). HEVs can therefore turn off rather than idle their engines when stopped and rely on their batteries to power the required ancillary loads. Because an electric motor, unlike an engine, can provide near-full torque before it begins to spin, the electric motor can then launch the vehicle and restart the engine after the vehicle is in motion.

Summary of efficiency improvements. Hybrid vehicle designs will include some or all of the following: use regenerative braking to recapture kinetic and potential energy; downsize the engine by using electric motors to assist with acceleration; provide electric launch and low-speed electric-only operation; load level the engine for efficient operation and battery charging; use advanced engine technologies; turn off the engine at idle; and use efficient electric accessories.

In addition to these hybrid functions, HEVs can benefit from efficiency improvements that could be applied to any vehicle. These include reducing vehicle mass (such as through use of lightweight materials), aerodynamic drag, and rolling resistance. Design changes to reduce ancillary loads can improve the efficiency of any vehicle, but are particularly important for HEVs, as high loads can increase the fuel consumption of an energy-efficient vehicle by as much as 50%. Similarly, less aggressive driving (avoiding rapid speed changes and excessive speeds) can improve the efficiency of any vehicle and make a particularly large impact on an HEV. Other areas of research into broadly applicable vehicle component improvements include work on advanced combustion and variable compression ratio engines, and six-speed as well as continuously variable transmissions.

Hybrid electric vehicle components. HEVs add three major components to conventional vehicle designs: energy storage, electric motors, and power electronics.

Energy storage system/batteries. Typically, nickel-metal hydride batteries are used in HEVs, although lithium-ion batteries are being considered for future use. The challenge is to develop low-cost batteries with high energy density (for extended use), high power density (for short bursts of high power to accelerate and accept high rates of regenerative energy), and long life that are safe for consumer use. A key issue is the state-of-charge window; that is, how much of the available energy can be used without significantly degrading the life of the battery. Ambient low and high temperatures affect battery life and performance. In addition, charging and discharging the battery generate heat that must be managed.

Electric motors and generators. Electric motors can convert electrical energy from the batteries into mechanical energy to help drive the wheels of the vehicle. An electric motor can typically also operate in reverse as a generator: taking mechanical energy to drive its shaft and create electrical energy, such as during regenerative braking. Typically, three-phase alternating-current (ac) motors are used because of their cost and reliability.

Power electronics. The electronics required for HEV operation include an inverter to switch between ac power for the motor and direct-current (dc) power for the energy storage systems, as well as converters to change the voltage of the battery system to a lower voltage for ancillary devices or to a higher voltage for more efficient motor operation. A 100-kW inverter operating at 90% efficiency generates 10 kW of heat that must be managed to ensure long component life.

Hybrid powertrain designs. Different HEV design approaches are possible with different component arrangements. The two classic HEV designs are known as "series" and "parallel" configurations (**Fig. 3**). A series configuration has no direct mechanical connection between the engine and the vehicle wheels. The engine is used simply to power a generator, which provides electricity that is used by an electric motor to drive the wheels or stored in the vehicle's batteries for later use. This configuration enables precise control of the engine near its point of highest efficiency. However, the electric drive components for a series vehicle must be very large to satisfy the peak driving demands. Two examples of series-configured vehicles are the Chevy Volt and Ford HySeries plug-in hybrid concepts (plug-in hybrid vehicles are discussed in the final section).

One disadvantage of a series configuration is the inefficiency associated with taking the mechanical energy from the engine crankshaft, converting it into electrical energy with the generator, and then converting it back to mechanical energy with the electric motor to turn the driveshaft. In a parallel configuration, the engine can directly power the wheels and the electric motor works in parallel to provide

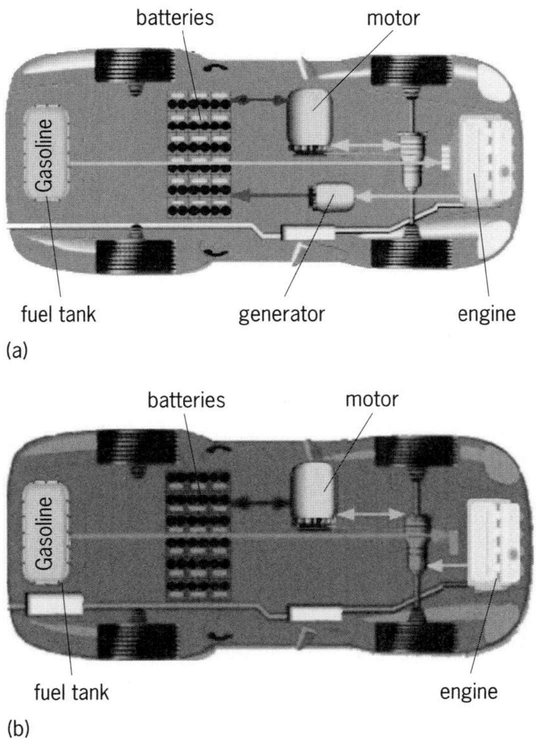

Fig. 3. Classic HEV designs. (*a*) Series HEV configuration, where the engine only provides power to an electric generator that either charges the batteries or powers an electric motor to turn the wheels. (*b*) Parallel HEV configuration, where either the engine or the electric motor can power the wheels. The engine can also charge the batteries.

or absorb supplemental power. Mechanically connecting the engine directly to the wheels avoids the double-conversion loss of the series configuration, but makes it more difficult to operate the engine at peak efficiency. Several commercial HEVs use more complicated variations of the classical design configurations to take better advantage of the benefits of each, including the power-split approach used in the Toyota and Ford HEVs and the two-mode configuration used in the Chevy Tahoe hybrid.

HEVs are often discussed with respect to their approximate placement on a continuum (low to high) from "micro" and "mild" to "full" hybrid. In general, hybrids moving up the spectrum use larger batteries and higher power motors (relative to the size of the vehicle's engine), and take advantage of more of the hybrid functions discussed previously. Hybrids on the lower end of the spectrum may also have their electric motors directly coupled to the engine, whereas those on the upper end can generally operate the drivetrain with the electric motor alone. For instance, three of GM's HEVs that could be placed into categories of micro, mild, and full hybrid are the Chevy Silverado, the Saturn Vue, and the Chevy Tahoe, respectively. These vehicles follow the trend of increasing relative electric component size and increasing number of hybrid functions, beginning with engine start/stop capability, then adding electric assist and sizable regenerative energy recapture, and

Fuel reduction of hybrid vehicles using 2007 EPA labels

		City label, mpg		Highway label, mpg		Combined label, mpg		HEV fuel reduction, %		
Hybrid electric vehicle	Standard vehicle	HEV	Std.	HEV	Std.	HEV	Std.	City label	HW label	Combined
Honda Insight	Civic 1.8L auto	48	25	45	36	46	29	48	20	37
Toyota Prius	Corolla 1.8L auto	48	26	45	35	46	29	46	22	37
Honda Civic	Civic 1.8L auto	40	25	45	36	42	29	38	20	31
Chevy Silverado	Silverado 2WD 5.3L	16	14	19	19	17	16	13	0	6
Ford Escape 2WD	Escape 2WD V6	31	18	29	23	30	20	42	21	33
Honda Accord	Accord 3L auto	24	18	32	26	27	21	25	19	22
Lexus RX400h 2WD	RX 350 2WD	28	18	25	23	26	20	36	8	23
Toyota Highlander 2WD	Highlander 2WD 3.3L	28	17	25	23	26	19	39	8	27
Mercury Mariner 4WD	Mariner 4WD V6	28	17	27	21	27	19	39	22	30
Lexus GS 450h	GS430	22	16	25	23	23	19	27	8	17
Toyota Camry	Camry V6 3.5L auto	33	19	34	28	34	23	42	18	32
Nissan Altima	Altima V6 auto	35	20	33	26	34	22	43	21	35
Saturn Vue	Vue 2WD 6 cyl. auto	23	18	29	25	26	20	22	14	23

finally adding full electrical operation and extensive engine load leveling capability.

Do HEVs save fuel? Toyota alone sold more than 1 million HEVs globally by mid-2007. More than two dozen hybrid models will be available to consumers soon. The average fuel consumption reduction of HEVs over the new EPA combined cycle label is 27% for vehicles sold in 2007 (see **table**). Fuel consumption is reduced by more than 40% for some vehicles in city driving, where significant kinetic energy is recaptured, the engine shuts off during idling, and the electric motor assists with acceleration. Fewer benefits are realized at relatively constant highway speeds.

What are the barriers? The most significant barrier to wider acceptance of HEVs is the additional cost for the batteries, power electronics, and electric motors. These components also add packaging and assembly challenges.

Plug-in HEVs, fuel-cell vehicles, and electric vehicles. Hybrid vehicle technology is not an entirely new concept. Hybrid and electric vehicles (EVs) were available in the late 1800s and early 1900s. The abundance of oil supplies at that time eventually contributed to conventional combustion-engine vehicles winning out as the dominant vehicle technology during the twentieth century. More recently, growing concern about the continued availability of petroleum reserves and greenhouse gases emitted from burning oil has led to renewed interest in hybrid and other alternative powertrain technologies.

Plug-in hybrid electric vehicles (PHEVs) are HEVs that include a charging plug and sufficient energy storage to be charged from an electrical outlet. PHEVs have all the benefits of HEVs and can operate from domestically produced electricity, which provides increased fuel reliability from greater fuel diversification. Two major designs are being considered today. One provides true all-electric driving that can operate solely by the electric powertrain. This requires that the battery pack, electric motors, and power electronics be sized for all vehicle operation. The other design, called a blended control strategy, uses the electric powertrain for most of the driving, but depends on an engine for rapid (high-power) acceleration. By foregoing equivalent all-electric driving, the design can use smaller battery packs, motors, and electronics. In either case, the battery pack energy may be sized to displace fuel use for 16–64 km (10–40 mi) of conventional HEV operation. This would have a substantial impact on oil consumption because the average trip length is about 7 km (4.4 mi), and about 50% of our daily driving (generally, multiple smaller trips) is less than 53 km (33 mi). PHEVs can substantially reduce our dependence on oil, provide more reliable personal transportation, use renewable energy to provide battery charging, and shift air pollution from urban centers during rush hour to more remote locations during nighttime vehicle charging. A battery pack that provides 64 km (40 mi) or less of driving range can be recharged overnight from a conventional 110- to 120-V outlet. A key benefit of PHEVs relative to an all-electric vehicle is that the vehicle need not carry a large battery pack, which is only fully used for longer but less-frequent trips.

Fuel cell vehicles (FCVs) are already being demonstrated, but must overcome several barriers before mass introduction. The required improvements include lowering the cost of the fuel cell stack, improving the effectiveness of on-board hydrogen storage, and developing an efficient hydrogen production and distribution infrastructure. FCVs will also be HEVs (or even PHEVs) to reduce the size and cost of the fuel cell stack and to take advantage of regenerative braking.

The simple design of EVs has many benefits. Many EVs have been built and demonstrated. Current challenges include the high cost of the energy storage system, the need for higher voltage charging stations, the development of safe battery packs with higher energy densities, and the time to efficiently charge the batteries.

Advances in HEV technology (energy storage, electric motors, power electronics, control strategies, etc.) will benefit PHEVs, FCVs, and EVs as those technologies become nearer term.

For background information *see* ALTERNATING CURRENT; ALTERNATING-CURRENT MOTOR; AUTOMOTIVE ENGINE; BATTERY; DIRECT CURRENT; DYNAMIC BRAKING; ELECTRIC VEHICLE; FLYWHEEL; FUEL CELL in the McGraw-Hill Encyclopedia of Science & Technology.

Robert Farrington; Jeffrey Gonder

Bibliography. T. Gillespie, *Fundamentals of Vehicle Dynamics*, Society of Automotive Engineers, Inc.,

1992; J. M. Miller, *Propulsion Systems for Hybrid Vehicles: IEEE Power & Energy Series 45*, Institution of Electrical Engineers, London, United Kingdom, 2004; E. H. Wakefield, *History of the Electric Automobile*, Society of Automotive Engineers, Inc., 1994.

Hydatellaceae

Flowering plants (angiosperms), with about 300,000–400,000 living species, are the largest and most diverse assemblage of land plants on our planet. They include many familiar and economically important plant groups such as the cereal grasses, legumes, and most fruits and vegetables. One of the major goals of systematic botany is to reconstruct the evolutionary history of this diverse lineage. Traditionally, hypotheses of evolutionary relationships have been based largely on morphological traits (for example, flower structure, stem anatomy), but current approaches also consider genetic evidence from one or a combination of the three genomes found in plants (plastid, mitochondrial, and nuclear). This type of work has provided many important insights into flowering plant evolution, and there has been substantial progress in reconstructing the plant tree of life. This knowledge has necessitated the production of a new classification scheme because some families and higher taxa were found not to be natural, and this framework now provides a solid evolutionary basis by which all aspects of comparative flowering plant biology can be focused.

Several major evolutionary lineages are currently recognized among living angiosperms (**Fig. 1**). These include three small lineages that split from the rest of the flowering plants before the major angiosperm radiation; these include the New Caledonian shrub *Amborella trichopoda*, water lilies (Nymphaeales), and woody Austrobaileyales. The other major lines include magnoliids (magnolias, laurels, and their relatives), eudicots (daisies, maples, oaks, poplars, etc.), and monocots (bananas, grasses, orchids, sedges, etc.). The discovery of additional major lineages of flowering plants, particularly among the well-studied lines near the very base of the flowering plant evolutionary tree, has not been expected because most flowering plant families have been sampled in molecular studies. The recent unexpected major realignment of the misunderstood and poorly known family Hydatellaceae, therefore, has taken the botanical world by surprise.

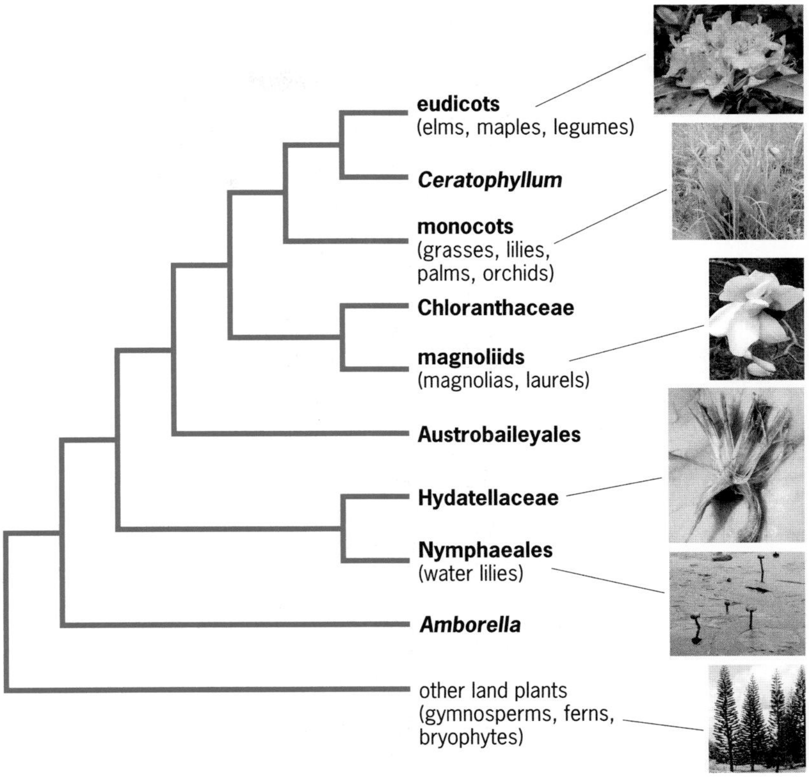

Fig. 1. Evolutionary tree showing the major flowering plant lineages and their relationships. Hydatellaceae and water lilies (Nymphaeales) are each other's closest relatives, and together they form a lineage that diverges near the base of the flowering plant evolutionary tree. (*Photographs by J. M. Saarela*)

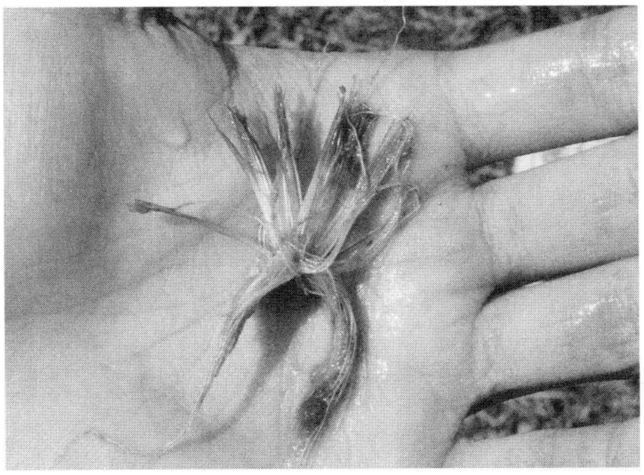

Fig. 2. *Hydatella inconspicua* (Hydatellaceae) from Lake Kai Iwi, New Zealand. (*Photograph by Justin Goh, School of Biological Sciences, University of Auckland, New Zealand*)

Family description. Hydatellaceae are a small family of minute (mature individuals are only a few centimeters in height; **Fig. 2**), nondescript, aquatic flowering plants that grow and flower in or at the edge of seasonal pools and swamps in Australia, New Zealand, and India. The family includes two genera (*Hydatella* and *Trithuria*, although these generic circumscriptions require careful study) and approximately 10 species characterized by narrow, linear leaves and highly reduced reproductive structures. Reproductive units in Hydatellaceae are unisexual or bisexual, and they include a series of sterile bracts (modified leaves associated with plant reproductive structures) that subtend tight aggregations of stamens (the male reproductive organ) and/or carpels (the female sex organ), each with no associated perianth (petals, sepals, or tepals). Current interpretations suggest that each reproductive unit in Hydatellaceae is a highly reduced inflorescence (a flower cluster segregated from any other flowers on the same plant, together with the stems and bracts associated with it) of simple, unisexual flowers, but it has also been hypothesized that the reproductive units might represent some sort of "prefloral" state that preceded the evolution of "standard" bisexual flowers, or that they might have arisen by reduction of

Fig. 3. **A species of water lily (*Nuphar* sp.) that is part of the lineage (Nymphaeales) most closely related to Hydatellaceae. (*Photograph by J. M. Saarela*)**

a more classic flower structure, perhaps in response to life in an aquatic habitat.

Traditional classification. In early plant classifications, *Hydatella* and *Trithuria* were thought to be monocots, a large lineage of plants often characterized as having parallel-veined leaves, floral parts in threes (that is, three petals, three sepals), and a single cotyledon (seed leaf). For over a century, these genera were included in the superficially similar monocot family Centrolepidaceae, which is closely related to grasses (Poaceae), sedges (Cyperaceae), and Australian rushes (Restionaceae). Like Hydatellaceae, the approximately 35 species of Centrolepidaceae are also small with linear leaves and highly reduced flowers and inflorescences, and they often co-occur with *Hydatella* and *Trithuria*. The two groups can be difficult to distinguish at normal magnification. In the 1970s, the German botanist U. Hamann published a careful morphological and anatomical study of Centrolepidaceae, and noted several major structural differences between *Hydatella* and *Trithuria* and the rest of the family and even other monocots, including differences in pollen, stomata (microscopic pores in leaves and stems that facilitate gas exchange), seeds, and ovules. Hamann concluded that if *Hydatella* and *Trithuria* are recognized as members of Centrolepidaceae, then that family would not be a natural unit. Consequently, he proposed a new plant family, Hydatellaceae, to accommodate *Hydatella* and *Trithuria*, and expressed hopes that this would emphasize their obscure characteristics and facilitate a search for their true evolutionary affinities. Nonetheless, in subsequent flowering plant classifications, Hydatellaceae continued to be treated as monocots—largely on the basis of their narrow, pointed leaves which resemble those of many grasses, sedges, and rushes—with multiple researchers noting major difficulties in identifying their nearest monocot relatives and calling for insight into their evolutionary affinities from molecular evidence.

DNA evidence. By examination of deoxyribonucleic acid (DNA) sequence data from multiple plastid and nuclear gene regions representing both Hydatellaceae genera, it was demonstrated that Hydatellaceae are not closely related to grasses or even other monocots. Unexpectedly, it was found that Hydatellaceae are in fact the closest living relatives of the aquatic water lily lineage (Nymphaeales) and are therefore part of a lineage that diverged near the base of the angiosperm phylogenetic tree (Fig. 1). The New Caledonian shrub species *Amborella trichopoda* is the only other lineage thought to have diverged from other flowering plants below this point, although the exact divergence order of *Amborella* and the Hydatellaceae-Nymphaeales lineage is not yet firmly established.

Morphological evidence. The surprising new home for Hydatellaceae illuminated by DNA evidence has prompted a careful reevaluation of its morphological characteristics, which previously (and incorrectly) had been used to place the family among monocots. Indeed, the inconspicuous, linear-leaved, and tiny-flowered Hydatellaceae are strikingly different in overall appearance compared to the often large-leaved and attractively flowered water lilies (**Fig. 3**), and it is likely because of this seeming disparity that the two groups had not previously been considered to be closely related. Upon close inspection, however, both lineages share several microstructural features, including several aspects of seed structure, stomata arrangement, pollen shape, and seed germination pattern, all of which provide additional evidence supporting a close relationship between these two lines. Surprisingly, these shared characters include many of those originally used to segregate Hydatellaceae from Centrolepidaceae. Such morphological evidence must be interpreted carefully and with caution as some of the shared characters also occur in other plant groups, including monocots, although they do not occur in consistent association. In cases like this, independent evidence from DNA can provide a new, informative, and relatively unbiased framework for interpreting morphology. This is a good example of the difficulties inherent in using morphological traits alone to uncover evolutionary relationships—it can be difficult to determine whether similar traits were inherited from a common ancestor, evolved independently in unrelated lineages, or were inherited initially but later lost (reversed in subsequent lineages).

Broader evolutionary implications. From a broader point of view, the new position for Hydatellaceae has important implications for understanding the early morphological, molecular, and ecological evolution of flowering plants as a whole. For example, the new placement for Hydatellaceae, with its simple, unisexual flowers, might provide insight into the origins of the "typical" bisexual flowers found in water lilies and many other angiosperms. Moreover, the addition of this branch near the base of the angiosperm phylogenetic tree might aid in clarifying the branching order of *Amborella* and the Hydatellaceae-Nymphaeales line and determining what the closest living relatives of the flowering plants are. Discovery of Hydatellaceae material in the fossil record could potentially provide new insight into the timing of the origins of the angiosperms.

For background information *see* DEOXYRIBONUCLEIC ACID (DNA); FLOWER; NYMPHAEALES; PLANT EVOLUTION; PLANT KINGDOM; PLANT MORPHOGENESIS; PLANT PHYLOGENY; PLANT TAXONOMY in the McGraw-Hill Encyclopedia of Science & Technology.

Jeffery M. Saarela

Bibliography. E. M. Friis and P. Crane, New home for tiny aquatics, *Nature*, 446:269–270, 2007; U. Hamann, Hydatellaceae—A new family of Monocotyledoneae, *New Zeal. J. Bot.*, 14:193–196, 1976; P. J. Rudall et al., Morphology of Hydatellaceae, an anomalous aquatic family recently recognized as an early-diverging angiosperm lineage, *Am. J. Bot.*, 94:1073–1092, 2007; J. M. Saarela et al., Hydatellaceae identified as a new branch near the base of the angiosperm phylogenetic tree, *Nature*, 446:312–315, 2007.

Hydroclimatology

As our understanding of the Earth's physical and biological systems has grown, it has become increasingly obvious that processes and patterns of the atmosphere, lithosphere, hydrosphere, cryosphere, and biosphere are dynamically linked over many space and time scales. This is very much the case for the Earth's climate and hydrologic (water) cycle. There is about 12,900 km^3 of water (in all three phases) in our atmosphere. While this is a relatively small portion of the total global freshwater (~0.04%), it has large turnover rates (~495,000 km^3 per year) because of the rapid cycling through evaporation, condensation, and precipitation and is transported effectively by atmospheric circulations. Atmospheric water also plays a key role in climate modification by absorbing, scattering, reflecting, and emitting radiation. This role is vastly different in the gaseous (water vapor), liquid (cloud droplets), and frozen (ice crystals) forms. Additionally, latent heat is absorbed and released during phase changes of water. Hurricanes (typhoons, tropical cyclones) and thunderstorms gain immense energy from the release of latent heat during condensation and freezing of moist air as it is lifted and cooled. Across the Earth's surface, large water and energy exchanges with the atmosphere occur through the processes of precipitation, evaporation, transpiration, sublimation, and the formation of dew and frost. Additionally, surface water has a large impact on the Earth's climate through albedo (reflectance). Snow and ice have a very high albedo, while oceans and lakes have a very low albedo. Consequently, the growth and shrinking of ice extent is dynamically linked with climate change. Hydroclimatology therefore focuses on the interaction and overlap between climatology and hydrology, including all of the atmospheric and surface components and fluxes of the hydrologic cycle (**Fig. 1**).

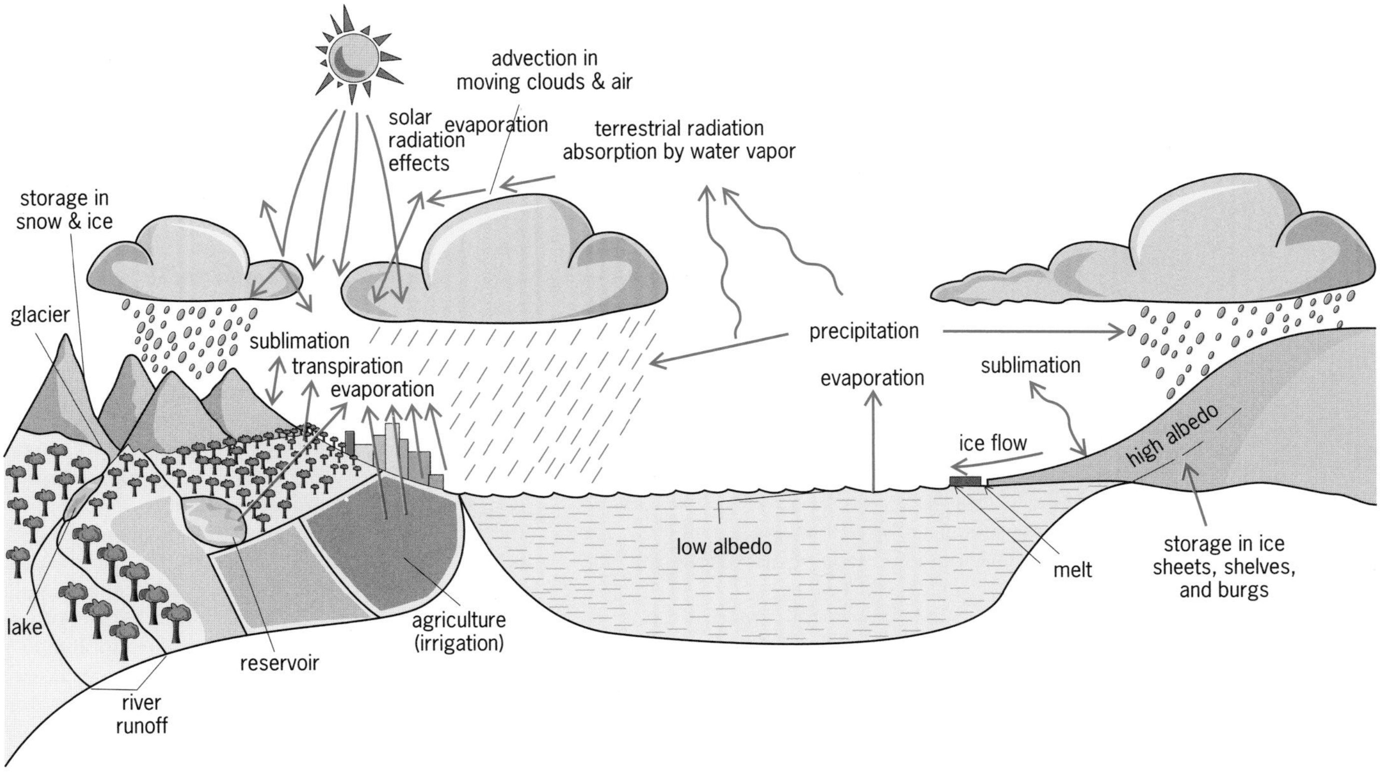

Fig. 1. Major components and processes in the hydroclimate system.

In addition to links between water and climate, hydroclimatology overlaps and interacts with other earth science disciplines. For example, understanding surface-atmosphere exchanges of water requires knowledge of related biophysical processes such as sap flow and transpiration, oceanic processes controlling surface-water temperature, and glaciological processes such as melt, sublimation, or wind-blown snow. Therefore, hydroclimatology is an interdisciplinary field, and much important hydroclimatological research may be classified under different fields of study. Nevertheless, the importance of hydroclimates to societies around the world, particularly to water resource availability (for drinking, irrigation or hydroelectric power generation) and natural hazards (floods or desertification), has produced an established subdiscipline within climatology and hydrology. Currently, many national and international research institutions, networks, and societies publish special journal editions, hold conferences, and compile datasets dedicated to building our knowledge of hydroclimates.

Measurement and modeling of hydroclimates. A large range of methods for measuring hydroclimate properties has been used, ranging from simple rain gauges and evaporation pans to precision chilled mirror dewpoint hygrometers. The spatial coverage and longevity of hydroclimate records is inversely proportional to the sophistication of the measurement, so the best record coverage is for basic variables such as daily rainfall total, daily minimum and maximum temperature, and changes in water levels. Precipitation, evaporation, and soil water storage are now commonly measured from tower-based meteorological instrumentation. The density of these networks is not globally uniform but concentrated on landmasses and in wealthier countries. Linking of regional networks and improvement to interpolation techniques has been used for near real-time hydroclimate information for applications such as assisting irrigators to manage their water resources efficiently. Fast-response water vapor analyzers (using spectroscopic measurements of water vapor concentration) are coupled with sonic anemometer measurements of vertical velocity fluctuations to monitor surface-atmosphere water vapor fluxes. This technique, known as the eddy covariance approach, determines vapor transport rates by turbulent eddies near the surface, with high temporal frequency. These instruments are currently deployed on thousands of research towers worldwide, with organized networks evolving to improve data consistency and produce synthesis analyses. A variety of instruments have also been deployed on vessels, vehicles, and aircraft, providing transects of hydrometeorological data.

Remote sensing has greatly extended the scope of hydroclimate measurement due to both the great spatial detail obtained from satellite imagery and the unique way water in different phases absorbs, transmits, and reflects radiation of different wavelengths. For example, water vapor strongly absorbs radiation in several regions of the infrared part of the electromagnetic spectrum. Since this radiation is reflected or emitted upward from the surface, satellite sensors that measure electromagnetic radiation in these spectral regions are used to determine total water vapor concentration. Furthermore, since water vapor at different levels of the atmosphere absorb preferentially in slightly different parts of the spectrum, atmospheric profiles (sounding) of water vapor can be retrieved. Cloud radar systems estimate cloud drop and raindrop density and location by analyzing backscatter of microwave radiation emitted from the Earth's surface and use multiple-angle views and the Doppler shift to estimate three-dimensional cloud motions. In addition, the distribution and forms of surface water can be monitored from space using a range of techniques.

Promising research developments include the dynamic combination of surface measurements from increasingly sophisticated observational networks (for example, FLUXNET) and remotely sensed data, allowing production of high spatiotemporal models of hydroclimatic variables such as evapotranspiration. Recent large-scale experimental approaches have included networks of surface stations and a range of remotely sensed data that is used to develop and validate coupled surface-atmosphere computer models and as initialization data for future projections.

Estimates of hydroclimate regimes prior to the instrument record have been derived from numerous paleoclimate proxies such as sediment and ice cores, pollen counts (palynology), and paleolimnology. Tree-ring records (dendrochronology) have been particularly useful for extracting hydroclimate estimates with annual resolution. Such records have been used to derive annual drought indices, such as the Palmer Drought Severity Index (PDSI), to assess the frequency and duration of past droughts and their climatic controls.

A range of modeling techniques has been developed to help investigate the relative roles of different processes in the hydroclimate system and ultimately to produce predictions of future hydroclimates. These range from scale physical models to computer simulations using numerical techniques and high-performance computers. Recent dynamic coupling of climate and hydrologic computer models has greatly enhanced applications and improved accuracy, particularly in hydrologic response to external climate forcing. For example, Y. A. Mohamed and coworkers [Hydrometeorology of the Nile: Results from a regional climate model, *Hydrol. Earth System Sci.*, 9:263–278, 2005] used a regional-scale, coupled climate/hydrologic model to help unravel the processes and responses of the Nile watershed, which has one of the longest hydroclimate records for model evaluation. Significant challenges remain in modeling, particularly parameterization of dynamic small-scale phenomena such as cloud microphysics. Small errors in these parameters can lead to large errors in estimating precipitation or percent cloud cover. Another problem is that many links between model components are either unknown or not understood well enough to be accurately incorporated. For example, recent observational research has shown that the scattering of light by clouds and aerosols improves the water-use efficiency of many plants, implying a dynamic feedback between hydroclimates and plants, as well as links with other aerosol-producing phenomena such as volcanoes, fires, or dust storms.

Hydroclimates and humans. Human interaction with the hydroclimate system is multifaceted, operates on multiple time and space scales, and involves direct and indirect impacts. People are highly vulnerable to hydroclimate changes. In the first 6 years of this century, droughts afflicted approximately 620 million people in 73 countries. Over the same period, floods caused more than 37,000 deaths and made 8.8 million people homeless. Hydroclimatology is critical to water resource availability and related resources such as agricultural products and hydroelectricity. Arid and semiarid regions are particularly vulnerable to changes in regional hydroclimatology because of their existence near the threshold of agricultural viabilities. Because of this, much research has aimed to understand the frequency, intensity, and duration of hydrologic extremes. Hydroclimate models are now being integrated with topographic, ground-water, agricultural, economic, and environmental impact models to produce comprehensive analysis tools for societal impact assessment.

Human activities also modify hydroclimatic regimes either intentionally or as unintended consequences. Intentional modification includes activities such as cloud seeding to increase rain or snow, damming and river control measures, irrigation, mulching, and frost protection. These occur mostly at the micro to local scale. Unintended hydroclimate by-products occur from activities such as deforestation, urbanization, and agricultural production at the local to regional scale. Population pressure and/or poor land-use management can enhance desertification in already marginal dryland regions. At the global scale, anthropogenic increases of atmospheric greenhouse gas concentrations have been linked with hydroclimatic changes in the past century with increasing confidence. Although not understood in detail, this effect is global in scope but produces varied and complex regional responses due to shifting patterns of atmospheric circulations.

Hydroclimate change and the future. Both observational and paleoclimate records show hydroclimates have changed significantly worldwide over a wide range of time scales. These changes have resulted from a large number of forcing mechanisms operating over different time and space scales such as changes in solar radiation, changes to atmospheric and oceanic circulations, and changes in local to regional topography. Hydroclimates respond to such forcing in complex ways that highlight the dynamic interaction among processes, including positive and negative feedback mechanisms and thresholds. For example, drought may be initiated by an increase in temperature because warmer air has a higher atmospheric demand for water and water is therefore removed more rapidly from plant leaves and soil pores. With depleted soil moisture and increased plant stress, further warming occurs by the increased (sensible heat) energy available to heat the air, producing a positive feedback mechanism. As thresholds are reached, such as the wilting-point of plants, abrupt shifts can occur in the hydroclimatic equilibrium. At the global scale, hydroclimate feedback occurs between temperature and atmospheric water vapor concentration (since water vapor is a greenhouse gas), temperature and surface ice cover (since open water and ice-free land surfaces absorb more solar radiation than ice), or between temperature and clouds (since a cloudier atmosphere increases the planetary albedo).

Relatively small changes in temperature have been found to result in large shifts in watershed hydrologic responses. For example, a general warming trend over the western United States in the past 50 years has resulted in an increase in the ratio of rain to snow and an advance in the timing of peak spring seasonal flows, both of which reduce winter-spring storage of water in the mountains and greatly alter seasonal river discharge patterns. Another important threshold is the melting point of water. Regions where, and at times when, temperatures range near 0°C can undergo large changes in watershed hydrology. These changes have large impacts on processes such as fluvial sediment flow as well as aquatic and riparian ecosystems. Such changes may be felt far downstream such as seasonal changes in water salinity in deltaic wetland ecosystems. The implications for water resource management for drinking, irrigation, or hydroelectric power are vast, complex, and potentially expensive. Increasing numbers of climate-change studies have focused on the attribution of hydroclimate changes observed over the past century. These increasingly highlight the role of anthropogenic changes to global greenhouse gas concentrations. In addition, global climate models predict significant further hydroclimatic changes in the next century. For example, the International Panel on Climate Change 2007 report predicts significant changes to a suite of hydroclimatic variables (**Fig. 2**), including potential changes in precipitation of up to ±30%, with subtropical regions generally getting drier and mid- to high-latitude regions as well as some tropical regions such as the tropical Pacific Ocean and western tropical South America getting wetter. Precipitation depletion, coupled with increasing evaporation rates across most of the Earth due primarily to tropospheric warming, produces widespread depletion of soil moisture, a critical biological resource. Limitations of current models and complexities in the hydroclimate system make it difficult to provide accurate predictions of future hydroclimates, and these become increasingly less accurate with longer projection time. However, continuing improvements in the biophysical understanding of the hydroclimate system, computational techniques and capacity, data for model validation and initialization, as well as downscaling of global- to regional-scale models mean that the ability to predict

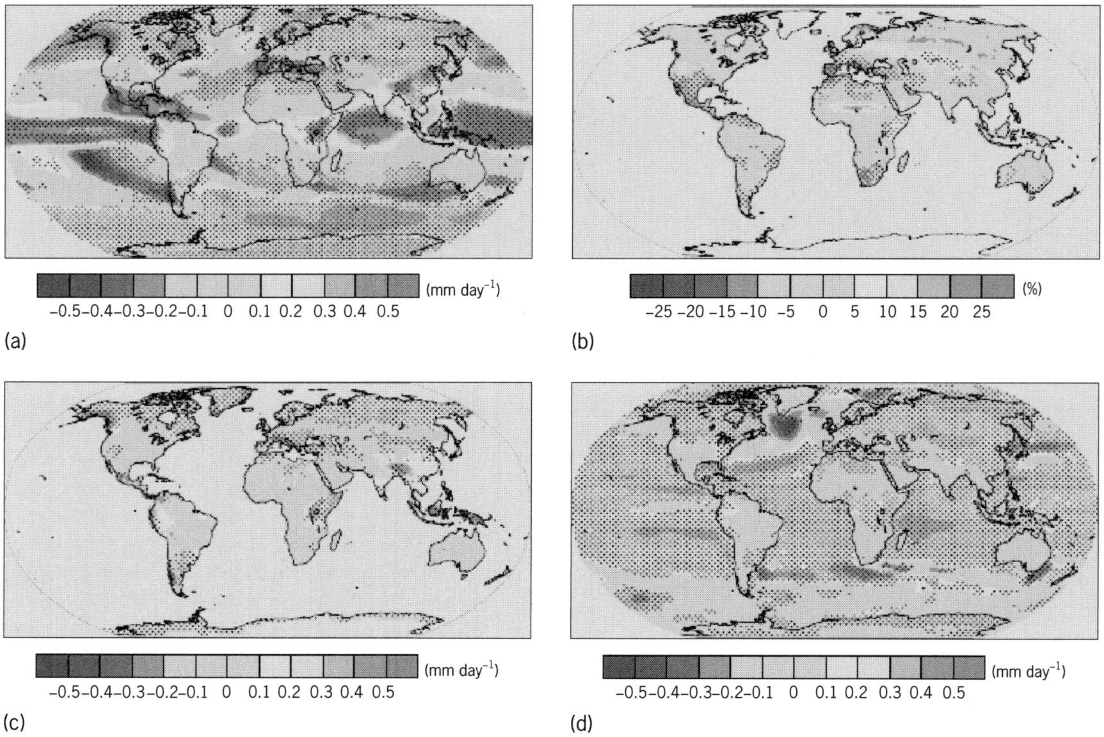

Fig. 2. Projected future changes in global hydroclimates showing multimodel mean changes in (*a*) precipitation (mm day^{-1}), (*b*) soil moisture content (%), (*c*) runoff (mm day^{-1}), and (d) evaporation (mm day^{-1}). Changes are annual means for the SRES A1B scenario for the period 2080–2099 relative to 1980–1999. Stippled regions indicate at least 80% of models agree on the sign of the mean change (*From IPCC 2007, Fig. 10.12*).

future hydroclimates at the basin scale is rapidly improving.

For background information *see* ALBEDO; ATMOSPHERIC GENERAL CIRCULATION; CLIMATOLOGY; CLOUD PHYSICS; HEAT BALANCE, TERRESTRIAL ATMOSPHERIC; HYDROLOGY; HYDROMETEOROLOGY; INFRARED RADIATION; METEOROLOGICAL SATELLITES; PALEOCLIMATOLOGY; PRECIPITATION (METEOROLOGY); RADAR METEOROLOGY; REMOTE SENSING; SATELLITE METEOROLOGY; TERRESTRIAL RADIATION; WATER RESOURCES in the McGraw-Hill Encyclopedia of Science & Technology. Andrew J. Oliphant

Bibliography. F. Beyrich and H-T. Mengelkamp, Evaporation over a heterogeneous land surface: EVA_GRIPS and the LITFASS-2003 experiment—An overview, *Bound.-Lay. Meteorol.*, 121:5–32, 2006; H. A. Cleugh et al., Regional evaporation estimates from flux tower and MODIS satellite data, *Remote Sens. Environ.*, 106:285–304, 2007; E. R. Cook, North American drought: reconstructions, causes, and consequences, *Earth Sci. Rev.*, 81:93–134, 2006; R. R. Heim, A review of twentieth-century drought indices used in the United States, *Bull. Am. Meteorol. Soc.*, 83:1149–1165, 2002; IPCC, *Climate Change 2007—The Physical Science Basis: Working Group I Contribution to the Fourth Assessment Report of the IPCC*, Cambridge University Press, 2007; D. R. Legates, Remote sensing in hydroclimatology: An introduction to a focus section of *The Professional Geographer*, 52:233–234, 2000; Q. Min, Impacts of aerosols and clouds on forest-atmosphere carbon exchange, *J. Geophys. Res. Lett.*, 110:D06203, 2005; N. W. T. Quinn et al., Model integration for assessing future hydroclimate impacts on water resources, agricultural production and environmental quality in the San Joaquin Basin, California, *Environ. Model. Softw.*, 19:305–316, 2004; S. K. Regonda et al., Seasonal cycle shifts in hydroclimatology over the western United States, *J. Climate*, 18:372–384, 2005; J. F. Reynolds, Global desertification: Building a science for dryland development, *Science*, 316:847–851, 2007.

Hydrogen storage materials

As concern regarding environmental issues such as pollution and global climate change escalates and energy independence grows in importance, substantial international research continues to be focused on hydrogen as an alternative to fossil fuels. Hydrogen, which in pure form exists as molecular H_2, is an attractive option for a variety of reasons. First, it occurs in vast amounts on Earth as a component of water (H_2O) and other chemical compounds, and it is the most abundant element in the universe. Second, the combustion of H_2 with oxygen releases energy with only water vapor and heat as by-products; no carbon dioxide (CO_2), a greenhouse gas, is produced. This aspect makes fuel cells operating with H_2 leading candidates to replace gasoline-powered internal combustion engines. Third, on a gravimetric basis (kilogram for kilogram), hydrogen in the form of H_2 contains about three times the energy of gasoline.

Similar to electricity, hydrogen is an energy carrier. To realize its benefits, hydrogen will need to be extracted from renewable sources such as water. For use in the transportation sector, where light-duty vehicles account for about 20% of energy use in the United States, hydrogen must be stored in a safe, efficient, and cost-effective way. The U.S. Department of Energy has set 2010 targets of 6 mass% for the gravimetric density (ratio of hydrogen mass to total mass) and 45 g/liter for the volumetric density (hydrogen mass per unit volume) for a vehicular hydrogen storage system. Materials for hydrogen storage must have substantially higher values to achieve these system targets after the additional weight of the storage tank and ancillary hardware is taken into account.

Challenges. Storing H_2 under compression is technically viable. With high-pressure tanks, system capacities of about 4 mass% and 20 g/liter have been achieved. Liquid H_2 (boiling point: −253°C, 20 K, or −424°F) can roughly double the volumetric density but requires a superinsulating cryogenic tank, and boil-off due to heat leaking into the tank is a major concern. Solid hydrides are of continuing interest as hydrogen storage media in part because they have a volumetric advantage; many hydrides, such as $LaNi_5H_7$, store more hydrogen per unit volume than liquid hydrogen. The challenges are the gravimetric density and the temperature and kinetics of H_2 release.

Recent developments. At present neither the compressed or liquid H_2 options nor any known material satisfies the Department of Energy targets. Significant progress has been made over the past few years, however, in several materials areas.

New hydrides. Light metal hydrides offer promising opportunities for maximizing gravimetric hydrogen density in addition to high volumetric density. Resurgent interest in complex hydrides—so named because they form metal-hydrogen complexes such as AlH_4^- (alanate), BH_4^- (borohydride), NH_2^- (amide), or NH^{2-} (imide)—can be traced to advances in sodium alanate ($NaAlH_4$). Adding dispersed particles of early transition metals such as titanium, typically introduced as $TiCl_3$, to $NaAlH_4$ lowered the dehydrogenation temperature of $NaAlH_4$ to below its melting temperature (190°C or 374°F) and permitted reversible hydrogenation. Although the gravimetric capacity of $NaAlH_4$ is lower than desired, its enthalpy ΔH for dehydrogenation is close to that considered ideal for hydrogen storage, $\Delta H \sim 30$–40 kJ/mole.

The enthalpy largely sets the thermodynamic limit for the operating temperature T and equilibrium pressure P of the system through the van't Hoff relation (1), where R is the gas constant and ΔS is

$$\ln(P[\text{bar}]) = -\frac{\Delta H}{RT} + \frac{\Delta S}{R} \qquad (1)$$

the entropy change during the reaction. (This equation and the signs of ΔH and ΔS are written for the dehydrogenation reaction; the signs reverse for hydrogenation.) For solid storage, ΔS is dominated by the entropy of hydrogen in the gas phase (130 J/K-mole H_2) and varies by less than 30% from that value for most systems. For ΔH below about 45 kJ/mole H_2, the equilibrium temperature in a 1-bar pressure of H_2 gas, given by Eq. (2), falls below 80°C (176°F), the

$$T(1\text{ bar}) = \frac{\Delta H}{\Delta S} \qquad (2)$$

operating temperature of a proton exchange membrane fuel cell. Hydrogen in most other stable complex hydrides, such as $LiBH_4$, is too strongly bound; the T(1 bar) temperatures are too high for practical use in transportation applications.

Recently, hydrogen storage has been demonstrated using the conversion of lithium nitride (Li_3N) first to lithium imide (Li_2NH) and then to lithium amide ($LiNH_2$) on absorption of hydrogen. Lithium hydride (LiH) is also produced in each step. Only the conversion between Li_2NH and $LiNH_2$ has a T (1 bar) value low enough to be considered practical, while offering a gravimetric capacity somewhat larger than that of $NaAlH_4$. Combining magnesium amide, $Mg(NH_2)_2$, with LiH produces the mixed imide compound $Li_2Mg(NH)_2$ on dehydrogenation, with a T (1 bar) value close to optimal, although the capacity is reduced. Hydrogen cycling in the Li-Mg-N-H system is further complicated by the appearance of another intermediate mixed Li-Mg-imide phase. Also, production of small quantities of ammonia (NH_3) along with H_2 is an issue for this and other nitrogen-containing hydrogen storage materials.

Several new quaternary hydrides with large hydrogen content have been found in the Li-B-N-H system. Of greatest interest for hydrogen storage are materials based on the $Li_4BN_3H_{10}$ phase, a complex hydride that contains one BH_4^- and three NH_2^- complexes per formula unit (see **illustration**). Optimum

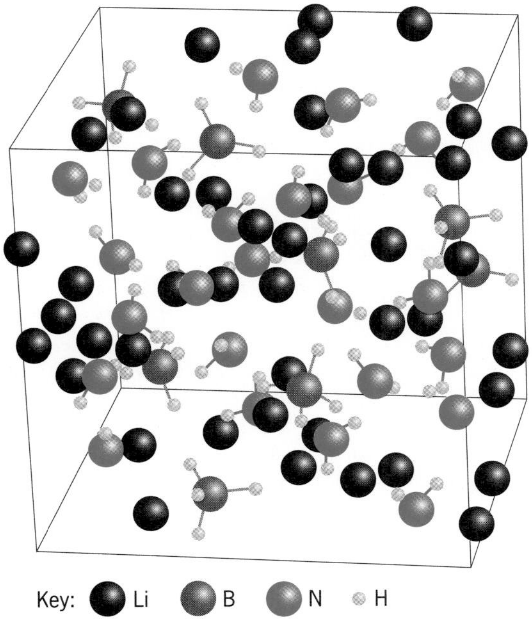

Crystal structure of $Li_4BN_3H_{10}$, illustrating BH_4 and NH_2 complexes.

hydrogen release, however, occurs near the composition $Li_3BN_2H_8$, corresponding to a 1:2 ratio of $LiBH_4$:$LiNH_2$. At this composition the material can decompose directly to the final Li_3BN_2 phase while releasing all of its hydrogen, whereas the 1:3 equilibrium composition contains excess nitrogen that is lost as ammonia gas. Attempts to rehydrogenate have been unsuccessful; experimental observations and theoretical calculations concur that ΔH for this system is likely below 20 kJ/mole H_2 and may in fact be exothermic; that is, the hydride may be thermodynamically unstable. Nevertheless, this system does suggest that other as yet undiscovered quaternary phases exist that could have promising hydrogen storage properties.

Destabilized hydrides. The high ΔH for complex hydrides can be moderated by reacting them with another compound, preferably another hydride, to form a mixed compound in the dehydrogenated state. A canonical example is the $LiBH_4$-MgH_2 system. Dehydrogenation of $LiBH_4$ alone produces LiH and elemental boron, characterized by high ΔH and limited reversibility. When MgH_2 is added to $LiBH_4$ in the ratio of 1:2, however, dehydrogenation instead produces MgB_2. The additional thermodynamic stability of MgB_2 in the dehydrogenated state in effect destabilizes $LiBH_4$ in the hydrogenated state: ΔH and T (1 bar) for the combined system are reduced compared to those of either $LiBH_4$ or MgH_2 individually. Furthermore, the combined system is fully reversible to the hydrogenated state. Several examples of destabilized hydride systems are given in the **table**. One advantage offered by this technique is that large numbers of such couples are possible, even though the number of known complex hydrides is limited. As yet, however, a reversible reaction couple having the desired combination of T (1 bar) and high gravimetric capacity has not been identified.

Reaction kinetics remains an important issue with most light metal hydride materials, including destabilized reaction couples. The chemical reactions involved with dehydrogenation and rehydrogenation typically involve mass transport between phases, which tends to be kinetically limited even when the reaction is thermodynamically favored. Temperatures well above the T (1 bar) values are often required to drive the reaction. Modest quantities of additives, often in the form of transition metal chlorides, are usually included with the reactants to try to enhance the kinetics and reduce the reaction temperature. The effectiveness of such additives, even when optimized, varies from a temperature reduction of as little as about 20°C (68°F) in some systems to reductions in excess of 100°C (212°F) for $NaAlH_4$ and $Li_3BN_2H_8$.

Cryogenic materials. H_2 molecules can bind to the surface of many materials, but the attractive forces are weak so that low temperatures, near that of liquid nitrogen (−196°C, 77 K, or −321°F), are required. The amount of H_2 adsorbed, however, generally increases with the surface area of the material, and if this is sufficiently large, the penalty of having to cool a tank containing the material might be offset by the improved hydrogen storage capacity. Carbon-based materials are of interest in view of their low mass and potentially low cost. Activated carbons, for example, are well known, can feature effective surface areas as large as 2500 m^2/g (that is, an ounce can have more than 17 acres of surface area), and can adsorb as much as 5 mass% H_2 at 77 K. Furthermore, there is a large and growing variety of carbon nanostructures, materials with different atomic-scale conformations which can lead to distinct physical properties. Among these are carbon nanotubes, fullerenes (C_{60} molecules or buckyballs), and aerogels; their hydrogen adsorption behavior is under active investigation. Intense research on other high-surface-area materials is also in progress. Metal-organic frameworks, composed of inorganic units (for example, ZnO_4) held together in a very porous crystalline lattice by organic links (for example, benzenedicarboxylate), were discovered by Omar Yaghi and coworkers several years ago. At least 12,000 different metal-organic frameworks have been documented, of which only a relative handful have been tested for hydrogen adsorption. So far surface areas as large as 5000 m^2/g (35 acres per ounce) and 5–7 mass% hydrogen adsorption at 77 K have been demonstrated. Other framework-type compounds are being synthesized by a number of research groups.

Selected solid hydride storage reactions

Reaction	Gravimetric capacity, mass%	Volumetric capacity, g/liter	ΔH, kJ/mole H_2	T (1 bar), °C*
$MgH_2 \leftrightarrow Mg + H_2$	7.7	111	74.5	279
$NaAlH_4 \leftrightarrow \frac{1}{3}Na_3AlH_6 + \frac{2}{3}Al + H_2$ ($NaAlH_4$ stage 1)	3.7	47	37	30
$\frac{1}{3}Na_3AlH_6 + \frac{2}{3}Al + H_2 \leftrightarrow NaH + Al + \frac{3}{2}H_2$ ($NaAlH_4$, stage 2)	1.9	32	47	100
$LiBH_4 \rightarrow LiH + B + \frac{3}{2}H_2$	13.9	92	61	457
$LiNH_2 + LiH \leftrightarrow Li_2NH + H_2$ (Li_3N, stage 2)	6.5	68	66	275
$Li_2NH + LiH \leftrightarrow Li_3N + H_2$ (Li_3N, stage 1)	5.5	68	95	430
$Mg(NH_2)_2 + 2LiH \leftrightarrow Li_2Mg(NH)_2 + 2H_2$	5.6	66	39	75
$3MgH_2 + MgCu_2 \leftrightarrow 2Mg_2Cu + 3H_2$	2.6	75	68	240
$2MgH_2 + Si \leftrightarrow Mg_2Si + 2H_2$	5.0	83	36	20
$2LiBH_4 + MgH_2 \leftrightarrow 2LiH + MgB_2 + 4H_2$	11.5	96	41	225
$6LiBH_4 + CaH_2 \leftrightarrow 6LiH + CaB_6 + 10H_2$	11.7	90	59	418
$Li_3BN_2H_8 \rightarrow Li_3BN_2 + 4H_2$	11.9	114	≤20 (<0?)	<0

*T(°F) = T(°C) × 1.8 + 32.

DFT modeling. A powerful technique for calculating energies of electrons in molecules and solids, density functional theory (DFT), has become a valuable tool in research on hydrogen storage materials. It has greatly expanded understanding of the properties of known hydrides, including the electronic structure, the hydrogen bonding character, and the enthalpy ΔH accompanying hydride formation. The electronic terms entering ΔH are usually the largest, but the vibrational contributions arising from the phonons, the quantized lattice vibrations of the materials involved in the process, can be significant. There is even a contribution at zero temperature (that is, 0 K, −273.15°C, or −459.67°F) called the zero-point energy (ZPE). By calculating the forces between ions with density functional theory for different displacements of ions in a material, the phonon energies can be derived, albeit at considerable computational expense. Noteworthy hallmarks of progress in the field are the increasingly frequent reports of such calculations. For example, it is now known that the ZPE component is about 25% of ΔH in $LaNi_5H_7$ and in the lithium amide–lithium imide reaction, serving to improve agreement between theory and experiment in each case.

Given the encouraging accord between calculated and measured properties for a variety of known systems, great promise lies in the imaginative use of density functional theory to synergistically guide the discovery and development of technologically viable hydrogen storage materials. Theoretical assessment of hundreds of destabilizing reactions has stimulated work on the experimental side. Hypothetical model hydrides including organometallic buckyballs (complexes of C_{60} molecules and transition-metal atoms like scandium), transition-metal–ethylene complexes, transition-metal–decorated polymers such as polyacetylene, and activated boron nitride nanotubes have been proposed. Synthesis of designer materials such as these and determination of their hydrogen-bearing characteristics is an open challenge whose resolution may lead to better pathways for hydrogen storage.

For background information *see* ABSOLUTE ZERO; CARBON NANOTUBES; CHEMICAL KINETICS; CRYOGENICS; ENTHALPY; FUEL CELL; FULLERENE; HYDROGEN; METAL HYDRIDES; PHONON; QUANTUM CHEMISTRY in the McGraw-Hill Encyclopedia of Science & Technology. Jan F. Herbst; Frederick E. Pinkerton

Bibliography. G. W. Crabtree, M. S. Dresselhaus, and M. V. Buchanan, The hydrogen economy, *Phys. Today*, 57(12):39–44, December 2004; M. Fichtner (organizer), Viewpoint set #42: Nanoscale materials for hydrogen storage, *Scripta Mater.*, 56(10):801–858, 2007; F. E. Pinkerton and B. G. Wicke, Bottling the hydrogen genie, *Ind. Physicist*, 10(1):20–23, February–March 2004; S. Satyapal, J. Petrovic, and G. Thomas, Gassing up with hydrogen, *Sci. Amer.*, 296(4):80–87, April 2007; L. Schlapbach (guest ed.), Hydrogen storage, *MRS Bull.*, vol. 27, no. 9, September 2002; L. Schlapbach and A. Züttel, Hydrogen-storage materials for mobile applications, *Nature*, 414:353–358, 2001; A. Züttel, Materials for hydrogen storage, *Mater. Today*, 6(9):24–33, September 2003.

Infant Australopithecus from Dikika

Australopithecus afarensis is an extinct species of the evolutionary lineage leading to modern humans. It is well known from numerous important fossil discoveries from sites in Ethiopia, Kenya, and Tanzania, dating to between 2.9 and 3.8 million years ago. Since anthropologists know a good deal about the anatomy of *A. afarensis*, this species provides the best evidence for reconstructing the evolutionary relationships and behavior of early human species. The recent exciting discovery of an infant skeleton of *A. afarensis* from eastern Africa has provided new insights into the biology of this important species. It is rare that skeletons of such antiquity survive, and the preservation and discovery of a nearly complete infant skeleton are unprecedented. The fossil skeleton (DIK-1-1), nicknamed "Selam," was discovered at Dikika in Ethiopia in 2000 by a research team led by Zeresenay Alemseged. It was recovered from the Hadar Formation, a thick series of sediments that has produced most of the fossil finds of *A. afarensis*, including the adult partial skeleton A.L. 288-1, known as "Lucy," found in 1974.

Fossil skeleton. The original DIK-1-1 find consisted of a sandstone block containing the partial cranium, brain endocast (a natural cast of the internal surface of the braincase derived from sediment that filled the cranium at the time of burial), and associated vertebrae, ribs, and scapula (shoulder blade) (**Figs. 1–3**). Subsequent field seasons at Dikika in 2002 and 2003 resulted in the recovery of additional skeletal elements, including limb and foot bones. The deposits from which DIK-1-1 was recovered are 3.31–3.35 million years old, similar in age to deposits that have provided other specimens of *A. afarensis* from the nearby site of Hadar. Since the initial discovery, the sandstone matrix has been painstakingly removed to expose the fossil bones, while maintaining the contact between the articulating pieces.

Compared to Lucy, the Dikika skeleton is more complete, probably because flood sediments rapidly buried the individual a short time after or at the time of death. The completeness of the skeleton means that anthropologists now have previously unknown skeletal elements to analyze and the opportunity to study the anatomy of bones from a single individual. Included in the unique skeletal elements is the hyoid bone of the throat, which is unknown for any early human species. Missing from the skeleton are the pelvis and vertebrae of the lower back.

Selam's deciduous (baby) teeth are erupted (that is, they have emerged from the gum line), and computed tomography (CT) scans reveal fully formed crowns of permanent teeth within the jaws for later root development and eruption. Based on the stage of eruption of the teeth, and assuming an apelike growth pattern, Selam is estimated to have been

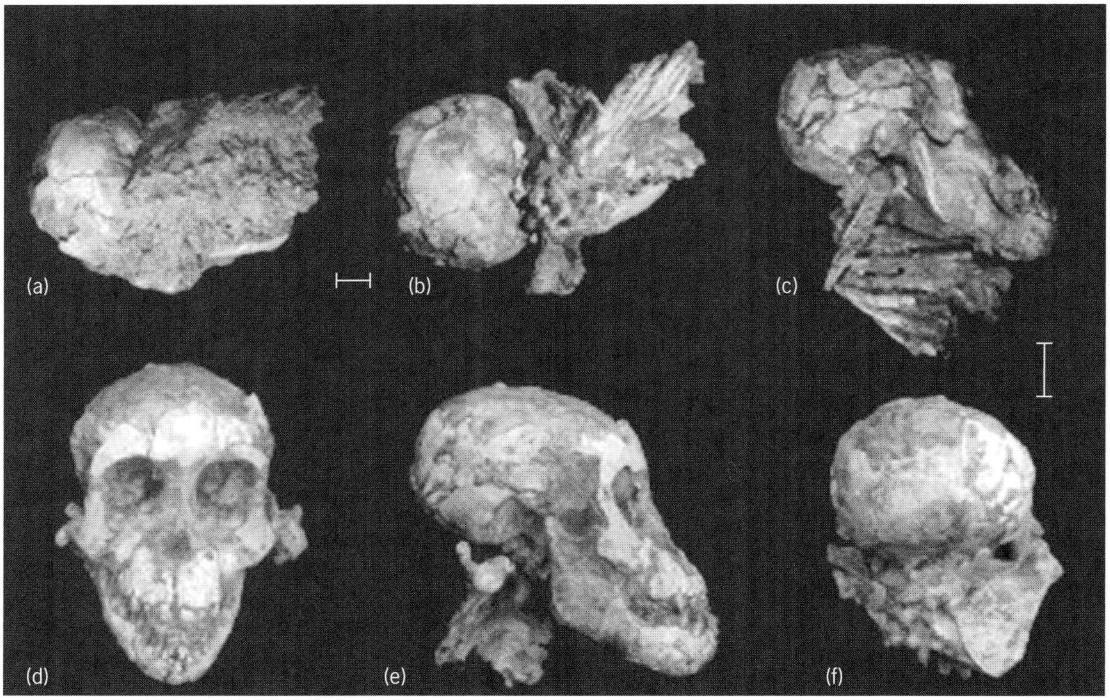

Fig. 1. DIK-1-1 skull, scapula, ribs, and vertebrae: (*a*) inferior view as discovered; (*b*) inferior view after some matrix removal; (*c*) side view showing skull, ribs, and scapula; (*d*) front view of face; (*e*) side view showing brain endocast, partial skull, and shoulder joint; (*f*) back view of skull, endocast, scapula, and ribs. Scale bars: 2 cm (0.8 in.). (*Copyright 2006 by Nature Publishing Group*)

approximately 3 years old at death. An apelike rate of growth is assumed because analyses have demonstrated that tooth formation and eruption occurred faster in *Australopithecus* than in recent humans. Selam's sex has been determined to be female because the unerupted teeth measured from CT scans are relatively small, which are more likely to come from female individuals.

There are a few previously known *Australopithecus* infant crania that can be compared to Selam. These include the *A. afarensis* A.L. 333-105 from Hadar and an infant skull of the closely related *A. africanus* from Taung, South Africa. Comparisons reveal that Selam is most similar to A.L. 333-105, particularly in the face. For example, A.L. 333-105 and Selam share nasal bones that are tall and narrow like those of great apes, whereas those of Taung are short and broad. Given the facial and cranial similarities with adults and infants of *A. afarensis*, as well as its derivation from the Hadar Formation, Selam has been attributed to this species.

Mode of locomotion. Even though the analysis of the Dikika skeleton is in its early stages, Selam's anatomy has implications for interpreting the locomotor behavior of *A. afarensis*, particularly the degree to which the species spent time climbing in trees. The presence of a bicondylar angle, the condition wherein the thighbones are angled toward one another at the knees, shows that she walked upright as in modern humans. This angle develops during the lifetime of an individual as a mechanical response to bipedal walking (that is, walking on two legs as in modern humans). Additionally, the heel bone is robust, which is evidence that body weight was transmitted through the foot in a fashion consistent with bipedal locomotion. On the other hand, the shoulder joint faces upward as in apes, rather than to the side, which facilitates over-the-head arm movements. This new information about the configuration of the shoulder can be added to the other apelike characteristics of the forelimb, such as curved fingers and long arms relative to legs, that are observed in Selam and other *A. afarensis* specimens. These characteristics suggest that the arms of *A. afarensis*, like those of apes, were adapted for climbing and hanging in trees.

Taken together, the anatomical characteristics of the legs, arms, and hands indicate that Selam and other members of *A. afarensis* were bipedal on the ground, but were also proficient at climbing in trees. An alternative interpretation is that *A. afarensis* was fully bipedal and terrestrial, and that the apelike forelimbs were merely retained from an arboreal (that is, tree-living) ancestor, but no longer used for activity in the trees. These opposing points of view are central to an ongoing debate about whether bipedality in *A. afarensis* was supplemented by some degree of arboreal behaviors. Selam's skeleton provides critical information to this debate by contributing new anatomical data. Specifically, the anatomy of her shoulder joint, like that of the arm and finger bones, is consistent with arboreal locomotion. Further study of the Dikika skeleton may be helpful in answering this question, particularly if the structure of additional skeletal elements can be shown to correspond directly to an individual's locomotor behavior during its lifetime, as do the curvature of the finger bones and the bicondylar angle of the thighbone.

Growth and development. The extraordinary preservation and the young chronological age of Selam's skeleton will continue to provide novel data about growth and development in *A. afarensis*. For the first time, growth in the cranium and postcranium (the skeleton without the skull) can be examined together in a single individual. Thus, the timing of developmental events, like the eruption pattern of the teeth and the growth of the limbs, can be better understood. Such information helps anthropologists reexamine the current hypothesis, based solely on dental eruption timing, that the growth pattern of *A. afarensis* is similar to that of apes. Models of brain growth can be presented as well. The estimated endocranial volume (a measure of the size of the internal cavity of the braincase that is used as a guide to brain size) for Selam of between 275 and 330 cm^3 is comparable to that of chimpanzees of similar age. When the endocranial volume of Selam is considered as a percentage of the volume of an adult *A. afarensis* (375–550 cm^3), it is apparent that her brain growth is 65–88% complete. Modern humans at the same age would have completed much less of their brain growth, owing to their absolutely larger adult brain volumes.

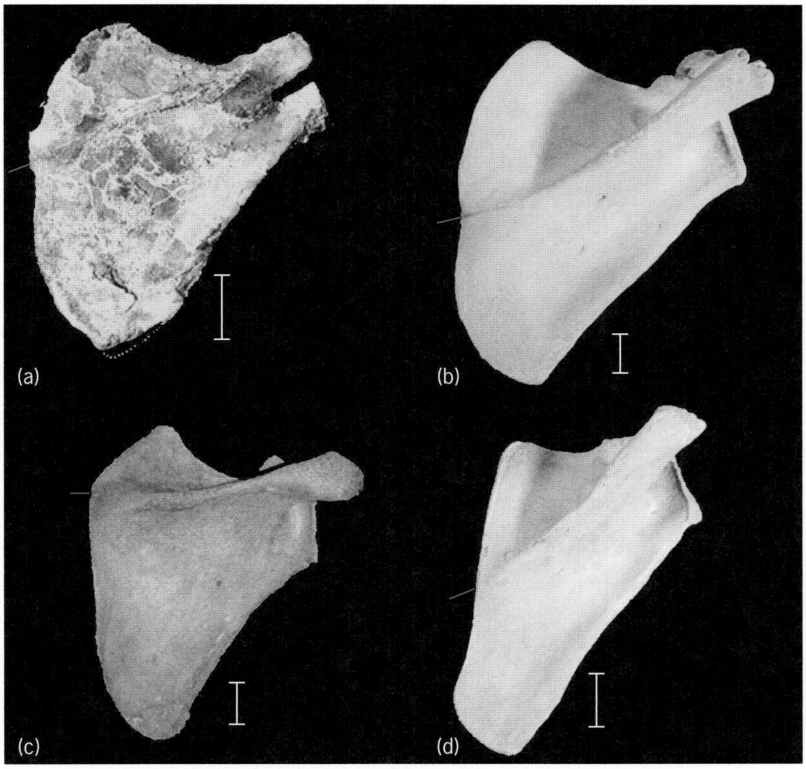

Fig. 3. Back view of right scapula of (*a*) DIK-1-1, (*b*) juvenile gorilla, (*c*) human, and (*d*) chimpanzee. Note that the line marking the placement and orientation of the spine of the scapula is more similar among the apes and DIK-1-1 than between DIK-1-1 and the human example. Scale bars: 1 cm (0.4 in.). (*Copyright 2006 by Nature Publishing Group*)

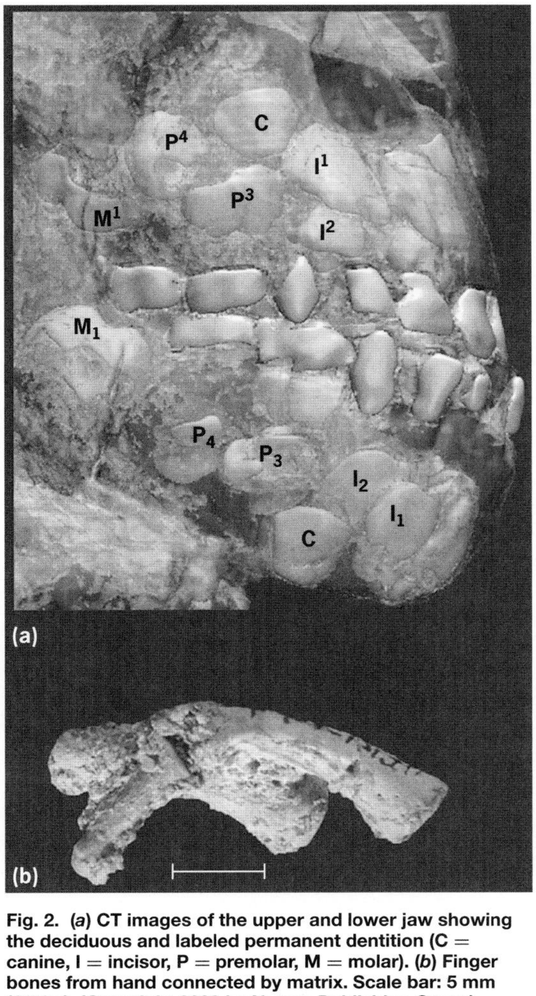

Fig. 2. (*a*) CT images of the upper and lower jaw showing the deciduous and labeled permanent dentition (C = canine, I = incisor, P = premolar, M = molar). (*b*) Finger bones from hand connected by matrix. Scale bar: 5 mm (0.2 in.). (*Copyright 2006 by Nature Publishing Group*)

With further anatomical analyses, Selam will be invaluable for inferring the locomotor behavior in *A. afarensis* and for helping to resolve the debate about whether the species was partly arboreal. The form of skeletal elements, such as the scapula and hyoid, that are rare or previously unknown in *A. afarensis* will improve the understanding of the anatomy of this extinct species. Selam also provides an important point of comparison to adult anatomy, which will facilitate studies of growth changes in the cranium and postcranium. Finally, the discovery of DIK-1-1 in the Afar region of Ethiopia underscores the importance of continued fieldwork for recovering new and significant fossils that can contribute to studies of early human evolution.

For background information *see* ANTHROPOLOGY; ANTHROPOMETRY; APES; AUSTRALOPITHECINE; FORENSIC ANTHROPOLOGY; FOSSIL HUMANS; FOSSIL PRIMATES; PHYSICAL ANTHROPOLOGY in the McGraw-Hill Encyclopedia of Science & Technology.

Elizabeth Harmon

Bibliography. Z. Alemseged et al., A juvenile early hominin skeleton from Dikika, Ethiopia, *Nature*, 443:296–301, 2006; B. H. Smith, Patterns of dental development in *Homo*, *Australopithecus*, *Pan*, and *Gorilla*, *Am. J. Phys. Anthropol.*, 94(3):307–325, 1994; J. T. Stern, Climbing to the top: A personal memoir of *Australopithecus afarensis*, *Evol. Anthropol.*, 9:113–133, 2000; C. V. Ward, Interpreting the posture and locomotion of *Australopithecus afarensis*: Where do we stand?, *Yearb. Phys. Anthropol.*, 45:185–215, 2003; B. Wood, A precious little bundle, *Nature*, 443:278–281, 2006; J. G. Wynn et al., Geological and palaeontological context of a Pliocene juvenile hominin at Dikika, Ethiopia, *Nature*, 443:332–336, 2006.

Inferring patterns of diversification

Episodes of prolific cladogenesis (branching of new species from common ancestral lineages), adaptive radiation (rapid speciation of an ancestral lineage to fill diverse ecological niches), species selection (differential success of lineages owing to species-intrinsic properties), key innovations, and mass extinctions are a few examples of evolutionary phenomena involving differential rates of *diversification* (*speciation* minus *extinction* rate). Although traditionally based on patterns of fossil diversity chronicled in the paleontological record, the study of diversification increasingly relies upon information from *phylogenetic analyses* of extant species (estimates of evolutionary relationships that collectively comprise the Tree of Life). Tremendous technical progress in the generation of molecular sequence data and parallel theoretical, methodological, and computational advances in the analysis of those data have wrought an exponential increase of ever more reliable phylogenetic trees. Depending on the nature of the data and methods of analysis, phylogenetic trees can provide two sources of information relevant to the inference of diversification rates (**Fig. 1**): *topological* distribution of species diversity across branches of the tree and *temporal* distribution of branching events through time. These phylogenetic observations can be compared to expectations generated under various null models to explore numerous evolutionary phenomena.

Stochastic models. Diversification is an inherently stochastic (random) process: over a given time interval, two lineages with the same underlying probability of diversification may nevertheless realize substantially different rates of diversification. It is therefore necessary to compare our phylogenetic observations (on the topological distribution of species diversity or temporal distribution of branching events) to expectations generated by an appropriate stochastic branching model. Several statistical models have been proposed (see **table**). Most are stochastic Markov processes that model lineage extinction, μ, and/or speciation, λ, as instantaneous events that occur in continuous time with equal and independent probability along any tip of the growing tree. (In general, a stochastic Markov process assumes a series of random events that lack "memory" such that the probability of the next event depends only on the most recent outcome.) During a brief interval Δt, every tip of the tree will speciate with probability $\lambda(t)\Delta t$ and go extinct with probability $\mu(t)\Delta t$, such that $\lambda(t)$ and $\mu(t)$ are lineage-specific rates of speciation and extinction. The Markov family of branching processes includes four common models of increasing complexity:

1. The *constant-rate pure-birth* model (or *Yule* model) assumes a constant speciation rate and a zero extinction rate.

2. The *generalized pure-birth* model (or *equal-rates Markov* model) allows the speciation rate to vary through time, but assumes a zero extinction rate.

3. The *constant-rate birth-death* model assumes constant and nonzero rates for both speciation and extinction, provided that the net diversification rate is positive.

4. The *generalized birth-death* model permits speciation and extinction rates to vary through time.

Although less commonly used, non-Markov models are perhaps more realistic, in that they allow the

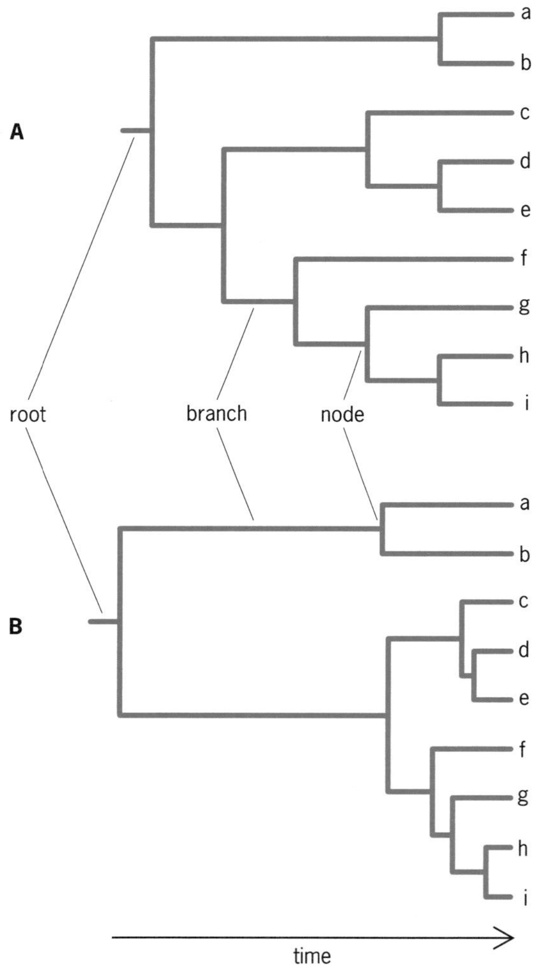

Fig. 1. Phylogenetic trees can provide two sources of information for the study of diversification rates. Both the upper and lower trees specify identical evolutionary relationships among species *a* through *i*. However, the lengths of branches may be arbitrary (as in A) or drawn to reflect the (absolute or relative) timing of branching events (as in B). Two corresponding classes of methods have been developed. *Topological* (also referred to as *tree-balance*) methods rely exclusively on the species diversity of lineages descended from a common node in the tree. By contrast, *temporal* (also referred to as *tree-shape*) methods exploit estimates of the waiting times between speciation events in the tree.

Common stochastic branching process models used in the study of diversification rates

Family	Model	Parameters	Comments
Markov	Generalized birth-death	$\lambda(t)$, $\mu(t)$	All Markov models assume speciation and extinction events are instantaneous, and that, even when $\lambda(t)$ and/or $\mu(t)$ may vary through time, they are nevertheless constant across every tip of the tree at any instant.
	Constant-rate birth-death	$\lambda(t) = \lambda$, $\mu(t) = \mu$, $\lambda \geq \mu$	Behavior of this model is well characterized mathematically.
	Generalized pure-birth	$\lambda(t)$, $\mu(t) = 0$	Also referred to as the equal-rates Markov model. Equivalent to the constant-rate pure-birth model when temporal information is ignored.
	Constant-rate pure-birth	$\lambda(t) = \lambda$, $\mu(t) = 0$	Also referred to as the Yule model. Equivalent to the generalized pure-birth model when temporal information is ignored. Most commonly used model owing to its tractable mathematical properties and convenient simplifying assumptions.
Non-Markov	Peripatric* pure-birth	$\lambda(t)$, τ	All non-Markov models relax the assumption that speciation and extinction events occur instantaneously. Consequently, expectations may differ substantially from Markov models. Explicitly models diversification by peripheral isolates speciation. More realistic, but seldom used owing to difficulty in reliable estimation of the refractory period parameter, τ.
	Allopatric* pure-birth	$\lambda(t)$, τ	Explicitly models diversification by allopatric/vicariant speciation. More realistic, but seldom used owing to difficulty in reliable estimation of the refractory period parameter, τ.

*Allopatric speciation = Differentiation of populations in geographical isolation to the point where they are recognized as separate species. Peripatric speciation = A special version of allopatric speciation that occurs when one of the isolated populations has relatively few individuals.

process of diversification to have memory, as might arise, for example, when speciation events are followed by a refractory period in which newly formed species reestablish their geographic range or effective population size.

Stochastic branching models are central to both hypothesis testing and parameter estimation. Although our current focus is on the use of these models for generating expected (null) distributions, numerous approaches have also been developed that allow various diversification-rate parameters to be estimated from phylogenetic trees with an absolute time scale. For example, maximum-likelihood approaches (a statistical technique for fitting a model to data, which simultaneously estimates values for the associated parameters) can be used to fit the observed distribution of branching times to a constant-rate birth-death model to estimate rates of speciation and extinction.

Fundamental questions. Although differential rates of diversification are associated with myriad evolutionary phenomena, the study of these diverse processes ultimately entails a limited number of fundamental inference problems for which phylogeny-based methods have been developed.

Detecting variation in diversification rates across lineages. This inference problem addresses the question, "Have lineages diversified under significantly different rates?" Detecting significant variation in rates of diversification has been used to investigate the prevalence of different modes of diversification (including adaptive radiation) in samples of estimated phylogenies, and also to evaluate whether a particular phylogenetic tree satisfies the assumptions of other inference methods (for example, for identifying diversification rate shifts through time; see below). In general, methods for detecting significant diversification rate variation involve statistics that variously summarize relevant aspects of the topological or temporal phylogenetic information. The value of the statistic is first calculated for the study tree. Monte Carlo simulation (a method that obtains a probabilistic approximation to the solution of a problem by using statistical sampling techniques) is then used to generate a null distribution of the statistic; this involves repeatedly simulating trees equal in size to the study tree under a stochastic model in which the probability of diversification is equal across all lineages (for example, the constant-rate pure-birth model). The statistic is calculated for each simulated tree, and resulting values collectively comprise a null distribution of the statistic against which the observed value can be compared to estimate the probability that all lineages in the study tree diversified under stochastically constant rates.

Locating shifts in diversification rate along branches. This second inference problem addresses the question, "Along which branches have significant changes in diversification rate occurred?" Locating significant shifts in diversification rate along branches of a tree has been used to investigate the influence of intrinsic traits (key innovations, such as morphological, behavioral, or physiological novelties) and incidence of extrinsic events (key opportunities, such as the dispersal of continental lineages to oceanic islands). Relatively few methods have been developed to address this inference problem. One method, the relative cladogenesis statistic, relies on temporal information to identify the set of k ancestral lineages that were contemporaneous at some arbitrary point in the past, t_k, that collectively gave rise to N extant species. The observed distribution of extant species diversity in each of the k ancestral lineages can be compared to the expected distribution generated under a stochastic branching model in which the underlying probability of diversification is equal among the set of lineages. Identification of an anomalously diverse lineage indicates that it diversified at a significantly higher rate than its contemporaries. The statistic can be evaluated over the entire duration of the tree (that is, by integrating t_k over the interval from the first to the last branching event) to assess the probability of a significant diversification rate shift along all internal branches.

Exploring the association between traits and rates of diversification. This third inference problem addresses the question, "Are rates of diversification significantly correlated with a particular organismal trait?" Identifying traits that are correlated with differential rates of diversification is critical to testing key innovation hypotheses, in which an evolutionary novelty is hypothesized to have promoted lineage diversification. Putative key innovations are typically morphological traits, but may also involve other biological attributes (associated with changes in ecology, physiology, behavior, etc.). Methods have been developed for both qualitative traits that occur in two or more discrete states and quantitative traits that exhibit a continuous number of states. These methods generally entail estimating the evolutionary history of the putative trait (**Fig. 2**). A statistic is then calculated that reflects the association between the inferred trait history and diversification rate. One such statistic, δ, provides a test of key innovations with two discrete states, i and j. The diversification rate is calculated for lineages in the study tree that possess alternate states of the trait, λ_i and λ_j, and the statistic is simply given by the difference between the state-specific rates, $\delta = (\lambda_i - \lambda_j)$. A null distribution of this statistic is then generated by repeatedly simulating branching times for nodes in the study topology using a stochastic branching model in which the diversification rate is not correlated with the evolutionary history of the trait. The δ statistic is calculated for each simulated tree, and these values collectively comprise the null distribution against which the observed value is compared to determine the probability that the trait is correlated with a significantly elevated diversification rate.

Identifying shifts in diversification rate through time. This fourth inference problem addresses the question, "Have rates of diversification significantly changed over time?" Detecting significant temporal shifts in diversification rate has been used to study the impact of extrinsic events that might simultaneously influence rates of diversification in all lineages of a study tree (for example, the onset of Pleistocene glaciations might uniformly have impacted rates of diversification in temperate groups). Establishing whether a particular tree has diversified under stochastically constant rates is also necessary to use other inference methods (for example, methods for estimating parameters of constant-rate branching process models assume that rates have not changed over time). Several methods have been developed to identify temporal shifts in diversification rate. For example, the constant-rate test relies on a statistic, γ, that summarizes the relative distance of branching events from the root of a tree. The γ statistic is first calculated for the study tree: values of $\gamma < 0$ indicate that branching events are clustered near the root (that is, rates have decreased through time), whereas values of $\gamma > 0$ indicate that branching events are concentrated toward the tips of the tree (that is, rates have increased through time). The observed value may then be compared to a null distribution of the γ statistic generated by Monte Carlo simulation under a stochastic branching model in which diversification rates are constant through time. All methods for detecting shifts in diversification rate through time require that rates of diversification across lineages are stochastically constant, which can be ascertained with the methods described previously.

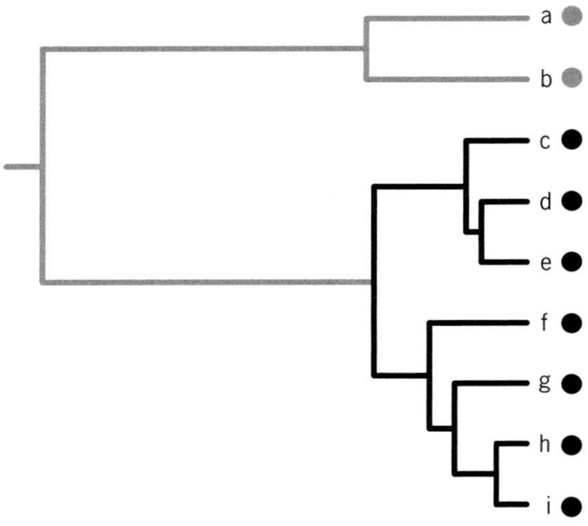

Fig. 2. Exploring association between organismal traits and rates of diversification. A hypothetical phylogenetic tree for species *a* through *i* includes estimates of the (absolute or relative) timing of branching events. The trait under consideration exhibits two discrete states, gray and black. Circles adjacent to the tips indicate the observed states in the extant species, and branches of the tree have been shaded to reflect the inferred evolutionary history of this trait. The key innovation hypothesis posits that the evolution of the black state promoted rates of diversification, which can be tested by means of a statistic that reflects the difference in state-specific rates of diversification.

Outlook. Our ability to address these fundamental questions about diversification is currently unprecedented. The profusion of phylogenetic research and the continuing development of phylogeny-based methods have profoundly transformed our capacity to discover the historical patterns and understand the processes of biological diversification.

For background information *see* BIODIVERSITY; EXTINCTION (BIOLOGY); MACROEVOLUTION; ORGANIC EVOLUTION; PHYLOGENY; SPECIATION; SPECIES

CONCEPT; STOCHASTIC PROCESS in the McGraw-Hill Encyclopedia of Science & Technology.

Brian R. Moore

Bibliography. B. R. Moore, K. M. A. Chan, and M. J. Donoghue, Detecting diversification rate variation in supertrees, pp. 487–533, in O. R. P. Bininda-Emonds (ed.), *Phylogenetic Supertrees: Combining Information to Reveal the Tree of Life*, Kluwer Academic, Dordrecht, the Netherlands, 2004; S. Nee, Birth-death models in macroevolution, *Annu. Rev. Ecol. Evol. System.*, 37:1–17, 2006; S. Nee, Inferring speciation rates from phylogenies, *Evolution*, 55:661–668, 2001; D. Rabosky, Likelihood methods for inferring temporal shifts in diversification rates, *Evolution*, 60:1152–1164, 2006; R. H. Ree, Detecting the historical signature of key innovations using stochastic models of character evolution and cladogenesis, *Evolution*, 59:257–265, 2005.

Integrated nanosensors

Integrated nanosensors are nanostructured systems in which several sensors of different types, including those sensitive to optical, magnetic, chemical, or biological stimuli, have been integrated on a single platform (**Fig. 1**). Integrated nanosensors are the subject of interdisciplinary research involving materials scientists, physicists, chemists, biologists, and engineers. The driving force behind the development of integrated nanosensors is the diverse range of applications (including the detection of explosives, gas-phase toxins, pathogens in food products, and so forth) and the development of sensitive biosensors (DNA, proteins, bacteria, neurons, and so forth).

Feynman's vision. It is generally accepted that a visionary discussion by Richard P. Feynman in 1959 of the problems and promise of miniaturization constituted the starting point for the new field that today is called nanotechnology. The spirit of this discussion was embodied in his statement: "I will not now discuss how we are going to do it, but only what is possible in principle—in other words, what is possible according to the laws of physics."

Nanoscale engineering is being applied to the miniaturization of the current generation of sensors and the creation of entirely new classes of sensors. The ability to manufacture sensor components—including power supply, sensing receptor, and transmitter—at a fraction of traditional sizes will allow sensors to be much smaller and thus much more easily incorporated into the environment for a broadening range of sensing applications. Nanoscale engineering could also improve the sensing element itself, which would be particularly important since shrinking the sensor size would also decrease the area of the sensor available for detection.

Use of nanoparticles. There are two kinds of nanosensors: nanoparticle sensors, which are intended to detect and study nanoparticles, and nanoparticle-based detector systems, which rely on the formation of nanoparticles that have been developed as sensing species. Nanoparticles are unique tools as sensors for three reasons.

First, nanoparticles are similar in size to many proteins. This is part of the reason they can operate well inside cells. The sensors, which can detect a wide variety of proteins, could serve as a tool for diagnosing diseases like cancer by identifying the malformed proteins made by sickly cells. For instance, gold nanoparticles with attached fluorescent dyes have been developed by V. Rotello and his team to detect specific proteins. Depending on its shape, a particular protein molecule stimulates certain sensors to release their dye and glow. By analyzing the pattern of glowing, the researchers can identify the protein.

Second, nanoparticles possess unique physical characteristics with sensitivities orders of magnitude better than conventional devices and provide such performance advantages as fast response and portability. For example, N. J. Halas and her collaborators fabricated nanoshells and nanorice, which consist of nonconducting cores that are covered by metallic shells. (Nanorice are prolate spheroidal nanoparticles that resemble grains of rice and have dielectric cores and metallic shells.) Nanoshells are about 10,000 times more effective at surface-enhanced Raman scattering (SERS) than traditional systems. Nanoshells provide an opportunity to design all-optical nanoscale sensors—essentially new molecular-level diagnostic instruments—that could detect as little as a few molecules of a target substance.

Third, nanoparticles have unique physical properties that do not exist in bulk materials. For example, the optical response of gold colloidal nanoparticles (5–20 nm in size) is characterized by a localized surface plasmon resonance around 520 nm, which is absent from the spectrum of bulk metals (**Fig. 2**).

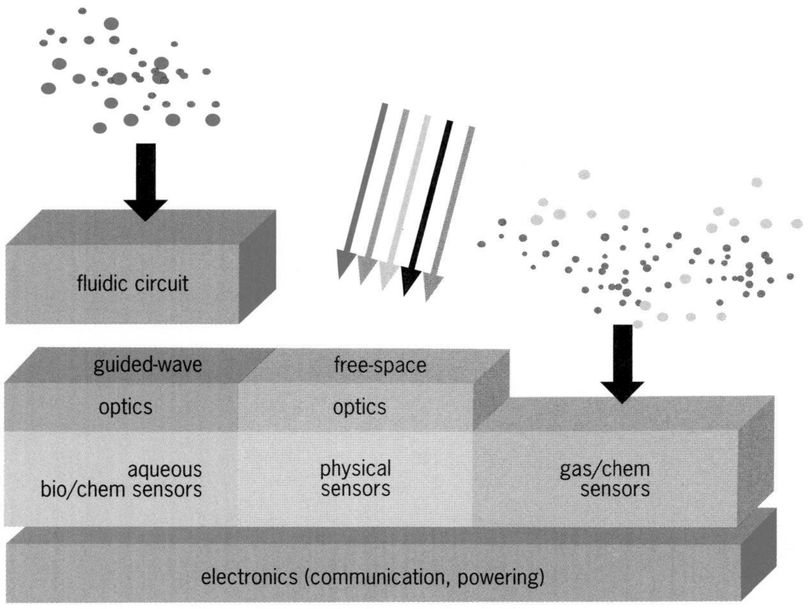

Fig. 1. Scheme of an integrated nanostructured supersensor. (*From http://nanosensors.ucsd.edu/Introduction.htm; courtesy of I. K. Schuller*)

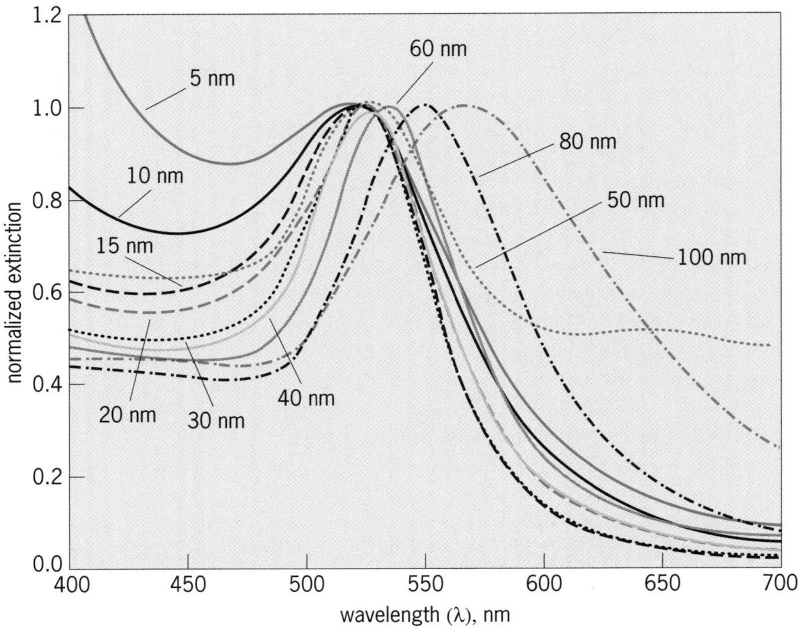

Fig. 2. **Extinction coefficient versus wavelength of gold colloidal nanoparticles in an aqueous medium. The different spectra correspond to different sizes of the gold nanoparticles. (*From A. P. Herrera et al., Synthesis and agglomeration of gold nanoparticles in reverse micelles, Nanotechnology, 16:S618–S625, IOP Publishing, 2005; courtesy of C. Rinaldi*)**

Intelligent sensors. Nanosensors allow for building integrated devices, providing an elemental base for intelligent sensors. Intelligent sensors are characterized as having significant powers to analyze, process, and store data. They can be utilized as autonomous systems or they can be spread out in large numbers to form networks. *See* AUTONOMOUS MICROSYSTEMS.

The concept of intelligent sensors, which currently are being developed, is an implementation of another of Feynman's visions. Among the most central and fundamental problems of biology in 1959, Feynman stated the question: "What is the sequence of bases in DNA?" It took more than 40 years to find an answer to this question on the basis of many technological innovations. A draft human genome sequence was presented in 2001 by the International Human Genome Sequencing Consortium. This sequence is "the biological example of writing information on a small scale" which inspired Feynman "to think of something that should be possible for technical applications. Biology is not simply writing information; it is doing something about it."

Use of advanced structures. State-of-the-art nanosensors are based on various advanced structures, such as quantum dots, nanopores (nanosensors for biomolecules), carbon-nanotube-based optical and electromechanical force sensors, transducers of biopotentials, and multianalyte biosensors. Quantum confinement effects in semiconductor nanostructures have made possible many approaches to the design of novel sensing materials. An attractive feature of nanostructured materials is the ability to tailor optical and electrical properties by changing simple parameters such as layer thickness and materials composition.

Quantum dots. For example, G. J. Brown and her collaborators developed infrared photodetector applications on the basis of semiconductor quantum dots. An indium gallium arsenide/indium gallium phosphide (InGaAs/InGaP) quantum-dot infrared photodetector (QDIP) is characterized by high sensitivity in the mid-infrared/infrared range.

One of the exciting prospects of nanotechnology is the use of quantum dots in biology. The emission ranges for selected quantum-dot core materials overlap the representative spectral ranges of biological interest (**Fig. 3**). These unique optical properties of quantum dots enable applications in various biological investigations. Detection of minute concentrations of nucleic acid sequences is important for

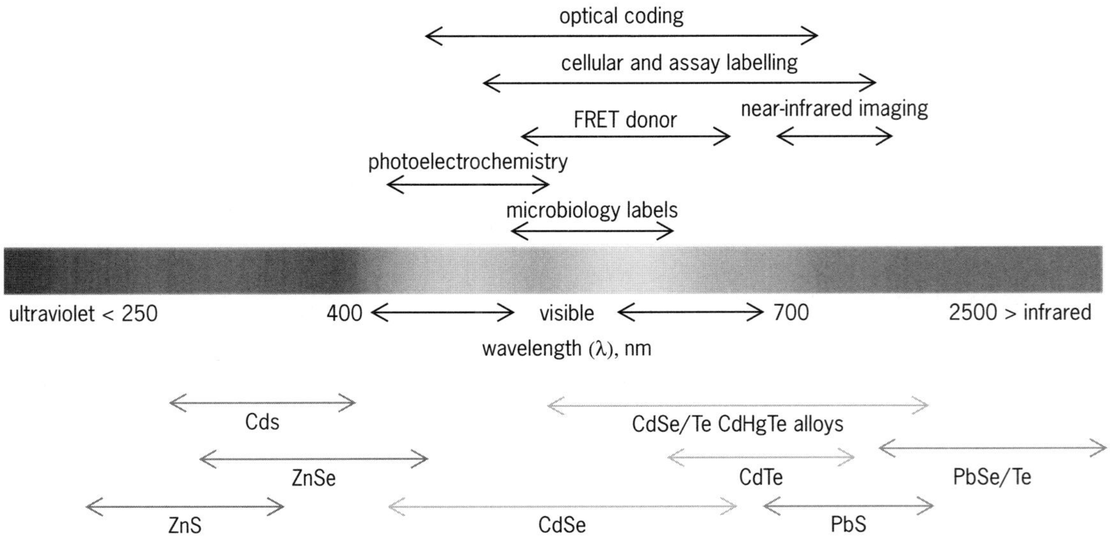

Fig. 3. **Ranges of emission wavelengths of selected quantum-dot core materials over the spectrum. Representative areas of biological interest are also presented. (*Reprinted by permission from Macmillan Publishers Ltd.: I. L. Medintz et al., Quantum dot bioconjugates for imaging, labelling and sensing, Nat. Mater., 4:435–446 © 2005*)**

medical diagnosis and for understanding biomolecular mechanisms. The use of quantum dots as DNA nanosensors promises to significantly enhance the sensitivity of fluorescence-based DNA detection.

The DNA nanosensor includes a quantum dot [a cadmum selenide–zinc sulphide (CdSe–ZnS) core-shell nanocrystal] conjugated with several streptavidins and two target-specific probes: a reporter probe labeled with a fluorophore acceptor (cyanine 5) and a capture probe labeled with biotin (**Fig. 4**). In the absence of a DNA target, the quantum dot and the reporter probe are unlinked and only the quantum-dot fluorescence is detectable. If a target DNA molecule is present in solution, it is sandwiched between the reporter probe and the capture probe. Several sandwiched complexes are then attached to the quantum dot through biotin-streptavidin binding and, consequently, the fluorophore is brought into close proximity to the quantum dot. This results in the fluorescence emission from the acceptors by means of fluorescence resonance energy transfer (FRET) after illumination of the quantum-dot donor. Finally, detection of the acceptor emission indicates the presence of DNA targets.

Artificial synapses. Interfacing biological matter with hardware systems at the microscale and nanoscale levels, which has been explored by G. R. Borghs and his collaborators, represents a fascinating research domain that holds promise for nervous system research. It should eventually result in integrated hybrid systems that would enable scientists to interact at a fundamental level with biological structures and gain insight into the mechanisms governing their functions. A functional neuroelectronic hybrid system can be seen as an artificial synapse between a biological entity (a neuronal cell or tissue) as the presynaptic element, and an integrated circuit as the postsynaptic element. This interface should enable the bidirectional transfer of information between the organic and inorganic components of the system. In the nervous system, information propagates between adjacent neuronal cells by means of electrical signals (action potentials) and chemical signals (neurotransmitters). Therefore, if one wants to integrate an electronic device into this information flow, this integration should include sensors and actuators that are able to detect and trigger action potentials or the release of chemical messengers (that is, neurotransmitters).

Nanowires. C. M. Lieber and his group have recently shown that nanowires can readily change their conductance upon binding of charged target biological molecules to receptors linked to their surfaces. Nanowires have the potential for very high sensitivity detection since the depletion or accumulation of charge carriers resulting from this surface binding can affect the entire cross-sectional conduction pathway of the nanowire. For some nanowires, such as hollow carbon nanotubes, every atom is on the surface and exposed to the environment; thus, even small changes in the charge environment can cause drastic changes in their electrical properties.

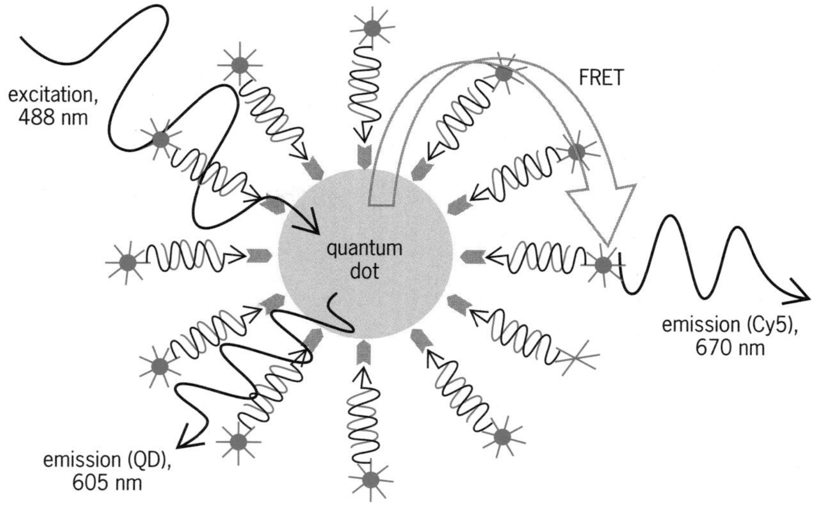

Fig. 4. Fluorescence emission from cyanine 5 (Cy5) due to fluorescence resonance energy transfer (FRET) between Cy5 acceptors and a quantum-dot donor in a nanosensor assembly. (*Reprinted by permission from Macmillan Publishers Ltd.: C.-Y. Zhang et al., Single-quantum-dot-based DNA nanosensor, Nat. Mater., 4:826–831 © 2005*)

Unique physical properties. Sensors with sizes at the nanoscale level possess unprecedented physical properties. C. Hierold and his collaborators, who have fabricated nanoscale force sensors based on single-walled carbon nanotubes, emphasize that the underlying physics at the nanoscale differs from the physics at the microscale: The functionality of nanosystems is determined by quantum-mechanical effects. As shown by R. A. Freitas, nanosensors can be expected to be closer to the theoretical limit of sensitivity (for their size) than macrosensors (for their size).

The progress in this direction has been tremendous: Whereas the development of semiconductor microchips has allowed for miniaturization of computers, the present efforts are related to building up a new elemental base at the nanoscale level and to the implementation of the novel principles of quantum computing.

Communication with nanodevices. An example of communication with nanodevices, which is a prerequisite of their integration, is a proposal for a wireless two-state quantum-dot device called a "cell" advanced by W. Porod, G. H. Bernstein, and their collaborators. A cell consists of five quantum dots and two electrons. Electrons can tunnel from one quantum dot to the next. The Coulomb repulsion between the two electrons causes them to move to opposite corners of the cell. This leads to two states of equal energy in the cell. If two cells are placed adjacent to each other and the first cell is forced into a certain state, the second cell will assume the same state, because doing so lowers its energy. The result is that a "1" has moved on to the next cell. By stringing many cells together, a nanoscale "pseudowire" can be produced to transport a signal.

Applications. Integrated nanosensors would have many applications in disparate fields such as process monitoring, robotics, and the environmental, medical, consumer, homeland security, and military areas. Applications of integrated nanosensors open great opportunities, for example, for monitoring the human body, due to their unprecedented sensitivity. According to the forecast of computing in 2020 as outlined by D. Butler, new computers—able to continuously measure most variables of a human body, at whatever scale needed—would take the form of networks of sensors with built-in facilities of data processing and transmission. This futuristic picture represents one of the ways in which Feynman's vision may be realized.

For background information *see* BIOTIN; CARBON NANOTUBES; FLUORESCENCE MICROSCOPE; HUMAN GENOME; MICROSENSOR; NANOPARTICLES; NANOTECHNOLOGY; PLASMON; PROTEIN; QUANTIZED ELECTRONIC STRUCTURE (QUEST); RAMAN EFFECT; SYNAPTIC TRANSMISSION in the McGraw-Hill Encyclopedia of Science & Technology.

Jozef T. Devreese; Yvan Bruynseraede

Bibliography. R. P. Feynman, There's plenty of room at the bottom, *Eng. Sci.*, 23:22–36, 1960; R. A. Freitas, Jr., *Nanomedicine*, vol. 1: *Basic Capabilities*, Landes Bioscience, Georgetown, Texas, 1999; G. L. Snider et al., Quantum-dot cellular automata: Review and recent experiments, *J. Appl. Phys.*, 85:4283–4285, 1999.

Integrity monitoring (satellite navigation)

Satellite navigation systems provide a number of significant benefits over traditional ground-based navigation aids. These benefits include global coverage, improved accuracy, and the ability to support area navigation (the ability to navigate along any arbitrary course as opposed to only along defined routes). The U.S. Global Positioning System (GPS) and the Russian Global Navigation Satellite System (GLONASS) are used today for a number of safety-critical applications, including navigation of aircraft and ships. Other satellite navigation systems are emerging and are expected to be in use for such applications within the next decade.

As defined in the *Federal Radionavigation Plan* (2005), integrity is "the measure of the trust that can be placed in the correctness of the information supplied by a navigation system. Integrity includes the ability of the system to provide timely warnings to users when the system should not be used for navigation." Many navigation applications have more stringent integrity requirements than can be met by stand-alone satellite navigation systems by themselves. For these applications, augmentations that provide external integrity monitoring may be employed to meet these more stringent requirements.

Requirements. Integrity requirements for safety-critical navigation applications are commonly specified using three parameters: (1) the alert limit, (2) the time-to-alert, and (3) the probability of hazardously misleading information. An alert limit is the maximum allowable position error before safety would be unacceptably compromised if the user were not promptly notified. The time-to-alert is the maximum allowable period from the onset of an out-of-tolerance condition until an alert is provided. The probability of hazardously misleading information is the maximum acceptable probability of occurrence of an out-of-tolerance condition without a timely alert.

Aircraft navigation is an application with especially stringent integrity requirements. For example, satellite navigation requirements developed within the International Civil Aviation Organization (ICAO) for aircraft Category I precision approaches include a 10–15-m (33–49-ft) vertical alert limit, a 6-s time-to-alert, and a maximum probability of hazardously misleading information of 2×10^{-7} per approach. In a Category I precision approach, the aircraft may descend to as low as 60 m (200 ft) before the pilot must either visually acquire the runway or conduct a missed approach. Less stringent integrity requirements are applicable to some other phases of flight including en route (the portion of flight between departure and approach), and more stringent integrity requirements are in development for other types of precision approach that provide navigation at lower heights above the runway.

Two basic techniques are commonly used to achieve the integrity requirements for safety-critical satellite navigation applications. Receiver techniques offer an inexpensive, effective solution for applications with alert limits of the order of several hundred meters or larger. Integrity requirements for applications with lower alert limits are typically met through the use of ground monitoring networks.

Receiver techniques. Satellite navigation receivers employ measurements of the transit time of signals broadcast from satellites to determine their three-dimensional position. The measured transit times are multiplied by the speed of light to obtain measurements of range to each of the visible satellites. Such measurements are referred to as pseudoranges when, as is common, the receiver employs an imprecise clock, because each transit time measurement includes a large bias due to the receiver clock error. With four satellites visible in the sky with favorable geometry, the receiver can utilize four pseudorange measurements to accurately estimate four parameters of interest: the receiver's three-dimensional position coordinates and the clock error. On rare occasions, failures in a satellite, the ground control network, or the user receiver can result in excessive position errors. Historically, satellite clock anomalies are the most common cause of extreme position errors, and notably result in a corruption of only one of the multiple simultaneous pseudorange measurements made by the receiver.

When pseudorange measurements are made to more than four visible satellites with favorable geometry, the receiver has more measurements than unknown parameters to estimate. The additional measurements enable the consistency of the five navigation solutions obtained by using four of the five satellites at one time to be compared to determine if any one of the measurements is erroneous.

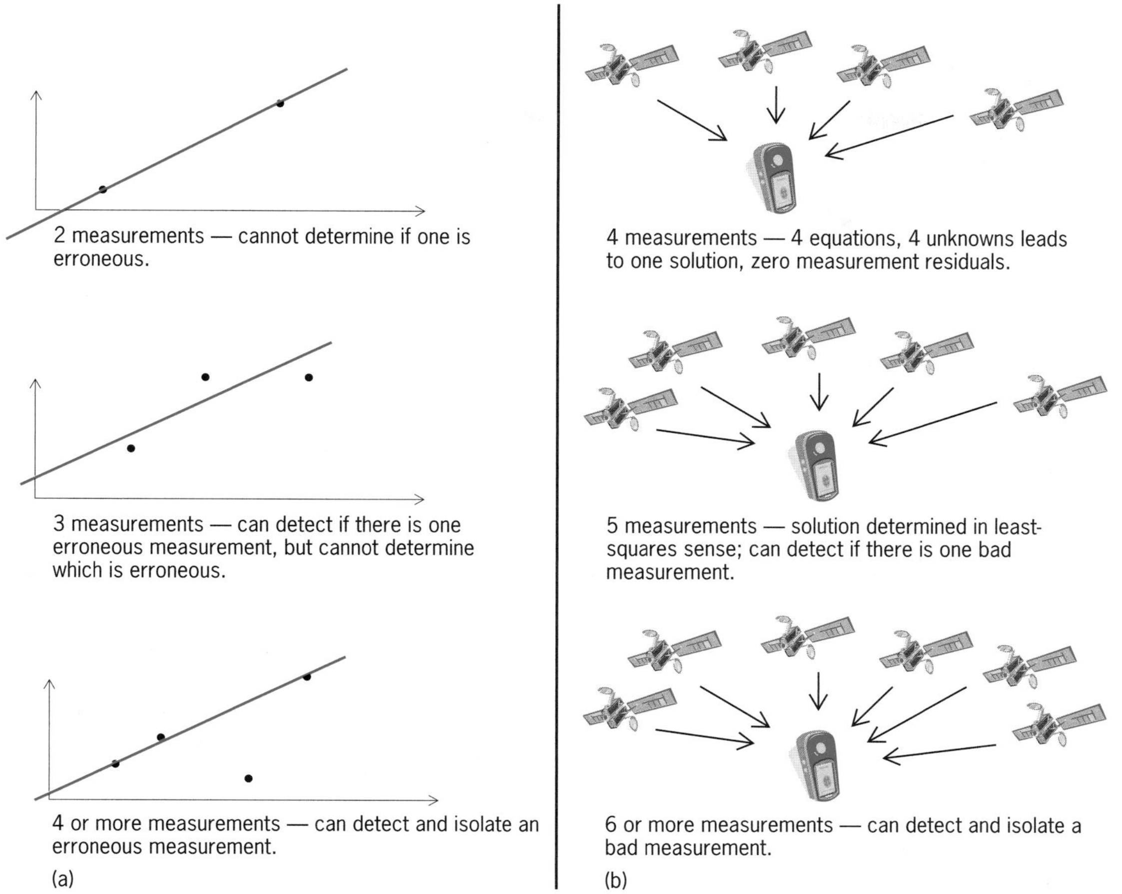

2 measurements — cannot determine if one is erroneous.
3 measurements — can detect if there is one erroneous measurement, but cannot determine which is erroneous.
4 or more measurements — can detect and isolate an erroneous measurement.
(a)
4 measurements — 4 equations, 4 unknowns leads to one solution, zero measurement residuals.
5 measurements — solution determined in least-squares sense; can detect if there is one bad measurement.
6 or more measurements — can detect and isolate a bad measurement.
(b)

Illustration of satellite navigation system receiver autonomous integrity monitoring concepts. (*a*) Two-dimensional problem: noisy measurements of a line. (*b*) Four-dimensional problem: estimating receiver position and clock error in satellite navigation.

The evaluation of measurement consistency in satellite navigation systems is frequently referred to as receiver autonomous integrity monitoring (RAIM). RAIM techniques were first developed in the 1980s for the GPS.

To illustrate the basic principles involved in RAIM, it is useful to consider an analogy to a lower-dimensional problem. Consider a scenario in which noisy measurements are used to determine the two parameters of a line, for example, its slope and y-intercept, in a two-dimensional space (see **illustration**). If two measurements are available, the two unknown parameters of the line may be estimated (two equations—two unknowns). In this instance, the two measurement residuals, or distances along the y axis from the measurements to the estimated line, are identically zero. It is not possible with only two measurements to determine if one of the measurements is erroneous. This situation is akin to estimating a receiver's three-dimensional position and clock error in satellite navigation with four visible satellites (four equations—four unknowns). With four pseudoranges, the measurement residuals are identically zero and it is not possible to determine if any of the pseudorange measurements are erroneous. Within a satellite navigation receiver, the measurement residual for each pseudorange may be determined as follows. First, the receiver's three-dimensional position coordinates and clock error are estimated using a least-squares technique. Then, the difference is formed between each pseudorange measurement and an estimate of that measurement made by computing the range from the estimated user position to the known satellite position adjusted by the product of the estimated receiver clock error and the speed of light.

With more than two measurements in our two-dimensional example, it is not generally possible to fit a line perfectly through all of the measurements. In this case, a least-squares solution may be used, where the two parameters of the line are chosen to minimize the sum of the squared measurement residuals. With three measurements, one can assess whether the measurements are consistent by examining the size of the measurement residuals. Large residuals are an indication that at least one of the measurements is erroneous. However, with only one redundant measurement (three measurements and two unknowns), it is not possible to tell which of the measurements are erroneous. This situation is akin to satellite navigation with five visible satellites: the receiver can determine if one of the measurements is erroneous, but not which one. With four or more measurements in the two-dimensional problem, or with six or more pseudoranges in satellite navigation, it is possible to not only tell if one of the measurements is erroneous, but also identify which one of the measurements is erroneous.

Ground monitoring networks. If one or more satellite navigation receivers within a network are situated at known locations, the errors present within a satellite navigation system can be accurately estimated in real time using precise satellite ephemeris information to determine whether measured pseudoranges are consistent with predicted pseudoranges for that location. Using a single reference station or a ground monitoring network along with a suitable data link, timely warnings can be provided to satellite navigation users when the system should not be used for navigation. Furthermore, the reference station or ground network can be used to provide pseudorange corrections to the end user to enable more accurate positioning to be achieved. This latter concept is often referred to as differential positioning. Differential GPS networks are currently operating in many parts of the world for air navigation, maritime navigation, and numerous other applications.

For background information *see* EPHEMERIS; LEAST-SQUARES METHOD; SATELLITE NAVIGATION SYSTEMS in the McGraw-Hill Encyclopedia of Science & Technology.

Christopher J. Hegarty

Bibliography. *Federal Radionavigation Plan*, U.S. Departments of Defense and Transportation, Washington, D.C., 2005; *Global Positioning System: Papers Published in NAVIGATION*, vol. V, The Institute of Navigation, Fairfax, Virginia, 1998; E. Kaplan and C. J. Hegarty (eds.), *Understanding GPS: Principles and Applications*, 2d ed., Artech House, Norwood, Massachusetts, 2006.

Ion channels (genetics)

Ion channels are proteins that span cellular membranes and mediate the passage of electrolytes across the hydrophobic barrier that maintains the integrity of the cell and interior compartments. They are found in both prokaryotes and eukaryotes and their structures are highly conserved. Ion channels come in a variety of types. Some are selective for cations, while others favor anions. Some are regulated by membrane voltage, while others are controlled by binding of ligands. It is now known that there are hundreds of genes encoding human ion channels, and some genes encode multiple functionally distinct variants as a result of alternative splicing [a process in gene expression that enables the production of multiple forms of messenger ribonucleic acid (mRNA) from a single RNA transcript, thus enabling the production of multiple forms of protein from one gene]. The regulated and coordinated actions of ion channels are required to generate and transmit the electrical signals necessary for a myriad of operations that are essential for normal physiological function.

Ion channel structure. Fundamentally, ion channels are protein-lined holes in membranes that permit the passage of ions into and out of cells or subcellular compartments, with "gates" that open and close the permeation pathway. In recent years, a number of elegant crystallography studies have shown precisely what these holes look like at the atomic level, how channels select for one ion over another, and how channels open and close.

Most channels consist of four or five independent protein subunits assembled symmetrically or nearly symmetrical around a central pore. In selective channels, the protein protrudes into the central pathway to form a narrow filter that interacts with the permeating ions, replacing water molecules in the ion hydration shell to allow passage of one ion selectively through the pore. Multiple crystal forms of related channels have revealed distinct structures that reflect tertiary rearrangements of the protein corresponding to gating states, that is, open and closed states, of the channel.

Ion channel protein families. Ion channel genes clearly fall into several related families (see **illustration**). Among the largest of these is the pore-loop (P-loop) cation channel superfamily, which includes all of the highly selective Na, K, and Ca channels. These channels are all formed as tetramers from identical or similar subunits [K channels, nonselective transient receptor potential (TRP) channels, nonselective cyclic nucleotide-gated channels, glutamate-activated channels], or as large monomers, each containing four pseudosymmetric domains (Ca channels and Na channels). All contain similar sequences in the structurally conserved core that lines the ion selectivity filter, including the so-called P-loop. The Cys-loop ligand-gated ion channel family includes acetylcholine receptor–, serotonin receptor–, γ-aminobutyrate (GABA)–, and glycine-activated channels, and all contain a basic structural unit of five subunits. Some proteins that clearly form ion channels do not appear to be part of a large family. Such outliers include the cystic fibrosis transmembrane regulator (CFTR), a member of the adenosine triphosphate (ATP) binding cassette (ABC) family of proteins that forms a Cl^- selective channel. Interestingly, another member of the ABC superfamily, the sulfonylurea receptor (SUR), does not form ion channels on its own, but is an obligate partner of one of the P-loop K channels and provides the regulatory domain of the ATP-sensitive K channel.

Ion channel nomenclature. Different ion channels have been recognized for decades. Traditionally, they have been named according to function, typically with reference to their ion selectivity and gating characteristics. Hence, voltage-gated K channels are potassium selective channels that open in response to membrane depolarization. Nicotinic acetylcholine receptor channels are nonselective channels that open in the presence of extracellular acetylcholine. However, the recognition of the underlying genes and proteins now permits a more systematic nomenclature based on structure. For example, based on the expressed proteins, one would name voltage-gated K channels "K_vx.y," where x is the subfamily of proteins and y is the individual member. Based on the underlying gene, "*KCNXy*" encodes a gene in the KCN structural family, where X is a letter indicating the gene subfamily and y is a number indicating the individual member. Thus, *KCNA1* encodes the protein K_v1.1, which underlies a voltage-activated K channel. It is important to realize that these various nomenclature systems are often used together or interchangeably.

Ion channel function. Ion channels that respond to voltage are the major molecular determinants of the fundamental electrical signal in excitable cells—the action potential. Following a trigger stimulus, activation of Na^+ and/or Ca^{2+} channels leads to membrane depolarization that is propagated throughout the cell. Subsequent activation of voltage-dependent K^+ channels repolarizes the cell and readies it for the next stimulus. The result of an action potential varies depending on the cell. In muscle cells, the calcium entering through voltage-dependent Ca channels during a single action potential results in the release of calcium from storage sites within the cells to cause contraction. In neurons, the action potential generally results in the release of a neurotransmitter that propagates the signal to the next cell.

Neurotransmitters and other chemical stimuli initiate or shape electrical activity by binding to ligand-gated channels. In the case of synaptic transmission in the central nervous system, neurotransmitters can be stimulatory, activating nonselective cation channels (such as the acetylcholine- or glutamate-activated channels) that cause the membrane to depolarize. Conversely, the inhibitory neurotransmitter GABA activates a Cl^- channel that inhibits membrane excitability. However, ligand-gated channels do not only respond to neurotransmitters. Others are regulated by nucleotides, G-protein [GTP (guanosine 5′-triphosphate)-binding protein] subunits, noxious chemicals, or pH.

As illustrated, ion channels have a significant impact on nearly every physiological function in the body. Thus, genetic mutations that alter function are likely to have pathophysiological significance. Below, we consider some recent examples of human gene mutations that have been linked with human disease.

Ion channels and diabetes. A network of ion channels is essential for the regulation of insulin secretion from the pancreatic islets of Langerhans. One type of ion channel, the ATP-sensitive K channel (K_{ATP}), senses changes in the available blood glucose—a rise in blood glucose (and the consequent rise in intracellular ATP) following a meal closes the normally active K_{ATP} channels. K_{ATP} channel closure leads to the increase of voltage-dependent Ca channel activity and Ca entry triggers the release of insulin. Once blood glucose returns to normal, K_{ATP} channels reopen, Ca channel activity decreases, and insulin secretion is silenced.

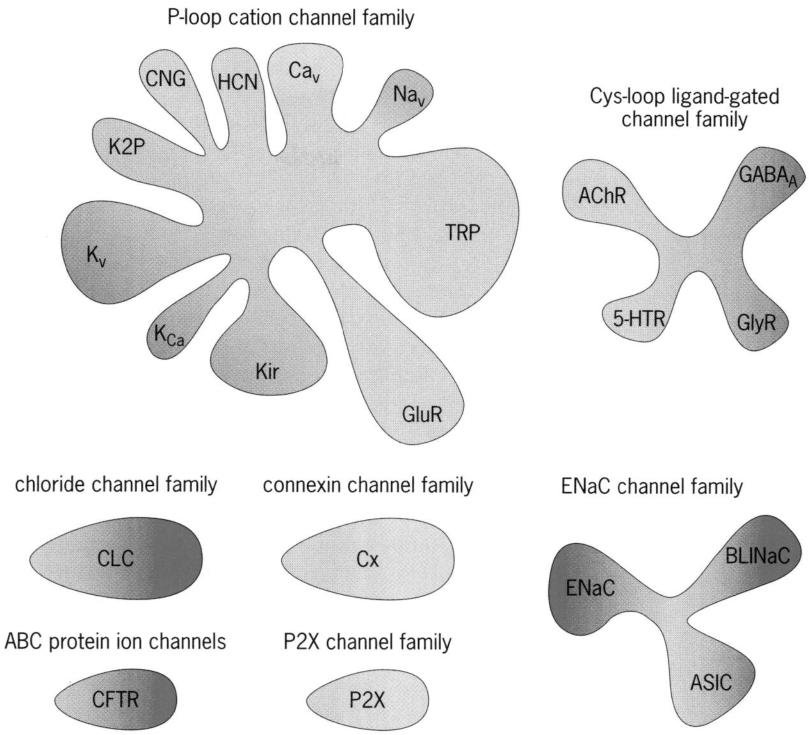

Ion channel gene families. The P-loop cation channel superfamily encodes many different ion channels, including nonselective channels (hyperpolarization-activated, HCN; transient receptor potential subfamily members, TRP; glutamate receptor channels, GluR) as well as highly selective potassium channels (voltage-gated, K_v; inward rectifier, Kir; two-pore channels, K2P; Ca-activated channels, K_{Ca}), Na channels (Na_v), and Ca channels (Ca_v). Similarly, the ligand-gated channel family encodes serotonin receptor (5-HTR)-, acetylcholine receptor (AChR)-, glycine-, and GABA-activated channels. The ENaC channel family encodes epithelial sodium-selective channels as well as nonselective cation channels. Other channel gene families include the CLC family of chloride channels, the ABC protein CFTR, and the connexin (CxR) channel families.

Given the essential role of ion channels in the regulation of insulin secretion, it is not surprising that mutations of K_{ATP} channel genes have now been identified in human patients afflicted with permanent neonatal diabetes mellitus, a disease characterized by high levels of glucose in the blood and dysregulated insulin secretion. Each of the identified mutations causes K_{ATP} channels to be relatively insensitive to changes in intracellular ATP and remain open. Because the K_{ATP} channels are always open, voltage-dependent Ca channels remain closed, insulin is not released, and blood glucose levels do not decrease.

The discovery that fundamental defects in islet excitability can underlie diabetes has led to two important paradigm shifts. First, patients with this type of early-onset diabetes have traditionally been treated with invasive and painful insulin injections; however, the recognition that the fundamental defect lies in the K_{ATP} channel has led to a revolution in the therapeutic intervention with patients receiving sulfonylurea drugs orally that inhibit K_{ATP} activity and restore insulin secretion. Second, the demonstration that diabetes can be caused by affecting islet excitability suggests that mutations of other pancreatic ion channel genes and environmental factors or deranged cell signaling pathways that impinge on islet excitability can have a similar effect.

Ion channels and cardiovascular diseases. A number of ion channels have recently been implicated in disorders of cardiac excitability. Genetic variations of both voltage-dependent K^+ and Na^+ channels have been identified in patients with an aberrant electrocardiogram finding—the long QT syndrome, which occurs when the cardiac action potential is prolonged and is known to be a risk factor for the development of life-threatening arrhythmias. By prolonging the duration of the action potential, the probability that a premature signal occurs and leads to ventricular tachycardia (excessively rapid heart rate) is increased. Interestingly, some of the same ion channels are often the unintended targets of drugs that are known to cause arrhythmias, illustrating the role that environmental factors may play in modulating ion channel function.

While mutations that cause long QT syndromes have been best characterized, the ion channel links of other cardiovascular diseases are now emerging. Inactivating mutation of the *KCNA5* gene that encodes a voltage-dependent K channel has been identified in a family afflicted with atrial fibrillation. The cardiovascular implication of ion channel mutations is not limited to the heart alone. Mutations of renal Na^+, K^+, and Cl^- channels have been identified in patients with abnormal blood pressure regulation, as in both Liddle's and Bartter's diseases.

Ion channels and neurological diseases. Electrical signals and ion channels are critical players in the nervous system. Ion channels convert chemical signals to electrical ones and propagate the signals to, from, and within the central nervous system. Given this fact, it is not surprising that ion channel mutations can underlie a number of neurological disorders. Epilepsy, multiple sclerosis, episodic ataxia (lack of muscular coordination), and migraine are among the conditions that have been linked with mutations in ion channel genes.

Ion channels may also be important therapeutic targets in the nervous system. Local anesthetics that block voltage-dependent Na channels have long been used therapeutically. By blocking action potential generation in neurons, these drugs can reduce the transmission of pain signals from sensory neurons. Are there other drugs targeting specific ion channels that might be more effective with fewer side effects? Evidence is mounting that members of the TRP channel family, for example TRPV1, may be the initial pain sensors, responding to the presence of noxious chemicals. Thus, blocking TRPV1 channels may provide a very specific anesthesia, blocking only pain signals, but leaving other sensory functions intact.

Future perspectives. The examples presented above demonstrate the wide range and potential implication of ion channels in human disease. Undoubtedly, future research will provide further examples. It seems certain that many other common diseases will be caused or affected by ion channel mutations.

For background information *see* BIOPOTENTIALS AND IONIC CURRENTS; CELL MEMBRANES; CELL PERMEABILITY; DIABETES; ION TRANSPORT; NERVOUS SYSTEM DISORDERS; NEUROBIOLOGY; PAIN; PROTEIN; SIGNAL TRANSDUCTION in the McGraw-Hill Encyclopedia of Science & Technology.

Colin G. Nichols; Thomas P. Flagg

Bibliography. F. M. Ashcroft, *Ion Channels and Disease*, Academic Press, San Diego, 2000; W. A. Catterall, K. G. Chandy, and G. A. Gutman (eds.), *IUPHAR Compendium of Voltage-Gated Ion Channels 2002*, International Union of Pharmacology, Leeds, United Kingdom, 2002; M. G. Hanna, Genetic neurological channelopathies, *Nature Clin. Pract. Neurol.*, 2:252–263, 2006; B. Hille, *Ion Channels of Excitable Membranes*, 3d ed., Sinauer, Sunderland, Massachusetts, 2001; C. G. Nichols, K_{ATP} channels as molecular sensors of cellular metabolism, *Nature*, 440:470–476, 2006; R. Roberts and R. Brugada, Genetics and arrhythmias, *Annu. Rev. Med.*, 54:257–267, 2003.

Laser marking

Laser marking is the process of directing a laser beam to etch the surface of a porous or nonporous material substrate to provide the contrast for legible characters, graphics, and machine-readable symbols, such as one- and two-dimensional barcodes. Laser marking devices date back to the early 1980s for marking and coding food products and metals. A laser marking system is an assembly of electrical, mechanical, and optical components, which includes a laser. There are many methods for directing the laser beam onto the material substrate to produce a mark. Mask, galvanometric scanning, and acoustic optic are the most common methods used.

A mask laser system uses a mask made of brass that is attached to the laser marking head to shape the laser beam image onto the material substrate. A mask laser marks a static image and is typically used to mark lot numbers and date codes onto food product packaging. A galvanometric scanning laser system uses two motorized mirrors to steer the laser beam to dynamically etch the material substrate (**Fig. 1**).

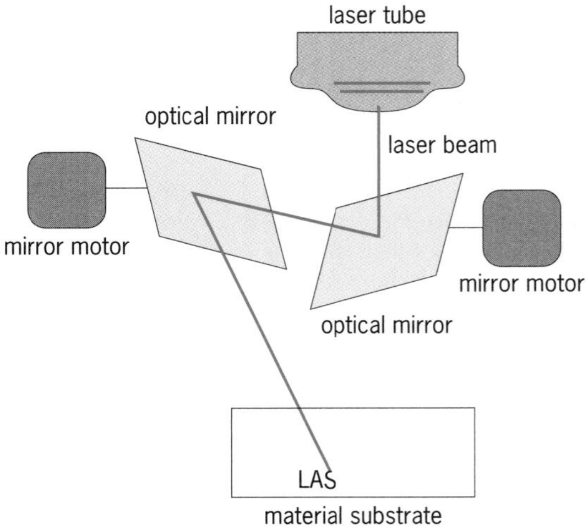

Fig. 1. Components of the galvanometric marking head.

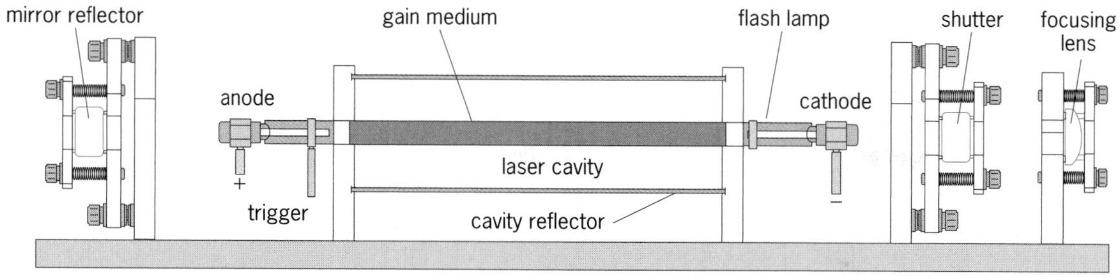

Fig. 2. Components of the laser.

An acoustic optic laser system steers the laser beam by acoustically moving the optical mirrors much like a speaker cone moves to generate sound waves in a sound-system speaker.

Galvanometric and acoustic optic methods of dynamically steering the laser beam have the benefit of producing unique laser marks on the material substrate, whereas a mask laser system is only capable of producing the same mark until the mask is manually changed.

The energy generated by the laser is in or near the optical portion of the electromagnetic spectrum. It includes the ultraviolet (100–400 nm), visible (400–700 nm), and infrared (700 nm–1 mm) radiation. A laser works by electrically stimulating the gain medium in the laser cavity to produce a light beam. The light is amplified by reflecting it back into the laser cavity using a mirror reflector and reflective shutter device (**Fig. 2**). The shutter controls the release of the laser light toward the focusing lens and onto the material substrate to etch. Closing the shutter allows the laser energy to recharge until it is needed.

Laser versus conventional marking. The benefit of laser marking methods over conventional marking and coding methods, such as inkjet printing or thermal printing, is the elimination of the cost of consumables associated with these printers. One key advantage of laser marking systems is their ability to produce an indelible mark on the substrate, which is imperative for applications requiring product traceability. A disadvantage of laser marking systems over conventional marking devices is the time it may take to etch the surface to produce a contrasting mark. A 4 × 4-in. (100 × 100-mm) zone for marking would take 1 s or less to print with an inkjet or thermal label printer, whereas a laser system could take at least 5–10 s to mark the same information. Some substrates do take longer to laser mark than others because of their material properties. In applications where end users are looking to increase production yield, marking speed is a large factor in deciding the best marking and coding technology.

A laser marking system capable of generating image data that changes "on-the-fly" requires software to drive the optical marking head to deflect and position the laser beam onto the substrate. Besides design software, other interfaces to a laser marking system include safety interlocking and mark cycle trigger inputs.

Modes. There are two modes in which a laser marking system may operate: static and dynamic. In most cases, a laser marking system will not be able to change between these modes without changing some of the laser system electronic hardware and software.

Static mode. The laser marking system is typically mounted in a stationary position. The material that is to be marked is statically positioned in the marking zone, which is underneath the laser marking head. The laser beam is emitted and directed by a mask or the marking head's galvanometric mirrors to produce the mark when a user activates the "start mark" cycle. At the end of the laser cycle, the marked material is removed from the marking zone and the cycle is then repeated if more material is to be marked.

Dynamic mode. The laser marking system is mounted in a stationary position. The material to be marked either is on a moving conveyor or is traversing past the laser marking head. A short laser pulse is emitted and directed by a mask, optical mirrors that are acoustically moved, or by using the marking head's galvanometric mirrors to direct vertical columns of laser light onto the material to render the image. The start of the mark cycle occurs when the marking laser system's product sensor is activated by seeing a product in the mark zone. A rotary line-speed encoder may be used to track the speed of the moving material to maintain the aspect ratio of the mark image should the speed of material moving past the laser marking head fluctuate. **Figure 3** shows the laser marking system operating in dynamic mode.

Marking. The beam of light that a laser emits is narrow and low-diverging, with the wavelength being well-defined. The wavelength of light is defined by the gain medium located inside the optical cavity. The gain medium may be gas, liquid, solid, or free electrons. Since the beam of laser light is narrow and focused, it is important that a stable gantry is used in the installation in order to position the laser marking head and to prevent surrounding vibration from interfering with the mark quality.

Marking lasers typically operate using two different wavelengths. For example, a CO_2 laser has a wavelength of 10.64 micrometers. This type of laser is ideal for marking soft materials such as paper, glass, or coated surfaces. Alternatively, a Nd:YAG laser has a wavelength of 1.064 μm and is ideal for marking

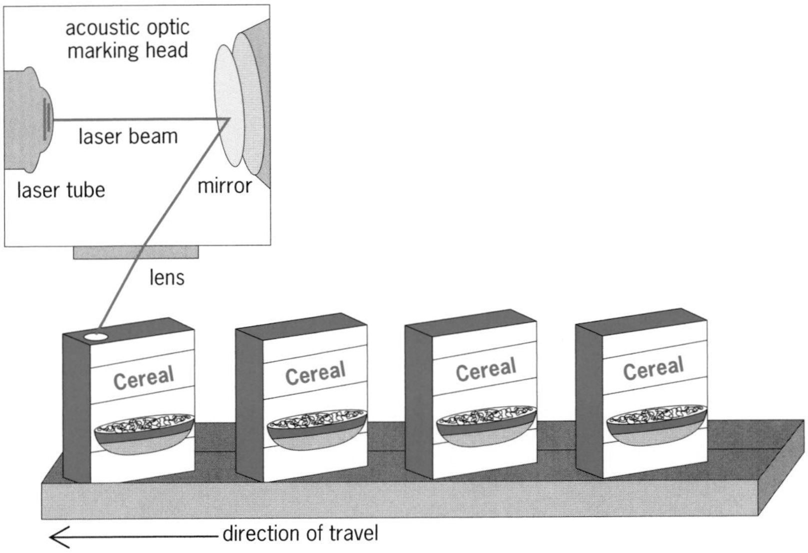

Fig. 3. Dynamic mode laser marking.

hard materials such as stainless steel and titanium. A laser system with a wavelength designed to mark soft materials is usually not effective in marking hard materials and vice versa.

To reduce the size of the marking device, fiber optics are now being introduced to deliver the laser light to the marking head. Another development that is allowing laser marking systems to be more efficient, powerful, and reliable is the use of a diode laser pump to excite the solid gain medium. In this technology, a very small, low-power and long-lasting light-emitting diode is used. It acts as a light source to excite the laser medium gain. The diode pump method is now replacing the conventional electric lamp method of exciting the gain medium and has the advantage of the reducing heat and energy requirements for operating the laser. For the end user, this means less maintenance and higher system uptime of the marking system.

Since producing contrasting marks is somewhat limited with laser marking systems, a line of coating materials are available that fuse or change color as the laser beam makes contact with the substrate. These materials range from ceramic coatings to polymer additives that were included during an extrusion or injection molding processes.

Outlook. Laser marking has a strong potential for marking and coding many products and materials because of its ease of use, ability to produce an indelible mark, and versatility of operating modes.

For background information *see* ACOUSTOOPTICS; ELECTROMAGNETIC RADIATION; LASER; OPTICAL MODULATORS; OPTICAL PULSES in the McGraw-Hill Encyclopedia of Science & Technology.

Dean Hornsby

Bibliography. American National Standard for Safe Use of Lasers (ANSI Z136.1-2000), 2000; Laser Institute of America, *Laser Safety Information Bulletin*, 1993; A. E. Siegman, *Lasers*, 1986; W. T. Silfvast, *Laser Fundamentals*, 2d ed., 2004.

Laser-ultrasonic cavitation

Transient illumination of light-absorbing nanoparticles using pulsed laser sources can produce rapid and highly localized heating. If a particle in a liquid medium absorbs a sufficient amount of laser energy, then vaporization of a layer of liquid blanketing the particle ensues and the hot vapor bubble goes through an expansion phase followed by a subsequent collapse. This phenomenon, termed optical cavitation, can be utilized for a number of biomedical applications. The localized increase in temperature and mechanical disruption associated with the bubble collapse have been used for the targeting of individual cells, producing selective damage in, for example, cancer cells and viruses infused with particles. Gas bubbles can also be stimulated to increase the echo generation in a nanoparticle-targeted region of tissue, thereby improving the contrast in diagnostic ultrasound imaging.

Researchers have also begun to explore applications in which the laser interaction with nanoparticles takes place in the presence of an applied acoustic field (**Fig. 1**). There are then two complementary physical processes producing favorable conditions for the formation of vapor bubbles. First, the liquid is subjected to a negative pressure (tension) during the rarefaction phase of the ultrasound. If the applied stress exceeds the so-called acoustic cavitation threshold pressure, the energy barrier to cavity formation is overcome. Next, the laser heating of the nanoparticle leads to the formation of a superheated vapor cavity whose presence serves to lower the cavitation threshold. The net effect is greater levels of cavitation activity at lower acoustic pressures. Depending on the application area, laser-illuminated nanoparticles can be used to promote and control acoustic cavitation, or ultrasound can be used to augment the effects of optical cavitation. In the former case, the laser-generated gas bubbles serve as nuclei for ultrasound cavitation, allowing for site-specific

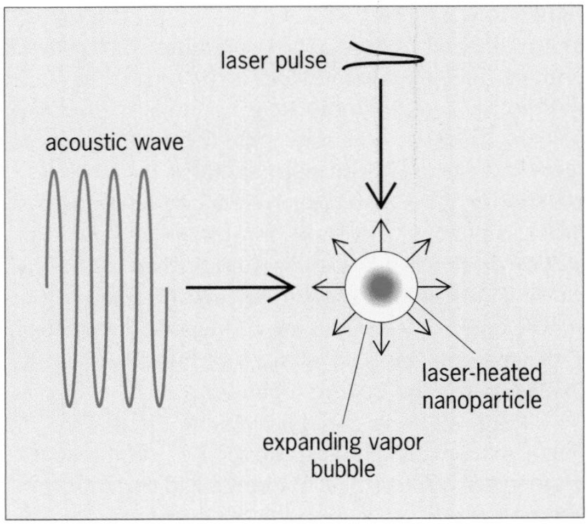

Fig. 1. Heating of a nanoparticle using a pulsed laser source in the presence of an acoustic field. The combined effects of the light and ultrasound lead to the creation of microbubbles.

initiation of a cavitation field under acoustic pressures that are less than would otherwise be required. In the latter case, an ultrasound field can be used to enhance bubble growth and related mechanical effects in the vicinity of laser-illuminated nanoparticles or to lower the optical energy deposition required to produce a given biological effect.

Synergistic effects, combining light and sound. Experimental evidence in vitro suggests that laser-excited nanoparticles are an effective means of nucleating acoustic cavitation. In the experimental arrangement illustrated in **Fig. 2***a*, a gel phantom (tissuelike medium) was infused with gold nanoparticles and placed in a water tank. A laser source emitting a nanosecond pulse and operating at a wavelength at which the gel medium was transparent and nanoparticles were absorbing was used to irradiate the phantom. A high-intensity focused ultrasound (HIFU) transducer was used to insonify the phantom with a 10-cycle tone burst. The light and ultrasound counterpropagated along the same axis. Cavitation activity was characterized by the detection of acoustic emission associated with bubble collapse using a second ultrasound transducer oriented perpendicular to, and confocal with, the HIFU transducer. The laser pulse was timed to fire when the center of the ultrasound tone burst arrived at the focal point of the transducer, near the peak negative pressure (Fig. 2*b*). Cavitation activity was examined as a function of both peak negative pressure at the focal point and optical energy in the laser pulse used to excite the nanoparticles. **Figure 3** shows some representative results illustrating the role of light and sound in reducing nucleation thresholds in the presence of nanoparticles. The abscissa has been normalized by P_{CA}, where P_{CA} is the minimum HIFU negative peak pressure required to produce detectable cavitation activity in the absence of an optical field. The ordinate gives cavitation activity, defined as the average power of the acoustic emissions from cavitation activity divided by the background noise power. Results are given for several different values of laser pulse energy (E), and E_{CA} is the minimum laser pulse energy required to generate detectable cavitation activity in the absence of an ultrasound field. From this plot, it can be seen that the use of a combination of optical and ultrasound fields offers a significant reduction in both the acoustic pressure and optical energy required to produce cavitation activity below those measured with sound (P_{CA}) and light (E_{CA}) alone. For example, detectable cavitation activity was observed at acoustic pressures as low as 0.2 P_{CA} using a pulse energy of only 0.025 E_{CA}. The synergistic effects produced by laser-illuminated nanoparticles and ultrasound potentially offer a predictable and controlled means of creating a targeted cavitation field in tissue at moderate acoustic pressure and optical energy levels.

Ultrasound therapy. A promising potential application of this technique is in the treatment of diseased tissue using high-intensity focused ultrasound. There is evidence that the therapeutic heating produced in tissue using ultrasound can be enhanced in the

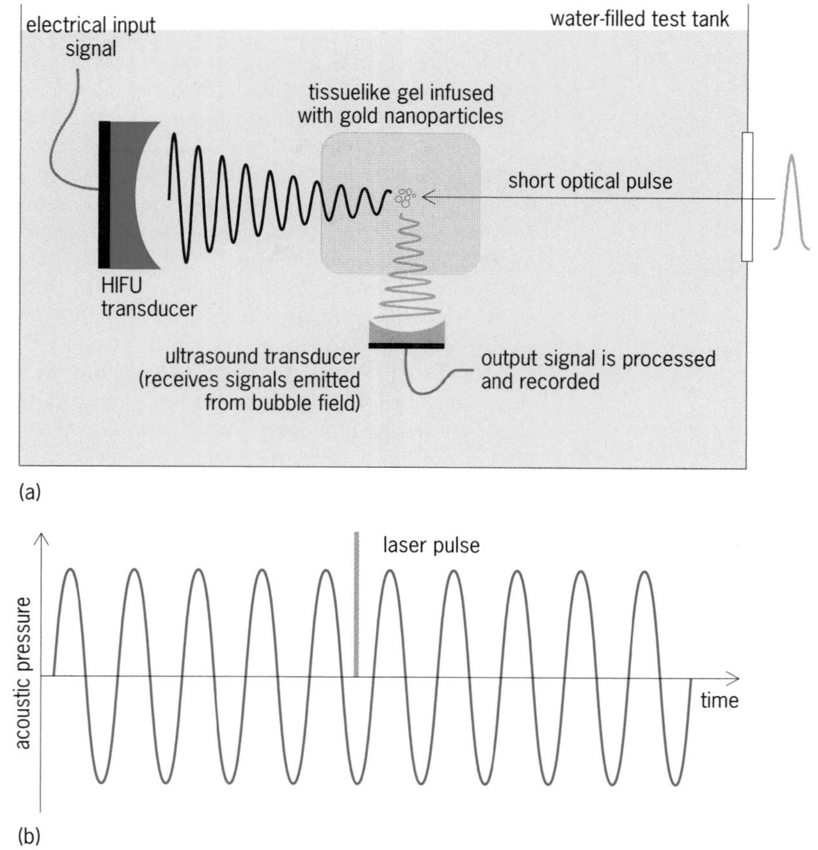

Fig. 2. Illumination by a nanosecond laser pulse of a nanoparticle-infused tissue phantom that is simultaneously insonified using a high-intensity focused ultrasound (HIFU) transducer. Cavitation activity is quantified by measuring the sound emitted by the bubble field using a second transducer. (*a*) Experimental arrangement. (*b*) Pressure at the ultrasound focus as the ultrasound pulse train passes. The laser is fired at the center of the pulse train near the peak negative pressure.

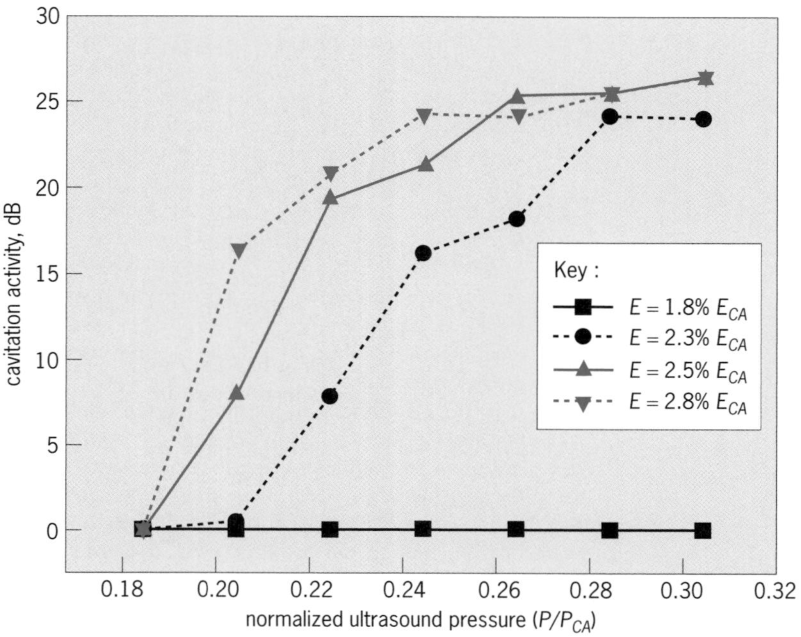

Fig. 3. Cavitation activity observed as a function of ultrasound focal pressure for several different values of laser energy. The combination of optical and ultrasound fields offers a significant reduction in both the acoustic pressure and optical energy required to produce cavitation activity below those measured with sound (P_{CA}) and light (E_{CA}) alone.

presence of inertial cavitation. In inertial cavitation, the explosive growth of a preexisting microscopic cavity occurs during the rarefaction phase of an applied ultrasound field. This is followed by violent bubble collapse and broad-bandwidth acoustic emission, along with the creation of large local forces and high-velocity fluid flow. The difficulties limiting the utility of cavitation for therapeutic applications are that very high pressure amplitudes are required to nucleate cavitation in many biological tissues and the nucleation process is quite unpredictable. Ultrasound contrast agents (gas-filled microspheres that can be injected into the body) can be used as nuclei and promote the cavitation process, but these agents have a relatively short lifetime due to diffusional instabilities and mechanical disruption by the ultrasound field. The vapor cavities produced by illuminating nanoparticles using a laser source offer an alternative means of producing controlled nucleation sites for cavitation. Advantages of this approach include the following:

1. The laser pulse can be precisely timed to coincide with the peak rarefaction pressure of the ultrasound pulse, allowing one to produce cavitation nuclei "on demand."

2. Nanoparticles are durable and can be reactivated multiple times.

3. By employing nanoparticles functionalized to specific cell types, one could achieve tissue-specific spatial targeting and thus lessen the need for precise ultrasound targeting in acoustically remote or aberrating media, such a breast tissue.

Laser therapy and imaging. In addition to HIFU therapy, researchers have explored a number of other biomedical applications of laser-illuminated nanoparticles including photoactivated release of drugs contained in nanoparticle-doped thermally activated polymers; tissue therapy (photothermolysis) and selective targeting of cells, bacteria, and viruses; and enhanced contrast imaging in tissue using photoacoustics. The addition of a secondary ultrasound field to "amplify" the laser-nanoparticle interaction offers potential benefits in applications where the ability to operate in a nonlinear regime, utilizing microbubbles and the associated mechanical effects, is critical. Safety considerations limit the amount of laser energy that can be deposited into tissue, and optical scattering and absorption result in a strong decay in energy with depth below the surface. In some potential biomedical applications, the energy that can be delivered to the nanoparticles may be insufficient to result in microbubble formation, or the microbubbles that are formed may not grow to a sufficient size to produce the desired therapeutic effect. Supplementing light with ultrasound potentially allows one to enhance bubble formation, reduce the required optical energy to acceptable levels, and effectively activate nanoparticle-infused tissue located at greater depths below the surface.

Consider, for example, therapeutic applications in which lasers are used to irradiate, and cause selective damage to, nanoparticle-targeted cells. Pulsed lasers allow for the extremely rapid deposition of energy and offer the possibility of inflicting highly localized damage in nanoparticle targeted cells with minimal damage to the surrounding tissue. There is evidence that the deposition of sufficient laser energy to cause explosive vaporization of the media surrounding the nanoparticles and subsequent bubble formation increases the severity of cell damage and efficacy of laser treatment. Ultrasound-enhanced laser therapy can potentially increase cavitation activity at a lower laser energy threshold, allowing for the technique to be applied in the treatment of tissue abnormalities or tumors buried well below the skin surface.

An additional nanoparticle-based laser technique that could potentially be aided by an applied ultrasound field is contrast-enhanced photoacoustic imaging. In photoacoustic imaging, a pulsed laser source is used to irradiate a region of tissue. Absorption of laser light by subsurface tissue constituents leads to local heating and acoustic-wave generation through the thermoelastic effect. The strength of the acoustic signal generated in a given region of tissue is directly proportional to the amount of light that is absorbed, and by measuring and processing the acoustic signals received at the tissue surface it is possible to create an image of the subsurface optical properties. A large increase in the photoacoustic response of nanoparticle-infused media has been observed when operating above the energy threshold for bubble formation, and this phenomenon can be used to increase the contrast in photoacoustic imaging or to detect the presence of early-stage tumor formation. A secondary ultrasound field will dramatically increase the photoacoustic emissions from the nanoparticles, providing substantial signal enhancement or reducing the requisite light levels.

Nanoparticles. Nanoparticles used to seed the bubble formation process should be nontoxic, be readily functionalized (through attachment to proteins or antibodies), and exhibit strong optical absorption at the excitation wavelength. Metallic nanoparticles have emerged as attractive candidates in this area. Metallic nanoparticles exhibit extremely strong absorption at the plasmon resonance frequency, where the optical field causes a coherent oscillation of electrons in the particles. The wavelength of peak optical absorption can be tuned by changing the size, shape, or composition of the particles. Nanoshells (consisting of a gold layer over a silica core), for example, have been developed that have absorption peaks in the near-infrared wavelength range where the penetration of light through tissue is maximal. Additional candidate nanostructures include quantum dots, nanorods, and nanotubes. Researchers also continue to make rapid progress in the development of molecularly targeted agents that can be conjugated to nanoparticles and delivered to cancer tumors, potentially serving as sites for bubble-mediated therapy and imaging. *See* INTEGRATED NANOSENSORS.

For background information *see* ACOUSTIC EMISSION; BIOMEDICAL ULTRASONICS; CAVITATION; LASER; NANOPARTICLES; NONLINEAR ACOUSTICS; PHOTOACOUSTIC SPECTROSCOPY; PLASMON; SONOCHEMISTRY; ULTRASONICS in the McGraw-Hill Encyclopedia of Science & Technology.

Todd W. Murray; Ronald A. Roy; R. Glynn Holt

Bibliography. C. H. Farny et al., Nucleating cavitation from laser-illuminated nano-particles, *Acoustic Res. Lett. Online*, 6(3):138–143, July 2005; M. Ferrari, Cancer nanotechnology: Opportunities and challenges, *Nat. Rev. Cancer*, 5:161–171, 2005; G. ter Haar and C. Coussios, High intensity focused ultrasound: Past, present and future, *Int. J. Hyperther.*, 23(2):85–87 (2007); T. G. Leighton, *The Acoustic Bubble*, Academic Press, 1994, paper 1997; V. P. Zharov et al., Photothermal nanotherapeutics and nanodiagnostics for selective killing of bacteria targeted with gold nanoparticles, *Biophys. J.*, 90:619–627, 2006.

Lead systems integrator

A lead systems integrator (LSI) is an organization (or set of organizations) that has been tasked to oversee the development and integration of system components for either a large, complex system or a system-of-systems (SoS). These large, complex systems and systems-of-systems are often referred to as "systems-of-interest."

The responsibilities of a lead systems integrator can vary considerably, depending on the type of system-of-interest being developed or upgraded, the system-of-interest's current state, and the desires of the acquisition agency. Lead systems integrators are more typically used in the development and integration of military systems. They are also used to support the development, integration, and evolution of other federal, state, and local government agency enterprise systems. Occasionally, they are used in the commercial marketplace to support the development, integration, and evolution of business enterprise systems. Lead systems integrators are selected for their technical expertise in the system-of-interest domain.

Based on information found in contractual statements of the work of lead systems integrators, most of the effort of lead systems integrators for new developments focuses on defining the system-of-interest concepts and overall architecture; conducting source-selection activities to acquire component systems to be integrated into the system-of-interest framework; managing changes to the system-of-interest requirements, architecture, and design; tracking component system evolutionary changes that may be occurring in parallel with the system-of-interest development activities; and integrating and testing the system-of-interest component systems. Because of the complexity and changing user needs associated with these systems-of-interest, they are often developed using an incremental or evolutionary development process that spans many years.

Activities similar to those for new developments are performed to evolve the system-of- interest over time to accommodate new user needs, an often large number of independently evolving, externally interoperating systems, and changing technologies. However, in the on-going evolution of the system-of-interest, there is less emphasis on source selection and more emphasis on negotiation with and coordination of the component system suppliers and external interoperators.

While a system-of-interest may be conceptually simple from a functional capability perspective, it can become very complex when dealing with such things as security, safety, information management, geographic distribution, and scalability for large numbers of users on the technical side and multiple stakeholders and vendors or suppliers on the management side. To better understand the activities of lead systems integrators, large, complex system and system-of-systems projects have been observed and engineers surveyed with respect to the types of issues they typically face.

Start-up. Once a lead systems integrator is under contract to develop a system-of-interest, it quickly begins to concurrently define the scope of the system-of-interest, plan the activities to be performed, analyze the requirements, and start developing the system-of-interest architecture and framework. As the scope, requirements, and architecture start to firm up, the lead systems integrator begins source selection activities to identify the desired component system suppliers. Then, as the suppliers start coming on board, the lead systems integrator must focus on team building, re-architecting, and feasibility assurance with the selected suppliers. Team building is critical since the lead systems integrator and the selected suppliers have often been competitors in the past and now must work together as an efficient, integrated team. Re-architecting is often necessary to make adjustments for the selected system components that may not be compatible with the initial system-of-interest architecture or other selected components. And feasibility assurance is conducted to better evaluate technical options and their associated risks. Many of the technical risks in a system-of- interest framework are due to incompatibilities between different existing system components or limitations of older system components with respect to today's technology.

Incremental acquisition and change management. As the system-of-interest development teams begin to coalesce, the lead systems integrator focuses on incremental acquisition activities for the development, integration, and testing of the required component systems for the system-of-interest. During this process, there are often continuous changes and new risks that must be managed. In addition, the lead systems integrator is continuously looking for opportunities to simplify the system-of-interest architecture and reduce effort, schedule, and risks. Key management issues include the following.

Number of stakeholders. The stakeholders in the development effort of any large, complex system or system-of-systems are numerous. They come from sponsoring and funding organizations as well as the various user communities that have high expectations for the planned system-of-interest. They also include the external interoperators, with whom memoranda of agreement and interface specifications must be developed and change-managed.

Number of development organizations. Because the system-of-interest components are often "owned" by an organization or supplier other than the sponsoring organization or lead systems integrator, there is often a separate development or manufacturing organization associated with each major component in the system-of-interest. In addition, there may be lower level suppliers associated with each major component, adding to the number of development organizations.

Number of decision "approvers." It has been reported that as the number of people involved in the decision-making process increases, the probability of getting a timely (or even any) decision often decreases. In the large system-of-interest development arena, the stakeholders and the component "owners," as well as the lead systems integrator, are often all involved in making key decisions.

Cross-cutting risks. These are risks that cut across organizational boundaries or major components (as opposed to component risks that can be managed by the component supplier). Key to a successful system-of-interest is negotiating solutions that are optimal for the system-of-interest, and not necessarily optimal for some of the components. This requires component stakeholders or suppliers to sometimes implement changes for the system-of-interest that are not optimal for the component.

Schedules. A key feature of systems-of-systems and large, complex systems is that many of the components within them are owned and managed by another organization. This means that system-of-interest timelines can be impacted by other "outside" goals and timelines. System-of-interest changes and enhancements may be done in conjunction with other planned component improvements. As a result, there may be long-lead enhancements that are not required by the system-of-interest but are more important to the component owner's product evolution, and will delay implementation of the system-of-interest features. Also, these other on-going changes (not required for the system-of-interest) may impact the stability of the component (including its architecture). While this problem may be perceived as primarily a scheduling issue, it can also impact the overall effort since component delivery delays can result in inefficient integration activities and significant rework.

Process scalability. As a lead systems integrator scales up its traditional systems engineering management processes for ever larger systems-of-interest, it often finds that there are new and unexpected issues. Typical management issues include the following:

1. Traditional planning and scheduling may lead to unacceptably long schedules, requiring the lead systems integrator to be more creative in both its technical and implementation approaches.

2. Planning and tracking activities must integrate inputs from a variety of different organizations, each with its own (and probably different) process.

3. Traditional oversight and coordination can spread key lead-systems-integrator personnel too thin. A key to success is determining what activities to oversee and coordinate and what activities to let the suppliers manage on their own.

4. More emphasis is required for contracting and supplier management. Incentives are often needed to better align the priorities and focus of the component supplier organizations. In addition, contracts must provide mechanisms to allow suppliers to participate more in the change-management process to help assess impacts and to develop efficient approaches to proposed changes.

5. Standardization of all processes is likely to be overly constraining and overly expensive. The lead systems integrator needs to decide what to standardize and what to let the suppliers control.

6. The decision-making process involves considerably more organizations. The lead systems integrator needs to find ways to streamline this process and to satisfice (that is, not everybody gets everything they want, but everybody gets something they are satisfied with) the various stakeholders.

7. Risk management for cross-cutting risks needs to cross organizational boundaries. It is important that risk management activities for cross-cutting risks do not select strategies that are optimal for one area of the system-of-interest but are to the detriment of other areas. The focus must be on the overall system-of-interest.

Evolution of processes. Since system-of-interest development efforts usually span many years and include many incremental or evolutionary developments, there are opportunities for the lead systems integrator to adapt and mature its processes to suit its environment. A key development is the attempt by lead systems integrators to blend traditional processes with those that are more agile. Lead systems integrators are more agile when dealing with risk, change, and opportunity management for future increments, but plan for stabilized evolutionary increments in the near term. Key to this approach is knowing when to plan, control, and stabilize and when to be more flexible, agile, and streamlined. The agile teams are responsible for performing acquisition intelligence, surveillance, and reconnaissance functions, and then rebaselining future increment solutions as necessary.

For background information *see* RISK ASSESSMENT AND MANAGEMENT; SYSTEM FAMILIES; SYSTEMS ENGINEERING; SYSTEMS INTEGRATION in the McGraw-Hill Encyclopedia of Science & Technology.

Jo Ann Lane; Barry Boehm

Bibliography. S. Blanchette, *U.S. Army Acquisition—The Program Executive Officer Perspective*, Spec. Rep. CMU/SEI-2005-SR-002, 2005; B. Boehm and J. Lane, 21st century processes for acquiring 21st century software-intensive systems of systems, *CrossTalk*, pp. 4–9, May 2006; A. J. Krygiel, *Behind the Wizard's Curtain*, CCRP Publication Series, National Defense Universities Press, 1999; J. Lane, *System of Systems Lead System Integrators: Where do They Spend Their Time and What Makes Them More/Less Efficient: Background for COSOSIMO*, University of Southern California Center for Systems and Software Engineering, USC-CSE-2005-508, 2005; R. Madachy, B. Boehm, and J. Lane, Assessing hybrid incremental processes for SISOS development, *Software Process Improvement and Practice*, 12(5):461–473, 2007.

Limb identity during development

The vast majority of vertebrate species have two pairs of bilaterally symmetric limbs that emerge during early development. The anterior limbs form at the junction between neck and torso and are known as forelimbs, while the posterior limbs form at the junction of torso and tail and are known as hindlimbs. The forelimbs and hindlimbs each acquire different shapes during development. Depending on the species, forelimbs and hindlimbs will, respectively, form arms and legs (humans and other primates), wings and legs (birds), pectoral fins and pelvic fins (fish), or forelegs and hindlegs (most amphibians, reptiles, and mammals). Despite the dramatic difference between, say, a hawk's wings (forelimbs) and talons (hindlimbs), the early stages of forelimb and hindlimb development are virtually identical in appearance, which has led researchers to search for the factors that control the differences in the development of the limbs.

Limb formation. Limbs start off as small buds arising on the flank of the embryo. The induction of limb buds is known to involve communication between two tissues: the flank mesenchyme (embryonic tissue of mesodermal origin), which forms the interior of the flank of the embryo; and the ectoderm, or surface layer, which overlies the flank mesenchyme. In the first stages of limb formation, before any limb bud is visible, certain regions of the flank mesenchyme start to secrete a growth factor, *Fgf10* (fibroblast growth factor 10). This induces the ectoderm to secrete other growth factors in return and initiates the physical outgrowth of the limb buds. Before limb outgrowth starts, these regions of the flank mesenchyme that initiate limb development are called the limb fields.

Classical transplantation experiments in chicken embryos have been performed to determine if the limb mesenchyme or ectoderm control the identity of the forming limb as a hindlimb or a forelimb. In these experiments, mesenchyme or ectoderm of the forelimb or hindlimb from various stages of embryonic development was transplanted into the opposing limb of early chick embryos. The eggs were then resealed, and the chicks allowed to develop until the physical differences between forelimb and hindlimb became visible. Transplants of limb ectoderm from one limb set to the other had no effect on limb development. However, when forelimb mesenchyme from the limb field was transplanted into the hindlimb, the transplanted tissue developed winglike feathers, indicating that the forelimb mesenchyme retained its forelimb identity despite having been placed into the hindlimb. Conversely, hindlimb mesenchyme transplanted into the forelimb produced scales typical of legs. These experiments showed that the factors that determine forelimb and hindlimb identity are in the mesenchyme, not the ectoderm, and they are already present in the early limb field.

Specific genes. With the advent of modern genetics, researchers working on limb development searched for genes that might prove to be the unknown hindlimb- or forelimb-specific factors. Different tissues of the embryo activate different genes to acquire their specific characteristics, and researchers looked for genes that are activated, or "expressed," differently in the hindlimb and the forelimb. Surprisingly, the vast majority of the genes known to determine the anterior/posterior, dorsal/ventral, and proximal/distal patterning of the limb are expressed identically in both forelimbs and hindlimbs. One set of genes, the *Hoxc* genes (a subset of the homeobox genes, which are involved in the development of limbs and many other body structures), was found to be expressed differently in forelimbs and hindlimbs, but these are not expressed until later stages of limb development and cannot be responsible for the early determination of limb identity. Research therefore focused on three genes: *Tbx4* and *Ptx1*, which are expressed only in the hindlimb, and *Tbx5*, which is expressed only in the forelimb (see **illustration**). All three genes are expressed in the limb field mesenchyme of their respective limbs and are maintained in the limb throughout much of development, making them ideal candidates to control limb-type identity and the eventual morphological shape of the limb.

Tbx4 and *Tbx5* were exciting candidate genes to control limb identity, as they belong to the T-box gene family, which is known to play many important roles during embryonic development. Members of this group of related genes are transcription factors, which means they bind directly to DNA and activate the expression of genes near to where they bind, giving them the capability to simultaneously control many other genes. Since *Tbx4* and *Tbx5* are quite similar, they have the potential to play similar roles in controlling limb development, but they are sufficiently different that they could produce the differences in morphology that characterize hindlimbs and forelimbs. Unfortunately, when either *Tbx4* or *Tbx5* was inactivated in mouse embryos,

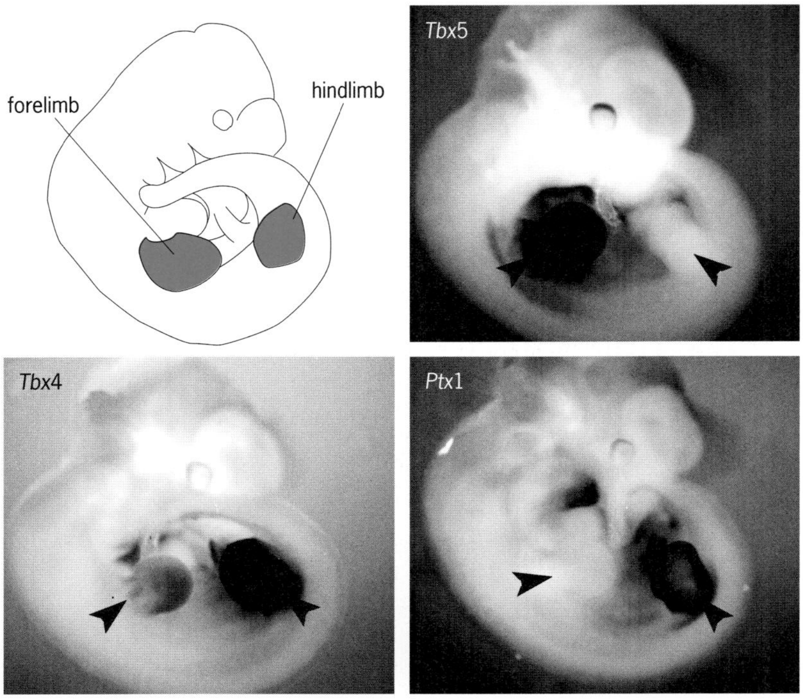

A cartoon of a midgestation mouse embryo shows the locations of the developing limbs, followed by photographs of mouse embryos that have been stained to show the expression of *Tbx5*, *Tbx4*, or *Ptx1*. The dark stain indicates the region in which each gene is expressed. Arrowheads indicate limbs.

the corresponding limb failed to form. This demonstrated that these genes were critical for the outgrowth of the limb, but made it impossible to assess whether the mutated limbs had forelimb or hindlimb characteristics.

The third candidate gene, *Ptx1*, is also a transcription factor, from the homeodomain family, and is also expressed in the hindlimbs from the limb field stage until late embryonic development. When *Ptx1* (also known as *Pitx1*) was mutated in mice, embryos with no functional *Ptx1* had hindlimbs that were short and had many characteristics of forelimbs, suggesting that *Ptx1* might be necessary for hindlimb identity. However, there was some disagreement as to whether the hindlimbs in *Ptx1* mutant embryos had truly been transformed into forelimbs, or if they were simply malformed hindlimbs. Furthermore, in these experiments, it was impossible to determine whether *Ptx1* could determine limb identity on its own or if it had to act in combination with *Tbx5*.

Since removal of *Tbx4*, *Tbx5*, and *Ptx1* was unable to resolve their respective roles in limb identity, several attempts have been made to "swap" these genes, that is, express *Tbx4* and *Ptx1* in the forelimb or *Tbx5* in the hindlimb, and observe any potential changes in limb identity. Initial experiments of this type were performed in chicks, where mutation of naturally occurring genes is currently unfeasible. Experimenters from several research groups were able to introduce experimental expression of *Tbx4* and *Tbx5* in their opposing limb types, but this resulted in mixed expression of both genes. Such embryos displayed a range of phenotypes, including limb truncations, unclassifiable malformed limbs, completely normal limbs, and most importantly limbs that appeared to have transformed from one limb type to the other. These results were described as indicating that the expression of *Tbx4* in forelimb or *Tbx5* in hindlimb is capable of transforming limb identity, and the research groups assumed that the low frequency of transformation and high frequency of nonspecific limb defects were due to the mixed expression of the two genes. In light of more recent evidence, it seems that this interpretation was incorrect. It is likely that the high levels of mixed gene expression perturbed limb development in a nonspecific way, and that the putative transformed limbs were simply malformed limbs that retained their original identity.

More recent "swap" experiments were performed in mouse embryos, where the endogenous T-box genes could be removed before the experimental genes were put in place. In these experiments, *Tbx5* was removed from the forelimb, and *Tbx5*, *Tbx4*, *Ptx1*, and combinations thereof were expressed experimentally in the forelimb. As had been previously shown, when none of these genes was expressed in the forelimb, the forelimb failed to form. However, when *Tbx5* was reintroduced experimentally, normal forelimbs developed, demonstrating that the experimental system itself did not upset limb development, since it was capable of recapitulating normal development. Using this system, *Tbx5* was replaced with *Tbx4*. These manipulated embryos continued to develop normal forelimbs, proving that either *Tbx4* or *Tbx5* was sufficient to cause limb bud outgrowth, but suggesting that they had no effect on the identity of the limb they were in. Conversely, when *Ptx1* alone was experimentally expressed in the forelimb, it was not sufficient to allow limb bud outgrowth. However, when *Ptx1* was experimentally expressed in the forelimb with either *Tbx4* or *Tbx5* (to allow outgrowth), the forelimb buds grew out with characteristics of hindlimbs. These transformed forelimbs expressed hindlimb-specific *Hoxc* genes, had hindlimb-like skeletal structures, and had hindlimb- like musculature. The results of these experiments demonstrate that, while *Tbx4* and *Tbx5* are necessary for the growth of the limb bud, they do not have any obvious effect on limb type and they can be substituted for one another. On the other hand, *Ptx1* is sufficient to change the forelimb to a hindlimb identity, and can therefore be considered the defining factor controlling limb type.

Similar conclusions were reached by two research groups that removed *Tbx4* and *Tbx5* while limbs were growing. In these experiments, *Tbx4* or *Tbx5* was expressed only very briefly during the initial stages of limb development, which was sufficient to permit approximately normal limb outgrowth, and then rapidly removed. Limb identity and shape was tracked until late stages of embryonic development. While these limbs did show malformations, no changes were seen in the expression of limb-specific gene expression, and the limbs formed all of the skeletal elements of their original limb identity. These results confirmed that *Tbx4* and *Tbx5* are dispensable for the maintenance of limb identity in later stages of limb development.

Outlook. While the basic steps of hindlimb versus forelimb identity now appear to have been established, many questions remain to be answered regarding the details and the mechanisms that control this process. The target genes that *Ptx1* controls are largely unknown, and virtually nothing is understood about how differential gene expression translates into differences in final forelimb or hindlimb shape. It is likely that the T-box genes will still play significant roles in these processes; *Ptx1* has been shown to bind DNA cooperatively with T-box genes, and thus it is possible that *Tbx4* and *Tbx5* assist *Ptx1* in binding its gene targets. Additionally, little is known about the genetic forces that position *Tbx4* and *Tbx5* expression at the correct places along the embryonic flank to permit limb outgrowth in the forelimb and hindlimb fields.

For background information *see* ANIMAL EVOLUTION; DEVELOPMENTAL BIOLOGY; DEVELOPMENTAL GENETICS; EMBRYONIC DIFFERENTIATION; GERM LAYERS; MORPHOGENESIS; PATTERN FORMATION (BIOLOGY); SKELETAL SYSTEM in the McGraw-Hill Encyclopedia of Science & Technology. L. A. Naiche

Bibliography. M. P. Logan, Finger or toe: The molecular basis of limb identity, *Development*, 130:6401–6410, 2003; L. A. Naiche et al., T-box genes in vertebrate development, *Annu. Rev. Genet.*, 39:219–239, 2005; L. Niswander, Pattern formation: Old models out on a limb, *Nature Rev. Genet.*, 4:133–143, 2003.

Mast cells

Mast cells represent a specific type of immune cells best known for their role in allergic disorders and anaphylactic (acute allergic or hypersensitive) reactions, but are now also considered as the sentinel cells that promote body defense to combat parasite and bacteria infection. Mast cells provide an abundant source of chemical mediators, cytokines (peptides released by some cells that affect the behavior of other cells, serving as intercellular signals), and growth factors, and can respond to many immunological and nonimmunological stimuli. These cells are thought to participate in many other inflammatory and pathological processes, such as arthritis, autoimmune diseases, fibrosis, angiogenesis, carcinogenesis, and inflammatory bowel disease, and in the regulation of cardiovascular function, etc., because increases in mast cell numbers and/or morphologic evidence of mast cell activation are frequently detected in such conditions. Among the mediators produced by mast cells, some are proinflammatory and can help to recruit and activate other immune cells to sites of reactions, whereas others are immunomodulatory and can dampen the reactions by suppressing the activities of other cells. Recently, it has been shown that mast cell–derived enzymes can degrade harmful biological molecules or external toxins. These findings reveal the two "faces" of mast cells, which not only aggravate allergy and other immunological disorders, but are also critical for host defense against pathogens and for maintaining tissue homeostasis.

Origin, development, and distribution. Mast cells are derived from bone marrow precursor cells that give rise to all types of blood cells. Paul Ehrlich was credited with the first description of the mast cell in animal tissues in his doctoral thesis in 1878. He observed prominent granules in some cells in the connective tissue when aniline dye was used to stain tissue sections and named these "granular cells of the connective tissue" as mastzellen (German for "well-fed cells"). Ehrlich also noticed that mast cell numbers were strikingly increased in human tissues with chronic inflammation and other disorders, including carcinomas.

Mast cells can be found in virtually all vascularized tissues. Although mast cells are of bone marrow origin, fully differentiated, mature mast cells are normally not found in the blood circulation. Mast cell precursors circulate in the blood and migrate to the peripheral tissues where they differentiate and become permanent residents. Mast cells are often found in close proximity to blood and lymphatic vessels, near or inside the nerve bundles, smooth muscle cells, mucus-producing glands, and hair follicles, or adjacent to epithelial surfaces. The highest density of mast cells can be found in the skin, conjunctiva (the mucous membrane covering the eyeball and lining the eyelids), and respiratory and gastrointestinal tracts where a barrier function is critical to fence off external insults. Allergic reactions frequently occur at these sites where mast cells are abundant.

Mast cell development is regulated by the interaction of different mast cell growth factors with their specific receptors. The stem cell factor (SCF)/c-Kit receptor interaction represents the most important pathway that governs rodent and human mast cell development. Other molecules such as nerve growth factor, various interleukins (IL-3, IL-4, IL-9, and IL-10), transforming growth factor-β, interferon-γ, and probably many others can also influence mast cell proliferation, differentiation, and survival.

Mature human mast cells are usually round or spindle-shaped in tissues with large oval nuclei. The most distinct features of mast cells are their prominent cytoplasmic granules with staining characteristics called metachromasia, that is, the ability to react and change the color of the staining dyes (**Fig. 1**).

Mast cell heterogeneity. Structural, phenotypic, and functional properties of mast cells are determined by species origin, stage of development, and anatomical location. Mast cell heterogeneity refers to the phenomenon that different mast cell populations exhibit different morphology (size, ultrastructure of the granules), histological staining characteristics, quantity of stored mediators, sensitivity to different stimuli or pharmacological inhibitors, or response to growth factors for proliferation and development.

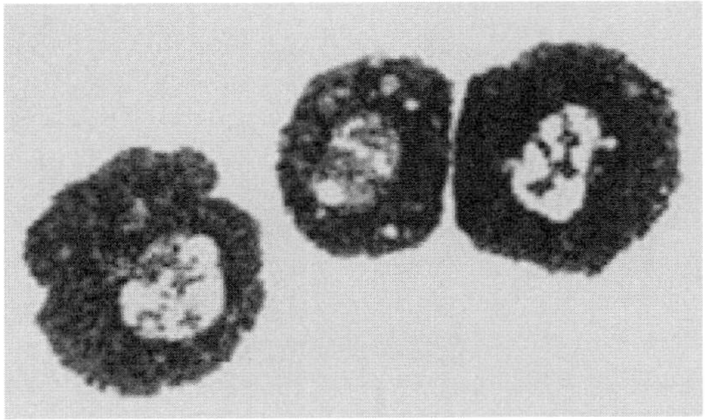

Fig. 1. Mast cells purified from mouse peritoneal cavity and stained with Giemsa exhibit cytoplasmic granules.

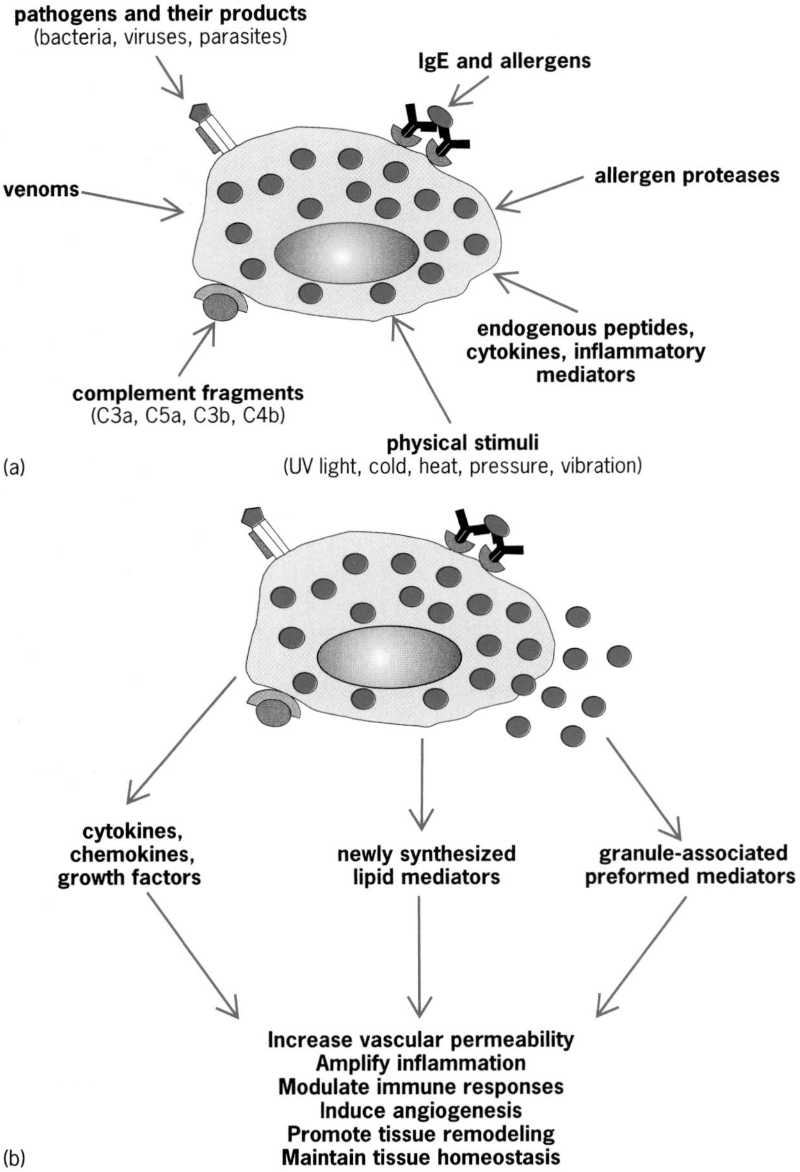

Fig. 2. Mast cell activation. (*a*) Multiple stimuli can activate mast cells. (*b*) Activated mast cells produce mediators and proteins that can influence the outcome of an immune response or its associated tissue changes.

It is generally agreed that mast cell heterogeneity is influenced by microenvironmental factors, such as growth factors, cytokines, and mediators, present in the surroundings of mast cells.

Mast cell activation. Mast cells can detect and respond to many internal or external stimuli via the many different types of signaling receptors expressed on their surface (**Fig. 2*a***). Some surface receptors also mediate the communication between mast cells and their surrounding cells. Once activated, mast cells can produce a wide array of biologically active mediators/products, which are generally divided into three categories: (1) preformed granule-associated mediators, including biogenic amines (such as histamine and serotonin), proteoglycans (such as heparin and chondroitin sulfate), and proteases (such as chymases, tryptases, and carboxypeptidases); (2) newly synthesized lipid mediators (such as leukotrienes and prostaglandins); and (3) cytokines, chemokines (small chemoattractant cytokines), and growth factors. Mast cell activation can lead to sequential release of all classes of mediators or differential release of specific classes of mediators, depending on the type and strength of the stimuli.

Allergy and anaphylaxis are the consequence of mast cells being "armed" by the binding of immunoglobulin E (IgE), a subtype of immunoglobulin produced during an immune response to a specific allergen, to the high-affinity IgE receptors expressed on mast cells. These IgE receptors are triggered by the subsequent binding of the specific allergen to the mast cell–bound IgE molecules. This type of mast cell activation is thought to be also critical for host defense against parasite infections. Aggregation of IgE receptors by IgE and antigen initiates a cascade of biochemical events inside the mast cells and triggers "degranulation," the immediate release of mediators stored inside the mast cell granules. These granule-associated mediators, together with the newly synthesized lipid mediators, are responsible for many signs and symptoms of acute allergic reactions and anaphylaxis. In some cases, those mediators can recruit other inflammatory cells, and a late-phase reaction (a second, sometimes more severe reaction) can develop a few hours later. IgE-dependent mast cell activation can also lead to activation of gene transcription and translation of many cytokines, chemokines, and growth factors. These cytokines and chemokines can further amplify the allergic reaction by recruiting more inflammatory cells to the site. The sustained release of mast cell mediators may lead to long-term chronic inflammation as well as structural alteration of the affected tissues (Fig. 2*b*).

In addition to IgE-dependent activation, mast cells can also be activated by a wide variety of other stimuli, including antigen-IgG immune complexes, products (C3a, C5a) of complement activation during some immune responses, bacteria, viruses, parasites, and pathogen-derived products. Other types of mast cell activators include the polybasic compounds (compound 48/80, mastoparan, polymyxin B), endogenous peptides (neuropeptides, endothelin, neurotrophins), venoms and venom components, cytokines and inflammatory mediators, physical stimuli, etc.

The functional versatility of mast cells is attributed to the wide spectrum of mediators that can be produced by mast cells, the many mechanisms by which mast cells can be activated, and the wide distributions of these cells in many tissues and organs. To define the precise contribution of mast cells in each of the biological and pathological processes is complicated by the phenotypic and functional plasticity of mast cells, which can secrete multiple mediators that possess opposing effects or that are also produced by other cell types. Thus, it has been a major challenge for many researchers in the mast cell field to predict and characterize the effects directly contributed by specific mast cell activation in vivo.

Experimental approaches to study mast cell functions. Our understanding of mast cell biology and functions was greatly facilitated by the establishment of techniques to derive and culture large numbers of mouse and human mast cells from stem cells that give rise to blood cells. Before the culturing techniques were developed, mast cells that were studied were restricted to mast cell populations purified from dispersed animal or human tissues, which yielded low numbers of cells or cells with compromised viability, together with the risk that some cell surface receptors might be lost during such purification procedures. Although many important discoveries have been made using the in vitro cultured mast cells, the in vivo relevance and biological importance of such in vitro observations need to be directly validated by in vivo studies. One useful animal model designed for such investigations involves the use of the mast cell-deficient c-*kit* mutant mice. These mice virtually lack tissue mast cells as a result of defective SCF/c-Kit interactions, which are critical for mast cell development and survival. The deficiency of mast cells in the c-*kit* mutant mice can be selectively repaired by adoptive transfer of genetically compatible, in vitro-derived mast cells. Using such an animal model, one can assess the extent to which the difference in the expression of biological responses observed in the mast cell-deficient mice compared to the wild-type mice reflects the role or requirement of mast cells in such responses. Using this approach, mast cells have been found to contribute to many inflammatory conditions, including parasite repulsion, experimental allergic encephalomyelitis (a mouse model of multiple sclerosis), experimental bullous pemphigoid (a chronic, autoimmune, subepidermal, blistering skin disease), autoimmune arthritis, and silicosis (a respiratory disease caused by inhalation of silica dust), and to promote atherosclerosis. Studies using this model have also shown that mast cells can dampen the toxicity of snake and insect venoms, mediate immune tolerance after tissue transplantation, and limit chronic inflammation elicited by ultraviolet irradiation or poison oak/ivy exposure.

Therapeutic approaches targeting mast cells. The knowledge inferred from animal studies combined with clinical observations can reveal useful therapeutic strategies that target mast cells or mast cell-specific products, either to suppress mast cell activation in allergy, asthma, anaphylaxis, and some autoimmune diseases or to enhance aspects of mast cell functions that are indeed beneficial to human health. Antihistamine drugs are used to block the effects of histamine released by mast cells on nerves and other cells. Drugs of the cromoglicate class are mast cell stabilizers, which block the release of inflammatory mediators from mast cells. In addition, leukotriene antagonists are used to block the effects of leukotriene mediators released by mast cells during allergic reactions. Recently, omalizumab, a monoclonal antibody against the IgE molecule, has been used as a new class of therapeutic agent for patients with moderate to severe persistent asthma by reducing exacerbations and steroid requirement in such patients.

For background information *see* ALLERGY; ANAPHYLAXIS; ASTHMA; CYTOKINE; HISTAMINE; HYPERSENSITIVITY; IMMUNOGLOBULIN; IMMUNOLOGY; INFLAMMATION; PARASITOLOGY in the McGraw-Hill Encyclopedia of Science & Technology. Mindy Tsai

Bibliography. S. C. Bischoff, Role of mast cells in allergic and non-allergic immune responses: Comparison of human and murine data, *Nat. Rev. Immunol.*, 7:93–104, 2007; S. J. Galli et al., Mast cells as "tunable" effector and immunoregulatory cells: Recent advances, *Annu. Rev. Immunol.*, 23:749–786, 2005; D. D. Metcalfe, D. Baram, and Y. A. Mekori, Mast cells, *Physiol. Rev.*, 77:1033–1079, 1997; M. Wills-Karp and G. K. K. Hershey, Immunological mechanisms of allergic disorders, pp. 1439–1479, in William E. Paul (ed.), *Fundamental Immunology*, Lippincott Williams & Wilkins, Philadelphia, 2003.

Mesozoic marine reptiles

The Mesozoic Era (an interval from 245 to 65 million years ago), commonly known as the Age of Dinosaurs, was a time in Earth's history when reptiles were the dominant large animals in most ecosystems. The roles played today by mammals such as lions, antelope, and elephants were filled in the Mesozoic by different types of dinosaurs, familiar, large reptiles closely related to crocodiles and birds. However, other groups of large reptiles also lived during this time, and most are less familiar. The Mesozoic oceans played host to a myriad of large reptiles, including ichthyosaurs, plesiosaurs, nothosaurs, placodonts, and mosasaurs. Each of these groups of reptiles contained many species, some among the largest creatures ever to live in the oceans, and none of them were dinosaurs. These animals may be thought of as analogous to the whales, seals, and otters of today. This article will discuss each of these groups in turn. The relationships of the various groups, and of the groups to mammals, are depicted in **Fig. 1**.

The Mesozoic Era is formally divided into three shorter time periods, termed the Triassic, Jurassic,

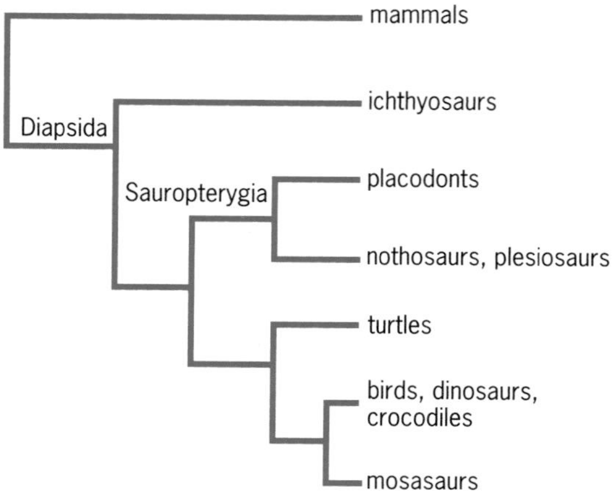

Fig. 1. Family relationships of animals discussed. Animals close together on the diagram—technically called a cladogram—are more closely related than those farther away. Most marine reptiles were basal diapsids, meaning that they were related to modern reptiles, but were not members of the two major groups alive today.

and Cretaceous periods (**Fig. 2**). The Earth during this time was quite different than it is today. The climate was much warmer; further, because there were no ice caps, temperatures from higher to lower latitudes were more uniform. Another consequence of the lack of ice was a relatively high sea level. Large areas of the continents were covered by ocean during some or all of the Mesozoic, including most of Europe and the middle of North America. Termed epicontinental seas, these shallow, nutrient-rich waters were rich in fish, cephalopods (squid and related animals), and many other invertebrates. As the Mesozoic progressed, many reptile groups evolved members that took advantage of this abundant sea life (Fig. 2). These reptiles were diapsids, a technical term that refers to the structure of the skull and applies to all living reptiles, including lizards, snakes, crocodiles, and turtles, and most fossil ones. (Birds are diapsids as well.) Most, but not all, Mesozoic marine reptiles first evolved in the Triassic. Some did not survive past the end of this period, while others persisted or even evolved into other forms. All surviving Mesozoic marine reptiles went extinct at the end-Cretaceous mass extinction, a huge die-off triggered probably by a giant meteor impact in the Yucatan region of Central America that also extinguished the dinosaurs and many other animal species.

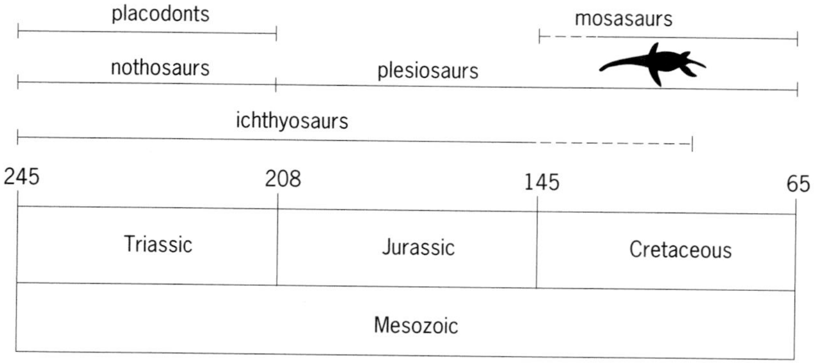

Fig. 2. Time scale and life spans of the groups discussed in this article. The Mesozoic is divided into three roughly equal periods: Triassic, Jurassic, and Cretaceous. Most marine reptiles first evolved in the Triassic, although different groups survived for different lengths of time. The end of the Cretaceous Period is marked by a mass extinction event, leading to the extinction of dinosaurs, marine reptiles, and many other species.

Ichthyosaurs. Ichthyosaurs (**Fig. 3*a***) were one of the first groups of marine reptiles to evolve in the Mesozoic, appearing early in the Triassic. Primitive, lizardlike members of this group have been found in Japan and China, and ichthyosaurs were distributed across the globe by the middle of the Triassic. Ichthyosaurs were highly pelagic animals, meaning that they swam in deep water away from shore and the sea bottom. They were very fishlike superficially, with short, finlike flippers and a sharklike tail that moved from side to side. However, they still breathed air and were most similar to present-day dolphins and porpoises. Ichthyosaurs were obligately aquatic (restricted to an aquatic mode of life) and could not leave the water at all. They gave birth to live young (rather than laying eggs), and several spectacular fossils of ichthyosaur mothers that died in the act of giving birth have been discovered in Germany. The ichthyosaurs have a long history, persisting until the Middle Cretaceous, and over this history several trends occur within the group. The vertebrae in the back become progressively shorter and more disklike, whereas the tail becomes less lizardlike and more fishlike. These adaptations resulted in more efficient swimming over distances. Some ichthyosaurs had the largest eyes of any vertebrate [25 cm (10 in.)], and possibly the largest of any known animal (only the giant squid has eyes as large). Squid and fish were the main prey of ichthyosaurs, and large eyes were a benefit to these visual hunters. Some ichthyosaurs were 15 m (49 ft) in length, the size of a large whale.

Placodonts. Placodonts (Fig. 3*b*) and the following two groups—nothosaurs and plesiosaurs—are all closely related, forming a clade (family group) called Sauropterygia, meaning "winged reptiles." Placodonts, however, did not have wing-shaped flippers. They were heavy, slow-moving animals that crawled across the sea floor using relatively unspecialized limbs. The body was short and stout, and in more derived species a hard carapace (shield) of bone covered the trunk. Placodonts look very much like turtles, and a close relationship between turtles and placodonts has been hypothesized by several workers. The current consensus, however, is that the resemblances between the two groups are superficial and that they evolved independently. Unlike turtles, all placodonts were durophagous, meaning that they ate tough-bodied prey such as clams, mussels, and other hard-shelled mollusks. The mouths of placodonts were filled with large, round, flat teeth, both on the jawbones and on the roof and floor of the mouth. These teeth formed a pavement useful for crushing, similar to the teeth in some rays. Placodonts were relatively small, with the largest less than 2 m (6.6 ft) in length. Placodonts arose early in the Triassic and died out at the end of this period, and relatively little is known about the biology of these strange animals.

Nothosaurs. The nothosaurs (Fig. 3*c*) are another entirely Triassic group, although some members gave rise to the plesiosaurs (see below) at the end of the Triassic. The term nothosaur actually covers a large group of animals of diverse body form. Several genera of small [<50 cm (20 in.)], lizardlike animals are known in great abundance from deposits in southern Germany and Italy. These animals lived in very shallow water and could probably leave the water as well. The large number of complete specimens of these animals has allowed detailed analysis of their growth and biology. Recently, more species of these small nothosaurs have been discovered in China, and some with preserved embryos within their body cavities, proving that they did not lay eggs but gave birth to live young. Live birth in such lizardlike, amphibious animals is surprising and is a strong argument for live birth in the more ocean-going members of the group. Another group of nothosaurs consisted of much larger animals that possessed shorter, more flipperlike limbs and a shortened body. The neck was relatively long in most taxa, and the head had an elongated, bizarre shape. The jaws were extremely long, extending back behind the face and the rest of the skull, so that the mouth was much longer than the rest of the head. Powerful muscles opened and closed these jaws, allowing quick snaps at rapidly moving fish and other prey. These nothosaurs were probably amphibious and gave birth to live young, although definitive evidence is lacking. These animals also became increasingly pelagic over their history, finally giving rise to the ocean-going plesiosaurs.

Plesiosaurs. Plesiosaurs (Fig. 3*e*) evolved from nothosaurs at the Triassic-Jurassic boundary and lived on until the end of the Mesozoic. All plesiosaurs shared a novel style of locomotion, a style unique among secondarily aquatic vertebrates (secondarily since all vertebrates originally had an aquatic ancestor) in that both sets of limbs were adapted as flippers. The trunk was short, stiff, and immobile, and the tail was short and blunt. The flippers were used in underwater flight, similar to those in aquatic turtles and birds. Recent research on plesiosaur flippers has demonstrated that their shape was tailored to the size of the prey pursued by different types of plesiosaurs. Different plesiosaurs hunted prey of widely different sizes, as indicated by the great variation of head size within the group. Most plesiosaurs had a small head and a long neck, yielding the familiar swanlike profile associated with the term plesiosaur. However, some plesiosaurs had short necks and very large heads, and these animals are given their own name: pliosaurs (Fig. 3*d*). Some pliosaurs have huge teeth, rivaling those of large theropod dinosaurs such as *Tyrannosaurus rex*. Until recently, conventional wisdom held that all pliosaurs were closely related. However, recent research has shown that the pliosaur body type—a short neck and a large head—evolved repeatedly within the clade. Animals with extremely long necks also evolved more than once. Small-headed plesiosaurs are known to have eaten ammonites (an extinct group of mollusks), squid, and fish based on preserved stomach contents, whereas pliosaurs ate much larger animals, probably including large fish, ichthyosaurs, and other plesiosaurs. Many plesiosaurs have been found with gastroliths, or stomach stones, preserved in their body cavities. These stones were probably used to grind up food after it had been swallowed whole. Plesiosaurs with incredibly long necks evolved during the Cretaceous Period. Called elasmosaurs, these reptiles were up to 15 m (49 ft) in length, and some possessed over 70 vertebrae in their necks. Movement and support of such a neck are not mechanically possible outside of the water, and it is thought that all plesiosaurs were obligately aquatic and gave birth to live young. However, definitive fossil evidence of this is still lacking.

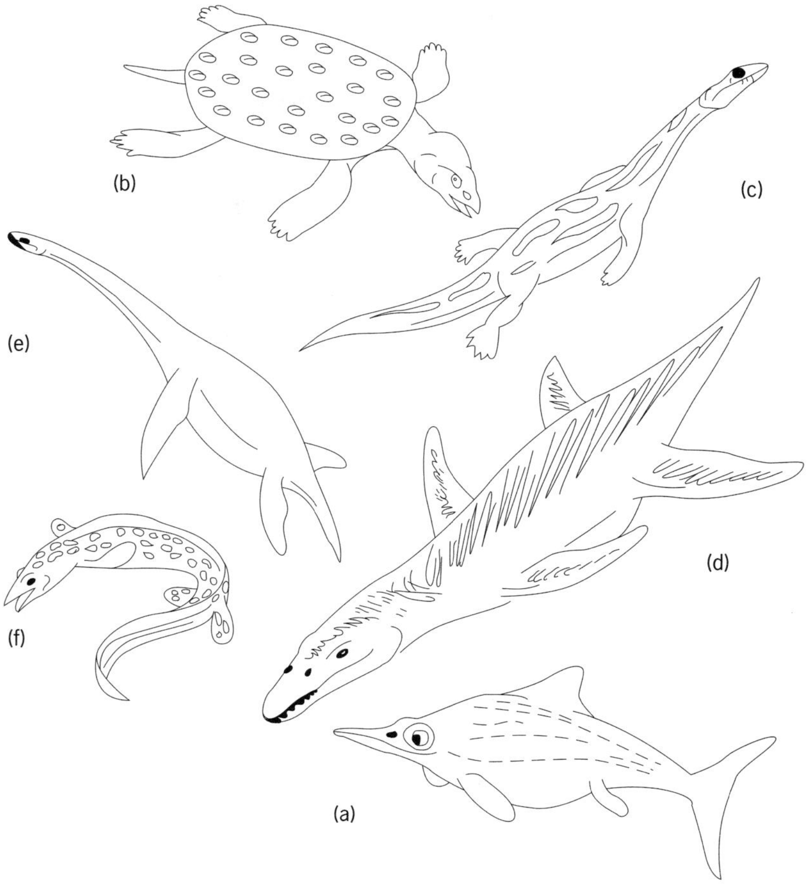

Fig. 3. Animals discussed in this article: (*a*) the ichthyosaur *Opthalmosaurus*, length 3.5 m (11.5 ft), Late Jurassic. (*b*) Placodont *Placochelys*, length 90 cm (3 ft), Middle Triassic. (*c*) Nothosaur *Nothosaurus*, length 3 m (10 ft), Middle Triassic. (*d*) Pliosaur *Liopleurodon*, length 12 m (40 ft), Late Jurassic. (*e*) Plesiosaur *Muraenosaurus*, length 6 m (20 ft), Late Jurassic. (*f*) Mosasaur *Platecarpus*, length 4.3 m (14 ft), Late Cretaceous.

Mosasaurs. Mosasaurs (Fig. 3 *f*) comprise the last group of Mesozoic marine reptiles. Unlike the other marine reptiles mentioned, mosasaurs were not basal diapsids (not early diapsids from which all others evolved). They are actually closely related to modern monitor lizards such as the Komodo dragon, and possibly to snakes. Some researchers believe that mosasaurs are ancestral to snakes, but this view has not gained wide acceptance. Mosasaurs evolved from terrestrial monitor lizards in the Early Cretaceous and were a very successful clade, giving rise to many species throughout the period. Ancestral mosasaurs, called aigialosaurs, were probably amphibious and were not radically adapted to life in the water. However, the limbs of true mosasaurs evolved quickly into flippers similar to those of modern whales, whereas the body was long and eel-like, and overall body size increased rapidly. Some mosasaurs were up to 10 m (33 ft) in length, and they fed on other marine animals. Little is known about the reproductive biology of mosasaurs, although large ones probably could not leave the water. Mosasaur fossils are extremely common in parts of Nebraska and Kansas, and mosasaurs had a worldwide distribution by the end-Cretaceous mass extinction.

Recent discoveries. Research on marine reptiles has advanced quickly over the past several years, particularly in our understanding of plesiosaur family relationships, prey-linked adaptation, and prey preference. The insight that pliosaurlike animals—with large heads, big teeth, and short necks—evolved several times is surprising because a spectrum of body traits evolved in concert with head size. Large-headed animals tend to have relatively long shoulder and hip bones, large hindlimbs, and flippers like those of falcons and other hunting birds (with wings in the birds being analogous to the flippers). However, details of the skull and vertebral column strongly support the idea that these traits evolved convergently. Why would all these features of body shape evolve together? Researchers hypothesize that all of these adaptations are tailored for the pursuit of large prey. An animal with a large mouth and big teeth can eat large animals, including other marine reptiles. Chasing down and dispatching these large animals requires speed, acceleration, and strength, and the convergent traits just described bestow these qualities. In contrast, plesiosaurs with long necks and small heads have relatively short shoulder and hip bones and flippers like those of albatrosses and seagulls (again with wings being analogous to flippers). Long-necked animals were probably adapted for cruising over long distances, moving through the water and catching small fish and squid with sweeps of their necks.

Recent fossil discoveries have also yielded some surprises. In Australia, a long-necked plesiosaur was found with a stomach full of clam shells, leading researchers to hypothesize that mollusks were the preferred prey of these plesiosaurs. However, other fossils from the same type of animals from North America do not contain clam shells, while squid, fish, and ammonite stomach contents are known from other fossils, so firm conclusions are elusive. Second, the reproductive biology of plesiosaurs is a frustrating problem because no pregnant plesiosaurs have been discovered yet. This lack of knowledge renders the Chinese fossils critically important. These recently discovered nothosaurs contain perfectly preserved embryos, therefore proving that they gave birth to live young. By extension, plesiosaurs probably gave birth to live young as well. Last, predation by marine reptiles on each other has been documented recently. A mosasaur skeleton in the Smithsonian Institution (in Washington, D.C.) contains partially digested plesiosaur bones as stomach contents, whereas a new plesiosaur skeleton from Wyoming contains ichthyosaur remains. Paleontologists are currently engaged in marine reptile fieldwork all over the world, and new discoveries will hopefully come to light in the near future.

For background information *see* ANIMAL EVOLUTION; DIAPSIDA; DINOSAUR; ICHTHYOPTERYGIA; MESOZOIC; NOTHOSAURIA; PALEONTOLOGY; PLACODONTIA; PLESIOSAURIA; REPTILIA; SAUROPTERYGIA in the McGraw-Hill Encyclopedia of Science & Technology. F. Robin O'Keefe

Bibliography. R. L. Carroll, *Vertebrate Paleontology and Evolution*, W. H. Freeman and Company, New York, 1987; R. Ellis, *Sea Dragons: Predators of the Prehistoric Ocean*, University of Kansas Press, Lawrence, Kansas, 2003; M. J. Everhart, *Oceans of Kansas: A Natural History of the Western Interior Sea*, Indiana University Press, Bloomington, 2005; C. McGowan, *Dinosaurs, Spitfires, and Sea Dragons*, Harvard University Press, Cambridge, Massachusetts, 1991; R. Motani, Rulers of the Jurassic seas, in *Special Edition: Dinosaurs and Other Monsters*, *Sci. Am.*, 14(2):4–11, 2004.

Metallographic preparation for electron backscattered diffraction

Electron backscattered diffraction (EBSD) is performed with the scanning electron microscope (SEM) to provide a wide range of analytical data such as from crystallographic orientation studies, phase identification, and grain (crystal) size measurements. A diffraction pattern can be obtained in less than a second, but image quality is improved by using a longer scan time. Grain mapping requires the development of diffraction patterns at each pixel in the field and is a much slower process. The quality of the diffraction pattern, which influences the confidence of the indexing of the diffraction pattern, depends upon removal of damage in the lattice due to specimen preparation. It has been claimed that removal of this damage can only be achieved using electrolytic polishing or ion-beam polishing. However, modern mechanical preparation methods, equipment, and consumables yield excellent quality diffraction patterns without the use of dangerous electrolytes and the problems and limitations associated with electropolishing and ion-beam polishing. If mechanical preparation results in quality polarized-light images of noncubic crystal structure elements and alloys (for example, Sb, Be, Hf, α-Ti, Zn, Zr), or color-tint etching of cubic or noncubic crystal structure elements or alloys produces high-quality color images, then the surface is free of harmful residual preparation damage and EBSD patterns with high pattern quality indexes will be obtained.

Polarized-light image quality is dependent upon eliminating preparation damage and the quality of

TABLE 1. Al–7.12% Si preparation method

Step	Surface	Abrasive/Size	Load, lb (N)	Speed rpm/direction	Time, min
1	CarbiMet® abrasive discs	220–320 (P240-P400) grit SiC, water cooled	5 (22N)	240–300 Contra	Until Plane
2	UltraPol® or TriDent® cloths	9-μm MetaDi Supreme® suspension	5 (22N)	120–150 Contra	5
3	TriDent® or TexMet® 1500 cloths	3-μm MetaDi Supreme suspension	5 (22N)	120–150 Contra	4
4	TriDent or TexMet® 1500 cloths	1-μm MetaDi paste	5 (22N)	120–150 Contra	3
5	MicroCloth®, VelTex®, or ChemoMet® cloths	MasterMet® colloidal silica	5 (22N) (7 lb/31N for ChemoMet)	120–150 Contra	3

the microscope optics. Consequently, one should always check the polarized light response of the metals to verify the preparation quality before EBSD. For cubic metals, the metals should be etched first with a general-purpose reagent to confirm the nature of the expected microstructure. The final polishing step should be repeated and a color-tint etch used to verify freedom from damage. EBSD is best performed with an as-polished, nonetched specimen due to the steep angle to the electron beam, as surface roughness can degrade the diffraction pattern. A well-prepared unetched specimen will exhibit a good grain-contrast image with a backscattered electron detector. This is a good test for freedom from surface damage.

Preparation. Specimen preparation methods for metals have been developed that yield excellent results and generally require less than 25 min to complete. Automated preparation equipment is recommended for accuracy and reproducibly. Manual (hand) preparation cannot produce the flatness, phase retention, and damage removal as easily as automated processing and is less reproducible.

Successful preparation requires that sectioning be done with equipment and consumables that minimize damage. Sectioning is a violent process that can introduce massive damage. The crystal structure does influence damage depth. Abrasive blades should be used that have been designed for metallography and recommended for the specific metal/alloy in question. A precision saw will yield even less damage, as the blades are much thinner and the applied loads are much lower. Cutting with machines and blades/wheels that introduce minimal damage is the most critical step in generating damage-free metallographic surfaces. The second critical rule for obtaining damage-free polished surfaces is grinding with the finest possible abrasive and surface to make all of the specimens in the holder co-planar and remove the sectioning damage. The proposed methods use flat, woven cloths or pads that minimize relief problems. To minimize damage, use less aggressive surfaces, such as silk, nylon, polyester, or polyurethane. The specimen preparation method must remove all scratches. If scratches are present, so too is damage below the scratch. Scratch depths produced in grinding and polishing are not uniform. A deep scratch will have deep deformation below it. The preparation method must remove the scratches and the underlying damage in order to obtain high-quality EBSD patterns.

Initial experiments. These covered a wide variety of metals and alloys prepared mechanically using three to five steps. The EBSD patterns shown were developed using the EDAX-TSL OIM™ EBSD system on a Philips (FEI) XL 30 SEM using a tungsten filament at 20 kV, spot size 5 (beam 140-nm diameter), 20-mm working distance, and an emission current of 73 μA ($\sim$0.84 nA on the specimen). The plane-of-polish was tilted 74° from horizontal. The TSL OIM data collection software system generated pattern quality indexes (PQI), and the results shown here are the average and 95% confidence limits for 25 randomly selected grains using unetched specimens. A second set of samples of high-purity metals were analyzed using the HKL Channel 5 EBSD system of Oxford Instruments with an FEI Quanta 200 SEM, using a tungsten filament, with a 70° tilt angle for the plane-of-polish, 20 keV, and a 10-mm working distance. The patterns were evaluated using the band contrast data, with the average and standard deviation calculated for a number of measurements. Several cast specimens had very large grains, so only a few EBSD patterns could be obtained. The silicon specimen was a single crystal, so all patterns were basically identical.

Aluminum alloys. The first example is Al–7.12% Si, prepared by the method in **Table 1**. The α-Al dendrites were sampled for EBSD patterns.

If a higher power electron source is used, such as lanthanum hexaboron (LaB_6) or field emission gun, the PQIs will be better than when using a tungsten filament. The average atomic weight of an Al–7.12% Si alloy is quite low, so the alloy's backscattered

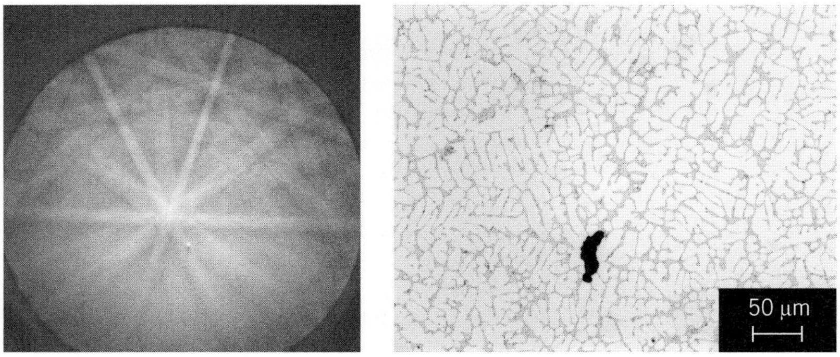

(a) (b)

Fig. 1. EBSD pattern for (*a*) α-Al in as-cast Al–7.12% Si; pattern quality index: 87 ± 4.2. Light micrograph of (*b*) as-cast Al–7.12% Si hypoeutectic alloy etched with 0.5% HF in water.

TABLE 2. Copper and copper alloy preparation method

Step	Surface	Abrasive/Size	Load, lb (N)	Speed rpm/direction	Time, min
1	CarbiMet waterproof paper	240 to 320 (P280 to P400) grit SiC water cooled	5–6 (22–27)	Contra	U.P.
2	UltraPol or TriDent cloths	9-μm MetaDi Supreme diamond suspension	5–6 (22–27)	Contra	5
3	TriDent or TexMet 1500 pads	3-μm MetaDi Supreme diamond suspension	5–6 (22–27)	Contra	4
4	TriDent or TexMet 1500 cloths	1-μm MetaDi Supreme diamond suspension	5–6 (22–27)	Contra	3
5	MicroCloth, VelTexor ChemoMet pads	0.05-μm MasterMet Colloidal silica suspension	5–6 (22–27) (7 lb/31 N for ChemoMet)	Contra	3

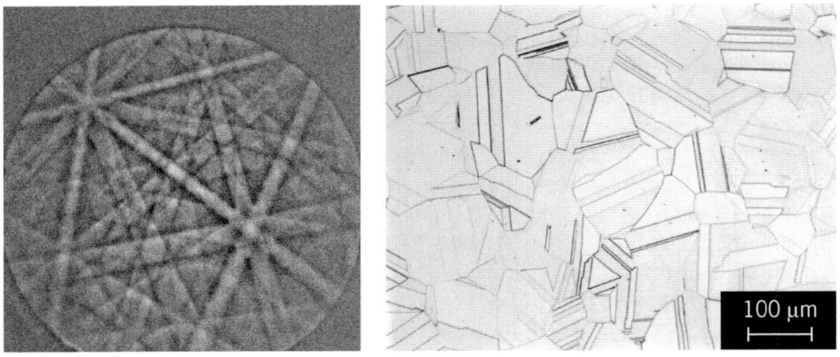

(a) (b)

Fig. 2. EBSD pattern for (*a*) cartridge brass, Cu–30% Zn; PQI: 221 ± 8.6. Microstructure of (*b*) wrought Cu–30% Zn etched with equal parts hydrogen peroxide (3%) and ammonium hydroxide.

electron generation efficiency is low, which makes it a very difficult subject for EBSD. As can be seen in **Fig. 1**, an excellent quality diffraction pattern was obtained from the α-Al dendrites. This demonstrates that mechanical preparation is capable of producing high quality EBSD patterns when properly performed.

Copper and copper alloys. Pure copper is extremely ductile and malleable. Copper and its alloys come in a wide range of compositions, including several variants of nearly pure copper for electrical applications that are very difficult to prepare damage-free. Rough sectioning and grinding practices can easily damage copper and its alloys and the depth of damage can be substantial. Scratch removal, particularly for pure copper and brass alloys, can be very difficult. If the scratches are not removed, there will be damage beneath. Following the preparation cycle with a brief vibratory polish using colloidal silica is very helpful for scratch and damage removal. Attack polishing additions have been used in the past to improve scratch removal but usually are not necessary using the contemporary methods followed by vibratory polishing.

Table 2 is a five-step method for preparing copper and its alloys. It is always helpful, particularly with alloys that are difficult to prepare damage-free, to etch the specimen after the fifth step and then repeat the fifth step. This reduces damage and gives better EBSD patterns. The specimen shown in **Fig. 2** was wrought cartridge brass, Cu–30% Zn, that was cold reduced 50% in thickness and then annealed at 704°C for 30 min producing a coarse twinned α-Cu matrix. This is a relatively difficult alloy to prepare free of scratches and surface damage and the EBSD pattern quality was superb.

EBSD patterns can be developed for both phases in a two-phase alloy, as long as preparation keeps both phases flat on the plane-of-polish. If relief is present, such that one phase is recessed below the surface, EBSD patterns will not be developed. As an example, a specimen of naval brass, an α-β brass consisting of Cu–39.7% Zn–0.8% Sn, was tested after etching, which attacked the β phase. EBSD patterns could be generated from the α phase but not from the recessed β phase. Repolishing and running the specimen unetched produced excellent results for both the α and β phases as shown in **Fig. 3**.

Table 3 summarizes the TSL OIM EBSD PQI results for the metals and alloys evaluated, many of which are difficult to prepare. These results clearly show that mechanical specimen preparation, if properly

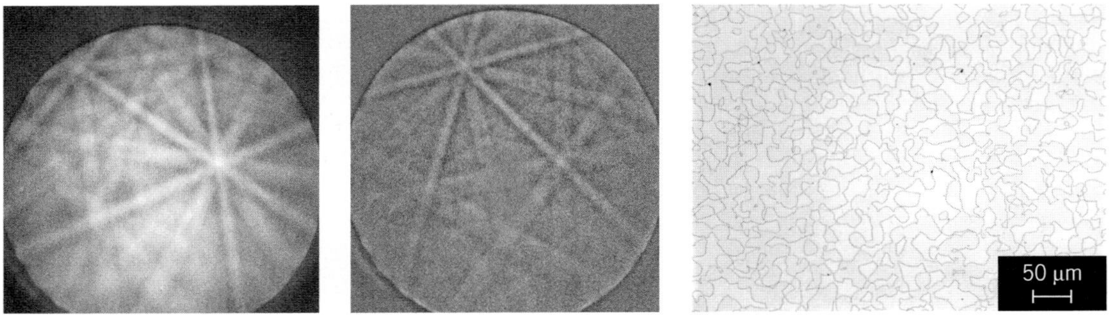

(a) (b) (c)

Fig. 3. EBSD patterns for Cu–39.7% Zn–0.8% Sn showing the (*a*) alpha and (*b*) beta phase with PQIs of 118.5 ± 8.7 for α-Cu and 150.4 ± 20.7 β-Cu. (*c*) Microstructure.

TABLE 3. Pattern quality index values using the TSL OIM EBSD system

Metal/alloy	PQI ± 95% CL	Metal/alloy	PQI ± 95% CL
α-Al in Al-7.12% Si	87 ± 4.2	α-Cu in Cu-30% Zn	221 ± 8.6
Cu-39.7% Zn-0.8% Sn	118.5 ± 8.7 for α	Cu-39.7% Zn-0.8% Sn	150.4 ± 20.7 for β
Elgiloy (Co-based)	221.4 ± 7.4	Pure Fe	249.6 ± 5.5
Si Core Fe B	199.9 ± 7.4	316 stainless steel	184.9 ± 8.5
2205 Duplex SS	248 ± 15.4 for α	2205 Duplex SS	207.9 ± 11 for γ
Ni-200	176.3 ± 17.6	HyMu 80 (Ni-base)	196.7 ± 7.2
Nitinol (Ni-Ti)	58.7 ± 4.3	CA625 Plus (Ni-base)	200.5 ± 6.5
Fine grain 718 (Ni-base)	80.7 ± 4.4	Pure Cr	259.8 ± 13.1
Pure Nb	166.2 ± 17.1	Pure V	125.9 ± 10.3
Pure Ta	169.7 ± 13.0	CP Ti ASTM F67 Gr2	119.1 ± 4.1
W in W-27 Cu	296.9 ± 20.1	Pure Bi	86.2 ± 1.8
Pure Pb	49.3 ± 3.0	Pure Ru	266.2 ± 21.8

performed, is fully capable of producing damage-free surfaces that yield acceptable EBSD patterns that can be indexed reliably. The Ni-based superalloys, Carpenter's Custom Age 625 Plus and the fine-grained 718, contained submicroscopic strengthening phases that make the EBSD analyses more difficult. The pure tantalum specimen was not fully dense.

Further experiments. A second set of experiments used the Oxford Instruments HKL Channel 5 EBSD system. Eighteen high-purity (generally >99.95%) specimens were prepared using methods similar to those above, but usually with five steps (four for Ti) and with longer times per step. High-purity specimens are the most difficult to prepare. These specimens varied from Mg (atomic number 12) to Bi (atomic number 83) and covered the range of metallic crystal structures, including body-centered cubic (6), face-centered cubic (4), hexagonal close-packed (5), diamond cubic (1), and rhombohedral/trigonal (2). **Table 4** lists the specimens prepared and analyzed.

Specimens of pure Sb, V, and Zr were susceptible to SiC embedment, even though the grit size was coarse, 240- and 320-grit. Hence, grinding was repeated after coating the paper with paraffin wax. Attack polishing was used, mainly with 30% conc. H_2O_2, for the last step for preparing Cr, Nb, Ti, W, and Zr. MasterMet colloidal silica was used for the last step, except for preparing Fe (MasterMet alumina was used) and Mg (water-free MasterPolish was used). Oil-based diamond suspensions (9-, 3- and 1-μm) were used to prepare the high-purity (99.999%) Mg. For the Bi and Pb pure specimens, grinding used four steps: 240-, 320-, 200- and 600-grit SiC paper coated with paraffin wax with low loads, followed by three polishing steps using 5-, 1-, and 0.3-μm alumina slurries and a final polish with MasterMet colloidal silica. All polishing steps used MicroCloth synthetic suede cloth. Although the Bi produced an excellent EBSD pattern, none was obtained with the pure Pb specimen. A 1-h vibratory polish with MasterMet colloidal silica using a MicroCloth pad was required to obtain a diffraction pattern for Pb. A 2-min chemical polish is normally used after mechanical polishing of Zr, so EBSD was conducted on a second specimen after chemical polishing. Surprisingly, no pattern could be obtained on the chemically polished specimen.

TABLE 4. Band contrast values using the Oxford Instruments HKL EBSD system

High-purity elements	Atomic number	Crystal structure	Band contrast (0–255)
Mg	12	hcp	161.2
Al	13	fcc	151.2
Si	14	Diamond cubic	205.8
Ti	22	hcp	134.0
V	23	bcc	102.2
Cr	24	bcc	88.27
Fe	26	bcc	105.4
Ni	28	fcc	85.0
Cu	29	fcc	122.6
Zn	30	hcp	170.8
Zr	40	hcp	77.3
Nb	41	bcc	145.6
Ru	44	hcp	66.0
Sb	51	Rhombohedral	180.2
Ta	73	bcc	122.8
W	74	bcc	91.6
Pb	82	fcc	108
Bi	83	Rhomb./trigonal	255

Band contrast results (normalized to 0 to 255) for the 18 high-purity specimens are summarized in Table 4. All of the EBSD patterns were indexed correctly.

Additional trials are being done using the specimens given the standard preparation cycle followed by a 1-h vibratory polish using colloidal silica. Of the seven specimens analyzed so far after vibratory polishing, six had higher band contrast values: Mg, Si, Ti, Ni, and Nb, while Bi had the same maximum band contrast value of 255.

For background information *see* ALUMINUM; ALUMINUM ALLOYS; COPPER; COPPER ALLOYS; CRYSTAL STRUCTURE; ELECTRON DIFFRACTION; ELECTRON MICROSCOPE; GRAIN BOUNDARIES; METALLOGRAPHY; METALS; POLARIZED LIGHT MICROSCOPE; SCANNING ELECTRON MICROSCOPE in the McGraw-Hill Encyclopedia of Science & Technology.

G. F. Vander Voort; E. P. Manilova

Bibliography. J. Pouchou, *L'Analyse EBSD: Principes et Applications* (in French), EDP Sciences, 2004; V. Randle and O. Engler, *Introduction to Texture Analysis: Macrotexture, Microtexture and Orientation Mapping*, Gordon and Breach, London, 2000; A. J. Schwartz, M. Kumar, and B. L. Adams (eds.), *Electron Backscatter Diffraction in*

Materials Science, Springer, 2000; G. F. Vander Voort, Color metallography, in vol. 9 of *ASM Handbook, Metallography and Microstructures*, G. F. Vander Voort (ed.), ASM International, 2004; G. F. Vander Voort, *Metallography: Principles and Practice*, ASM International, 1999; G. F. Vander Voort, The SEM as a metallographic tool, in *Applied Metallography*, ed. by G. F. Vander Voort, Van Nostrand Reinhold, 1986; G. F. Vander Voort et al., *Buehler's Guide to Materials Preparation*, 2004.

Microbial diversity in caves

Caves remain one of the last unexplored frontiers on Earth. Yet, unlike space or the deep ocean, cave exploration is available to the common man or woman; with training and minimal equipment (a helmet and three sources of light), it is possible to walk where no human has walked before. Nonetheless, the dangers of cave exploration are very real, and not every individual looks forward to the idea of 100-m (330-ft) pits, or crawls so tight that one must exhale to fit. Therefore, the exploration of these passages, as both a human and scientific endeavor, is well behind that of the other terrestrial areas of our planet. As scientists begin to examine this environment, they are making discoveries that are changing our ideas on how caves form and the role that microbial species play in this unique environment.

Microbial ecology and geomicrobiology. Life on Earth has been microscopic for much of its 3.7-billion-year history, predominantly comprising the microscopic bacteria and archaea. Despite the relatively simple structure of these organisms, their mechanisms of energy acquisition for growth are anything but basic; they have developed metabolic mechanisms to trap energy from sources as diverse as sunlight (photosynthesis) to the energy gradients between minerals (chemolithotrophy). As a result of such activity over billions of years, these tiny organisms have had a profound impact on our global environment, from the deposition of massive oil and gas deposits to the generation of the oxygen that we breathe. Today, microscopic organisms continue to play a critical role in our biosphere, recycling nutrients, degrading toxins, and fixing carbon dioxide gas.

Historically, the role of microbial species in geologic transformation was overlooked, by our ability to describe many mineral transformations through purely inorganic, chemical processes. Nonetheless, as analytical techniques became more refined, it was apparent that microbial species were intimately involved in many of these transformations. Likewise, as molecular methods increased our ability to study microbial species in the natural environment, we realized that these organisms were indeed playing a critical role in certain geologic processes, from mineral deposition and ore enrichment to chemical weathering and clay formation. With the emergence of a compelling argument for microbial interactions in geologic processes, geologists, biologists, and chemists began to work together to better understand the role that microbial species play in geological settings; this marriage between the geological, biological, and chemical sciences emerged as the science of geomicrobiology.

Assessing microbial activity in caves is not a new idea, dating back to work in caves during the early 1900s. Despite this, as well as the extremely starved nature of the cave environment and our (then) limited appreciation for the mechanisms of microbial energy acquisition, no microbial species were thought to subsist in the cave environment. To explain the microbial species that were seen, researchers suggested that they were brought in by air currents or via the activities of cave-visiting trogloxene species (that is, terrestrial animals, such as bats, that use the cave habitat for specific purposes, on a sporadic basis, and that exhibit no cave-related adaptations). Even so, unusual formations and strange mineral deposits became increasingly difficult to describe by purely inorganic chemistry, and researchers began to recognize that these features were biogenic. With this recognition, the field of cave geomicrobiology emerged.

Mineral-depositing microbes. While there are hundreds of minerals found within cave environments, when most people think of caves they imagine a dark passage decorated with dripstone formations, such as stalactites, stalagmites, and soda straws (speleothems). Just a few hundred years ago, scientists believed that a cave formation was a life form: a simple, vegetative organism comprising rock. With a better grasp of chemistry, we now know that water above the cave picks up carbon dioxide from the soil, making a weak carbonic acid. As this acid interacts with the calcium carbonate rock (limestone) in which the cave is formed, it becomes saturated with this mineral. Then, as the water enters the cave, the carbon dioxide is able to evaporate, reducing the acidity of the water and causing the calcium carbonate to precipitate, building up into delicate crystal formations. With this understanding of the chemical nature of such processes, investigators wondered if microorganisms could play a role in precipitation of these minerals.

In the 1960s, it was demonstrated that microbial species could precipitate carbonate minerals when fed a diet of calcium salt, and that this ability was a general phenomenon of soil species. In order to determine if such activity occurred in caves, microbiologists found that microbes isolated from stalactites demonstrated a greater capacity to precipitate carbonates, both in the number of species able to carry out this process and in the quantity of carbonate produced. Despite this, we still do not have a clear understanding of how or why microbial species precipitate carbonate minerals, as this activity entombs the microorganism in a mineral matrix that ultimately kills it. We do know that the calcium found in limestone is toxic to microorganisms, and the precipitation of carbonate minerals removes this toxic ion. Therefore, the ability of a microorganism to precipitate calcium carbonate may be a survival strategy, allowing it to survive the toxic, calcium-laden cave environment, albeit one that ultimately leads to its demise. The implication that the greenhouse gas, carbon dioxide, plays an important role in this process, becoming locked away in the mineral structure, ensures that work will continue in elucidating the role of these species in speleothem formation.

Rock-eating microbes. While we know that microbial species play an important role in the precipitation of minerals within caves, we also suspect that they play an important role in dissolving them as well. The reasons for this dissolution are as diverse as the species that carry them out, with some species dissolving the rock to mobilize essential nutrients and energy, while others produce toxic, acid by-products that are buffered by the dissolving rock to allow continued growth. The change in structure of cave rock can also vary, although often the removal of mineral cements can lead to a softening of the rock for many centimeters, producing a crumbling and powdery mass that cave explorers call "punk rock." We are still in the process of determining the mechanisms by which microbial species produce punk rock; it is known, though, that microbial species will pursue whatever energy sources are available, even that of the limestone, when other nutrient and energy sources are limiting. Of course, the idea that anything can make a living by eating rock at first appears to be absurd; however, when this rock is analyzed for chemical changes, it appears that the microbial species have removed certain ions essential for growth, including magnesium and phosphorus, while enriching ions that are associated with energy production, such as iron and manganese. While these data suggest a potential goal for rock dissolution, it provides perhaps the first clue to the metabolic processes that microbial species employ to survive in these nutrient-deprived environments.

While microbial activity on surfaces can subtly enlarge cave passages, geologists used to believe that caves themselves could only form through the dissolution of limestone by weakly acidic rainwater. In this process, caves form from the top down, with the characteristic speleothems. However, this belief was challenged, especially after the explorations of enormous cave systems, such as the 120-mi (192-km) Lechuguilla Cave within the Guadalupe Mountains of New Mexico. This cave contained massive quantities of gypsum, many meters thick, which is highly soluble and easily washed away under previously understood cave-forming processes. These clues led investigators to suspect that these caves were formed from the bottom up, through the dissolution activity of sulfuric acid–rich ground water; the reaction of sulfuric acid with limestone leads to the precipitation of gypsum. When the chemical makeup of this gypsum was analyzed, the elements of sulfur were found to be isotopically light, a chemical marker of biological activity. While Lechuguilla Cave may have formed millions of years ago and any such activity is now extinct, cave explorers found other cave systems that were still actively growing. Within these cave systems, large numbers of microorganisms were using the chemical gradients of sulfur ions to produce the sulfuric acid that could be seen directly dissolving limestone rock. Over time, this could lead to cave formation in a dramatic display of microbial activity, with the formation of cave systems that extend over 100 mi (160 km) in length and contain some of the largest cave chambers on Earth.

Subterranean microbial life. The recognition of microbial species in subterranean environments has greatly expanded our knowledge of life on Earth, and has demonstrated that microbial species can make a living at the extremes of starvation, even as minimal as the hydrogen released by volcanic activity or radioactive decay. All the more surprising is the rate at which these low energy levels are converted into biomass for cell growth, with some estimates that subterranean species may divide only once every 30,000 years. Our understanding of cave geomicrobiology is increasing our knowledge of microbial activity on subterranean surfaces, moving from our one-time bias that microbial species would not be able to make a living in the dark, starved, and inhospitable environment of a cave to the recognition of intricate mechanisms of energy and nutrient acquisition that support growth—a growth that has profoundly impacted the structure, chemistry, and even aesthetic nature of the cave environment. Nonetheless, cave geomicrobiology is only just beginning to emerge as a new field, with many outstanding questions regarding the nature of microbial species in caves, their diversity, their metabolic activities, and even the mechanisms under which they evolve. The identification and characterization of these microbial species may have profound effects on human society, from issues related to global warming and drug discovery, to understanding how life could subsist in other subterranean environments on Earth and elsewhere.

For background information *see* ARCHAEBACTERIA; BACTERIA; CAVE; EUKARYOTAE; GYPSUM; LIMESTONE; MICROBIAL ECOLOGY; PROKARYOTAE; STALACTITES AND STALAGMITES in the McGraw-Hill Encyclopedia of Science & Technology.

Hazel A. Barton

Bibliography. H. A. Barton, Introduction to cave geomicrobiology: A review for the non-specialist, *J. Cave Karst Stud.*, 68:43–54, 2006; H. A. Barton and V. Jurado, What's up down there: Microbial diversity in starved cave environments, *Microbe*, 2:132–138, 2007; N. R. Pace, A molecular view of microbial diversity and the biosphere, *Science*, 276:734–740, 1997; A. N. Palmer, Origin and morphology of limestone caves, *Geol. Soc. Amer. Bull.*, 103:1–21, 1991.

Micromachined ultrasonic transducers

The name "micromachined ultrasonic transducer (MUT)" is used for devices that are made using microfabrication techniques to generate and detect ultrasound. Although this general term also applies to transducers made by shaping bulk piezoelectric materials using micromachining techniques, the more specific definition refers to devices with tiny membranes that are driven by either capacitive or piezoelectric actuation. The advent of both capacitive and piezoelectric micromachined ultrasonic transducers, cMUTs and pMUTs, respectively, has been made possible by advances in integrated circuit fabrication technology. This article focuses particularly on cMUTs.

Since the discovery of the piezoelectric effect by the brothers Pierre and Paul-Jacques Curie in 1880, piezoelectric crystals have been the materials of choice to build ultrasonic transducers. Although the idea of using an electrostatically actuated membrane as an ultrasonic transducer dates as far back as the early 1900s, a practical device was not developed until the 1990s. The eminent scientist Paul Langevin considered using the electrostatic transducer to detect submarines during World War I. He concluded that electric field strengths of the order of a million volts per centimeter (10^8 V/m or 100 V/μm) would need to be maintained in the device to generate electrostatic forces as large as a kilogram per square centimeter (or an atmosphere). At the time, such a device was impractical. Today, high electric fields can be generated by applying only a few tens of volts between a transducer's electrodes that are separated by a submicrometer gap, which is conveniently realizable by using microfabrication techniques.

Micromachined ultrasonic transducers offer many manufacturing and performance advantages. Batch manufacturing provides significant cost reduction compared to existing piezoelectric transducer fabrication technology, which relies on meticulous and labor-intensive steps. Tight parameter control in integrated circuit fabrication processes results in uniform quality. The microlithographic definition of membrane dimensions and the ability to control the thickness of the thin films in the device structure enable the realization of devices with a wide range of operating frequencies (10 kHz–100 MHz). Micromachined transducers can also be conveniently integrated with supporting electronic circuits on the same substrate or by flip-chip bonding. In immersion applications, micromachined ultrasonic transducers demonstrate sensitivities comparable to piezoelectric transducers, but over a much wider frequency spectrum.

Structure and operating principle. A cMUT can be built in any practical shape and size by connecting the basic building blocks (cells) in parallel. A cMUT cell is simply a thin membrane suspended over a thin gap (**Fig. 1**). An insulation layer can also be formed optionally on the bottom electrode. The membrane's shape, size, and thickness are the major factors that determine the frequency of operation. When a dc voltage is applied between the top electrode, usually located on top of the membrane, and the bottom electrode, the membrane is attracted toward the bottom electrode by the electrostatic force, and induced stress within the membrane balances the attraction. Driving the membrane with an ac voltage superimposed on the bias generates ultrasound. If the biased membrane is subjected to ultrasound, a current is produced due to the capacitance change under constant bias voltage.

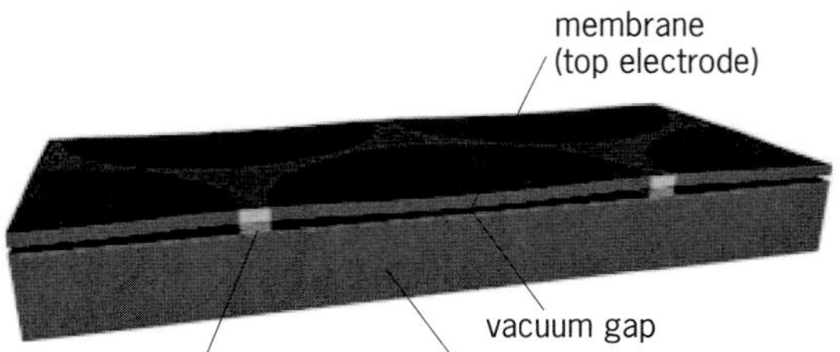

Fig. 1. Cross section of a cMUT.

Fabrication processes. Micromachining takes advantage of the fabrication techniques perfected by the integrated circuit industry in the last two decades. These well-developed techniques have helped to fabricate various micro-electro-mechanical systems (MEMS) including cMUTs.

The first generation of cMUTs was fabricated based on a sacrificial release method. In this approach, several layers of different materials are deposited and patterned on a substrate and then the sacrificial layer is etched to form a movable membrane.

A newer fabrication technique is based on a direct wafer bonding technique (**Fig. 2**). This technique is fascinating because it contradicts everyday experience. If two flat and smooth surfaces are brought into close enough proximity, the molecules from the two surfaces start to attract each other (due to van der Waals forces). This attraction enables the wafers to fuse together, without any intermediate "gluing" material. Galileo Galilei (1564–1642) predicted this effect—an adhesion between two plane and smooth surfaces—long ago. The first technical application of this technique can be found in the nineteenth century, when polished glass parts were bonded together for various optical applications (optical contacting). As a micromachining method, the direct wafer bonding technique has been used for about 2 decades.

For the realization of a cMUT structure in its simplest form (as shown in Fig. 1) using the wafer bonding technique, seven key fabrication steps are required: First, an electrically conductive silicon wafer is thermally oxidized to form an insulation layer (Fig. 2*a*). Then, standard optical lithography and etching methods are used to pattern this insulation layer (Fig. 2*b*). In a second thermal oxidation step, a thin oxide layer is grown at the bottom of each cMUT cell (Fig. 2*c*). Now this wafer can be brought in contact with another wafer to perform direct wafer bonding. Normally, so-called silicon-on-insulator (SOI) wafers are used, which consist of three different layers: a thin single-crystal silicon layer (0.5–30 μm) on top of a thin buried oxide layer and the bulk of the wafer, a thicker ($\sim$500 μm) silicon layer. Because air inside the gaps of the cMUT

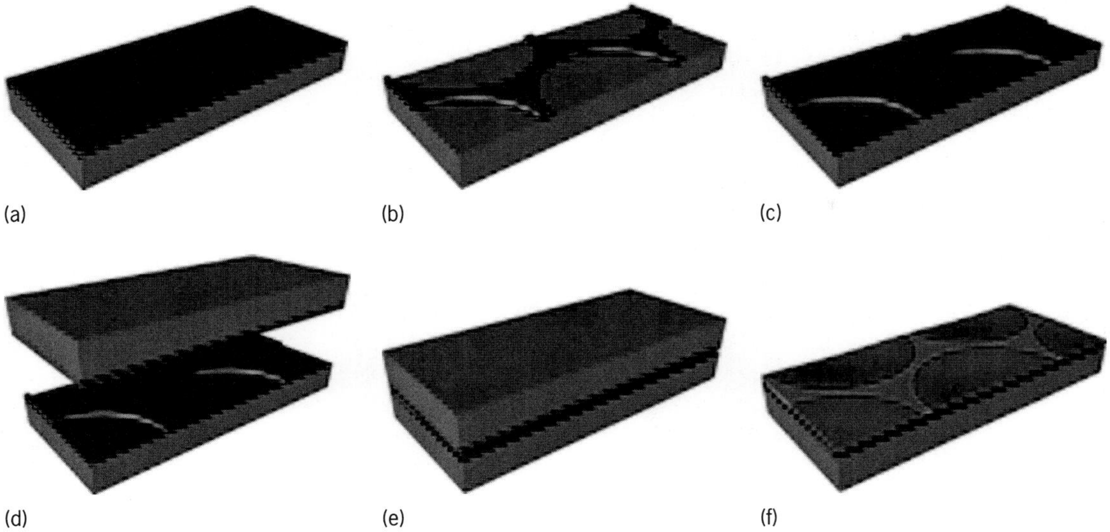

Fig. 2. **Fabrication process based on direct wafer bonding. (*a*) Thermal oxidation of an electrically conductive silicon wafer. (*b*) Standard lithography step and patterning of the silicon dioxide in a liquid etch step. (*c*) Second oxidation to grow insulation layer at the bottom of the cMUT cells. (*d*) Direct bonding to a silicon-on-insulator (SOI) wafer in a vacuum chamber. (*e*) Annealing of the bonded wafers to improve bond strength. (*f*) Removal of silicon and insulation layer from SOI wafer. Single-crystal silicon cMUT membrane on top of vacuum gaps remains.**

would increase damping and, therefore, limit the vibration amplitude of the membranes, the bonding is done in a vacuum environment (Fig. 2*d*). For stronger bonds the wafers are then annealed at temperatures exceeding 900°C (1650°F; Fig. 2*e*). After this thermal treatment, the SOI wafer layers, except the thin silicon layer, are removed via standard grinding and etching techniques (Fig. 2*f*). Therefore, the direct wafer bonding method for cMUT fabrication fulfils the main purpose of transferring the thin single-crystal silicon layer (which becomes the cMUT membrane) from the SOI wafer to the oxidized and patterned silicon wafer (which becomes the substrate and the insulation layer).

This fabrication process has several advantages over the sacrificial release method. It allows the fabrication of cMUT cells with high precision and reproducibility, with almost any cell shape, and with a wide range of membrane diameters and thicknesses.

Applications. Ultrasonic waves can be generated in liquids, gases, and solids. Since its invention, the main research focus for cMUT technology has been mostly on generating ultrasound in air and liquids, motivated by applications such as nondestructive testing and medical imaging.

Underwater imaging applications. Underwater acoustic imaging systems provide images of underwater objects when water turbidity precludes the use of optical means of viewing. The wide frequency range of operation of cMUTs makes them a suitable choice to implement both conventional low-frequency (for example, 10 kHz) sonar systems and high-frequency (for example, 500 kHz–3 MHz) imaging systems for use with crewless underwater vehicles or as handheld units for divers.

Medical imaging applications. Three applications in the field of medical imaging are expected to be greatly impacted by the introduction of cMUT technology. These are real-time three-dimensional (3D) imaging based on large two-dimensional (2D) transducer arrays; intracavital imaging using miniature devices; and real-time, high-resolution imaging using high-frequency arrays. Some of the devices demonstrated so far include a 128 × 128-element 2D array (**Fig.** 3*a*), a 2-mm-diameter (0.08-in.) forward-viewing intravascular imaging array (Fig. 3*b*), a 16 × 16-element 2D endoscopic imaging array with integrated front-end electronics (Fig. 3*c*), and 64- and 128-element 1D linear arrays with operating frequencies up to 70 MHz (Fig. 3*d*).

Gas-coupled applications. In recent years, there has been an increasing research interest in cMUTs for various gas-coupled ultrasound applications. Well-established state-of-the-art transducers, based on piezoelectric bulk-wave resonators, are limited in their applicability for gas-coupled ultrasound. Although the acoustic impedance mismatch between the gaseous media and the piezoelectric transducer, which results in high insertion loss and narrow bandwidth, can be significantly improved by using various matching layers, there are other requirements that need attention. There are limitations in terms of high-temperature environments and for low-gas-pressure conditions. CMUTs, especially those fabricated using wafer bonding, not only have the potential to overcome these limitations, but can also provide performance improvements with regard to a wide design space for frequency and bandwidth, transducer shapes, sizes and array types, wide dynamic range, and robustness. Some promising examples of air-coupled ultrasound applications of cMUTs are nondestructive evaluation with noncontact ultrasound, ultrasonic gas flowmeters, low-pressure anemometry, metrology for an integrated inertial guidance

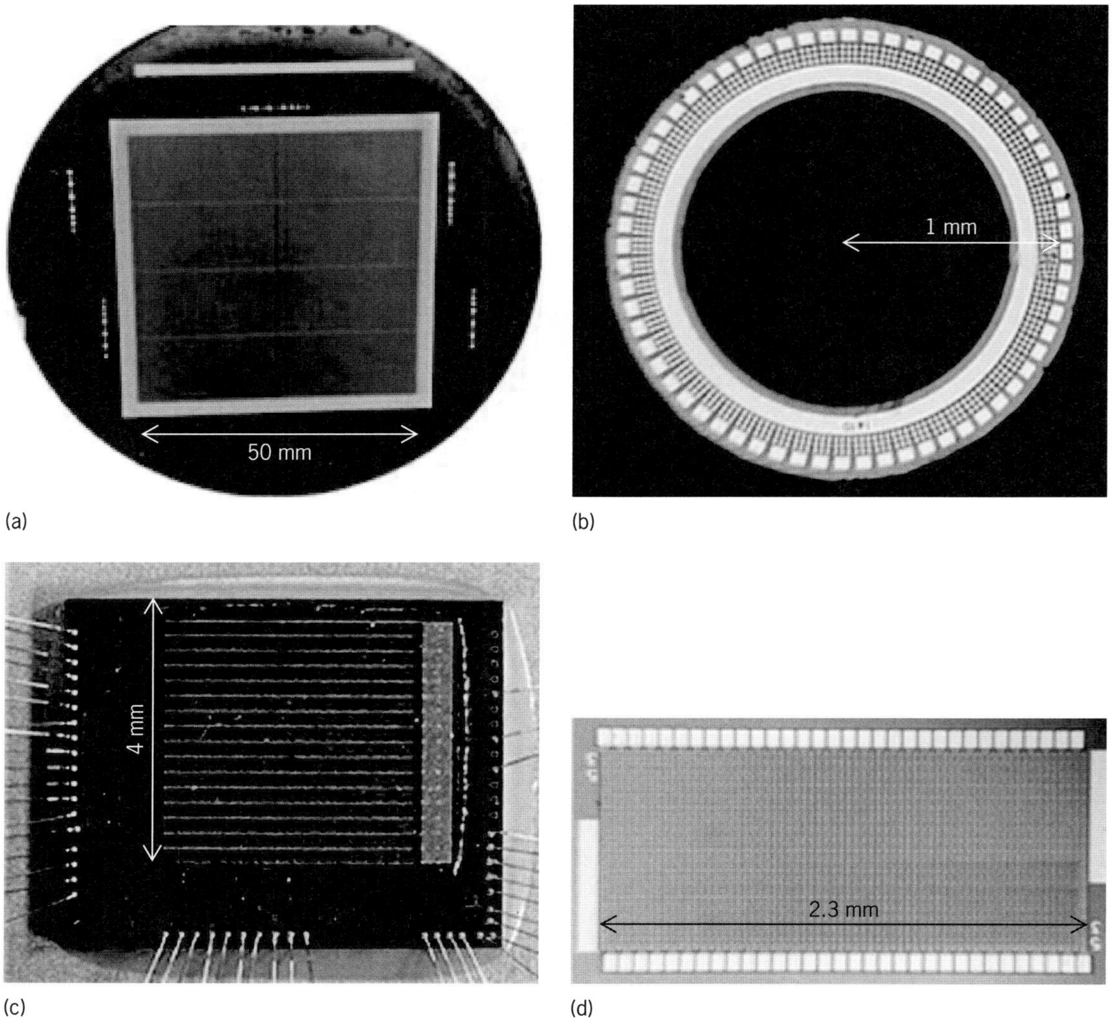

Fig. 3. Some cMUT arrays for medical imaging. (*a*) A 128 × 128-element 2D cMUT array for high-resolution 3D imaging. (*b*) A 64-element, forward-viewing cMUT ring array for 3D intravascular-intracardiac imaging. (*c*) A 16 × 16-element 2D cMUT array integrated with front-end electronic circuits for endoscopic 3D imaging. (*d*) A 64-element, high-frequency, 1D array for high-resolution imaging.

system, and chemical-biological sensor applications. *See* AIR-COUPLED ULTRASONIC TESTING.

Solid-coupled applications. The third group of applications is focused on generating ultrasonic waves in solids. One example, where cMUTs already have been shown to be suitable, is the generation of flexural plate waves, also known as Lamb waves, in thin solid plates. These Lamb waves can be used for sensor and acoustic filter applications. Another example, which is currently under investigation, is launching bulk waves in solid rods, using the electrostatic transducer principle. This will be the electrostatic equivalent to the well-known piezoelectric-based Langevin or sandwich transducer.

In summary, cMUT technology based on the well-established integrated circuit manufacturing techniques can potentially enable many applications that have not been possible, mostly due to limitations in the existing transducer technologies based on piezoelectric materials.

For background information *see* BIOMEDICAL ULTRASONICS; ELECTRONIC PACKAGING; INTEGRATED CIRCUITS; MEDICAL IMAGING; MEDICAL ULTRASONIC TOMOGRAPHY; MICRO-ELECTRO-MECHANICAL SYSTEMS (MEMS); MICROPHONE; MICROLITHOGRAPHY; NONDESTRUCTIVE EVALUATION; PIEZOELECTRICITY; TRANSDUCER; ULTRASONICS in the McGraw-Hill Encyclopedia of Science & Technology.

Butrus T. Khuri-Yakub; Ömer Oralkan; Mario Kupnik

Bibliography. A. S. Ergun et al., Capacitive micromachined ultrasonic transducers: Fabrication technology, *IEEE Trans. Ultrason. Ferroelec. Freq. Control*, 52:2242–2258, 2005; A. S. Ergun, G. G. Yaralioglu, and B. T. Khuri-Yakub, Capacitive micromachined ultrasonic transducers: Theory and technology, *J. Aerospace Eng.*, 16:76–84, 2003; Ö. Oralkan et al., Capacitive micromachined ultrasonic transducers: Next-generation arrays for acoustic imaging?, *IEEE Trans. Ultrasonics Ferroelec. Freq. Control*, 49:1596–1610, 2002.

Model-based systems

Engineers who design systems using text specification documents focus their work upon the completed system to meet performance, time, and budget goals. Consistency and integrity are difficult to maintain within text documents for a single complex system, and they are more difficult to maintain as several systems are combined into higher-level systems, especially when these systems are maintained over decades, and evolve technically and in performance through updates. This system design approach frequently results in major changes during the system integration and test phase, and in time and budget overruns.

Engineers who build systems using system specification documents within a model-based systems environment go a step further and aggregate all of the data. They interrelate all of the data to ensure consistency and integrity. After the model is constructed, the various system specification documents are prepared, all from the same database. The consistency and integrity of the model are ensured, and therefore the consistency and integrity of the various specification documents are ensured.

This article attempts to define the term "model-based systems" relative to such an environment. The intent is to expose the complexity of the enabling problem by outlining "what" is needed, "why" it is needed, and "how" needs are being addressed by international standards writing teams.

Technical terminology. Many preschool children encounter their first systems engineering problem in a story about "The Three Little Pigs." The problem is to build a house. Immediately, a cognitive model of a single house forms in the child's mind. The available construction materials are straw, sticks, or bricks. This single cognitive model then splits into three alternative solution models. Next the potential of a destructive wind is introduced with a house-survival and inhabitant-safety requirement. Additionally, the three stakeholders have different degrees of construction cost tolerance and acceptable risk. The models now become clouded with uncertainty. Convergence to a design solution requires learning to gain insight and understanding. A decision-making meeting takes place to evaluate proposals. Each stakeholder attempts to convey his or her solution model to the others. They differ and none communicate well enough for learning and understanding to drive the respective cognitive models to consensus for decision making. They leave, proceed to build, and end with two failures and one success.

A model-based systems environment would have computer-aided models for construction cost-and-risk analysis, complemented by small-scale physical or equivalent computer models for alternative house design investigations. These models would be executed to evaluate the emergent behaviors of different house designs for a range of building material options, expected wind conditions, construction costs, and acceptable risks. The model-based systems environment would allow designers to be creative as they propose and evaluate design alternatives, database results, and recall and reuse archived work. Design documentation would be generated from the commonly used database for communication, mutual learning, and peer review.

The failure to model leads to a weakened ability to develop the predictive capability called insight and understanding. This in turn weakens the ability to communicate. Weak communication weakens learning, and this in turn weakens consensus building.

Systems engineering context. In this generic systems engineering context, cognitive and scaled physical models are foundational to complex systems communication, understanding, and decision making. Only since Isaac Newton's contributions have engineers communicated their cognitive models using the language of applied mathematics, and only in the last half-century could their equations be represented for digital solution in computer-sensible languages, that is, languages that computers can read, process, and respond to.

The phrase "model-based systems" is used freely across domains of study and levels of abstraction. It conveys the direct message that the realized system is based upon a process of modeling, and the implicit message that modeling is mathematical, with computer methods providing digital solutions and system data management.

Within the engineering community, attempts to realize visions of model-based systems of engineering have been ongoing since the 1970s with varying degrees of success. Most attempts encountered problems associated with information interfacing issues. By the end of the 1970s it became obvious that resolution required internationally approved standards.

On July 11, 1984, the First International Standards Organization (ISO) TC 184/SC4 meeting was held at the National Institute of Standards and Technology. Participating countries included Canada, France, Germany, Switzerland, the United Kingdom, and the United States. The purpose of this meeting was to create an international standard that enabled the capture of information comprising a computerized product model in a neutral form without the loss of completeness and integrity, throughout the life cycle of a product. Today, the breadth, depth, and maturity levels of these and related efforts have the potential of enabling model-based systems engineering environments to meet expanding user needs.

Communication of views. Common to all model-based system views is the need to communicate decomposition, modeling, and inference-based decision.

Decomposition. Decomposition is the art of viewing a quantity in terms of a combination of simpler quantities. Here, "art" refers to the ability to select simplifications that are easily learned and communicated, while also providing predictive insights that are robust relative to modest change and uncertainty. Each of the components must also be checkable and have well-defined interfaces, so that a model of the whole can be assembled with emergent behavior that can be detected and verified.

Modeling. Modeling is the process of simplification designed to clearly show the basic structure or workings of an object, system, or concept. A model is a representation of a thing, an idea, or a reality that has an execution engine to query the model. The human mind, an electronic computer, a physical test facility, and so forth, may accomplish the query operation. Results are reproducible and can be verified. They can be presented for communication and documentation in a text language, a graphic language such as a schematic diagram, a computer-generated display, and so forth.

Inference-based decision. Inference-based decisions and insights are derived by the process of making logical judgments based on well-known relationships and model results. These decisions must be verified by testing, with robustness measured to quantify the sensitivity of predictions to change. The measures provide justification for reuse-redo decisions, and form the basis for change messaging directed only to groups doing work that may need to initiate a change response-action.

Underlying model execution and results collection is the need to join and interface the data that flow between models. Even for simple systems, it is a major effort to ensure information consistency as data flow through the networked set of executable models. Complicating the problem further is the absolute need to maintain the consistency of information shared by all engineers and managers across all system life-cycle stages.

Support environment. Irrespective of application, the support environment for model-based systems must be learnable, scalable, and checkable. With the need to interface an ever-expanding set of capabilities, users encounter issues related to learning and results checking. Creators and maintainers additionally encounter issues related to scaling.

Scaling is fundamentally an interface issue. If interfaces are difficult to work with, new capability cannot be easily added, enabled, and checked. Furthermore, unexpected emergent behaviors of the whole can occur; these too must be detectable and checkable. If interfacing difficulties are allowed to perpetuate, unwanted emergent behavior is guaranteed to grow and become progressively more difficult to fix. The ability to scale further ends. Parallel to this issue is the need to also maintain database information integrity, consistency, metadata, and traceability as usage expands.

Semantic foundations. Resolution demands a computer-sensible semantic foundation based on international standards. This is needed at the environment specification level; model representation, exchange, and sharing level; model management level; and model-usage level.

Environment specification. At the environment specification level, a semantic foundation for model-based systems engineering exists. It was created by teams of experts in writing standards from the ISO, the Object Management Group (OMG), and the International Council on Systems Engineering (INCOSE). It collects and defines the core functional elements for any systems engineering process. The work provides the basis agreed to by the ISO, the OMG, and the INCOSE for work on AP233 (ISO 10303-233: Application protocol: System engineering and design) and SysML (System Modeling Language).

Within the context of systems engineering, a "system" must exhibit observable and reproducible properties and have a boundary that separates it from all other things in the environment. It is essential to know what is inside the system and what is outside of it.

Within the same context, the term "system of systems" can be defined, and in like manner "model of models." Realizations of both carry requirements related to system or model scalability. These critically important issues relate to interface control, modularity, openness, commonality, and rapid upgrade and certification of all component systems or models. These issues are nontrivial to resolve at both the engineering level and the socioeconomic level, where suppliers are protecting market share while customers are seeking a more competitive supplier base.

Representation, exchange, and sharing. At the model representation, exchange, and sharing level, a semantic foundation is provided by the STEP (Standards for the representation and Exchange of Product data), ISO 10303.

At the highest abstraction level, STEP standards consist of application protocols (APs) that service focused domains of study for which software vendors supply competitive products. Each application protocol collects, in a computer-sensible manner, the foundational concepts of a particular domain along with their associated attributes and relationships. Information is made computer sensible by use of the EXPRESS data modeling language. The EXPRESS models of the application protocols define neutral read-write data formats for the representation, exchange, and sharing of product data information. Software user groups never see or interact with this deep detail; however, they use the STEP-enabled, standards-based information interfacing capabilities.

Since these neutral formats are standards-based, they provide a time-stable information representation format consistent with the exchange needs of multiorganizational collaboration, long-term data archival, and reuse. Neutral formats also provide an access route from outdated software to new versions or best-of-breed capabilities.

Management. At the model-management level there is the need to configuration manage models and model data associated with system analysis, data associated with system design, and version and traceability relationships between design and analysis. The STEP standards have a "product data management" capability that services these needs. This capability includes the system metadata (who, when, where—what, why, how), versioning, context, authorizations, relationships, and so forth.

Usage (semantics). At the model-usage level there is a need for semantic standards to communicate concepts and system characterization data. At this level of abstraction, professional societies and organizational groups establish relevant semantic standards.

Currently there is a gap between what is actually made and what can be made computer sensible. Many groups are addressing the associated problem through ontology technology via the World Wide Web Consortium's (W3C's) Web Ontology Language (OWL) standards.

OWL can actually be used to explicitly represent the meaning of terms in vocabularies and the relationships between those terms. However, it is hard to find references to major projects with a model-based systems engineering environment at a capability and maturity level where relationships are captured and used for model-based reasoning. The use of OWL for taxonomy capture today opens the door for more capability tomorrow. Examples of ongoing work supporting a managed collection of process-plant life-cycle data are the EPISTLE Reference Data Library and the controlled natural language Gellish.

Enabling an integrated collection of interoperable models with well-defined interfaces to build a need-based model of models is a nontrivial task. This is basically an information interface control problem. It can make very effective use of the STEP standards and advances in ontology technology.

Usage (graphical presentation). Another consideration at the model-usage level is model data preparation and presentation. SysML is a graphical modeling language for representing systems and product architectures, as well as their behavior and functionalities. It builds on the experience gained in the software engineering discipline of building software architectures in UML (Unified Modeling Language). SysML has the potential to provide a major contribution to the learnable-checkable problem. It could provide a high degree of consistency across tools in the input data preparation process and the output presentation process, and hence in the process of peer review of work in progress—by different groups using different SysML-based tool sets. Market forces will eventually dictate the degree to which this vision is realized.

For background information *see* COMPUTER-BASED SYSTEMS; MODEL THEORY; MODELING LANGUAGES; SYSTEM FAMILIES; SYSTEMS ARCHITECTURE; SYSTEMS ENGINEERING; SYSTEMS INTEGRATION in the McGraw-Hill Encyclopedia of Science & Technology.

Harold P. Frisch

Bibliography. J. F. Andary, H. P. Frisch, and D. W. Oliver, Model based systems engineering, in A. P. Sage and W. B. Rouse (eds.), *Handbook of Systems Engineering and Management*, Wiley, 2008; M. P. Gallaher, A. C. O'Connor, and T. Phelps, *Economic Impact Assessment of the International Standard for the Exchange of Product Data (STEP) in Transportation Equipment Industries, Final Report*, Planning Rep. 02-5, NIST, 2002; S. J. Kemmerer (ed.), *STEP: The Grand Experience*, NIST Spec. Publ. 939, 1999; J. T. Pollock and R. Hodgson, *Adaptive Information*, Wiley, 2004.

Molecular modeling for drug design

Molecular modeling is a collective term that refers to theoretical methods and computational techniques to model and mimic the behavior of molecules, such as visualizing two- and three-dimensional (2D and 3D) structures, organizing compounds and their properties into databases, providing tools for analyzing molecular properties, or simulating the behavior of molecules on an atomic level. Drug design is an iterative process that begins when a compound is identified to display an interesting biological profile and ends when its activity profile and the chemical synthesis are optimized. Today, molecular modeling permeates all aspects of drug design. Scientists have used computer models of new chemical entities to help define activity profiles, geometries, and reactivities.

The vast majority of drugs are small molecules designed to bind, interact, and modulate the activity of specific biological receptors. Receptors are often proteins that bind and interact with other molecules to perform the numerous functions required for the maintenance of life. Because of genetic abnormalities, physiologic stressors, or some combination thereof, the function of specific receptors may become altered and alterations manifested as minor physical symptoms, as in the case of a runny nose due to allergies, or as life-threatening and debilitating events, such as sepsis or depression. The role of drugs is to correct the functioning of these receptors.

One of the basic tenets of drug design is that the biological activity of a drug is dependent on the 3D placement of its specific functional groups. As early as 100 years ago, Emil Fischer recognized the lock-and-key principle, whereby a drug and its receptor are complementary both spatially and chemically. The application of molecular modeling for drug design is split between those methods dealing with biological systems where no structural information is known at the atomic level (the unknown receptor) and those systems where a 3D structure is known from x-ray crystallography or nuclear magnetic resonance (NMR) spectroscopy. The Garland R. Marshall group at Washington University has focused most of its efforts over the last 30 years on the common problem encountered where one has little structural information. The objective is to derive an abstract model called pharmacophore, which was first defined by Paul Ehrlich in 1909. A pharmacophore is a set of structural features common to all active molecules that exhibit a particular biological activity. Once a pharmacophore has been developed, it can be used to find or suggest other active molecules. This approach to drug design is often referred to as ligand-based drug design. Others, such as Peter Goodford and Tak Kuntz, have taken the lead in developing approaches to therapeutic targets where the structure of the target was available at atomic resolution. The seminal work of Goodford and colleagues on designing inhibitors of the 2,3-diphosphophorylglycerate (DPG) binding site on hemoglobin for treating sickle-cell disease stimulated many others to obtain crystal structures of their therapeutic target. This approach to drug design is often referred to as receptor-based drug design. Advances in molecular biology have provided the means of cloning and expressing proteins in sufficient quantities to screen a variety of conditions for crystallization. Thus, it is almost expected that a crystal structure is available for any therapeutic target of interest. Unfortunately, many therapeutic targets, such as G-protein–coupled receptors that are the target of around half of all modern medicinal drugs, are still significant challenges to structural biology.

Molecular modeling methods are now widely used in the pharmaceutical industry, based on the idea that you can see exactly how your molecule interacts with its receptor target. Molecular modeling has contributed greatly to help identify compounds that would be expected to interact favorably with the receptor, or compounds that contain the required functional groups of the pharmacophore. Once such a compound has been identified, a program of chemical modification is undertaken to enhance its potency and selectivity, to ensure that neither it nor any of its metabolites is toxic, and to provide appropriate transport characteristics to enable it to pass through cell membranes and reach its target. Molecular modeling techniques play a significant role in suggesting what modifications to make and in understanding the experimental biological profiles. One of the most dramatic examples of computer-aided drug design is the development of superoxide dismutase mimetics by Dennis Riley. By understanding the redox chemistry of manganese superoxide reductase, Riley was able to design a novel pentaazacrowns scaffold complexed with manganese that catalyzes the conversion of superoxide to hydrogen peroxide at diffusion-controlled rates. This is the first example of a synthetic enzyme with a catalytic rate equal to or better than nature's.

Ligand-based drug design. Molecular modeling techniques can help us to deduce a pharmacophore. Typical features are hydrogen-bond donors and acceptors, positively and negatively charged groups, and hydrophobic groups. A 3D pharmacophore specifies the spatial relationships between these groups. These relationships are often expressed as distances or distance ranges but may also include other geometrical measures such as angles and planes. The 3D pharmacophore may also contain features that are designed to mimic the presence of the receptor. These are commonly represented as exclusion spheres, which indicate locations within the pharmacophore where no part of a ligand is permitted to be positioned. Some of these features are shown in **Fig. 1**. Once a pharmacophore is extracted from a set of ligands, it can be used as a model for the design of other molecules that can accomplish the same activity.

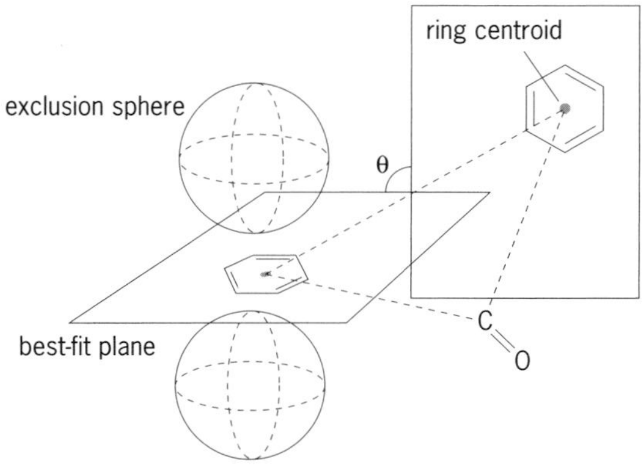

Fig. 1. Features that can be incorporated into pharmacophores.

By 1979, the Marshall group had developed a systematic computational approach for generating pharmacophore hypotheses, the active-analog approach. The basic premise was that each compound tested against a biological assay was a 3D question for the receptor. Each molecule was, in general, flexible and could present any number of possible 3D arrays of interactive chemical groups. By computationally analyzing the sets of possible pharmacophoric patterns associated with each active molecule, one could find those pharmacophoric patterns common to a set of actives. At first sight, it would appear that a systematic search over 20–30 molecules would greatly magnify the combinatorial explosion associated with a systematic conformational analysis. However, one can significantly reduce the scale of the problem by using information about molecules whose conformational space has already been considered. Thus, it is only necessary to consider those conformations that would enable the current molecule's pharmacophoric groups to be positioned in the same locations that have already been found for previous molecules. Marshall and colleagues showed that it is possible to determine what torsion angles of the rotatable bonds will enable conformations consistent with the previous results obtained. It is best to choose the most conformationally restricted molecules first, as these will have a reduced conformational space.

A well-known example is the pharmacophore for the angiotension-converting enzyme (ACE), which is involved in regulating blood pressure. Four typical ACE inhibitors are shown in **Fig. 2**, including captopril, which is widely used to treat hypertension. Three groups within this class of inhibitors (for example, captopril) are required for activity: a terminal carboxyl group, an amido carbonyl group, and a zinc-binding group. The problem is to determine conformations in which the inhibitors can position these three pharmacophoric groups in the same relative positions in space. To derive an ACE pharmacophore, four points were defined for each molecule. The derivation of these four points for captopril is shown in Fig. 2. Note that one of the points corresponds to the presumed location of the enzyme's zinc atom. Five distances were defined between the four points. The number of rotatable bonds in each molecule varied between 3 and 9 and the molecules were considered in order of increasing number of rotatable bonds. The entire conformational space was explored for the first most inflexible molecule. For each conformation, a point was registered in a five-dimensional hyperspace that corresponds to that particular combination of the five distances. When the second molecule was considered, only those torsion angles that would enable these distances to be achieved were permitted to the rotatable bonds. As more molecules were examined, the common region in the five-dimensional hyperspace was reduced. Marshall and colleagues analyzed 28 ACE inhibitors of diverse chemical structure available by 1987 and 2 inactive compounds with appropriate chemical functionality. Based on these data, a unique conformation for the core portion of each molecule interacting with a hypothetical ACE active site was deduced. The two inactive compounds were geometrically incapable of appropriate interaction. This pharmacophore model was the basis for the discovery of a large family of ACE inhibitors. These safe, effective, and orally available anti-hypertensive drugs have been used for the past 2 decades.

Inhibitors of the angiotension-converting enzyme served as a test case for the active-analog approach in which one tries to deduce the receptor-bound conformation of a series of active analogs, based on the assumption of a common binding site. The crystal structure of the complex of lisinopril with ACE was finally determined in 2003. The common backbone conformation of ACE inhibitors and the location of the zinc atom, hydrogen-bond donor, and cationic site of the enzyme determined by the active-analog approach essentially overlap that seen in the crystal structure of the complex. At least for this case, the assumption regarding the relative stability of groups in catalysis or recognition is valid.

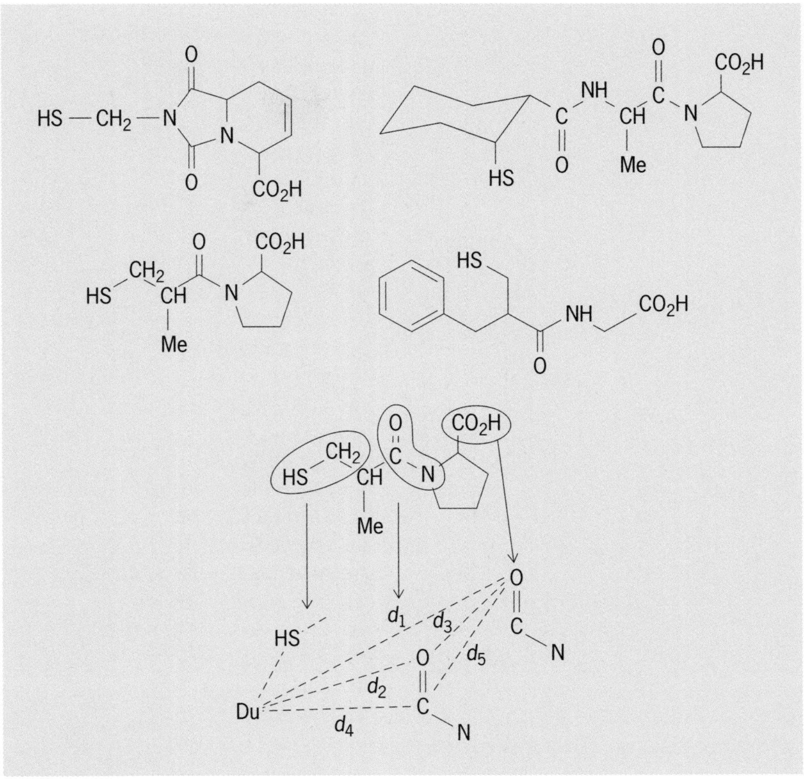

Fig. 2. A pharmacophore based on inhibitors of the angiotension-converting enzyme.

Receptor-based drug design. As determinations of crystal structures have become more commonplace, efforts to design ligands to compliment a cavity on a molecular surface have become more sophisticated. This approach is usually implemented by searching a database that may contain an extremely large number of compounds. Certainly, the pioneering effects of Goodford and colleagues to design compounds binding to the DPG site on hemoglobin were dominated by chemical intuition, physical models, and very primitive modeling systems. The development of molecular-modeling software DOCK by I. Kuntz was a major innovation, and other software (AutoDock, DREAM, and so forth) are readily available over the Internet for exploring possible complex formation between a possible ligand molecule and a target receptor. Goodford has developed the use of probe atoms and chemical groups in a grid to map the binding site and identify optimal binding subsites. This led to experimentally determining subsite binding with the subsequent assembly of fragments by crystallography or NMR. The Washington University group led by Chris Ho has developed a number of receptor-based design tools. Recently, Ho developed a new-generation software package, RACHEL, that has been commercialized by Tripos. The ability to include synthetic feasibility and generate candidates with druglike properties has become a dominant theme in drug design.

An impressive example of the application of receptor-based drug design methods was the design of an inhibitor of the human immunodeficiency virus (HIV) protease by a group of scientists at DuPont Merck in 1994. This enzyme is crucial to the replication of the HIV (acquired immune deficiency syndrome) virus and so inhibitors may have therapeutic value as anti-AIDS treatments. The starting point was a series of x-ray crystal structures of the enzyme bound with a number of inhibitors. Their objective was to discover potent, novel compounds that were orally available. Many of the previously reported inhibitors of this enzyme possessed substantial peptide character and were biologically unstable, poorly absorbed, and rapidly metabolized. The x-ray structures of the HIV protease revealed several key features that were subsequently incorporated into the designed inhibitor. The enzyme is a dimer with C_2 symmetry. It is a member of the aspartyl protease family with the two active-site aspartate residues lying at the bottom of the active site. There is one particularly important feature that is routinely observed in crystal structures of HIV protease with inhibitors bound in the active site. This is the presence of a tightly bound water molecule near the top of the active site. It accepts two hydrogen bonds from the backbone amide hydrogens and donates two hydrogen bonds to the carbonyl oxygens of the inhibitor.

A flow chart showing the various phases leading to the final compound is reproduced in **Fig. 3**. The first step was a 3D search of a subset of the Cambridge Structural Database. The pharmacophore for this search is composed of two hydrophobic groups and a hydrogen-bond donor or acceptor. The hydrophobic groups were intended to bind in two hydrophobic

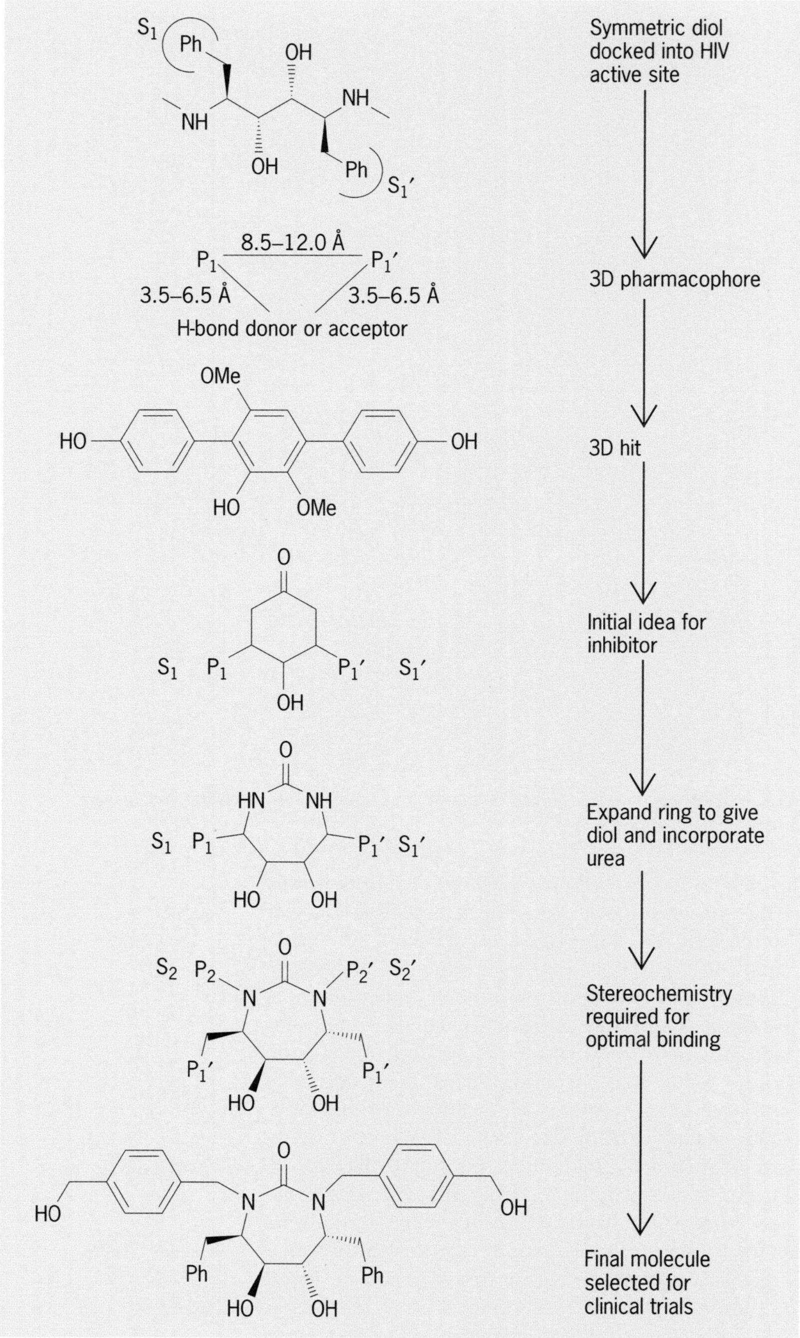

Fig. 3. Flow chart showing the design of novel orally active HIV-1 protease inhibitor.

pockets and the hydrogen-bond donor or acceptor to bind to the catalytic aspartate residues. The search yielded the hit shown in the figure. This molecule not only contained the desired elements of the pharmacophore, but it also had an oxygen atom that could displace the bound water molecule. Displacement of the water was expected to be energetically favorable due to the increase in entropy. The benzene ring in the original compound was changed to a cyclohexanone, allowing the substituents to be positioned in a more appropriate orientation. The DuPont Merck group had previously explored a series of peptide-based diols that were potent inhibitors but with poor oral bioavailability. They were keen to retain the diol functionality, so the next step was to expand the ring to a seven-membered diol. The ketone was then changed to a cyclic urea, to strengthen the hydrogen bonds to the flaps and to aid the synthesis. Additional modeling studies based upon the x-ray structure were done to predict the optimal stereochemistry and the conformation required for optimal interaction with the enzyme. The results of these studies showed that the 4*R*, 5*S*, 6*S*, 7*R* configuration was most appropriate. Nitrogen substituents were predicted to bind the S_2 and S_2' pockets of the enzyme, so various analogs were synthesized to enhance the potency while maintaining the desired pharmacological properties.

Prediction of affinity. In order to prioritize synthesis and testing of a compound, an estimate of the binding affinity to compare with the synthetic efforts is of practical use. Molecular modeling has contributed greatly to predict binding affinity, starting from the 3D structure of a ligand-receptor complex. Unfortunately, even for systems where the crystal structure of a complex of the compound with the receptor is already available, accurate prediction of affinity de novo is still challenging. The holy grail of rational drug design is to accurately predict the biological activity of a compound.

Outlook. Molecular modeling is no longer merely a promising technique in drug design. It is a practical and realistic way of helping the medicinal chemist. On its own, it is unlikely to lead to pharmaceutical novelties, but it has become a significant tool, an aid to thought, and a guide to synthesis. Drugs still must be synthesized and tested, but molecular modeling can contribute a clear molecular rationale and above all spur the imagination.

For background information *see* ACQUIRED IMMUNE DEFICIENCY SYNDROME (AIDS); ANGIOTENSIN RECEPTOR BLOCKERS; CHEMICAL BONDING; COMPUTATIONAL CHEMISTRY; CONFORMATIONAL ANALYSIS; LIGAND; MOLECULAR BIOLOGY; MOLECULAR MECHANICS; MOLECULAR SIMULATION; PROTEIN; STEREOCHEMISTRY; X-RAY CRYSTALLOGRAPHY in the McGraw-Hill Encyclopedia of Science & Technology.

Ye Che; Zhexin Xiang

Bibliography. P. J. Goodford, Drug design by the method of receptor fit, *J. Med. Chem.*, 27:557–564, 1984; I. D. Kuntz et al., Geometric approach to macromolecule-ligand interactions, *J. Mol. Biol.*, 161:269–288, 1982; A. R. Leach, *Molecular Modeling: Principles and Applications*, 2d ed., 2001; G. R. Marshall et al., The conformational parameter in drug design: The active analog approach, *Computer-Assisted Drug Design*, pp. 205–226, 1979; T. I. Oprea et al., *Cheminformatics in Drug Design*, 2005.

Multihull ships

Multihull ships consist of multiple hulls performing essentially the same flotation function as that of a conventional single-hull ship (monohull). The most familiar, and the oldest example, of a multihull vessel is a Pacific islander outrigger canoe (or proa). When

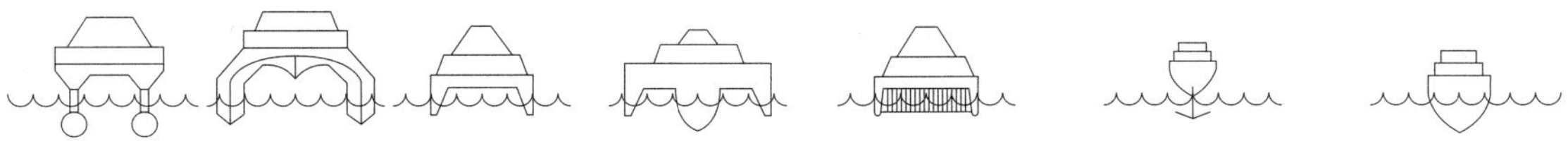

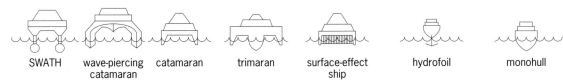

Fig. 1. Typical hull-form cross sections for various types of ships.

compared to a conventional western monohull canoe, the additional hull or hulls on the outrigger primarily provides additional stability in ocean waves.

Supporting lift. In general, ship multihull geometries can be categorized by how their hulls are used to support the ship, either by buoyant lift, dynamic lift, or powered lift—or combinations of these (**Fig. 1**). The monohull uses hydrostatic lift, relying on buoyancy to generate the supporting lift force. The hydrofoil uses dynamic lift, using the speed of the wing-shaped foil in the water to generate the lift force. Planing monohulls use high-powered propulsion to go at speeds fast enough to create a dynamic lift force at the surface of the water. Multihull ships use combinations of these effects to provide the supporting lift force. Ships that use multiple submerged buoyant lift bodies for support include small-waterplane-area twin-hull (SWATH) and Sea SLICE (sometimes spelled SeaSlice) designs. Common displacement multihull ships that use buoyant (hydrostatic) lift methods include catamarans and trimarans, though quadramarans have also been developed. A modification to the buoyant lift catamaran is the wave-piercing catamaran, which uses the two hulls in a long, slender shape to pierce through waves. Multihull ships that make use of powered lift are surface-effect ships (SES), which use a captured bubble of air between the two side hulls to reduce resistance. A hybrid buoyant-powered lift catamaran hullform, basically a catamaran-SES hybrid, has been developed called the Harley hullform. A hybrid hydrofoil-assisted catamaran, with a wing-shaped body extending into the water from the middle of the hullform between the side hulls, has also been developed and tested by the U.S. Navy. Multihull designs typically have proprietary and closely held geometry and performance characteristics that are not available in the public domain, so detailed characterizations are difficult to obtain and estimate.

Geometry. Catamarans and multihulls, other than trimarans, generally consist of equal-sized hulls. A trimaran typically has a larger center hull and two equal sized outer hulls, placed symmetrically about a larger center hull in the transverse direction, but at various points along the longitudinal direction, depending upon design. The SWATH ship typically has two large submerged hulls placed symmetrically below the water, with hydrodynamically efficient shaped support struts connecting the submerged hulls to the main hull. The Sea SLICE design has four underwater bodies instead of two. The submerged hulls provide hydrostatic support, while the struts have a reduced waterplane area presented to the free surface of the water, reducing residual (wave-making) resistance. The wave-piercing catamaran uses typical catamaran geometry, but shapes the two hulls such that they pierce waves, rather than riding over them, during operation at high speeds, reducing residual resistance.

Functions. Multihull ship functions run the breadth of ship uses, from small sailboats, such as the Hobie Cats, to larger multihull sailboats built for the 1988 America's Cup and The Race (an around-the-world catamaran race that started on December 31, 2000) competitions, all the way to motor-powered high-speed cargo transport and military applications. Many powered vessels are being designed and operated by commercial and military organizations that take advantage of many different characteristics of

Fig. 2. The High Speed Vessel (HSV-2) *Swift*, a catamaran chartered by the U.S. Navy Military Sealift Command (MSC).

Fig. 3. United Kingdom research vessel *RV Triton*, a trimaran. (© *2007, SPG Media Limited*)

Fig. 4. The research vessel *R/V Kilo Moana*, a SWATH ship. (*School of Ocean and Earth Science and Technology, University of Hawaii*)

multihulls. High-speed and luxury super catamarans, those over 60 ft (18 m) in length, are becoming common, as well. Ferries use the large deck area for storage and transportation of passengers and vehicles. High-speed ferries further take advantage of improved resistance characteristics and stability at high speed to reduce transportation times. Military vessels have begun using multihulls, particularly catamarans, such as the U.S. Navy Military Sealift Command (MSC) chartered High Speed Vessel (HSV-2) *Swift* (**Fig. 2**), the Joint High Speed Vessel (JHSV), the Littoral Combat Ship Experimental [LCS(X)] Seafighter, and trimarans, such as the United Kingdom research vessel *RV Triton* (**Fig. 3**) and a U.S. Navy Littoral Combat Ship (LCS) design variant, for their high speed, lower power operational capability, as well as their large, open deck areas used for military equipment and operations. Research vessels (R/V; **Fig. 4**) and some military surveillance ships use multihulls for their exceptional stability for long voyages or for required stability for their sensors and instrumentation. One special U.S. Navy experimental ship was developed as a unique SWATH design with a very low radar signature, the *Sea Shadow* (**Fig. 5**).

Fig. 5. *Sea Shadow*, a U.S. Navy SWATH ship (*U.S. Navy*)

Ship resistance. Multihull ships share displacement among the hulls, which leads to a wetted surface that is generally larger than a comparable displacement monohull. This increases the frictional resistance coefficient, which adds a component that requires more power for equivalent low-speed operations. The slender hullform of the multiple hulls reduces the residual (sometimes termed residuary) resistance coefficient, which is related to wave-making, wave-breaking, and eddy-current losses, therefore reducing the ship resistance at higher speeds. Ship slenderness is characterized by length over volumetric displacement ($L/V^{1/3}$). Since displacement is shared among hulls, the draft is also smaller. Overall, the combination of frictional and residual resistance is reduced through design using a high slenderness, which results in high length-to-hull-depth (L/D) ratio and high beam-to-draft (B/T) ratio, which in turn leads to the multihull ship being able to make higher speeds with less power than an equivalent monohull design. In addition, the arrangement of the hulls can be adjusted to account for wave interference effects, allowing for further manipulation of resistance characteristics during the design process. For SWATH designs, the higher frictional resistance component of the submerged hulls is compensated for by the reduced residual resistance of the small waterplane area of the strut assembly, which reduces wave-making resistance. The trade-off is that SWATH ships are not used for high-speed operations, but for their improved stability characteristics.

Sea keeping. The multiple hulls that comprise the multihull ship are each generally long and slender. Overall, the ships have a wide effective beam for the overall ship as a manifestation of having to connect multiple hulls into a single unit. Both transverse and longitudinal stability are generally very good for multihull ships because of this property. Multiple hulls make up for lack of transverse stability of a typical high-L/D-ratio monohull form by providing a much larger effective beam across the hulls. Sailing multihull ships use the wide beam width to support transverse moments from wind and waves, whereas conventional monohull sailboats use keels and ballast/appendages to provide additional transverse stability. These multihull sailing ships have high stability, especially in the transverse direction, and are unlikely to capsize, but can pitch-pole (tumble end-over-end) in storms or at high speed in high waves. As noted above, high-performance sailing catamarans have been used in the 1988 America's Cup and The Race. Dynamic stability of multihill ships is generally good, though high-speed operations of powered craft can lead to ride characteristics that require some sort of active control system in order to prevent local velocities and accelerations that can lead to equipment operation problems and potential crew and passenger seasickness. The stability and survivability of damaged multihull ships is usually improved over that of monohull designs, since the compartments are generally smaller in the hulls, which reduces the effects of flooding in hull compartments, and physical separation of the hulls and equipment improves built-in redundancy and therefore results in higher reliability in operation.

Internal design. Internal ship arrangements are challenging for multihulls due to their long, slender shape. Engines and machinery can be large, and allowing for installation, maintenance, and operation in the arrangements can be difficult. For the trimaran, the sizing of the center hull and side hulls depends on the trade-off of hull interior arrangement needs and desired resistance and stability characteristics in the longitudinal placement for the outer hulls fore and aft. The external arrangement and size of topside area can make the topside and deckhouse arrangements somewhat easier than those of conventional monohull ships, allowing room for aircraft landing and combat and weapon systems, as well as for personnel accommodations.

Production. Production of multihull ships can be quite challenging. The strength of the connecting structure between the hulls is key to being able to construct a safe and capable multihull ship. The transverse moments across hulls require either extra structural steel or the use of lightweight, but strong, composite materials. Indeed, it is the advent of the ability to construct ships using these advanced composites that has made it possible to produce many of the creative combinations of hull forms that exist today. The slender shape of the hulls leads to challenges with fabrication as well, since the welding, joining, and testing has to be accomplished in tight spaces with little additional clearance. The docking, drydocking, and berthing of multihull ships are different enough from typical monohulls to result in additional challenges in production and operation.

Transport factor. One relatively new metric to compare the operational utility of high-speed ships is the transport factor (TF), which is a measure of merit that combines displacement, speed, and installed horsepower. The TF is defined as (1.6878/550)*2240*displacement (in long tons)*speed (in knots)/installed horsepower. [In metric units, the TF is defined as 9.80665*displacement (in metric tons)*speed (in m/s)/installed power (in kilowatts).] TF is then plotted versus speed or nondimensional speed (Froude number) to give an indication of how ships compare in their ability to efficiently provide high-speed transport. High-TF vessels are better at carrying large payloads at high speeds, and many multihull ships compare quite favorably to comparable monohulls.

Internal hull configurations. Although the term "multihull" implies multiple hull forms arranged externally to accomplish desired floatation, ships can be constructed with hulls in many different geometries, with multiple hulls arranged internally to one another. For internal configurations, multiple hulls are used primarily to improve safety or provide arrangement flexibility. In the case of multihull submarines, concentric hulls are constructed to increase reserve buoyancy, as in many ex-Soviet nuclear submarines. In the case of surface vessels, many petroleum tankers and cargo ships that require additional safety margins for protection either of the cargo or the environment use multiple hulls to prevent damage to the hull structure from allowing the environment into the ship or the cargo from coming out of the ship.

Prospects. The future of multihull ships is promising, as the need for multiple ways to come up with new designs that can carry cargo at high speeds with good stability and safety is increasing. Advances in composite material design, ship production techniques, active control systems for ride stabilization, and higher-density propulsion systems should provide for new technologies to allow continued innovative design creation and production.

For background information *see* AIR-CUSHION VEHICLE; FERRY; HYDROFOIL CRAFT; MERCHANT SHIP; SHIP DESIGN; SHIP POWERING, MANEUVERING, AND SEAKEEPING; SUBMARINE in the McGraw-Hill Encyclopedia of Science & Technology. Clifford Whitcomb

Bibliography. T. C. Gillmer and B. Johnson, *Introduction to Naval Architecture*, United States Naval Institute, Annapolis, Maryland, 1982; C. Kennell, Design trends in high-speed transport, *Mar. Technol.*, 35(3):127–134, July 1998; C. Kennell, On the nature of the transport factor component TF_{ship}, *Mar. Technol.*, 38(2):106–111, April 2001; V. Dubrovsky and A. Lyakhovitsky, *Multi-Hull Ships*, Backbone Publishing, 2001.

Multispectral satellite observations of severe storms

Currently, the principle tool for monitoring and studying severe convective storms is Doppler weather radar. Since the early days of weather satellites and in parallel with the advances in radar meteorology, scientists have been evaluating the capabilities of various satellite-borne instruments to provide additional, or alternative, information about storm severity. The major limitation of this method is that traditional satellite instruments observe only the uppermost layers (cloud tops) and do not "see" inside the cloud, and therefore do not have the capability to reveal the internal structure of the storm. Thus, some basic questions remain as to whether there are any storm-top characteristics or features that could provide information about the nature of the storm, its intensity, and potential severity, and whether there is any information in the cloud tops that could be used for storm warnings.

Given the fact that the entire cumulonimbus cloud is the result of complex processes inside the storm and their interactions with the atmosphere surrounding the storm, the appearance of the cloud reflects some history of the storm formation, internal structure, and intensity. Therefore, storm-top observations may provide evidence of processes that have formed it, offering a record of storm characteristics that can be used for storm monitoring and warnings, along with other observational tools. In areas lacking sufficient radar coverage, weather satellites represent the principle tool for storm-activity monitoring.

Satellites and spectral bands. Among the numerous satellites, several have contributed substantially

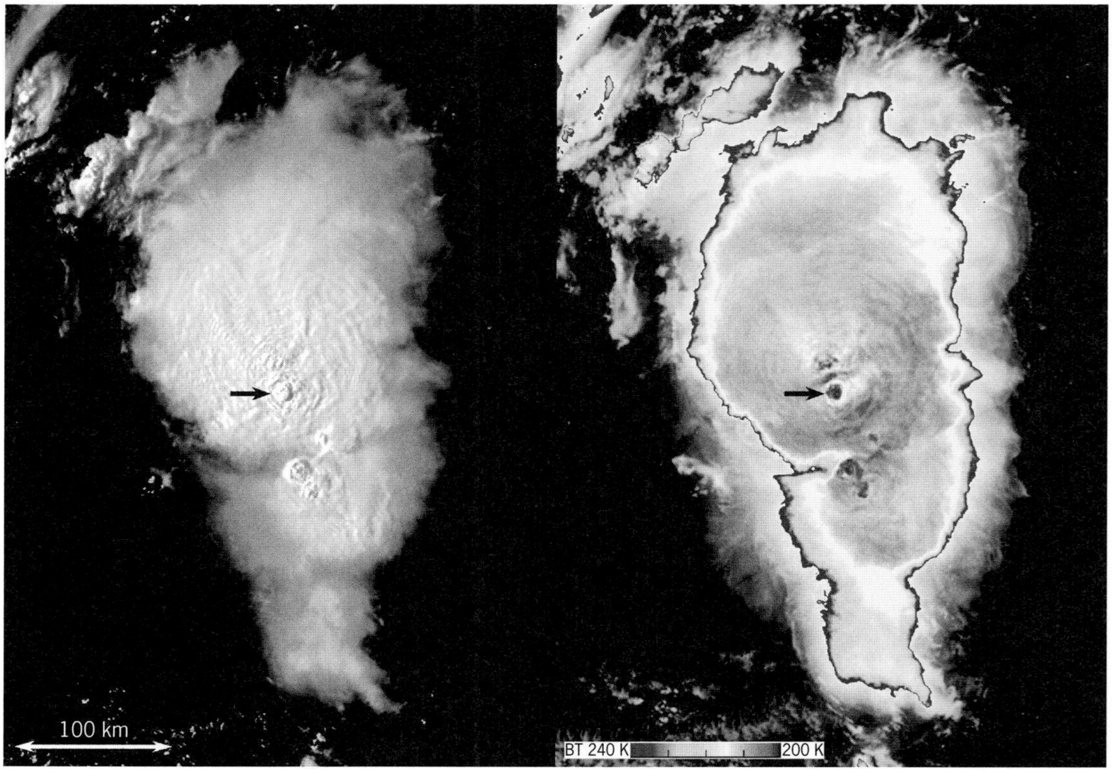

(a) (b)

Fig. 1. Example of an overshooting top as seen in the AVHRR (Advanced Very High Resolution Radiometer) imagery, June 25, 2006, 16:08 UTC, NOAA15, Czech Republic. (*a*) Composite image of the AVHRR bands 1, 2, and 4. (*b*) Enhanced AVHRR band-4 image. The overshooting top is indicated by an arrow. (*Data source: NOAA CLASS*)

to the present understanding of severe convective storm tops as viewed from space. Among them are, namely, the NOAA (National Oceanic and Atmospheric Administration) GOES (Geostationary Operational Environmental Satellites) and POES (Polar Operational Environmental Satellites) satellites, European geostationary satellites Meteosat and MSG (Meteosat Second Generation) operated by EUMETSAT (European Organization for Exploitation of Meteorological Satellites), and the multinational polar-orbiting satellites, Terra and Aqua, managed by NASA (National Aeronautics and Space Administration). All these satellites are equipped with multispectral radiometers for monitoring the Earth and its clouds in various spectral bands, ranging from the visible (0.4–0.7 μm) to the thermal infrared (IR) region up to about 14.5 μm. Some satellites observe at longer wavelengths, in the microwave range. Bands from the visible up to $\sim$4.5 μm include the reflected solar radiance, so they provide information about the reflectivity (or reflectance) of the observed targets during daylight. From $\sim$4.5 μm and above, the bands contain information only on the thermal emitted radiance, which is usually converted into the equivalent blackbody brightness temperature (BT). The spectral bands between 3.5 and 4.5 μm contain both solar-reflected and thermal-emitted components. Detailed descriptions of the bands and technical details of the individual instruments that house these bands are beyond the scope of this article.

Anvil characteristics, overshooting tops. Storm updrafts—the vertical currents that drive convective storms—are a primary source of ice particles that form the storm anvil. Anvil characteristics are very closely related to the properties of the feeding updraft. For example, the stronger the updraft, the faster the growth rate of the anvil. Therefore, it is not surprising that the earliest work specifically addressed this topic, characterizing the anvils using thermal IR spectral bands (mean and minimum BT, size and shape of clouds with a BT below a certain threshold, and rate of change of these quantities). Simultaneously, significant overshooting tops (**Fig. 1**) have been included among the characteristics related to storm severity (for example, the stronger the updraft, the higher the overshoots). These characteristics can be detected in both visible bands (because of illumination effects and the shadows they cast on the surrounding lower anvil top) and in thermal IR bands (as they penetrate the tropopause they continue to cool; thus the lowest storm cloud-top temperatures are typically associated with these). However, the lowest observed BT and detection of overshooting tops are strongly related to the image resolution. The better the instrument resolution, the finer the details it can detect.

Cold-U/V shapes. Among the first storm-top features that have been correlated with severe weather were the so-called enhanced V shapes. These can be found in enhanced thermal IR imagery and resemble in shape either a "V" or "U," located at the outer upwind part of the storm anvil, with a warmer area inside (**Fig. 2*a***). Recently, these have been more accurately referred to as cold-U/V shapes. The outline of these features (U or V) seems to be related to storm-relative winds at anvil levels, as well as to the satellite viewing geometry (the more oblique the viewing angle, the more "closed" the feature appears, thus resembling a "V" in shape). The explanation for these is very closely related to the mechanism generating the embedded close-in warm area (CWA) inside the cold-U/V shape. This is attributed to the result of wake effects downwind the overshooting top area. It is the magnitude of this phenomenon (difference between the lowest BT within the cold-U/V and highest BT inside the CWA) and some other characteristics of the cold-U/V shape that have been postulated as possible satellite-based storm severity indicators.

Cold ring-shaped storms. A similar feature, only recently documented, is referred to as the cold ring shape (alternatively as the cold doughnut shape, Fig. 2*b*). In this case, the cold area at the outer part of the anvil completely encircles a central warm spot (CWS), which is the only difference from the cold-U/V shaped storms. However, not enough cold ring-shaped storms have been studied to investigate a link with severe weather. It is assumed that the only difference between cold-U/V and cold ring-shaped storms might be in the strength of the upper-level winds that drive the appearance of storm anvils, while the mechanism (wake effect) creating the CWA and CWS might be the same. An alternative mechanism for the CWS may be that the cloud summit in the area of overshooting tops comes into thermal balance with the warmer, lower stratosphere into which this elevated dome rises.

Jumping cirrus. Another mechanism that possibly explains the inner warm area of the cold-U/V shape is the so-called jumping cirrus, documented by T. T. Fujita based on aircraft observations at anvil levels. In this mechanism, cirrus cloud particles are ejected above the anvil top after the collapse of significant overshooting tops and then carried downwind by the ambient flow. Since the cirrus particles are blown higher into the warmer, lower stratosphere, they can be heated and may partially mask the colder anvil top underneath. The vertical characteristics of these cirrus clouds also have been documented based on stereoscopic observations, enabled by simultaneous observations from two geostationary satellites displaced longitudinally in orbit.

Cloud-top microphysics. The introduction of the 3.7-μm spectral band on polar-orbiting satellites in 1978 and of a similar band centered at 3.9 μm on a geostationary satellite in 1995 have enabled more detailed microphysical studies of cloud tops than ever before. These observations have revealed that some convective storms exhibit a significant increase in their cloud-top reflectivity at these wavelengths, which in most cases has been attributed to very small ice particles generated by strong updrafts. However, since some of these small ice particles may be formed by processes other than updrafts, increased 3.7/3.9-μm reflectivity may not always be a manifestation of storm severity. On the contrary, observations have shown that some strong storms do not exhibit increased 3.7/3.9-μm reflectivity. The 3.7/3.9-μm reflectivity is also a basis of the temperature-particle effective radius profiles method, developed by D. Rosenfeld.

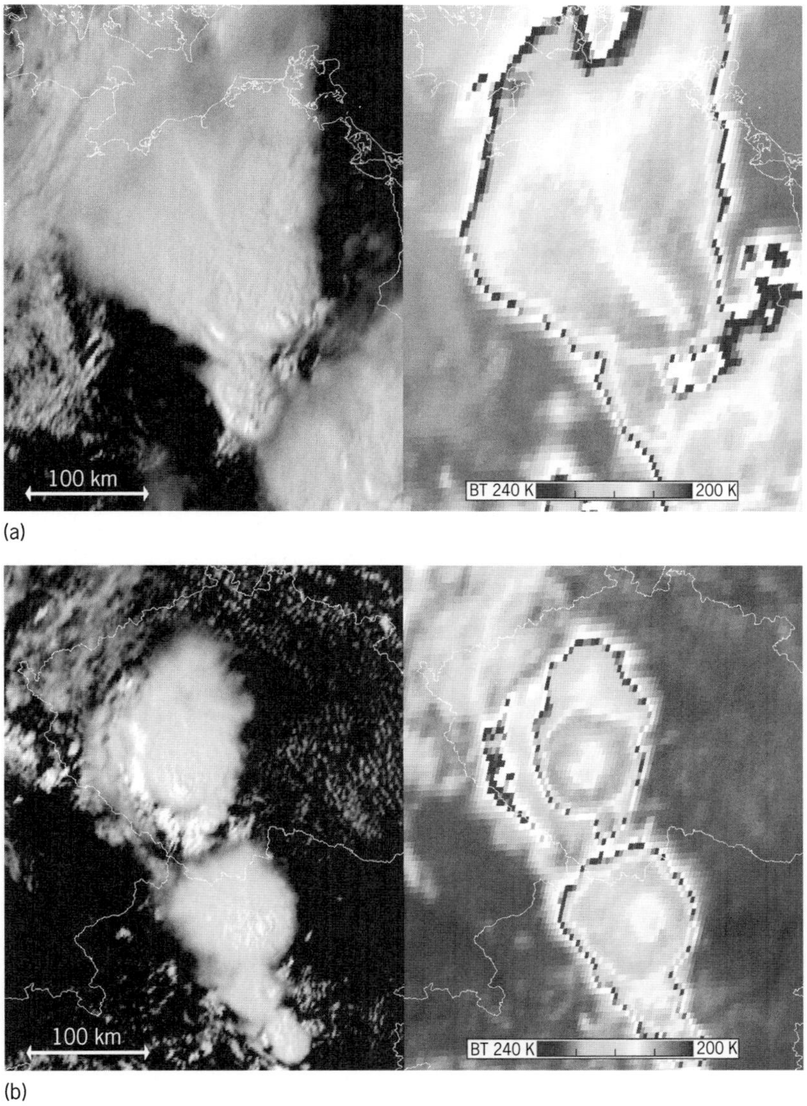

Fig. 2. Two basic cloud-top forms of severe convective storms in thermal IR spectral band. (*a*) Cold-U shape (May 26, 2007, 15:00 UTC, MSG-2, Germany). (*b*) Cold-ring shape (June 25, 2006, 13:45 UTC, MSG-1, Czech Republic and Austria). Left frames show the storms in visible band, right frames in enhanced thermal IR band. (*Data source: EUMETSAT and CHMI*)

Above-anvil plumes. Among the features revealed by the 3.7-μm studies is a special one resembling plume spreading above the anvil top. This feature often appears to be streaming from an almost point-sized source (on the order of a single image pixel, about 1 × 1 km), located downwind of the coldest tops, usually above the CWA. After being discovered in the 3.7-μm band, the plumes were later documented in all the other spectral bands (**Fig. 3**). Most recently, model results of P. K. Wang suggest a gravity-wave-breaking mechanism as the likely source of the above-anvil plumes. Moreover, it now appears that some of the older, documented cases attributed to jumping cirrus could instead be classified as plumes. At least in some cases, warm plumes might be the explanation behind the cold-U/V shaped storms by means of masking the colder anvil top underneath.

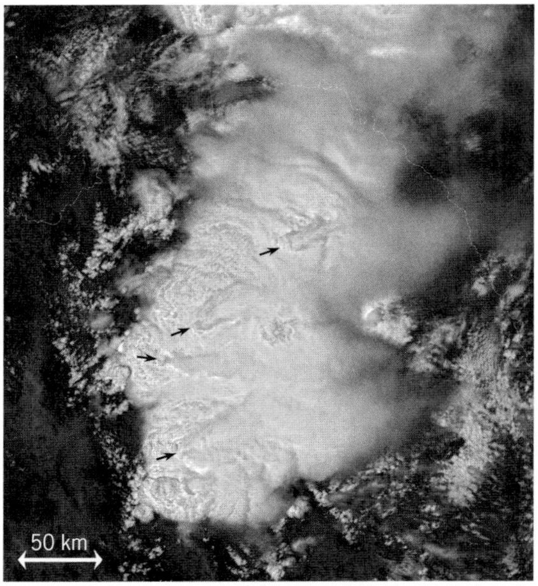

Fig. 3. Plumes above tops of convective storms [May 7, 2007, 20:05 UTC, MODIS/Aqua, Mexico (south of Mexico/Texas border)] for 250-m resolution band 1. The individual plumes are marked by arrows. (*Data source: NASA LAADS*)

Lower-stratospheric water vapor. Another feature that appears to be closely related to the above-discussed plumes is moisture transported into the lower stratosphere by storm activity (a similar process as the one generating ice plumes). One possible technique for detecting this moisture is of findings BT differences between an IR window band (10–12.5 μm) and a water-vapor absorption band (WV, 6.2-μm). With no water vapor above the cloud tops, the IR window band and WV band will have nearly the same BT, but occasionally satellites detect warm, lower-stratospheric gaseous moisture with a background of a very cold cloud top. In the IR window band, this gaseous moisture is transparent, but the WV band adds extra radiance, making the storm top appear warmer than in the IR window band. The most recent observations have documented this phenomenon to occur not only in the tropics, but also at the midlatitudes.

Outlook. Present research goals include studies of how the above phenomena are actually interrelated. This is being achieved by means of multispectral observations of storm tops, where combinations of various spectral bands, each providing somewhat different information, are studied to determine the mutual links between the individually observed cloud-top features. A better understanding of these should lead to an improved conceptual model of the processes occurring at the upper parts of convective storms. Providing great help in these efforts will be advanced instruments flown aboard the latest research satellites, such as the cloud profiling radar (CPR) aboard the CloudSat satellite and the cloud-aerosol lidar on the CALIPSO satellite. With the help of these instruments, it should be possible to learn more about the nature and significance of the cloud-top features found at or above the tops of severe convective storms.

For background information *see* BLACKBODY; CLOUD; CLOUD PHYSICS; MESOMETEOROLOGY; METEOROLOGICAL RADAR; REMOTE SENSING; SATELLITE METEOROLOGY; STORM DETECTION; THUNDERSTORM in the McGraw-Hill Encyclopedia of Science & Technology.

Martin Setvák

Bibliography. T. T. Fujita, Principle of stereoscopic height computations and their applications to stratospheric cirrus over severe thunderstorms, *J. Meteor. Soc. Jpn.*, 60:355–368, 1982; G. M. Heymsfield, R. Fulton, and J. D. Spinhirne, Aircraft overflight measurements of Midwest severe storms: Implications on geosynchronous satelite interpretations, *Mon. Weather Rev.*, 119:436–456, 1991; I. M. Lensky and D. Rosenfeld, The time-space exchangeability of satellite retrieved relations between cloud top temperature and particle effective radius, *Atmos. Chem. Phys.*, 6:2887–2894, 2006; R. A. Mack, A. F. Hasler, and R. F. Adler, Thunderstorm cloud top observations using satellite stereoscopy, *Mon. Weather Rev.*, 111:1949–1964, 1983; D. W. McCann, The enhanced-V: A satellite observable severe storm signature, *Mon. Weather Rev.*, 111:887–894, 1983; A. J. Negri, Cloud-top structure of tornadic storms on 10 April 1979 from rapid scan and stereo satellite observations, *Bull. Am. Meteor. Soc.*, 63:1151–1159, 1982; P. K. Wang, The thermodynamic structure atop a penetrating convective thunderstorm, *Atmos. Res.*, 83:254–262, 2006.

Mycorestoration

Mycorestoration is the utilization of fungi to rehabilitate stressed or contaminated environments. Much like our own immune systems, ecosystems become strained by excessive pollutants (natural or anthropogenic), high populations of infectious bacteria, viruses, or protozoa, and the stripping away of needed habitat. Fungi shoulder a vital responsibility in managing these stresses through sustaining a robust soil foundation, increasing primary productivity, managing infectious outbreaks, and recycling every compound the world can create. Unfortunately, human beings have acted as viruses to Earth's immune system for 150 years through massive deforestation, resource harvesting, industrial production, and incredible population growth. Our previous view of an endlessly endowing world forever supporting our infinitely growing economy, as well as digesting all the waste, is fortunately beginning to change. With that change in perspective, we are recognizing nature's allies in ecorestoration, the fungi being our cavalry in this battle.

Fungi are ubiquitous in all environments throughout the world, providing several ecological services. One main role of fungi is decomposition, being the natural recycling centers of planet Earth, converting every biological and inert compound into usable biomass and nutrients for plants, animals

Five categories of mycorestoration

Process	Definition
Mycoremediation	Using fungal decomposers to degrade, transform, or lock up contaminants and recalcitrant substances
Mycoforestry	Restoring forests that are stressed by logging, mining, pathogens, or overdominant tree species using mutualistic and saprophytic fungi
Mycofiltration	Removing pollutants, sediment, bacteria, or protozoa from runoff, ground water, and streams by using mutualistic and saprophytic fungi
Mycopesticides	Controlling specific insect pest populations with very specific fungal parasites
Mycoagriculture	Improving agricultural soil quality, crop production, and agricultural waste recycling with fungi

(especially insects), and other microbes. Their role as mycorrhizae, mutualists associated with over 95% of terrestrial plant roots, greatly increases plant productivity, as well as transforming healthy soil into a gigantic filtration system. Both the decomposers and mutualists work alongside parasitic fungi to turn desolate landscapes into diverse forest and grassland communities, even benefiting vegetable gardeners and other farmers. Some fungi even kill insect pests, leaving almost no organism untouched by this diverse kingdom. All this function and diversity has been recognized by mycologists and can now be exploited through five different processes (**table**).

Mycoremediation. Products manufactured by humans are subjected to attack by tenacious organisms looking for food or a cozy residence. Fungi can grow on and deteriorate any food, building material, and clothing that humans can invent. This leads to the development of durable synthetic products, chemicals, preservatives, or toxins to ensure our efforts will not dissolve before our eyes. However, most products outlive their usefulness, and nondegradable wastes accumulate exponentially. Previous chemical or mechanical methods have been successful in degrading some wastes, but most, such as plastics, are just sequestered in landfills. In the past 50 years, though, innovators have found that fungi can be used to break down our most recalcitrant refuse and chemicals by a process called mycoremediation.

Based on earlier success with fungal production of antibiotics such as penicillin, and food chemicals such as citric acid and alcohol, complex bioreactors using yeasts and filamentous fungi were developed to break down pharmaceuticals, phenols, toxic oils, and heavy dyes from industrial wastewater streams. They have the ability to mineralize, release, and store various elements or ions and to accumulate toxic materials. The ease of growth and fast reproductive rates of fungi such as *Aspergillus* lent to their efficacy in a laboratory setting; however, scaling up processes in natural settings has met with mixed success. Further investigation with higher fungi, particularly the wood-decay fungi, helped move this powerful tool into the field.

Wood-decay fungi make their living by digesting the sturdiest compounds known to the natural world (**Fig. 1**). Wood consists of complex carbohydrates such as cellulose and hemicellulose, all bound tightly together by complex phenolic biopolymers such as lignin. Naturally, in order to degrade such a recalcitrant diet, these fungi have powerful enzymatic systems that can also be exploited by humans. Our lab has recently shown that phenolic resin polymer, a "plastic" found in plywood, brake linings, and rotary telephones, can be broken down by a fungus. Gusse and Volk (2007) have demonstrated the ability of fungi to break down a laundry list of durable pollutants. Field studies have even shown their ability to degrade these compounds in soil, air, and water, as well as in bioreactors.

Mycoforestry. For most people and industries, the view of our forests has been that of an endless abundance of any material we ever need. This view continues with forestry practices mimicking agriculture by replanting large monocultures of fast-growing timber species with limited regard for sustainable harvesting practices, the importance of biodiversity, or the organisms that maintain the rich splendor of our vast forests. Mycoforestry is the recognition and utilization of fungi in their natural role as the "master gardeners" of forest ecosystems.

Fungi serve forest communities in a number of ways. The decomposers convert leaves, wood, animal waste, and bodies to essential nutrients usable by other microbes and plants. This process also creates an invaluable soil structure consisting of nondigestible organic material, a teeming network of threadlike fungal hyphae (long, branching filaments composing the mycelium of a fungus) [**Fig. 2**], and other microbes. This community then joins the mycorrhizal community to support plants and animals by providing a harvest of minerals unreachable by plants, as well as water cleansed of infectious microbes and otherwise noxious chemicals. Other fungi in the forest also function as parasites that attack mature trees, thinning dominant trees within stands and creating openings for new

Fig. 1. Wood-decay fungi (*Trichaptum biforme*) covering the cut end of a log. In this species the growth is mostly in the outer region of the tree, the sapwood.

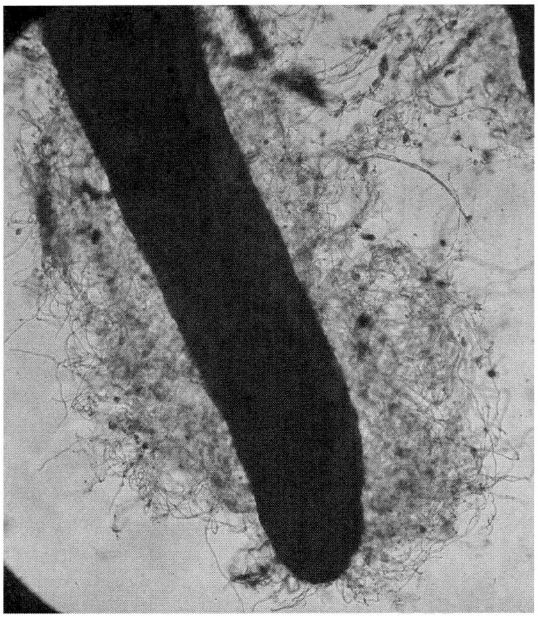

Fig. 2. Mycorrhizal fungal hyphae forming a dense net around a root tip. The hyphae emanate into the soil to scavenge for mineral nutrients and water.

saplings. Many of these fungi also serve as food for bacteria, other fungi, and animals, even humans. These ecological groups of fungi can be used in mycoforestry to restore or improve forest communities that have been clear-cut, damaged by logging roads or mining, or deprived of natural cycles such as fire.

Mycofiltration. The thriving networks of fungi in forest soils (**Fig. 3**) are a perfect example of a tight-knit biological filter using materials that other organisms cannot, and producing wastes on which other organisms thrive. It has been said that 1 cubic inch of soil contains enough hyphae to stretch over 8 mi (13 km) when lined up end-to-end. Moreover, the fungi of the forest could have an absorptive surface area that is 10–100 times greater than that of leaves in the forest. This means that fungal hyphae with their degradative abilities and stimulation of organisms present in soil are the most complex, powerful water purification system on Earth. They just need to be set up in the correct situations to manage detoxification communities for solving chemical problems.

Fig. 3. Dense layer of hyphae covering the surface of forest soil. The millions of hyphae only 10 micrometers thick become so intertwined that they are visible to the naked eye.

Mycofiltration is the use of fungi grown on various substrates to create membranes called biofilms that can filter microbial pathogens such as bacteria, fungi, or protozoa, and toxic chemicals such as antibiotics or fertilizers, as well as particulate runoff. Dense growths of decomposer fungi such as *Pleurotus* (oyster mushrooms) or *Stropharia* (wine caps) can be grown on wastes such as straw, wood chips, or grass clippings. The fungal product is then buried along roadsides, sensitive watersheds, factory discharges, manure-rich farm lands, or other stressed habitats to "strain out" harsh inputs. Decomposer fungi tend to thrive in these environments; however, mycorrhizal fungi also assist to purify these same environments by processing compounds for their plant mutualists. Both of these fungi can also stimulate microbial communities that digest nutrients not processed by the fungi.

Mycopesticides. Fungi have coexisted with insects as long as our planet has been a shining biological example to the universe. As decomposers, they are in contact with vast numbers of insects that eat rotting plant material. Some ants even feed leaves to fungal gardens that they use for food. On the darker end of the spectrum is the Entomophthorales, an entire order of fungi that parasitize various insects. Microscopic spores floating through the air land inconspicuously on the tough endoskeletons of ants, termites, beetles, and many other insects, where they germinate and burrow into the juicy protein-rich insides of their host. They even invade crucial parts of their brains, programming them to scale the tallest nearby structures and produce spores. Some fungi such as *Cordyceps* use the energy from digesting insects to create a fruiting body twice the size of the insect (**Fig. 4**). The fruiting body then disseminates millions of spores into the environment, intent on finding their next victim (Volk, 2007).

Mycologists have found these phenomena useful against insect pests intent on destroying building materials, food stocks, and greenhouse plants. Strict insect pathogenic genera such as *Beauveria, Metarhizium*, or *Cordyceps*, as well as other more promiscuous fungi such as *Aspergillus* and *Trichoderma*, are being researched and even patented as biocontrols against termites and ants as well as white pine weevil (a highly destructive beetle pest of pine and spruce). There are problems with developing rapidly growing fungal strains with prolific, wide-reaching spore dispersal mechanisms, but current research looks promising. It turns out that the insects we attempt to kill are as tricky as the fungi we use to fight them.

Fig. 4. *Cordyceps* fruiting bodies projecting their brilliant colors and millions of spores from the leftover carcass of an underground insect. (*Photo: Sean Westmoreland*)

Mycoagriculture. Agriculture and mushroom production have always been as closely linked as crops and soil, but recognition of how to use these cohabitations to our advantage has been largely ignored in large-scale agriculture. Fungi are the major controllers of all crops and wastes produced in agricultural settings—from the mycorrhizae that assist in crop production to decomposers that degrade field stubble, postharvest plant wastes, and the manure made by animals that feed on crops. However, as large-scale farming practices become the norm, reliance on natural nutrient cycling processes has been exchanged for large machinery and "a better world through chemistry."

As the traditional small farmers suffer from depleted topsoil, dependence on chemicals that potentially make their families ill, and an overindustrialized way of life, fungi may come to the rescue. Mycorrhizal inoculants are sold to assist with root enhancement for increased nutrient acquisition, and mushroom compost is used to enhance soil fertility. Some farmers even inoculate edible mushrooms on waste plant material in the rows between their vegetable plants. Research into the ability of edible mushroom saprophytes (organisms, especially fungi, that live on dead or decaying organic matter) such as *Hypholoma* (brick caps) and *Stropharia* to increase nutrient availability, as well as to increase overall plant and mushroom production, has demonstrated that enhancing age-old natural processes can benefit both the soil and the plants.

Outlook. The fungi are an amazing kingdom of organisms that naturally already do a lot to make our lives better. Now it is just a matter of exploiting these fungi to do even more to help solve environmental problems we created in the past.

For background information *see* BIODEGRADATION; FOREST ECOSYSTEM; FUNGAL BIOTECHNOLOGY; FUNGAL ECOLOGY; FUNGI; MYCOLOGY; MYCORRHIZAE; PESTICIDE in the McGraw-Hill Encyclopedia of Science & Technology.

Adam C. Gusse; Thomas J. Volk

Bibliography. A. C. Gusse, P. A. Miller, and T. J. Volk, White rot fungi demonstrate first biodegradation of phenolic resin, *Environ. Sci. Tech.*, 40:4196–4199, 2006; A. C. Gusse and T. J. Volk, Fungal bioconversion, in *McGraw-Hill Yearbook of Science & Technology*, McGraw-Hill, pp. 90–93, 2007; H. Singh, *Mycoremediation: Fungal Bioremediation*, Wiley-Interscience, 2006; P. Stamets, *Mycelium Running: How Mushrooms Can Help Save the World*, Ten Speed Press, Berkeley, California, 2005; T. J. Volk, TomVolk.net, 2007.

Nanomaterials in the forest products industry

Nanotechnology is the study and engineering of matter at the dimensions of 1–100 nm where physical, chemical, or biological properties are fundamentally different from those of the bulk material. The nanotechnology paradigm is to modify bulk properties and functionality by controlled manipulations at the nanoscale. Nanotechnology research has dramatically grown within the past 10 years because of recent developments in nanoscale characterization techniques, processes, and understanding of material behavior at the nanoscale. By expanding our understanding and control of matter at such levels, new avenues in product development can be opened. Nano-based science has applications across nearly all economic sectors and allows the development of new technologies with broad commercial potential within the forest products industry.

Forest products industry. Wood has been widely used as an engineering material for thousands of years because of its availability and unusual ability to provide high mechanical strength and high strength-to-weight ratio while retaining its toughness. Wood has also been used as a material source for a variety of composite products based on wood chips and flakes, wood sawdust (wood flour), wood fiber (individual wood cells), and recently cellulose nanomaterials. The forest products industry produces many items, such as solid wood lumber, wood-based composites (plywood, oriented strandboard, fiberboard, three-dimensional engineered fiberboard), engineered structural members (I beam, finger-jointed wood), wood-plastic composites (decking), paper products (paper, filters,

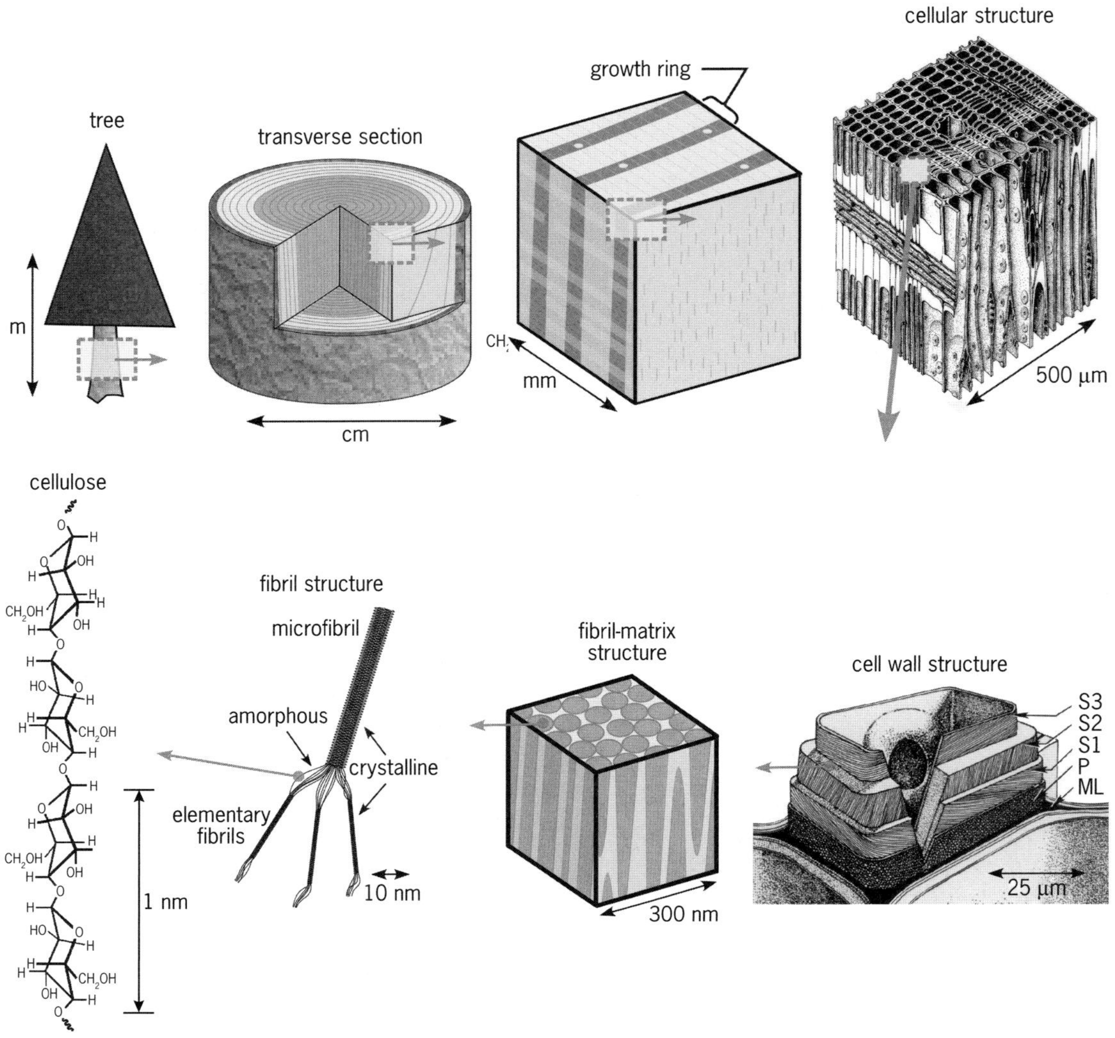

Fig. 1. Hierarchical structure of wood, showing the size scale of each structural feature within wood. The linear polymer chains of cellulose (30% of wood by weight) arrange to form cellulose fibrils, which are the base strengthening component within wood. (*Adapted from R. J. Moon, C. R. Frihart, and T. Wegner, Nanotechnology applications in the forest products industry, Forest Prod. J., 56(5):4–10, 2006*)

corrugated containers), and products that are used in conjunction with wood (adhesives, coatings/paints, stains, stabilizers).

Nanotechnology offers the potential to transform the forest products industry in virtually all aspects, including production processes (raw materials, engineered wood and wood-based materials), improved energy efficiencies, new applications for composite and paper products, and composites made from cellulose nanomaterials. Additionally, there are opportunities to use new products in conjunction with wood-based materials to improve a particular function. For example, an array of built-in nanosensors within wood- and paper-based products can be used to monitor force, moisture level, temperature, pressure, chemical concentrations, and attack by wood decay fungi. More traditionally, nanofillers can be used within wood or wood composites, offering new opportunities to improve durability (with regard to wear, ultraviolet or biological decay, fire, and dimensional stability) or to improve chemical bonding with paints, sealants, and adhesives. Also, by building functionality onto the surfaces of cellulose nanomaterials, new opportunities for products such as pharmaceuticals, chemical sensors, self-sterilizing surfaces, and electronic wood-based devices may be achieved.

Structure of wood. Nanoscale manipulation of wood occurs near its smallest structural scale (**Fig. 1**). Wood is a hierarchical structured composite, in which several unique structures span several length scales: growth ring structure (each ring represents one growth year), cellular tissue structure, multilayer cell wall structure, fibril-matrix structure within each cell wall layer, cellulose microfibril structure, and the structure of the three main polymer components (cellulose, hemicellulose, and lignin). Bulk properties of wood result from the culmination of interactions within and between each structural scale. The last three structural scales mentioned above are the realm of nanotechnology research of wood, which occurs in two general categories: (1) controlled manipulation of the three main polymer components through various treatment processes (chemical or genetic modification) to modify bulk properties, and (2) removal of the polymer components and processes to make new materials based on these starting materials (such as cellulose nanowhiskers).

Cellulose nanowhiskers. Cellulose is the world's most abundant biopolymer and is present in virtually all plants. Its main function is to act as a reinforcement material. Cellulose is a linear chain of ringed glucose monomers (10,000 to 15,000) linked together. Multiple cellulose chains arrange to form cellulose microfibrils having regions that are disordered (amorphous regions) or highly ordered (crystalline regions). Microfibrils have diameters of 3–20 nm and lengths that can reach several micrometers, depending on the source of the cellulose (plant, wood, or bacteria). Cellulose nanocrystals are obtained by acid hydrolysis of microcrystalline cellulose, in which the more chemically reactive amorphous regions within microcrystalline cellulose are dissolved. The remaining crystalline cellulose is nano-sized and has a rod or whisker shape (length: 100–300 nm, diameter: 3–5 nm, for wood cellulose source) [**Fig. 2**]. The tensile modulus (the ratio of stress to elastic strain in tension) of these crystalline particles is estimated to be 150 GPa [(10^9 kg/m^2)], which is greater than that of Kevlar® (130 GPa; Kevlar® brand fiber is an innovative technology combining high strength with light weight).

Not all cellulose nanocrystals are the same; aspect ratio and surface chemistry are different depending on the cellulose source (wood, plants, bacteria) and the production process (type of acid used during the acid hydrolysis, or any additional chemical reaction used to intentionally modify the cellulose nanocrystal surface chemistry). Research is being conducted on several aspects of cellulose nanocrystal processing science, including (1) characterization of nanowhisker properties from various sources and various production routes, (2) development of new low-cost, large-scale production routes (enzymes, cow-digested fiber, and as a by-product of other industrial wood processes), and (3) new chemical treatments for modifying the nanowhisker surface chemistry.

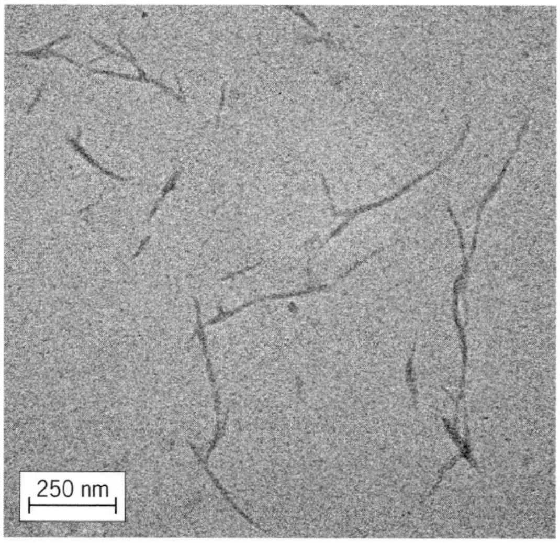

Fig. 2. Transmission electron microscope image of a cellulose nanocrystal. (*Courtesy of James Beecher, USDA Forest Service, Forest Products Laboratory, Madison, Wisconsin*)

Strength and stiffness of reinforcement materials*

Material	Tensile strength, GPa	Modulus, GPa
Cellulose nanocrystals	7.5	150
Glass fiber	4.8	86
Steel wire	4.1	207
Kevlar	3.8	130
Graphite whisker	21	410
Carbon nanotubes	11–73	270–970

*Data taken from <http://woodscience.oregonstate.edu/faculty/simonsen>.

Cellulose nanowhiskers offer several advantages as a reinforcement particle. They have a high aspect ratio, high stiffness, and a unique surface chemistry. The reinforcement efficiency is enhanced by this high surface area and aspect ratio, resulting in better reinforcement with less material than is possible with larger sized particles (such as wood fibers). The specialized surface chemistry offers new applications that exploit this chemistry (for example, dispersion in new matrix polymers, self-assembly, chemical sensors). Additionally, cellulose nanowhiskers can be considered as a green technology because they are naturally produced, are biodegradable, and are likely to have lower health risks if ingested or inhaled than those associated with other reinforcement materials. Cellulose nanowhiskers also have comparably low production costs. These advantages make up for some of the shortcomings in properties as compared to other reinforcement nanoparticles, such as carbon nanotubes (see **table**). Interestingly, some researchers are coating carbon nanotubes with cellulose nanowhiskers to change the surface chemistry of the carbon nanotubes while retaining their mechanical properties.

Cellulose nanowhisker composites. The addition of cellulose nanowhiskers to bulk composites, panels, or thin films has been demonstrated to improve composite thermal stability, mechanical strength, toughness, and flexibility. The arrangement of cellulose nanowhiskers within composites and films has a major effect on the final properties. Recent research has focused on new chemical treatments to tailor the surface chemistry of the nanowhiskers for controlling the degree of dispersion within the matrix material and to control the bonding strength between the nanocrystalline particles and the matrix material, both of which strongly influence the resulting composite properties. A major challenge is in minimizing nanocrystal particle agglomeration (that is, grouping together large numbers of particles) so that the desired level of reinforcement particles can be added while still achieving a uniform dispersion within the polymer matrix. Researchers are also investigating the optimum nanocrystal structural organization within composites for a desired property and developing assembly techniques necessary to produce these desired structures. Electric fields, magnetic fields, and shear deformation have been used to improve the nanowhisker alignment. For more complex nanowhisker arrangements, techniques using self-assembly and nanomotors are being investigated.

Some potential consumer applications for cellulose nanowhiskers will be in the production of biodegradable, lightweight, and high-strength composite panels in the electronics, automotive, and aerospace industries. Additionally, cellulose nanowhiskers have been used as the reinforcement network structure in the development of robust, flexible, durable, lightweight, and dimensionally stable films. These reinforcement films are seeing applications in razor-thin flexible display screens, polymeric fuel cell membranes, and barrier applications. For barrier applications, the surface chemistry, high surface area, high aspect ratio, and small pore size of the nanowhisker network assembly are used to filter a number of toxic industrial chemicals. By modifying the surface chemistry and the spacing of the nanowhiskers, different toxins can be selectively filtered.

Outlook. Nanotechnology has the means to revolutionize the forest products industry, both by providing new opportunities for current product improvements and by developing new applications of cellulose in a wider consumer products realm that is outside the scope of the traditional forest products industry. Cellulose nanowhiskers are a unique particle that can be manipulated for the development of specified properties in various applications such as reinforcement for polymers and specialized barrier films.

For background information *see* CELL WALLS (PLANT); CELLULOSE; NANOSTRUCTURE; NANOTECHNOLOGY; WOOD ANATOMY; WOOD COMPOSITES; WOOD ENGINEERING DESIGN; WOOD PROCESSING; WOOD PRODUCTS; WOOD PROPERTIES in the McGraw-Hill Encyclopedia of Science & Technology.

Robert J. Moon

Bibliography. S. Beck-Candanedo, M. Roman, and D. G. Gray, Effect of reaction conditions on the properties and behavior of wood cellulose nanocrystal suspensions, *Biomacromolecules*, 6:1048–1054, 2005; R. J. Moon, C. R. Frihart, and T. Wegner, Nanotechnology applications in the forest products industry, *Forest Prod. J.*, 56(5):4–10, 2006; M. A. S. A. Samir, F. Alloin, and A. Dufesne, Review of recent research into cellulosic whiskers, their properties and their applications in nanocomposite field, *Biomacromolecules*, 6:612–626, 2005.

Neanderthal extinction

Neanderthals were humans that lived across Europe, the Middle East, and into central Asia and southern Siberia between 300,000 and 24,000 years ago. The prevalent view has been that these people were "archaic" and somehow brutish and apelike. With the arrival in Europe of "modern" human populations from tropical Africa around 40,000 years ago, the demise of these "backward" people was imminent. Present-day humans are all descended from the African colonizers and the Neanderthals went extinct, leaving no genetic trace among the people of today. The story is still considered accurate by a sector of the paleoanthropological community. In recent years, however, new research has raised important questions regarding the impact of moderns on Neanderthals, the degree of genetic mixing that may have taken place, and ultimately the causes of the extinction of the Neanderthals.

Effect of cold climates. The Neanderthals were tough and had a body adapted to their lifestyle. They evolved somewhere in Eurasia, isolated from tropical African humans, after 500,000 years ago. The robust body and relatively short limbs have been used to support the idea that these features had to do with adaptation to cold climate (see **Fig. 1** for an image of a typical cranium). It is well established in many mammals that reduced limb lengths and large body mass are ways of reducing heat loss in cold environments, but there are other ecological factors that can also have a bearing. In the case of Neanderthals, which evolved alongside a fauna including elephants, rhinos, and other large herbivores and carnivores, a tough and muscular body would have been an advantage. Neanderthals lived in semi-open country, on forest edges, and in places where there was sufficient cover to allow close approach to prey animals. These were ambushed at close quarters with thrusting spears. The discovery that the right arm had better developed musculature than the left has offered support in favor of this method of hunting.

Experimental studies on living humans with body shapes similar to Neanderthals and tropical modern humans have similarly shown no real heat conservation advantage in the Neanderthal body shape. The reality is that when the climate of Europe became intensely cold during the glaciations, a common feature of long periods when the Neanderthals lived,

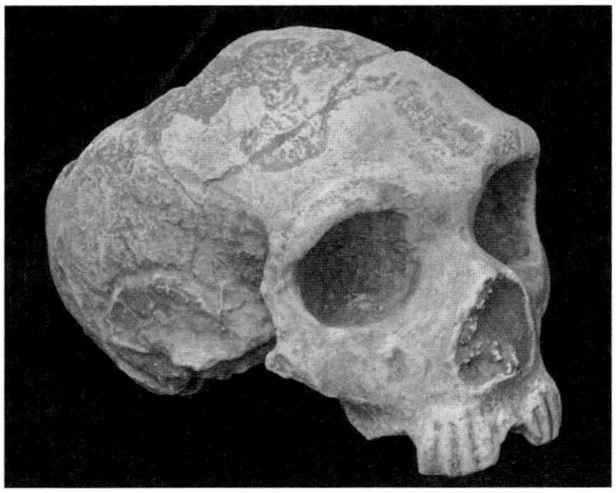

Fig. 1. Neanderthal cranium, with pronounced brow ridges, large braincase, and receding forehead, from Forbes' Quarry, Gibraltar. This skull, discovered in 1848, preceded the 1856 Neander Valley (Germany) discovery by 8 years. (*Photo courtesy of Clive Finlayson/Gibraltar Museum*)

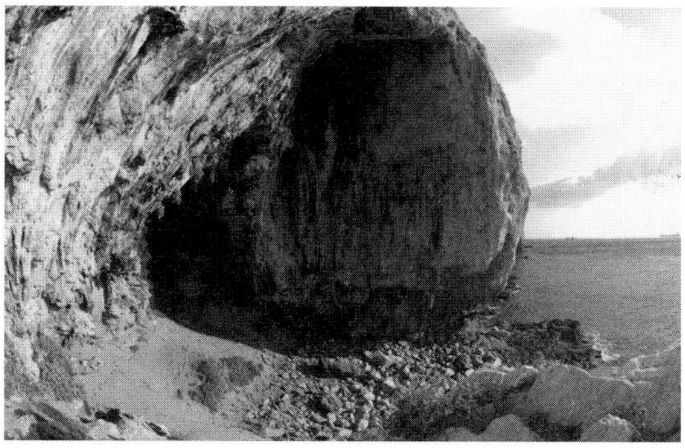

Fig. 2. Gorham's Cave, Gibraltar. Situated on the coast on the eastern side of the Rock of Gibraltar, the site was occupied by Neanderthals between 120,000 and 24,000 years ago. This is the last known site of Neanderthal occupation. (*Photo courtesy of Clive Finlayson/Gibraltar Museum*)

these people lived mostly in the milder regions in the south and west of the continent. These observations have altered our view of the Neanderthal, not as a human that survived the glacial cold but rather of one that was forced into refugia every time conditions deteriorated, bouncing back when climate improved. The pattern was typical of many temperate climate animals that lived alongside the Neanderthals.

Arrival of modern humans. The theory that modern humans caused the extinction of the Neanderthals has no evidence in support. The coincidence in the timing of the arrival of modern humans in Europe and the extinction of the Neanderthals has been cited repeatedly in favor. Were it accurate, this evidence could equally be interpreted to mean that the progression of moderns across Europe was only possible as the Neanderthals became extinct, for other reasons, leaving vacant territory for the new arrivals.

The recent discoveries in Gorham's Cave, Gibraltar, at the southernmost tip of Europe have shown that Neanderthals survived in isolated refugia as recently as 24,000 years ago (**Fig. 2**). Thus, 16,000 years after the first entry of moderns into Europe, there were still pockets of Neanderthals around. Not only was the modern arrival as the cause of extinction unlikely, but the purported speed of the wave of competition was unrealistic. Instead, the period between 40,000 and 24,000 years ago in Europe has to be seen as one in which human populations (Neanderthal survivors and modern pioneers) were scattered across the land at low density, given that this was the buildup to the last Ice Age, without one causing much of an impact on the other.

Genetic mixing. The above observation raises doubts regarding the degree of genetic mixing that may have taken place between Neanderthals and modern humans. Recent work from Portugal and Romania has revealed fossil early modern humans that appear to retain archaic features, and this led to the revival of the idea that the two populations may have interbred. The view runs contrary to the genetic evidence that seems to indicate that no such interaction took place, not at least to the degree that would have left traces among people today. An alternative explanation for these new fossil discoveries is that there were a number of populations scattered across Eurasia in different stages of morphological evolution, reflecting differing degrees of isolation. On the one hand, there were the very robust Neanderthals; at the other end, there were the gracile moderns arriving from the tropics. In between, there may have been a number of populations with intermediate morphologies. The issue remains unresolved.

Exchange of ideas. Another highly controversial issue in recent years has been the question of whether Neanderthals were less intelligent than modern humans. The prevalent view is that moderns had something special that Neanderthals did not have and was reflected by their ability to paint on cave walls and decorate themselves with pendants made from mammal teeth and shells. Linked to this was the greater range of behavior patterns observed among moderns when compared to Neanderthals, including the use of bone and ivory and the production of blades from flint.

This debate reached its culmination with the controversy about acculturation (modification of the culture of a group or individual as a result of contact with a different culture). Certain archeological sites, especially in southwestern France and northern Spain, produced stone tool technologies (Châtelperronian) that appeared analogous to those made by moderns and, in some cases, were accompanied by ornaments. These technologies were, however, associated with Neanderthals. The discussion developed between those who saw Neanderthals as capable of independently developing the skills to produce these artifacts and those who saw these as a case of Neanderthals mimicking the newly arrived moderns and their technology (Aurignacian). Since this debate started, the sites that associated modern humans with the Aurignacian have been shown not to be reliable. Vogelherd, in Germany, was a type locality that had Aurignacian in association with modern human skeletons. However, it has now been shown that the skeletons were younger than the industry. Ironically, the makers of the Aurignacian culture, supposedly reflecting the first wave of modern human advance across Europe, are currently unknown.

Reasons for extinction. When we look at the broad picture across Europe after the last period of major global warming, 120,000 years ago, what we find is a human population (the Neanderthals) that was constantly being restricted to refugia every time the climate deteriorated. With climatic improvements, the population recovered, but each time another period of deterioration hit the continent the Neanderthals were decimated. This interpretation could not have been foreseen by those who regarded the Neanderthals as cold-adapted and therefore unaffected by harsh climate.

The cause of the repeated population fragmentation was not climate directly, but rather climate's effect on the landscape. Each time the climate got worse, tundra advanced south and steppe west. Europe became a world of treeless landscapes in which a bulky human, used to ambushing prey from within cover, could not get near enough to hunt successfully. The evidence points to attempts to cope.

The Châtelperronian technologies of the French Neanderthals would seem to reflect a change in lifestyle, but the rate of change was too fast to allow their bodies to change.

Instead, a population of moderns with a more gracile build, which has been shown recently to have been expert at endurance running, reached the steppes and plains of central Asia and eastern Europe where it encountered an untapped larder of large mammals, from woolly mammoth to steppe bison, living in treeless landscapes. These people developed portable tool kits and projectile technology. In a world of expanding treeless landscapes, these moderns found a door that led them west into Europe and east toward Siberia and, eventually, North America.

Meanwhile, Neanderthals were managing to survive in their classic landscapes of semi-open vegetation with scattered woods and bushland. These were restricted to the south and west where climate was less severe. It is here that the last populations held out, but their numbers were so low that extinction was inevitable. The most recent evidence has revealed that the last populations living in Gibraltar were hit badly by a series of harsh climatic events in which drought seems to have been a key factor causing their disappearance.

Rather than seeking a single cause to the Neanderthal extinction, however, we should see the process as a protracted one that lasted tens of thousands of years. The last populations in Gibraltar went extinct because of local climatic effects, but it is equally plausible that others disappeared because of inbreeding, disease, or localized competition from other humans.

For background information *see* EARLY MODERN HUMANS; EXTINCTION (BIOLOGY); FOSSIL HUMANS; MOLECULAR ANTHROPOLOGY; NEANDERTALS; PALEOCLIMATOLOGY; PHYSICAL ANTHROPOLOGY; PREHISTORIC TECHNOLOGY in the McGraw-Hill Encyclopedia of Science & Technology. Clive Finlayson

Bibliography. C. Finlayson, *Neanderthals and Modern Humans: An Ecological and Evolutionary Perspective*, Cambridge University Press, 2004; C. Finlayson et al., Late survival of Neanderthals at the southernmost extreme of Europe, *Nature*, 443:850–853, 2006; C. Finlayson and J. S. Carrión, Rapid ecological turnover and its impact on Neanderthal and other human populations, *Trends Ecol. Evol.*, 22:213–222, 2007; J. R. Stewart, The ecology and adaptation of Neanderthals during the non-analogue environment of Oxygen Isotope Stage 3, *Quatern. Int.*, 137:35–46, 2005; T. H. van Andel and W. Davies (eds.), *Neanderthals and Modern Humans in the European Landscape during the Last Glaciation*, MacDonald Institute Monographs, 2004.

Negative refraction

Refraction is one of the most fundamental phenomena in nature. It gives rise to such well-known effects as the apparent bending of objects partly immersed in water, rainbows, mirages, green flashes, and haloes. Refraction is also utilized in many existing optical instruments, including microscopes, telescopes, and eyeglasses. All these phenomena and applications rely on conventional or "positive" refraction. What would the world look like if the sign of refraction were reversed?

The law of refraction predicts that an electromagnetic wave, crossing the interface between two materials with refractive indices n_1 and n_2, changes its trajectory, depending on the difference in the refractive indices, such that $n_1 \sin\theta_1 = n_2 \sin\theta_2$, where θ_1 and θ_2 are the angles from the normal of the incident and refracted waves. The direction of the refracted wave depends on the sign of n_2 (assuming $n_1 > 0$). The refraction is referred to as positive when $n_2 > 0$ (**Fig. 1***a*) and as negative when $n_2 < 0$ (Fig. 1*b*). While positive refraction is a well-known phenomenon, a negative index of refraction leads to many unusual and often surprising effects. For example, Fig. 1*c* and *d* show calculated images of a metal rod in a glass filled with regular water and in a glass filled with negative-index water.

Left-handed world. The refractive index is one of the basic characteristics of electromagnetic wave propagation in continuous media. It is closely related

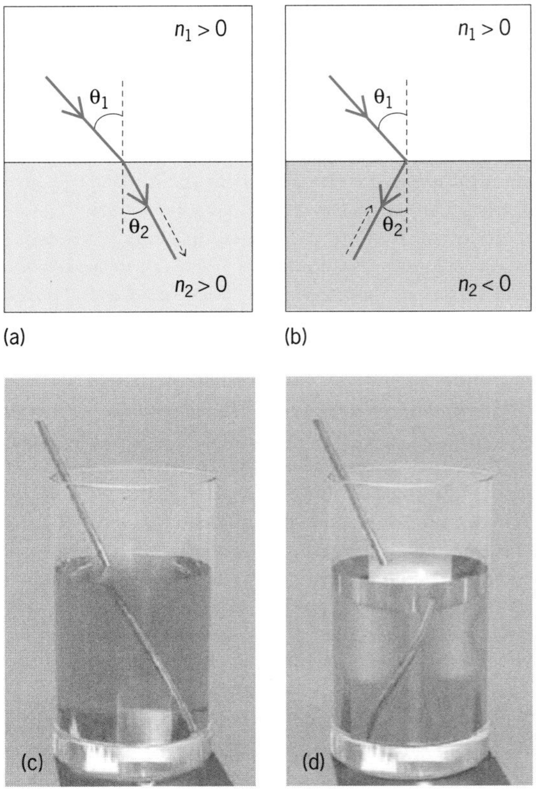

Fig. 1. Refraction: Diagrams of (*a*) positive refraction and (*b*) negative refraction; and calculated images of a metal rod (*c*) in a glass filled with regular water ($n = 1.3$), and (*d*) in a glass filled with "negative-index water" ($n = -1.3$). In parts *a* and *b*, solid lines with arrows indicate the direction of the energy flows, broken lines with arrows show the direction of the wave vectors. (*Parts c and d from G. Dolling et al., Photorealistic images of objects in effective negative-index materials, Opt. Express, 14:1842–1849, 2006*)

to two fundamental physical parameters that characterize material properties, the dielectric permittivity ε and the magnetic permeability μ, through the equation $n = \pm\sqrt{\varepsilon\mu}$. While nearly all transparent conventional materials have positive ε and μ, corresponding to positive n, there are no fundamental physical reasons prohibiting materials from possessing simultaneously negative ε and μ, and as a result negative n. Although not found in nature, such materials were recently created artificially and were named "metamaterials."

A detailed theoretical study of electromagnetic wave propagation in materials with simultaneously negative ε and μ was performed by Victor Veselago in 1967. Maxwell's equations, which relate the electric field E, the magnetic field H, and the wave vector $\boldsymbol{k}$, predict that E, H, and $\boldsymbol{k}$ form a "left-handed" set and the sign of the refractive index is negative if both ε and μ are negative, and a "right-handed" set if both ε and μ are positive. The former class of materials is often referred to as left-handed materials or negative-index materials (NIMs), while the latter class is referred to as right-handed materials or positive-index materials (PIMs). At the same time, the direction of the Poynting vector S, which defines the direction of the energy flow, is the same in positive-index and negative-index materials. Thus, the Poynting vector is antiparallel to the $\boldsymbol{k}$-vector in negative-index materials and is parallel to the $\boldsymbol{k}$-vector in positive-index materials. The opposite directionality of the $\boldsymbol{k}$-vector and the Poynting vector is often taken as the most general definition of negative-index materials. Therefore, the negative refraction illustrated in Fig. 1 is a direct result of the opposite directionality of $\boldsymbol{k}$ and S and of the continuity of the tangential components of the wave vector at the interface between the two media.

Although the term "left-handed materials" was originally coined to describe materials with simultaneously negative ε, μ, and n, currently it is used in a broader context to include other optical structures that possess antiparallel $\boldsymbol{k}$-vectors and Poynting vectors and support negative refraction. Examples of such materials include photonic crystals, anisotropic waveguides, organic and uniaxial gyrotropic crystals and a thin film on a metal substrate, and nanotransmission lines.

Negative refraction has been demonstrated at microwave frequencies in a metamaterial wedge and in the visible frequency range at the interface between a bimetal Au-Si_3N_4-Ag waveguide and a conventional Ag-Si_3N_4-Ag slot waveguide using plasmons. Negative refraction at optical frequencies was demonstrated in photonic crystals. Although many unusual phenomena associated with the negative index of refraction can be observed in photonic crystals, the main limitation of photonic crystals is that the size of their characteristic features is comparable to the wavelength of light. On the contrary, optical metamaterals with feature sizes much smaller than the wavelength of light are predicted to enable many truly remarkable phenomena. However, currently optical metamaterials are available only in the form of subwavelength thin films, thus permitting the measurement of a phase advance but not of negative refraction per se.

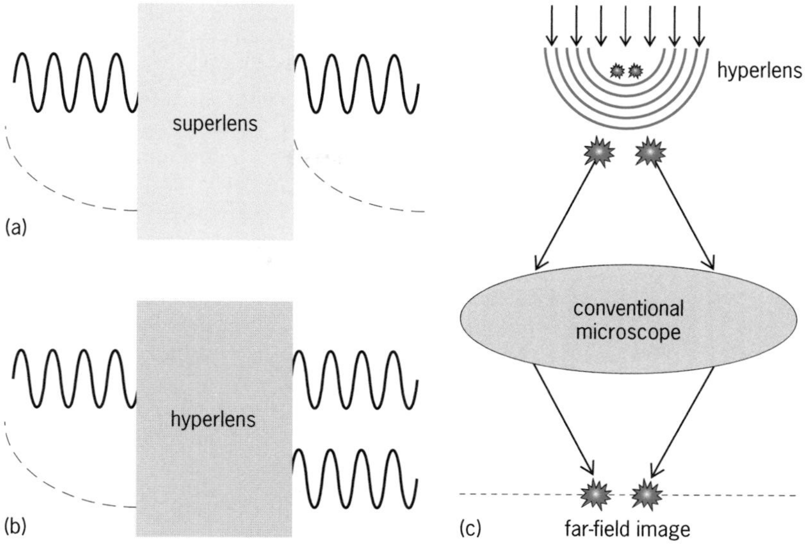

Fig. 2. Schematics of (*a*) superlens, (*b*) hyperlens, and (*c*) imaging system using a hyperlens. In parts *a* and *b*, solid lines correspond to the propagating field components, broken lines correspond to the evanescent field components.

Besides negative refraction, negative-index materials have been predicted to give rise to a wide variety of extraordinary linear and nonlinear optical phenomena, including reversed Cerenkov radiation, the reversed Doppler effect, backward phase-matched second-harmonic generation and optical parametric amplification, lasing without a cavity, bistability, and gap solitons in PIM-NIM couplers with no external feedback.

Superresolution: from "super" to "hyper" lens. A very unusual property of negative-index materials gives rise to the possibility of imaging using a flat slab of negative-index material with $n = -1$ surrounded by a conventional medium with $n = 1$. Moreover, under the appropriate conditions this slab not only focuses propagating field components but also recovers the evanescent field components, which decay exponentially with distance from the source (**Fig. 2***a*), through the excitation of a plasmon resonance on the surfaces of the negative-index material. These evanescent field components, which are responsible for imaging of the high-frequency and correspondingly small-scale features of the object, cannot be restored by conventional lenses, inevitably limiting their resolution. Thus, at least in the ideal (lossless) case, an imaging system based on a slab of negative-index material, named a "superlens" by John B. Pendry, has the potential for significantly improving resolution in the image plane. Unfortunately, a superlensing effect is extremely sensitive to losses in the negative-index-material slab. While the superlens is likely to be useful in numerous near-field applications, including biomedical imaging and nanolithography, superresolution in the far field is challenging.

Recently, a promising solution, a hyperlens, was proposed independently by Nader Engheta and

Evgenii Narimanov. Instead of reamplifying and refocusing the evanescent field components as Pendry's superlens does, a hyperlens converts those evanescent waves into propagating waves (Fig. 2*b*). Once all the components are propagating waves, they can easily be imaged by a conventional lens (microscope) in the far field (Fig. 2*c*). The only remaining limitation of a hyperlens is that the object plane must be situated very close to the hyperlens surface.

Optical metamaterials. Many of the predicted extraordinary properties of negative-index materials would not have been possible without rapid progress in the design and fabrication of optical metamaterials. As mentioned above, no materials in nature possess negative ε and μ in the same range of frequencies. While dielectric permittivity of some existing materials is negative at certain frequencies, no isotropic materials with negative μ are known. Moreover, magnetism is usually weak at optical frequencies so that $\mu \approx 1$.

On the contrary, metamaterials are built of artificial or "meta" atoms, which are resonant structures such as split-ring resonators and paired metal nanorods or nanostrips. The meta-atoms can be engineered and arranged such that their ε, μ, and n are positive, negative, or even zero at any selected frequency. The first optical metamaterials with a negative index of refraction have been demonstrated using paired nanorods, and independently by another group using paired dielectric voids in metal.

While these first experiments confirmed the possibility of the realization of a negative index of refraction at optical frequencies, the negative-index materials were realized only in the form of subwavelength films and possessed significant losses. Some essential requirements for practical negative-index-material designs include minimized losses or a large ratio of the real and imaginary parts of n, often taken as a figure of merit; a broad bandwidth over which both ε and μ are negative; optimized impedance matching; and realization of three-dimensional negative-index materials.

Using a self-supporting fishnet structure consisting of rectangular dielectric voids in parallel metal films, a figure of merit of 3 has been demonstrated at a wavelength $\lambda = 1.4\ \mu$m. This structure represents the current state-of-the-art for optical negative-index materials. Recently, the first three-layered negative-index material with a figure of merit of 2.5 at $\lambda = 1.41\ \mu$m was also reported.

Refractive index engineering. While one of the original motivations behind the development of metamaterials was the demonstration of negative-index materials, metamaterial technology has stimulated rapid progress in an entirely new and exciting branch of modern optics, transformation optics, which is based on the idea of mapping a coordinate transformation to a set of material parameters, ε and μ. Metamaterials allow precise control over these material parameters and, more generally, enable refractive index engineering. Such mapping turned out to be particularly useful for cloaking applications and facilitated the first experimental demonstration of cloaking of a copper cylinder at microwave frequencies. In that experiment the object was concealed by a cylindrical metamaterial cloak built using split-ring resonators. The coordinate transformation and a schematic of the first nonmagnetic cloak operating at optical frequencies, as proposed theoretically, are illustrated in **Fig. 3*a***. Figure 3*b* shows the results of numerical simulations of cloaking of an ideal metallic cylinder. Currently, in both microwave and optical cloak designs, the effect has been achieved at only one frequency. Obviously, broadband cloaking would be desirable for most practical applications, and further research is therefore required.

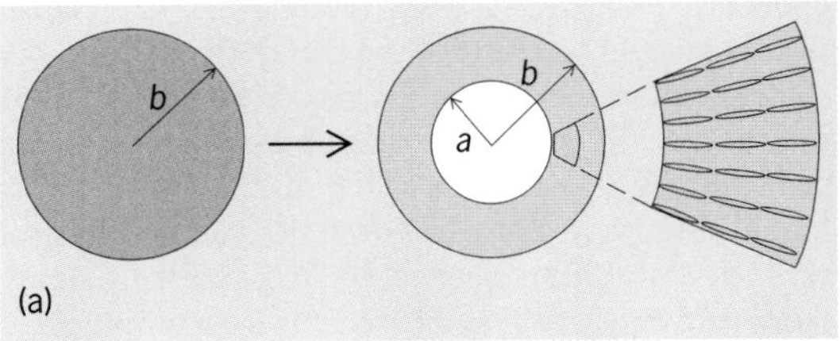

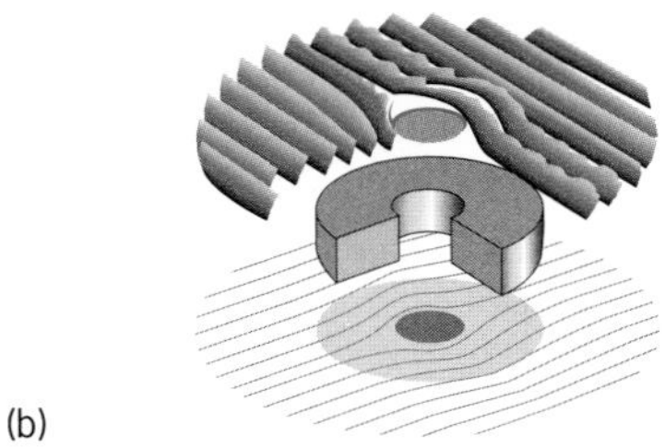

Fig. 3. Cloaking. (*a*) The transformation of a cylindrical region $r < b$ into a concentric cylindrical shell $a < r < b$ and an enlarged section of a nonmagnetic optical cloak built with metal wires embedded in a dielectric host. (*b*) Numerical simulations of an ideal metallic cylinder with radius $r = a$ illuminated with TM (tranverse-magnetic)-polarized wave with the cloak turned "on."

Finally, cloaking is only one realization of the great potential of transformation optics in conjunction with metamaterials. Other promising applications include field concentrators and a variety of reflectionless devices. Metamaterials are bringing new degrees of freedom for designing structures with almost any desired optical properties, thus presenting enormous opportunities for a wide range of applications relying on refractive index engineering.

For background information *see* MAGNETISM; MAXWELL'S EQUATIONS; PREMITIVITY; PLASMON; POYNTING'S VECTOR; REFRACTION OF WAVES in the McGraw-Hill Encyclopedia of Science & Technology.

[The authors gratefully acknowledge the support of the Army Research Office through Grants W911NF-07-1-0343 and 50342-PH-MUR.]

Natalia M. Litchinitser; Vladimir M. Shalaev

Bibliography. G. V. Eleftheriades and K. G. Balmain, *Negative Refraction Metamaterials: Fundamental Principles and Applications*, Wiley-IEEE Press, 2005; V. M. Shalaev, Optical negative-index metamaterials, *Nature Photonics*, 1:41–48, 2007; E. Shamonina and L. Solymar, Metamaterials: How the subject started, *Metamaterials*, 1:12–18, 2007; V. Veselago et al., Negative refraction index materials, *J. Comput.*

Theor. Nanosci., 3:189–218, 2006; V. G. Veselago and E. E. Narimanov, The left hand of brightness: Past, present and future of negative index materials, *Nat. Mater.*, 5:759–762, 2006.

New coatings for wood

There has never been a broader array of coatings for wood available on the market than now. In the past, coatings for wood, such as stains, primers, and top coats, were primarily oil-based. However, as new air-quality regulations have mandated lower volatile organic content (VOC) in architectural coatings throughout the United States and now proposed in Canada, manufacturers have had to reformulate their products to meet the new regulations. [Volatile organic compounds are organic chemicals that produce vapors readily at room temperature and normal atmospheric pressure, including gasoline and solvents such as toluene, xylene, and tetrachloroethylene, which form photochemical oxidants (including ground-level ozone) that affect health, damage materials, and cause crop and forest losses; many are also hazardous air pollutants.]

There are two primary approaches to reformulating a low-VOC coating for wood. In the majority of these reformulations, the solvent portion of the product has been replaced with water. In a smaller percentage of these reformulations, the amount of solid ingredients has increased significantly to produce a high-solids coating. Given the special characteristics of wood, each of these approaches has presented challenges to the coatings formulator and, ultimately, the user of these coatings.

Characteristics of wood. Wood is one of the world's most common materials of construction. As such, it is a renewable resource, easy to use, and durable for centuries when properly maintained. It can come from hardwood, softwood, or tropical wood species. However, regardless of origin, once the tree is harvested from the forest, it becomes vulnerable to attack by a host of enemies. Degradation can come from water, sunlight, insects, and microorganisms.

By far, water is wood's worst enemy. Because wood comprises about 50% cellulose and 25% hemicellulose, it is subject to swelling and shrinking as it gets wet and dries. Continued wet/dry cycles create a continuous movement of wood that causes stress between the wet surface and the dry interior. This stress causes cracking, warping, bowing, twisting, and cupping of wood, resulting in structural problems. This excessive moisture also invites microorganisms such as mold and mildew to grow, causing aesthetic problems. Continued exposure to moisture will lead to rot and destruction of the wood itself.

Thus, understanding the characteristics of wood is critically important for the coatings formulator. The very nature of wood's reaction to water is what makes an oil- based coating an easier product to use on wood and a water-based coating more difficult to use on wood. When water-based coatings are applied to wood, they will usually swell the grain of the wood, causing grain raising (**Fig. 1**). This usually results in the need for sanding, especially on fine furniture and cabinetry. Because they tend to dry faster as the water soaks into the wood, water-based stains are subject to lapping, which is seen as a darker area at the overlap of two brush strokes applied side by side. Other problems can occur in exterior water-based coatings for wood such as poor water resistance and adhesion failure compared to oil-based coatings. In order to understand why these problems occur, it is necessary to know the basics of how water-based coatings are formulated and how they differ from their oil-based counterparts.

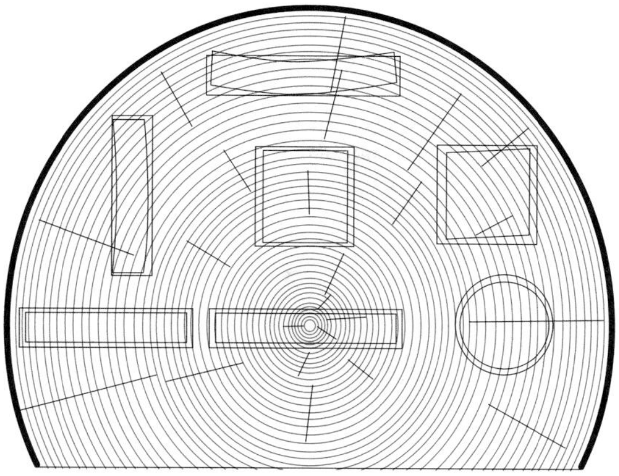

Fig. 1. Mixed grain patterns cause differences within wood that result in cupping and warping. Where the wood is cut from the tree determines how the wood will warp when exposed to water. (*From Wood Handbook: Wood as an Engineering Material, USDA Forest Service*)

Conventional oil-based coatings. Conventional oil-based coatings usually have several basic categories of ingredients that can be broken down into four main groups: solvent, binder, pigments/fillers, and driers/additives. The solvent acts as a carrier for the other ingredients and usually comprises one or more petroleum distillates such as mineral spirits, mineral oil, or xylene. It is this component of oil-based coatings that contributes to the depletion of the atmospheric ozone layer, and its content is now regulated by governmental agencies.

Binders can be as simple as drying oils such as linseed, tung, teak, or soybean oil or more highly formulated chemicals such as alkyds (a class of adhesive resins made from unsaturated acids and glycerol), polyurethane, epoxies, silicones, and fluorinated polymers. These generally deliver the bulk of the protection properties to the wood. Binders can be used by themselves or in combination.

Pigments and fillers impart color and opacity to a coating. Pigments generally are composed of iron oxides that result in basic brown, red, and yellow tones, but they can be as sophisticated as highly formulated organic molecules that impart stronger, intense colors like deep greens, reds, and blues. The most common white pigment is titanium dioxide. Fillers are usually made up of mined materials such as clay, calcium carbonate, mica, talc, or diatomaceous earth (yellow, white, or light-gray, siliceous, porous deposit made of the opaline shells of diatoms). They

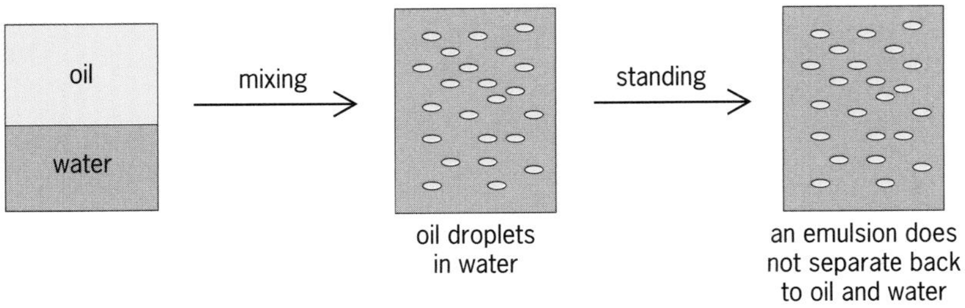

Fig. 2. An emulsion is a system containing two liquids (such as oil and water), one of which is incompletely miscible in the other and is dispersed as discrete droplets or particles into the other liquid. However, upon standing, the droplets remain dispersed.

are designed to fill voids between the pigments and the binder and occupy space in dried coating film.

Driers are used in very small amounts and in combination to initiate the drying of the coating after it is applied to the wood. These usually comprise organic metal compounds containing cobalt, manganese, zirconium, calcium, nickel, and others. They work to accelerate the cross-linking reaction of the binder, resulting in a fully developed polymer coating on the wood. When they are omitted from a formula, the coating will not dry properly. Other additives such as defoamers may be added to aid in speedy manufacture and good application characteristics.

Conventional oil-based coatings generally provide good protection to wood because of their water-resistant nature. Since oil and water do not mix, it follows that the oil-derived components that make up an oil-based coating would provide protection from water. In addition, the solvent carrier helps those ingredients penetrate well below the surface of the wood, giving it good adhesion to the wood. Since there is no water in the coating, there is no wood swelling, grain raising, or lapping.

High-solids, low-VOC coatings are essentially concentrated versions of the conventional oil-based coating. Many of the ingredients are the same, but they usually contain about a third of the solvent found in the conventional oil-based coating. As a result, they use enhanced drier combinations to facilitate drying a higher amount of solids left on the wood. However, overapplication of these products leads to a tacky residue on the wood surface. For exterior products, high-solids coatings can also cause higher rates of mold and mildew growth on the film if not properly formulated.

Water-based coatings and basic emulsion chemistry. The basic approach to designing water-based counterparts to oil-based coatings for wood is to remove as much solvent as possible, replace it with water, and modify the binder such that it can be carried in water. Since the binders are not naturally water-loving (hydrophilic), chemical modifications and other additives had to be developed to make this possible. Thus, emulsion chemistry has evolved to address these needs. Because oil and water do not mix, they separate when they are put together and shaken. However, an emulsion is a system containing two liquids, one of which is incompletely miscible in the other and is dispersed as discrete droplets or particles into the other liquid. However, upon standing, the droplets remain dispersed (**Fig. 2**).

Thus, what stabilizes these oil-like materials in water? Surface-active agents, known as surfactants, surround oil droplets and stabilize them in the water phase (**Fig. 3**). Because light is refracted by the particles, the emulsion looks milky.

The chemistry of the surfactant molecule(s) is a complicated subject. However, in simplistic terms, a surfactant acts as a chemical bridge between the oil and water phases of the emulsion. One part of the molecule has an affinity for the oily-like binder phase, whereas another part has an affinity for the water phase of the emulsion. The surfactants surround the oil phase and form discrete droplets called micelles that can be measured (**Fig. 4**). These particles are typically less than 1 micrometer in diameter; moreover, the smaller they are, the more translucent the emulsion.

While surfactants are generally necessary to emulsify the oil-like binders in a coating formula to make it water-based, they have a few drawbacks that make them problematic for coating wood. As the emulsion is applied to the wood, the droplets break and deposit the ingredients at the surface. The water phase evaporates, the oily phase penetrates into the wood, and the surfactant is deposited onto the surface. It becomes trapped into the film polymer matrix as

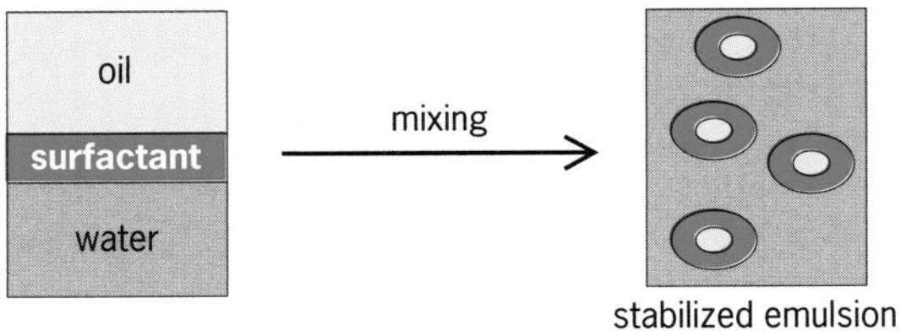

Fig. 3. Surfactants (surface-active agents) surround oil droplets to stabilize them in the water phase.

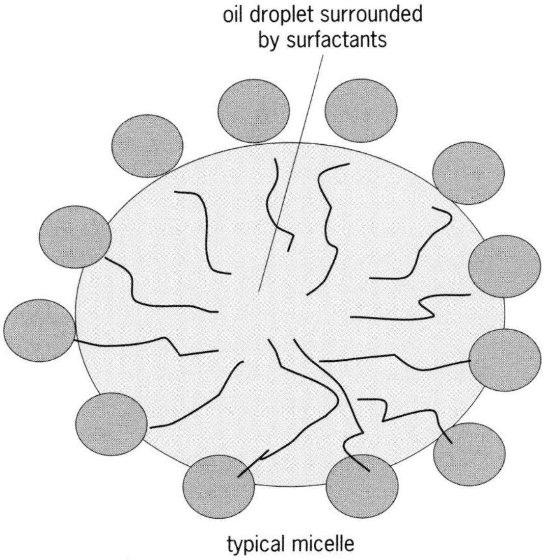

Fig. 4. Surfactants surround the oil phase and form discrete droplets called micelles.

the coating dries. The deposited surfactants still retain their affinity for water after application on the surface. This makes the wood more water-sensitive instead of imparting the desired protection features. This represents the fundamental difference between oil-based and water-based coatings for wood and why the formulator must test the performance of water-based coatings when they are applied to wood. Since the wood swells even more in the presence of surfactants and water, many parameters such as grain raising, lapping, water resistance, and adhesion can be negatively impacted.

Fortunately, the major binder manufacturers have recognized this problem and new binders are being developed every day with lower surfactant content. As more testing is done on water-based acrylics, polyurethane, and other binder emulsion systems that are incorporated into making a coating for wood, better lower-VOC coatings that perform well and meet the latest air-quality regulations are offered in the marketplace. It does behoove consumers or contractors to do their homework and ask a lot of questions of the coating manufacturer before making their purchase and coating their wood.

For background information *see* EMULSION; LACQUER; PAINT AND COATINGS; PRIMER (SURFACE COATING); SHELLAC; SURFACE COATING; SURFACTANT; VARNISH; WOOD DEGRADATION; WOOD PROPERTIES in the McGraw-Hill Encyclopedia of Science & Technology.

Victoria Scarborough

Bibliography. M. J. Rosen, *Surfactants and Interfacial Phenomena*, Wiley, New York, 2004; R. M. Rowell (ed.), Moisture properties, chap. 4 in *Handbook of Wood Chemistry and Wood Composites*, CRC Press, New York, 1995; USDA Forest Service, *Wood Handbook: Wood as an Engineering Material*, General Tech. Rep. 113, USDA Forest Service, Forest Products Laboratory, Madison, Wisconsin, 1999; Z. W. Wicks et al., Organic Coatings: Science and Technology, vol. 1: *Film Formation, Components, and Appearance*, Wiley, New York, 1992.

New species of Borneo

More than 400 new species of plants and animals have been discovered on the island of Borneo over the last decade. These include many new plant and insect species, but also several new fish, amphibian, and reptile species. In addition, remarkable discoveries on the genetic origin of three large mammal species have been made.

Island of Borneo. The natural richness of Borneo has attracted scientists and explorers for hundreds of years, yet it takes much effort to uncover the natural secrets of the island. The interior of Borneo is still a remote area that is difficult to reach. Most of it is very mountainous, and few inland roads exist. Upstream river travel from the coast is only possible up to the locations where the many large and impassable rapids block further boat traffic. From here on, the explorer has no other choice than entering the rugged mountain terrain by foot.

However, with so many parts of the globe already explored, more scientists are interested in coming to the wilderness of Borneo and traveling to some of the last unexplored places on earth, which is why so many new species have been discovered recently.

The largest part of Borneo is Indonesian territory, usually referred to as Kalimantan. The northern part belongs to Malaysia (the states of Sarawak and Sabah) and Brunei Darussalam (**Fig. 1**).

Borneo is the second largest island of the Malay Archipelago and straddles the equator. The island has a humid tropical climate with high rainfall. Although some seasonal fluctuations exist, there is no significant dry season without much rain. The soils of most parts of Borneo are rather poor, which is one of the reasons that the human population level has never been high here. Until a few decades ago, most of Borneo was covered with dense tropical rainforests, which have developed uninterruptedly for millions of years. Consequently, an enormous number of plant and animal species could evolve, making Borneo one of the most important centers of biodiversity in the world.

Discovery of new species. Not all new species are immediately recognized as such when found in the forest: New species are discovered in various ways.

Scientists exploring a forest or a stream might come across a plant or animal that they immediately distinguish as a species that has not yet been scientifically described and classified. This can happen only to scientists who have an extensive knowledge of the local flora or fauna, and usually involves specialists of a certain group of plants or animals. For instance, a new tree species of *Coelostegia* was discovered inadvertently during a routine inventory of a montane forest plot, and a new frog species was discovered when rocks were turned over in a stream. Specimens of a potential new species are collected and need to be verified with descriptions of existing species before being classified and published as a species new to science.

Occasionally, scientists come across an animal or plant part for sale at a local market, or given to them by local people, that came from an unknown species. This happened, for instance, with a new species of catfish that was found at a market.

New species are frequently discovered through examination of earlier collected specimens. This often happens when taxonomists work on the revision of a certain taxon, for which they need to precisely check many specimens from various collections. Scientists doing inventories of the flora or fauna of a certain area will collect as many different specimens as possible, and only when checking their collection afterward find that it includes one or more previously unknown species. For instance, *Schumannianthus monophyllus*, a new plant species, was collected in the 1950s, but only published as a new species in 2006.

The latest techniques of DNA analysis have enabled scientists to determine how long ago two different varieties of a species became genetically separated. In this way, it was discovered that the Bornean

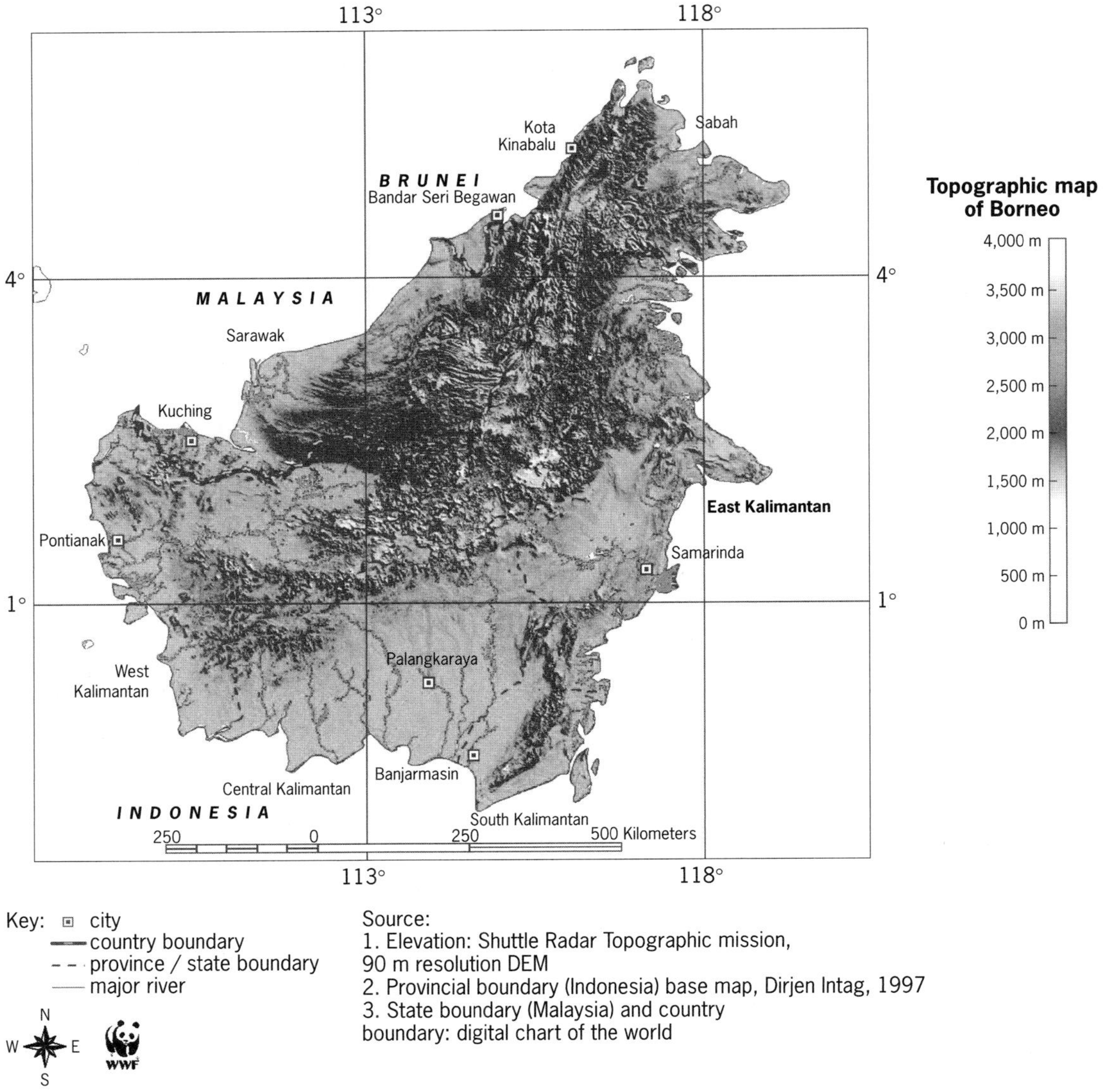

Fig. 1. Topographic base map of Borneo. (*Courtesy of WWF-Indonesia*)

orangutan and the Sumatran orangutan separated at least 1.1 million years ago, and have since then followed different evolutionary paths. The accepted separation distance for species is 1–2 million years, and the Bornean orangutan (*Pongo pygmaeus*) is thus now considered a new and different species. This technique has so far only been applied to a few species, but has already shown significant results.

Plants. Hundreds of new plant species of Borneo have been described during the last decade. Scientific descriptions and classifications of individual, new plant discoveries are usually published in botanical journals and receive relatively little attention. Once revisions and compilations of plant groups have been issued in major publications, this new information attracts the attention of a wider audience.

Orchids are among the most popular plants, and are sought after by many collectors as well as scientists. Searching for Bornean orchids often involves climbing since many species are epiphytes, growing nonparasitically on tree branches. More than 80 new orchid species have been discovered during the last ten years.

Wild gingers (Zingiberaceae) are common at the forest floors of most rainforests of Borneo. Apart from a few species found almost everywhere in the lowlands, there are also many species with a restricted distribution. During the last decade, more than 55 new ginger species were discovered.

Rattans, the climbing palms of Asia, were thought to be well researched in Borneo since many species are manufactured into handicraft, tools, and furniture. However, several new species from high elevations were discovered after extensive research on Mount Kinabalu, in the state of Sabah.

Occasionally, even new tree species are discovered. The aforementioned new *Coelostegia* species is a rather big tree with conspicuous bluish and thorny large fruits (**Fig. 2**). It escaped discovery for so long only because of its restricted range in the mountains. Other newly discovered trees include six species from the laurel family (Lauraceae), five new

Elaeocarpus species, and even a new tropical oak species, *Lithocarpus palungensis*.

Invertebrates. Invertebrates are by far the largest and most complicated group of animals, and new species are still being discovered throughout the world almost on a monthly basis. Certain orders such as Bornean stick insects and butterflies have been reasonably well covered, while other orders still need extensive research. During the last few years, more than 260 new insect and spider species were described. Many of these include beetles (Coleoptera), flies (Diptera), true bugs (Heteroptera), and spiders (Arachnida).

Four new species of freshwater crabs as well as two new species of freshwater prawns were discovered from the rivers of Borneo.

The most spectacular new insect discovery is that of the giant cockroach. This species was found during a scientific expedition to a large limestone landscape in the easternmost part of Borneo. The giant cockroach has a length of 10 cm (4 in.) and is believed to be the largest cockroach in the world.

Fish. Considerable progress has been made on the research of fish species of Borneo over the last decade. No less than 14 new species of catfish were discovered in recent years. Some of these, the labyrinth catfishes (Clariidae), are capable of spending time outside the water and can even travel short distances over land. These fishes are equipped with a labyrinth organ (an accessory respiration organ) arising from their gills, which enables them to take in oxygen from the air.

A revision of the suckerfishes of Borneo [the taxa *Gastromyzon* (**Fig. 3**) and *Neogastromyzon*] combined with additional field research revealed 19 new species. Suckerfish have enlarged fins that enable them to cling onto rocks in fast-flowing rivers (most common in the interior of Borneo) where they feed on algae and insects.

Siamese fighting fish of the genus *Betta* are well known amongst aquarium lovers. In 2006, six new species were described from the rivers of Borneo.

Perhaps the most interesting fish discovery is the new species of miniature fish, *Paedocypris micromegethes*. This species has an adult length of 8.8 mm (0.35 in.), making it one of the smallest fish species in the world. It was found in a blackwater stream (slow-moving water in which acids are leeched from decaying vegetation to create very transparent, tea-colored water) in a peat swamp forest and is restricted to this type of habitat. Since most peat swamp forests have been destroyed or are under heavy environmental pressure, species such as this miniature fish are highly endangered.

Amphibians. The frogs and toads of Borneo were thought to be well known by now. However, during the last decade, no less than five new species were discovered. These comprise species from restricted locations from the highlands as well as tree frog species, which are generally difficult to detect.

Reptiles. The snake species of Borneo are relatively well known after years of time-consuming inventories. These animals are usually found by coincidence, and two remarkable new snake species, both from west Kalimantan, were discovered during the last decade. One is a new *Enhydris* species, a 0.5-m (1.6-ft) long mud snake, which can change its color like a chameleon. The other species belongs to the *Hydrophis* taxon. This group was thought to consist mainly of marine species; however, the new species was surprisingly found about 1000 km (620 miles) upriver.

Fig. 2. Fruits of *Coelostegia montana*, a new tree species from the mountains of East Kalimantan, Borneo. (*Photo by Stephan Wulffraat, courtesy of WWF-Indonesia*)

Six new lizard species were discovered, including two species of snake-resembling legless lizards (family Dibamidae) that were discovered inside a rotten log and under the surface.

Mammals. The latest discovery of a mammal species new to science involves the Bornean clouded leopard (*Neofelis diardi*) [**Fig. 4**]. It was always assumed that the clouded leopards of Borneo and of the mainland are the same species. However, DNA analysis established that this species became separated from the mainland species at least 1.4 million years ago, and additional analysis of morphology has further proved significant differences.

A few years ago, two photographs were taken by an automatic camera trap of a mysterious animal that resembled a red civet or marten. The published photographs led to wide speculations on the identity of

Fig. 3. *Gastromyzon* zebrinus, a new species of suckerfish from West Kalimantan, Borneo. (*Photo and copyright by Tan Heok Hui*)

Fig. 4. Bornean clouded leopard (*Neofelis diardi*). (*Copyright by WWF–Canon/Alain Compost*)

this animal. Unfortunately, not a single specimen has been collected so far, a necessity for properly identifying this species.

A major discovery was made in 2003 when it was found through DNA analysis that the elephants of Borneo became separated from the other Asian elephants at least 300,000 years ago. This established for the first time that the Bornean elephants were not introduced, but are indigenous to Borneo.

Outlook. Large primary forest areas of Borneo have disappeared in the last decades due to extensive logging and conversion into oil palm and other plantations. Devastating forest fires have destroyed millions of hectares (1 hectare = 10,000 m^2) of forest land. The enormous biological richness of Borneo is clearly demonstrated by all of the new species discovered, even in the twenty-first century. All possible effort should now be taken to save the remaining forests of Borneo because otherwise many species will be lost forever.

For background information *see* BIODIVERSITY; CONSERVATION OF RESOURCES; EAST INDIES; ECOLOGICAL COMMUNITIES; ISLAND BIOGEOGRAPHY; POPULATION ECOLOGY; RAINFOREST; SPECIATION; ZOOGEOGRAPHY in the McGraw-Hill Encyclopedia of Science & Technology.

Stephan Wulffraat

Bibliography. WWF-Indonesia, Biodiscoveries: Borneo's botanical secret, World Wide Fund for Nature-Indonesia, Jakarta, 2006; WWF-International, Borneo's clouded leopard identified as a new cat species, World Wide Fund for Nature, Gland, Switzerland, 2007; WWF-Indonesia, Borneo's lost world: Newly discovered species on Borneo, World Wide Fund for Nature-Indonesia, Jakarta, 2005; WWF-Indonesia, The search continues: The latest remarkable species discoveries from Borneo July 2005 to September 2006, World Wide Fund for Nature-Indonesia, Jakarta, 2006.

Nobel prizes

The Nobel prizes for 2006 included the following awards for scientific disciplines.

Chemistry. The chemistry prize was awarded to Roger D. Kornberg of Stanford University for determining the molecular basis of transcription in eukaryotic cells, a more complex process than transcription in bacteria, which lack a cell nucleus.

To do this, Kornberg used genetically modified yeast cells, which are simple eukaryotes, and a technique known as x-ray crystallography to determine the atomic structure of the crystallized proteins. X-ray crystallography does not produce an actual image of the atoms in a molecule. Instead, it maps the distribution of electrons in the molecule from which a representation of a three-dimensional structure is computer-generated.

Transcription is a chemical reaction whereby an amino acid sequence (gene) from deoxyribonucleic acid (DNA) is copied as a strand of messenger ribonucleic acid (mRNA). After transcription, mRNA is transported to the ribosome in the cytoplasm, where the genetic information is used to synthesize proteins. In eukaryotic cells, transcription takes place within the cell nucleus where the enzyme RNA polymerase II, aided by the general transcription factors and regulated by a molecular complex known as Mediator, copies genes.

The biochemical mechanism of transcription is important not only because it is necessary for synthesizing all the proteins needed to build and maintain an organism, but also for understanding the cause of, and devising treatments for, diseases involving protein mutations, such as cancer.

In 2001, Kornberg published x-ray diffraction results showing yeast RNA polymerase, DNA, and the beginning of an RNA chain. In essence, he captured the process of transcription in action. Kornberg obtained this "picture" by leaving one of RNA's four bases (adenine, cytosine, guanine, or uracil), thereby stopping the transcription process.

In earlier publications, Kornberg and colleagues reported the discovery of Mediator. And in later publications, Kornberg's laboratory has shown the structures of RNA polymerases in various stages of the transcription process as well as DNA, RNA, and the general transcription factors.

Kornberg's laboratory continues working on the structural biology and biochemistry of transcription and the mechanism of transcription control.

For background information *see* AMINO ACIDS; BIOCHEMISTRY; CELL (BIOLOGY); CELL NUCLEUS; DEOXYRIBONUCLEIC ACID (DNA); ENZYME; EUKARYOTAE; GENE; GENE ACTION; NUCLEIC ACID; PROTEIN; RIBONUCLEIC ACID (RNA); TRANSCRIPTION; X-RAY CRYSTALLOGRAPHY in the McGraw-Hill Encyclopedia of Science & Technology.

Physiology or medicine. Andrew Z. Fire (Stanford University, California) and Craig C. Mello (University of Massachusetts Medical School, Worcester) shared the Nobel Prize for Physiology or Medicine for their discovery that double-stranded ribonucleic acid (RNA) can trigger the suppression of gene activity in a homology-dependent manner, a process termed RNA interference (RNAi).

In 1990, scientists performing experiments in which multiple copies of a gene responsible for the darkening of the color of the flowers of Petunia were introduced into its genome observed the development of white or patchy (variegated) flowers as well as darker flowers. Apparently the white or variegated plants had recognized the "foreign" gene and silenced it along with the homologous endogenous gene, a process named cosuppression or homology-dependent gene silencing. Subsequent experiments showed that RNA transcribed from the introduced genes (transgenes) initiated the silencing. In 1998 Fire, Mello, and colleagues reported their discovery that the injection of double-stranded RNA (dsRNA) into the genome of the roundworm (nematode) *Caenorhabditis elegans* led to the sequence-specific breakdown of cytoplasmic messenger (that is, posttranslational) RNA containing the same sequence as the dsRNA trigger. This so-called RNA interference was related to the cosuppression seen earlier in Petunia. Their work further showed that only double-stranded RNA could produce the effect; that only relatively few molecules were needed indicting that the dsRNA was acting catalytically; and that the effect could spread between tissues suggesting that it could be transmitted between cells.

Since these initial discoveries, research has revealed that not only introduced dsRNA but also dsRNA synthesized within the cell can have RNAi-like effects by reducing or silencing gene activity, thus constituting an essential control system for gene expression in many cellular processes from development of the organism to its physiological functioning. Furthermore, RNAi has been shown to protect against RNA virus infections in a variety of organisms including plants and invertebrates, although its role in viral immunity in vertebrates including humans remains to be elucidated. It is also believed to aid in maintaining the stability of the genome by silencing mobile elements.

RNAi is now a widely used tool in broad areas of basic and applied research. In fact, the function of almost any gene can be investigated since the specificity of the RNAi effect allows researchers to suppress the expression of specific genes and follow the resulting phenotypic effect.

The discovery of RNA interference is considered one of the major advances in biomedical knowledge in recent decades with much promise for new applications in medicine.

For background information *see* DEOXYRIBONUCLEIC ACID (DNA); GENE; RIBONUCLEIC ACID (RNA) in the McGraw-Hill Encyclopedia of Science & Technology.

Physics. John C . Mather (NASA Goddard Space Flight Center, Greenbelt, Maryland) and George F. Smoot (University of California, Berkeley) shared the Nobel Prize in Physics for their discovery of the blackbody form and anisotropy of the cosmic microwave background radiation.

One of the most significant discoveries in modern cosmology was the detection of the microwave background radiation by A. Penzias and R. Wilson in 1963. It is widely accepted to be the leftover heat from the big bang, and indeed its discovery gave credence to this theory. According to the hot big bang model, the universe started in an extremely hot, dense state composed of electrons, neutrinos, photons, positrons, and quarks. With the expansion of the universe, much of the matter annihilated with antimatter to form photons, which then cooled with further expansion. According to thermodynamic theory, the spectrum of the radiation from the big bang should be similar to that emitted by a blackbody (Planck spectrum).

Observations made with NASA's *Cosmic Background Explorer* (*COBE*) satellite, launched in 1989, provided strong confirmation of the thermodynamic predictions. The observed spectrum closely agreed with the predicted blackbody spectrum, thereby excluding virtually any explanation for the background microwave radiation other than the big bang theory. Other aspects of the model, such as the assumption of homogeneity of the radiation, were also confirmed. An important observation of *COBE* was the extremely small variations in temperature in varying directions (anisotropy). These pointed to the beginnings of aggregation of matter in the early universe, a necessary precursor to the development of stars and galaxies.

COBE had been proposed back in the 1970s, but setbacks such as the loss of the *Challenger* space shuttle delayed its launch. John Mather was credited with being the coordinator of the entire project as well as having principal responsibility for the experiment that demonstrated the blackbody form of the microwave background radiation. George Smoot was responsible for the measurement of the variations in temperature. *COBE*'s four-year mission has been followed by other spacecraft, including NASA's *Wilkinson Microwave Anisotropy Probe* (*WMAP*), which measured microwave background radiation with greater accuracy and thereby provided extensive information on, for example, the age and composition of the universe.

For background information *see* BIG BANG THEORY; COSMOLOGY; COSMIC BACKGROUND RADIATION; INFLATIONARY UNIVERSE COSMOLOGY; WILKINSON MICROWAVE ANISOTROPY PROBE in the McGraw-Hill Encyclopedia of Science & Technology.

Odontogriphus: earliest mollusk

The origin and early evolutionary history of mollusks, one of the most diverse and fascinating groups of animals alive today, is traditionally based on the study of shells found in fossil deposits dating from the Early Cambrian, about 540 million years ago. Different hypotheses concerning the supposed morphology of the ancestral mollusk have emerged in the last 20 years, one preponderant view being that early forms were microscopic and possessed a simple mineralized shell. The restudy of a collection of half-billion-year-old fossils from the Burgess Shale (a fossil deposit featuring exceptionally preserved Middle Cambrian marine biota located in southern British Columbia) opens new insights concerning early molluscan evolution.

General description of mollusks. Mollusks, including the familiar snails, bivalves, and octopods, represent the second most diversified phylum on Earth, with up to 250,000 described living species—only the arthropods boast greater variety. Most mollusks are aquatic, but some have also conquered terrestrial habitats; they can be grazers, suspension feeders, parasites, and carnivores. Extant members of the phylum are grouped into seven classes, with the gastropods (snails) representing almost half of all known living species. Recent mollusks are typically characterized by the presence of a shell (or shells), sclerites as spines or scales, one or more pairs of gills (ctenidia), a muscular foot, and a feeding apparatus called a radula. Various of these features are modified or secondarily lost in some groups.

Molluscan roots. The oldest convincing molluscan shells are millimetric in size and are found in small shelly fossil (SSF) assemblages (mineralized microfossil skeletons that mark the first appearance of skeletal hard parts) dating from the Early Cambrian, around 540 million years ago. These SSF assemblages represent the first compelling evidence for a rapid and massive diversification of animal life (referred to as the Cambrian explosion). However, the "mineralization" of many groups during the Cambrian explosion probably involved only a small fraction of animal diversity. Reconstructing the early history of mollusks based on shells alone is therefore bound to be incomplete and misleading.

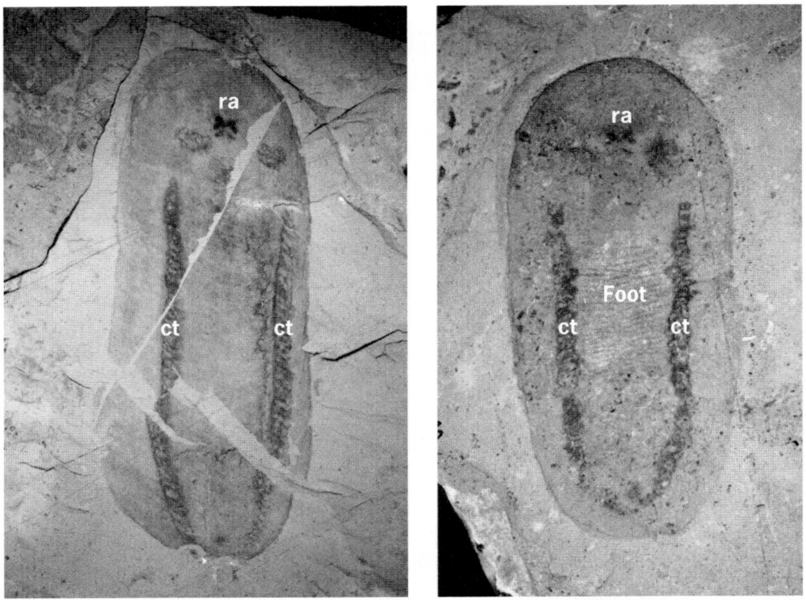

(a) (b)

Fig. 1. *Odontogriphus omalus* **from the Burgess Shale. Type specimen numbers are from the Royal Ontario Museum. (*a*) ROM57720. Specimen length 107 mm (4.2 in.). (*b*) ROM57723. Specimen length 71 mm (2.8 in.). Terms: ra, radula; ct, ctenidia (gills).**

Burgess Shale. In the pantheon of paleontological marvels, the 505-million-year-old Middle Cambrian Burgess Shale has acquired an almost mythical status. This site, discovered in 1909 by Charles Walcott, is remarkable for the stunning preservation of nonmineralized parts such as guts, eyes, appendages, and soft body outlines. Without the Burgess Shale and other comparable sites in the Early Cambrian of China (for example, Chengjiang), our understanding of the origins and diversification of animals would be extremely impoverished, for almost 90% of the genera in the Burgess Shale would not have left any remains under normal conditions of fossilization.

Odontogriphus, a Cambrian chimera. *Odontogriphus* is truly emblematic of the Burgess Shale fauna. The singular original specimen was collected by Walcott himself almost 100 years ago, but was not recognized and formally described as a distinct organism until 1976. Like many other animals from the Burgess Shale, *Odontogriphus* was considered a chimera (displaying a variety of characters from different groups of animals) and remained in a phylogenetic limbo for decades.

Despite their often superb preservation, carcasses of Burgess Shale organisms were affected by many processes, in particular by decay (the study of such processes, from the time of death of an organism to burial and fossilization, is known as taphonomy). A large sample size of soft-bodied organisms in the Burgess Shale will always show a range of preservational quality, from perfectly connected tissues to almost unrecognizable fragmented shapes. Walcott's original specimen of *Odontogriphus* shows obvious signs of decay and is incomplete, and this poor taphonomic condition most likely led to the uncertain placement of this organism in the panoply of phyla.

Recently, the discovery of nearly 200 new *Odontogriphus* specimens by researchers from the Royal Ontario Museum provides a much more complete view of the animal (**Fig. 1**). The body is elongated, flat, smooth, and semicircular in outline. The largest specimens were up to 12 cm (4.7 in.) in length and 4 cm (1.6 in.) in width. The outer body covering was probably thick, but there is no evidence of a shell or spines. A ventral muscular foot was surrounded on each side and at the rear by up to 100 gill-like structures (ctenidia) [Fig. 1*b*]. Internal anatomy included a complete gut with a prominent stomach, straight intestine, and subterminal anus. One of the most important features is located in the mouth region near the anterior end of the animal (**Fig. 2***a*,*b*). In Walcott's specimen, this part of the body was not particularly clear, but the new specimens possess a pair of chevronlike tooth rows. Each row is composed of two teeth, which are mirror images of each other. The teeth have pointed denticles facing backward. We interpret this mouth apparatus as homologous to a radula—a unique feature of the mollusks (Fig. 2*c*). In mollusks, the radula is composed of chitinous teeth in rows along a tonguelike structure and functions like a rasping (scraping) tool in feeding. In *Odontogriphus*, the radulalike structure must have grown in the same way as in extant mollusks by

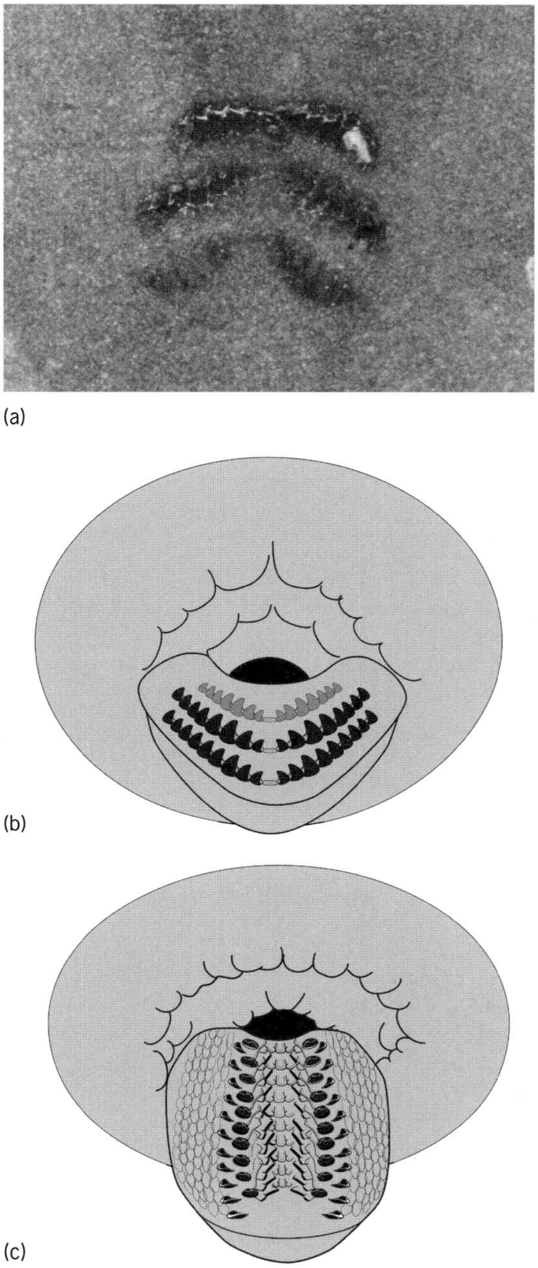

Fig. 2. Radula. (*a*) *Odontogriphus omalus* ROM57716. Width of the radula = 5.2 mm (0.2 in.). (*b*) Reconstruction of the radula of *Odontogriphus*. (*c*) Reconstruction of the radula of a modern chiton (Polyplacophora). (*Illustrations b and c by Marianne Collins*)

posterior addition of a new row of teeth and shedding of the first row. Replacement teeth in fossil *Odontogriphus* are evident in the presence of a third and sometimes a fourth row of teeth behind the two first rows. These additional rows are smaller and fainter than the first two and were evidently not yet completely formed. The presence of an isolated tooth in the intestine of one specimen suggests that the front teeth were ingested after shedding, as in extant forms.

Affinities and ecology. *Odontogriphus* shares an affinity with *Wiwaxia* (**Fig. 3**), an animal known from the Burgess Shale and other Cambrian fossil deposits. Both animals have an almost identical radula and a similar body outline. Unlike *Odontogriphus*, *Wiwaxia* was covered with sclerites (scalelike elements attached to the dorsal and lateral sides like armor). *Wiwaxia* has variously been interpreted as a member of a modern group of polychaete worms, a member of a new phylum, or a segmented worm of unknown affinity, but was evidently related to the mollusks because of the radula. Another animal resembling *Odontogriphus* is *Kimberella*, a fossil from the Ediacaran Period (about 550 million years ago) of Russia and Australia. *Kimberella* was soft-bodied and ovoid in shape, and may have had a radula (based on rasping traces associated with bodies of the animal on the same rock surfaces). These traces, called *Radulichnus*, form a series of chevronlike structures similar to traces made today by grazing mollusks on bacteria and algae growing on hard surfaces. *Kimberella* had no shell or sclerites, but the dorsal surface was probably relatively tough. Because of the radular form, *Odontogriphus*, *Wiwaxia*, and *Kimberella* were probably grazers feeding on bacterial mats. In the Burgess Shale, many specimens of *Odontogriphus* are found associated with *Morania*, a cyanobacterium that formed matlike films (Fig. 3).

Early evolution of mollusks revisited. Early mollusks were large as evidenced by *Odontogriphus*, *Wiwaxia*, and *Kimberella* and possessed a radula. They were probably originally without a shell, and it is possible that acquisition of sclerites in *Wiwaxia* was a secondary evolutionary event prior to the invention of shells. A new animal called *Orthrozanclus* has recently been described from the Burgess Shale. It has sclerites and one frontal shell and resembles both *Wiwaxia* and another group of organisms called the halkieriids. The latter group possesses sclerites and two shells (one at the front, one at the back). *Orthrozanclus*, *Wiwaxia*,

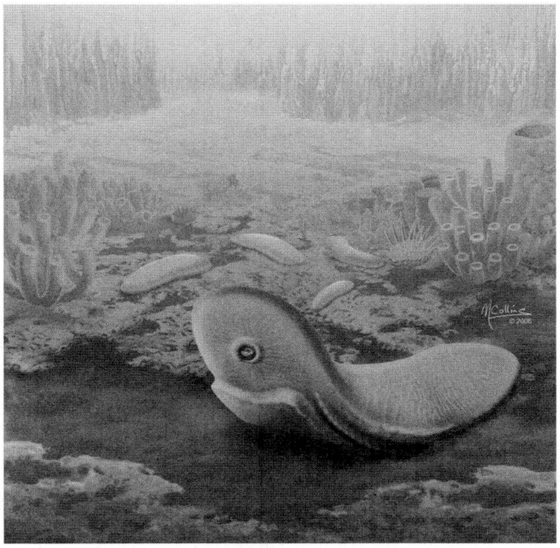

Fig. 3. Reconstruction of a seabed assemblage of *Odontogriphus* (foreground) and *Wiwaxia* (background, with body-covering sclerites) grazing on bacterial mats. (*Illustration by Marianne Collins*)

and the halkieriids are therefore probably related. The position of these three as early ancestral (stem-group) mollusks has been proposed, with *Odontogriphus* occupying a basal position because of its simple body plan. An alternative position of this group, between the mollusks and the annelids + brachiopods, has also been suggested; the rationale for this is that the shells in halkieriids are reminiscent of shells in brachiopods, and that the sclerites in *Wiwaxia* are reminiscent of setae (slender, usually rigid chitinous bristles) in annelids. However, the presence of a radula in *Odontogriphus* and *Wiwaxia* and the lack of obvious segmentation in these animals firmly suggest a molluscan affinity. The discovery of new species of fossils in Ediacaran and Cambrian deposits would certainly contribute to refining or rejecting this hypothesis.

For background information *see* ANIMAL EVOLUTION; BURGESS SHALE; CAMBRIAN; EDIACARAN BIOTA; FOSSIL; MOLLUSCA; PALEONTOLOGY; TAPHONOMY in the McGraw-Hill Encyclopedia of Science & Technology. Jean-Bernard Caron; Amélie Scheltema; Christoffer Schander; David Rudkin

Bibliography. J. B. Caron et al., A soft-bodied mollusc with radula from the Middle Cambrian Burgess Shale, *Nature*, 442(7099):159–163, 2006; S. Conway Morris and J. B. Caron, Halwaxiids and the early evolution of the lophotrochozoans, *Science*, 315:1255–1258, 2007; A. H. Scheltema, K. Kerth, and A. M. Kuzirian, Original molluscan radula: Comparisons among Aplacophora, Polyplacophora, Gastropoda, and the Cambrian fossil *Wiwaxia corrugata*, *J. Morphol.*, 257(2):219–244, 2003.

Optical properties of deep-earth minerals

Earth's lower mantle contains vast amounts of rock, extending from just beneath the 660-km seismic discontinuity, all the way to the core-mantle boundary at 2900 km depth. Lower-mantle materials are essentially semiconductors, insulating enough to promote convective heat transfer from the core, but heat conduction is gaining recognition as an important control on geodynamic processes of the mantle at high pressure and temperature conditions. The optical properties of minerals, determined by light absorption experiments, provide information on the radiative component thermal conductivity (κ_{rad}). The radiative component heat transfer is generally dependent on temperature, pressure, and the electronic structure of iron doping.

Structure and composition of low-mantle minerals. The lower mantle constitutes roughly half of the planet's mass and is thought to consist of about 80% silicate perovskite [$(Mg,Fe)SiO_3$] and 20% ferropericlase [$(Mg,Fe)O$], each with 10–20% iron depending on the temperature and depth. Silicate perovskite exhibits an orthorhombic deviation from the ideal cubic-perovskite symmetry due to tilting of the corner-linked SiO_6 octahedra (B site). Magnesium and ferrous iron (Fe^{2+}) occupy the dodecahedra (A site), and small amounts of aluminum and ferric iron (Fe^{3+}) may occupy either A or B sites. At conditions near the core-mantle boundary, historically named the D″ (D-double-prime) layer, silicate perovskite may transform to a so-called post-perovskite phase, isostructural with calcium iridiate ($CaIrO_3$). Ferropericlase, the nonsilicate oxide, has the *B*1 (rocksalt) structure throughout lower-mantle pressure-temperature (*P-T*) conditions.

Role of iron in minerals. The presence of Fe in perovskite and ferropericlase strongly affects their physical properties, compared with iron-free end members magnesium-silicate ($MgSiO_3$) and periclase (MgO). In addition to density, sound velocity, and rheology, iron can influence transport properties such as diffusion and conductivity. Optical properties and electrical and thermal conductivity depend dramatically on Fe composition because iron belongs to *d*-block elements, called transition metals, which have *d*-shell electrons in their valence orbitals. As a consequence, iron in compounds can exist in a number of oxidation states. Ferric iron is stable in most surface environments, while the Earth's mantle contains both ferrous and ferric iron, depending on the activity of oxygen in the mantle (that is, oxidation fugacity) and on the bulk mineral's composition (such as Al content). Since processes in the bio- and geosphere involve changes in redox state (Fe^{3+}/Fe^{2+} ratio), there are a number of broad consequences ranging from biogeochemical control to oxygenation of the atmosphere and oxidation state of the mantle. If both ferric and ferrous irons are present in the deep mantle as suspected, electrical conductivity can increase substantially because of electron hopping between iron ions with different oxidation states.

The spin state of iron is another important factor affecting the physical properties of iron-containing minerals. Depending on the total spin of a system (high = spin unpaired or low = spin paired), the electronic structure of minerals changes, causing modifications of thermoelastic, magnetic, and optical properties. The spin state depends upon the energy balance between the crystal-field splitting and spin-pairing energy, and this balance depends on thermodynamic conditions (for example, pressure). Electronic spin transitions, predicted theoretically almost 20 years ago, have recently been discovered experimentally under high static pressures of 40–120 GPa in both silicate perovskite and ferropericlase (**Fig. 1**).

Optical properties of minerals at high pressure and temperature. Optical properties of lower-mantle minerals synthesized at various conditions have now been studied over a wide spectral range, including mid- and near-infrared, visible, and ultraviolet (UV) [2000–35000 cm^{-1}] and at high pressures and temperatures in situ. The iron-free materials enstatite and periclase are insulators with a wide band gap, so they are transparent in this spectral range. In Fe-bearing minerals, the major absorption band is believed to be caused by a crystal-field transition of high-spin Fe^{2+} ion. This transition is split into several components by the Jahn-Teller effect, in which the number of components and the symmetry of the ground and excited states are determined by a local symmetry of the iron ion environment. In ferropericlase, the $^5T_{2g} \rightarrow {}^5E_g$ crystal-field transition of Fe^{2+} ion in an octahedral site is split into two components. In silicate perovskite, $^5E_g \rightarrow {}^5T_{2g}$ crystal-field transition of Fe^{2+} ion in a dodecahedral is split into three components. These transitions are in near-infrared spectral range (7000–12500 cm^{-1}) and are expected to increase gradually in frequency with pressure. An abrupt change of absorption is expected at the spin transition; the crystal-field transitions were predicted to move to visible and ultraviolet, leaving the near IR transparent. Crystal-field bands of Fe^{3+} are of low intensity in the spectral range of interest because they are spin-forbidden. Other than crystal-field transitions, contributions include Fe^{2+}-Fe^{3+} intervalence transitions (near 15,000 cm^{-1} in near-infrared–visible range) and Fe-O charge transfer (in ultraviolet).

Until quite recently, optical properties of mantle mineral at pressures comparable to the lower mantle (24–133 GPa) were unknown. The effects of temperature were estimated using corresponding measurements at ambient pressure. The effects of pressure were assumed to be small and were mainly attributed to those of the crystal-field band absorption. This approach becomes problematic if the absorption mechanism alters because of pressure-induced changes in electronic structure or if mechanisms of absorption other than those observed at ambient pressure come to play under pressure.

Ultrahigh-pressure experiments (including those at high temperature) have recently become available with the development of the diamond-anvil cell. This technique involves using two opposed brilliant-cut diamond anvils to contain a very small sample cavity (**Fig. 2**) with hard gasket support (such as rhenium foil) for side support. These measurements deal with very small samples, typically ($40 \times 40 \times 15\ \mu m^3$) of laboratory-grown single crystals (Fig. 2) immersed in a transparent soft material serving as a pressure medium. The results of the experiments provide information about the wavelength-dependent absorption coefficient, $\alpha(\lambda,T)(cm^{-1}) = \ln(10)/l$, where l is the path length in the material corresponding to the transmitted signal attenuation in 10 times.

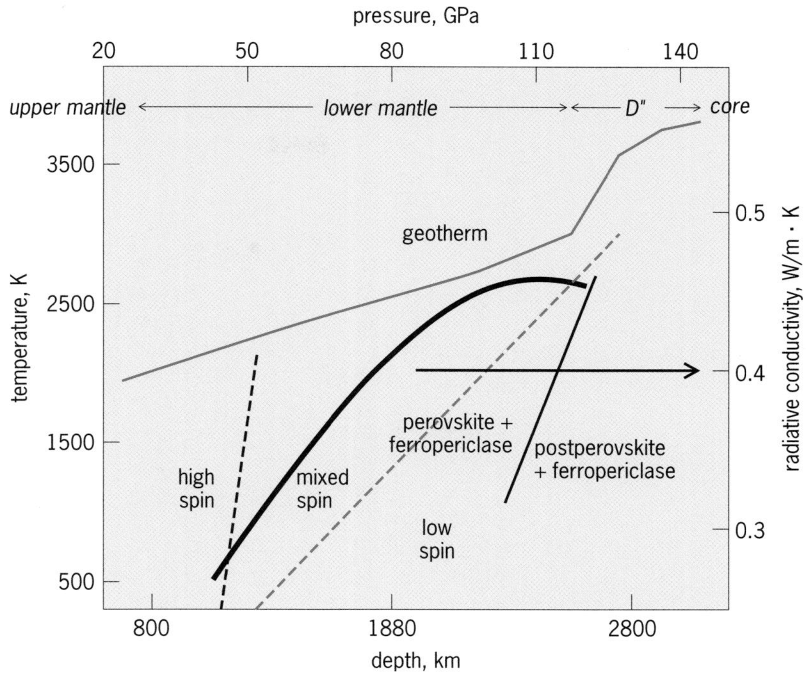

Fig. 1. Radiative conductivity of the lower Earth's mantle superimposed to the phase-composition diagram of constituting minerals and the Earth's geotherm.

Pressure-dependent optical absorption in ferropericlase and silicate perovskite reveals a variety of unexpected phenomena that include substantial changes in absorption in the spectral range of interest (**Fig. 3**). Contrary to the predictions, these changes are mainly related to an increase of absorption of Fe-O (in perovskite and ferropericlase) and Fe-Fe (in ferropericlase) charge-transfer bands, effects that were largely overlooked in the past. The changes in absorption related to the spin transitions are notable, but the effects are much smaller. Moreover, these changes become even less pronounced at high temperature because of a smearing out of the spin transition.

The results suggest that the absorption of mantle rocks is mainly governed by the iron concentration: Fe^{2+} in ferropericlase and Fe^{3+} in silicate perovskite. In the case of ferropericlase, the effect is critically

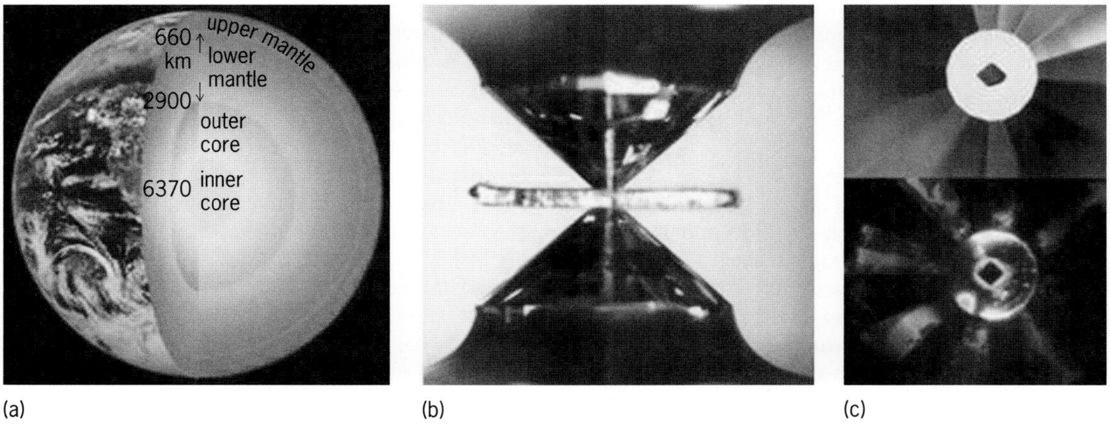

Fig. 2. *(a)* **Cross section of the Earth's interior showing the lower mantle region.** *(b)* **Photograph of a diamond anvil cell from the side. Each anvil is about 2.5 mm in height.** *(c)* **View through the diamonds of a ferropericlase crystal [$Mg_{0.75}Fe_{0.25}$]. Darkening in white light upon compression from room pressure (upper image) to 62 GPa (lower image).**

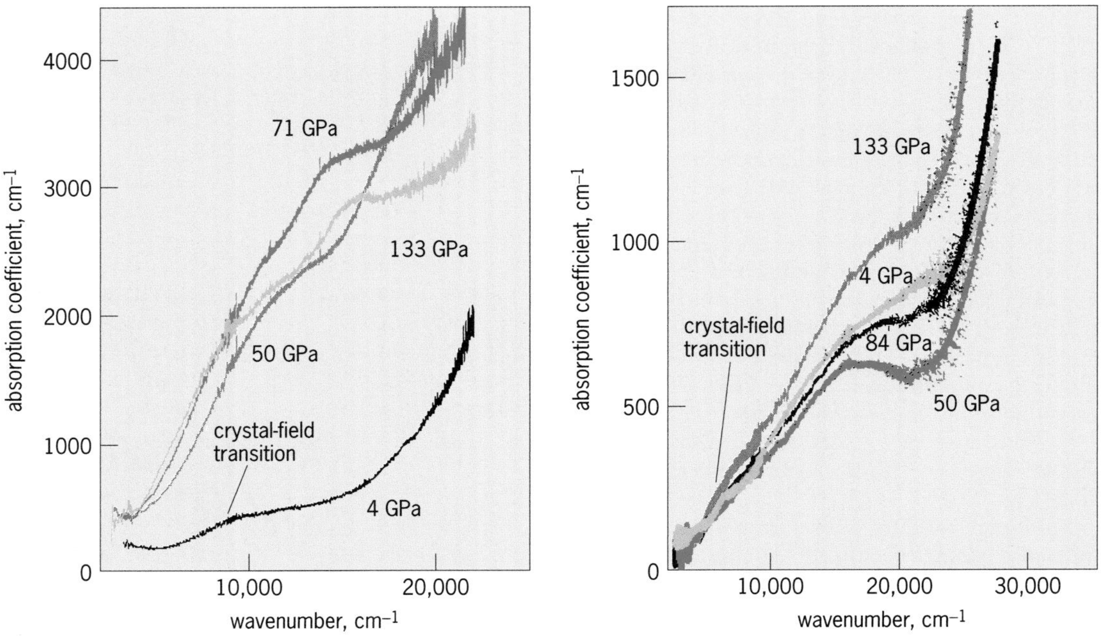

Fig. 3. Optical absorption of the lower-mantle minerals under pressures. *(a)* $Mg_{0.85}Fe_{0.15}O$ ferropericlase. *(b)* $(Mg_{0.9}Fe_{0.1})SiO_3$ perovskite.

dependent on Fe concentration, as the absorption increases drastically near the percolation limit (~12% Fe). In the case of silicate perovskite, the presence of Fe^{3+} increases intervalence Fe^{2+}-Fe^{3+} and Fe^{3+}-O charge-transfer bands.

Since the absorption coefficient of the mantle materials depends so critically on the iron concentration and the oxidation state, this information becomes increasingly important. According to the current paradigm, silicate perovskite contains about 10% of iron (50% may be Fe^{3+}) and ferropericlase about 20% (Fe^{3+} is low). If this is correct, the experimental data show that silicate perovskite is much less absorptive than ferropericlase.

Radiative thermal conductivity of the deep mantle. Radiative heat transfer is a diffusive process that occurs through emission and absorption of light by hot mantle rock. Grain size plays an important role in determining the radiative heat transport. If the $l\alpha$ product (l = grain-size, α = absorption coefficient) is large (high absorption regime), the thermal conductivity can be expressed as

$$k_{\text{rad}}(T) = \frac{4}{3}\int_0^{\infty} \frac{1}{\ln 10 \times \nu^2 \times \alpha(\nu, T)} \times \frac{\partial[n(\nu, T)^2 I(\nu, T)]}{\partial T} d\nu$$

where $I(\nu,T)$ is the Planck function and $n(\nu,T)$ is the refraction index. The grain size of the lower mantle rock is not well constrained, but it is likely in the 0.1–1-cm range. Thus, based on the absorption data (Fig. 3) we conclude that in the spectral range of interest, the high-absorption regime is realized. Silicate perovskite is much less absorptive and its content (80%) is well above the percolation limit, so the radiative conductivity is determined by this fraction. Calculated under these assumptions, radiative thermal conductivity is shown in Fig. 1.

The experimentally determined radiative conductivity is substantially smaller than a previous estimation based on extrapolations of the optical properties of minerals to higher pressures. Substantial contribution from the UV-visible absorption tails to the radiative conductivity, which was neglected previously, is responsible for the discrepancy. Since the absorption of silicate perovskite in the UV-visible tails is mostly due to the O-Fe^{+3} charge transfer, and the contribution from the Fe^{2+}- Fe^{3+} intervalence charge transfer is also substantial, the amount of the ferric iron in the mantle (that is, redox state) may be controlling the radiative heat conductivity (Fig. 3). According to our estimates, the radiative contribution does not exceed 0.54 W/(m·K) on top of the D″ layer. The value of k in the thermal boundary layer strongly influences its thinness and stability, where low values of k typically imply a less stable and thinner boundary layer. Moreover, a decrease in k also implies a decrease in plume temperature. These results should be taken into account when exercising model calculations of the mantle dynamics. Substantial efforts are still required to estimate the lattice (vibrational) contribution to thermal conductivity.

For background information *see* CONDUCTION (HEAT); EARTH INTERIOR; ELECTRON-HOLE RECOMBINATION; ELECTRON SPIN; FUGACITY; HIGH-PRESSURE MINERAL SYNTHESIS; IRON; JAHN-TELLER EFFECT; MINERAL; OXIDATION-REDUCTION; OXIDE AND HYDROXIDE MINERALS; PEROVSKITE; SILICATE SYSTEMS; TRANSITION ELEMENTS in the McGraw-Hill Encyclopedia of Science & Technology.

Alexander Goncharov; Steven D. Jacobsen; Viktor Struzhkin; Pierre Beck

Bibliography. J. Badro et al., Transitions in rerovskite: Possible nonconvecting layers in the lower mantle, *Science*, 305:383–386, 2004; R. G. Burns, *Mineralogical Applications of Crystal Field Theory*, 2d ed., Cambridge University Press, United Kingdom, 1993; A. F. Goncharov, V. V. Struzhkin, and S. D. Jacobsen, Reduced radiative conductivity of low-spin (Mg,Fe)O in the lower mantle, *Science*, 312:1205–1208, 2006; A. M. Hofmeister, Mantle values of thermal conductivity and the geotherm from phonon lifetimes, *Science*, 283:1699–1706, 1999; H. Keppler and J. R. Smyth, Optical and near infrared spectra of ringwoodite to 21.5 GPa: Implications for radiative heat transport in the mantle, *Am. Mineral.*, 90:1209–1212, 2005; H. K. Mao and P. M. Bell, Electrical conductivity and the red shift of absorption in olivine and spinel at high pressure, *Science*, 176:403–406, 1972; C. McCammon, Perovskite as a possible sink for ferric iron in the lower mantle, *Nature*, 387:694–696, 1997; D. M. Sherman, The high-pressure electronic structure of magnesiowüstite (Mg,Fe)O: Applications to the physics and chemistry of the lower mantle, *J. Geophys. Res.*, 96:14299–14312, 1991.

Organ regeneration

There is a great shortage of organs for transplantation, generating a need for alternative treatments for organ loss. Induced organ regeneration is being increasingly used as a treatment modality for patients who have lost function of an organ.

Organ regeneration is the restoration of the anatomical structure and physiological function of an organ following its loss by trauma or chronic disease. As currently practiced, organ regeneration is induced by investigators or clinicians at the site of a wound. Suitable wounds for induction of regeneration are those that have been inflicted accidentally leading to massive loss of organ mass, or wounds that have been inflicted deliberately on an organ that has lost its function due to a disease process. In contrast, organ repair leads to closure of a wound, as does regeneration, but does not lead to restoration of the original organ. Spontaneous regeneration occurs without aid and is observed in the early mammalian fetus and in certain amphibians. Induced regeneration is aided by the investigator or clinician; it is the focus of this article.

The field of induced organ regeneration in the adult mammal is relatively young, and the information currently available is centered on skin, peripheral nerves, the conjunctiva of the eye, and the urethra.

Nonregenerative tissues. The spontaneous response of organs to injury can be interpreted by referring to the healing response of the "tissue triad"—the epithelium, basement membrane, and stroma. These three tissue types are anatomically grouped together in most organs. They comprise anatomical structures that can be used as reference structures to compare the response of different organs to injury. The epithelium (epidermis in skin, myelin sheath in nerves) regenerates spontaneously and, while doing so, induces regeneration of the attached basement membrane. The stroma (dermis in skin, endoneurium in nerves) is nonregenerative, that is, it does not regenerate spontaneously in adults. It follows that regeneration of the stroma is the major problem in studies of induced organ regeneration. Once the stroma has been induced to regenerate, epithelium and basement membrane may regenerate spontaneously.

Regeneration viewed as chemical synthesis. Use of the language of chemistry is becoming an increasingly common approach to describe the regeneration of tissues and organs. "Reactants" are cells of a specific type (such as keratinocytes in skin or Schwann cells in peripheral nerves), solutions of macromolecules that regulate cell behavior (for example, cytokines such as platelet-derived growth factor) by binding on receptors, or solid surfaces (for example, scaffolds or more specifically regeneration templates) that provide cell regulation by binding of cells on specific ligands located on the active solid surface. The products of the synthetic process are the desired regenerates, described morphologically and functionally in reference to the native tissue that has been regenerated. Two classes of experimental "reactors" for organ synthesis are currently used: cell culture (in vitro reactor) and the freshly inflicted wound (in vivo reactor). Reactants are introduced into the reactor, and the resulting tissue products are analyzed to find out how much, or how well, each of the tissue types of the triad has been synthesized. This approach simplifies greatly the description of the experimental protocol used in various laboratories. It also has the advantage of eliciting readily the minimum number of reactants that are required to be added by the investigator to achieve synthesis of a given tissue type or an entire organ.

Analysis of a large number of protocols used by a variety of independent investigators of skin and peripheral nerve regeneration has shown that these two organs may be grown without adding solutions of cytokines or stromal cells (for example, fibroblasts) to the reactor. Both of these reactants are supplied in sufficient quantities and in the appropriate timing by the inflammatory process of the wound, which acts as the reactor. Such intrinsic regulation of the synthetic processes is not available in vitro (cell culture).

Simple conclusions have been drawn from the analysis of protocols used by several independent investigators. In studies conducted in vivo (that is, in a healing wound) only an appropriate solid regulator (regeneration template) need be added to the wound to induce synthesis of the stroma. Once the stroma has been synthesized inside the wound, epithelia from the edges of the wound spontaneously migrate over it, restoring the epidermis and synthesizing the basement membrane. Although this process of sequential tissue synthesis has the benefit of simplicity, it is delayed by the relatively slow process of epithelialization and may not be the clinically preferred mode for restoring the organ.

A faster method used clinically for skin restoration places on the surface of the newly synthesized stroma several small epithelial autografts, which can be harvested with minimal scarring due to the intrinsic ability of epithelia to regenerate spontaneously. In another rapid method, the template is seeded with autologous epithelial cells (cells from the same organism) prior to application onto the wound. In this process, stroma and epithelia are synthesized simultaneously (rather than sequentially, as in the method described above), leading to regeneration both of dermis and epidermis within about 3 weeks.

Induced organ regeneration in the adult mammal. Regeneration templates are being increasingly used to induce regeneration of organs that have failed.

Skin has been partly regenerated in the guinea pig, swine, and humans. Originally referred to as artificial skin, the newly synthesized skin was later recognized as the product of a regenerative process. A detailed comparison of the structural and functional similarities and differences among normal skin, scar, and regenerated skin has been made. Among its characteristics, regenerated skin is mechanically competent, fully endowed with blood vessels, and sensitive to touch as well as to heat or cold. The regenerated dermal-epidermal junction, with its extensive formations of epidermal folds (rete ridges) and capillary loops, leaves no doubt that the de novo regenerated skin organ is clearly not scar (see **illustration**). However, regenerated skin currently differs from intact skin by the absence of skin appendages (hair follicles, sweat glands, etc.).

Peripheral nerves have been regenerated in rats and other rodents, as well as in humans, across unprecedented distances separating the two stumps that result from a complete cut (transection) of the nerve trunk. Regeneration was most complete using a number of templates that appear to share certain critical structural features (see below). The regenerated nerves have a structure and function that are very close, though currently not identical, to that of native nerves.

The conjunctiva has been also regenerated in a rodent (rabbit) model following the complete excision of the conjunctival stroma and grafting of a template that was similar to that used for dermis regeneration.

The urethra has been regenerated in humans following resection of segments of the urethra and their replacement with collagen matrices derived from bladder submucosa that were seeded with autologous cells. The bladder was regenerated in a dog model, following removal of about 50% of the native bladder, using a collagen-based matrix that had been seeded with allogenic urothelial cells and with allogenic muscle cells.

Significant progress has been reported in inducing regeneration of bone, heart valves, and the spinal cord.

Clinical applications of organ regeneration. An FDA-approved device for dermis regeneration, Integra™, is based on a collagen-glycosaminoglycan (CG) scaffold. It is currently used to treat patients with massive burns, patients undergoing plastic or reconstructive surgery of the skin, as well as patients with chronic skin wounds. Another device that has been used to treat chronic skin wounds is Apligraf™. Paralysis of the limbs resulting from trauma to peripheral nerves is being treated with NeuraGen™ tubes. Open grafts to treat urethral or bladder defects have been used for several years.

Mechanisms of organ regeneration. Over the years, studies have shown that induction of regeneration requires the use of a template or other substance that effectively blocks contraction. Contraction is a spontaneous healing process by which the perimeter of a wound shrinks over several days, leading to closure of the wound. Studies of the mechanism of biological activity of regeneration templates have shown that these scaffolds block contraction by two mechanisms. In the first mechanism, an active scaffold reduces to about 20% the number of contractile fibroblasts (myofibroblasts) that normally are recruited during wound contraction. In the second, myofibroblasts bind avidly on the surface of the scaffold, thereby losing their contiguity

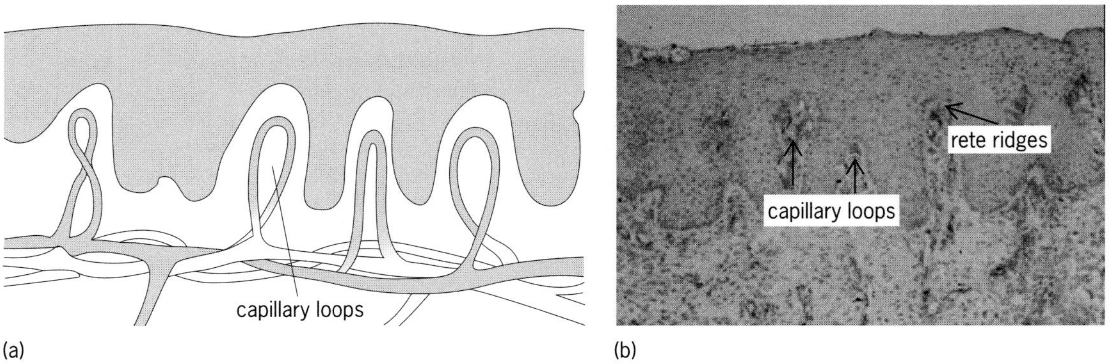

Comparison of (*a*) normal and (*b*) regenerated skin. The dermal-epidermal junction of regenerated skin shows capillary loops inside the rete ridges. Neither rete ridges nor capillary loops form in scar. Presence of rete ridges is one of two major features that distinguish intact and regenerated skin from scar. Another distinguishing feature is presence of quasirandomly oriented collagen fibers in the intact and regenerated dermis, rather than a highly aligned stroma, as in dermal scar. (*C. C. Compton et al., Organized skin structure is regenerated in vivo from collagen-GAG matrices seeded with autologous keratinocytes, J. Invest. Dermatol.,110:908–916, 1998*)

with each other as well as losing the orientation of their contractile axes in the plane of the wound. Both effects, occurring almost simultaneously, reduce the macroscopic mechanical force generated by the population of contractile cells, leading to significant blocking of contraction. A hypothetical mechanism based on mechanical splinting action by a stiff scaffold has been rejected by the data.

The structural determinants of regeneration templates have been partly identified. They include specific ligands for binding of contractile fibroblasts as well as a critical scaffold degradation rate that matches the rate of tissue synthesis (synchronous synthesis). There is also some evidence for a required three-dimensional scaffold configuration that matches the configuration of the stroma being synthesized at the anatomical site (isomorphous replacement). Biological activity has so far been identified in scaffolds based on graft copolymers of type I collagen and a glycosaminoglycan. Undoubtedly, there are several other approaches of induced regeneration that will probably be identified in future studies.

For background information *see* BIOMEDICAL CHEMICAL ENGINEERING; CONNECTIVE TISSUE; EPITHELIUM; NERVOUS SYSTEM (VERTEBRATE); REGENERATION (BIOLOGY); SKIN in the McGraw-Hill Encyclopedia of Science & Technology. I. V. Yannas

Bibliography. A. Atala, Regeneration of urologic tissues and organs, *Adv. Biochem. Engin./Biotechnol.*, 91:181–210, 2005; I. V. Yannas (ed.), *Regenerative Medicine*, vols. I and II, 2005; I. V. Yannas, *Tissue and Organ Regeneration in Adults*, 2001.

Organic synthesis in water

Water is the most abundant molecule on Earth and the universal solvent in which the chemistry of the life processes has developed over billions of years. Its unique chemical and physical properties have fascinated scientists for a long time. The connection between water's deceptively simple molecular structure and its influence as a solvent on biological reactions remains one of the greatest scientific challenges. In sharp contrast to the general academic interest in water, the modern chemist rarely uses or even considers water as a medium for synthetic organic reactions. Students working on all levels of organic chemistry are trained to use anhydrous reaction conditions. The diligent synthetic chemist keeps glassware, syringes, reagents, and solvents free from traces of water. The reason behind the lack of interest in using water as a reaction solvent can be traced back to the beginning of the twentieth century when new synthetic methods based on organometallic reagents were introduced. These methods were extremely powerful but often required inert conditions because of the high reactivity of the carbon-metal bond. Throughout the better part of the last century, synthetic methods were developed for use primarily in anhydrous organic media. It was not until the early 1980s that the use of water was reevaluated when it was found that the rate and selectivity of the important and useful Diels-Alder reaction could be greatly enhanced in water compared to the same reaction in an organic solvent. The interest in water as solvent was further invigorated in the 1990s with the introduction of the concept of "green" chemistry, which addresses the environmental impact of chemicals and chemical processes. Water, being cheap, safe, nontoxic, and environmentally benign, was soon recognized as perhaps the ultimate green solvent. In view of the potential reward of replacing hazardous organic solvents with water, researchers took up the academic challenge of developing new synthetic methods that were compatible with the aqueous medium. This article highlights two recent important developments in this area: the discovery of metal-based catalysts that operate in water, and new insights into the influence of aqueous biphasic reaction conditions on reaction rates.

Metal catalysis. The discovery of efficient catalysts for various reaction types is an important entry into faster and cleaner reactions in any medium. It is essential to the success of aqueous synthesis that catalysts developed are as efficient and selective as those used in organic solvents are. Unfortunately, switching the solvent to aqueous reaction media is not a straightforward task. Thus, the most important recent development in aqueous synthesis may have been the discovery of water-tolerant metal catalysts for several important reaction types, such as alkylation, cross coupling, metathesis, and Michael addition. Moreover, by complexing various chiral molecules with the active metal center, catalysts have been obtained that also induce selectivity in reactions that generate enantiomeric products, that is, molecular mirror images. Enantioselective catalysts are of great importance, for example, in the pharmaceutical industry as drugs must often be prepared in an enantiomerically pure form. The difficult transition to synthesis in water is illustrated by the fact that the first catalytic enantioselective reaction in water was reported as late as 1998. Seminal work in the field of aqueous catalysis was performed by S. Kobayashi and colleagues in Japan, who showed that some metals of the periodic table were more compatible with the aqueous medium than others. It was found that lanthanide metals, such as lanthanum (La) and ytterbium (Yb), were tolerant to water and retained catalytic activity in an aqueous aldol reaction. Subsequent work has shown that lanthanide-based catalysts may be useful for several types of reactions in water. Various enantioselective reactions in water have been achieved by complexing lanthanides as well as some other metals with chiral ligands. For example, S. Kobayashi and coworkers developed catalysts for the enantioselective Mannich-type addition reaction in water as a route to optically active α-amino ketones or esters, which are useful compounds in the preparation of nitrogen-containing biologically active compounds. The reaction proceeded with excellent enantioselectivity in pure water. As shown in reaction (1), the reac-

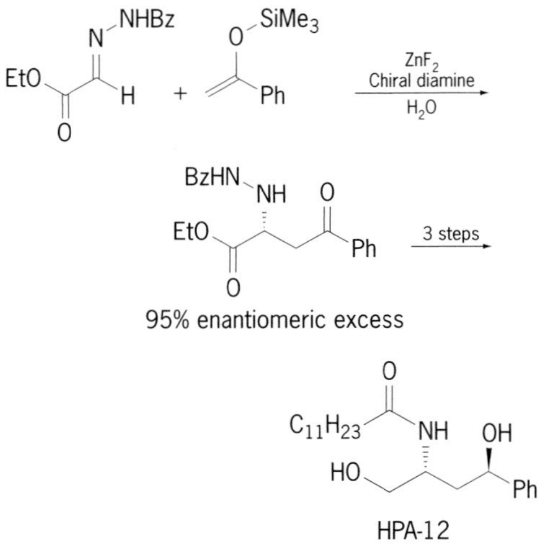

(1)

tion was applied to an enantioselective synthesis of HPA-12, which is an attractive compound that specifically inhibits intracellular transport of sphingolipids.

In 2005, the Nobel prize in chemistry was awarded to Yves Chauvin, Robert H. Grubbs, and Richard R. Schrock for the discovery of catalysts for metathesis reactions. In these types of reactions, bonds between carbon atoms are first broken and then reformed in useful new ways by using a specific ruthenium catalyst. Metathesis reactions are now widely used in the chemical industry. To address the emerging interest in using water as a process solvent, Grubbs and associates have recently reported metathesis catalysts that have been modified to have high solubility and activity in water. An example of such a modification is shown in structures below, where the original

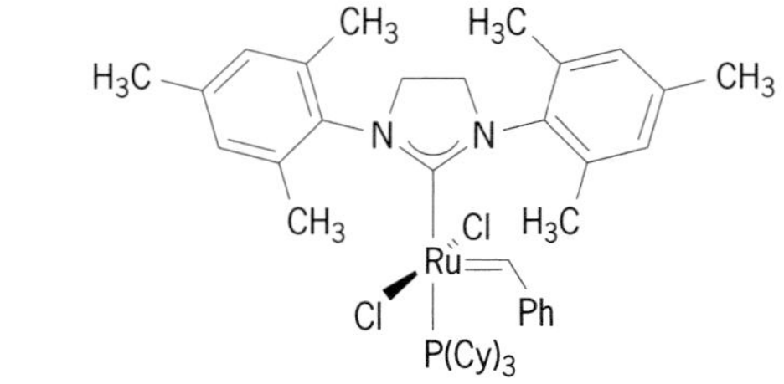

(1)

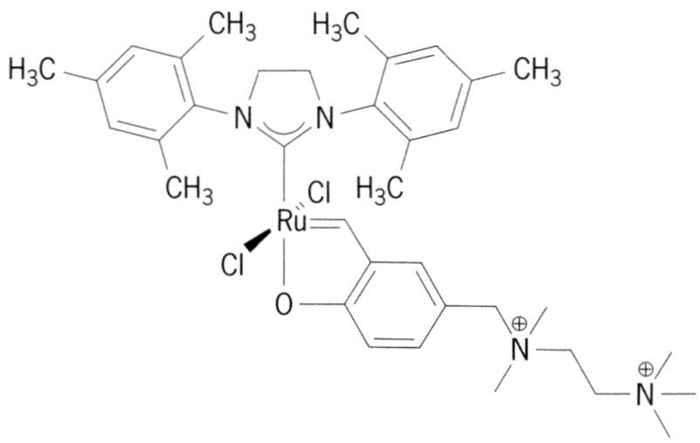

(2)

metathesis catalyst (**1**) has been modified (**2**) to be compatible with the use of water as the reaction solvent.

Aqueous biphasic reactivity. Many organic compounds have limited solubility in water. Because the common perception has been that solubilizing reacting molecules is a prerequisite for an efficient reaction, strategies have been developed to increase aqueous solubility of hydrophobic substrates, most commonly by using an organic cosolvent or other additives. While such modifications may be efficient, the practical advantages of using water, such as low cost, simplicity of reaction, and simple product isolation, diminishes with their use. New insight into the complex relationship between solubility and reactivity in water was recently reported by Nobel Laureate K. B. Sharpless and coworkers, where several reactions using aqueous biphasic ["on-water" (the insoluble reactants initially float on the water)] conditions were greatly accelerated compared to those run in organic solvent alone or even in the absence of any solvent. Reaction (2) shows how the reaction of

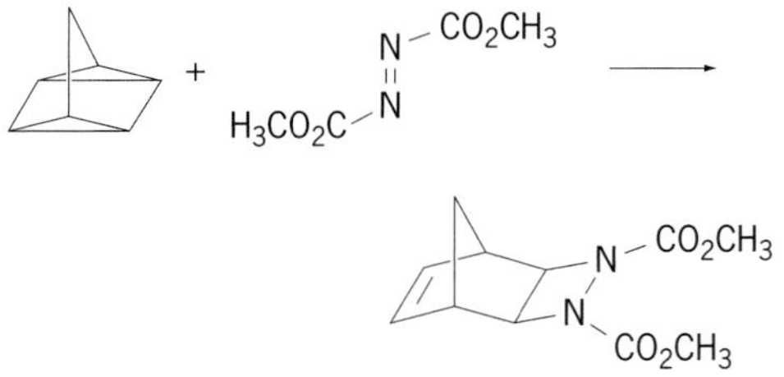

(2)

Solvent	Time, h
Toluene	>120
Dichloromethane	72
None	48
Methanol	18
Water	0.17

quadricyclane with dimethyl azodicarboxylate yields 1,2-azetidine as a single product. The time to completion for the reaction is only 10 min at 23°C, compared with 48 h for the solvent-free reaction, 18 h in methanol, and more than 120 h in toluene.

Beyond the short reaction times, a practical advantage of on-water reactions that should be of great interest to the process chemist is the potentially straightforward isolation by phase separation or filtration of the water insoluble product. Current theoretical work by Nobel Laureate R. A. Marcus and Y. Jung focuses on understanding the mechanism behind the rate-accelerating effects. The on-water protocol is also being extended to other reaction types, such as the important Wittig reaction reported by M. Bergdahl and coworkers to be accelerated using on-water conditions. Finally, U. M. Lindström and coworkers have recently made important advances in the area of aqueous biphasic catalysis, that is, running on-water reactions in the presence of highly water- soluble and water-compatible metal catalysts.

For background information *see* ASYMMETRIC SYNTHESIS; CATALYSIS; DIELS-ALDER REACTION; GREEN CHEMISTRY; LANTHANUM; ORGANIC SYNTHESIS; SOLVENT; SPHINGOLIPID; YTTERBIUM in the McGraw-Hill Encyclopedia of Science & Technology.

U. Marcus Lindström

Bibliography. M. Bergdahl et al., Wittig reactions in water media employing stabilized ylides with aldehydes, *J. Org. Chem.*, 72:5244–5259, 2007; J. J. Jordan and R. H. Grubbs, Small-molecule *n*-heterocyclic carbene containing olefin metathesis catalysts for use in water, *Angew. Chem. Int. Ed.*, 46:5152-5155, 2007; J. E. Klijn and J. B. F. N. Engberts, Fast reactions "on water," *Nature*, 435:746-747, 2005; S. Kobayashi and C. Ogawa, New entries to water-compatible Lewis acids, *Chem. Eur. J.*, 12:5954-5960, 2006; C. J. Li and T. H. Chan, *Organic Reactions in Aqueous Media*, Wiley, 2007; C. J. Li and L. Chen, Organic reactions in water, *Chem. Soc. Rev.*, 35:68-82, 2006; U. M. Lindström, *Organic Reactions in Water*, Blackwell, 2007; U. M. Lindström et al., Amino acid induced rate acceleration in aqueous biphasic Lewis acid catalyzed Michael addition reactions, *Angew. Chem. Int. Ed.*, 46:4543-4546, 2007.

Paleogenomics

Genomics is the study of the organization, structure, and function of the total complement of genes and other DNA within the cell(s) of an organism. This term usually refers to the haploid (possessing a single set of unpaired chromosomes) nuclear DNA content in eukaryotes. Genomics can be differentiated from genetics (along a continuum), with the former concerned about the function and interaction of multiple genes, noncoding DNA [such as introns (segments of DNA that do not encode part of the gene's protein and are spliced out of messenger RNA in processing) and intergenic DNA (DNA sequences located between gene clusters)], and genome architecture. The latter is concerned mainly with a single gene and its products. Following this scheme, paleogenomics focuses on recovering and understanding the genomic information in long extinct species. Strictly defined, paleogenomic research must include original fossil remains as its primary data source. Therefore, the term "paleogenomics" should not be applied to comparative genomic studies that infer ancestral patterns or character states from living species alone. Investigations into the genomes of recently extinct organisms are more properly termed "archeogenomics" or "ancient DNA studies." Because the field of genomics is focused on living organisms, it is concerned with only a fraction of the genomic novelty that has evolved. Insights from paleogenomic research are therefore indispensable for a thorough understanding of the nature and evolution of the eukaryote genome.

Paleogenomics has many overlapping goals with ancient DNA studies, and the successes already achieved in ancient DNA studies highlight the potential for paleogenomic research. In the first study examining the genetics of an extinct species, the quagga (*Equus quagga*) was shown to be more closely related to zebras than to horses, highlighting the approach's capacity to provide insight into evolutionary relationships. Also of interest is the identification of single nucleotide polymorphisms (SNPs, that is, single base-pair differences between two copies of a DNA sequence from two individuals), which allow researchers to answer questions about the population genetics of extinct species, such as interbreeding or population divergence. Divergence estimates are especially important because they help establish and test molecular clocks. Uniquely, paleogenomics may be able to answer whether the mutation rate around which molecular clocks are built is constant or variable. These are fundamental phylogenetic problems important to nearly all evolutionary biologists. Paleogenomic approaches may also be used to study disease in early populations by analyzing epidemiology in extinct cultures as well as the coevolution among animals and their parasites. Moreover, ecological and symbiotic interactions of long-extinct species can be revealed by analyzing genomic remains in coprolites (fossil feces) and fossilized gut contents, as has been done for mastodons and bees. Broader architectural questions that are beyond the scope of genetics will also drive the field of paleogenomics. For example, the spread of mobile genetic elements (which constitute much of the genome in most eukaryote species), whole genome duplications, genome size, and synteny (the chromosomal order of genes) are all topics of interest.

In traditional paleontology, an actualistic approach (uniformitarianism) is used to describe and analyze the characteristics of extinct organisms by making comparisons and analogies with living species. There are three primary approaches to paleogenomic research (sequencing-based, paleohistology-based, and evolutionary-development-based), all of which use the same actualistic method employed in other paleontological sciences.

Sequencing-based approach. The aim of this approach is to recover the actual DNA sequences that comprised the genome of extinct organisms. Sources of ancient DNA include mummified and frozen remains, museum specimens, and remains preserved in peat bogs, sinkholes, caves, and sediment, as well as organisms preserved in amber inclusions. Anoxia (absence of oxygen), low temperature, neutral pH, and desiccation create the most favorable taphonomic (postmortem and burial) conditions for preservation of genomic material (DNA).

Early results were insightful and promising; however, apart from some premature claims, it is uncertain whether DNA can be amplified and sequenced from species many millions of years old. This is the approach's greatest weakness. However, the advent of massively parallel sequencing methods, which yield up to 25 million base pairs of sequence in one 4-h run (**Fig. 1**), has dramatically improved the prospects for sequence-based paleogenomics. For example, megabases from the nuclear and mitochondrial genomes of mammoth (*Mammuthus primigenius*) have revealed a high degree of sequence identity with elephants. Researchers have also sequenced megabases from the nuclear genome of Neanderthals (*Homo neanderthalensis*) and Pleistocene cave bears (*Ursus spelaeus*). The entire mitochondrial genomes of multiple moa species have also been sequenced. Some of these studies have taken a promising "metagenomic" approach by sequencing many anonymous DNA samples from fossils and

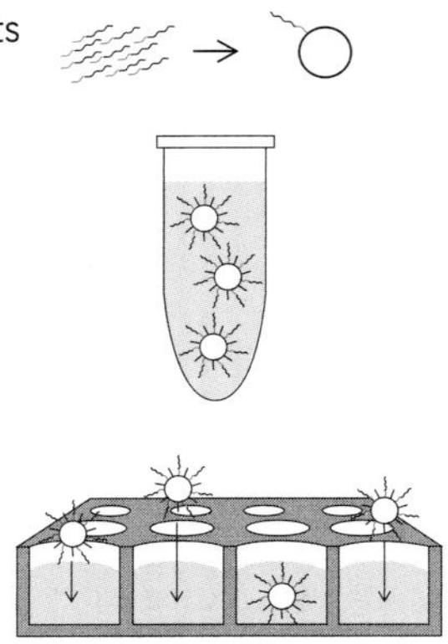
sample preparation, fragments of DNA attached to beads
emulsion-based PCR creates millions of copies per bead
beads deposited into fiber-optic slide, sequencing reaction and detection

Fig. 1. Diagram of the massively parallel picoliter-scale sequencing method, which accommodates 1.6 million reads, each approximately 100 base pairs in length.

using bioinformatic tools to isolate and compare the sequences of interest.

Although the results of direct-sequencing studies are exciting, degradation and artifacts are common, and contamination is a major hurdle. Incorporating incorrect base pairs during the amplification of ancient DNA using the polymerase chain reaction (PCR) often results in miscoding lesions, a common artifact. The authenticity of DNA amplification and sequencing from fossils and other ancient remains can be maximized by using sterile excavation protocols and isolating laboratory space and equipment from other DNA-based experiments. Increasingly, the information gleaned from comparative genome studies, as well as bioinformatic analyses of diagenetic damage, such as miscoding lesions, can be used to help authenticate paleogenomic data. Repeated independent laboratory investigation remains the most convincing form of confirmation.

Paleohistology-based approach. A strong positive relationship between cell size and nucleus size has been recognized since the nineteenth century in a wide variety of tissue types across multicellular plants and animals. In the early 1970s, this relationship was formalized as the nucleotypic theory. It not only relates genome size (and hence nucleus size) to cell size, but also states that the genome influences phenotype independent of its information content.

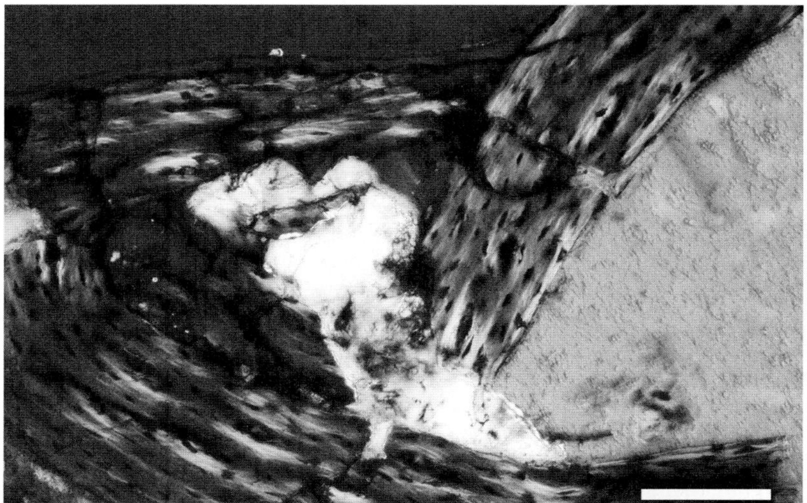

Fig. 2. Histological micrograph of a *Rhamphorhynchus* (pterosaur from the Jurassic Period) radius in cross section. Scale bar: 100 micrometers.

With regard to paleogenomics, the nucleotypic relationship could be used to infer genome characteristics from the cell size of extinct organisms (**Fig. 2**). Despite the vast potential of this approach, only a handful of studies have been conducted: on fossil bone from lungfish and amphibians, conodont epithelial cells, angiosperm stomata, and fossil bone from birds and other dinosaurs. At its core, this approach relies on inferring genomic characters from fossilized cells or other microscopic features using relationships derived from living animals as the basis for analysis. The strength of this approach is that genomes can be investigated for species extinct many millions of years, thereby independently complementing comparative and evolutionary studies on the evolution of genome architecture. For example, the size of the pockets in which bones cells resided (lacunae) during life suggests that the genome size and abundance of mobile genetic elements in carnivorous dinosaurs remained stable for some 230 million years, through the surviving lineage of modern birds. The limit to this approach is that only gross aspects of the genome [such as size, ploidy (the number of complete chromosome sets in a nucleus), and percent of the genome composed of mobile elements] can be inferred with confidence.

The paleohistology approach of investigating genome size relies (as does the evolutionary development approach discussed below) on exploring correlations between characters within species. Because species are not independent from one another, normal statistical analyses overestimate the degrees of freedom and may give grossly fallacious results. The so-called comparative method resolves this predicament by using the branching pattern (topology) of a phylogeny to make hierarchical contrasts among characters. The expected variances, which are proportional to the branch lengths originating from the most recent common ancestors, are used to scale the contrasts. Statistical analyses can then be performed on the contrasts. Inferences of genome characters can be made using the phylogenetically corrected regression as well as the phylogenetic and histological information pertaining to the extinct organism.

Evolutionary-development-based approach. Another approach is to track morphological innovations in the fossil record that are the product of known developmental genes in living relatives. This method brings fossil data into developmental genetic studies to make and test hypotheses about the origin and evolution of the genes responsible for novel morphology. For example, fossil data have been used in studies of genes involved with biomineralization in echinoderms and digit identity in birds. The strength of this approach is that genes that determine morphology, and are, therefore, likely important for survival, can be studied over great spans of evolutionary time. The weakness, from a genomics perspective, is that such studies tend to focus on only one or a few genes. Furthermore, morphological development often proceeds by complex interacting gene networks, and it may be difficult to pin down the genetic changes responsible for shifts in morphology seen in the fossil record.

Outlook. Much about genome evolution remains a mystery. For instance, within the genus *Xenopus* (the clawed frog), the ploidy level spans between 2n and 12n, although how and why this range of genome sizes evolved is unknown. Future paleogenomic research may be able to address these and other conundrums. The best of such studies will synthesize analyses from all the approaches detailed above, in addition to comparative genomics, to robustly evaluate hypotheses about the origins of genomic novelty.

For background information *see* ANIMAL EVOLUTION; DEOXYRIBONUCLEIC ACID (DNA); DINOSAUR; EXTINCTION (BIOLOGY); GENOMICS; ORGANIC EVOLUTION; PALEONTOLOGY; PHYLOGENY; TAPHONOMY in the McGraw-Hill Encyclopedia of Science & Technology. Chris Organ

Bibliography. D. Birnbaum et al., "Paleogenomics": Looking in the past to the future, *J. Exp. Zool.*, 288:21–22, 2000; D. J. Bottjer et al., Paleogenomics of echinoderms, *Science*, 314:956–959, 2006; M. B. Hebsgaard et al., Geologically ancient DNA: Fact or artifact?, *Trends Microbiol.*, 13:212–220, 2005; C. L. Organ et al., Origin of avian genome size and structure in nonavian dinosaurs, *Nature*, 446:180–184, 2007; H. N. Poinar et al., Metagenomics to paleogenomics: Large-scale sequencing of mammoth DNA, *Science*, 311:392–394, 2006.

Parasitic plants

Although not generally recognized, parasitic organisms make up a large percentage of the Earth's total biodiversity. The word "parasite" generally conjures up images of organisms such as microbes and worms, but this life-form has evolved repeatedly in many groups, including flowering plants (angiosperms). A parasitic angiosperm is one that attaches to a host root or stem via a modified root called the haustorium, thereby deriving some or all of its water and nutrients. Some parasitic plants are green and photosynthetic, but they still obtain water and solutes from their host. These are called hemiparasites (somewhat of a misnomer in that they are fully parasitic). Other parasitic plants have lost their ability to conduct photosynthesis and are thus fully heterotrophic (that is, obtaining nourishment from exogenous organic matter). These are called holoparasites, and they obtain carbohydrates from host phloem as well as water from the host xylem. Here a plant is considered parasitic only if it forms a direct haustorial connection to a host plant. This excludes plants such as Indian pipe (*Monotropa*) that attach to mycorrhizal fungi and are technically known as mycoheterotrophs.

Because of their curious nature, the natural history of parasitic plants has been studied for centuries. Work from the mid-1800s onward focused mainly on the morphological and anatomical features of these plants, with particular attention to adaptations associated with their parasitic habit. Since then, data from physiology, biochemistry, genetics, ecology, pathology, and phylogeny have resulted in a virtual explosion of information about these plants. It should be pointed out, however, that the vast majority of parasitic plants do not harm economically valuable crops. Indeed, of the 268 genera and 4558 species of parasitic plants (see **table**), only about 25 genera can be considered crop pests.

Recently, great advancements have been made in the area of the evolutionary biology of parasitic plants. Current information on parasitic angiosperms can be organized according to two interrelated disciplines: molecular evolution and molecular phylogeny. Parasitic plants have provided excellent models to study genetic changes at the molecular level because they represent natural genetic mutants that differ from typical land plants in fundamental ways. Indeed, the holoparasites stretch the definition of plant itself, having lost one of the key traits associated with this lineage: photosynthesis. Accompanying these changes at the molecular level are changes in morphological features as well. Reduction and complete loss of typical plant organs has compromised previous efforts to classify several parasitic plant families. More recently, molecular data have helped rectify this situation, but these studies have also produced some real surprises.

Molecular phylogenetics. Phylogenetics is primarily concerned with the branching events that took place during the evolutionary history of a group of organisms. With regard to parasitic plants, a basic question is: "How many times did parasitism evolve among angiosperms?" Previous morphology-based classifications provided many hypotheses, but different researchers often had conflicting ideas about the relationships of these plants, particularly the holoparasites. Thus, until recently, no definitive answer to this question could be given because well-supported phylogenetic trees were not available for all groups. As shown in the table, all parasitic angiosperms can now be placed in the global angiosperm phylogeny. For the most part, these placements are robust, being supported by analyses using genes from the nucleus, plastid, and mitochondrion. These results indicate that parasitism arose independently in 12 lineages (clades) of flowering plants.

Rafflesiaceae are a notable family for which evidence about their evolutionary relationships has been elusive. This family is considered here in the strict sense, that is, including only the large-flowered genera *Rafflesia*, *Rhizanthes*, and *Sapria*. *Rafflesia* is commonly known as the "Queen of the Parasites" because some members have the largest flowers among all angiosperms [exceeding 1 m (3.3 ft) in diameter] (see **illustration**). The placement of these amazing parasites in the global angiosperm phylogeny has been extremely difficult because of morphological reductions in vegetative parts—there are no stems, leaves, or roots, only an endophyte—the vegetative portion of the parasite that lives

Parasitic plants

Clade	Example genera	No. genera/species	Family	Order
1	*Cassytha*	1/16	Lauraceae	Laurales
2	*Hydnora, Prosopanche*	2/15	Hydnoraceae	Piperales
3	*Olax, Ximenia*	14/104*	"Olacaceae"	Santalales
3	*Misodendrum*	1/10	Misodendraceae	Santalales
3	*Arjona, Quinchamalium, Schoepfia*	3/50	Schoepfiaceae	Santalales
3	*Amyema, Nuytsia, Tristerix*	73/906	Loranthaceae	Santalales
3	*Agonandra, Opilia*	11/34	Opiliaceae	Santalales
3	*Comandra, Santalum*	37/531	"Santalaceae"	Santalales
3	*Arceuthobium, Phoradendron, Viscum*	7/546	Viscaceae	Santalales
3	*Balanophora, Helosis, Thonningia*	17/43	Balanophoraceae	Santalales
4	*Cynomorium*	1/2	Cynomoriaceae	Saxifragales
5	*Krameria*	1/18	Krameriaceae	Zygophyllales
6	*Rafflesia, Rhizanthes, Sapria*	3/30	Rafflesiaceae	Malpighiales
7	*Bdallophyton, Cytinus*	2/7	Cytinaceae	Malvales
8	*Apodanthes, Berlinianche, Pilostyles*	3/23	Apodanthaceae	Cucurbitales
9	*Mitrastemon*	1/2	Mitrastemonaceae	Ericales
10	*Lennoa, Pholisma*	2/5	Boraginaceae	Euasterid
11	*Cuscuta*	1/170	Convolvulaceae	Solanales
12	*Orobanche, Pedicularis, Striga*	88/2046	Orobanchaceae	Lamiales
	Total parasitic plants	268/4558*		

*Not counting 13/64 autotrophic (nonparasitic) "Olacaceae."

inside the host vine *Tetrastigma* (Vitaceae). Molecular phylogenetic methods have placed Rafflesiaceae in Malpighiales and recently more specifically near the spurge family Euphorbiaceae. This result is surprising because spurges typically have small flowers, thus indicating that the stem lineage leading to *Rafflesia* underwent a massive increase in flower size (ca. 79-fold).

With 163 genera and over 2200 species, the sandalwood order (Santalales) is one of the largest clades of parasitic plants. One of the most important innovations to evolve in Santalales is stem parasitism, which includes mistletoes.

The order Lamiales is a clade containing the family Orobanchaceae. Among flowering plants, parasites in the genera *Orobanche* (broomrape) and *Striga* (witchweed) inflict the most damage upon crops and are thus the subject of intensive efforts to reduce their impact on world agriculture.

***Rafflesia pricei*. Mature open flower, about 30 cm (12 in.) in diameter, from Tambunan, Sabah, Malaysia (Borneo). (*Photograph by D. L. Nickrent*)**

Molecular evolution. Molecular evolution is concerned with the rates and patterns of change in macromolecules (DNA, RNA, and protein) and mechanisms underlying those changes. This is an active field intimately associated with molecular phylogeny. This is especially true with model-based phylogeny inference methods for which mathematical algorithms aim to accurately model molecular processes such as base substitutions. As mentioned above, parasitic plants represent natural genetic mutants that provide a window into understanding how evolutionary processes work at the molecular level. The use of parasitic plants as genetic models will be demonstrated by discussing three areas: evolutionary rate increases, plastid genome evolution, and horizontal gene transfer.

One of the earliest concepts developed when molecular data were applied to evolutionary questions concerns the molecular clock, that is, the number of DNA base substitutions (for example, changing an adenine to a guanine) accumulated per unit time (that is, the rate). Initially thought to be constant across different organismal lineages, later work showed that this strict molecular clock could not be applied to all organisms and all genes. Studies of some animals and plants suggested that differences in accumulated substitutions could be best explained by the generation time: Slowly reproducing organisms had fewer substitutions than rapidly reproducing ones. When DNA sequences of ribosomal genes of holoparasitic angiosperms such as *Balanophora* and *Rafflesia* were obtained, it was found that they exhibited substitution rates as high as 3.5 times that seen in nonparasitic, autotrophic plants. Because these parasites did not have rapid generation times and because some rapid generation plants such as *Arabidopsis* did not show the same rate increases, factors other than generation time were discussed.

Parasitic plants have also been important in the study of plastid genome evolution. Plastids are derived from cyanobacteria that became incorporated into eukaryotic cells hundreds of millions of years ago. This endosymbiotic event was followed by the massive loss of genes from their genome (plastome), most of which now reside in the host cell nucleus. Today, most land plant plastids are streamlined organelles with ~100 genes for which order is highly conserved. The primary function of the plastid is photosynthesis and, as expected, some of the genes needed for this process are coded by the plastome (for example, photosystem I and II genes, *rbcL*, etc.). However, several thousand genes are needed for the processes that take place in the plastid; thus, the majority of genes are coded by the nucleus and the products (proteins) are imported into the plastid. What then happens to the plastid and plastome of a holoparasitic angiosperm that does not photosynthesize? Answers to this question began emerging following study of beechdrops (*Epifagus*, Orobanchaceae), a root holoparasite of beech trees (*Fagus*) in North America. The complete plastome sequence of *Epifagus* showed that it is about half the size of a typical angiosperm plastome, the difference mainly attributable to the loss of all photosynthetic and respiratory genes. More recently, the complete plastome sequences of additional parasitic plants (*Cuscuta* and *Pholisma*) have been completed and they too have lost these genes.

The loss of genes in parasitic plant plastomes may be viewed as a continuation of a process that began soon after the initial endosymbiotic event. How far then can this process proceed? In other words, can the plastome be lost entirely? All plastids examined to date contain a plastome, albeit some are small. Whereas the plastome of *Epifagus* is 70 kilobases (kb) long, its relative *Conopholis* has been reduced to 40 kb. Although a complete plastome sequence is not available, Southern blot data suggest a plastome as small as 20 kb in *Cytinus* (Cytinaceae). A number of hypotheses have been advanced as to why nonphotosynthetic plastids retain plastomes [for example, the plastome has a gene (or genes) essential for survival, but which cannot be transferred to the nucleus], but overall none of them provide explanations that are satisfactory for all organisms. To date, only one lineage of holoparasite has eluded polymerase chain reaction (PCR) amplification of any plastome genes, namely Rafflesiaceae, again highlighting that one explanation may not serve all examples. Given the high substitution rates in holoparasites such as *Rafflesia*, molecular methods such as PCR and Southern blotting are compromised because the target genes may be too divergent to accept primers and probes designed from more "typical" plants. Ultimately, complete genome sequencing will provide the answer to the fate of the plastome in *Rafflesia*.

Parasitic plants have played a key role in uncovering another biological phenomenon that has only recently received wide attention: horizontal (or lateral) gene transfer (HGT) [where genetic material (DNA) is exchanged among organisms that are distantly related]. The most prevalent form of genetic inheritance is vertical transmission, that is, a parental generation providing genes to the next (filial) generation, and so on. Indeed, this form of inheritance provides the cornerstone for Darwinian evolution. However, HGT between unrelated species can occur in bacteria and fungi, although its frequency, mechanism, and significance in eukaryotes are not understood. It should be pointed out that HGT here does not refer to transmission of mobile genetic elements, but to larger segments of DNA (for example, whole genes) from the mitochondrion and nucleus. Widespread transfer of mitochondrial genes in flowering plants and a gymnosperm has been demonstrated, and this same phenomenon was documented for a parasitic plant. Ironically, the transfers are not unidirectional (host to parasite), but movement of parasite genes into hosts has been documented as well. These involved the discovery of the mitochondrial gene *atp1* from *Cuscuta* in *Plantago* and mitochondrial *nad1* and *matR* genes from Santalales in the fern *Botrychium*. The mechanism of HGT is still a matter of speculation; however, the role played by parasitic plants in providing interspecific conduits is being investigated.

For background information *see* AGRICULTURAL SCIENCE (PLANT); CELL PLASTIDS; PARASITOLOGY; PLANT CELL; PLANT GROWTH; PLANT KINGDOM; PLANT PATHOLOGY; PLANT PHYLOGENY; RAFFLESIALES in the McGraw-Hill Encyclopedia of Science & Technology. Daniel L. Nickrent

Bibliography. C. C. Davis et al., Floral gigantism in Rafflesiaceae, *Science*, 315:1812, 2007; J. Kuijt, *The Biology of Parasitic Flowering Plants*, University of California Press, Berkeley, 1969; D. L. Nickrent et al., Phylogenetic inference in Rafflesiales: The influence of rate heterogeneity and horizontal gene transfer, *BMC Evol. Biol.*, 4:40, 2004; D. L. Nickrent, J. P. Der, and F. E. Anderson, Discovery of the photosynthetic relatives of the "Maltese mushroom" *Cynomorium, BMC Evol. Biol.*, 5:38, 2005; D. L. Nickrent et al., Molecular phylogenetic and evolutionary studies of parasitic plants, pp. 211–241, in D. Soltis, P. Soltis, and J. Doyle (eds.), *Molecular Systematics of Plants II: DNA Sequencing*, Kluwer Academic, Boston, 1998; M. C. Press and J. D. Graves (eds.), *Parasitic Plants*, Chapman and Hall, London, 1995.

Pathological gambling

By definition, gambling is the process of placing something of value (usually money) at risk in the hope of gaining something of greater value. Excessive forms of gambling have been described as compulsive, problematic, or pathological, among other terms. Pathological gambling (PG) is the diagnostic term adopted by the American Psychiatric Association. Along with kleptomania, pyromania, intermittent explosive disorder (aggressive outbursts that can be violent or destructive and that are out of proportion to any possible provocation or stimulus), and trichotillomania (compulsive hair pulling), PG is classified as an "Impulse control disorder not elsewhere classified."

A diagnosis of PG requires the presence of five or more inclusionary criteria reflecting persistent and recurrent maladaptive gambling behaviors and must not be better accounted for by a manic episode. The inclusionary criteria include a preoccupation with gambling; needing to gamble larger amounts (tolerance); repeated unsuccessful attempts to control, cut back, or stop gambling; and restlessness or irritability during such attempts (withdrawal). Other criteria target gambling as a way of escape or to relieve a dysphoric (distressed) mood and the phenomenon of "chasing," trying to win back gambling losses through more gambling. Lying to conceal the extent of gambling, committing illegal acts to finance gambling, jeopardizing or losing significant relationships or opportunities, and relying on others to relieve a desperate financial situation caused by gambling are also captured in the inclusionary criteria for PG. In addition, individuals with PG often experience problems in legal, financial, and marital domains.

Similar to the inclusionary criteria for substance dependence, those for PG target aspects of tolerance, withdrawal, interference in major areas of life functioning, and repeated unsuccessful attempts to cut back or quit. Other diagnostic criteria are more unique to PG: for example, chasing. Unlike the categorization of substance-use disorders (SUDs), which have separate diagnostic categories for various degrees of the severity of substance use (that is, substance abuse), there are no diagnostic criteria for forms of excessive gambling that do not meet the criteria for PG, although the term problem gambling is often used in this context.

Psychiatric comorbidity. PG has been conceptualized as lying along an impulsive-compulsive spectrum. Because the clinical features of PG share similarities with those for substance dependence, PG has also been described as a behavioral addiction, or a nondrug addiction. Beyond shared diagnostic criteria, these disorders are associated with impulsivity, demonstrate "telescoping" (faster rate of progression from initial to problematic behavior in women compared with men), and show lower prevalences in older adults and higher prevalences in adolescents and young adults. Consistently, high rates of co-occurrence have been found between PG and SUDs. Other co-occurring psychiatric disorders with PG include antisocial personality, mood, anxiety, and psychotic disorders. Population-based studies have typically not found an increased association between obsessive-compulsive disorder (OCD) and PG, despite a proposed classification of PG as an OCD-spectrum disorder.

Personality features. PG and SUDs share several personality features. Both groups exhibit high levels of impulsiveness and sensation-seeking. PG subjects also score high on measures of compulsivity. Unlike OCD, compulsive features in PG appear more closely related to impaired control over mental activities and urges or worries about losing control over motor behaviors rather than compulsions such as checking and hand washing that are typically seen in OCD.

Neuropsychology and neurocognition. Relatively few studies have investigated neuropsychological and neurocognitive functioning in PG. Reported findings indicate impairments in executive functioning (as measured by various analytical ability tests such as the Embedded Figures Test, the Wisconsin Card Sorting Test, and the Porteus Mazes Test), risk–reward decision-making [as measured with delay discounting (see below) and gambling tasks], and cognitive control (as assessed by the Stroop Color-Word Interference Test).

Genetics. Familial factors, including genetics, contribute to the risk for PG, and common genetic contributions to PG and SUDs exist. Family studies of pathological gamblers show high rates of first-degree relatives with SUDs. In the Vietnam Era Twin (VET) registry, the genetic liability for PG was estimated to lie between 35% and 54%. VET data also indicate that a significant portion of the liability for PG is accounted for by that for alcohol dependence, and similar findings have been reported for the relationships between PG and both antisocial behaviors and major depression. Molecular-genetics studies of PG have investigated the roles of specific candidate genes related to norepinephrine, serotonin, and dopamine neurotransmitter systems. In part, these genes have been studied in PG given their implication in SUDs and compulsive disorders. The frequencies of some dopamine receptor gene alleles have been found to be elevated in PG as opposed to controls, and allelic variants have also been associated with PG. A gender-related difference was found in the promoter region of the serotonin transporter gene (5-HTTLPR); a less functional short variant was found more frequently in PG males, but not in females. However, these molecular-genetic studies should be interpreted cautiously given methodological limitations, including small sample sizes, characterization of participants, and inconsistent consideration of allelic differences attributable to racial differences.

Neurobiology. Emerging data suggest a number of neurobiological features that determine in part PG.

Dopamine. Dopamine (DA) may mediate rewarding and reinforcing aspects of gambling. Decreased levels of DA have been observed in the cerebrospinal fluid (CSF) of pathological gamblers, whereas amphetamine, a drug that directly influences DA neurotransmission, selectively primes ("cross-primes") motivation for gambling in PG. Additional support for DA implication in PG stems from genetic studies (as discussed above) and associations between PG and DA agonist use in the treatment of Parkinson's disease.

Serotonin. Serotonin (5-HT), a neurotransmitter influencing impulsivity, is implicated in PG. In animals, forebrain 5-HT depletion facilitates impulsive choice, and the indirect 5-HT agonist fenfluramine decreases impulsive behavior. In humans, individuals with impulsive characteristics and early onset alcoholism have on average low levels of the 5-HT metabolite 5-hydroxyindoleacetic acid (5-HIAA). Low levels of 5-HIAA are seen in CSF samples of PG subjects as compared to controls [after controlling for CSF collection (tapping) time, which was increased in the PG group]. The administration of metachlorophenylpiperazine (m-CPP), which acts as a partial agonist with high affinity for 5-HT receptors, induces a behavioral "high" and also increases prolactin levels in individuals with PG, with between-group differences seen in PG and control subjects.

Stress response systems. Stress response pathway studies in PG suggest noradrenergic dysfunction, with higher peripheral norepinephrine levels and higher blood levels of epinephrine in PG subjects compared to controls during gambling and on gambling-concentrated days. Cortisol levels rise during casino gambling in both recreational and pathological gambling groups. These data are consistent with a role for the stress response and noradrenergic systems in PG, possibly relating to arousal and excitement.

Endogenous opioids. Opioidergic pathways, implicated in the mediation of hedonic and rewarding and reinforcing behaviors, contribute to PG. Problem gamblers as opposed to nonproblem gamblers show lower levels of β-endorphin during casino gambling sessions. Opioid antagonists have shown efficacy in placebo-controlled trials in PG (see below).

Neuroimaging. Neuroimaging studies have implicated frontostriatal circuitry in PG. The ventromedial prefrontal cortex (vmPFC) contributes to both reward processing and prediction. Subjects with vmPFC lesions demonstrate deficits in planning, such as making frequent decisions leading to negative consequences. These subjects perform more poorly on the Iowa Gambling Task (IGT, a simulated test for examining monetary decision behavior under uncertainty) than do control subjects. PG subjects perform similarly to subjects with vmPFC lesions with regard to IGT performance. PG subjects also more commonly choose lower monetary rewards that are immediate over higher monetary rewards that are delayed, a phenomenon termed "delay discounting." Decreased activation of the vmPFC has been observed in PG subjects in response to gambling cues, during performance of the Stroop Test, and during simulated gambling. The degree of vmPFC activation during simulated gambling correlates inversely with gambling severity in PG subjects. Together, these data suggest an important role for the vmPFC in the etiology of PG with regard to decision-making and risk-reward assessment.

Other components of frontostriatal circuitry have been implicated in PG. During a simulated gambling task, PG subjects show decreased ventral-striatal activation as opposed to control subjects, and this activation correlates inversely with gambling severity in PG subjects. Additional support for this pathway in PG etiology is found during neuroimaging studies of craving, where PG subjects demonstrate diminished activation of the ventral striatum when compared to controls.

Treatment. With prevalence estimated at 1–2% and annual costs of over $5 billion in the United States alone, PG is costly for society, affected individuals, and their families. Those afflicted may experience unemployment, debt, incarceration, family dysfunction, and suicidiality. Few systematic studies have investigated the efficacies and tolerabilities of specific treatments for PG. Currently, no drugs are approved by the U.S. Food and Drug Administration for the treatment of PG. Several placebo-controlled, randomized clinical trials have been conducted, although results are often limited by small sample sizes, short durations, and study designs that frequently exclude individuals with co-occurring disorders.

Selective serotonin reuptake inhibitors (SSRIs) have shown mixed results in PG, with some but not other studies showing benefit of the use of active drug over placebo. Lithium has demonstrated superiority to placebo in reducing gambling and manic symptoms in individuals with PG and co-occurring bipolar disorder. Opioid antagonists such as naltrexone and nalmefene, perhaps due to their modulation of the mesolimbic dopamine system, have demonstrated efficacy superior to placebo. Similar to their effect in SUDs, opioid antagonists may help reduce appetitive urges or cravings.

Behavioral interventions for PG have also been examined. Gamblers Anonymous (GA), a 12-step self-help group modeled on Alcoholics Anonymous, appears helpful, particularly in conjunction with professional treatment. However, a majority of individuals who participate in GA discontinue after attending only a few meetings. Motivational interviewing, motivational enhancement, and cognitive behavioral therapy have also received initial empirical support. More research is needed to define optimal treatment strategies for individuals with PG. Specific therapies targeting specific groups seem warranted, particularly given the frequent co-occurrence of other psychiatric disorders with PG. An improved understanding of the neurobiology of PG and its relationship to clinical characteristics and treatment response will help develop improved prevention and treatment strategies.

For background information *see* ADDICTIVE DISORDERS; ALCOHOLISM; ANXIETY DISORDERS; BRAIN; MOTIVATION; NEUROBIOLOGY; NEUROTIC DISORDERS; OBSESSIVE-COMPULSIVE DISORDER; PSYCHOLOGY in the McGraw-Hill Encyclopedia of Science & Technology.

Jennifer D. Bellegarde; Marc N. Potenza

Bibliography. J. A. Brewer, J. E. Grant, and M. N. Potenza, The treatment of pathological gambling, *Addic. Disorders Treat.*, in press, 2007; J. A. Brewer and M. N. Potenza, The neurobiology and genetics of impulse control disorders: Relationships to drug addictions, *Biochem. Pharmacol.*, in press, 2007; A. E. Goudriaan et al., Pathological gambling: A comprehensive review of biobehavioral findings, *Neurosci. Biobehav. Rev.*, 28(2):123–141, 2004; J. E. Grant, J. A. Brewer, and M. N. Potenza, The neurobiology of substance and behavioral addictions, *CNS Spectrums*, 11(12):924–930, 2006; M. N. Potenza, Should addictive disorders include non-substance-related conditions?, *Addiction*, 101(suppl. 1):142–151, 2006.

Phloem loading

Phloem loading is the first step in the long-distance transport of nutrients in plants. Loading also provides the driving force for this transport. Why is long-distance transport needed? Although vascular plants produce their own organic molecules by photosynthesis, not all organs are self-sustaining. Any tissue that uses or stores more carbohydrate than it produces is a sink, including roots, fruits, tubers, and immature leaves. The sources of carbohydrate are generally mature leaves. This nutrient transport, from source to sink, takes place in the phloem, a highly specialized tissue that links all the organs of the plant and is found in close association with the water-conducting tissue, the xylem.

The vascular tissues—xylem and phloem—form the plumbing system of the plant, lifting water from the soil to the uppermost leaves, and sending the products of photosynthesis to the sinks. In leaves, the vascular tissues, along with some additional supporting cell types, constitute the veins (**Fig. 1***a*). The large veins are visible to the naked eye, but they constitute only a small percentage of the network. Smaller (minor) veins ramify throughout the mesophyll—the photosynthetic tissue—delivering water in the xylem and picking up sugars and other compounds in the phloem for export (Fig. 1*b*).

What types of molecules are transported in the phloem? All plants transport sucrose, and in many this is the sole carbohydrate found in phloem sap. In other plants, the phloem transports raffinose and stachyose, which are trisaccharides and tetrasaccharides, respectively; in others, the phloem transports sugar alcohols. Reducing sugars such as glucose and fructose are not found. In addition to carbohydrate, the phloem also transports ions and a wide variety of organic molecules such as amino acids, organic acids, and even macromolecules (proteins and RNA). There is increasing evidence that some of these macromolecules are regulatory and are able to influence gene expression at sites far from where they were produced.

The distances that all these nutrients and macromolecules travel in the phloem can be extreme—consider the degree of separation between the branches of a redwood tree and its roots—yet this transport occurs without the aid of a heart or any other type of pump. Instead, plants take advantage of their hard cell walls to create pressure in the phloem, more at the source end than in the sinks, and it is this pressure differential that drives the bulk flow of solution, just as water flows through a garden hose. The pressure at the source end (1.5 MPa = 218 psi) can be over five times that of an automobile tire.

Cell types. How are such pressures generated? To answer this question, we must first consider the two functional cell types in the phloem: sieve elements (SEs) and companion cells (CCs) [**Fig. 2**]. The SE is one of the most specialized cell types in nature. SEs are highly elongated cells joined end-to-end to form long tubes. As the SEs develop, large pores form in the adjacent end walls, creating channels for the flow of solution through them. These end walls look like sieves—hence, the name sieve elements. The developing SEs also lose their nuclei, vacuolar membranes, ribosomes, and many other organelles, which would otherwise impede flow. Since protein turnover occurs in the SEs, along with other life processes, they are clearly not dead. They are kept alive by adjacent, metabolically active CCs. The two cell types are connected by narrow pores called plasmodesmata and are often referred to as the SE-CC complex. Although long-distance transport occurs only in the sieve tubes, both the SEs and the CCs are under pressure. The pressure is established by one of three known phloem-loading mechanisms, operating through the symplast or apoplast.

The symplast and the apoplast. Almost all plant cells, not just phloem cells, are joined to one another by plasmodesmata. These fine channels are lined by plasma membrane, which is therefore continuous between the cells. In this sense, all the cells joined by

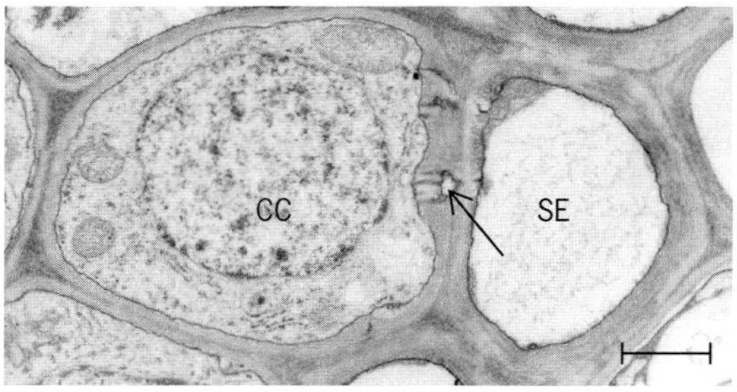

Fig. 2. **A sieve element (SE) and companion cell (CC) in a minor vein of an *Amborella trichopoda* leaf, photographed in transverse section in an electron microscope. Note that the SE is almost empty; the small amount of cytoplasm is confined to the cell periphery. The CC is dense with cytoplasm. The two cells are joined by a plasmodesma (arrow). In minor veins, CCs are larger than SEs; in the rest of the plant, they are smaller. The sieve plate joining this SE to the next was not captured in the section. Magnification bar = 1.0 μm. (*Electron micrograph by Robert Turgeon and Richard Medville*)**

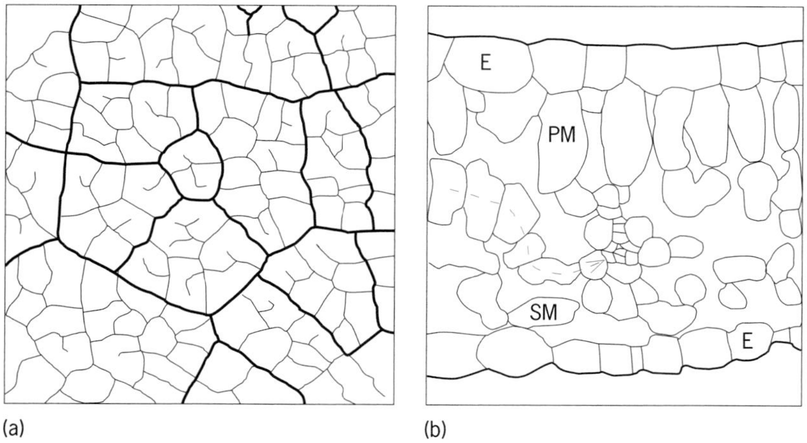

(a) (b)

Fig. 1. **Leaf tracings. (*a*) A cleared leaf illustrating different vein orders. No mesophyll cell is more than six or seven cells away from a small vein. (*b*) Transverse section of a leaf. Sucrose and other products of photosynthesis diffuse from palisade mesophyll (PM) and spongy mesophyll (SM) cells to the minor veins (dashed line). Veins are vascular bundles with both xylem and phloem for water and photoassimilate conduction, respectively. E = epidermis.**

plasmodesmata, including the phloem, are part of an interconnected unit called the "symplast." Molecules can pass from one cell to another within the symplast without crossing plasma membranes, as long as they are smaller than the size exclusion limit of the plasmodesmata. The size exclusion limit varies considerably in different cell types, but it is often around 1 kDa. Limits that are smaller or much larger are known in certain cases. Sucrose has a molecular mass of 342 Da and therefore passes through all but the narrowest plasmodesmata. The apoplast is the space outside the cells, specifically the walls and the dead xylem elements that carry water. It seems odd that sugars and other nutrients would leave the confines of the symplast and enter the apoplast in their travels, but they sometimes do for a very good reason (see below).

Apoplastic phloem loading. Ernst Münch developed the concept that long-distance transport is driven by a difference in pressure between cells of the source and sink phloem. He published this idea in 1928. Over the next decades, a number of laboratories demonstrated that the sugar content of the phloem is much higher than that in surrounding cells, although this turns out not to be true for all species. Since water follows solute into cells by osmosis, and the walls are rigid, the pressure is highest in cells with the highest solute content. The discovery that the phloem is pressurized supported Münch's hypothesis.

The presence of so much sugar in the phloem indicates that a metabolically active pump is operating since energy is needed to establish and maintain differences in the concentrations of neutral solutes. This explains why sucrose enters the apoplast on its way to the phloem: the sucrose pump is on the plasma membranes of the SE-CC complex.

To dissect the transport pathway in detail, let us start with synthesis of sucrose in mesophyll cells (**Fig. 3*a***). Sucrose diffuses through plasmodesmata from one mesophyll cell to the next toward the minor veins. This is a symplastic pathway. No mesophyll cell is more than six or seven cells from a minor vein. Within the vein, sucrose effluxes into the apoplast in the vicinity of the phloem. The efflux step lowers the sucrose concentration in the terminal cell of the symplastic pathway, allowing diffusion to continue. Once in the apoplast, the sucrose is carried into either the SE or the CC by specific transporter proteins.

Although energy is used in the transport step, it is not used directly. Rather, this is a "secondary active transport" phenomenon that involves another type of protein, a proton pump (**Fig. 4**). Proton pumps on the plasma membrane pump H^+ outward, into the apoplast. They do so against a concentration gradient, using metabolic energy in the form of adenosine triphosphate (ATP). A pH gradient is produced as a result of this energy expenditure; the pH of the apoplast around the phloem is approximately 5, whereas the pH of phloem sap is much higher, approximately 8. This 1000× difference in proton concentration (since each whole number change in the

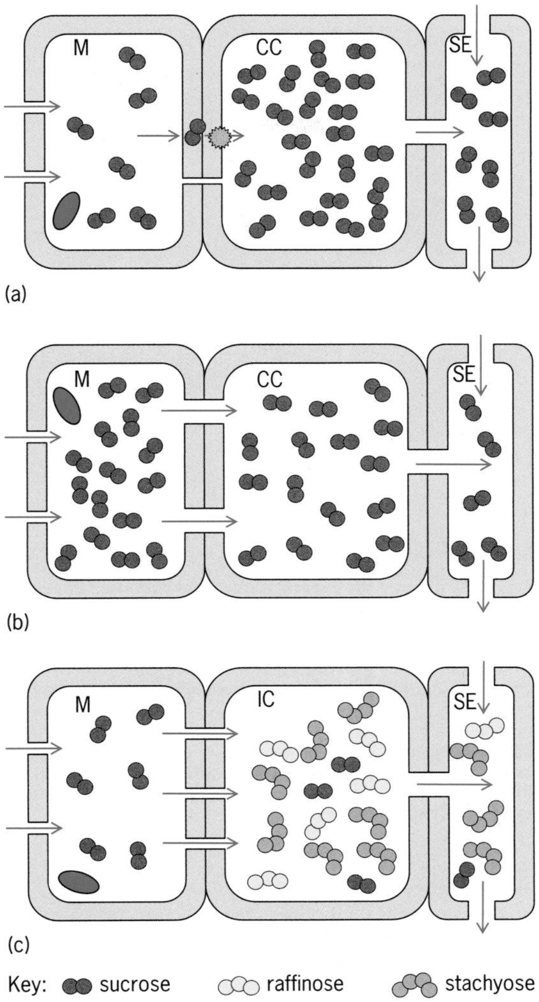

Fig. 3. Three phloem-loading mechanisms. (*a*) Apoplastic loading. Sucrose molecules synthesized in mesophyll cells (M) enter the apoplast near the SE-CC complex and are carried into the CC and/or SE by a transporter on the plasma membrane (star). Once inside the SE, one element of a long sieve tube, they are carried away by bulk flow of solution. (*b*) Diffusion. A high concentration of sucrose in the cytoplasm of mesophyll cells acts as the driving force for diffusion into the CC and from there into the SE. (*c*) Polymer trapping. Sucrose diffuses into a unique type of CC called an intermediary cell (IC) through especially abundant plasmodesmata. Inside the IC, the sucrose is converted to raffinose and stachyose, preventing backdiffusion. Ovals represent chloroplasts (cell plastids occurring in the green parts of plants, containing chlorophyll pigments, and functioning in photosynthesis).

pH indicates a 10-fold change in the proton concentration) is a form of stored energy. The sucrose transporter recognizes both protons and sucrose and, by coupling transport of the two, carries sucrose into the SE-CC complex against its concentration gradient. Water is drawn in osmotically, pressurizing the entire complex.

A second advantage to this loading strategy is that it is highly specific: only sucrose molecules are pumped into the SE-CC complex because the sucrose transporter does not recognize other sugars well. Less is known about the loading of other types of molecules and ions, but it is very likely that there is a spectrum of transporter types on the SE-CC complex,

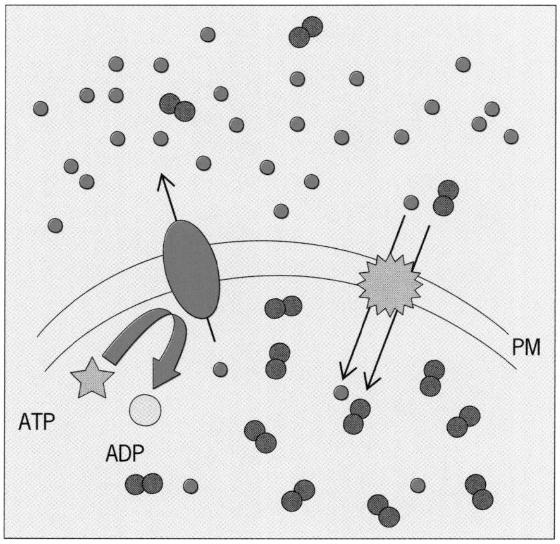

Key: sucrose H+

Fig. 4. Diagrammatic representation of secondary active sucrose transport. Protons are energetically pumped by an ATPase protein (large oval) across the plasma membrane (PM) of the cell into the extracellular space, the apoplast. [ATPase is an enzyme that hydrolyzes adenosine triphosphate (ATP) into adenosine diphosphate (ADP) and phosphate.] The expenditure of this energy establishes a proton gradient from the outside to the inside of the cell. A specific sucrose-proton cotransporter (star) carries protons and sucrose into the cell. In this step, the protons travel down a concentration gradient, allowing sucrose molecules to be pumped in against a concentration gradient.

all using the same pool of protons in the apoplast as an energy source. Sucrose uptake from the apoplast can be visualized by floating leaf tissue on a solution of [^{14}C]sucrose, freeze-drying the tissue, and pressing it against a photographic film (**Fig. 5**). The film shows the pattern of the minor veins where radiolabeled sucrose accumulates.

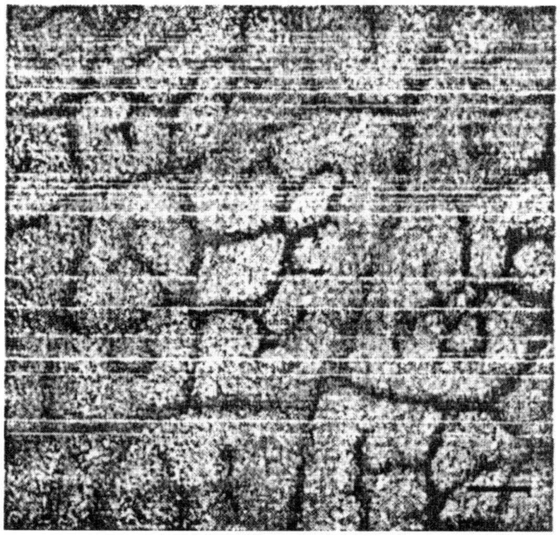

Fig. 5. Autoradiograph of *Veronica spicata* leaf tissue. The surface of the leaf was abraded to allow penetration of solutions. Tissue was then floated on [^{14}C]sucrose for 1 h, washed, rapidly frozen, freeze-dried, and pressed against a photographic film. Upon development of the film, it is evident that the minor veins accumulated radiolabeled sucrose from the apoplast. Magnification bar = 0.5 mm. (*Autoradiograph provided by Emilie Rennie*)

Symplastic phloem loading. If loading involves an apoplastic step, as described above, one might expect to find that the plasmodesmata between the SE-CC complex and surrounding cells are few in number and possibly very narrow; otherwise, the loaded sucrose would diffuse through them back into the mesophyll cells. Indeed, in many plants, the plasmodesmata at this location are sparse. Their precise size exclusion limit has not been measured.

However, contrary to these expectations, some plants have highly abundant plasmodesmata at all the cell-cell interfaces from mesophyll to minor vein phloem. Could these plasmodesmata be involved in phloem loading? This seemed unlikely at first. True active transport requires the presence of a membrane and of transporter proteins that mediate flux across it. Although internally complex, plasmodesmata are essentially long pores lined with plasma membrane; there is no membrane across the mouth of the pore on either end. The laws of thermodynamics dictate that, in open pores, net movement of neutral molecules is down a concentration gradient, not in the opposite direction.

One theoretical solution to this problem is that the sucrose gradient might, in fact, go downhill from the mesophyll to the phloem. This would allow sucrose to simply diffuse into the SE-CC complex (Fig. 3*b*). As long as the sucrose concentration is high enough in the mesophyll cells, there will be enough sucrose in the minor vein phloem to generate the pressure differential needed for bulk flow in the sieve tubes. Recall that phloem loading, in the sense of forcing sucrose into the phloem to a concentration higher than in the mesophyll cells, was not discovered until well after Münch published his model. Therefore, this notion of passive transport from mesophyll to phloem is probably Münch's original concept. Is there evidence for such a transport scheme? In one species, *Salix babylonica* (weeping willow), the conditions for passive entry into the phloem seem to be met; the concentration of sucrose in mesophyll cells is as high as it is in the phloem, and an inhibitor of sucrose transporters has no effect on phloem transport. How widespread this phenomenon might be has yet to be discovered.

However, there is another symplastic mechanism that has been better characterized. In select families of plants (for example, the cucurbits, such as melon and cucumber), the plasmodesmata in the minor vein CCs are especially numerous and asymmetrically branched. All the species that have such CCs (called intermediary cells) translocate their carbohydrate primarily as raffinose and stachyose (Fig. 3*c*). The synthesis of these sugars is part of the phloem-loading mechanism.

In these plants, sucrose from the mesophyll diffuses into the intermediary cells through the abundant plasmodesmata in their walls. Inside the intermediary cells, most of the sucrose is converted to raffinose and stachyose. This keeps the sucrose concentration low so that diffusion will continue. It is thought that the intermediary cell plasmodesmata are especially narrow, allowing sucrose passage but excluding the larger sugars, raffinose and stachyose.

Thus, the plasmodesmata act as one-way valves, allowing sugar in, but preventing escape in the opposite direction. Larger plasmodesmata between the intermediary cell and SE allow all sugars to pass through and to be carried away to the sinks. This mechanism is known as polymer trapping. Interestingly, plants with intermediary cells also have "ordinary" companion cells in the minor veins, so it seems likely that polymer trapping evolved as an addition to the apoplastic loading mechanism, which persists to a certain degree in these plants.

Distribution of phloem loading mechanisms. At present, it is not clear why different strategies of phloem loading have evolved. Apoplastic loading is most common and is especially prevalent in herbaceous groups. Most crop plants, and perhaps all monocots, load this way. The polymer trap mechanism is more common in the tropics than in temperate regions and is not particularly slanted toward any growth form. Plants with abundant minor vein plasmodesmata, at least those without intermediary cells, are almost all trees. They are found all over the world, except in the Arctic, but trees are not found in the Arctic for many reasons unassociated with phloem loading. Further work is needed to determine how the loading types are distributed, especially those with symplastic mechanisms. This knowledge is needed before the evolution of the different mechanisms can be understood.

For background information *see* CARBOHYDRATE; CELL PERMEABILITY; CELL WALLS (PLANT); LEAF; PHLOEM; PHOTOSYNTHESIS; PLANT MINERAL NUTRITION; PLANT TISSUE SYSTEMS; PLANT TRANSPORT OF SOLUTES; PRIMARY VASCULAR SYSTEM (PLANT); TRANSFER CELLS; XYLEM in the McGraw-Hill Encyclopedia of Science & Technology. Robert Turgeon

Bibliography. X-Y. Chen and J-Y. Kim, Transport of macromolecules through plasmodesmata and the phloem, *Physiol. Plant.*, 126:560–571, 2006; S. Lalonde, D. Wipf, and W. B. Frommer, Transport mechanisms for organic forms of carbon and nitrogen between source and sink, *Annu. Rev. Plant Biol.*, 55:341–372, 2004; N. Sauer, Molecular physiology of higher plant sucrose transporters, *FEBS Lett.*, 581:2309–2317, 2007; A. Schulz, Role of plasmodesmata in solute loading and unloading, pp. 135–161, in K. J. Oparka (ed.), *Plasmodesmata*, Blackwell, Oxford, 2005; R. Turgeon, Phloem loading: How leaves gain their independence, *Bioscience*, 56:15–24, 2006.

Phototropin

Growth and development of plants are dependent on sunlight. Consequently, plants have evolved a sophisticated array of photoreceptor systems to control their growth. Phototropins are plant photoreceptors that are specifically receptive to UV-A/blue light (320–500 nm) [UV-A is one of the types of ultraviolet radiation produced by the Sun, in the wavelength range 320–400 nm; blue light has an approximate wavelength range of 400–500 nm]. They control a range of responses concerned with optimizing the photosynthetic efficiency of plants, thereby promoting their growth, particularly under weak light conditions.

Biological functions. The effects of blue light on plant development have been studied for almost two centuries. One of the responses identified is phototropism, the light-driven directional growth movement of plants. Typically, stems show positive phototropism (movement toward the light), whereas roots exhibit negative phototropic movement (away from the source of light). Despite extensive research, the molecular identity of the blue light-absorbing photoreceptor responsible for phototropism remained elusive until only recently. Its identity finally became known through genetic analysis methods using the model plant *Arabidopsis thaliana*. *Arabidopsis* contains two phototropins referred to as phot1 and phot2. Mutants of *Arabidopsis* lacking both phototropins lose their phototropic responsiveness (**Fig. 1***a*).

Originally phototropins were thought to be involved only in phototropism (hence their name). However, subsequent genetic and physiological analyses have shown that they control other photoresponses as well. For instance, phototropins control the movement of chloroplasts (cell plastids occurring in the green parts of plants, containing chlorophyll pigments, and functioning in photosynthesis and protein synthesis) in response to different light intensities (Fig. 1*b*). Both phot1 and phot2 act to induce chloroplast accumulation movement to the upper cell surface under weak light conditions to maximize light capture for photosynthesis. Phot2 alone induces chloroplast avoidance movement to the cell sidewalls in bright light to increase mutual shading and prevent photodamage of the photosynthetic machinery. Phototropins also induce the opening of stomatal pores in the leaf and stem epidermis, allowing plants to regulate water loss and CO_2 uptake. Additionally, the positioning and expansion of leaves are regulated by the phototropins (Fig. 1*c* and *d*). Collectively, these responses serve to enhance the photosynthetic performance of plants and maximize their growth potential. Many plant species are able to track the movement of the Sun by a process known as solar tracking or heliotropism, and this is likely mediated by the phototropins. Phot1 activity is also important for the rapid but transient growth inhibition of young seedlings upon their emergence from the soil.

Structure and activity. Phototropins, like photoreceptor pigments associated with mammalian vision, comprise many amino acids (900–1000) that form the main structure of the protein (the apoprotein) to which an accessory chemical cofactor is bound that can absorb light and impart color (the chromophore). Together, the apoprotein and chromophore constitute what is known as the photoreceptor holoprotein. Specifically, the phototropins bind a vitamin-B-related chromophore known as flavin mononucleotide (FMN), giving the protein a yellow color and enabling it to absorb light of blue and UV-A wavelengths.

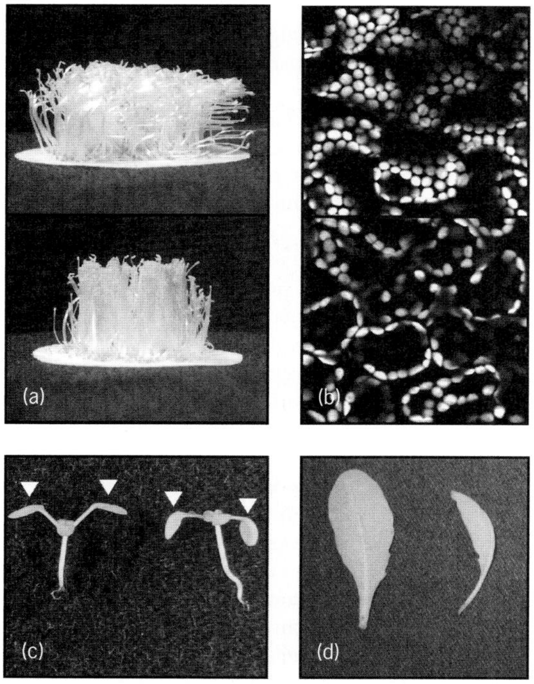

Fig. 1. Biological functions of phototropins. (*a*) Phototropic response of wild-type *Arabidopsis* seedlings (top panel) and a phototropin-deficient mutant (bottom panel) irradiated with blue light from the right. (*b*) Chloroplast positioning in leaf cells of *Arabidopsis*. Microscope images represent cells viewed from above. Accumulation movement of chloroplasts to the cell surface (top panel) occurs when leaves are irradiated from above with low-intensity blue light. Avoidance movement of chloroplasts (bottom panel) to the cell sidewalls occurs when leaves are irradiated with high-intensity blue light. (*c*) Leaf positioning in wild-type (left) and phototropin-deficient *Arabidopsis* plants (right) as indicated by the white arrows in response to irradiation from above. (*d*) Rosette leaves from mature wild-type (left) and phototropin-deficient *Arabidopsis* (right). The leaf from the mutant is unexpanded and curled and therefore lies on its side.

The amino acid structure of plant phototropins can be separated into two segments: an N-terminal photosensory region and a C-terminal effector region that includes a classic serine/threonine kinase motif (a kinase is any enzyme that catalyzes phosphorylation reactions) [**Fig. 2**]. The N-terminal region is composed of two so-called LOV domains, each of which binds the blue light–absorbing cofactor FMN as a chromophore. LOV domains exhibit homology to motifs found in a diverse range of eukaryotic and prokaryotic proteins involved in sensing light, oxygen, or voltage—hence the acronym LOV. X-ray crystallography has shown that the LOV domain consists of five antiparallel β-sheets (a type of secondary structure in which two or more polypeptide chains run alongside each other and are linked in a regular manner by hydrogen bonds between the main chain C═O and N—H groups), an α-helical connector, and two additional α-helices, binding the FMN inside an enclosed structure.

Bacterially expressed LOV domains are yellow in color owing to their bound flavin cofactor and are photochemically active as monitored by absorbance spectroscopy (**Fig. 3**). In the dark, LOV domains bind FMN noncovalently to form a spectral species, LOV_{447}, absorbing maximally near 447 nm (Fig. 3*b*). Irradiation of the domain induces the formation of a covalent bond between the C(4a) carbon of the FMN and the sulfur atom of a nearby, conserved cysteine residue within the domain. Formation of the cysteinyl adduct (a chemical addition product) occurs within microseconds of illumination and produces a spectral species, LOV_{390}, absorbing maximally near 390 nm. Formation of LOV_{390} is fully reversible in darkness, and represents the active signaling state that leads to photoreceptor activation. Why the phototropins contain two LOV photosensors is still unclear. Photochemically active LOV2 is necessary for phototropin function, whereas the photoreactivity of LOV1 has been shown to be dispensable. However, LOV1 has been proposed to mediate receptor dimerization (a chemical reaction in which two identical molecular entities react to form a single dimer) and/or modulate the photochemical reactivity of LOV2.

Mode of receptor activation. The current view of phototropin receptor activation is that LOV2 functions as a repressor of the C-terminal kinase domain in the dark, and that this mode of repression is alleviated upon photoexcitation, resulting in receptor autophosphorylation [addition of a phosphate to a protein kinase (possibly affecting its activity) by virtue of its own enzymic activity]. Photoexcitation leads to displacement of an α-helix from the surface of the LOV2 core. Unfolding of this α-helix, designated Jα, results in activation of the C-terminal kinase domain (Fig. 2). Rearrangements within the central

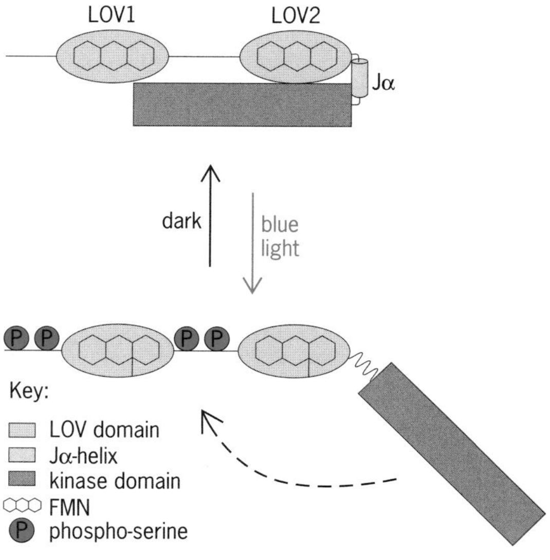

Fig. 2. Phototropin structure and mode of activation. Domain structures and abbreviations used are indicated. In the dark, phototropin is unphosphorylated and inactive. Absorption of light invokes a photochemical reaction within the LOV domains. Photoexcitation of the main light sensor LOV2 causes disordering of the Jα-helix and activation of the C-terminal kinase domain, which consequently leads to autophosphorylation of the photoreceptor. The relative positions of known phosphorylation sites are indicated.

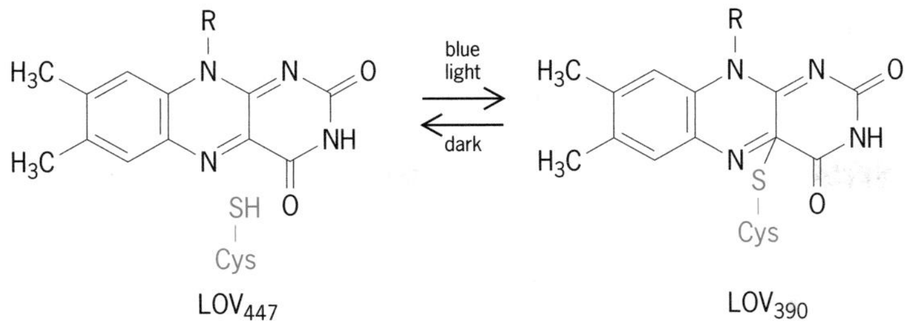

Fig. 3. LOV-domain photochemical reactivity. Schematic representation of LOV-domain photochemistry. Details of the reaction are described in the text.

β-sheet scaffold have also been reported to play a role in propagating the photochemical signal generated within LOV2 that leads to surface protein changes that are necessary for activation of the C-terminal kinase domain and autophosphorylation of specific serine residues. Light-activated phototropin can return to its nonphosphorylated state upon incubation in darkness. This recovery process involves dephosphorylation of the receptor by an as-yet unidentified protein phosphatase (Fig. 2).

Although sites of phototropin autophosphorylation have been mapped upstream of LOV2, there is still no information as to their functional consequences (Fig. 2). A truncated version of phot2 composed of only LOV2 and the C-terminal kinase domain is functionally active in plants, implying that autophosphorylation does not play a role in receptor signaling. Furthermore, the kinase domain of phot2 has been reported to stimulate phototropin responsiveness in the absence of light, consistent with a model in which the N-terminal photosensory region, including LOV2, acts as a dark-state repressor (Fig. 2).

Phototropins are typically associated with the plasma membrane, but a small fraction of the receptor pool is rapidly internalized (within minutes) upon blue light irradiation. One consequence of autophosphorylation may therefore be to promote receptor dissociation from the plasma membrane to desensitize the photosensory system, a process analogous to other receptor kinases. Another consequence of phototropin autophosphorylation is to mediate 14-3-3 binding. 14-3-3 proteins (which are given numerical designations based on column fractionation and electrophoretic mobility) are key regulators of protein function in eukaryotes and preferentially bind to phosphoserine/threonine-containing motifs. However, the biological significance of 14-3-3 binding to phototropins is not known and awaits further investigation.

Receptor signaling. So far, no substrate for phototropin kinase activity has been identified other than the receptors themselves. Nonetheless, a number of phototropin-interacting proteins have been isolated. Nonphototropic hypocotyl 3 (NPH3) is a novel protein that directly interacts with phot1. NPH3 is thought to serve as a protein scaffold to assemble components of a phototropin receptor complex. *Arabidopsis* mutants lacking NPH3 fail to show phototropism, demonstrating that NPH3 is essential for this response. In addition, NPH3 is required for optimal leaf positioning and leaf flattening in *Arabidopsis*. A protein closely related to NPH3, known as root phototropism 2 (RPT2), also interacts with phot1 and is required for both phototropism and stomatal opening by blue light.

Positive phototropism ultimately results from an increase in growth on the shaded side of the stem owing to an accumulation of the plant growth hormone auxin. As light passes through the stem, it becomes progressively diffracted, thereby generating a gradient of phototropin activation across the organ, with the highest level of activity occurring on the irradiated side (**Fig. 4**). Formation of this biochemical gradient underlies the directionality of the phototropic response. Illumination with unilateral light leads to a gradient of auxin across the growing region of the stem, with an accumulation of auxin on the shaded side of the stem. Currently, little information is available regarding how a differential stimulation of phototropin activity across the stem results in this subsequent auxin gradient.

Phototropin signaling is undoubtedly complex given the range of responses that these photoreceptors mediate. Increases in cytosolic calcium concentrations appear to serve as a signaling messenger downstream from receptor activation, particularly in the rapid growth-inhibition response that occurs in seedlings emerging from the soil. Moreover, cytoskeletal changes are important for both blue light–induced chloroplast movements and stomatal opening.

Neochrome. In some plant species, including the fern *Adiantum capillus-veneris*, phototropism and chloroplast movement are induced by red light as well as blue. *Adiantum* contains a novel dual

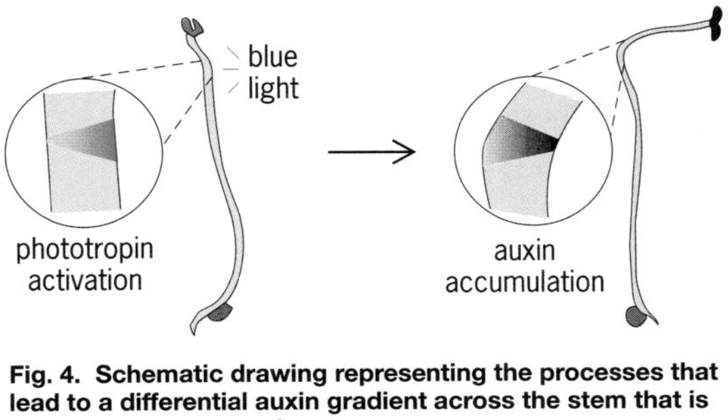

Fig. 4. Schematic drawing representing the processes that lead to a differential auxin gradient across the stem that is required for phototropism.

red/blue light-sensing photoreceptor known as neochrome, comprising a red light-absorbing phytochrome photosensory domain fused to the N-terminus of an entire phototropin receptor (a phytochrome is a protein plant pigment that serves to direct the course of plant growth and development in response variously to the presence or absence of light, to photoperiod, and to light quality). The presence of such a hybrid photoreceptor is proposed to enhance light sensitivity and aid the prevalence of species such as ferns in low light conditions typically found under the canopy of dense forests. Similarly, photoactivation of red light-absorbing phytochromes is known to enhance blue light-induced phototropism in *Arabidopsis*. However, this phototropic enhancement results from an interaction between separate phytochrome and phototropin receptor systems that likely involves phytochrome kinase substrate (PKS) proteins (phytochrome signaling components that bind phototropins).

Additional LOV-sensor proteins. LOV-photosensory motifs are not only restricted to the phototropins. *Arabidopsis* contains three other single LOV-containing proteins known as the ZTL/ADO family that play important roles in regulating the targeted degradation of components associated with circadian clock function and flowering. Furthermore, single LOV-containing proteins are found in other organisms besides plants. For example, white collar-1 (WC-1) is a photoreceptor that mediates phototropism in addition to other photoresponses in *Neurospora crassa* and related fungi. Surprisingly, LOV-sensing motifs are present in a large number of otherwise very different bacterial proteins. Such proteins are found in Eubacteria, Cyanobacteria, and even Archaea. The presence of LOV-containing proteins throughout various kingdoms of life therefore demonstrates that this photosensory module has been conserved throughout evolution. The number and diversity of the LOV-containing proteins found throughout nature will undoubtedly increase as more genome sequences become available.

For background information *see* AUXIN; PHOTOMORPHOGENESIS; PHOTOSYNTHESIS; PHYTOCHROME; PLANT GROWTH; PLANT MORPHOGENESIS; PLANT MOVEMENTS; PLANT ORGANS; PLANT PHYSIOLOGY in the McGraw-Hill Encyclopedia of Science & Technology.

John M. Christie

Bibliography. W. R. Briggs, Blue-UV-A receptors: Historical overview, pp. 171–197, in E. Shäfer and F. Nagy (eds.), *Photomorphogenesis in Plants*, 3d ed., Springer, Dordrecht, 2006; W. R. Briggs, The LOV domain: A chromophore module servicing multiple photoreceptors, *J. Biomed. Sci.*, 14:499–504, 2007; W. R. Briggs and J. M. Christie, Phototropins 1 and 2: Versatile plant blue-light receptors, *Trends Plant Sci.*, 7:204–210, 2002; R. B. Celaya and E. Liscum, Phototropins and associated signaling: Providing the power of movement in higher plants, *Photochem. Photobiol.*, 81:73–80, 2005; J. M. Christie, Phototropin blue-light receptors, *Annu. Rev. Plant Biol.*, 59:21–45, 2007.

Plant metabolomics

Metabolomics is the general term for the study of the global metabolite composition of a biological system, the metabolome, under a given set of conditions. The investigation of the metabolome of a cell can be defined by a range of terms: metabolite profiling, metabolite fingerprinting, metabonomics, and metabolomics. Here, the term metabolite profiling is used to describe the rapid, but accurate quantification and identification of metabolites using a single analytical technique. Metabolite profiling is becoming increasingly important in all research areas. Its application ranges from diagnostics through gene function identification to systems biology.

During the twentieth century, analysis of metabolites focused on a handful of compounds that were of possible interest in a given experiment. While this approach yielded important insights into the pathway of attention to the researcher, it simultaneously was limited to the pathway under study and often was very labor-intensive. Estimations of the number of naturally occurring compounds are currently of the order of 200,000. Whereas only a subset of metabolites are present in one single organism, the estimated numbers for different eukaryotes currently range from 4000 to 20,000 metabolites. The past view on single-metabolite systems was thus rather narrow, given that in these experiments not even 1% of the metabolites were covered.

The connection of chromatographic methods to mass spectrometry, while already proposed years ago, has only recently become routinely applicable in plant biology. This achievement enabled the multivariate and comprehensive analysis of complex sample mixtures with hundreds of metabolites of known and unknown chemical structure.

Metabolite profiling protocols have only been recently developed, and thus a great deal of innovation and method improvement is still in progress. Systems can employ different separation techniques and/or various detection components. Separation techniques include liquid chromatography (LC), gas chromatography (GC), and capillary electrophoresis (CE). On the detector side, different types of mass spectrometry (MS), nuclear magnetic resonance (NMR), and ultraviolet light spectroscopy (UVIS) systems exist.

The complex nature of biological compounds does not allow the measurement of all metabolites of one organism by a single comprehensive method, but rather requires the analysis by more than one technique. Hence, the selection of a suitable system depends on the biological question and, in general, represents a compromise between selectivity, speed, and sensitivity. While NMR systems are rapid, selective, and nondestructive, they have low sensitivity and are costly. In an LC/NMR study of plant extracts, 2700 putative compounds could be detected; however, less than 50 were identified. Nevertheless, NMR is of unparalleled importance in structure elucidation and for in vivo determination of metabolite levels. CE/MS and LC/MS systems have been shown to be highly sensitive and are able to separate compounds with high molecular weights, but these setups have problems with chromatographic reproducibility and resolution, as well as relatively high separation time, selectivity, and costs. Further advances might overcome these problems and the problems in the identification of a tremendous number of unknown compounds (indicated by unknown peaks in the chromatogram). GC/MS has proven to be a highly reliable and robust platform, providing good separation and high sensitivity at comparatively low cost (**Fig. 1**). For these reasons, it is now widely and routinely applied to study metabolic phenotypes in plants, microbes, and mammals. Nevertheless, there are drawbacks in the study of the metabolome by GC/MS. In order to visualize the metabolic content, the compounds have to be volatile. This requirement is accomplished by chemical derivatization (chemically modifying a compound to produce a new compound with similar chemical structure, called a derivative, having properties more amenable to separation analysis), but at the cost of additional time and chemical modification.

One major challenge of metabolomics lies in the handling of vast amounts of data points. Recent developments address the housing of mass spectral tags (MSTs) in public databases and the storage of metabolomic experimental data in open repositories, similar to available databases for transcript profiling. Still, today no standardization in describing known and unknown MSTs exists, and thus the exact identification of the same MSTs in different laboratories remains a hurdle.

Applications of metabolite profiling. As noted, metabolite profiling is becoming increasingly important for many research areas. In diagnostics, metabolite profiling has been used to study the mode of action of various herbicides. For example, barley seedlings were treated with LC50 dosages (50% lethal concentration dosages; that is, dosages where the concentration of chemical is expected to produce death in 50% of the organisms exposed to that concentration) of various herbicides and were then comparatively profiled with untreated plants. In combination with bioinformatic tools, this allowed the identification of herbicidal inhibitors of acetyl-CoA carboxylase (ACC) and acetolactate synthase (ALS), which are important enzymes in biosynthetic pathways.

Loss of function can facilitate the identification of gene function. The Keio knockout population of *Escherichia coli*, which consists of approximately 4000 strains, as well as other loss-of- function populations are currently being profiled at the metabolome level to study the effect of gene loss. Gain-of-function analysis by metabolite profiling similarly provides very powerful information. A transgenomic approach where every single gene of *E. coli* and yeast were independently expressed in *Arabidopsis thaliana* highlights this. For example, the expression of a threonine aldolase (an enzyme converting threonine into glycine, an important amino acid in the metabolic network of photorespiration) resulted in an expected increase in threonine, but also revealed an increase in the metabolites of the methionine pathway (homoserine and methionine) and a decrease in isoleucine biosynthesis, thus revealing novel information of the network of amino acid metabolism. Thus, gene function annotation can be facilitated by using metabolite profiling. Lately, the analysis of stress responses by metabolic analysis could reveal new insight into the reaction of a plant to cold stress. Following 1-h cold stress, *Arabidopsis* plants responded with an early increase of maltose. In a more recent study, the effect of nitrogen limitation in tomato plants was investigated. Nitrate starvation has a dramatic effect on leaf metabolism and resulted in decreases of amino and organic acids, whereas levels of carbohydrates increased; these findings confirm and extend studies of nitrate nutrition over many years.

The term systems biology is variously defined; here it is used in the context of the comprehensive multidimensional analysis of the inventory of a cell. Systems biology approaches are now starting

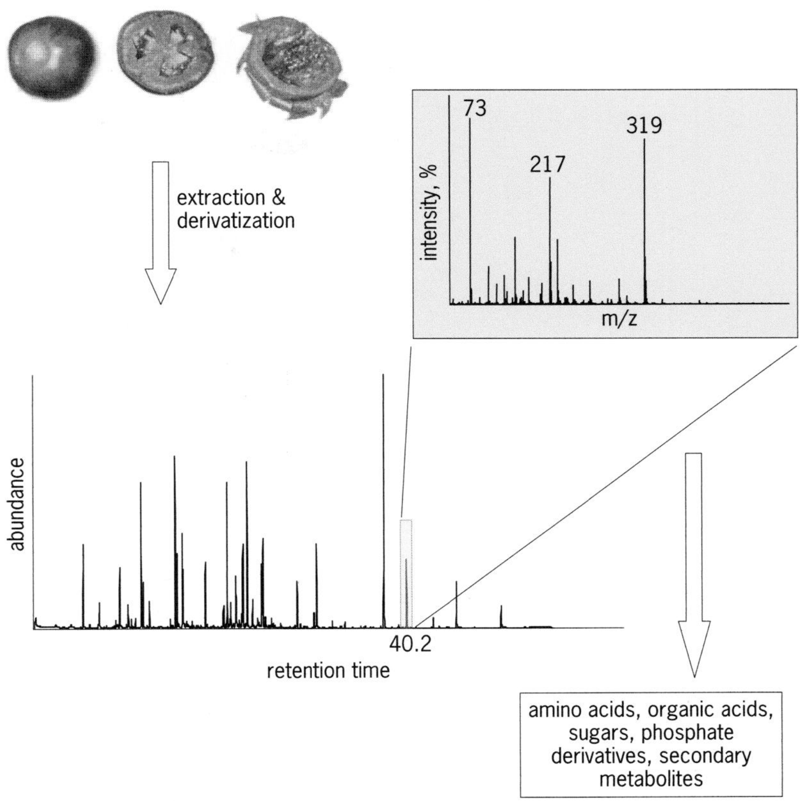

Fig. 1. Principle of GC/MS-based metabolite profiling. From a given biological sample (here, pericarp tissue of a tomato), metabolic content is extracted using methanol/water extraction. The metabolic extract is then derivatized using methyl-alkylation reagents to form volatile components that are thermally stable and thus ideal for GC/MS analysis. A small aliquot is injected into the GC/MS system. Compounds will be separated by their different chemical and physical properties during chromatography. From the chromatographic column, eluting metabolites will finally be detected in the mass spectrometer. The distinct mass spectral tag (the information of time, mass spectral fragmentation, and abundance) allows consequently the identification and quantification of the eluting substances.

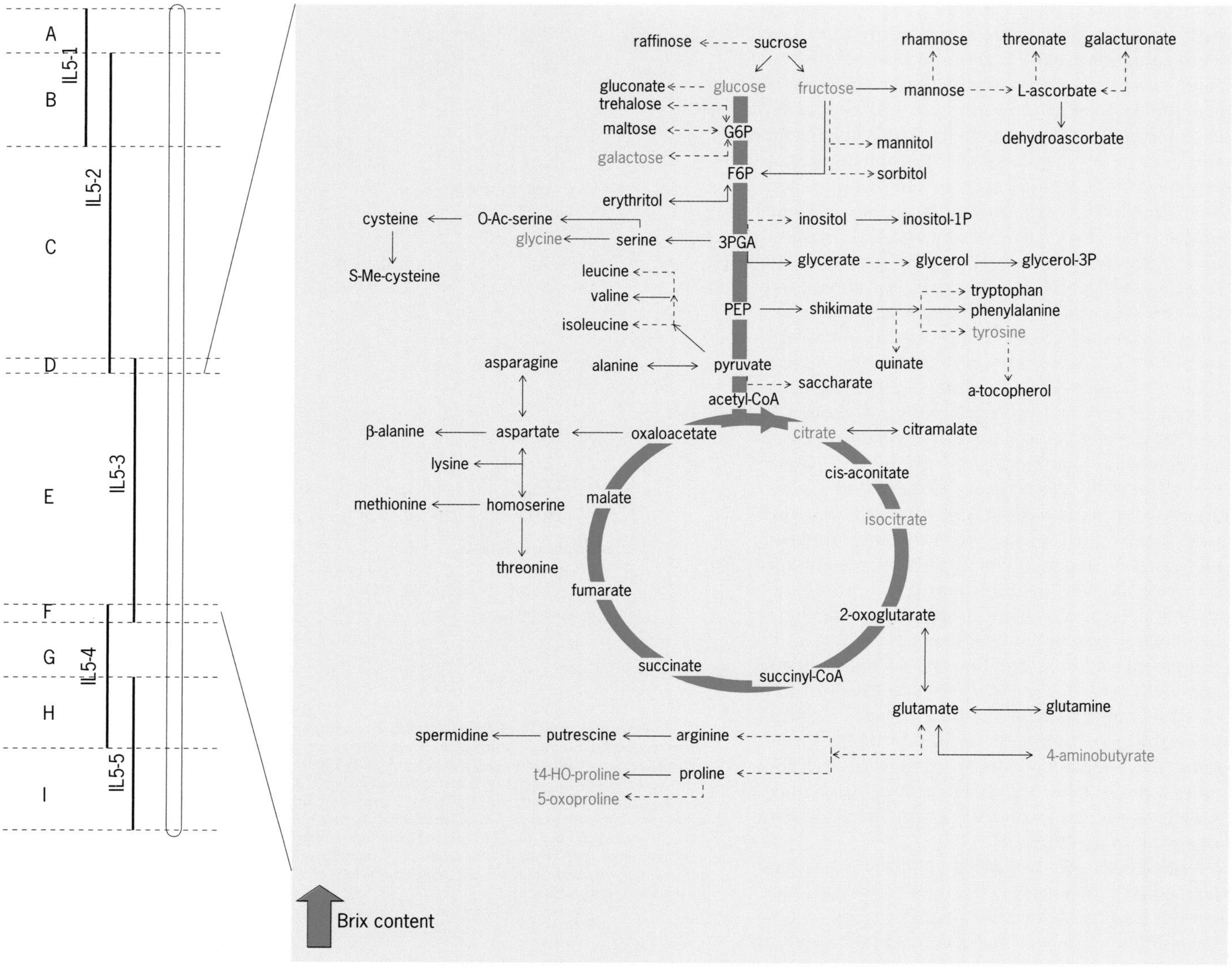

Fig. 2. A genomic region for organoleptic (relating to sensory properties of a product, such as taste, color, odor, and feel) quality in tomatoes. A tomato introgression population with well-defined genomic segments from a wild donor allowed the mapping of a genomic region responsible for metabolic and organoleptic quality in tomatoes. Sugars and organic acids are increased in comparison to a tomato elite variety by the introgression of a section of chromosome from a wild relative in this defined region (IL5-3). These metabolites are known to contribute positively to the taste and Brix (total soluble solids) content. Independent Brix measurements were in agreement with the metabolic results and showed an significant increase (up arrow). Metabolite-assisted breeding (MetAB) will facilitate the process of introducing new favorable traits into modern crops.

to include metabolomic analysis more frequently. Metabolite-transcript correlations from a large data set collected throughout development in wild-type and transgenic potato tubers, which were engineered to have enhanced sucrose metabolism, allowed the identification of candidate genes for biotechnology approaches. In this study, the transcript levels of approximately 280 transcripts were compared to changes in metabolite levels in paired samples. It was found that 517 metabolite-transcript correlations ($P < 0.01$) out of over 26,616 were identified to be significant. Most of the correlations were known, but some strong correlations between genes and nutritional important metabolites were new and lead to the identification of novel candidate genes for metabolic engineering for a broad number of traits. In another study, the metabolomic and transcript networks were investigated after response to sulfur starvation. The investigation of the network topology revealed hubs of homeostasis control points and activation of anthocyanin (plant color pigment) biosynthesis, as well as an auxin regulatory circuit in response to sulfur stress.

These examples show the power of metabolite profiling in the understanding of biological systems and proves to be a tool, in conjunction with other "-omics" approaches (for example, genomics and proteomics), that can help us gain more insight into the control and regulation of metabolic networks in living organisms. Metabolite profiling can also be extended to study and explore the phenotypic relevance of genomic regions by utilizing introgression lines (hybrids of two plant populations that have been backcrossed to introduce new genes into a wild population) or other mapping populations. A tomato introgression population consisting of isogenic (genetically identical) lines harboring a single defined genomic segment of a tomato wild species in the background of a tomato elite variety allowed the identification of genomic segments responsible for sugar, amino acid, organic acid, and vitamin content. The strategic breeding process or metabolite-assisted breeding (MetAB) uses metabolic information in combination with genome annotation (**Fig. 2**). This approach opens up an advance in breeding of metabolic quality as it supports breeding for metabolites.

Outlook. Future developments in plant metabolomics will include the identification of the tremendous number of unknown compounds. This will be achieved using two approaches: (1) improved mass spectrometry technologies such as Fourier transform ion cyclotron resonance mass spectrometry (FTIRC-MS), and (2) multilaboratory standard addition experiments using thousands of commercially and custom-synthesized substances. While the latter relies on the availability of substances as well as the labor and costs, the first approach using FTIRC-MS technology allows high-resolution elucidation of chemical compounds down to the chemical mass formulas and thus leading to the chemical structures.

While in the past plant metabolomics was mainly focused on the development and improvement of metabolite profiling protocols, the metabolomic approach is now being increasingly employed to address biological questions. It provides a useful tool for gene function annotation and understanding of biological networks and pathways. Metabolite profiling enables the comprehensive analyses of whole pathways. Still much effort is needed for the identification of unknown metabolites and the housing of the compiled data in public databases. Open databases will lead to a better understanding of biochemical pathways under different environmental conditions and genetic constraints. This will be even more insightful than transcript expression databases.

For background information *see* AGRICULTURAL SCIENCE (PLANT); ELECTROPHORESIS; GAS CHROMATOGRAPHY; LIQUID CHROMATOGRAPHY; MASS SPECTROMETRY; NUCLEAR MAGNETIC RESONANCE (NMR); PLANT METABOLISM in the McGraw-Hill Encyclopedia of Science & Technology. Nicolas Schauer

Bibliography. A. R. Fernie et al., Innovation—Metabolite profiling: From diagnostics to systems biology, *Nat. Rev. Mol. Cell Biol.*, 5:763–769, 2004; N. Schauer and A. R. Fernie, Plant metabolomics: Towards biological function and mechanism, *Trends Plant Sci.*, 11:508–516, 2006; N. Schauer et al., Comprehensive metabolic profiling and phenotyping of interspecific introgression lines for tomato improvement, *Nat. Biotechnol.*, 24:447–454, 2006; D. Steinhauser and J. Kopka, Methods, applications and concepts of metabolite profiling: Primary metabolism, *EXS*, 97:171–194, 2007; L. W. Sumner et al., Methods, applications, and concepts of metabolite profiling: Secondary metabolism, *EXS*, 97:195–212, 2007.

Plant pathogen forensics

Forensic science is defined as "the identification, collection, preservation, analysis, and interpretation of physical evidence that is relevant to a criminal act or process of interest." When a criminal act involves the use of a microbial agent, such as a pathogenic virus, bacterium, or fungus, a microbial forensic investigation will focus on tracing the source of that agent and collecting evidence for use in the attribution of the event and the prosecution of those responsible.

Why is it important to have capability in plant pathogen forensics? As with humans and animals, plants can be attacked by pathogens, including viruses, bacteria, fungi, and other microbes. Usually, dissemination of these agents is natural or unintentional, their movement facilitated by wind, water, agricultural equipment, seeds or propagative plant parts, insect vectors, animals, or farm workers. However, there is a possibility of deliberate or inadvertent introduction of a plant pathogen to an agricultural target. In the United States and elsewhere, endemic, introduced, and emerging pathogens pose a real threat to crops, forests, and range plants. Vulnerability arises from the fact that the vast land areas involved are not consistently monitored, and diseases may not be noticed quickly. The preferential use, in crop production, of plant genotypes having attractive commercial attributes results in plant homogeneity in the field, increasing the risk that the consequences of a new or more aggressive plant pathogen could be serious. Because many plant pathogens are easy to collect and are not directly harmful to humans, they pose little threat to the handler and can be grown, transported, and disseminated with little technical expertise or equipment. Increasing accessibility and simplicity of technological information related to pathogens increase the likelihood that rogue states or individuals may use such knowledge for criminal use.

Vulnerability of the U.S. food supply. Vulnerability of plants used as food arises at many points from the field to the consumer. Intentional pathogen introduction to an agricultural target, such as in the Midwestern grain belt, would cause less public alarm than direct targeting of humans; indeed, the effects may not be apparent for some time. Even after discovery, there is little concern that the U.S. population would suffer from nutritional shortages. However, deliberate plant pathogen introductions could significantly impact the availability of certain foods, as well as their quality and quantity. Furthermore, a few plant pathogens can be opportunistic human pathogens, primarily affecting the immunocompromised, the very young, or the elderly. Mycotoxins produced by some plant pathogenic fungi can be harmful to humans, and other species produce allergenic spores. The most significant impacts of deliberate plant pathogen introductions, however, are likely to be economic, and could include higher food prices and the costs of growing crops that are less desirable. Additional costs for disease management, both short term (crop destruction, pesticide application, or redirecting the use of the crop) and long term (development of resistant varieties), also will affect the economy. Imposition of quarantines and embargoes on U.S. agricultural products by global trading partners not only affects producers, but has downstream effects on the service industries that harvest, store, package, transport, add value to, and market our agricultural products. Perhaps even more important, there could be a loss of trading partners and markets throughout the world. Knowledge of the targeting of the food supply by those intending harm would surely lead to a loss of public trust in our food and in the ability of government to ensure its safety. Ultimately, the rural communities that rely on agricultural production may be destabilized and grower livelihoods threatened.

Is forensic plant pathology something new? Plant pathologists routinely examine outbreaks of plant diseases and look for clues about the origin of the pathogen, evidence of new pathogen strains or biotypes, unusual distribution patterns, and other evidence that might allow the identification of the source(s) of infection and the prevention or minimization of future outbreaks. Because most plant pathogens are introduced through natural means, it is rare for law enforcement personnel to be involved. The primary goals of a plant disease diagnostician are to identify the pathogen to the level necessary for effective and rapid disease management or containment procedures. Faster diagnosis usually translates into less crop damage and lower recovery costs.

However, growing concerns in today's world necessitate new awareness, vigilance, and preparedness to ensure the security of our agricultural systems. An important element of preparedness is the strong capability for criminal investigation and attribution. Like a traditional plant disease diagnostician, a forensic scientist investigating a potential crime scene involving plant disease will need to identify the pathogen, take note of the pattern of symptoms in the field, and consider other factors that influence the occurrence and severity of disease. However, several additional, unique challenges face the forensic investigator. The first critical question is whether a crime has occurred—whether the pathogen was introduced naturally or intentionally. Indeed, if the disease is of natural origin, there is no need for law enforcement intervention. However, as noted above, any number of unusual features of the disease could introduce uncertainty about natural causes and justify investigation.

Elements of a plant pathogen forensic investigation. On the scene of an agricultural crime, whether it is in a farmer's field, greenhouse, orchard, forest, or pasture, law enforcement personnel are faced with many decisions regarding the preservation of evidence, sample collection, handling and processing, chain-of-custody requirements (a record of the time and nature of each step, as well as those having responsibility for the samples at each point), appropriate laboratory assays, and data interpretation. Although plants and their pathogens have much in common with pathogens from animal and human crime scenes, few forensic investigators have significant experience in the agricultural realm. Thus, cooperation among law enforcement and the plant pathology community will be important for an optimal outcome. Regardless of the crime target (plant, animal, or human), certain types of information and tools are needed to formulate a reliable picture of the series of events and components that comprise the crime. These include:

1. Established and validated sampling strategies
2. Epidemiological models
3. Validated pathogen identification and typing methods
4. Knowledge of pathogen genetics and evolution
5. Understanding of the environment and background communities
6. Knowledge of gene expression and modification for both pathogen and host
7. Specific standards for discrimination or matching of microbial samples
8. Databases and analysis strategies

Technologies needed for plant pathogen forensics. Although the subdiscipline of plant pathogen forensics is in its infancy, much information and many tools currently being applied to traditional plant disease diagnostics and management can be applied in new ways to assist the forensics investigator. In addition, technologies and strategies from a wide array of other related disciplines can be brought to bear on new forensic needs. **Figure 1** illustrates some of the core disciplines that provide critical support to the plant pathogen forensic investigator. Adaptation of existing methodologies must address the importance of standardization, validation, reliability, and analytic proficiency as well as the need to know the limitations and confidence levels of the tests being used.

Intentional pathogen introduction. Answering the question of whether an outbreak is natural or human-influenced may be extremely challenging. However, certain features can suggest criminal activity. The answers to the following questions can assist the forensic investigator in the early stages of an investigation:

1. Are pathogens or pathogen strains unfamiliar or unexpected in this location or for this crop?
2. Is the disease unusually severe or at an abnormally high incidence?
3. Were weather and wind conditions prior to the outbreak conducive to disease development?
4. Has the disease appeared at an unexpected time or season?
5. Are multiple pathogens present in the same plant(s)?
6. Do simultaneous clusters of disease occur in noncontiguous areas?

7. Are patterns of field incidence and spread unusual?

8. If the pathogen is insect-transmitted, were the insects present at the time the plants became infected, and were these vectors inoculative (carrying or containing a pathogen and able to introduce it into a host plant)?

Field site as crime scene. If criminal intent is suspected, law enforcement personnel arriving at the scene must gather information critical to the investigation and to ultimate attribution. The geographical distribution of the disease should be determined as quickly as possible. Tools used for disease mapping include aerial photography, remote sensing, satellite images, or transect walking by human observers equipped with Global Positioning System (GPS) devices and photomagnifying computer accessories. The mapped pattern of disease can be compared with geographical patterns of the same disease during natural outbreaks.

Based on the disease pattern of diseased areas, the investigator must choose sites from which to collect samples, which are likely to include diseased plants or plant parts, soil, water, and other evidence. Control samples from apparently uninfected fields of the same crop (and possibly of other nearby crops) should be included. Standard operating procedures that stress the avoidance of sample contamination and the need to guard the chain of custody of the evidence must be established.

It is important to decide in advance whether sampling should be random or targeted. Where symptoms are obvious, sampling primarily from those areas would be logical, although it will be important to sample adjacent or nearby fields as well; if plants in those fields are not symptomatic, it is still possible that they are infected. Samples should be taken both from field edges nearest the field of interest and randomly.

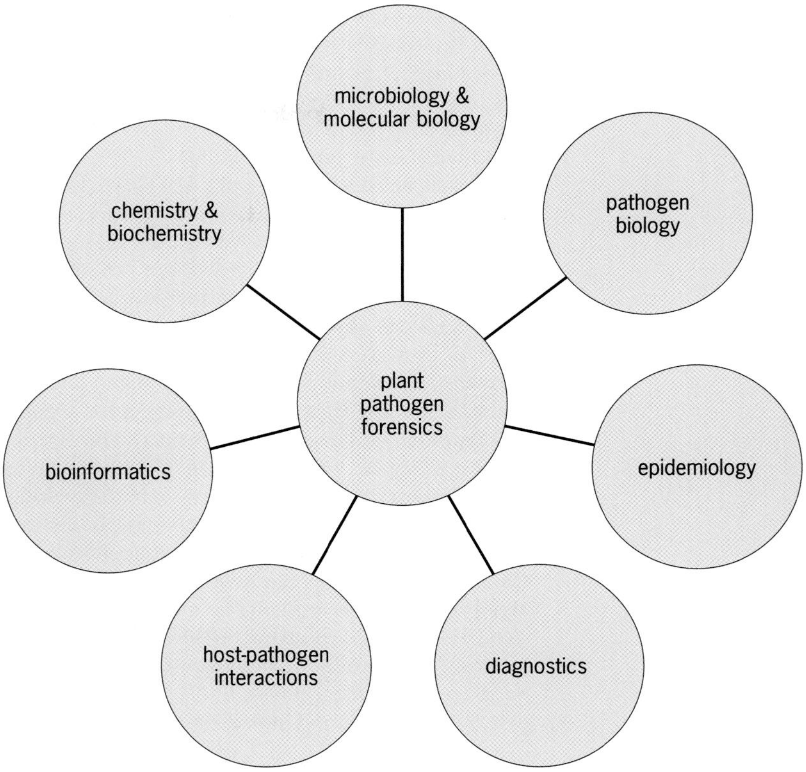

Fig. 1. Some scientific disciplines relevant to plant pathogen forensics.

For plant diseases involving insect vectors, symptoms often begin at points along a field edge where the insects have entered the field. The vectors may fly or be carried field to field and plant to plant

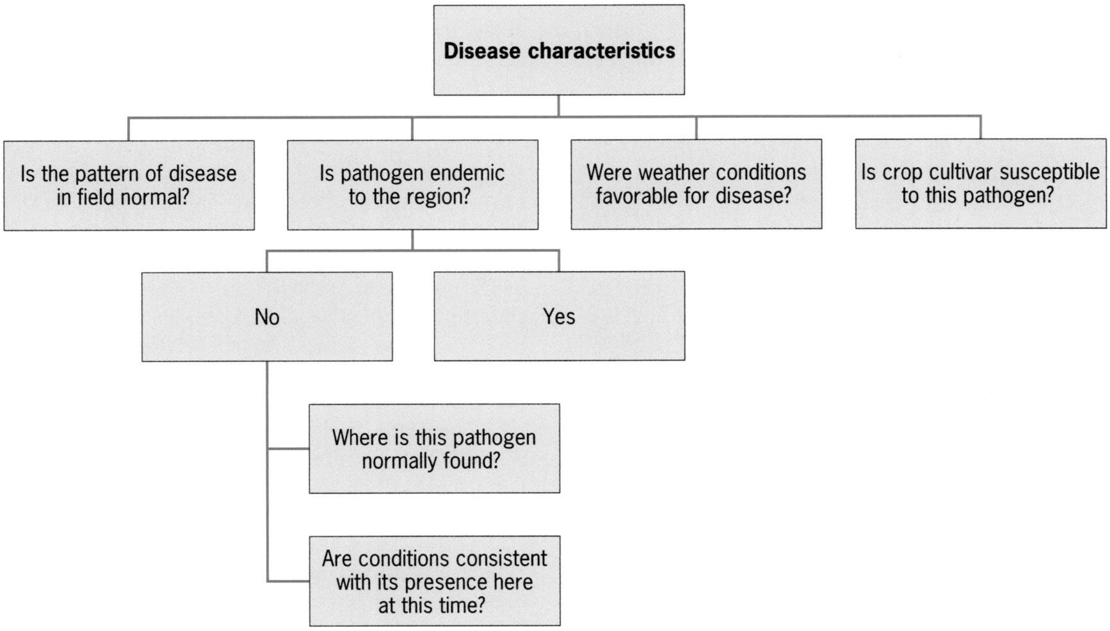

Fig. 2. Example of a decision tree for information gathering related to the source of the plant pathogen.

on wind currents, easily traversing several miles. In contrast, disease can appear first in the field interior if "volunteer" plants, infected in the previous season, provide the inoculum source. Disease incidence patterns that differ from normal patterns could be considered suspicious.

Forensic activities at the scene of a suspected crop-related crime. In addition to preservation of the crime scene, the forensic investigator will make observations and will collect evidence of various types, including photographs, notes, and samples. Making wise decisions about where, how, and what to collect are crucial to a successful evaluation. A "decision tree," or series of questions arranged in a simple grid, for use by a forensic investigator arriving at a potential crime scene may facilitate crime scene activities (**Fig. 2**). The questions will guide the investigator in determining, first, if a crime may have been committed. If the answer to that question is positive, the decision tree will guide in the collection of evidence. Decision trees can be created for several different incident components including disease characteristics, pathogen characteristics, environmental features, and agronomic practices. Examples of relevant issues include:

1. What background information is relevant to the investigation?
   (a) Weather/climate information
   (b) Waterways and irrigation systems—locations, use history, water samples
   (c) Planting history in the affected field (dating back several years)
   (d) Crop history (cultivar, planting date, fertilizers and/or pesticides applied and dates of application, cultivation practices used, nature and frequency of farmer monitoring, dates of first symptoms, etc.)
   (e) Natural and human-made debris at field site (trash, equipment or tools, empty containers, etc.)
   (f) Other evidence of human visits (footprints, fingerprints, tire tracks, broken fences or gates, trampled crops, residue on plants or surfaces, odors, etc.)
   (g) Interviews with farmers, landowners, neighbors, competitors, etc.
2. What samples are needed?
   (a) Which parts of the plant (roots, leaves, fruits, stems, flowers)?
   (b) Soil (in the root zone, at field borders)?
   (c) Water (from nearby streams or ponds, irrigation ditches or pipes)?
   (d) Weeds within the field or at the field borders (possible alternate hosts)?
   (e) Insects (particularly potential vectors of plant pathogens, primarily leafhoppers, aphids, whiteflies, mites, and beetles)?
3. How should sampling be done?
   (a) Should collecting be focused on areas having obvious symptoms, or should it be performed in a random pattern?
   (b) How many samples are needed?
   (c) What types of containers, swabs, and other implements are appropriate?
   (d) What sampling strategies will best preserve the integrity of the samples' molecular signatures?
   (e) How can the samples be protected from contamination?
4. What happens to the samples after collection?
   (a) What "chain of custody" information is required?
   (b) At what temperature should samples be stored and transported?
   (c) How can sample degradation be minimized and informative "signatures" preserved during transport and storage?

Looking to the future. Strong capability in microbial forensics must be part of the preparedness for ensuring the biosecurity of agricultural commodities as they move from the field to the consumer. The emerging discipline of plant pathogen forensics will require collaboration and cooperation among professionals in law enforcement and the security community; plant pathologists in academia, government, and private industry; and crop producers and consultants. Our efforts must include identification and prioritization of threatening plant pathogens as well as adaptation of existing technologies and the design of new and better tools for forensic applications. Educating a new generation of plant pathologists trained to understand the needs and issues of both the law enforcement and the agricultural communities will be critical to success. In addition, those "on the ground," including producers, crop consultants, extension personnel, and plant disease diagnosticians, must receive training to recognize and react appropriately when intentional pathogen introduction is a possibility. These efforts will ensure that U.S. capabilities in agricultural microbial forensics are strong, balanced, and science-based, thereby increasing the ability to respond appropriately, and perhaps also to serve as a deterrent to those who might otherwise intentionally use microbes against our nation's plant resources.

For background information *see* AGRICULTURAL SCIENCE (PLANT); BIOTECHNOLOGY; ECOLOGY; FARM CROPS; FORENSIC BIOLOGY; PATHOTOXIN; PLANT PATHOLOGY in the McGraw-Hill Encyclopedia of Science & Technology. Jacqueline Fletcher

Bibliography. B. Budowle et al., Microbial forensics, pp. 1–26, in R. G. Breeze, B. Budowle, and S. E. Schutzer (eds.), *Microbial Forensics*, Elsevier Academic Press, 2005; J. Fletcher et al., Plant pathogen forensics: Capabilities, needs, and recommendations, *Microbiol. Mol. Biol. Rev.*, 70:450–471, 2006; R. S. Murch, Microbial forensics: Building a national capacity to investigate bioterrorism, *Biosecur. Bioterr. Biodef. Strategic Pract. Sci.*, 1:117–122, 2003; M. Wheelis, R. Casagrande, and L.V. Madden, Biological attack on agriculture: Low-tech, high impact bioterrorism, *Bioscience*, 52:569–576, 2005; S. M. Whitby, *Biological Warfare against*

Crops, Palgrave, Basingstoke, United Kingdom, 2002.

Polymeric molecular magnets

As length scales shrink and demands on magnetic materials increase, engineers are increasingly looking toward molecule-based materials to develop the next generation of magnetic devices. These materials show promise for specialized applications, including magneto-optic recording, magnetic shielding, components of lightweight transformers, generators and motors, photomagnetic switches, sensitive molecular sensors, and spintronic devices. The appeal of molecule-based magnets is a result of their highly tunable nature, transparency, low density, moderate solubility in organic solvents, and facile preparation under mild conditions. This is in contrast to the current technologically useful magnetic atom-based (inorganic) materials, which have greater densities and require high temperatures and/or pressures to prepare them. Through molecular synthesis, organic chemistry provides the ability to fine-tune structures and properties to a much greater extent than is possible for inorganic materials. The possibility of combining multiple physical properties, such as optical/magnetic, conducting/magnetic, and so forth, has prompted the quest for hybrid materials, many of which consist of polymeric molecular magnets.

Three major classes of molecule-based magnets have been introduced: (1) true ionic molecular systems in which discrete cations and anions contain unpaired electrons, (2) single-molecule magnets that constitute clusters of varying sizes and geometries, and (3) polymeric molecular magnets (coordination polymers) where transition- or lanthanide-metal cations are linked via neutral or anionic ligands. Some of the most highly used ligands for the preparation of polymeric molecular magnets are shown in **Fig. 1**. For most of these materials, a random orientation of their unpaired electrons results in a paramagnetic state at room temperature. However, below a certain critical temperature, the spins often align in ordered ferromagnetic, ferrimagnetic, or antiferromagnetic configurations.

Synthesis and topologies. As shown in **Fig. 2**, coordination polymers are typically assembled using a building-block approach, where metal ions are referred to as nodes and neutral or anionic ligands are referred to as molecular linkers. The coordination

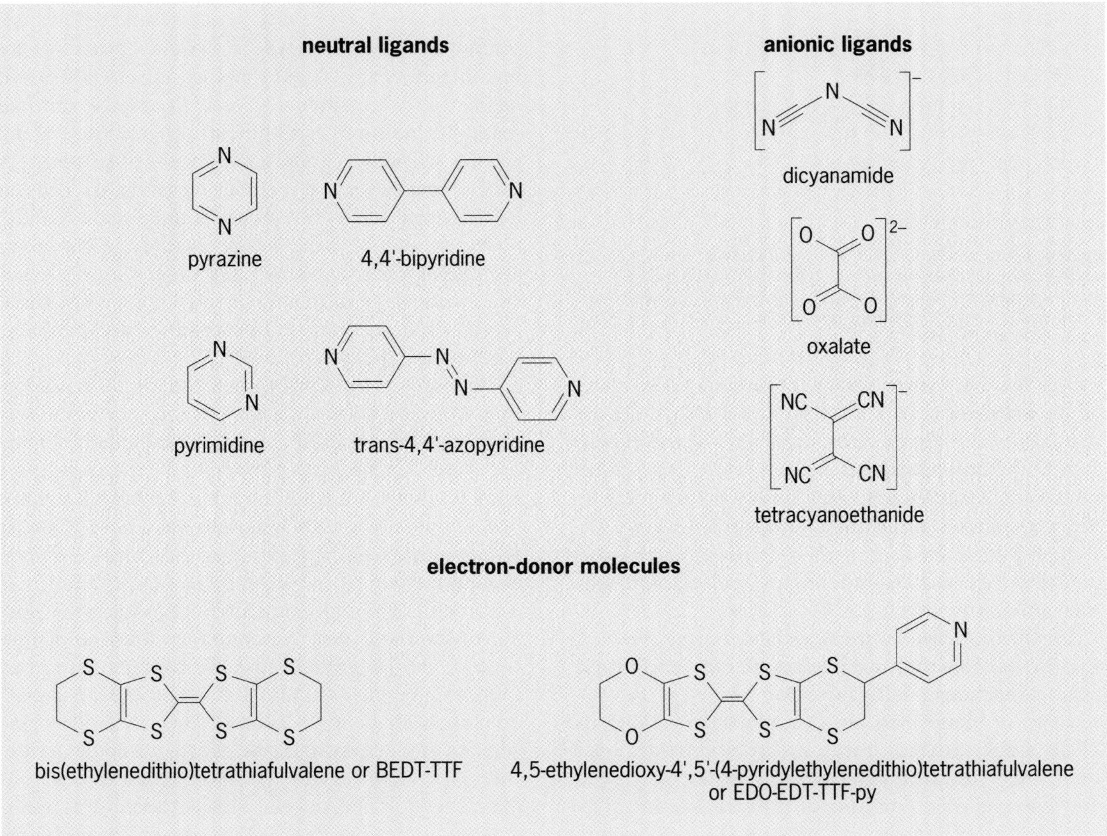

Fig. 1. The molecular linkers that hold polymeric molecular magnets together are typically neutral or anionic ligands. These molecules act as Lewis bases that form coordination bonds with Lewis acid transition-metal cations. Hybrid materials that possess multiple properties (such as conductivity, magnetism, chirality) can be formed through the use of tetrathiafulvalene-based electron-donor molecules.

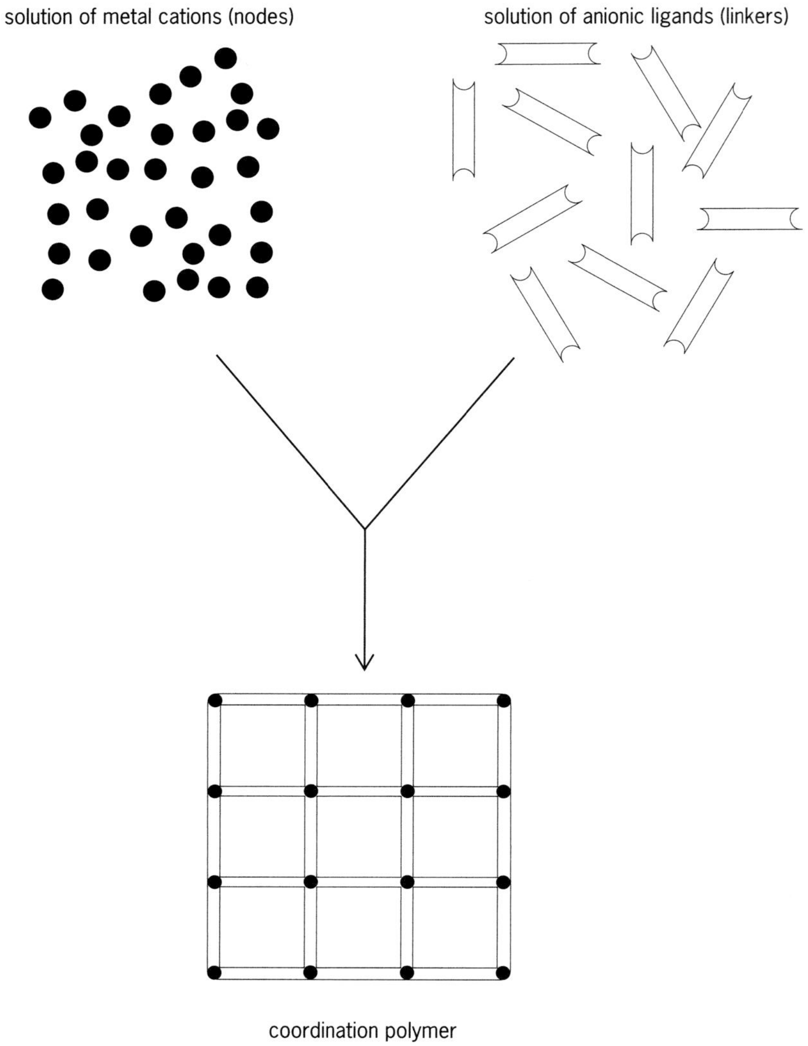

Fig. 2. Polymeric molecular magnets are typically formed through solution techniques that involve combining solutions that contain cationic metal nodes with solutions of anionic or neutral ligands that act as molecular linkers. Crystals are typically grown over a period of days or weeks as the molecular components self-assemble into 1D, 2D, or 3D networks. The formation of a 2D layer is illustrated.

bonds that hold the polymers together are the result of an interaction between the metal cation (Lewis acid) and the lone electron pairs on a ligand (Lewis base). Various methods can be used to crystallize coordination polymers, with the most common involving saturation, diffusion, and solvothermal techniques. Although less commonly used, microwave and electrochemical methods can be effective in specific situations.

Various dimensionalities can be achieved by varying the coordination number, geometry, and ligand arrangement around the metal ion, as well as the multiplicity of Lewis base sites on the ligands. Various chain motifs (such as linear, zigzag, and helical) can be formed. Ladder-type structures are an intermediate between one-dimensional (1D) chains and two-dimensional (2D) layers. A variety of 2D grids (including square, rectangular, and honeycomb) can also be prepared. There are a large number of possible three-dimensional (3D) networks that are frequently named after their inorganic atom-based counterparts that adopt similar connectivities (such as diamondoid, rutile, and perovskite). Some applications require robust, porous 3D coordination polymers. Often, such materials can be formed by incorporating a solvent molecule during the crystallization process and subsequently removing it. Because nature frequently prefers tightly packed structures, interpenetration of multiple lattices is a somewhat common feature of 3D networks. The charge of polymeric molecular materials can also be tuned such that anionic, cationic, and neutral polymers can be prepared.

Dicyanamide-based magnets. The dicyanamide ion, $[N(CN)_2]^-$, or dca, attracted much attention when the binary class of compounds, $M(\text{dca})_2$ ($M =$ V, Cr, Mn, Fe, Co, Ni, and Cu) were discovered. Depending on the choice of metal (and number of unpaired *d* electrons), canted antiferromagnets ($M =$ V, Cr, Mn, and Fe) can be produced with critical temperatures (T_c) as high as 47 K, while ferromagnets ($M =$ Co and Ni) with T_c of up to 21 K can also form. Hysteretic effects are also observed where coercive fields in excess of 1 tesla can be achieved for the Fe analog.

It is also possible to incorporate neutral organic ligands into the structures, thus providing a means to modulate or tune the resulting magnetic behaviors. For instance, $Mn(\text{dca})_2(\text{pyrazine})$ [pyrazine = 1,4-diazine] possesses a unique 3D double-interpenetrating ReO_3-like structure that antiferromagnetically orders below 2.5 K. Neutron diffraction experiments revealed that the ordered magnetic moment is collinear and oriented along the *ac* diagonal of the structure. Substitution of other metals, such as Fe, Co, Ni, and Cu, leads to isostructural networks where long-range magnetic ordering occurs well below 2 K. In addition to the use of pyrazine as a linker, other organic bases (for example, pyrimidine, 4,4′-bipyridine, *trans*-4,4′-azopyridine, and many others) can be used as molecular building blocks.

Anionic networks and hybrid materials. With an eye toward applications, several research groups have been developing anionic polymeric molecular magnets that can be incorporated into charge-transfer salts to form multifunctional materials that have both conducting and magnetic moieties. Anionic dicyanamide-based coordination polymers are especially attractive for developing structure and property relationships because many dicyanamide binding modes are known. Through the use of templating cations, various anionic structural motifs have been obtained. As shown in **Fig. 3**, these motifs include 1D (chains, ladders, and triple ladders), 2D (square, triangular, and hexagonal), and 3D (triple rutile, cubic, and $LiSbO_3$) structures. We have recently reported that several of the anionic $Mn(\text{dicyanamide})_3^-$ coordination polymers exhibit long-range magnetic ordering.

Bimetallic oxalate coordination polymers can form layered structures with long-range ferromagnetic ordering. These anionic layers have successfully been used as the charge-compensating layers in charge-transfer salts. For example, the (BEDT-TTF)3[MnCr(ox)$_3$] [BEDT-TTF is bis(ethylenedithio) tetrathiafulvalene; ox is oxalate] salt was the first molecular ferromagnetic metal. In this material, the partially oxidized BEDT-TTF molecules form conductive sheets that are interleaved with polymeric bimetallic oxalate bridged layers.

A second approach for coupling magnetic properties with secondary functionality, such as conductivity or spin-crossover, in molecular materials is to establish coordination bonds directly between the electron-donor molecules and the spin-containing transition elements. To accomplish this, N-donor functionalities, such as pyridine or bipyridine, have been covalently tethered to electron-donor molecules through covalent bonds. This approach has led to the crystallization of paramagnetic transition-metal complexes such as Mn(hfac)$_2$(EDO-EDT-TTF-py)$_2$ [hfac = hexafluoroacetylacetonato, EDO-EDT-TTF-py = 4,5-ethylenedioxy-4′,5′-(4-pyridylethylenedithio)tetrathiafulvalene]. However, the ultimate goal of coupling these systems into coordination polymers with cooperative magnetic behavior remains elusive.

Hydrogen-bond networks. In low-dimensional (1D and 2D) systems, weak interactions between chains or layers often determine the solid-state structure. Hydrogen bonds and π-π interactions are two commonly used structure-directing entities. Hydrogen bonds are responsible for the formation of ice, and the importance of such bonds in biological systems is well documented. However, the role of hydrogen bonds in magnetic materials is relatively unexplored. The strength of a hydrogen bond is governed by the magnitude of the electrostatic interaction that can be produced. To this end, H···F and H···O types of hydrogen bonds are highly desirable.

We are currently studying the chemical and magnetic exchange properties of the bifluoride ion, HF_2^- (that is, F···H···F$^-$), which is known to possess the strongest hydrogen bond. We recently reported the result of the first attempt at using this structural unit, where the aqueous reaction of $Cu(BF_4)_2$, NH_4HF_2, and pyrazine resulted in the formation of a 3D pseudocubic coordination polymer with the chemical formula [Cu(HF$_2$)(pyrazine)$_2$]BF$_4$. The framework contains both Cu-pyrazine-Cu and Cu-F···H···F-Cu bridges. Heat capacity and muon-spin relaxation studies revealed that a transition to long-range magnetic order occurred below 1.54 K. While the low critical temperature is not technologically useful for device applications, the work illustrates the feasibility of using hydrogen bonds as superexchange mediators in magnetic materials. Furthermore, thermogravimetric analyses of this compound demonstrates the rigidity of the H···F bonds in the Cu-F···H···F-Cu bridge, as a temperature greater than 200°C is required to disassemble the network.

Photomagnetic and spin-crossover materials. $M_3[M'(CN)_6]_2$·yH$_2$O and RM[M' (CN)$_6$].yH$_2$O (M = divalent metal ion, M' = trivalent metal ion) are metallocyanide coordination polymers based on the face-centered-cubic array of Prussian blue. The M' ion is directly bonded to the carbon end of the strong-field cyanide ligand, which produces a low-spin (LS) configuration. In contrast, M is in a high-spin (HS) state. Depending on the identities and combinations of M and M', a plethora of magnetic behaviors can be observed including ferrimagnetism, ferromagnetism, and photomagnetism. Freshly prepared samples of K_xCo[Fe(CN)$_6$]y·zH$_2$O show paramagnetic behavior, however, when samples of this material are exposed to red light, a change in magnetic response occurs and the material becomes ferrimagnetic with a T_c as high as 16 K. Application of blue light or warming above ~150 K reverses the effect. The effect is driven by a photo-assisted charge transfer from the diamagnetic spin state, Fe^{2+} ($S = 0$)/Co^{3+} ($S = 0$), to a paramagnetic spin state, Fe^{3+} ($S = 1/2$)/Co^{2+} ($S = 3/2$), accompanied by a dramatic increase in magnetization and hysteresis, which are useful properties of any magnet.

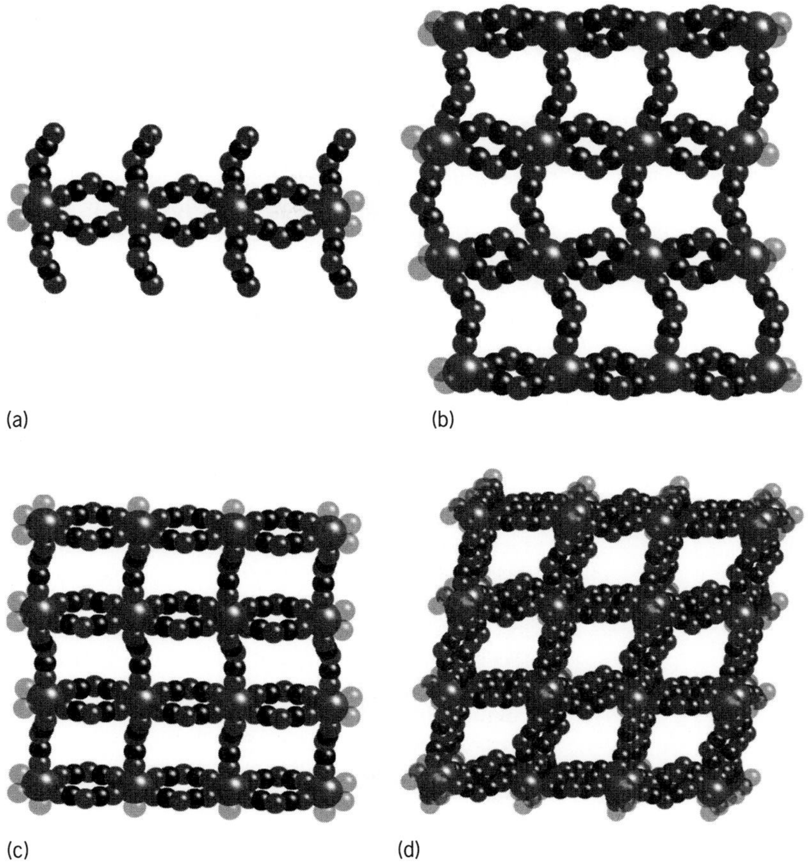

Fig. 3. Dimensionalities of anionic dicyanamide polymeric molecular magnets. Metal nodes are illustrated as large balls and the 5-atom dicyanamide ligands. The nitrile nitrogen atoms of adjacent dicyanamide ligands that propagate the polymer are illustrated as light grey balls. *(a)* 1D chain as observed in (PPh$_4$)$_2$Co(dca)$_4$. *(b)* Triple ladder structure as observed in (PPh$_4$)$_3$Cu$_4$(dca)$_{11}$. Note that the outer legs of the ladders contain square pyramidal coordination of the copper ions. This prevents propagation of the polymer perpendicular to the ladder direction. (c) 2D square later as observed in (PPh$_4$)Mn(dca)$_3$. In contrast to (PPh$_4$)$_3$Cu$_4$(dca)$_{11}$ all copper ions are octahedrally coordinated. *(d)* 3D cubic network as observed in (SPh$_3$)Mn(dca)$_3$. The various lattice types are a result of cation templation and the relative ratio of nodes to linkers.

Magnetic coordination polymers derived from Fe(II) are especially interesting because of their possible spin-crossover behavior. Such materials have potential application as molecular switches where stimuli, such as light, pressure, guest sorption, and temperature, can cause the material to change its electronic configuration from high- to low-spin configuration. Spin-crossover behavior has been observed in polymeric molecular materials with 1D, 2D, and 3D topologies. The $Fe_2(azpy)_4(NCS)_4 \cdot (guest)$ [azpy = *trans*-4,4′-azopyridine] is a particularly interesting example that exhibits guest-dependent spin-crossover behavior.

Room-temperature magnets. Two classes of ferrimagnetic 3D coordination polymers that contain vanadium ions, namely $V(TCNE)_x \cdot yCH_2Cl_2$ (TCNE = tetracyanoethanide; $x \sim 2$) and $V[Cr(CN)_6] \cdot yH_2O$, have been shown to have T_c that exceed room temperature. Their pyrophoric nature has limited the technological application of these materials and the lack of detailed structural information has made it difficult to assess the origin of their high-T_c behavior. While a closely related (although 2D) material, $[Fe(TCNE)_2(CH_3CN)_2][FeCl_4]$, has recently been reported, its T_c of 90 K is much less than the vanadium compound. The x-ray structure determination of this material indicates that each Fe^{2+} center is coordinated to four different TCNE ligands. This important information likely reveals some commonalities with the structurally elusive $V(TCNE)_x$ compound.

Outlook. Polymeric molecular magnets are a promising class of compounds that have the potential to address many of the requirements of contemporary materials research. As outlined above, these coordination polymers have many significant advantages over traditional inorganic materials. As advances are made in nanoscience and crystal engineering, these materials will likely find specialized applications. Materials chemists are learning how to control the structure (dense, porous, flexible, etc.) and properties (photomagnetism, spin-crossover, etc.) of these materials, which will be important for the development of sensitive sensors, spintronic devices, and so on.

For background information *see* ANTIFERROMAGNETISM; COORDINATION CHEMISTRY; CRITICAL PHENOMENA; CRYSTAL STRUCTURE; FERRIMAGNETISM; FERROMAGNETISM; HYDROGEN BOND; INORGANIC AND ORGANOMETALLIC POLYMERS; MAGNETIC FERROELECTRICS; MAGNETISM; NEUTRON DIFFRACTION; POLYMER in the McGraw-Hill Encyclopedia of Science & Technology.

John A. Schlueter; Jamie L. Manson

Bibliography. S. R. Batten and K. S. Murray, Structure and magnetism of coordination polymers containing dicyanamide and tricyanomethanide, *Coord. Chem. Rev.*, 246:103–130, 2003; O. Kahn, *Molecular Magnetism*, VCH Publishers, 1993; J. S. Miller and M. Drillon (eds.), *Magnetism: Molecules to Materials*, Wiley-VCH, vol. 1–5, 2001–2007; J. A. Real, A. B. Gaspar, and M. C. Munoz, Thermal, pressure and light switchable spin-crossover materials, *Dalton Trans.*, 250:2127–2157, 2006.

Post-Paleozoic ecological complexity

The greatest documented mass extinction in the history of life occurred at the end of the Paleozoic Era about 252 million years ago (mya). What happened then to the existing ecological complexity? At this time, 56% of marine genera and as much as 90% of species were decimated. The fossil record is most complete for skeletonized marine benthic (bottom-dwelling) animals. Geographically widespread, marine higher taxa with drastically different lifestyles were reduced to few surviving species or were extinguished entirely. Paleontologists have known of mass extinctions for almost 150 years, from counts of fossil taxa in successive geologic strata. Gradual shifts among dominant higher taxa and guilds (different ways of life) are also known, at least at a coarse global scale, from these counts and from analogies with living forms. However, tracking how the ecological world became complex has proven to be a daunting task.

Even the idea of complexity in ecology is many-sided. As ecology first developed from natural history, ecologists began to classify patterns of assemblages or communities, asking how many of which organisms, and what different ways of life, gave a sufficient accounting of complexity. One idea of an ecologically complex pattern is that something exists beyond a random assortment of individuals. Complex communities may be thought of as resilient systems that can be recognized in aggregate form in different places. Another tradition of studying complexity focuses on reciprocal interactions among organisms: competition, predation, symbioses, and other phenomena as processes. This facet of complexity involves dynamic interactions among species that may partly determine relative abundances. For ecologists, these different facets are regularly measured by indices of community diversity: counts of taxa (richness) and individuals (abundance), and the relative commonness or rarity of different taxa (evenness). Characteristic patterns of species abundance are also documented in models of relative abundance distributions (RADs) that reflect frequencies of individuals among species. These RADs allow a mathematical description of all the measured information of richness and abundance in a community. Most RADs are linked conceptually with varied processes of species interactions, allocation of resources and niche space, and therefore different facets of complexity. All of these indices and models can be validated empirically in living communities, but what of paleontology and the fossil record?

Certainly, the number of species on Earth has increased over time, given the shared genetic code that testifies to a single origin and biological commonality of life. This increase must imply growing complexity in community structure, as living species do interact and partition resources in the habitats they occupy. Yet interactions among fossil species can only be inferred indirectly. Modes of life of fossil

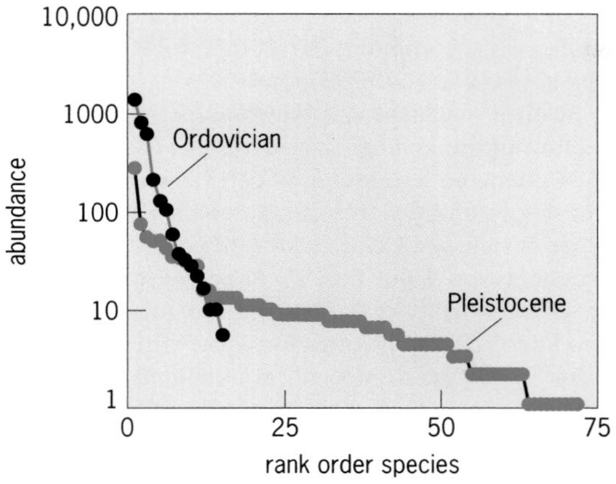

Fig. 1. Two contrasting relative abundance distributions from Ordovician and Pleistocene assemblages of skeletonized bottom-dwelling marine animals, representing early Paleozoic and late post-Paleozoic time intervals, respectively. Abundances are shown on a log scale.

organisms are inferred from their descendents, by functional reconstruction based on analogous anatomy (skeleton, muscle) or physiology of the living organisms, or less frequently by fossilized gut contents, tracks, or other evidence. An incomplete fossil record preserves only scattered glimpses of ecological history, so any broad ecological patterns must be assembled by fitting together a puzzle with many missing pieces for each past time interval, and comparing these with the much more complete puzzle of the living world.

Moreover, the diversification of multicellular life is not linear, but periodically set back by catastrophic mass extinctions. It is only in the past few years that cooperatively maintained databases of species' identities and abundances in thousands of fossil communities of known ages have become available for comparative analyses. Using this wealth of data, paleontologists are now using ecological diversity indices to track the fossil history of diversity structure in communities through time. A new method devised to compare RADs among fossil communities has led to a startling realization: the Permo-Triassic mass extinction that so greatly reduced global taxonomic richness also triggered a post-Paleozoic shift to more complex marine communities.

What to measure. There are a number of ways to measure ecological complexity in the fossil record. Counts of families and genera that lived in successive geologic periods each demonstrate increasing global richness through the Phanerozoic, the period of time between the end of the Precambrian and today, encompassing over a half-billion years of Earth history. This eon includes the Paleozoic Era, extending from about 540 to 252 mya. The Paleozoic is divided into six periods. From oldest to youngest they are Cambrian, Ordovician, Silurian, Devonian, Carboniferous, and Permian. The post-Paleozoic includes the Mesozoic (Triassic, Jurassic, and Cretaceous periods) and Cenozoic Eras. Alpha diversity (local taxonomic richness) of marine benthic animals at the genus level increases by a factor between 2.5 and 3.7 from the mid-Paleozoic to the Late Cenozoic, and this increase appears robust even after corrections for sampling and preservational biases. Despite temporary reductions during mass extinction intervals, these overall measures of taxonomic richness imply a general increase in complexity, of communities with more species and guilds, based on analogies with living organisms and other evidence. However, these measures do not take into account, in most cases, changes in relative abundance at local scales.

Relative abundance is usually summarized by an evenness index (a standardized statistic of abundance among taxa). Low evenness is indicative of dominance by one or a few taxa, whereas high evenness indicates that taxa are more equally abundant. Living communities that have more even abundances are more evenly branched in terms of a hierarchy of functional niches; their structure is more complex than those dominated by a few niche types, implicating multiple ecological structuring factors. Statistical methods for estimating alpha diversity across fossil assemblages are needed to account for differences in the numbers of fossil specimens counted in

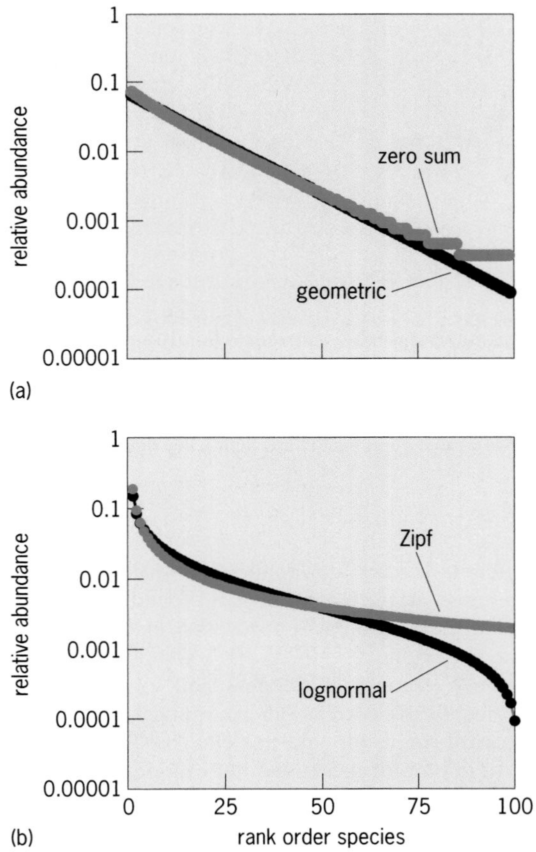

Fig. 2. (*a*) Simple and (*b*) complex model relative abundance distributions (RADs) are shown as the proportions of individuals belonging to each species in a community, with species ordered from most to least abundant. Distributions were calculated with the same richness (S = 100) and evenness (Pielou's J = 0.80, where Pielou's J is a measure of how evenly individuals are distributed among species in a community). Abundances are shown on a log scale.

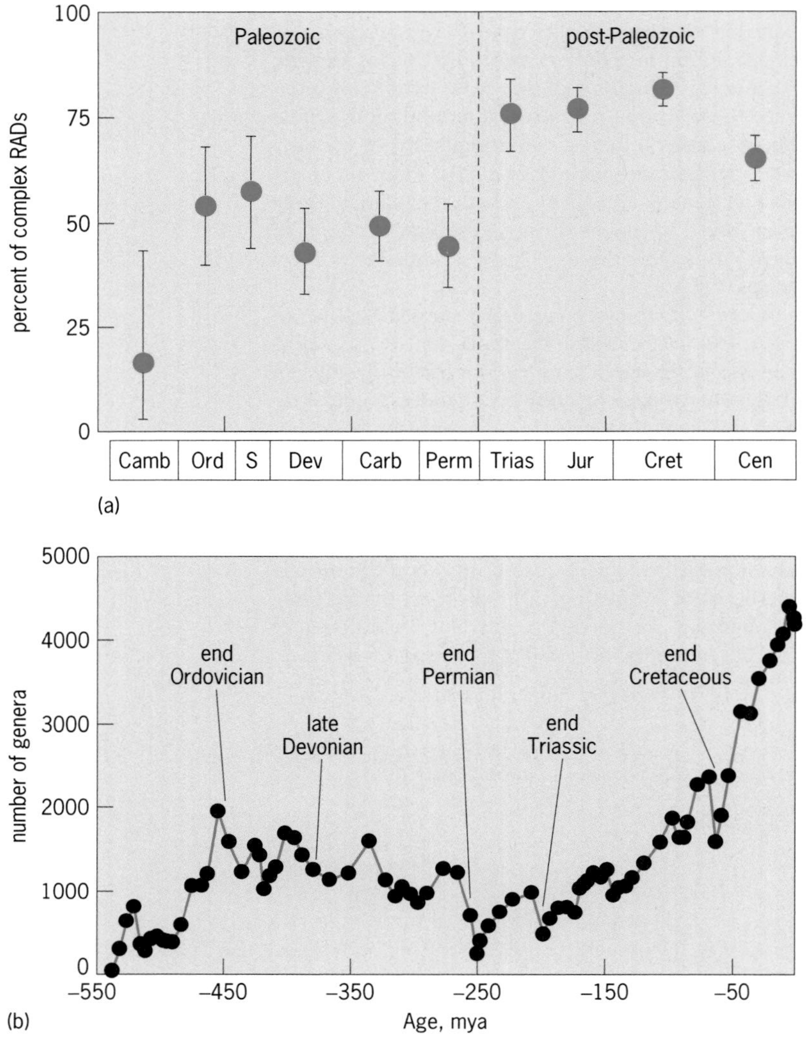

Fig. 3. **(*a*) Percentages of complex RADs for fossil assemblages in the Paleozoic and post-Paleozoic are shown. All statistically significant RADs are summarized within successive geologic time periods as mean values (circles) and error bars indicating one unit of support in the likelihood analysis. Simple RADs are the inverse of this plot; support bars for simple and complex RADs overlap in the Paleozoic, but are entirely separate in the post-Paleozoic. (*b*) Global genus richness of marine animals through the Phanerozoic is shown at the stage level. The five largest mass extinctions during this time are also indicated.**

different studies. After applying these methods, evenness at the genus level is found also to be significantly higher in Cenozoic than in Paleozoic marine communities.

While taxonomic richness and evenness have strong positive correlations according to many mathematical simulation and modeling studies, comparisons of empirical data sets of living organisms indicate that the sign and strength of these correlations are actually quite inconsistent. Communities with identical richness and evenness may differ in ecological structure and complexity. Neither diversity measure yields a comprehensive picture of the statistical distribution of abundances (RADs) underlying a biological community, which is fundamental to all elements of diversity measurement. Therefore, a new method was developed to compare RADs of species among fossil communities through the Phanerozoic, and across mass extinction intervals.

Relative abundance distributions. A more refined picture of the ecological structure of Paleozoic and post-Paleozoic communities can be seen in different shapes of RADs from an Ordovician fossil assemblage of marine benthic invertebrates versus a Pleistocene assemblage (**Fig. 1**). [The Pleistocene, part of the Cenozoic, extends from 1.8 mya to 10,000 years ago.] The most common species in the Ordovician assemblage are strongly dominant in abundance with evenly proportional steps between successively less abundant species, whereas the flatter shape of the Pleistocene assemblage reflects its greater evenness.

In what may be termed "simple" RADs, proportional abundance declines more or less exponentially from one species to the next (**Fig. 2*a***). For the geometric model distribution, species added to the community preempt portions of the available resources nonrandomly; remaining portions are utilized by successive species without increasing the total amount of resources. For the zero-sum multinomial distribution (in which gains or losses in species' abundances are balanced by other losses or gains, thereby summing to zero), preemption of resources and species interactions are subsidiary to factors affecting rates of immigration, origination, and extinction. This distribution emerges from a neutral model of communities, where the total number of individuals in ecologically similar species stays approximately constant and the survival or reproduction of any one individual does not depend on interactions with other individuals. In some "complex" RADs, proportions drop sharply among the most common species, then less rapidly among most remaining species with similar abundances, leaving a tail of rare species (Fig. 2*b*). For Zipf distributions (where the probability of occurrence falls steeply and then tapers off) or lognormal distributions (in which the logarithm of the parameter has a normal distribution), added species may either add new niches or facilitate opportunities for additional species; proportional allocation of resources or niche space to added species is not distinguished from random distribution. Lognormal distributions may also be formed by amalgamations of ecologically distinct groups of species, each with its own underlying distribution. The simple distributions imply fewer interactions, either by dominant species preempting a few critical resources or by ecological interactions among species playing less important roles than other "interaction-neutral" factors. Complex distributions imply a greater number of interactions among species, complex niche partitioning, niche facilitation (one species benefits from the presence of another), or multiple combinations of niche breadths and ecological factors.

The most likely statistical fit among each of these four model RADs was determined for 1176 Phanerozoic fossil assemblages, each with 10 or more species and 100 or more specimens. Sample sizes were explicitly accommodated, and a method of penalized likelihood (where a "penalty" term is added to the log-likelihood of the data to circumvent difficulties) was used to select RADs with statistically significant evidence ratios. By combining results in successive geologic time intervals, it is clear that proportions of marine fossil assemblages with simple RADs decrease markedly at the end of the Paleozoic Era (**Fig.** 3*a*).

Mass extinction. What is most remarkable is that the end-Permian mass extinction and subsequent recovery of benthic faunas permanently changed the structure—and complexity—of marine ecosystems. Statistical likelihood tests support the position of this shift at the Permian-Triassic boundary (~252 mya); none of the remaining largest mass extinctions had such a profound effect (Fig. 3*b*). Nor do Phanerozoic trends in sediment type (carbonate versus siliciclastic) or higher diversity correspond with a shift in ecological complexity. Multiple forcing factors have been implicated for the end-Permian extinction: global warming, severely elevated atmospheric and oceanic concentrations of CO_2, toxic sulfides, anoxia (absence of oxygen), and massive runoff of terrestrial organic detritus. As benthic ecosystems slowly reorganized, reefs temporarily disappeared and many previously dominant, nonmotile filter-feeders and benthic detritivores [bottom-dwelling consumers of dead organic material (detritus)] never fully recovered, while other higher taxa in guilds that had been less common eventually radiated.

What possible ecological factors seem to correspond with this lasting shift in ecosystem complexity? In the post-Paleozoic, motile active organisms such as gastropods, infaunal (living within sediments) bivalves, many arthropods, echinoids, and fishes became more common than previously dominant, nonmotile epifauna (organisms living on top of the ocean floor). Groups lacking metabolic control over diffusive gas exchange had suffered greater extinction rates than those with active exercise metabolism and buffered physiologies. New guilds including deep-burrowing organisms exploited resources more fully below the sediment-water interface, and shell-crushing predators and skeletally defended prey diversified in a pattern of sustained escalation. There is little doubt that life got more complicated, and stayed that way, for marine benthic animals in the post-Paleozoic.

For background information *see* BIODIVERSITY; ECOLOGICAL COMMUNITIES; ECOSYSTEM; EXTINCTION (BIOLOGY); GEOLOGIC TIME SCALE; MARINE ECOLOGY; PALEOECOLOGY; POPULATION ECOLOGY; TAXONOMY in the McGraw- Hill Encyclopedia of Science & Technology. Scott Lidgard

Bibliography. A. M. Bush and R. K. Bambach, Did alpha diversity increase during the Phanerozoic? Lifting the veils of taphonomic, latitudinal, and environmental biases, *J. Geol.*, 112:625–642, 2004; A. M. Bush, R. K. Bambach, and G. M. Daley, Changes in theoretical ecospace utilization in marine fossil assemblages between the mid-Paleozoic and late Cenozoic, *Paleobiology*, 33:76–97, 2007; D. H. Erwin, *Extinction: How Life on Earth Nearly Ended 250 Million Years Ago*, Princeton University Press, 2006; A. H. Knoll et al., Paleophysiology and end-Permian mass extinction, *Earth and Planetary Science Letters*, 256:295–313, 2007; P. J. Wagner, M. A. Kosnik, and S. Lidgard, Abundance distributions imply elevated complexity of post-Paleozoic marine ecosystems, *Science*, 314:1289–1292, 2006.

Preimplantation genetic diagnosis and therapy

Preimplantation genetic diagnosis (PGD) is a reproductive technology that allows couples at genetic risk to avoid the birth of affected offspring and to avoid facing prenatal genetic diagnosis and termination of pregnancy. Principally different from prenatal diagnosis, instead of preventing the birth of an affected child, preimplantation genetic diagnosis allows establishment of an unaffected embryo for the pregnancy from the onset and birth of an unaffected offspring. The detection and consequent avoidance of transfer of embryos with chromosomal abnormalities are also becoming an integral part of assisted reproductive technology (ART).

Preimplantation genetic diagnosis is performed by three major approaches: (1) testing of eggs by removing the first and second polar bodies (PB1 and PB2), which are by-products of an egg's division during meiosis and are naturally extruded during maturation and fertilization of the oocyte prior to embryo formation; (2) embryo biopsy at day 3; and (3) embryo biopsy at day 5 (**Fig. 1**). PGD through testing eggs by PB1 and PB2 removal was first introduced in 1990 in Chicago, by testing blastomeres in 1990 in London, and by blastocyst biopsy in 2004 in Sidney. The biopsied materials from the eggs and embryos are tested for single-gene defects, using polymerase chain reaction (PCR) analysis, or for chromosomal anomalies, using fluorescent in-situ hybridization (FISH) analysis. With improved reliability of DNA microarray technology, a combined testing of causative genes, chromosomal errors, and health-related genetic variability will become possible in an integrated analysis.

Although each of the above procedures has certain advantages and disadvantages and may be used depending on indication or patient preferences, accurate diagnosis may require the use of a combination of two or three different methods, especially when there is more than one indication. Also for some single indications, as well as detection and avoidance of the transfer of embryos with numerical chromosomal errors, the reliability of selection will be higher with a sequential testing of the egg and embryo. This will allow detecting the paternally derived chromosomal anomalies usually missed in egg testing, mosaic embryos originating from maternally derived meiotic errors, and uniparental

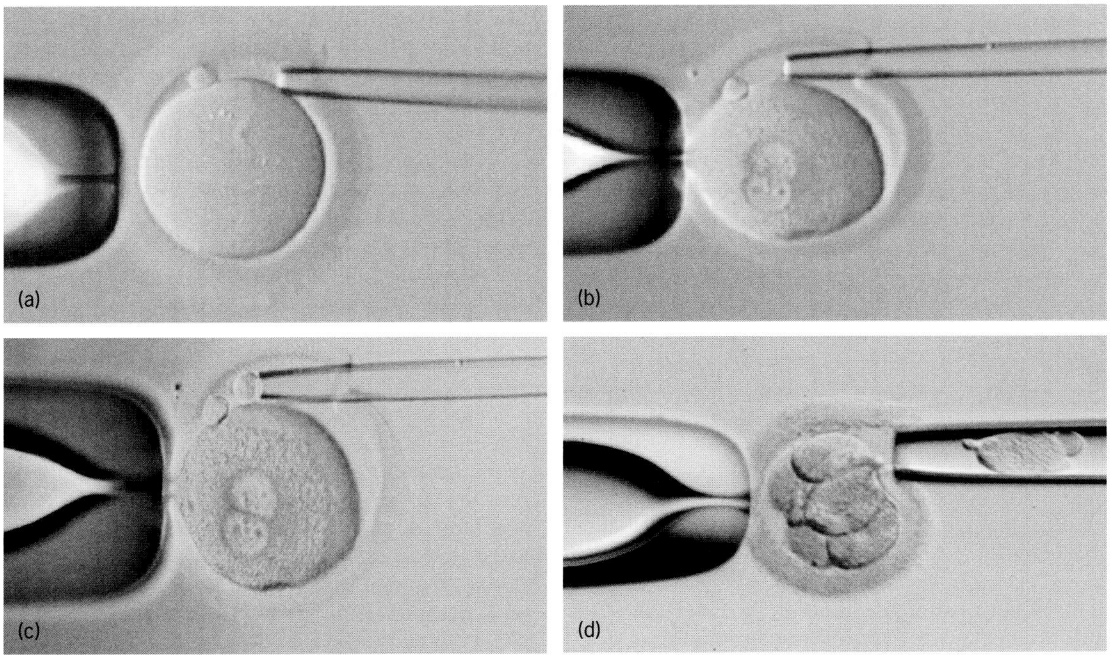

Fig. 1. Preimplantation genetic diagnosis methods. (*a*) First polar body (PB1) removal: performed after egg maturation and used in PGD for single-gene defects. (*b*) Second polar body (PB2) removal: performed after introcytoplasmic cell sperm insertion and used in PGD for single gene defects. (*c*) Simultaneous PB1 and PB2 removal: performed for PGD for chromosomal disorders. (*d*) Embryo biopsy: performed on day 3 of embryo development when the embryo contains 6 to 8 cells.

disomics (chromosome duplication where both homologous chromosomes originate from the same parent) missed in FISH analysis and embryo biopsy. Thus, a double or even triple biopsy procedure may be required for an accurate preimplantation genetic diagnosis.

Notably, preimplantation genetic diagnosis has been demonstrated to have no short-term or long-term detrimental effects. This is also in agreement with the available outcomes of thousands of preimplantation genetic diagnosis cycles for single-gene and chromosomal disorders, which have shown a comparable prevalence of congenital abnormalities after preimplantation genetic diagnosis and in the general population.

Application to genetic practices. Indications for preimplantation genetic diagnosis, similar to those for prenatal diagnosis, initially were applied if couples at genetic risk could not accept pregnancy termination associated with prenatal diagnosis. Preimplantation genetic diagnosis then gradually expanded to include the risk for common diseases with genetic predisposition and even nongenetic conditions. In fact, this technology could be applied to those conditions that may or may not appear during the lifetime of the offspring, such as common late-onset disorders (for example, Huntington's disease or breast cancer). Further, application of preimplantation genetic diagnosis as a means for the treatment of affected siblings has become a reality, already resulting in successful stem cell therapy of children with congenital and acquired disorders, for which there is still no other alternative radical treatment.

Preimplantation genetic diagnosis is presently applied to more than 100 different conditions, with the most frequent indications being cystic fibrosis, hemoglobin disorders, and dynamic mutations, such as muscular dystrophy and fragile X-linked mental retardation. The choice between the use of prenatal diagnosis and preimplantation genetic diagnosis mainly depends on the patient's view on pregnancy termination, which is strongly influenced by social and religious factors.

As many population groups object to the presently used methods of embryo biopsy, a pre-embryonic genetic diagnosis has been introduced, which principally moves the predictive genotyping to an even earlier stage than a traditional preimplantation genetic diagnosis. This makes preimplantation genetic diagnosis more ethically acceptable as it overcomes a patient's negative reaction to the micromanipulation and potential discarding of the embryo. Pre-embryonic genetic diagnosis has recently been applied to maternally derived gene defects and chromosomal abnormalities, and may be done in the future for paternally derived abnormalities. The first pre-embryonic diagnosis was applied for Sandhoff disease (a rare, genetic, lipid storage disorder resulting in the progressive deterioration of the central nervous system) by removing and testing PB1 and PB2 prior to the fusion of the maternal and paternal genomes. Only those eggs predicted to have a normal maternal contribution were allowed to fuse with the paternal genome and progress to the embryo stage, thereby avoiding the formation and possible discarding of any affected embryo.

One of the most frequent indications for preimplantation genetic diagnosis is the risk for X-linked disorders, previously performed by gender determination. Again, the advantage of egg testing in preimplantation genetic diagnosis for this group of diseases is that the embryo biopsy is avoided, as X-linked disorders are maternally derived and require no further embryo testing.

Preimplantation genetic diagnosis is also of special value for couples in which one of the partners is affected. Here, it concentrates on detecting the maternal contribution to the embryo and then selects for transfer only the embryos originating from the mutation-free eggs; if the mother is affected, the embryo biopsy is the only option.

Although the knowledge of sequence information for inherited diseases is a general requirement for undertaking preimplantation genetic diagnosis, it may also be performed by linkage analysis, irrespective of the availability of specific sequence information. This is a more universal approach, allowing tracking of the gene defect inheritance without actual testing for the gene itself.

As mentioned, preimplantation genetic diagnosis application is no longer limited to conditions presented at birth, but has become applicable for late-onset disorders with genetic predispositions. This, though, has never been an indication to prenatal diagnosis because they may not obligatorily be present during the potential progeny lifetime to justify pregnancy termination. Preimplantation genetic diagnosis has a distinct advantage because only the embryos free of genetic predisposition are selected, obviating the concerns over pregnancy termination. The present experience on these diseases includes preimplantation genetic diagnosis for congenital malformations and cancer predisposition, such as ataxia-telangiectasia, familial adenomatous polyposis coli, von Hippel-Lindau syndrome, retinoblastoma, neurofibromatosis type I and II, and familial posterior fossa brain tumor. Although preimplantation genetic diagnosis for this group of disorders raises important ethical issues, it is obvious that many couples at risk would not otherwise forgo pregnancy without preimplantation genetic diagnosis, which, in contrast to prenatal diagnosis, allows patients to have an unaffected offspring, rather than preventing the birth of a child with inherited predisposition to disease.

Application for stem cell therapy. Nondisease testing is presently another controversial preimplantation genetic diagnosis application, which will be more widely used in the future because of the expected expansion of stem cell therapy for an increasing number of human diseases where potential human leukocyte antigen (HLA)—identical donor progeny is required.

The first application of preimplantation genetic diagnosis for stem cell therapy was performed with regard to Fanconi anemia complementation group C (FA-C). In this work, together with the selection of embryos free of FA-C, only those HLAs compatible to the sibling with FA-C were replaced, resulting in birth of an unaffected child who became an HLA-compatible donor for the sibling requiring bone marrow transplantation. This first case resulted in a successful hematopoietic reconstitution in the affected sibling, opening the prospect of the application of preimplantation genetic diagnosis for stem cell therapy not only in inherited diseases but also in sporadic bone marrow failures, not requiring preimplantation genetic diagnosis, performed with a sole purpose of ensuring the birth of the potential HLA-compatible donor offspring. The first preimplantation HLA typing and successful stem cell therapy for a sporadic disorder was achieved in a patient with Blackfan-Diamond anemia (a rare blood disorder caused by a failure of the bone marrow to generate enough red blood cells), who is no longer dependent on blood transfusion.

Preimplantation HLA typing is becoming the most attractive indication, performed with or without testing for causative genes. The present experience includes hundreds of cases, performed for thalassemia, Fanconi anemia, Wiscott-Aldrich syndrome, X-linked adrenoleukodystrophy, X-linked hyper IgM syndrome, X-linked hypohidrotic ectodermal dysplasia with immune deficiency, incontinentia pigmenti, leukemias, and inherited and sporadic forms of Blackfan-Diamond anemia. Also, because most patients requesting preimplantation HLA typing are of advanced reproductive age, testing for numerical chromosomal anomalies is useful for avoiding the risk of delivering a child with chromosomal disorder and for improving reproductive outcome.

Moreover, because preimplantation genetic diagnosis provides a unique opportunity for establishment of disease-specific embryonic stem cell lines, it opens the possibilities for investigating the primary disease mechanisms in the genetically abnormal cells in the attempt to develop treatment regimens for genetic disorders. It may also become possible to derive embryonic stem cells from the same embryos selected for transfer, so transplantation treatment may be started before the birth of the HLA-matched unaffected child.

Application in ART. Approximately every second embryo obtained from poor-prognosis in-vitro fertilization (IVF) patients is chromosomally abnormal. Further, only one-fifth of these chromosomally abnormal embryos survive implantation, the remaining being eliminated at the start or during implantation. Obviously, the incidental transfer of these embryos in the absence of chromosomal testing should lead to implantation failures or may compromise the pregnancy outcome, leading to spontaneous abortions. Since this type of abnormality can be detected by preimplantation genetic diagnosis (**Fig. 2**), it should contribute to the improvement of pregnancy outcome of IVF patients. Hence, it is not surprising

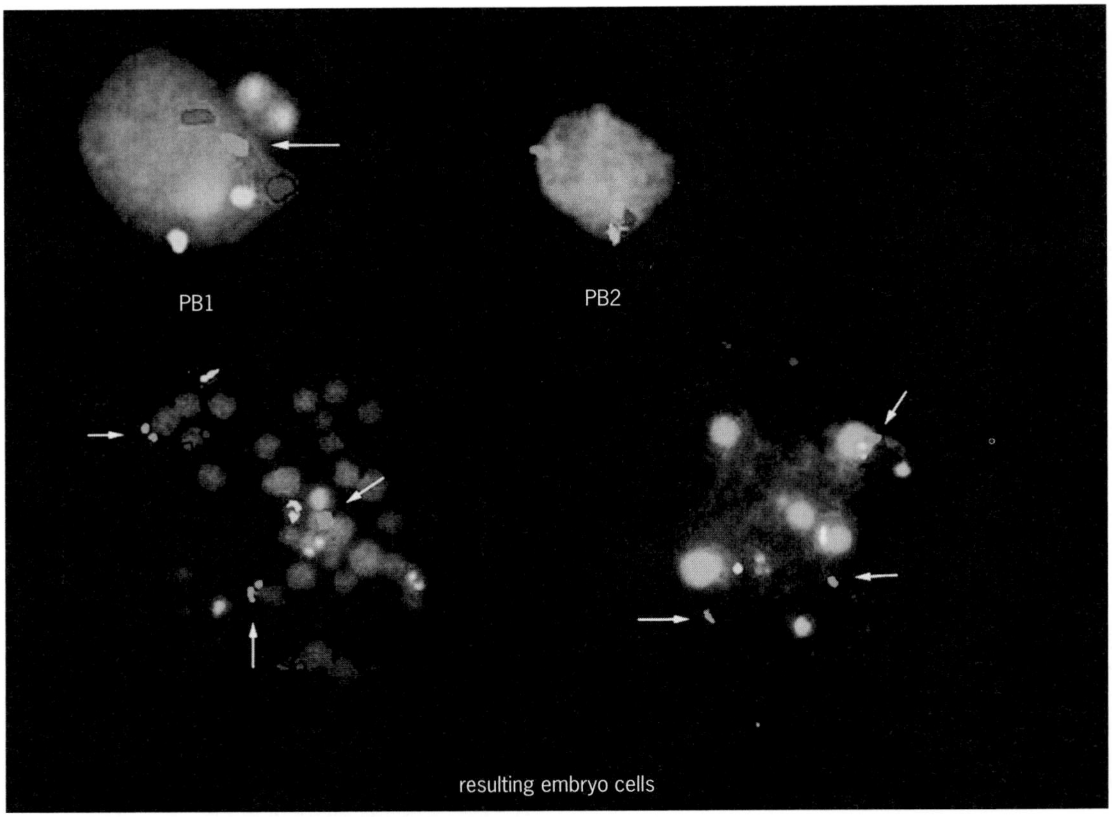

Fig. 2. PGD for chromosomal disorders. (Top) Egg testing by PB1 and PB2 demonstrating the error in PB1 (single green signal instead of two, shown by arrow), suggesting trisomy 21 in the embryo. (Bottom) Two cells from the resulting embryos are shown: one in metaphase (bottom left) and the other in interphase (bottom right).

that preimplantation genetic diagnosis for chromosomal disorders has been applied in over 20,000 IVF cases, resulting in improved implantation and pregnancy rates in patients of advanced reproductive age, those with repeated IVF failures and recurrent spontaneous abortions, and those with chromosome translocations.

The positive impact of such testing is particularly obvious from the comparison of reproductive outcome in the same patients with and without preimplantation genetic diagnosis, as previous reproductive experience of the patients may serve as the best control for the preimplantation genetic diagnosis impact. Analysis demonstrated significant improvements after preimplantation genetic diagnosis, including a fivefold improvement of implantation, a threefold reduction of spontaneous abortion, and a twofold increase of take home baby rate (that is, the rate of live births per cycle) after preimplantation genetic diagnosis. These improvements were even more striking in patients carrying chromosomal rearrangements. Thus, the current practice of selection of embryos in assisted reproductive technology may soon be upgraded to include the detection and avoidance of embryos with chromosomal anomalies.

While it may be predicted that preimplantation genetic diagnosis will become a routine for IVF patients of advanced reproductive age, it may appear of higher value for younger IVF patients as well. This may contribute to improving the overall standards of assisted reproductive practices by substituting the traditionally practiced selection of embryos for transfer by the selection of embryos free of aneuploidy (the occurrence of one or more extra or missing chromosomes) with higher potential to result in successful pregnancy.

For background information *see* BIOTECHNOLOGY; CONGENITAL ANOMALIES; GENE AMPLIFICATION; HUMAN GENETICS; INFERTILITY; PREGNANCY; PRENATAL DIAGNOSIS; REPRODUCTIVE TECHNOLOGY; STEM CELLS in the McGraw-Hill Encyclopedia of Science & Technology. Anver Kuliev; Yury Verlinsky

Bibliography. S. Munne and L. Gianaroli, Chromosomal status of human embryo, in K. Elder and J. Cohen (eds.), *Human Preimplantation Embryo Selection*, pp. 209–233, Informa Healthcare, London, 2007; Preimplantation Genetic Diagnosis International Society (PGDIS), Guidelines for good practice in PGD, *Reproductive BioMedicine Online*, 9:430–434, 2004; Y. Verlinsky and A. Kuliev, *Atlas of Preimplantation Genetic Diagnosis*, Taylor & Francis, London/New York, 2005; Y. Verlinsky and A. Kuliev, *Practical Preimplantation Genetic Diagnosis*, Springer, Berlin/New York, 2006.

Race-car aerodynamics

In recent years motor racing has become one of the most popular of sports, attracting record numbers of followers. In some racing categories the vehicles resemble production sedans while in others they look more like fighter airplanes, and there is also a great variety of tracks that range from paved to unpaved and from straight to oval or regular road courses. In all forms of racing, however, aerodynamics eventually surfaced as a significant design parameter, and nowadays all race-car designs have some level of aerodynamic element.

The complexity of race-car aerodynamics is comparable to airplane aerodynamics and is not limited to drag reduction. The generation of aerodynamic downforce (force directed downward, or negative lift) and its effect on lateral stability result in a major enhancement in race-car performance, particularly when high-speed turns are involved. In the process of designing and refining current race-car shapes, all available aerospace-type design tools are used. Because of effects such as flow separation, vortex flows, and boundary-layer transitions, the flow over most types of race cars is not easily predictable. Owing to the competitive nature of this sport and the short design cycles, engineering decisions must rely on information gathered from track and wind-tunnel testing, and even computational fluid dynamics.

Although the foundations of aerodynamics were formulated over the past 200 years, not all its principles were immediately utilized by race-car designers. Naturally, the desire for low drag was recognized first, and early designers focused mainly on streamlining their race cars. Although there was some experimentation with the addition of wings to influence the vertical load on the vehicle during the late 1920s, this major innovation was completely ignored for the following 35 years. Once designers realized the significance of aerodynamic downforce and its effect on vehicle performance, fixtures such as inverted wings or even underbody diffusers were added. The benefits of aerodynamic downforce and the improved performance are basically a result of increasing the tire adhesion by simply pushing the tires more toward the ground. Because of this additional load, larger friction (traction) levels can be achieved, and the vehicle can turn, accelerate, and brake more quickly. Furthermore, by controlling the fore/aft downforce ratio, vehicle handling can be easily modified to meet the needs of a particular race track.

The foremost and simplest approach to generate downforce was of course to add inverted wings to existing race cars. But almost immediately it was realized that the vehicle body may be used to generate downforce as well. The main advantage of this approach is that even small values of negative pressure under the vehicle can result in a sizable aerodynamic downforce because of the large planview area of the vehicle. **Figure 1** shows a modern Indy (Indianapolis Motor Speedway) car as tested in a wind tunnel. The large front and rear wings are immediately visible. However, underbody diffusers and vortex generators are hidden below the car and cannot easily be detected by the competition. The principles of these and some other downforce generating methods will be discussed.

Race-car wings. Airplane wing design matured by the middle of the twentieth century and it was only natural that race-car designers borrowed successful airplane wing profiles to use on their vehicles. This approach, however, was not entirely successful due to the inherent differences between these two applications. A race-car lifting surface design is different from a typical airplane wing design for the following reasons:

1. Race-car (front) wings operate very close to the ground, resulting in a significant increase in downforce. This increase is a manifestation of a phenomenon known as the wing-in-ground effect, which, interestingly, is favorable for the performance of both ordinary airfoils creating lift and inverted airfoils creating downforce. Of course, the effect does not come freely because a similar increase in drag is measured. Since many race cars use front wings mounted close to the ground, this principle is widely utilized in race-car design.

2. In most forms of motor racing a large rear wing is used. In the case of open-wheel race cars such as Indy cars these wings have very small aspect ratio (span/chord ratio), contrary to the much higher aspect ratio of airplane wings. The first result of the smaller aspect ratio was a significantly higher drag, but with the fringe benefit of delaying wing stall (the sudden drop of lift). This penalty could be reduced by adding very large end plates, seen on most race

Fig. 1. Typical Indy (Indianapolis Motor Speedway) car model as tested in a wind tunnel. A rolling belt on the floor is used to simulate the moving road. The wheels are mounted separately and rotated by the belt, while the 40%-scale race car's body positioned is changed from above. (*Mark Page; Swift Engineering*)

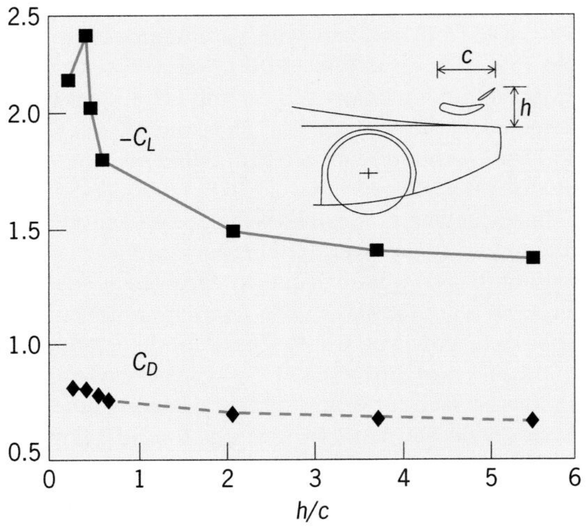

Fig. 2. Effect of rear-wing vertical position on the lift and drag coefficients, C_L and C_D, of a prototype race car. The rear-wing vertical position is expressed as h/c, where the wing height h and chord c are defined in the inset diagram. The rear-wing negative incidence $\alpha_w = 12°$, and the lift and drag coefficients are based on the planview area of the vehicle. (*From J. Katz, Race-Car Aerodynamics, 2d ed., Robert Bentley Inc., Cambridge, Massachusetts, 2006*)

cars, which indeed improve the lift-to-drag ratio. A second problem resulted from basing early designs on existing high-lift airfoil shapes, borrowed from airplanes having several elements (flaps and slots). But as noted, these airfoils were developed for airplanes having very wide wings (high aspect ratio), and therefore their performance was not optimized for race-car application. Recently, quite different, custom-designed airfoil shapes have been used to address this problem.

3. The third major difference between aircraft and race-car wings is the strong interaction between the lifting surface and the other body components. As an example, the data for a prototype race car with large underbody diffusers is presented in **Fig. 2**. In this case the wing height (h) was varied up to a height where the interaction is minimal. Clearly, the combined downforce increases as the wing approaches the vehicle's rear deck. At a very close proximity the flow separates between the rear deck and the wing and the downforce is reduced. The horizontal positioning (such as fore-aft) of the wing also has a strong effect on the vehicle's aerodynamics (usually downforce increases as the wing is shifted backward), but racing regulations state that the wing trailing edge cannot extend behind the vehicle body (from top view). The very large change in the downforce of this prototype car is due to the increased underbody diffuser flow, but the effect remains clear with sedan or even open-wheel race cars as well.

Creating downforce with the vehicle's body. Once the potential of using aerodynamic downforce to win races was realized, designers began experimenting with methods other than simply attaching inverted wings. One approach is quite similar to the previously mentioned wing-in-ground-effect model. Colin Chapman, designer of the famous Lotus 78, developed this concept to fit Formula One (F1) race-car geometry. In his design the vehicle's side pods had an inverted airfoil shape (in ground effect) and the two sides of the car were sealed by sliding "skirts." These side seals created a "two-dimensional" environment for the small-aspect-ratio inverted-wing- shaped side pods (resulting in air speeds much higher than the vehicle's speed). The concept (as shown in the lower inset to **Fig. 3**) worked very well, resulting in large suction forces under the car (in accord with the Bernoulli principle that higher flow speeds result in lower pressure). The "skirted car" was highly successful and the Lotus 78 won the world championship in 1977. By the end of the 1980s this method was used in many forms of racing, resulting in downforce values exceeding the weight of the vehicle. The sliding seals, however, were not trouble-free. Irregularities in the road surface occasionally resulted in seal failure and the immediate loss of downforce, with catastrophic consequences. [The effect of increasing the gap between the ground and the seal on the downforce is shown in Fig. 3; a 20-mm (0.8-in.) gap could result in loss of 50% of the downforce.] This problem led to the banning of all sliding seals in F1 cars by 1983. In the following years, this ruling was mandated in most other forms of racing as well, and the only parts of the vehicle allowed to be in contact with the ground were the tires.

Another approach that worked well is based on controlling the low pressure under the car, independent of the vehicle's speed. This approach resulted in the so-called suction cars. The first was the 1969 Chaparral 2J. This car used auxiliary engines to drive two large suction fans behind the vehicle. The whole periphery around the car underbody and the ground was sealed, and the fans were used to suck the leaking air through the seals to maintain the controllable low pressure. Another benefit from this design was that the ejected underbody flow (backward) reduced the flow separation behind the vehicle, and therefore

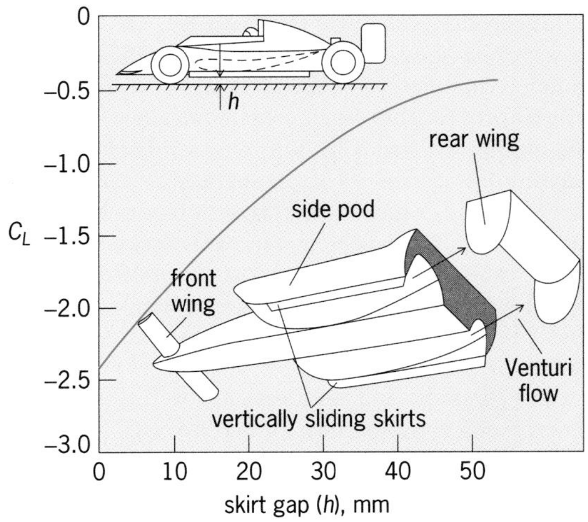

Fig. 3. Creation of downforce with side skirts. Graph shows effect of side-skirt-to-ground gap h on a vehicle's total downforce coefficient, $-C_L$. The gap h is defined in the upper inset diagram, and the underbody of the race car is diagrammed in the lower inset. 1 mm = 0.04 in. (*From J. Katz, Aerodynamics of race cars, Annu. Rev. Fluid Mech., 38:27–64, 2006, Annual Reviews*)

the vehicle's drag was reduced. The downforce was controlled by the auxiliary motors and did not increase with the square of speed, making the car quite comfortable (no stiff suspension) and competitive. The design was immediately successful. However, this success was not well received by the competition, and regulation almost immediately outlawed such designs.

Underbody diffusers (tunnels). Once the sliding skirts were banned the suction under the car was significantly reduced (Fig. 3). A logical evolution of this concept led to underbody "tunnels" formed under the sidepods, which sometimes were called diffusers. The integration of this concept into an actual race-car underbody is depicted in the upper part of **Fig. 4**. Flow visualizations clearly show the existence of the side vortices responsible for reattaching the flow in the tunnels (diffusers). Surface pressures measured along the tunnel centerlines are shown in Fig. 4 as well, and a sharp suction peak at the tunnel entrance is evident. In this study several diffuser angles were used and the resulting downforce and drag coefficients for the complete vehicle are shown in the table inserted in the figure. For this particular geometry, diffuser angles larger than 14° caused the flow to separate from the diffuser walls with the result of less downforce. The significance of the pressure peak at the diffuser entrance for race-car application is that the location of the vehicle's center of pressure could be controlled by the fore-aft shifting of the diffuser entrance. Of course, the downforce usually increases with reduced ground clearance, an effect that continues down to very small ground clearance values.

Add-ons. Simple modifications can be added to an existing car to increase downforce. One of the simplest add-ons is the vortex generator. These devices can take the form of small triangular plates or resemble miniature wings, and they are not always visible. A simple option is to add such vortex generators at the front of the underbody where their long vortex trails can induce low pressure under the vehicle. This principle is widely used for open-wheel race cars (such as Indy or F1 cars).

The discussion of vehicle-body-related downforce would not be complete without mentioning some of the widely used add-ons such as spoilers and dive plates. One of the earliest types of "spoilers" mounted on the rear deck of sedan-type vehicles is quite effective and widely used. Current stock cars use them, and the rear downforce that they generate has been shown to increase with increasing angle (measured from the horizontal plane). A 60° rear spoiler causes a change of about −0.20 in the lift coefficient C_L. Spoilers under the chin of a sedan-type vehicle were tested in the 1970s, and showed a positive effect on front downforce. Apart from reducing the pressure below the front underbody of the car, they have a positive effect on the flow across front-mounted radiators.

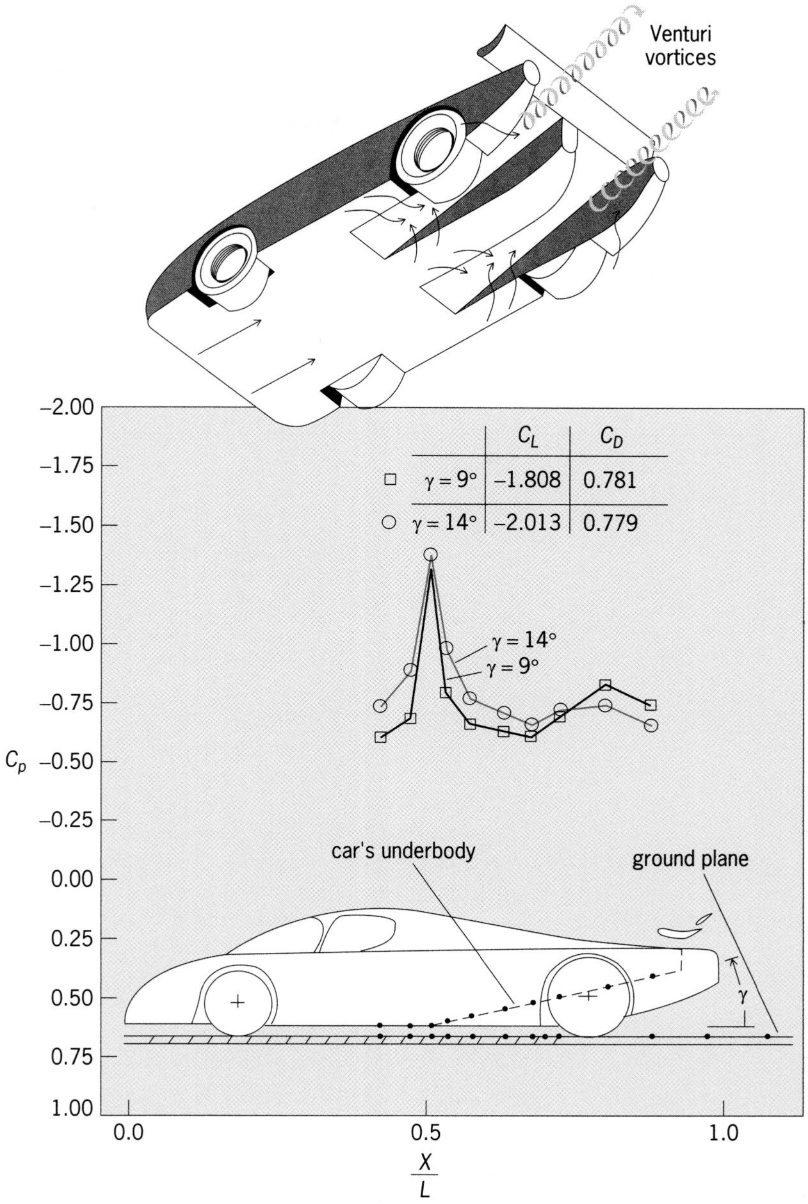

Fig. 4. **Creation of downforce with underbody diffusers (tunnels). Upper drawing shows race-car underbody. Graph shows effect of underbody diffuser angle γ on the diffuser centerline pressure distribution. Here, C_p is a nondimensional pressure coefficient, and negative values mean pressures below the ambient level. Resulting downforce coefficients, $-C_L$, and drag coefficients C_D for the complete vehicle are shown in the table. Lower diagram defines the diffuser angle γ, and locates measurement points, whose fore-aft positions are graphed in terms of X/L, where X is distance from front of car and L is length of car. Rear-wing negative incidence $\alpha_w = 12°$. (*From J. Katz, Race-Car Aerodynamics, 2d ed., Robert Bentley Inc., Cambridge, Massachusetts, 2006*)**

For background information *see* AERODYNAMIC FORCE; AERODYNAMICS; AILERON; AIR-CUSHION VEHICLE; AIRFOIL; AUTOMOBILE; BERNOULLI'S THEOREM; VORTEX; WING in the McGraw-Hill Encyclopedia of Science & Technology. Joseph Katz

Bibliography. W. H. Hucho, *Aerodynamics of Road Vehicles*, 4th ed., SAE International, Warrendale, Pennsylvania, 1998; J. Katz, *Race-Car Aerodynamics*, 2d ed., Robert Bentley Inc., Cambridge, Massachusetts, 2006; J. Katz, Aerodynamics of race cars, *Annu. Rev. Fluid Mech.*, 38:27–64, 2006; W. F. Milliken and D. L. Milliken, *Race Car Vehicle Dynamics*, SAE International, Warrendale, Pennsylvania, 1995.

Reactive power compensation technologies

Reactive power compensation is defined as the management of reactive power to improve the performance of alternating-current (ac) power systems. In general, the problem of reactive power compensation is related to load and voltage support. In load support the objectives are to increase the value of the system power factor, to balance the real power drawn from the ac supply, to enhance voltage regulation, and to eliminate current harmonic components produced by large and fluctuating nonlinear industrial loads. Voltage support is generally required to reduce voltage fluctuation at a given terminal of a transmission line. Reactive power compensation in transmission systems also improves the stability of the ac system by increasing the maximum active power that can be transmitted. *See* VOLTAGE STABILITY.

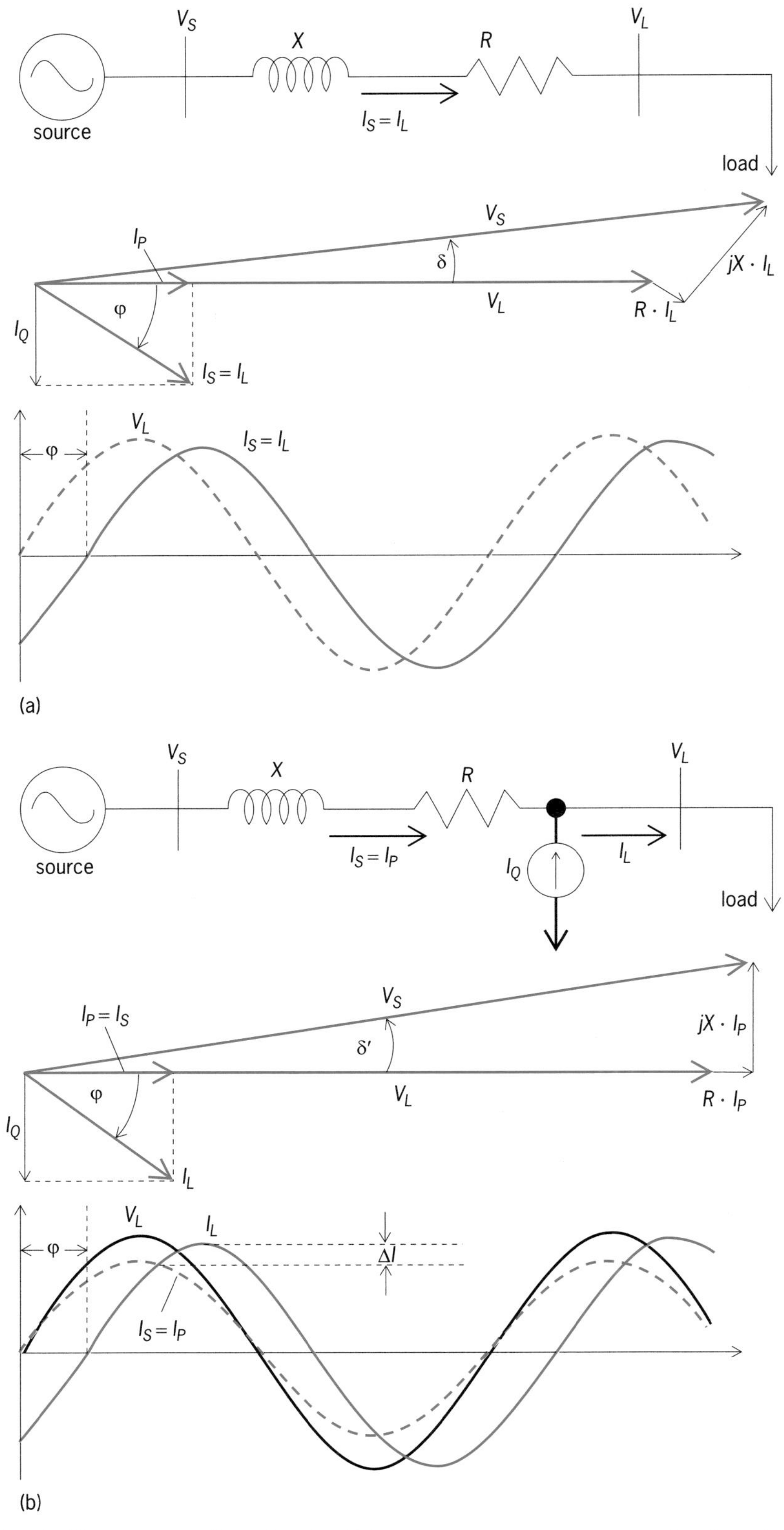

Fig. 1. Shunt compensation principles. (*a*) System without reactive compensation ($I_S = I_L$). (*b*) System that employs shunt compensation with a current source ($I_S < I_L$). A circuit diagram, a phasor diagram, and current and voltage waveforms are shown in each part of the figure.

Operating principles. Series and shunt reactive power compensation are used to modify the natural electrical characteristics of ac power systems. Series compensation modifies the transmission or distribution system parameters, while shunt compensation changes the equivalent impedance of the load. In both cases, the reactive power that flows through the system can be effectively controlled, improving the performance of the overall ac power system.

Shunt reactive power compensation. The principles and theoretical operation of shunt reactive power compensation will now be explained. A basic ac system consists of a power source whose voltage is V_S, a power line with impedance $R + jX$, and a typical inductive load with a voltage V_L (**Fig. 1**). In the system without compensation, the source current I_S and load current I_L are the same, and because the load is normally inductive, this current lags the load voltage V_L. This situation is represented by the phase angle φ between V_L and I_L (or I_S). As a result, the power source must supply the total load current, maintaining a high level of source current from the generator and through power lines, which means more power losses and lower power transmission capability (Fig. 1*a*). The load current I_L may be divided into two components: I_P, which is in phase with V_L and produces the real power (active power), and I_Q, which lags the voltage V_L by 90° and produces reactive power (imaginary power). Then the source V_S could generate only the real component I_P, and the I_Q component could be produced near the load with a reactive power compensator.

If the reactive power (imaginary power) is supplied near the load, then the source current is reduced or minimized, reducing power losses and improving voltage regulation at the load terminals. This shunt compensation can be done in three ways: with a capacitor, with a current source (Fig. 1*b*), or with a voltage source. As a result, the system voltage regulation is improved and the value of current required from the source is reduced.

Series reactive power compensation. Reactive power compensation can also be of the series type. Typical series compensation systems use capacitors to decrease the equivalent reactance of a power line at rated frequency (**Fig. 2**). In a self-regulated manner the series capacitor C generates reactive power that balances a fraction of the line reactance. (This

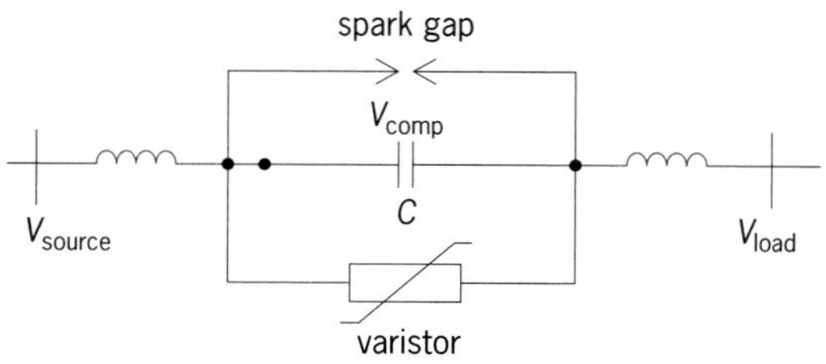

Fig. 2. Series capacitor compensator with protections.

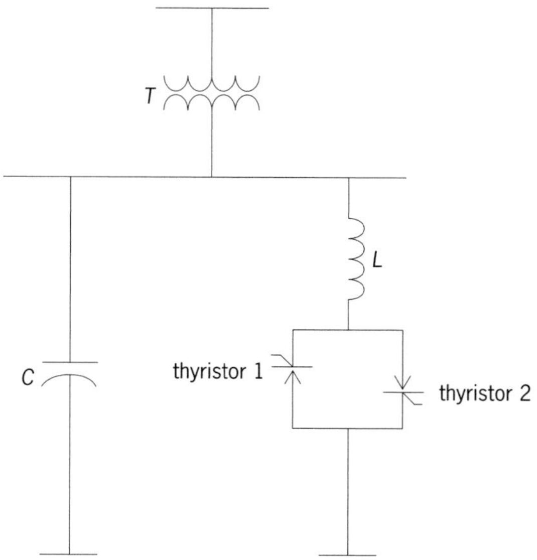

Fig. 3. Thyristor-controlled reactor (TCR) topology.

happens because the voltage drop in the capacitor and the inductance connected in series are 180° opposed.) The spark gap and varistor in Fig. 2 are used to avoid destruction of the capacitor C in overvoltage situations.

Like shunt compensation, series compensation can be implemented with current-source or voltage-source devices. However, the compensation strategy is different when compared with shunt compensation. In this case, a voltage V_{comp} is added between the line and the load to change the angle of V_{load}, which is now the voltage at the load side. With the appropriate magnitude adjustment of V_{comp}, unity power factor can again be reached at V_{load}.

Technologies. Traditionally, fixed or mechanically switched capacitors or inductors, or rotating synchronous condensers, have been used for reactive power compensation. However, in the last few decades, two new families of reactive power generators, using power electronic technologies, have been developed: thyristor-based static compensators and self-commutated static converters.

Thyristor-based static compensators. As in the case of synchronous condensers, the aim of achieving fine control over the entire reactive power range has been fulfilled with the development of thyristor-based compensators, and with the advantage of faster response times and reduced costs. They consist of standard reactive power elements (reactors and capacitors) that are controlled with thyristors to provide rapid and variable reactive power. These compensators can be grouped into two basic categories: the thyristor-switched capacitor (TSC) and the thyristor-controlled reactor (TCR).

In the TSC, the capacitor bank is split up into small steps, which are individually switched in and out using bidirectional thyristor switches. Each single-phase branch consists of two major parts, the capacitor and a pair of thyristor switches. The capacitor may be switched with a minimum of transients if the thyristor is turned on at the instant when the capacitor voltage and the network voltage have the same value. Despite the attractive theoretical simplicity of the switched-capacitor scheme, its popularity has been hindered by a number of practical disadvantages: Reactive power compensation is not continuous, and each capacitor bank requires a separate thyristor switch. Therefore the construction is not economical.

The TCR uses a bidirectional switch, implemented with a pair of oppositely connected thyristors, in series with an inductor L and a shunt capacitor C (**Fig. 3**). Since phase-angle control is used, a continuous range of reactive power consumption is obtained. By increasing the thyristor gating angle from 90° to 180°, the fundamental component of the reactor current is reduced. This is equivalent to increasing the inductance, reducing the reactive power absorbed by the reactor. The main disadvantage of this configuration is the generation of harmonic currents, which forces the implementation of more complicated topologies (with passive filters, using delta connections or twelve-pulse configurations, which are not shown in Fig. 3).

Thyristors are also used for series compensation. A thyristor-controlled series compensator (TCSC) provides a technology that addresses specific dynamic

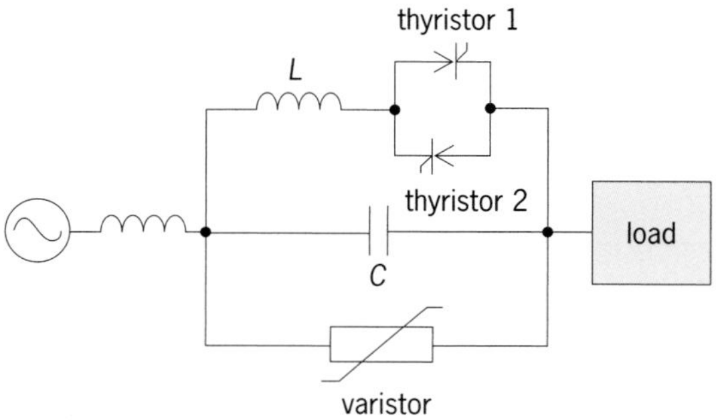

Fig. 4. Thyristor-controlled series compensator (TCSC).

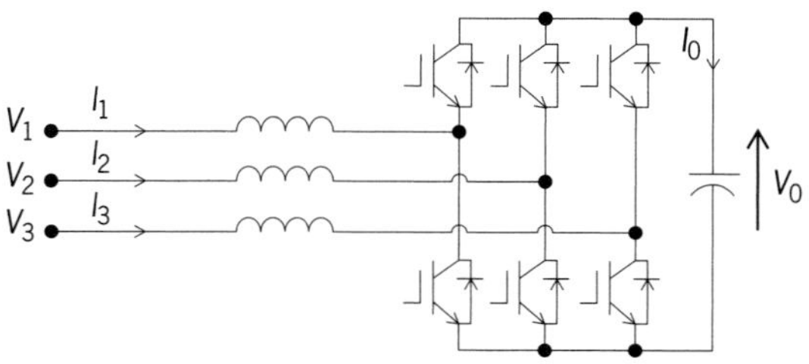

Fig. 5. Self-commutated, voltage-source converter using IGBT (insulated-gate bipolar transistor) valves.

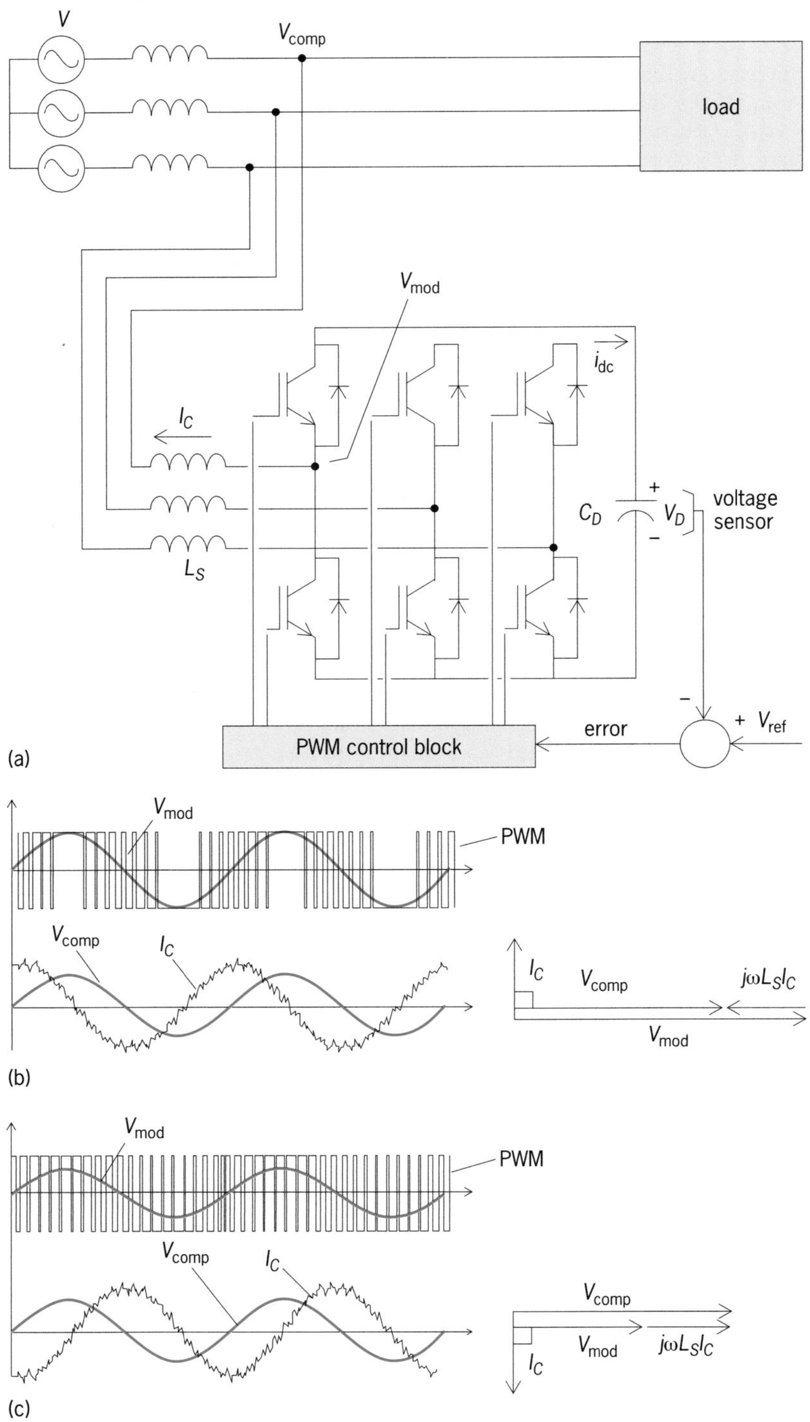

Fig. 6. Current and voltage waveforms of a voltage-source, self-commutated shunt reactive power compensator. (*a*) Compensator topology. (*b*) Current and voltage waveforms for leading compensation ($V_{mod} > V_{comp}$). (*c*) Current and voltage waveforms for lagging compensation ($V_{mod} < V_{comp}$). In each case, a phasor diagram is also shown.

problems in transmission systems (**Fig. 4**). TCSCs are an excellent damping tool when interconnecting large electrical systems. Additionally, they reduce the problem of subsynchronous resonance, a phenomenon that involves an interaction between large thermal generating units and series-compensated transmission systems.

Self-commutated converters. With the development of gate-commutated semiconductor devices [gate turnoffs (GTOs) and insulated-gate bipolar transistors (IGBTs)] attention has focused on self-commutated reactive power compensators. Several approaches are possible, including current-source and voltage-source converters. The current-source inverter uses a reactor supplied with a regulated dc current, while the voltage-source uses a capacitor with regulated dc voltage. The self-commutated, voltage-source converter is the one more commonly used as a reactive power compensator (**Fig. 5**).

The main advantages of self-commutated reactive power compensators are the significant reduction in size, cost, and harmonics generation. Self-commutated compensators are used to stabilize transmission systems, improve voltage regulation, correct power factor, and also correct load unbalances. Moreover, they can be used for the implementation of shunt and series compensators. **Figure 6** shows a shunt reactive power compensator, implemented with a self-commutated, voltage-source converter. The control of the reactive power is done by adjusting the amplitude of the fundamental component of the output voltage V_{mod}, which is modified using pulse-width modulation (PWM) [Fig. 6(*a*)]. When V_{mod} is larger than the voltage V_{comp}, the reactive power compensator generates reactive power [Fig. 6(*b*)], and when V_{mod} is smaller than V_{comp}, the compensator absorbs reactive power [Fig. 6(*c*)]. Its principle of operation is similar to the synchronous machine. The compensation current can be leading or lagging, depending on the relative amplitudes of V_{comp} and V_{mod}. The capacitor voltage V_D, connected to the dc link of the converter, is kept constant and equal to a reference value V_{ref} with a special feedback control loop.

New technologies. Based on self-commutated power electronics converters, reactive power compensators have been developed not only to compensate reactive power but also to improve voltage regulation, flicker, harmonics, real and reactive power, transmission-line impedance, and phase-shift angle. Even though the final effect is to improve power system performance, the control variable in all cases is basically the reactive power. Using self-commutated converters, the following high-performance power system controllers have been implemented: the static synchronous compensator (STATCOM), the static synchronous series compensator (SSSC), the dynamic voltage restorer (DVR), the unified power flow controller (UPFC), the interline power flow controller (IPFC), and the superconducting magnetic energy storage (SMES).

The STATCOM is based on a solid-state voltage source, implemented with an inverter and connected in parallel to the power system through a coupling reactor, in analogy with a synchronous machine, generating a balanced set of three sinusoidal voltages at the fundamental frequency, with controllable amplitude and phase-shift angle. This equipment, however, has no overload capability.

The SSSC is a voltage-source converter that injects a voltage in series to the line, 90° phase-shifted with the load current, operating as a controllable series capacitor. The basic difference, as compared with series capacitor, is that the voltage injected by an SSSC is not related to the line current and can be independently controlled.

The DVR is a device connected in series with the power system and is used to keep the load voltage constant, independent of the source voltage fluctuations. When voltage sags or swells are present at the load terminals, the DVR responds by injecting three ac voltages in series with the incoming three-phase network voltages, compensating for the difference between faulted and prefault voltages. Each phase of the injected voltages can be controlled separately in magnitude and angle.

The UPFC consists of two switching converters operated from a common dc link provided by a dc storage capacitor. One is connected in series with the line, and the other in parallel. This arrangement functions as an ideal ac-to-ac power converter in which the real power can freely flow in either direction between the ac terminals of the two inverters and each inverter can independently generate (or absorb) reactive power at its own ac output terminal.

The IPFC consists of two series voltage-source converters whose dc capacitors are coupled, allowing active power to circulate between different power lines. When operating below its rated capacity, the IPFC is in regulation mode, allowing the regulation of the P and Q flows on one line, and the P flow on the other line.

The SMES system is a device for storing and instantaneously discharging large quantities of power. It stores energy in the magnetic field created by the flow of dc current in a coil of superconducting material that has been cryogenically cooled. These systems have been in use for several years to improve industrial power quality and to provide a premium-quality service for individual customers vulnerable to voltage fluctuations. The SMES recharges within minutes and can repeat the charge-discharge sequence thousands of times without any degradation of the magnet.

Applications. Northern States Power Company (NSP) of Minnesota operates a reactive power compensator in its 500-kV power transmission network between Winnipeg and Minnesota. The purpose of this device, located at the Forbes substation in Minnesota, is to increase the power interchange capability on existing transmission lines. This solution was chosen instead of building a new line, as it was found superior with respect to increased utilization as well as reduced environmental impact. With the reactive power compensator in operation, the power transmission capability was increased to about 200 MW.

Today it is possible to travel by railroad between London and Paris in just over 2 h, at a maximum speed of 300 km/h (186 mi/h). The railway power system is designed for power loads in the range of 10 MW. The traction feeding system is a modern 50-Hz, 2 25-kV supply incorporating an autotransformer scheme to keep the voltage drop along the traction lines low. Power step-down from the grid is direct, via transformers connected between two phases. A major feature of this power system is the reactive power compensator support. The primary purpose of this compensator is to balance the unsymmetrical load and to support the railway voltage in the case of a feeder station trip when two sections have to be fed from one station. The second purpose is to ensure a low tariff for the active power by maintaining unity power factor during normal operation.

The Sullivan substation, in northeastern Tennessee, is supplied by a 500-kV bulk power network and by four 161-kV lines that are interconnected through a 1200-MVA transformer bank. Seven distributors and one large industrial customer are served from this substation. The STATCOM is implemented with a 48-pulse, two-level voltage-source inverter that combines eight six-pulse, three-phase inverter bridges, each with a nominal rating of 12.5 MVA. The STATCOM regulates the 161-kV bus voltage during daily load increases to minimize the activation of the tap-changing mechanism on the transformer bank, which interconnects the two power systems. The use of this reactive power compensator to regulate the bus voltage has resulted in the reduction in use of the tap changer from about 250 times per month to 2 to 5 times per month. Tap-changing mechanisms are prone to very costly failure. Without the STATCOM, the transmission company would be compelled to either install a second transformer bank or construct a fifth 161-kV line into the area; both are costly alternatives.

The convertible static compensator (CSC) at the Marcy substation, a versatile and reconfigurable device based on FACTS (flexible alternating current transmission systems) technology, was designed, developed, tested, and commissioned in the New York 345-kV transmission system. The CSC consists of two 100-MVA voltage-source converters that can be reconfigured and operated as STATCOMs, SSSCs, UPFCs, or IPFCs. The CSC provides voltage control, improved power-flow transfers, and superior power flow control.

For background information *see* ALTERNATING CURRENT; ALTERNATING-CURRENT CIRCUIT THEORY; CONVERTER; ELECTRIC POWER SYSTEMS; ENERGY STORAGE; PULSE MODULATION; REACTIVE POWER; SEMICONDUCTOR RECTIFIER; STATIC VAR COMPENSATOR; SUBSYNCHRONOUS RESONANCE; SUPERCONDUCTING DEVICES; SYNCHRONOUS MOTOR; VARISTOR in the McGraw-Hill Encyclopedia of Science & Technology.

Juan W. Dixon; Luis A. Morán

Bibliography. J. Dixon et al., Reactive power compensation technologies: State-of-the-art review, *Proc. IEEE*, 93:2144–2164, 2005; T. J. E. Miller, *Reactive Power Control in Electric Systems*, Wiley, 1982; E. Wanner, R. Mathys, and M. Hausler, Compensation systems for industry, *Brown Boveri Rev.*, 70:330–340, 1983.

Reference values and reference intervals

Results of clinical laboratory tests and other observations are interpreted by comparing them with *reference values* obtained from relevant individuals whose state of health is known. Instead of reporting numerous reference values, they are usually subjected to statistical analysis, followed by cutting off the extremes of the distribution by setting *reference limits* and reporting a *reference interval* containing a stated proportion (often the central 95%) of the reference values.

Strategy. The reference value strategy or philosophy has largely replaced the use of normal values, whose practice suffered from many weaknesses: The term "normal" has several and sometimes conflicting meanings—healthy, usual, Gaussian statistical distribution. The individuals from which the normal values were derived and their relevance were not reported, and physiological factors strongly influencing laboratory results (for example, posture, food intake, and use of a tourniquet in specimen collection) were not taken into account.

Clinical observations are interpreted by comparing them with other data. The basic idea of the reference value strategy is to provide values from subjects who are relevant controls for the patients examined and to ensure that these controls are well described. In the case of a laboratory result, an *observed value* is compared with one or several sets of reference values. A didactic way of putting it is as follows: Diseases can be regarded as experiments of nature and, as in the experimental sciences, the positive and negative controls should be sufficiently well described to enable critical evaluation and repetition of the experiments.

Terminology and definitions. National and international organizations have issued recommendations concerning the production and use of reference values. The nomenclature used is shown in the **illustration**.

Reference values are the values observed on reference individuals, selected using defined criteria. The individual's state of health should be defined (diseases are also regarded as "states of health"). Usually all subjects satisfying the criteria (for example,

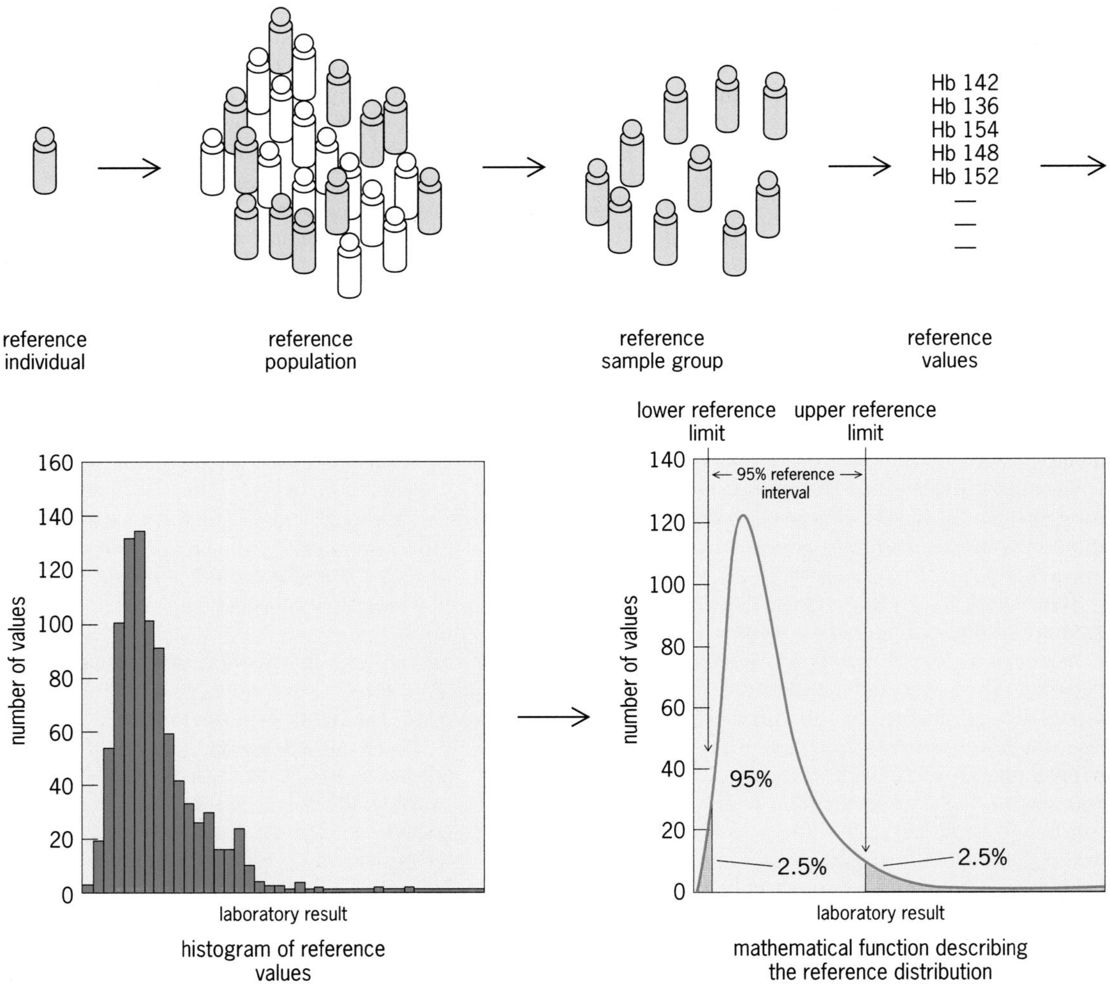

The reference values measured on the reference sample group are arranged as a histogram, the distribution of which is imitated by a suitable mathematical function where the reference limits are easy to calculate (parametric method). Here, the usual Gauss function did not fit; however, when the reference values were replaced with their logarithms, it fitted acceptably. Using this function, the reference limits were computed to eliminate 2.5% of the lowest and highest values. The central 95% reference interval is defined by these limits.

males aged 20–30) are not included, only a representative sample (*reference sample group*) of the criteria-satisfying *reference population*. The values needing evaluation (for example, those obtained on patients) are called *observed values*.

All the reference values can be supplied in the form of a table or histogram. However, a less cumbersome way is to subject the values to data reduction by statistical treatment. It is customary to start by arranging the data in the form of a histogram (or a curve fitted to it) showing the *reference distribution*. Then rare values at both ends of the distribution are omitted by setting *reference limits*. These limits bound a *reference interval*. It has become customary (but not obligatory) to set the limits so that 2.5% of the extreme values at both ends are cut off. The resulting interval then becomes "the central 0.95 (or 95%) interval." If the data have a Gaussian distribution, the interval corresponds to the mean ±1.96 (roughly 2) standard deviations. If nonparametric statistics are used, the central reference interval is obtained by eliminating an equal number of the highest and lowest values.

The calculated reference limits are not exact, and *confidence intervals* can be computed for them describing how exactly they were determined. The larger the number of reference values, the more exact are the limits. The confidence intervals are rarely reported, but it is good to bear in mind that the reference limits are not razor-sharp.

Common mistakes. The reader is discouraged from using nonrecommended terms such as "normal reference values" and "reference range." The former term demonstrates ignorance of the basic principles, while the latter does not comply with mathematical terminology, where a range is described by one figure (for example, 17), whereas an interval is expressed with two (for example, 12–29). Another mistake is to assume that the terms "reference values" or "intervals" imply good health. If derived from individuals selected with criteria of "good" health, the terms mentioned should be preceded by a qualifyer such as "health-related." Finally, the term "reference value(s)" is often used instead of reference interval. The former term implies that all values are presented; if parts of them are eliminated, other terms have to be used, usually reference interval.

Reference individuals. To obtain suitable reference individuals, one has to define the critera used in selecting them; however, just as an investigator is free to devise his controls, there are no strict rules defining the criteria. Various criteria for inclusion and exclusion of individuals are used, and they can be applied a priori or a posteriori; either preselected individuals are examined or results from suitable individuals are picked out from larger sets of data.

Values from healthy persons are obviously needed since both therapy and preventive medicine aim at achieving such values. However, what is conceived as health varies in different situations, for example, depending on whether total body or partial (for example, kidney) functions are studied, the age of the subject, and so forth. Health is thus a goal-oriented and relative concept. Therefore, when presenting "health- related reference values," one should make available to the user the health criteria applied and the purpose for which the values were collected.

In the oldest age groups, it is difficult to find persons who satisfy even mild criteria of health. One suggestion is to collect values from such individuals, to store them, say for 10 years, and to use the values of those who survived. In this case, the goal is obviously to promote survival.

Special sets of reference values are those from the general population or the usual values of hospital patients. It would be desirable to have separate reference values for ambulant and bedridden subjects, for regular and emergency situations, and sometimes for different ethnic and genetic groups.

Values typical of disease or physiological states such as pregnancy are also needed. As incipient disease may be indistinguishable from health, it is important to define the stage of the disease which the values represent.

Gender, pregnancy, age, ethnic origin, and so forth, have effects that may lead to partitioning of the reference individuals and values into groups, but subdivision is worthwhile only if the values of the groups clearly differ.

Finally, the subject's own previous values are of great interest. Typical individual variations have been collected and expressed as *reference change*. An unusually great deviation suggests a change in the physiological state or development of a disease (for example, myocardial infarction).

Specimen collection and its analysis. Numerous physiological and pharmacological factors influence the concentration of body constituents. Variations are also caused by the way the specimen is collected, treated before analysis, and analyzed. Evidently, the analytical and specimen-collection techniques producing the reference and observed values should be as identical as possible, including quality control.

A change in posture from upright to supine or the use of a tourniquet causes considerable changes in the composition of blood. Intake of food or pharmacological agents (including alcohol, coffee, and contraceptive pills) and time of the day also influence numerous components. The handling of the specimen also has effects.

The significance of these *preanalytical factors* can be reduced by standardizing the conditions of specimen collection. Recommended procedures have been published. For instance, before coming for a blood test, the subject is told to refrain from eating, drinking, or smoking after 10 p.m. the preceding night, to avoid exercise (walking long distances) in the morning, to arrive at 8 a.m. in the clinic, and then to sit for 15 min before venipuncture, which is performed without use of a tourniquet, etc. If the standard procedure has not been followed, this should be reported, and the user taught the possible consequences of the noncompliance. This especially concerns tests done on emergency patients.

Observed values in relation to reference values. Traditionally, the observed value is compared with a reference interval, often printed on answer forms. More advanced systems provide an interval corresponding to the age and sex of the patient, and provide an alert when the observed value falls outside the interval.

However, the use of the central health-related reference interval has drawbacks. The values falling outside the interval are easily taken to be pathological, although in fact they are only rare. The more tests that are performed on the patient, the greater becomes the chance that one or several observed values fall outside the interval, and this often leads to unnecessary examinations. One way of avoiding this situation is to use multivariate analysis: Instead of constructing reference intervals for single analyses, analogous multidimensional intervals are computed for packages of different tests.

Instead of using the reference interval, one can use fractiles or percentiles (percentile = 100 × fractile, where a fractile is the value at or below which a given fraction of the data lies) to indicate how the observed value is located with respect to the reference distribution. A hemoglobin value could be reported as 143 g/L (11%), meaning that 11% of the reference values are smaller than (or equal to) 143 g/L. The usual reference limits would then correspond to the percentiles 2.5 and 97.5.

Decision limits. The reference limits only indicate which values falling outside them are rare; they definitely do not distinguish between healthy and sick. Limits considered to substantiate a diagnosis, indicating a risk, etc., are called *decision limits*. Few have been published, but doctors subconsciously use them. For instance, low hemoglobin values may be judged using the following limits: (1) The value is somewhat low—check next time. (2) The patient suffers from anemia. (3) Start treatment for anemia. (4) An immediate transfusion is needed.

Apparently, efforts should be made to produce useful decision limits.

For background information *see* ANALYSIS OF VARIANCE; BIOMETRICS; DATA REDUCTION; EXPERIMENT; LITERATURE OF SCIENCE AND TECHNOLOGY; MEDICAL INFORMATION SYSTEMS; MEDICINE; STATISTICS in the McGraw-Hill Encyclopedia of Science & Technology.

Ralph Gräsbeck

Bibliography. R. Gräsbeck, The evolution of the reference value concept, *Clin. Chem. Lab. Med.*, 42:692–697, 2004; R. Gräsbeck and T. Alström (eds.), *Reference Values in Laboratory Medicine*, Wiley, Chichester, United Kingdom, 1981; National Committee for Clinical Laboratory Standards, *How To Define and Determine Reference Intervals in the Clinical Laboratory: Approved Guideline*, 2d ed., NCCLS C28-A2, Wayne, Pennsylvania, 2000; H. E. Solberg, Establishment and use of reference values, in C. A. Burtis, E. R. Ashwood, and D. E. Bruns (eds.), *Thietz Textbook of Clinical Chemistry and Molecular Diagnosis*, 4th ed., Elsevier Saunders, St. Louis, 2006; H. E. Solberg and R. Gräsbeck, Reference values, *Adv. Clin. Chem.*, 27:1–79, 1989.

Reversible express lanes

A most unique toll road, Tampa's Crosstown Expressway reversible express lanes (REL) opened to motorists in July 2006. It is developed, owned, and operated by Tampa-Hillsborough County Expressway Authority. REL is a common-sense transportation solution that addresses urban congestion by combining the innovations of concrete segmental bridges, reversible express lanes, cashless open-road tolling, and full electronic controls. The revolutionary "six lanes in six feet" freeway was constructed within the existing right-of-way of the Lee Roy Selmon Crosstown Expressway (**Fig. 1**). It provides three lanes toward Tampa in the morning peak and three lanes out of Tampa and into the rapidly growing suburb of Brandon in the afternoon peak. During midday, a central segment is closed and the Tampa and Brandon segments operate independently on a direction that optimizes local traffic circulation. Only cars and buses are allowed on the REL. A $1.50 toll was charged in 2007, but entry is unimpeded because tolls are collected electronically via in-vehicle transponders or with license-plate recognition. REL has provided a major reduction in congestion. Before-speeds of less than 15 mi/h (24 km/h) in the peak hours rose to free-flow speeds of about 60 mi/h (100 km/h), which translates to up to a full hour of round-trip travel-time savings for many commuters. In addition, REL was constructed at a record low cost per mile, had minimal environmental impacts, created a minimal disruption to adjacent traffic, and spurred development growth in both Tampa and Brandon. Actual traffic volumes have exceeded forecasts by 25%.

The growth of traffic from 13.1 million tolling transactions in 1982 to 30.2 million transactions in 2002 resulted in severe congestion for thousands of daily Crosstown Expressway motorists. The Expressway is also a classic commuter toll road, with directional percentage splits of more than 75/25 during the peak hours. In the morning, more than 75% of the traffic is Tampa bound; the reverse is true in the afternoon. Almost 80% of all of the daily traffic occurs during the morning and afternoon commuting peak periods.

The Authority's solution to relieve peak-hour congestion was to build 10 mi (16 km) of reversible express toll lanes between Interstate 75 and downtown Tampa. Like many urban areas, purchasing the necessary additional land in this corridor for typical highway widening was neither physically nor financially feasible. To minimize the footprint, most of the project was constructed as a concrete segmental bridge using only 6 ft (1.8 m) of space within the existing median (Fig. 1). This resulted in an aesthetically pleasing structure, lower project costs, and reduced impacts to the community and the environment. The shape of the box that supports the deck and transfers loads to the pier limits the view of the underside of the bridge to only half of the structure, providing light and limiting the structure's visual impact. The resultant perception is that of an overpass instead of a "double-decker" structure.

Technological innovations include cashless three-lane-wide open road tolling at free-flow speeds supplemented by a unique approach to video toll collection for motorists without transponders as well as a centralized traffic management center with state-of-the-art software to control the reversible lane operations and provide multiple safeguards to preclude vehicles from entering the expressway in the wrong direction.

Terminal gateways. The Brandon and downtown gateways to REL were planned by engaging the public in the design process. This resulted in highly positive community acceptance and support at both ends of the project.

In addition to their value as transportation projects, these gateways were major investments in urban architecture, landscaping, and public facilities that have been a catalyst for new private investment. They are a case study on the integration of major highway infrastructure into existing communities. They are also a case study of flexible traffic operations since REL is able to operate in six modes, including all eastbound, all westbound, and four combinations of directional operation of the Brandon and Tampa gateways.

The Brandon Parkway end of REL includes scenic landscaping, a winding off-road recreational trail for walking and cycling, and numerous sites for resting, relaxing, and enjoying the environment. The Parkway has become the prime location for construction, with over $100 million of new restaurants, shopping, residential, and private leisure activities. During off-peak travel periods, the Parkway functions as a set of internal circulation roads, facilitating local trips to shopping areas, public services, and restaurants.

In downtown Tampa, REL descends to Meridian Avenue. It has transformed a former narrow two-lane street through an aging industrial district into a modern six-lane urban thoroughfare. Representing a $50 million investment in downtown Tampa, the gateway includes custom-designed urban architecture and offers a visually exciting and pedestrian-friendly environment, which has been the primary catalyst for almost $1 billion of new residential and commercial development.

Traffic improvement. Before opening REL, the traffic on the existing four-lane divided toll facility was at level of service (LOS) F during the peak hours of operation. LOS is a grading scheme for representing the quality of traffic operations; it ranges from A (best) to F (worst). Of the total 115,000 average trips during a weekday, more than 75,000 occurred between I-75 and downtown Tampa on the east end of the highway. The trip time from the east averaged 30–40 min during the morning commute.

REL opened on a limited basis in mid-2006 and fully in January 2007. Since then, it has provided motorists a trip time of 10 min or less for their morning and afternoon commute into and out of Tampa. This has yielded a time savings of 20–30 min for each of the peak-hour directions, thus delivering a time savings of up to 1 h per day at a cost of $3.

Fig. 1. Elevated bridge on a 6-ft-wide (1.8-m) pier provides the equivalent of six lanes of peak commuting period capacity.

Travel time was not only substantially shortened, it became more reliable due to the safe conditions resulting from the express lane design and the elimination of vehicle conflicts caused by large trucks and numerous entrance and exit ramps. The reduced trip time also is responsible for enhanced public transit service from suburban Brandon to downtown Tampa. Within weeks of opening the REL, public transit ridership was up by over 40% on two express bus routes.

It is noteworthy that the REL is ahead of traffic forecasts. The forecast entries for the first year of operation was 12,500 vehicles per day. In February 2007, REL carried 15,960 vehicles.

Tolling system innovations. REL is the first transportation project in Florida to employ a totally cashless toll collection method known as "open road tolling," and it is the first implementation of free-flow tolling in a configuration wider than two lanes for the SUNPASS™ statewide electronic toll collection system (**Fig. 2**). In addition, video toll collection was added to ensure open access to all users, with or without a transponder.

The toll-by-plate program creates a unique video toll account (VTA) for occasional users who may call a special toll-free number in advance of using the REL, or up to 72 h after use, to register for a VTA. By providing their license plate number and a credit card, motorists may register for either a limited time use of the facility or for an on-going VTA, which only requires a minimum $5 balance in a prepaid account.

The toll system has been made more customer friendly by changing the overall philosophy of identifying violators. Under normal business practices for electronic tolls in the United States, violators are identified as those vehicles without a transponder or an ETC account. By providing multiple payment options, motorists can enter and pay later. A violation

Fig. 2. Electronic tolling structure (gantry) collects tolls at free traffic flow speeds.

is registered only when "failure-to-pay" occurs. Not only is this a more user-friendly approach to toll collection, it also has reduced mistaken violations and increased net revenues for the agency, while allowing the organization to focus its violation enforcement resources on those who intentionally and repeatedly refuse to pay tolls. *See* ELECTRONIC TOLL COLLECTION.

Construction and cost. The three-lane post-tensioned and steel-reinforced concrete bridge was constructed in 9-ft (2.7-m) segments (**Fig. 3**) at an off-site casting yard, delivered to the Crosstown Expressway, and then assembled in the median of the existing roadway, virtually eliminating any impacts to adjacent land uses, the surrounding community, or the environment.

Fig. 3. Match casting, where each new segment is cast against the previously completed one to ensure that all the pieces fit together properly. The curved shape of segments was one of the bridge's aesthetic features.

The construction started with the installation of cast-in-place piers in the median. Subsequently, a steel truss designed for REL was placed between the piers to temporarily support the segments while they were being assembled, allowing much of the work to be performed from above, therefore minimizing impacts to the traffic on the existing Crosstown Expressway lanes below. All segments were match-cast at the casting yard, so the on-site assembly was rapid, the resultant geometry flawless, and assembly was expedient.

Weighing about 70 tons each, the 59-ft-wide (18-m) segments were delivered to the expressway on 13-axle flatbed trucks designed for the project. The segments were then assembled during off-peak times. After the segments were lowered onto the truss, they were pulled together with post-tensioned steel cables inside the bridge.

Concrete segmental bridge construction is most efficient for longer structures and the efficiency increases as the length of the project increases. With more than 3000 segments, REL took advantage of the "cookie cutter" approach to bridge development. The total contract cost for the project was approximately $300 million in year 2004 terms. This includes all of the planning, design, right-of-way, construction, and construction management and inspection for the reversible express lanes and the two gateways. The cost also includes the electronic control and safety systems required to operate the lanes and the new three-story traffic management center.

The actual contract price for the 17.5 lane miles of bridge structure was just over $100 million. At approximately $120 million, the deck cost for the segmental bridge portion of the project was approximately $65 per square foot, far below the average cost for structures in Florida during the past 20 years. The average cost per lane mile for the reversible bridge is approximately $7 million and is among the lowest for bridges constructed in the United States.

The bulk of the construction funding for the REL was provided through a combination of revenue bonds and loans. One of the most interesting financing components was a unique loan from the State of Florida. In 1999, based on an endorsement from the Florida Transportation Commission which called the REL project ". . . a unique demonstration of innovative ideas, new technology and the beneficial impact of transportation on economic development and urban revitalization," Florida loaned the Expressway Authority $25 million in order to accelerate construction.

Worldwide applicability. Several of the concepts employed on REL have direct application to urban transportation needs worldwide. The concept of increasing the capacity of transportation corridors through innovative design and maximizing the use of existing public rights-of-way is directly applicable to traffic congestion problems in all urban areas (tolled or not). The tolling technology, payment, and enforcement programs are applicable to other express toll lanes, high-occupancy-toll (HOT) lanes, and open road tolling facilities everywhere.

[Comments, technical assistance, and photographs provided by FIGG Engineering Group are acknowledged.]

For background information *see* HIGHWAY ENGINEERING; PRECAST CONCRETE; TRANSPORTATION ENGINEERING; HIGHWAY BRIDGE in the McGraw-Hill Encyclopedia of Science & Technology.

Panos D. Prevedouros; Martin Stone

Bibliography. L. Figg and W. D. Pate, Precast concrete segmental bridges—America's beautiful and affordable icons, *PCI J.*, July-August, 2004; C. S. Papacostas and P. D. Prevedouros, *Transportation Engineering and Planning*, 3d ed., Prentice Hall, 2001; J. M. Rodriguez, Taking the high road, *Civil Engineering*, ASCE, July 2004; M. Stone and E. Suarez, Reversible elevated open road toll lanes in Tampa, *Tollways*, International Bridge Tunnel and Turnpike Association, Washington, D.C., Winter, 2007.

Sea urchin: genomic insights

The sea urchin, despite its unusual body plan, is a close relative of the vertebrates. It belongs to the phylum Echinodermata, which also includes sea cucumbers and sea stars. Echinoderms together with their sister group Hemichordata (wormlike creatures) and the Chordata (vertebrates and close relatives) form the deuterostome clade (see **Fig. 1**). Initially, sea urchins were characterized as deuterostomes based on developmental characteristics, such as gut formation—a classification that is just over 100 years old. [During embryonic development, animals in which the blastopore (the first opening of the early digestive tract) becomes the mouth are called protostomes; those in which the mouth develops after the anus are called deuterostomes.] The sea urchin (see **Fig. 2**) is the first representative of the echinoderms to have its genome sequenced, which may shine light on what are the common molecular characteristics of deuterostomes. This will further allow conclusions to be drawn about the origin of features that are assumed to be specific to chordates or vertebrates, and will reveal inventions in the echinoderm lineage that indicate the unique characteristics of the sea urchin.

An initial survey of the genes likely to function in the sea urchin immune system reveals striking differences in comparison to the genes involved in vertebrate immunity and shows a dramatic increase in "passive," or innate, immunity. Nevertheless, homologs of key genes involved in the vertebrate adaptive immune system are contained in the sea urchin genome, suggesting that the foundation on which adaptive immunity evolved was present in the last common ancestor. Similarly, the repertoire of sensory receptors indicates a common origin for those of echinoderms and vertebrates, with different expansions occurring in each lineage. On the other hand, the genes coding for proteins that form the matrix of the sea urchin skeleton are seemingly unrelated to those of vertebrates, arguing for an independent origin of calcified structures. However, the way that skeletogenic cells are specified in deuterostomes suggests that these structures may still be closely related. As these examples demonstrate, most features of extant deuterostomes have a common, clearly recognizable genetic basis in the last common ancestor. As such, most independent developments may be viewed as variations on a common theme, underlining the close relationship of all deuterostomes.

Sea urchin immunity. The immune system defends an organism against intrusion of pathogens, such as harmful bacteria, viruses, or parasites. The vertebrate immune system, arguably the best studied example, is generally divided into two parts: innate and adaptive immunity. Innate immunity refers to defense mechanisms that are part of the biological makeup of each organism and available at birth. Much more complex is the adaptive immune system, which each individual builds through ongoing immune challenges. The cornerstone of acquired immunity are cells that have specific affinity for antigens: T cells (thymus-derived lymphocytes) and antibody-producing B cells (bone-marrow-derived lymphocytes). Exposure to pathogens activates these cells and leads to their proliferation. Following immune challenge, they are maintained at

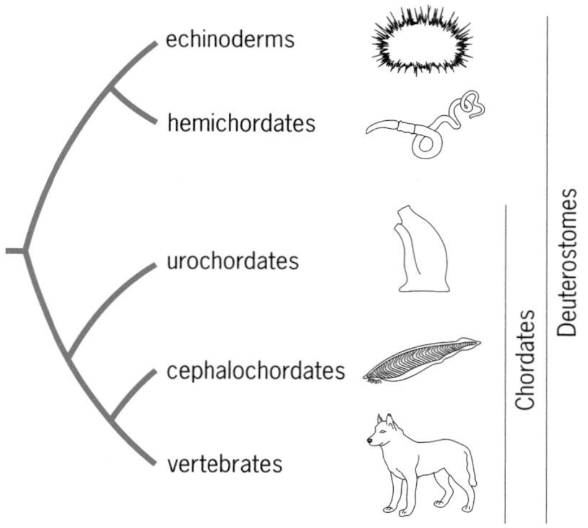

Fig. 1. Simplified phylogenetic tree of the deuterostomes.

Fig. 2. Purple sea urchin (*Strongylocentrotus purpuratus*), the subject of a recent gene sequencing project. (*Photo courtesy of Laura Francis*)

low levels and reexposure elicits a fast and specific response.

The immune systems of invertebrates have generally been regarded as fairly primitive because nothing resembling acquired immunity is known. However, a lack of acquired immunity does not imply a lack of intricate immune functions, as examination of the sea urchin genome, and evidence from other organisms, indicates. A striking finding is the vastly expanded set of genes involved in innate immunity in the sea urchin genome. These include the Toll-like and scavenger receptors, both expressed in immune cells of vertebrates to recognize common features of pathogens, like surface lipoproteins. Only a few genes of each receptor type are commonly found in vertebrate genomes, but the sea urchin genome contains more than 200 of each. Together they account for about 2% of all identified genes—a significant fraction. The high number of these receptors is accompanied by a moderate expansion of downstream adaptor proteins, which connect the receptors to the cellular machinery that elicits a response. This may indicate a possible diversification of these responses. Thus, the absence of a vertebratelike acquired immune system may be offset by a more elaborate innate immune system. This seems to be an emerging theme from invertebrates. In flies, for example, immune diversity is generated by alternative splicing of a single innate immune receptor, creating thousands of different isoforms.

Perhaps the most unexpected finding in the sea urchin genome is the identification of homologs of genes crucial in establishing the huge diversity of the adaptive immune system in vertebrates. T cells and B cells are each equipped with one single type of receptor or antibody. The gene coding for this receptor/antibody is custom made for each cell and is assembled by recombining DNA fragments from a pool of alternative pieces into one gene. Two recombination enzymes facilitate this cut-and-paste process, which is assumed to be specific to vertebrates. Other genes that are involved in this recombination process, mainly DNA repair enzymes, are also present in the sea urchin genome. Although the sea urchin genome contains genes that, like antibodies, fall into the immunoglobulin class, no genetic locus has been identified that would facilitate a similar recombination process. Thus, although clear homologs of most key factors are present, they may have functions other than establishing diversity of an adaptive immune system.

These findings indicate that the last common ancestor of all deuterostomes must have employed Toll-like receptors and immunoglobulins in its immune system. Its genome already contained recombination enzymes that were not yet tied into immunity. From there, sea urchins and vertebrates took different paths: the first expanded its innate immune repertoire; the latter achieved diversity through acquisition of a recombinant system.

Sea urchin sensory receptors. The sea urchin is not famous for an outstanding ability to perceive and react to external stimuli, mostly due to the lack of obvious sensory organs and its rather passive lifestyle. Over the last few decades, however, responses of sea urchins to various light and chemical stimuli have been described in detail, although how these are sensed at the molecular level is still unknown. The sequencing of the sea urchin genome now allows us to identify its set of sensory receptors.

Sensory receptors fall into a class of so-called G-protein–coupled receptors. These are integral membrane proteins that are associated with guanosine triphosphate (GTP)–binding proteins, which transduce the signal by activating further proteins inside the cell. Not all G-protein–coupled receptors are sensory receptors, but all are employed in transducing signals across membranes. More than 900 genes coding for such receptors have been identified, about half of which show a high similarity to vertebrate taste and smell receptors. Thus, sensory receptors are one of the largest gene families in the sea urchin genome. For a few receptors, expression has been observed in the tube feet and the pedicellariae (pincerlike structures scattered over the surface of echinoderms). Although much remains to be discovered, it seems that the sea urchin senses its environment with sensory cells covering its entire surface, and accomplishes this much better than had ever been suspected.

The sensory receptors are clustered in the sea urchin genome, as rapidly expanding gene families commonly are. They also show a high degree of self similarity, which indicates their close relationship and (on a genomic time scale) their recent origin. This and the fact that vertebrate receptors also evolve rapidly make it almost impossible to find pairs of homologous genes. Thus, although it is clear that sensory receptors in both lineages are closely related to each other—and more closely related than either group is to sensory receptors of protostomes (for example, arthropods, mollusks, and annelids)—evolution took a different turn in each lineage.

Sea urchin skeletogenesis. The sea urchin possesses a skeleton that surrounds the body cavity and supports the spines. The skeletons of both vertebrates and echinoderms largely consist of calcium carbonate, which suggests they have a common origin. However, the absence of skeletons in hemichordates and other deuterostome phyla, as well as the absence of any fossils showing skeletons in early deuterostomes, casts doubt on this scenario.

In skeletogenic tissues such as bone, the calcium carbonate is embedded in a matrix of proteins. These proteins include the collagens, which give bone and cartilage tensile strength. Collagens are present in the skeletal structures of echinoderms and vertebrates, although vertebrates have a larger and slightly divergent set. This expansion has been assumed to underlie the diversification of bone and cartilaginous tissue, of which vertebrates produce several different kinds. Other matrix proteins have a high affinity to the calcified mineral, the noncompressive part of bones. Between echinoderms and vertebrates, the mineral-binding proteins differ considerably, supporting the hypothesis that the two calcified structures evolved independently.

Despite this significant difference in protein composition, circumstantial evidence for a common origin may be derived from how these tissues are formed. In both taxa, homologous regulatory genes control the specification of skeletogenic cells. Although it is uncertain whether the last common ancestor had calcified support structures, this program could have been used to specify cartilagelike tissues. Current specification programs may be derivatives of this ancient progenitor. Thus, the different deuterostome lineages were equipped with a common "genetic toolkit" which they could expand upon to produce biomineralized structures.

Outlook. As the above examples illustrate, the sea urchin, despite its derived morphology, shows a number of surprising similarities to vertebrates and, to this end, humans. Identification of most similarities is based on comparisons of gene sets and sequence composition. As of today, only a few instances of common regulatory mechanisms, as in the specification of skeletogenic cells, are known. However, comparing such processes will help to reveal how changes in sequence lead to altered biological function and new body plans. Much remains to be discovered, but knowledge of the sea urchin genome is already helping to sketch the contours of what constitutes a deuterostome.

For background information *see* ANIMAL EVOLUTION; CHORDATA; ECHINODERMATA; ECHINOIDEA; GENETIC MAPPING; GENOMICS; HEMICHORDATA; IMMUNITY; INVERTEBRATE EMBRYOLOGY; SEA URCHIN; SENSE ORGAN; SKELETAL SYSTEM in the McGraw-Hill Encyclopedia of Science & Technology.

Stefan C. Materna

Bibliography. M. D. Cooper and M. N. Alder, The evolution of adaptive immunity, *Cell*, 124:815–822, 2006; T. Hibino et al., The immune gene repertoire encoded in the purple sea urchin genome, *Dev. Biol.*, 300:349–365, 2006; Sea Urchin Sequencing Consortium, The genome of the sea urchin *Strongylocentrotus purpuratus*, *Science*, 314:941–952, 2006.

Shape-memory polymers

Shape-memory polymers are an emerging class of actively moving polymers. They are able to respond to a specific stimulus by changing their shape. Shape-memory polymers are different from shape-changing materials. Both polymer concepts are based on elastic polymer networks equipped with molecular switches or stimuli-sensitive domains. The triggering of these switches results in the stimulated movement of the polymers. Shape-changing polymers are able to change their shape as long as they are exposed to a certain stimulus. Once the stimulus is terminated, they recover their original shape. This process of stimulated deformation and recovery can be repeated several times, while the geometry of shape changes is invariant from any alteration and is usually somehow related to the original shape. This is in contrast to shape-memory polymers whose temporary shape can be varied in each cycle. Most known shape-memory polymers are dual-shape materials. They are able to change from a shape A into shape B. While shape B is a permanent shape obtained during the polymer processing, shape A is a temporary shape. The temporary shape is created in a process called programming, in which the sample is mechanically deformed and subsequently fixed. The deformation leading to the temporary shape defines the change of shape. When exposed to an external stimulus, shape-memory polymers recover their permanent shape. This process of programming and recovery can be repeated several times with different temporary shapes in subsequent cycles.

Polymer architecture. The shape-memory effect is not an intrinsic material property and does not require specific repeating units. It results from a combination of the programming process as described above combined with a suitable polymer architecture and morphology. This allows the adjustment of intrinsic material properties (for example, mechanical properties) by varying molecular parameters such as the type or ratio of comonomers. A suitable polymer architecture enabling the shape-memory effect consists of netpoints and molecular switches that are sensitive to an external stimulus. Sensitivity to heat or light has been achieved for the shape-memory polymers reported so far. Indirect actuation of the thermally induced shape-memory effect has been realized by irradiation with infrared radiation, application of electrical current, exposure to alternating magnetic fields, or immersion in water.

In such a polymer network, netpoints are crosslinking chain segments that determine the permanent shape (**Fig. 1**). The netpoints can be of a chemical (covalent bonds) or physical

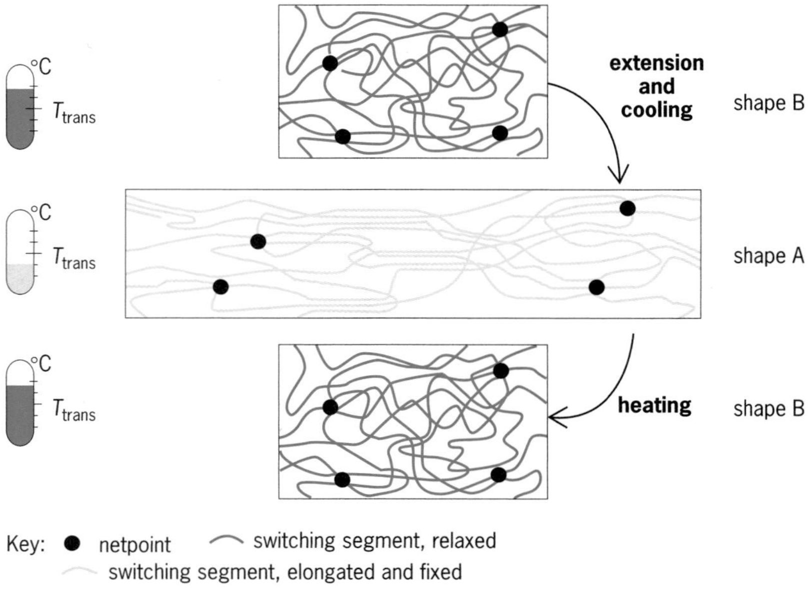

Fig. 1. Molecular mechanism of the thermally induced shape-memory effect of a polymer network T_{trans} is the thermal transition temperature related to the switching phase. ***(Adapted from A. Lendlein and S. Kelch, Shape-memory polymers, Angew. Chem. Int. Ed., 41:2034–2057, 2002)***

(intermolecular interactions) nature. Physical crosslinking is realized in polymers whose morphology consists of two segregated domains. Here the domains with the highest thermal transition T_{perm} act as netpoints, while the chain segments associated with the domains having the second highest thermal transition T_{trans} act as molecular switches (switching segments).

Molecular switches have the function of fixing the deformed shape temporarily under the conditions relevant for the particular application. This is realized by introducing additional reversible crosslinks. Like the permanent netpoints, these additional crosslinks can be formed by physical interactions or covalent bonds.

Physical crosslinking is realized by crystallization or vitrification of the domains related to T_{trans}, resulting in shape-memory polymers displaying a melting transition (T_m) or a glass transition (T_g). These switching domains are formed either by chain segments linked to two netpoints that contribute to the elastic behavior or by side chains that are only connected to one netpoint. In both cases, aggregation prevents the recoiling of the chain segments.

Functional groups that form covalent bonds that are reversibly controlled by an external stimulus act as molecular switches. The introduction of photoreversible functional groups extends shape-memory technology to using light as a stimulus.

Programming and recovery. Shape-memory properties are quantified in cyclic stimuli-specific mechanical tests. Each cycle consists of programming the test specimen and recovering its permanent shape. Different test protocols have been developed for programming and recovery in which the elongation ε is measured while the tensile stress σ is held constant or vice versa. From these measureands, the shape-memory effect is quantified by determining the shape fixity ratio (R_f) for the programming and the shape recovery ratio (R_r) for the recovery process. Quantification of the thermally induced shape-memory effect includes the determination of the switching temperature T_{switch}.

In the programming procedure, the test specimen is elongated from ε_0 to its maximum strain ε_m, while the temporary crosslinks are disabled. The stress is maintained for some time to allow the relaxation of the polymer chain, and then the temporary crosslinks are formed, fixing the temporary shape. The programming of this temporary shape can be stress or strain controlled. In strain-controlled programming, the sample is held at constant ε_m during formation of the temporary crosslinks, and afterward the sample holder (tensile tester) is driven back to its original position, while the stress-free condition is passed at the $\varepsilon_u(N)$ position. In stress-controlled programming, the sample stress is held constant at σ_m (maximum stress) after reaching ε_m, while the temporary crosslinks are fixed. Afterward, the stress is reduced until a stress-free condition is obtained. Here, the fixed deformation can be larger than ε_m. In thermally induced shape-memory polymers, the programming procedure can be additionally varied between cold stretching of the sample ($T < T_{trans}$) or a temporary heating ($T > T_{trans}$) and subsequent cooling.

After programming, the recovery is continued in stress- or strain-controlled mode. In the stress-controlled experiment, σ of the sample is held at 0 MPa during deactivation of the temporary crosslinks while the sample recovers its original shape. Here ε is measured. In the strain-controlled experiment, ε is held constant while the stimulus is applied. Here, the stress during recovery is measured. Cyclic programming and recovery allow the determination of R_f and R_r. The ability to fix the temporary deformation during the programming process is quantified by $R_f(N)$. It is the ratio of the strain in the stress-free state of the Nth cycle $\varepsilon_u(N)$ and the maximum strain ε_m [Eq. (1)]. R_r quantifies the ability of a material

$$R_f(N) = \frac{\varepsilon_u(N)}{\varepsilon_m} \qquad (1)$$

to memorize its permanent shape. It compares the change in strain of a cycle after the shape-memory effect has passed $\varepsilon_m - \varepsilon_p(N)$ with the change in strain in the course of programming of the Nth cycle given by $\varepsilon_m - \varepsilon_p(N-1)$ [Eq. (2)].

$$R_r(N) = \frac{\varepsilon_m - \varepsilon_p(N)}{\varepsilon_m - \varepsilon_p(N-1)} \qquad (2)$$

Recently, triple-shape materials were introduced, enabling two shape changes and having two transition temperatures $T_{trans,A}$ and $T_{trans,B}$. For programming, such a polymer network is heated to T_{high}, where the material is in the elastic state and is deformed. Cooling under external stress to T_{mid} ($T_{trans,A} < T_{mid} < T_{trans,B}$) results in formation of physical crosslinks related to $T_{trans,B}$. When the external stress is released, shape B is obvious. Subsequent deformation of this shape at T_{mid} and cooling to T_{low} under external stress creates shape A. Reheating to T_{mid} recovers shape B and subsequent heating shape C.

Potential applications. Thermally induced shape-memory polymers have been developed from various polymeric building blocks. A trend is the development of biodegradable shape-memory polymers for biomedical applications. By incorporating additional building blocks, the mechanical parameters of

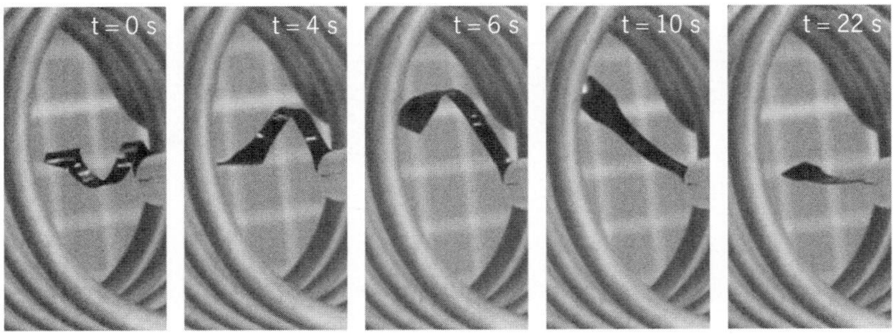

Fig. 2. Magnetically induced shape-memory effect of a thermoplastic shape-memory composite from nanoparticles consisting of iron(III) oxide in a silica matrix and a poly(etherurethane) exposed to an alternating magnetic field (f = 258 kHz, H = 30 kA·m^{-1}). *(From R. Mohr et al., Initiation of shape-memory effect by inductive heating of magnetic nanoparticles in thermoplastic polymers, PNAS, 103(10):3540–3545, 2006)*

the permanent and the temporary shape can be adjusted.

Besides optimizing shape-memory polymer architectures, recent shape-memory research has focused on making them sensitive to stimuli other than heat. This is achieved by the indirect thermal actuation of the shape-memory effect or the introduction of molecular switches that are sensitive to other stimuli.

Indirect actuation of the thermally induced shape-memory effect can be realized by incorporating magnetic nanoparticles into shape-memory polymers. These composites are programmed like conventional thermally induced shape-memory polymers, but instead of increasing the environmental temperature, the sample temperature is increased by inductive heating of the incorporated particles in an alternating magnetic field, resulting in the recovery of its permanent shape. Alternatively, the inductive heating of the particles can be used for the programming of the temporary shape. In **Fig. 2**, the uncoiling of a corkscrewlike spiral composite of an aliphatic poly(etherurethane) with 10 wt% silica-coated iron(III)oxide particles is shown as an example of the magnetically induced, indirect actuation of the shape-memory effect. R_r of indirectly heated shape-memory polymers is comparable to samples that were actuated by increasing the environmental temperature. Potential applications of magnetic-field driven shape-memory composites are seen where adjustments in a noncontact mode are required (for example, in medical devices or smart implants). Here, biodegradable shape-memory polymer matrixes are favored.

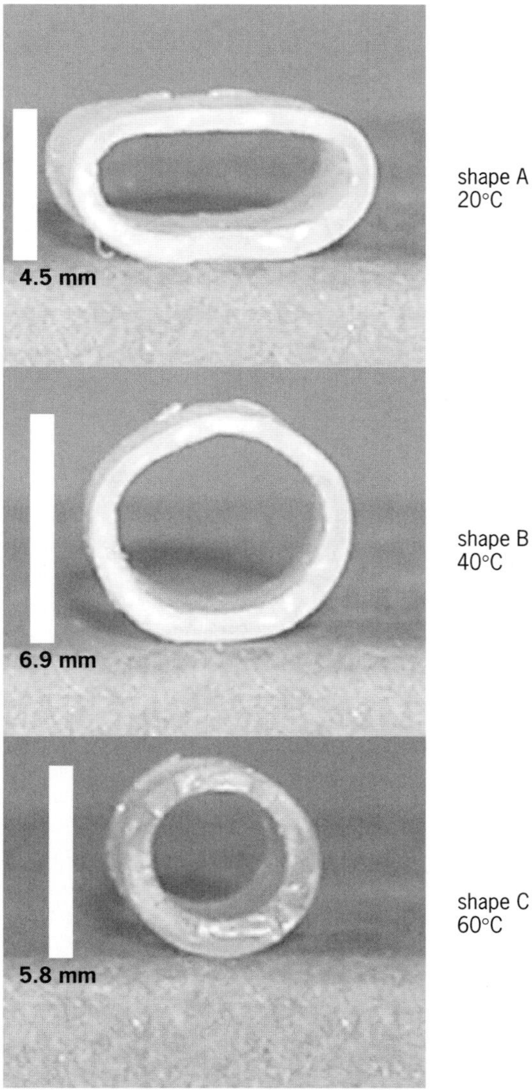

Fig. 3. Series of photographs illustrating the triple-shape effect of an intelligent tube. The triple shape material is a polymer network containing poly(ε-caprolactone) segments and poly(ethylene glycol) side chains. The series shows the recovery of shapes B and C by subsequent heating to 40 and 60°C, beginning from shape A, which was obtained as a result of the two-step programming process. (*From I. Bellin et al., Polymeric triple-shape materials, PNAS, 103(48):18043–18047, 2006*)

Incorporating photosensitive molecular switches, such as cinnamic acid (CA) or cinnamyliden acetic acid (CAA), has enabled the development of light-induced shape-memory polymers. Here, the programming and the recovery of the shape-memory effect is independent of any temperature effect. Instead of increasing the environmental temperature, light of suitable wavelengths is used for programming and recovery. Upon irradiation with light of suitable wavelengths, the photosensitive functional groups undergo a [2 + 2] cycloaddition reaction with each other, resulting in covalent crosslinks. When irradiated with light of different suitable wavelengths, these crosslinks are cleaved again. In the programming cycle, the polymer is stretched and then irradiated with ultraviolet light of $\lambda > 260$ nm, creating new covalent bonds and resulting in the fixation of the strained polymer chain segments in their uncoiled conformation. Upon irradiation with light having wavelengths $\lambda < 260$ nm, these newly formed covalent bonds are cleaved and the permanent shape is recovered. Photosensitive groups have been incorporated in two alternative polymer network structures: a graft and an interpenetrating polymer network. In both cases, the permanent shape is determined by the netpoints that are crosslinking the chain segments of the amorphous polymer networks.

Recently, triple-shape materials were introduced that are able to perform two shape changes after two subsequent temperature increases. Triple-shape polymers are polymer networks with at least two pronounced phase-separated domains that act as physical crosslinks. These domains are related to individual transition temperatures T_{trans}, which can be either a glass transition or a melting temperature, depending on the type of chain segment. Temporary shapes A and B are fixed by physical crosslinks built during a two-step programming procedure, while shape C is determined by the covalent crosslinks during network formation. In this process the highest thermal transition determines shape B and the second highest transition is associated to shape A. The triple-shape effect of an intelligent tube is visualized in **Fig. 3** to demonstrate that the two shape changes do not have to necessarily be unidirectional. One potential application for such materials is for stents, which are inserted into the human body in compressed shape A. When the device is placed at the desired position, it could be expanded into shape B, and, if required, removal can be facilitated by the realization of shape C.

Outlook. Shape-memory polymer research is progressing rapidly. While the fundamental research is focusing on implementing stimuli other than heat to actuate shape-memory polymers, or to actuate them remotely, the technology platform of available shape-memory polymers is progressing from laboratory demonstration objects to highly sophisticated applications, including fabrics, intelligent packaging, self-repairing automobile bodies, switches and sensors, and medical devices. The application requirements can be very complex in these areas, so there is a strong demand for actively moving materials that are able to perform complex movements. These requirements could be fulfilled by materials that perform more than one predetermined shift.

For background information *see* COPOLYMER; GLASS TRANSITION; MACROMOLECULAR ENGINEERING; MATERIALS SCIENCE AND ENGINEERING; POLYMER; STRESS AND STRAIN; SUPRAMOLECULAR CHEMISTRY in the McGraw-Hill Encyclopedia of Science & Technology. Marc Behl; Andreas Lendlein

Bibliography. M. Behl and A. Lendlein, Actively moving polymers, *Soft Matter*, 3:58–67, 2007; M. Behl and A. Lendlein, Shape-memory polymers, *Mater. Today*, 10:20–28, 2007; M. Behl, R. Langer, and A. Lendlein, Intelligent materials: Shape-memory polymers, in M. Shahinpoor and H. J. Schneider (eds.), *Intelligent Materials*, RSC London, United Kingdom, 2007; I. Bellin et al., Polymeric triple-shape materials, *PNAS*, 103:18043–18047, 2006; N. Y. Choi, S. Kelch, and A. Lendlein, Synthesis, shape-memory functionality and hydrolytical degradation studies on polymer networks from poly(rac-lactide)-b-poly(rac-lactide) dimethacrylates, *Advanced Eng. Mat.*, 8(5):439–445, 2006; A. Lendlein and S. Kelch, Shape-memory polymers, *Angew. Chem. Int. Ed.*, 41:2034–2057, 2002; A. Lendlein and S. Kelch, Shape-memory polymers as stimuli-sensitive implant materials, *Clin. Hemorheol. Microcirc.*, 32(2):105–116, 2005; A. Lendlein et al., Light-induced shape-memory polymers, *Nature*, 434:879–882, 2005; R. Mohr et al., Initiation of shape-memory effect by inductive heating of magnetic nanoparticles in thermoplastic polymers, *PNAS*, 103(10):3540–3545, 2006.

Sloan Digital Sky Survey

The Sloan Digital Sky Survey (SDSS) is an astronomical imaging and redshift survey, begun in 2000 and continuing through 2008, designed to map a large volume of the universe. Discoveries in the SDSS have contributed to the understanding of the solar system, the Milky Way Galaxy and other galaxies, and the early universe.

Planning for the SDSS began in the 1980s, when astronomers realized that the efficiency of charge-coupled devices (CCDs) relative to photographic plates would allow a much deeper wide-area map of the sky than previously possible. The survey is a collaboration among 25 universities and research institutions, including around 200 astronomers. The telescope structure was installed at Apache Point Observatory in New Mexico in the fall of 1995 and had first light in May 1998. The SDSS began in earnest in 2000 and will continue observing until July 2008. Much of its data are released to other astronomers and the public in general through its Web site.

Telescope, camera, and spectrograph. The SDSS telescope (**Fig. 1**) has a 2.5-m (98-in.) primary mirror, a modest size by modern standards. However, its optical design (called Ritchey-Chrétien) delivers a remarkably undistorted image over a large angular patch of sky (3°, or six times the diameter of the full moon). Depending on the behavior of the atmosphere at the telescope, the effective resolution of SDSS images is usually about 1.4 arcseconds.

The imaging camera was the widest field-of-view CCD instrument used in astronomy when it was built, consisting of thirty 2048 × 2048-pixel devices. It operates in an extremely efficient mode known as drift scanning, which allows it to image a large swath of sky in one continuous exposure (of 8 or more hours duration) rather than in many individual exposures. This mode allows almost all of the available observing time to be spent with the camera shutter open. The camera takes nearly simultaneous images in five wavelength bands called u, g, r, i, and z (with wavelength centers from 350 to 900 nm), thus measuring the colors of each object it detects. With this camera, the SDSS has imaged about 10,000 square degrees, or about one-quarter of the entire sky, and detected 300 million stars and galaxies in this area.

Fig. 1. Sloan Digital Sky Survey (SDSS) 2.5-m (98-in.) telescope at Apache Point Observatory, New Mexico. The secondary mirror is visible in the center of the aperture. The square structure surrounding the primary and secondary mirrors shelters them from scattered light and wind. (*Photo courtesy of Fermilab; Sloan Digital Sky Survey*)

In addition to the imager, the telescope also uses a multiobject spectrograph to study individual stars, galaxies, and quasars in detail. Fiber optics simultaneously captures the light from each of 640 objects in the field of view of the telescope and redirects it to a set of spectrographs that measures their optical spectra at a moderate resolution, about 0.3 nm. The survey has taken over 1 million spectra in this way, primarily of galaxies, but also of many stars in our own galaxy and very distant quasars.

The most important use of the spectra is to infer the velocities of objects from their Doppler shifts. The Doppler shift of the light from a receding galaxy causes features in the galaxy spectrum to appear at longer wavelengths than they would normally be observed (**Fig. 2**). Furthermore, Hubble's law states that the speed of recession v of a galaxy is related to its distance D by the equation

$$v = H_0 D$$

where the most recent estimates indicate that H_0 equals 22 kilometers per second per million light-years (72 kilometers per second per megaparsec). Thus, a spectroscopic survey like the SDSS can then be used to make a three-dimensional map of the universe (**Fig. 3**). The spectra can also be used to classify the type of galaxy and infer some physical properties (such as age, mass, and composition).

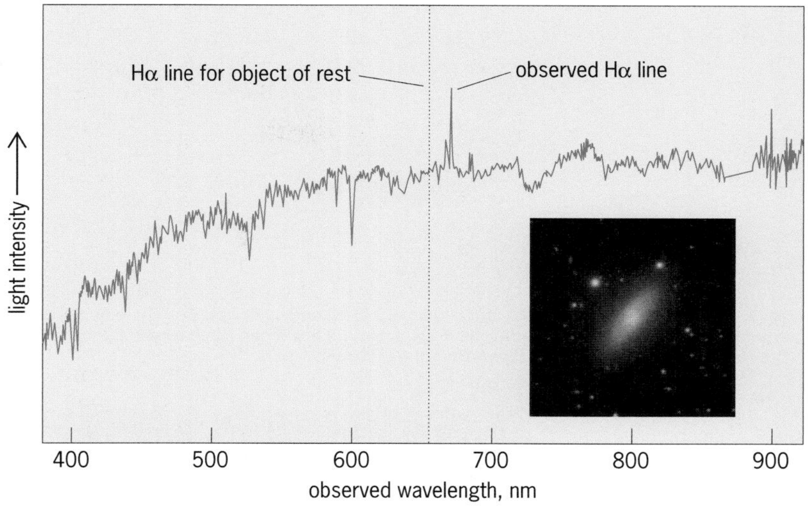

Fig. 2. Spectrum of a galaxy (image inset) receding at around 6000 km s^{-1} (3700 mi s^{-1}), or about 2% of the speed of light. The vertical line shows the wavelength at which one expects to observe the emission line Hα for an object at rest (656.3 nm). The actual emission line for this galaxy is observed at around 670 nm, a 2% longer wavelength. By Hubble's law, this galaxy is 270 million light-years away. The other emission and absorption features in this spectrum can be used to deduce the typical ages and total mass of the stars in this galaxy. (*Courtesy of Sloan Digital Sky Survey*)

Discoveries and measurements. The primary motivation of the SDSS was to measure the large-scale distribution of galaxies in the universe, but because it was the first large, digital imaging survey, it has made a number of other contributions to astronomy as well. The major contributions of the SDSS have included the following.

Solar system objects. The five bands in the SDSS are not imaged perfectly simultaneously, but a few minutes apart. Objects in the solar system move by a few arcseconds in this interval, and therefore can be recognized by their motion. In this manner, more than 60,000 asteroids have been discovered. By studying the population of asteroid orbits and their colors, the SDSS has confirmed the existence of asteroid "families"; the members of each probably have a common origin.

Milky Way Galaxy and its neighbors. With the five-band SDSS imaging survey, astronomers are exploring the structure of the Milky Way Galaxy by mapping the distribution of stars. With an optical survey like the SDSS, the thin disk of stars and central bulge of our Galaxy are totally obscured by dust; on the other hand, the SDSS has revolutionized our understanding of the outer stellar halo. **Figure 4** shows a map of the stellar distribution produced by the SDSS, showing some of the prominent structures that have been discovered. Most apparent is the Sagittarius stream, associated with a disrupting dwarf galaxy. While the Sagittarius stream extends across over 70° of the sky, more compact objects have also been found. Beginning with Willman I and the Ursa Major dwarf spheroidal galaxy in 2005, and followed up the following year with a host of other objects, the population of known Milky Way satellite galaxies has approximately doubled due to the SDSS. **Figure 5** shows an example of one of these galaxies, Canes Venatici I.

Star formation, stellar mass, and dark matter in galaxies. The high-quality imaging and spectroscopy of the SDSS has led to a more precise catalog of the stellar masses and rates of star formation of galaxies as a function of galaxy type (**Fig. 6**). The majority of galaxies with high stellar mass are "red-and-dead" elliptical galaxies that ended their star formation long ago. The majority of low-mass galaxies are spiral galaxies currently forming stars. This general pattern had been known

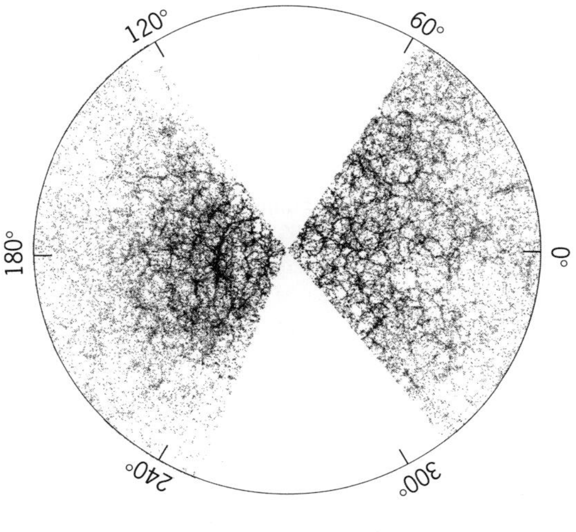

Fig. 3. Distribution of galaxies within 4° of the celestial equator from the SDSS. In this polar plot, the radius indicates the distance from the Milky Way, and the angle is the right ascension. The diameter shown is 8 × 10^9 light-years. Blank areas have not been surveyed by the SDSS. (*Courtesy of Sloan Digital Sky Survey*)

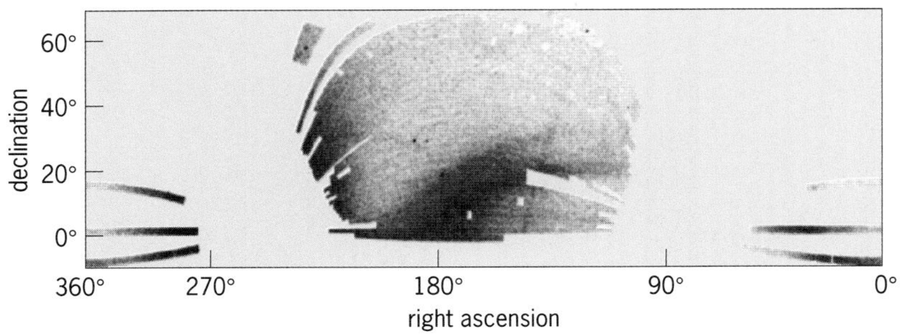

Fig. 4. Map of stars in the SDSS. Grayscale signifies density of stars on the sky. Several streams of stars are noticeable. Areas in background are those unobserved by the survey. (*Courtesy of Vasily Belokurov; Sloan Digital Sky Survey*)

Fig. 5. Image of Canes Venatici I, a recently discovered satellite of the Milky Way Galaxy. (*Photo courtesy of Vasily Belokurov; Sloan Digital Sky Survey*)

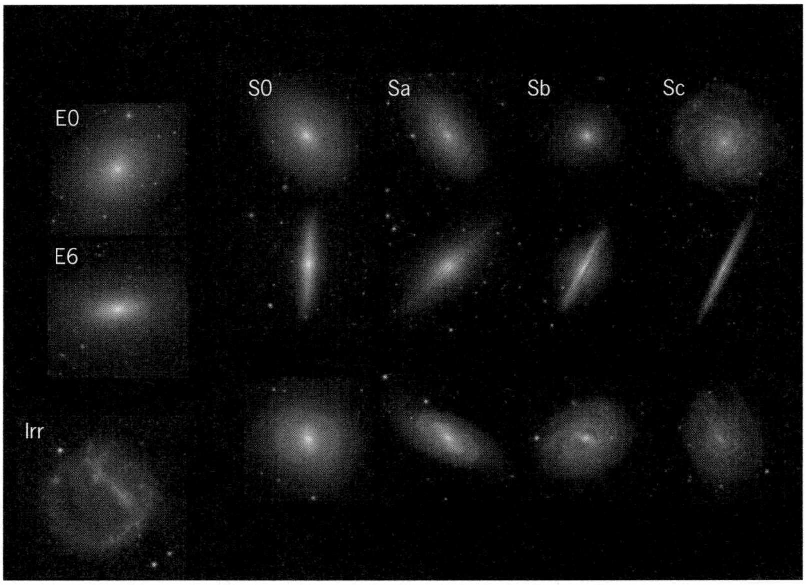

Fig. 6. Hubble sequence of galaxies, as observed by the SDSS. Elliptical galaxies (types marked E0 and E6) are old, spheroidal, and very massive. Spiral galaxies are on average less massive. While some spiral galaxies (S0 and Sa) tend to have very little current star formation, most (Sb and Sc) tend to have stars forming in their outer disks. (For the spiral galaxies, top row shows plan views, middle row shows side views, and bottom row shows plan views of barred examples.) Irregular galaxies (Irr) tend to be asymmetric, have very strong star formation, and be low in mass. (*Photo courtesy of Sloan Digital Sky Survey*)

before the SDSS, but the SDSS and other modern surveys have quantified it far more precisely. The SDSS has also been able to measure the average total mass content of galaxies using a technique known as weak lensing. Analyzing the distortions that the mass in foreground galaxies induces on the apparent shapes of background galaxies, one can infer the average amount of mass in each foreground galaxy. In agreement with other experiments, the SDSS finds that galaxies have far more total mass than can be explained by the ordinary matter detected in them, indicating the presence of "dark matter." In this manner, the SDSS has yielded a more precise, better quantified, and more complete census of the contents of galaxies than previously available.

Large-scale structure. The big bang theory of cosmology predicts what the large-scale, three-dimensional distribution of galaxies looks like. For example, theory predicts that around every galaxy, there should be a very diffuse shell (450 million light-years in radius) of excess galaxies. These very large shells are the remnants of sound waves that traveled during early epochs, expanding around dense locations, each of which now has a galaxy. The SDSS has mapped a large enough volume to detect these shells, known as the baryon acoustic oscillations, and is using them as a very robust tool for measuring size scales in the universe. Using these sorts of measurements of the large-scale structure in the universe, the SDSS has helped to determine the values of cosmological parameters, such as Hubble's constant, H_0, and the average density of mass in the universe.

Extremely distant objects. The SDSS has discovered the most distant quasars yet observed. These quasars are extremely luminous and are almost certainly associated with black holes millions of times the mass of the Sun, located at the centers of galaxies. The distances implied by their Doppler shifts indicate that the observed light was emitted when the universe was less than 10^9 years old. In these spectra, astronomers have found signs of the Gunn-Peterson effect, which is the absorption of light by neutral hydrogen gas. This discovery reveals that at around 10^9 years of age the universe had not quite yet had enough activity (from young stars and quasars) to ionize all of the hydrogen. Thus, the observed SDSS quasars are among the first objects to form in the early stages of the universe.

For background information *see* ASTEROID; ASTRONOMICAL IMAGING; ASTRONOMICAL SPECTROSCOPY; BIG BANG THEORY; COSMOLOGY; DARK MATTER; DOPPLER EFFECT; GALAXY, EXTERNAL; GRAVITATIONAL LENS; HUBBLE CONSTANT; LOCAL GROUP; MILKY WAY GALAXY; QUASAR; REDSHIFT; TELESCOPE; UNIVERSE in the McGraw-Hill Encyclopedia of Science & Technology.

Michael Blanton

Bibliography. A. Dekel and J. P. Ostriker (eds.), *Formation of Structure in the Universe*, Cambridge University Press, 1999; W. L. Freedman (ed.), *Measuring and Modeling the Universe*, Carnegie Observatories Astrophysics Series, vol. 2, Cambridge University Press, 2004; F. Levin, *Calibrating the Cosmos: How Cosmology Explains Our Big Bang*

Universe, Springer-Verlag, New York, Astronomers' Universe Series, 2006; L. S. Sparke and J. S. Gallagher, III, *Galaxies in the Universe: An Introduction*, 2d ed., Cambridge University Press, 2007.

Slow earthquakes

Slow earthquakes occur by the same mechanism as ordinary earthquakes, as both are caused by shear slip on faults. However, slow earthquakes take a long time to occur relative to ordinary earthquakes. The slower a fault slips, the less efficiently it generates seismic waves. Thus, the distinguishing characteristic of slow earthquakes is that the waves they generate are weak, particularly at high frequencies, compared to waves from ordinary earthquakes. Highly sensitive seismic and geodetic earthquake monitoring networks installed over the past decade have led to a rapidly expanding collection of newly discovered, unusual seismic phenomena, including silent earthquakes, low-frequency earthquakes (LFEs), very low frequency earthquakes (VLFs), and deep nonvolcanic tremor.

Silent earthquakes. Silent earthquakes are extreme examples of slow earthquakes, in which a fault slips so slowly that no seismic waves at all are generated. Because they generate no waves, silent earthquakes do not register on seismographs and are difficult to detect. They can be detected using geodetic measurements of ground deformation. Silent earthquakes have been observed along the San Andreas Fault using borehole strain meters, in Japan and Cascadia (Pacific Northwest) using permanent Global Positioning System (GPS) networks, and in Japan using borehole accelerometers. Long-term aseismic transients, lasting years, have been documented in several areas of Japan including the Bungo Channel and the Tokai region near Tokyo. Recently reported silent earthquakes range as large as magnitude 7.5.

Low-frequency earthquakes. Low-frequency earthquakes (LFEs) were discovered in southwest Japan and are small earthquakes of magnitude 2 or less. Several studies published in recent years have shown that LFEs are small, slow earthquakes with durations of about 1 s that are the result of shear slip on the plate interface. They occur on the down-dip extension of the locked part of the subduction megathrust and overwhelmingly during episodes of deep tremor (see below).

Very low frequency earthquakes. Very low frequency earthquakes (VLFs) were also discovered in Japan. VLFs are similar to LFEs. However, they are somewhat larger, up to magnitude 4, and somewhat slower, with characteristic durations of approximately 10 s. Waveform modeling indicates that they occur as slow slip on the subduction plate boundary. They are also colocated with LFEs on the deep extension of plate boundary and occur during episodes of deep tremor.

Deep tremor. Deep nonvolcanic tremor is a weak shaking of the Earth that was discovered in Japan. Since then, deep tremor has been found in many other places, including Cascadia and central California. Tremor differs from ordinary earthquakes. Rather than occurring impulsively and ending quickly (in seconds), tremor shakes the Earth for hours, days, or even weeks at a time. It is called nonvolcanic because there is a different kind of tremor scientists have known about for decades that occurs in volcanoes and is caused by fluids moving through a volcano's subterranean plumbing system. It is natural to make a connection between the two kinds of tremor, which is quite reasonable because the tremor source is located where metamorphic dehydration reactions are expected to liberate water from the subducting slab. Most studies of tremor favor fluid-flow coupling directly into ground motion as the source of tremor, but D. R. Shelly and coworkers have shown that the wiggles that tremor makes on seismographs match the wiggles of low-frequency earthquakes. Thus, under Shikoku, Japan, at least, deep tremor episodes can be regarded as a swarm of hundreds of thousands of tiny slow earthquakes. A key question is whether this result generalizes to other areas. It seems likely that it does.

Episodic tremor and slip. The seismic signature of deep tremor looks quite similar in the different areas it has been discovered (**Fig. 1**). It would be quite surprising if such similar signals were generated by different mechanisms. Low-frequency earthquakes have not been observed in Cascadia, possibly because there is no borehole seismic network comparable to Japan's Hi-net (high sensitivity seismograph network), but there is a near-perfect spatial and temporal correlation of tremor and much larger silent earthquakes in Cascadia. This same correlation between tremor and silent earthquakes was substantiated in Japan as well. The two phenomena are so closely intertwined that they are commonly referred to as ETS (episodic tremor and slip), where the "S" refers to the slip in the silent earthquake. The location of the tremor source is the same as that of

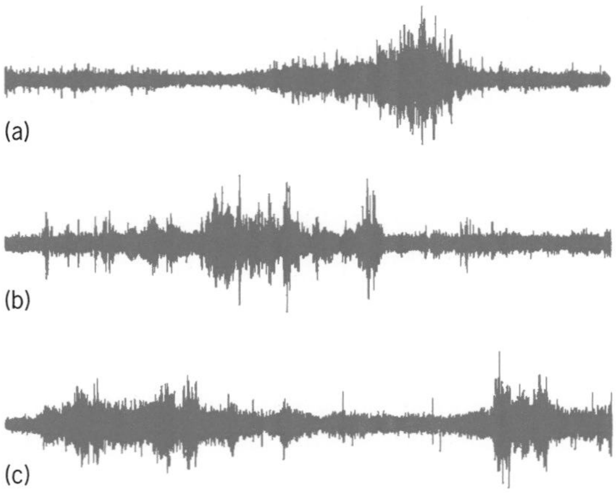

Fig. 1. Twelve-minute records of ground motion during episodes of deep, (*a*) nonvolcanic tremor in Shikoku, Japan, (*b*) Washington State, and (*c*) central California. Tremor is the large amplitude oscillation in each case. The remarkably similar signature of tremor in these different areas suggests that the underlying tremor mechanism is the same.

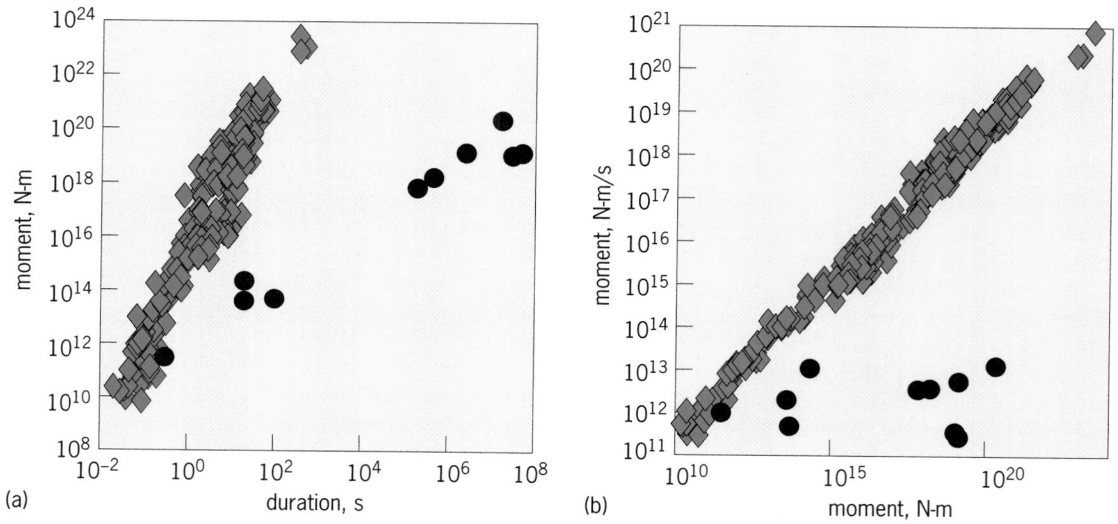

Fig. 2. (*a*) Characteristic duration of earthquakes versus seismic moment, which is a fundamental measure of earthquake size on logarithmic axes. Diamonds indicate measurements from hundreds of "ordinary" earthquakes. Circles indicate measurements of recently discovered slow earthquakes of various types. Ordinary earthquakes grow dramatically in size as their duration increase. Slow earthquakes grow only gradually in size with increasing duration. (*b*) Growth rate versus earthquake size (moment). Ordinary earthquakes grow explosively with increasing size, but slow earthquakes grow at a nearly constant rate, independent of size.

the silent earthquakes they accompany, though some have argued that the depth may differ. The location is quite interesting and strategic. The adjoining, shallower parts of these faults generate dangerous magnitude 8+ earthquakes. Because of its location and sense of motion, deep tremor should drive the dangerous part of the fault towards failure. So tremor has the potential to trigger a large earthquake. For this reason, it is extremely interesting and bears careful study.

In each of these cases, the aseismic transients arise due to the same mechanism, slow slip on the deep extension of faults. Moreover, these different events occur not just at the same place, but also at the same time. The simplest interpretation is that deep tremor, low-frequency earthquakes, very low frequency earthquakes, and silent earthquakes all represent different aspects of the same process. That is, they are a different manifestation of slow fault slip. What makes this even more interesting is that these phenomena are not just slow, they are slow in a very systematic way; namely, the larger the slow earthquakes are, the more slowly they occur.

Scaling law for slow earthquakes. Seismologists use a quantity known as the seismic moment as a fundamental measure of earthquake size. Measured in this way, the size of an ordinary earthquake increases as the duration cubed. This is known as a scaling law, a rule that describes how earthquake behavior changes with earthquake size. The scaling law for ordinary earthquakes is that once earthquakes get started, they grow explosively in size and power.

Slow earthquakes of various types occur over a wide range of geologic environments from the subduction zones of Japan, Mexico, and Cascadia, to the slopes of Kilauea volcano in Hawaii, to the San Andreas Fault of California. They all obey a common scaling law that differs from the scaling law for ordinary earthquakes. Slow earthquakes do not grow explosively. Whether they are large or small, slow earthquakes grow at a constant rate. That is, the seismic moment grows linearly with their duration (**Fig. 2**). That observation led to the various recently discovered slow earthquake types being defined as a new category of earthquake.

What puts the brakes on slow earthquakes? The slow earthquake scaling law raises the fundamental question of why slow earthquakes behave this way. The speed at which regular earthquakes grow is controlled by the stresses transmitted by seismic waves, and for that reason their rupture speed is comparable to wave speeds of several kilometers per second. Ordinary earthquakes are "wavelike" in this sense. Clearly slow earthquakes are different, so something else must control their growth. An important aspect of these slow earthquakes is that their slip is small relative to their overall extent. This leads to (at least) two possibilities.

If the slip is approximately constant regardless of slow earthquake size, then the area of faulting must grow at a constant rate. This means that the rupture speed must decrease as the fault area increases. This behavior is characteristic of diffusive phenomena, but what could be diffusing? An obvious possibility is fluids. There is strong evidence for fluids at the locations where many of these slow earthquakes occur. However, D. R. Shelly and coworkers found that the tremor source migrated at speeds of up to 45 km/h, which is much too fast for fluid diffusion. Instead, stress may be the diffusing quantity.

Another possibility is that slip in slow earthquakes is not constant, instead just extremely small relative to ordinary earthquakes. In this scenario, the rupture velocity must decrease even more rapidly with increasing earthquake size than in the diffusive model. While this explains the observed behavior, it leaves open the question of what it is that is applying the speed limit.

Outlook. Clearly, there are many important unanswered questions about slow earthquakes, and perhaps recent progress on understanding them raises more questions than answers. It does frame questions in important new ways that should lead to a deeper understanding of these mysterious phenomena and perhaps normal earthquakes as well.

For background information *see* EARTHQUAKE; FAULT AND FAULT STRUCTURES; GEOPHYSICS; PLATE TECTONICS; SEISMOLOGY; SUBDUCTION ZONES in the McGraw-Hill Encyclopedia of Science & Technology.

Gregory C. Beroza

Bibliography. S. Ide et al., A scaling law for slow earthquakes, *Nature*, 447:76–79, 2007; S. Ide, D. R. Shelly, and G. C. Beroza, The mechanism of deep low frequency earthquakes: Further evidence that deep non-volcanic tremor is generated by shear slip on the plate interface, *Geophys. Res. Lett.*, 34:L03308, 2007; Y. Ito et al., Slow earthquakes coincident with episodic tremors and slow slip events, *Science*, 315:503–506, 2007; K. Obara, Nonvolcanic deep tremor associated with subduction in southwest Japan, *Science*, 296:1679–1681, 2002; G. Rogers and H. Dragert, Episodic tremor and slip on the Cascadia subduction zone: The chatter of silent slip, *Science*, 300:1942–1943, 2003; D. Shelly et al., Low-frequency earthquakes in Shikoku, Japan, and their relationship to episodic tremor and slip, *Nature*, 442:188–191, 2006; D. R. Shelly, G. C. Beroza, and S. Ide, Non-volcanic tremor and low frequency earthquake swarms, *Nature*, 446:305–307, 2007.

Software-defined radio

Software-defined radio (SDR) provides the capability to implement a variety of radios on a single device. In theory, software-defined radios have an operating frequency, bandwidth, and modulation type that are programmed via software or programmable hardware. Limitations of current hardware generally limit the operating frequency range. Software-defined radio has become practical through the development of very powerful, inexpensive, programmable digital processing hardware such as field-programmable gate arrays (FPGAs) and digital signal processors (DSPs), as well as high-speed analog-to-digital (A/D) and digital-to-analog (D/A) converters. Although not strictly software, field-programmable gate arrays are commonly considered part of SDR architectures since software is typically not fast enough to implement radios without some hardware assistance. Radios differ widely in capability, and the ability of a software-defined radio to implement a particular type of radio is limited by the capabilities of its electronic circuitry.

Software-defined radio is a very attractive way to protect an investment in radio hardware, enabling upgrades, multiple radio types, and new standards to be implemented with a software download. The ability to implement different modulations and operate in different frequency bands allows more efficient spectrum use. Even high-volume consumer devices such as cell phones are moving to SDR-like architectures. As applications of software-defined radio increase, economies of scale are making it increasingly attractive.

Radio overview. Radio has been around for more than a century. The term radio refers to the techniques used to transmit and receive electromagnetic waves in free space. The approximate frequency range is 30 kHz to 300 GHz. Radio technology is used in radio, television, radar, cellular telephony, and other applications. Early radio transmitted voice and music, which are acoustic analog audio-frequency waveforms. AM, FM, and shortwave radio are a few examples. Modern radio communications is evolving to be primarily digital, even for analog sources such as voice, music, and video. There are increasing amounts of inherently digital information (computer databases, e-mail) that are being transmitted via radio. Digital information and transmission are more precise, easier to correct for information loss, and more suitable for modern digital processing. Interestingly, the earliest radios were used for digital information (telegraph).

Radio basics. Radio architectures vary, but the basic principles developed years ago are still in use. The **illustration** is a diagram of a superheterodyne radio receiver. (A transmitter has a similar

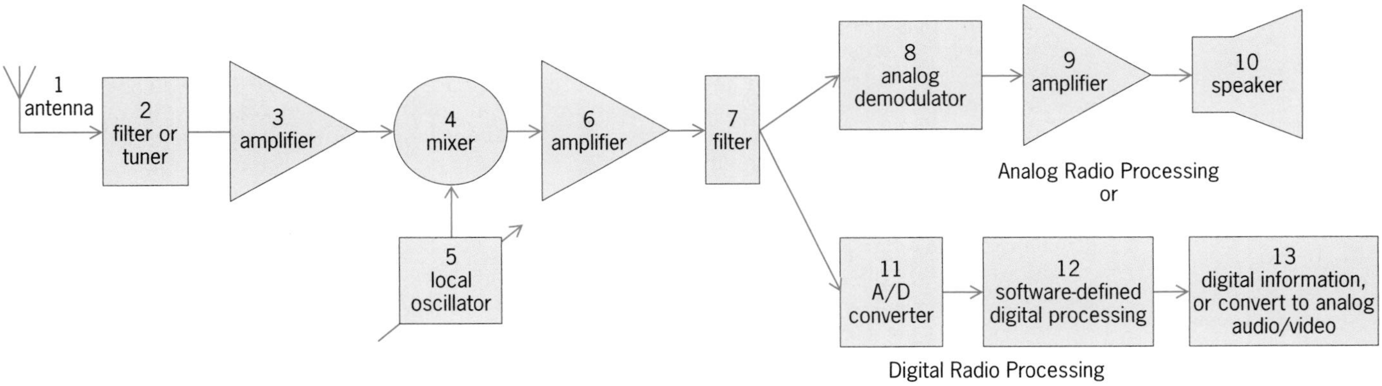

Superheterodyne radio receiver.

architecture, and will not be discussed.) An antenna (1) receives electromagnetic energy and feeds it to a filter or tuner (2). The tuner or filter attenuates energy outside the frequency band of interest to minimize interference. A tuner is used when the frequency band is variable; a filter is used when the frequency band is fixed. The signal, which has lost a lot of energy in its path from a remote transmitter, is amplified by block (3) to minimize added noise in later stages. The amplified signal, at the original frequency, is input to a mixer (4). The mixer "multiplies" the signal input with that of a local oscillator (5), resulting in output frequencies at the sum and difference frequency. For instance, in AM radio, a signal at 880 kHz is mixed with a local oscillator set to 1335 kHz. The mixer output will have frequencies of $1335 + 880 = 2215$ kHz and $1335 - 880 = 455$ kHz. The signal is amplified by block (6) and filtered at 455 kHz by block (7). The signal at 2215 kHz is rejected by the filter. The filtered signal at intermediate frequency (IF) can be either demodulated with analog circuitry (block 8), as in an AM or FM radio, or converted to digital samples (block 11) and processed digitally. In older digital microwave and digital satellite systems, this digital processing was done with dedicated, fixed-function circuitry. However, more modern systems are implemented on programmable platforms such as field-programmable gate arrays and digital signal processors. The digital processing block (12) in the illustration is the heart of a software-defined radio.

Architecture. In current SDR architectures, the radio circuitry used from the antenna (block 1 in the illustration) to the IF output stage (block 7) has limited programmability. The SDR architecture that affords the greatest flexibility is one where the signal is sampled by the analog-to-digital converter directly at the antenna. This is not practical today except in limited cases due to limitations in available converter and digital processing technology. However, it is a promising approach for the future.

It is possible to design filters whose bandwidth and center frequency are selectable via software. Amplifiers, local oscillators, and mixers can be designed to operate over a very broad frequency range. Antennas in general operate over limited frequency bands. Tuners and filters have higher performance if their operating frequency is limited or fixed. Designing a radio to operate over a very wide frequency range and to meet a range of radio requirements significantly increases the cost of the radio and compromises its performance. Therefore, most software-defined radios are designed for a limited selection of frequency bands and set of radio applications. However, technology is advancing, and more universal and flexible software-defined radios are likely to be introduced in the future.

Radio architectures used in software-defined radio vary in the following ways:

IF stages. There can be one or more IF stages. The purpose of additional IF stages is to do a better job of removing interference by moving the processing (IF) frequencies to places where known interferers do not exist.

Zero IF. A direct conversion receiver mixes the signal directly to baseband. This is an increasingly popular architecture and is used commonly in cell phones. Conversion to baseband requires quadrature mixing and sampling to ensure that all of the waveform information above and below the carrier frequency is retained.

Direct sampling. As mentioned above, this architecture samples the signal at the original frequency. This is not practical today except for low-frequency operation.

Digital downconversion. Either the downconversion to baseband can be done digitally, as is depicted in the illustration, or the IF signal can be mixed to baseband using a quadrature mixer and then sampled. The industry is moving toward digital downcoversion, and eventually direct sampling, as analog-to-digital converters and digital processing get faster and cheaper.

Digital processing. Once the signal has been converted to digital samples by the analog-to-digital converter (block 11 in the illustration), the processing can be performed digitally by field-programmable gate arrays, digital signal processors, or other processors. Field-programmable gate arrays are generally used at the interface to the analog-to-digital converter, where the sample rate is higher. The sample rate can be reduced by filtering and sample rate conversion, and the lower-rate samples passed on to a programmable processor. In many cases, all of the digital processing is done in field-programmable gate arrays. A major benefit of software-defined radio is the ability to demodulate a variety of waveform types. So long as the signal is sampled at a high enough rate, with enough precision, any waveform can be processed and demodulated Although currently impractical from a cost standpoint, a single SDR platform could conceivably process WiFi, WiMAX, UMTS, GSM, Zigbee, cordless telephones, and Bluetooth.

Modulation types. Digital modulations such as frequency-shift keying (FSK), phase-shift keying (PSK), minimum-shift keying (MSK), differential phase-shift keying (DPSK), quadrature amplitude modulation (QAM), orthogonal frequency-division multiplexing (OFDM), and direct sequence spread spectrum (DSSS) have all been implemented on SDR platforms. Modulation schemes are chosen to best match system requirements for complexity, data rate, signal integrity, and so forth.

It is possible to implement older analog modulations such as AM and FM in a digital processor. Additional processing to remove noise and interference is possible with digital processing that could not be done in analog. Many modern AM and FM radios use all-digital processing for demodulation. In fact, direct digital FM and AM broadcasting has recently been authorized by the Federal Communications Commission (FCC) using a technology called in-band on-channel (IBOC).

Examples. Software-defined radio has been implemented in a number of systems.

Joint Tactical Radio System (JTRS). The Joint Tactical Radio System (JTRS) is an ambitious program from the U.S. Department of Defense (DoD) to standardize a platform that can serve a variety of communications needs. The major thrust of the JTRS is to protect the DoD investment in tactical radios by developing a platform that can evolve and adapt to new requirements, yet maintain compatibility with existing radios. It is designed to support a wide variety of legacy communication standards that will allow it to interoperate with radios in the field. In addition, the JTRS can support advanced new digital waveforms developed by the DoD. At the heart of the JTRS is the software communications architecture (SCA), a framework that provides a common waveform implementation language. Implementation differences are abstracted through radio primitives, the equivalent of software drivers in the PC world. The goal is for waveforms to be easily ported across JTRS family hardware, and for multiple hardware platforms from different manufacturers to easily support the same waveform.

GNU radio. GNU radio is an open-source software project aimed at collaboration in development of software-based radios. Standard hardware components for radio-frequency and FPGA circuitry are available from third parties. Applications under development include passive radar, radio astronomy, Global Positioning System (GPS), distributed sensor networks, ad-hoc mesh networks, RFID (radio-frequency identification) detector/reader, multiple-in multiple-out (MIMO) radio, and Digital Radio Mondiale (a new standard digital radio system for frequencies below 30 MHz).

Cellular base stations. Although cellular base stations are specialized for cellular systems, many of their manufacturers have evolved to SDR architectures to enable upgrades and support for a variety of standards.

For background information *see* ANALOG-TO-DIGITAL CONVERTER; DIGITAL-TO-ANALOG CONVERTER; FIELD-PROGRAMMABLE GATE ARRAYS; RADIO; RADIO RECEIVER; SIGNAL PROCESSING in the McGraw-Hill Encyclopedia of Science & Technology.

Michael D. Rauchwerk

Bibliography. J. H. Reed, *Software Radio, a Modern Approach to Radio Engineering*, Prentice Hall, 2002; W. Tuttlebee (ed.), *Software Defined Radio: Origins, Drivers and International Perspectives*, Wiley, 2002.

Solitons in electrical networks

Solitons are a unique class of pulse-shaped waves that propagate in nonlinear dispersive media, that is, media whose response to a disturbance is not proportional to the disturbance's amplitude (nonlinearity), and in which the speed of propagation of a wave depends on its frequency (dispersion). In the course of propagation, the soliton's wave energy remains confined within its fixed pulse shape without dispersing, and the soliton exhibits other remarkable nonlinear dynamics. The first reported soliton (by John Scott Russell in 1834) was a monopulse water wave in a narrow canal where the shallow water possessed both nonlinearity and dispersion. Other examples of solitons include vibrations in nonlinear spring-mass lattices, acoustic waves in plasmas, optical pulses in fiber-optic cables, and fluxons in superconducting Josephson junctions. In long Josephson junctions, the Josephson phase satisfies the sine-Gordon equation which has soliton solutions. Such soliton solutions are called fluxons as each soliton corresponds to the situation in which the supercurrent circulates around the center of the soliton and a single magnetic flux quantum is trapped in it. Indeed, solitons are found throughout nature in situations where a fine balance between nonlinearity and dispersion is achieved. Dispersion alone would flatten out a pulse-shaped wave during its course of propagation as different Fourier components of the wave would travel at different speeds. Nonlinearity, however, can counteract dispersion by making the pulse steepen and even topple over and break, as would be observed in the water waves approaching the seashore.

Optical solitons in fiber-optic cables are of particular technological significance. Because of their ability to propagate over long distances without dispersion of energy, the light-wave solitons can carry a large amount of digital information in long-distance communication. Laboratory experiments have demonstrated that optical solitons in a single optical fiber can transmit 1 terabit (10^{12} bits) of data per second over 1000 km (600 mi).

In electronics, the nonlinear transmission line (NLTL) serves as a nonlinear dispersive medium that propagates voltage solitons. These electrical solitons on the NLTL have been actively investigated since the 1960s, particularly in the microwave domain. Unlike optical solitons, electrical solitons are not suitable for long-distance communication because of the high resistive loss present in the NLTL. Instead, they are utilized primarily for sharp pulse generation, which is of considerable interest in modern electronics. In ultrafast time-domain metrology, the short duration of a pulse directly translates to high temporal resolution, and, hence, narrow electrical pulses can be used to sample rapidly varying signals or as probe signals for high-precision time-domain reflectometry (TDR). In addition to high-resolution metrology, periodic sharp pulse trains can be utilized for impulse radar ranging or in ultrawideband (UWB) communication systems.

Nonlinear transmission line. The NLTL can be constructed from a linear transmission line (two conductors running in parallel) by periodically loading it with voltage-dependent capacitors (varactors) such as reversed-biased *pn* junction diodes, Schottky diodes, or metal-oxide-semiconductor (MOS) capacitors (**Fig. 1***a*). Alternatively, the NLTL can be obtained by forming an artificial *LC* (inductance-capacitance) transmission line with varactors and discrete inductors (Fig. 1*b*). The NLTL is a nonlinear dispersive medium. The nonlinearity originates from the varactors, whose capacitance changes with the applied voltage. The dispersion arises from the structural periodicity, that is, the periodic lumped loading of varactors. A common feature

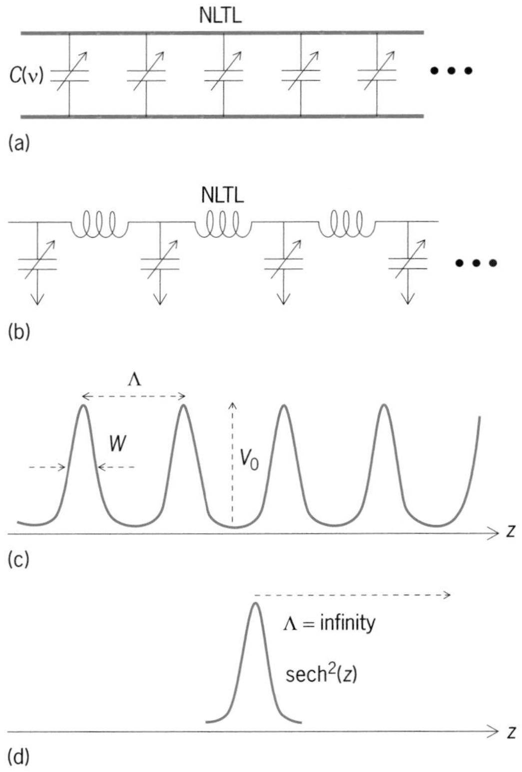

Fig. 1. Nonlinear transmission line (NTLT). (*a*) NTLT consisting of a linear transmission line periodically loaded with voltage-dependent capacitors (varactors), *C*(*v*). (*b*) Artificial nonlinear *LC* transmission line formed with varactors and discrete inductors. (*c*) Cnoidal wave traveling on the NTLT with amplitude V_0, pulse spacing Λ, and pulse width *W*; *z* = distance along line. (*d*) Monopulse soliton traveling on the NTLT with profile sech²(*z*). (*After D. S. Ricketts, X. Li, and D. Ham, Electrical soliton oscillator, IEEE Trans. Microw. Theory Techniques, 54:373–382, 2006, and D. S. Ricketts, X. Li, and D. Ham, Electrical soliton modelocking, IEEE LEOS Newslett., 20(3):4–11, June 2006*)

of any periodic structure like the NLTL is the existence of a cutoff frequency, beyond which no Fourier component can propagate, and below which different Fourier components travel at different speeds (that is, dispersion).

The NLTL was first introduced in 1960 by Rolf Landauer for parametric amplification. It was later shown by several scientists that the NLTL represented a discrete version of the Korteweg-de Vries (KdV) equation, which was derived by Diederik J. Korteweg and Gustav de Vries in 1895 to model the dynamics of shallow-water-wave solitons, and is the governing nonlinear differential equation for these solitons, as well as mechanical lattice solitons and plasma solitons. As such, the properties of voltage solitons on the NLTL are common among all the KdV solitons. In the 1970s and 1980s, the NLTL was at first used as a convenient experimental platform for studying KdV soliton dynamics since voltage signals are easily probed with oscilloscopes. Researchers soon learned, however, that this structure could be used for engineering applications due to the narrow soliton pulses it produced.

Electrical solitons on the NLTL. The general traveling-wave solution on the NLTL obtained by solving the KdV equation is a periodic train of pulses known as a cnoidal wave (Fig. 1*c*). For a given NLTL, an infinite number of cnoidal waves are possible via different interdependent combinations of the amplitude V_0, pulse spacing Λ, and pulse width *W*. Initial or boundary conditions will determine a specific cnoidal wave that can propagate on the NLTL. By letting the pulse spacing go to infinity, the monopulse soliton solution is obtained, which has a profile of the square of the hyperbolic secant (sech^2; Fig. 1*d*). The sech^2 waveform was first given by Joseph Boussinesq in 1871 and Lord Rayleigh in 1876 to account for Russell's shallow-water-wave soliton. For a sufficiently high amplitude that results in considerable nonlinearity, the cnoidal wave can be well approximated as a periodic train of sech^2 pulses. For simplicity, cnoidal waves will be referred to as a train of solitons, although, strictly speaking, "soliton" refers only to the monopulse solution.

Electrical solitons or general KdV solitons exhibit fascinating properties. The wave velocity and pulse width are closely related to the pulse amplitude. The higher the amplitude, the more prominent the nonlinear effect, and hence the steeper and narrower the soliton pulses. Since the capacitance of the varactors decreases as the applied voltages increase, solitons with higher amplitudes also experience smaller capacitance, which directly translates to smaller *LC* delays and a greater wave velocity. Due to this amplitude-dependent speed, if a taller soliton is placed behind a shorter one (**Fig. 2***a*, top), the taller one will catch up with the shorter one and move ahead of it after a collision. During the collision (Fig. 2*a*, middle), the two solitons interact very strongly and experience a significant amplitude modulation (a nonlinear collision). After the collision (Fig. 2*a*, bottom), the two solitons return to their original shapes, but have acquired a permanent time (phase) shift shown by the difference in d_1 and d_2. It is this particlelike character (strong interaction while retaining their identities) that led Norman J. Zabusky and Martin D. Kruskal in 1965 to coin the name "soliton" in the spirit of calling individual particles with names ending with "-on," for example, photon, electron, neutron, and proton.

Nonsoliton waves can also propagate on the NLTL, but only by changing their shape to form into a soliton or multiple solitons with a dispersive tail. The evolution of nonsoliton waves into solitons can be analyzed by using the inverse scattering method, a powerful mathematical tool developed by Clifford S. Gardner, John M. Greene, Kruskal, and Robert M. Miura in 1967. Phenomenologically speaking, a nonsoliton input close to a soliton profile will be sharpened into the soliton (Fig. 2*b*), while one significantly different from a soliton shape will break up into multiple solitons of different amplitudes (Fig. 2*c*). In either case, the input pulse width gets compressed traveling down the line. Once a soliton or solitons are formed, they propagate without further sharpening or breakup. Since 1967 this transient electrical soliton-forming process has occupied an

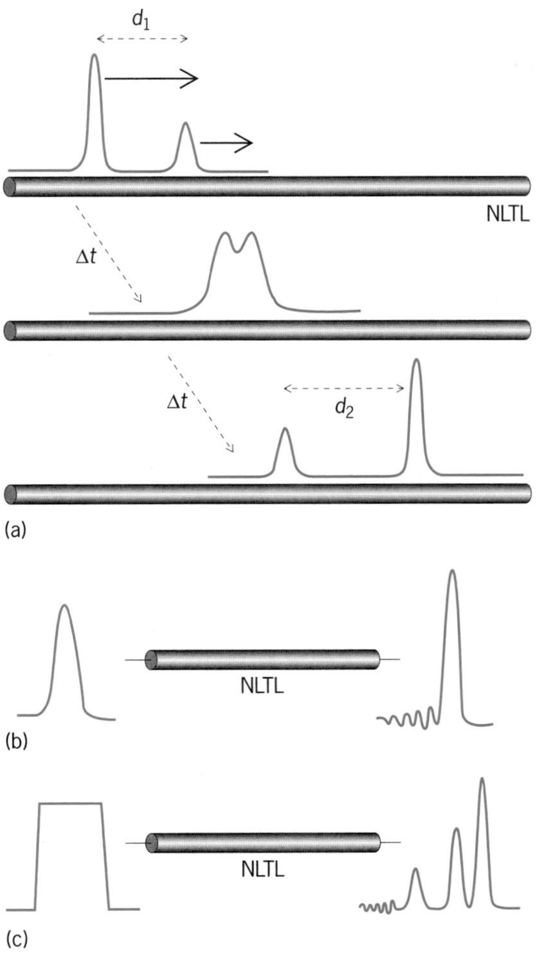

Fig. 2. Evolution of waveforms on an NLTL. (*a*) Collision of two solitons. Solitons are shown before collision with separation d_1, during the collsion, and after the collision with separation d_2, with time intervals Δt between successive frames. (*b*) Sharpening of a nonsoliton input that is close to a soliton profile into a soliton as it travels through the NTLT. (*c*) Breakup of a square wave into multiple solitons of different amplitudes as it travels through an NTLT. (*After D. S. Ricketts, X. Li, and D. Ham, Electrical soliton oscillator, IEEE Trans. Microw. Theory Techniques, 54:373–382, 2006, and D. S. Ricketts, X. Li, and D. Ham, Electrical soliton modelocking, IEEE LEOS Newslett., 20(3):4–11, June 2006*)

especially significant place in modern electronics for sharp pulse generation.

NLTL as a two-port system for pulse compression. The generation of picosecond pulses from electrical solitons propagating on an NLTL was proposed by Russell B. Riley in 1961. The active studies of narrow pulse generation on the NLTL during the past 40 years have culminated in an advanced monolithic NLTL that can achieve pulse widths down to the low picosecond range (<10 ps). In these developments, the NLTL has been used as a two-port system (input and output ports), which utilizes the transient soliton-forming process to compress an external high-frequency input signal into narrow soliton output pulses. The advantage of using the NLTL for sharp pulse generation is that the pulse rise time is no longer limited by the speed of the transistors in the active circuitry, but by the *RC* time constant of the varactors, which are usually much faster than the transistors. In particular, Schottky diode varactors can be as fast as 2 THz.

Electrical pulse compression by a factor of 2 can be achieved with a homogeneous (constant cutoff frequency) NLTL. For higher-order pulse compression, NLTLs with higher cutoff frequencies (smaller *LC* time constants) must be used to allow the propagation of the high-frequency Fourier components of the compressed pulse. As noted above, however, this procedure would inevitably result in the pulse breaking into multiple soliton pulses due to the significant difference between the shapes of the input pulse and the sharp solitons supported by the NLTL. In 1988, M. Tan, C. Y. Su, and W. J. Anklam introduced a tapered, nonlinear transmission line consisting of several cascaded homogeneous sections of NLTL with increasing cutoff frequency (that is, decreasing *LC* time constants and weaker dispersion). A pulse passing through the cascaded sections will be successively compressed, and the *LC* time constants are chosen so that only a single soliton will arise during the course of compression. Pulse compression by a factor of 7 has been demonstrated with such tapered NLTLs. Mark J. W. Rodwell and D. W. van der Weide adapted Tan's technique and integrated it on a monolithic substrate with advanced gallium arsenide (GaAs) technologies to produce the fastest electrical pulses ever generated from an all-electrical system. Rodwell's student, Michael G. Case, achieved a pulse width of 5 ps, while van der Weide demonstrated an edge rise time of 480 fs. These ultrafast NLTLs have been used in numerous applications—in particular, ultrafast metrology, where a sampling bandwidth of over 700 GHz has been achieved.

The disadvantage of this two-port topology is that it requires an external high-frequency input. In addition, due to the absence of feedback, the output waveform is sensitive to the quality and shape of the input signal.

Electrical soliton oscillator. One meaningful extension of two-port NLTL operation would be to construct a one-port (output port only) self-sustained electrical soliton oscillator by properly combining the NLTL with an amplifier that would provide positive active feedback (**Fig. 3**). Such an oscillator would self-start by growing from ambient noise to produce a periodic train of soliton pulses in a steady state, and hence would function as a self-contained soliton generator not requiring an external high-frequency input.

In the 1990s, G. J. Ballantyne and his colleagues demonstrated an NLTL-based soliton oscillator with a linear amplifier. Its oscillations, however, were not always reproducible, lacking robustness and controllability. The difficulty arises because the NLTL's soliton dynamics do not easily lend themselves to standard amplification techniques. In the case of a linear amplifier, for example, any small perturbation from noise, nonideal termination, or the dispersive tail is nondiscriminatively amplified and continues to grow into a soliton. These solitons with various

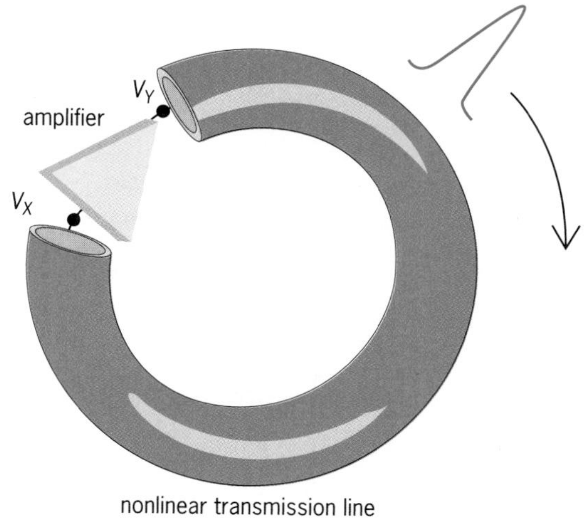

Fig. 3. **Electrical soliton oscillator. The amplifier input and output voltages are V_X and V_Y. (*After D. S. Ricketts, X. Li, and D. Ham, Electrical soliton modelocking, IEEE LEOS Newslett., 20(3):4–11, June 2006*)**

amplitudes would circulate in the oscillatory loop of Fig. 3 at different speeds and continually collide with one another, resulting in amplitude and pulse repetition rate variations in oscillations, often tending toward chaotic states.

The first robust self-sustained electrical soliton oscillator was developed in 2005 by David S. Ricketts, Xiaofeng Li, and Donhee Ham. The oscillator employs a special nonlinear amplifier with adaptive bias control to "tame" the unruly dynamics of solitons on the NLTL. Initially, the amplifier is biased in the gain region, which allows startup from ambient noise. As the oscillation grows and forms into pulses, the dc component of the output waveform steadily increases. The amplifier uses this increase in the dc component to lower its bias point such that the small perturbations are attenuated while the large signal pulses still receive enough gain without significant distortion. This threshold-dependent gain-attenuation mechanism is a technique widely employed in mode-locked lasers in optics, where it is known as saturable absorption, but was originally introduced in electronics during the vacuum-tube era by C. Chapin Cutler in 1955 for his linear pulse oscillator. The proof-of-concept prototype implemented in a CMOS integrated circuit produces a train of solitons with a 1.14-GHz repetition rate and a pulse width of 293 ps. Although the pulse widths achieved are not record numbers as compared to currently available electrical pulse generation techniques, including two-port pulse compression with an NLTL, the soliton oscillator, especially its NLTL, can be quickly scaled down and optimized to generate much narrower pulses.

In summary, electrical solitons on NLTLs have been actively investigated in the past 40 years for sharp electrical pulse generation applications. Two major topologies have been developed. The record-holding ultrafast GaAs NLTL used as a two-port passive network can produce pulses as narrow as a few picoseconds. The recently developed electrical soliton oscillator provides a new approach to robust self-sustained all-electrical soliton pulse generation.

For background information *see* FOURIER SERIES AND TRANSFORMS; HYPERBOLIC FUNCTION; INVERSE SCATTERING THEORY; NONLINEAR PHYSICS; OPTICAL PULSES; RADAR; SOLITON; TRANSMISSION LINES; ULTRAWIDEBAND (UWB) SYSTEMS; VARACTOR in the McGraw-Hill Encyclopedia of Science & Technology.

Xiaofeng Li; David S. Ricketts; Donhee Ham

Bibliography. T. H. Lee, Device physics: Electrical solitons come of age, *Nature*, 440:36–37, 2006; M. Remoissenet, *Waves Called Solitons: Concepts and Experiments*, 3d ed., Springer, 1999; D. S. Ricketts et al., Electrical soliton oscillator, *IEEE Trans. Microw. Theory Techniques*, 54(1):373–382, January 2006; M. J. W. Rodwell et al., Active and nonlinear wave propagation devices in ultrafast electronics and optoelectronics, *Proc. IEEE*, 82:1037–1059, 1994.

Space elevator

Classical theories on the strength of solids, based on fracture mechanics or maximum stress, assume a continuum. Even if this robust hypothesis has been demonstrated to work at the nanoscale for elastic calculations, it has to be revised for computing the strength of nanostructures or nanostructured materials. Accordingly, quantized strength theories recently have been formulated and validated by atomistic and quantum-mechanical calculations or nanotensile tests. As an example, we will discuss the implications on the strength prediction for the carbon nanotube-based space elevator's megacable.

Space elevator concept. A space elevator consists of a cable attached to the Earth's surface for carrying payloads into space. If the cable is long enough (that is, around 150,000 km, a value that can be reduced by a counterweight) the centrifugal forces exceed the gravitational forces, so that the cable would work under tension. The **illustration** shows the conceptual scheme and artistic views of the cable, as could be seen from the Earth or space. The cable could be anchored to either a mobile or a fixed-base station (platform). Mobile platforms can be controlled to avoid high winds, storms, and space debris. However, fixed-base stations have the advantage of being cheaper. The elevator would carry payloads into space by climbing a planar ribbon cable using a pairs of rollers (climber) to hold the cable with friction. A laser beam could be used to sustain and control the energy required to power the climber. Note that the climber will be naturally accelerated after having reached the geosynchronous orbit and could accumulate energy or exchange it with another climber.

A space elevator would revolutionize the method for carrying payloads into space at low cost, but its design is very challenging. In October 2006, teams from around the United States gathered in Las Cruces, New Mexico to compete for $400,000 in prize money as part of NASA's Centennial Challenges—the space agency's program of contests

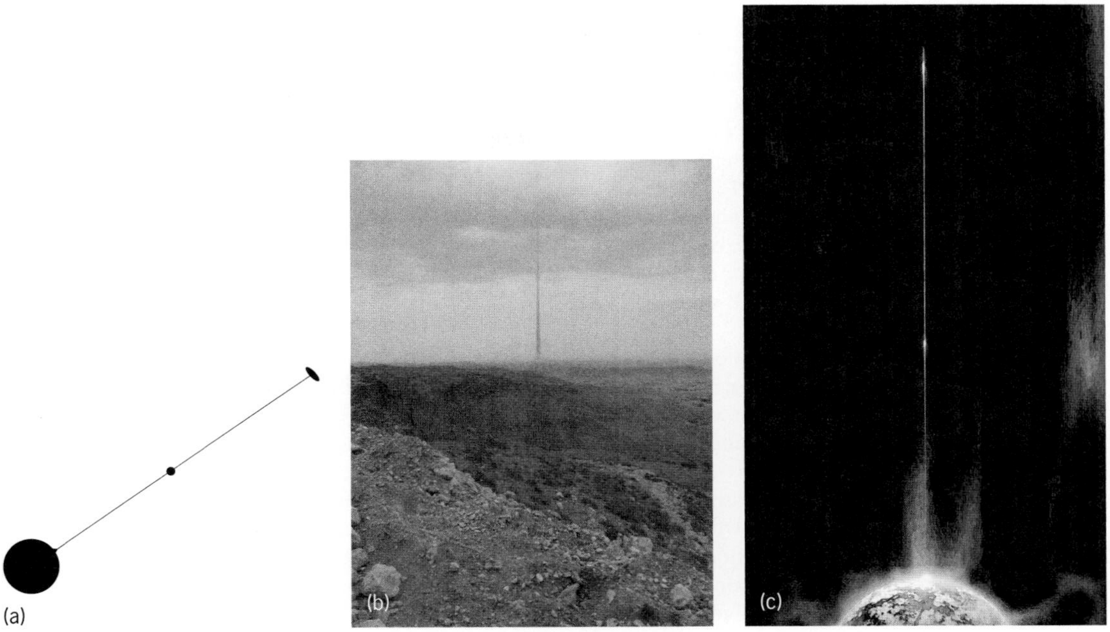

The space elevator (*a*) concept with artistic views from (*b*) Earth and (*c*) space. (*Courtesy of StudioAta, Torino, Italy*)

to stimulate innovation and competition in solar system exploration—in which one of the contests was to create a prototype space-elevator cable. The most critical component in the space-elevator design is undoubtedly the cable, which requires a material with very high strength and low density. The fact that the contest produced no winner in the "tether" category is a telling example that the current expectations from nanotube-based cables are unrealistic.

Classical theories on strength of solids need to be revised at the nanoscale. Theories on the strength of solids are of great interest in the materials science community. Fracture, as a critical transformation of the solid state, is predicted to take place when a given parameter reaches a critical value. The main two classical treatments consider this parameter to be the maximum stress or the energy release rate (or, equivalently, the stress-intensity factor). While the maximum stress is based on the concept of force, the energy release rate comes from classical fracture mechanics and is based on the continuous energy balance during crack growth.

For a linear elastic structure containing a crack of infinitesimal length (that is, a nanocrack), both theories equally predict physically unreasonable strengths—vanishing or infinite, respectively—clearly showing (even contrasting) paradoxes.

Solution of the paradoxes. Removing the hypothesis of continuous crack growth, both theories based on stress or on energy can be reconciled, giving extremely consistent strength calculations. Such predictions, and thus their deviations with respect to the continuous counterparts, recently have not only been confirmed by atomistic and quantum-mechanical simulations at the nanoscale, but also at larger size scales, as well as for fatigue crack growth.

Consider, for example, the failure stress of a two-dimensional lattice such as a single-wall nanotube with a given atomic size (the fracture quantum) containing an elliptical nanohole. The quantized approach has recently shown that the two classical strength predictions are only reasonable for large defects. By unifying their results and extending its validity to small defects, the quantized approach has found that even an atomistic defect is capable of dramatically reducing the strength of a single nanotube.

For a nanostructured cable composed of nanotubes, assuming a mean field approach and imposing the force equilibrium, the cable strength is calculated as the mean strength of the nanotubes composing the cable. Crucially, one must consider nanotubes with a realistic distribution of defects, which are statistically unavoidable in the manufacture of nanotubes by any current means.

Cable strength. The most relevant example of a nanotube bundle is the space-elevator megacable. Considering a cable with constant cross-sectional area and a vanishing tension at the planet surface, the maximum stress–density ratio, reached at geosynchronous orbit, is for the Earth equal to 63 GPa/1300 kg/m^3, corresponding to 63 GPa if low carbon density is assumed for the cable. Such a large failure stress has been experimentally measured during tensile tests of ropes composed of single-walled carbon nanotubes or multiwalled carbon nanotubes, both expected to have an ideal strength of ~100 GPa. For steel, the maximum stress expected in the cable would be ~383 GPa, whereas for Kevlar ~70 GPa, both much higher than their strengths.

An optimized cable design must consider a uniform tensile stress profile rather than a constant cross-sectional area. Accordingly, the cable could be built of any material by simply using a large enough taper ratio—the ratio between the maximum (at the geosynchronous orbit) and minimum (at the Earth's surface) cross-sectional area. The taper ratio is an exponential function of the density to strength ratio. For steel or Kevlar, a giant and unrealistic taper ratio of $\sim 10^{33}$ or $\sim 2.6 \times 10^8$, respectively, would be required. For carbon nanotubes, it theoretically is only ~1.9. Thus, the feasibility of the space elevator seems to have become plausible only because of the discovery of carbon nanotubes. The cable would represent the largest engineering structure hierarchically designed from the nano- (single nanotube with length of the order of a hundred nanometers) to the megascale (space elevator cable with a length of the order of a hundred megameters).

Basing the design of the megacable, or any large-scale structure, on the theoretical strength of a single carbon nanotube, as in the current view, is naïve. For example, just one atomic vacancy introduced in a defect-free carbon nanotube would reduce its strength by a factor of ~20%. In contrast, quantized theories can help in designing flaw-tolerant megacables using appropriate taper ratios, which can compensate for the strength reduction inherent in nanotube cables. The current paradigm, which considers the strength of a large-scale nanotube-based cable to be equivalent to that of a defect-free nanotube, is not flaw-tolerant. In the case of the space elevator cable, this means that taper ratios about 1–3 orders of magnitude larger are required. The strength of ~10 GPa, as often experimentally observed in individual carbon nanotubes, would dramatically increase the taper ratio up to ~613.

Conclusion. Is the space elevator out of order? Our opinion, which goes against the trend, is at present yes. However, our proposed flaw-tolerant concept could be the key for a future terrestrial space elevator design. Moreover, a lunar space elevator, due to the lower gravity of the Moon with respect to that of the Earth, could perhaps be realized with existing engineering materials, even if strong size effects on material strength are unavoidably expected.

Human beings have already demonstrated and built a megastructure, namely the Great Wall of China, having a length of 6352 km. But the global strength of the megacable will be imposed by its weakest link, in contrast to the local strength of the Great Wall.

However, never say never.

For background information *see* CARBON NANOTUBES; MATERIALS SCIENCE AND ENGINEERING; METAL, MECHANICAL PROPERTIES OF; STRENGTH OF MATERIALS; STRESS AND STRAIN in the McGraw-Hill Encyclopedia of Science & Technology.

Nicola M. Pugno

Bibliography. B. C. Edwards and E. A. Westling, *The Space Elevator: A Revolutionary Earth-to-Space Transportation System*, 2003; A. A. Griffith, *Phil. Trans. Roy. Soc.*, 221A:163–199, 1920; V. V. Novozhilov, *Prik. Mat. Mek.*, 33:212–222, 1968; J. Pearson, The orbital tower: A spacecraft launcher using the Earth's rotational energy, *Acta Astronaut.*, 2:785–799, 1975; N. Pugno, Dynamic quantized fracture mechanics, *Int. J. Fract.*, 140:159–168, 2006; N. Pungo, New quantized failure criteria: application to nanotubes and nanowires, *Int. J. Fract.*, 141:313–323, 2006; N. Pugno, On the strength of the carbon nanotube-based space elevator cable: from nanomechanics to megamechanics, *J. Phys. Condens. Matter*, 18:S1971–S1990, 2006; N. Pugno, Space elevator, *Nano Today*, 2:44–27, 2007; N. Pugno, The role of defects in the design of the space elevator cable: From nanotube to megatube. *Acta Mater.*, 55:5269–5279, 2007; N. Pugno and R. Ruoff, Quantized fracture mechanics, *Phil. Mag.*, 84:2829–2845, 2004.

Space flight

Space flight in 2006 continued its recent trend of slow but steady growth in human and robotic activities, moving forward in its two dominant themes: commercial utilization of low and geosynchronous orbits, and expansion of human presence in space toward exploration farther outward from Earth's boundaries. Based on the number of launches to orbit plus the number of launched satellite payloads, the utilization of space, which had reached its lowest level since 1961 in 2004 and remained on that level in 2005, in 2006 showed signs of reversing this trend. After staying 2 consecutive years at 55 total space launch attempts worldwide, in 2006 the number climbed to 66 (including four that failed).

As the U.S. space budget continued to stay its course on a relatively stable level with some upward adjustment for inflation, a major focus was the renewed flight operations of the space shuttle in support of the *International Space Station* (*ISS*). But Russian launch services continued to dominate human flights, as in the previous year. International space activities extended their trends of reduced public spending and modest launch services. Of the four failed launches (up from three in 2005, two in 2004) two were Russian (a Proton-M and a Dnepr), one Indian (GSLV), and one U.S.-commercial (Falcon 1). The National Aeronautics and Space Administration (NASA) went into the second year after implementing the Vision for Space Exploration, mandated by President George W. Bush on January 14, 2004, America's long-term plan for returning astronauts to the Moon to prepare for voyages to Mars and other destinations in the solar system. The year included three space shuttle flights, progress in preliminary design and procurement for America's next-generation spacecraft, and a number of scientific milestones. The second shuttle mission to the *ISS* since the *Columbia* loss in 2003 was followed by two more, which picked up again the continuing assembly of the space station with delivery of large components and supporting spacewalks by crewmembers. Launched in January 2006, the *New Horizons* spacecraft set out on a long voyage to study the dwarf planet Pluto. Until 2006, when the International Astronomical Union voted to change the body's designation from planet to dwarf planet, Pluto was the last unvisited planet in the solar system. Also in January, NASA's *Stardust* sample return capsule returned successfully to Earth, carrying interstellar dust specimens and comet particles from an encounter with comet Wild-2. The *Mars Reconnaissance Orbiter* (*MRO*) reached the Red Planet in March, initiating science operations in October. On the Mars surface, the twin rovers *Spirit* and *Opportunity* continued their exploration after exceeding their expected design life, amazing scientists with their findings. The *Cassini* spacecraft made more history at Saturn, discovering potential liquid water on the moon Enceladus, the *Spitzer Space Telescope* took the first temperature measurements on a planet outside the solar system, and the *Hubble Space Telescope* found evidence for dark energy in the universe already at its beginning.

The commercial space launch market was slightly below the 2005 level, which had shown improvement after a 2004 decline, begun in 2003. Of the 62 successful launches worldwide, about 21 (34%) were commercial launches (compared to 23 out of 52, or 44% in 2005), carrying 33 commercial payloads (compared to 36 in 2005). In the civil science satellite area, worldwide launches totaled 12, double the number of the preceding year. Among several privately-owned companies entering into the commercial space transportation enterprise, public attention was particularly paid to Virgin Galactic, which had achieved the first privately funded (suborbital) space flights reaching 100 km (62 mi) altitude in 2004. Early pioneers of future space tourism have spent 7–10 days each on the *ISS*, flying under contract with the Russian Federal Space Agency on Soyuz spacecraft—Dennis Tito (United States, 2001), Mark Shuttleworth (South Africa, 2002), Gregory Olsen (United States, 2005), and Anousheh Ansari (Iran-United States, 2006, the first woman "space tourist").

Despite chronic shortage of state funding, Russia's space program showed continued dependable participation in the buildup of the *ISS*. Europe's space activities in 2006 remained at the level of the previous year's total of five flights of the Ariane 5 heavy-lift launch vehicle, bringing the number of successes of this vehicle to 30. China launched six Long March rockets, one more than in 2005, and Japan also reached a launch rate of six, four more than in the preceding year.

The total number of people launched into space since 1958 (counting repeaters) in 2006 rose to 1015, including 108 women, respectively 453 individuals (44 women), in a total of 254 missions. Some significant space events in 2006 are listed in **Table 1**, and the launches and attempts are enumerated by country in **Table 2**.

International Space Station

During 2006, the *International Space Station* (*ISS*) marked the sixth anniversary of continuous crewed operations (November), during which NASA and *ISS* partner scientists have gathered vital information on the station that will help with future long-duration missions of the new exploration program, as the station has a unique microgravity environment that cannot be duplicated on Earth.

There was continuing debate about the provision of assured crew return capability after the Russian obligation to supply Soyuz lifeboats to the station expired in April 2006. Of great significance to the continuation of *ISS* assembly and operation proved to be Russia's shouldering the burden of providing crew rotation and consumables resupply flights to the station, after the loss of *Columbia* in 2003 brought shuttle operations to a standstill that lasted until the return-to-flight (RTF) mission of STS-114/*Discovery* in July 2005 and the subsequent 1-year delay until the second shuttle RTF flight, performed by STS-121/*Discovery* in July 2006.

There was a reduction in resupply mission to the station, which now could be supported only by Russian crewless automated Progress cargo ships, and station crew size was reduced from three to a two-person "caretaker" crew per expedition (also known as Increment), except for brief 10-day stays by visiting cosmonaut/researchers or commercial "tourists" arriving and departing on the third seat of Soyuz spacecraft. Operations procedures had to be revised accordingly, and such vital areas as onboard systems maintenance and spares provision had to be replanned carefully to continue crewed occupancy and a viable science research program on board despite the sudden constriction in logistics.

During 2006, as *ISS* entered its seventh year of operations as a staffed facility, three-person operation of the station was resumed in July with mission STS-121/*Discovery*. The first shuttle flight test of new inspection and protection techniques and systems, it also delivered the first European long-duration crewmember, Germany's Thomas Reiter, to join the Expedition 13 crew of Russian Commander Pavel V. Vinogradov and U.S. Flight Engineer Jeffrey Williams, who had arrived in March 2006 on *Soyuz TMA-8*/12S, replacing the Expedition 12 crew consisting of U.S. Commander William McArthur and Russian Flight Engineer Valery Tokarev. By end-2006, *ISS* assembly, about 52% complete, had grown to a total orbital mass of 470,000 lb (213 metric tons), length of 52 m (171 ft), height of 27.4 m (90 ft), width of 73 m (240 ft), and a habitable volume of about 425 m^3 (15,000 ft^3). Its solar arrays now had a surface area of 1784 m^2 (19,200 ft^2). Some 83 crewmembers of several nations had conducted a total of 81 spacewalks dedicated to assembly and servicing (53 from the station, 28 from a docked shuttle), for a total of 498 h 3 min.

Following the recommendations of an independent advisory panel of biological and physical research scientists called Remap (for "Research Maximization and Prioritization"), NASA in 2002 had established the formal position of a "Science Officer" for one crewmember aboard the *ISS*, responsible for expanding scientific endeavors on the station. In 2006, McArthur, Williams, and Michael "L-A" Lopez-Alegria served as Science Officers.

By end-2006, 59 carriers had been launched to the ISS: 20 shuttles, 2 heavy Protons [carrying the Functional Cargo Block (FGB) module "*Zarya*," and the service module "*Zvezda*"], and 37 Soyuz rockets

TABLE 1. Some significant space events in 2006

Designation	Date	Country	Event
New Horizons	January 19	United States	Launch on a Atlas 5 as humankind's first mission to the dwarf planet Pluto, with flyby at Jupiter in Feb. 2007, to arrive at Pluto in summer 2015.
Akari (*Astro-F*)	February 21	Japan	Infrared observatory, launched by JAXA on an M-V rocket into a 739-km (462-mi) high polar orbit. Mass: 925 kg (2035 lb).
MRO	March 10	United States	Arrival of NASA's *Mars Reconnaissance Orbiter* (*MRO*) at the Red Planet and start of maneuvers into final science orbit at 250 km (155 mi) × 316 km (196 mi).
Soyuz TMA-8/*ISS-12S*	March 29	Russia	Launch of the seventh *International Space Station* (*ISS*) crew rotation flight on a Soyuz-FG, bringing the Expedition 13 crew of Pavel Vinogradov and Jeffrey Williams, plus 8-day visitor Marcos Cezar Pontes.
Venus Express	April 11	Europe	Arrival of ESA's first Venus exploration probe at the target planet, after a 153-day cruise, to study Venus from a 250 × 66,000-km (155 × 41,000-mi), 24-hour orbit.
Progress M-56/21P	April 24	Russia	Crewless logistics cargo/resupply mission to the *ISS*, on a Soyuz-U rocket.
CloudSat/*CALIPSO*	April 28	United States	Launch of twin environmental satellites by NASA on a Delta 2 into a polar orbit at 705 km (438 mi) for observation of cloud and aerosol effects on climate and weather.
Progress M-57/22P	June 24	Russia	Crewless logistics cargo/resupply mission to the *ISS*, on a Soyuz-U rocket.
STS-121 (*Discovery*)	July 4	United States	The long delayed second shuttle "Return to Flight" Mission (ULF1.1) to the *ISS*, with cargo for the *ISS*, ESA Astronaut Thomas Reiter for the *ISS*, and 3 EVAs during a 13 d 6 h 37 m mission.
STS-115 (*Atlantis*)	September 9	United States	*ISS* Mission 12A with new truss elements (P3/P4) to resume station assembly and deployment of new solar arrays, with 3 EVAs, for a mission duration of 11 d 19 h 6 m.
Soyuz TMA-9/*ISS-13S*	September 18	Russia	Launch of the eighth crew rotation flight to *ISS* on a Soyuz-FG, bringing the Expedition 14 crew of Michael Lopez-Alegria and Mikrail Tyurin, plus first female space tourist Anousheh Ansari (10 d).
Progress M-58/23P	October 23	Russia	Crewless logistics cargo/resupply mission to the *ISS*, on a Soyuz-U rocket.
STEREO-A/*STEREO-B*	October 25	United States	Launch of twin "Solar-Terrestrial Relations Observatory" satellites by NASA on a Delta 2 into solar orbit, *STEREO-A* ahead of Earth, *STEREO-B* trailing behind Earth.
STS-116 (*Discovery*)	December 9	United States	*ISS* Mission 12A.1 with the P5 truss element for assembly, *ISS* crewmember Sunita Williams to replace Thomas Reiter, four EVAs. Total mission duration: 12 d 20 h 45 m.

(23 crewless Progress cargo ships, the DC-1 docking module, and 13 crewed Soyuz spaceships).

Soyuz TMA-8. *Soyuz TMA-8*/ISS-12S (March 29–September 28) lifted off on a Soyuz-FG launch vehicle at the Baikonur Cosmodrome in Kazakhstan, carrying Expedition 13 crewmembers Vinogradov and Williams, as well as Brazilian visiting crewmember 10 (VC10) and "taxi cosmonaut" Marcos Cezar Pontes, officially known as a space flight participant (SFP). The seventh *ISS* crew rotation flight by a Soyuz, it arrived at the station on March 31, docking smoothly at the FGB nadir port. Hatch opening and crew transfer were nominal. Eight days later (April 8), the previous crew return vehicle, *Soyuz TMA-7*/11S, undocked from the service module aft port for a safe landing in Kazakhstan with McArthur, Tokarev, and Pontes. Expedition 12's total mission elapsed time from launch to landing was 189 days 19 min in space (~187 days on board the station).

TABLE 2. Successful launches in 2006 (Earth-orbit and beyond)

Country of launch	Number of launches (and attempts)
Russia	23 (25)
United States (NASA/DOD/Commercial)	22 (23)
People's Republic of China	6 (6)
Japan	6 (6)
Europe (ESA/Arianespace)	5 (5)
India	— (1)
TOTAL	**62** (66)

Progress M-56. Designated ISS-21P, the first of three crewless cargo ships to the *ISS* in 2006 lifted off on a Soyuz-U rocket at Baikonur on April 24 and docked at the station on April 26. As all Progress transports, it carried about 2 tons of resupply for the station, including maneuver propellants, water, food, science payloads, equipment and spares.

Progress M-57. ISS-22P was the next crewless cargo ship, launched in Baikonur on a Soyuz-U on

June 24 and arriving at the station with fresh supplies two days later.

Soyuz TMA-9. *Soyuz TMA-9*/ISS-13S (September 18, 2006–April 21, 2007) was launched in Kazakhstan. The eighth crew rotation flight by a Soyuz, it carried Expedition 14 crewmembers Commander Lopez-Alegria and Flight Engineer 1, Mikhail Tyurin plus Iranian-American "space tourist" Ansari, flying under contract with the Russian Federal Space Agency. TMA-9 docked to the *ISS* on September 20, replacing the previous crew return vehicle, *Soyuz TMA-8*/12S, which undocked on September 28 with the Expedition 13 crew of Vinogradov and Williams (182 d 22 h 43 m in space) and Ansari (10 d 21 h 4 m 17 s). Flight Engineer 2 Reiter remained on the *ISS*.

Progress M-58. ISS-23P, the third automated logistics transport in 2006, lifted off on its Soyuz-U on October 23, docking at the *ISS* on October 26 with 2 tons of resupply cargo.

United States Space Activities

Launch activities in the United States in 2006 showed a solid increase from the previous year. There were 22 NASA, Department of Defense (DOD), and commercial launches, out of 23 attempts (2005: 16 of 16 attempts, 2004: 19 of 19; 2003: 26 of 27 [loss of Columbia]).

Space shuttle. After the loss of Orbiter *Columbia* on the first (and only) shuttle mission in 2003, and a return to the skies in summer 2005 by *Discovery*, the second RTF test flight was delayed until well into 2006. During the stand-down all resupply and crew rotation flights to the *ISS* were accomplished solely by Russian Soyuz and Progress vehicles.

STS-121. After two launch scrubs on July 1 and 2 due to weather concerns, *Discovery*, on its thirty-second flight, lifted off on the second RTF mission (ISS-ULF1.1) on July 4, 2006. The crew transferred the loaded Italian-built Multi-Purpose Logistics Module (MPLM) "Leonardo" to the station, performed three successful EVAs, transferred cargo between the space vehicles, and reberthed the MPLM with return cargo in the shuttle cargo bay. *Discovery* undocked and returned to Earth, touching down on July 17 after 202 orbits.

STS-115. Several launch scrubs preceded the successful liftoff of *Atlantis* (ISS Mission 12A) on September 9. The orbiter docked at the station on September 11. All objectives were successfully accomplished, in particular resupply of the station and resumption of station assembly after a hiatus of 4 years by delivering the 16-metric-ton (17.5-ton) P3/P4 integrated truss structure with the second set of solar arrays (of four). Three spacewalks were performed. *Atlantis* returned to Earth on September 21 after 186 orbits and a perfect mission.

STS-116. Shuttle *Discovery* launched on December 9 on ISS Mission 12A.1, one of the most challenging shuttle missions in NASA history, carrying an all-rookie crew. The crew successfully achieved all objectives of the 12-day mission: Delivery and installation of the P5 truss segment (serving as a spacer at the end of the new P4 truss), exchange of an *ISS* crewmember for another, reconfiguring and activating the new electrical power system and thermal control system, and transferring extra oxygen for EVA prebreathing and nitrogen to storage tanks on the outside of the "Quest" Airlock. Sunita Williams officially replaced Reiter as Expedition 14 Flight Engineer 2, and Reiter returned on *Discovery* after 168 days on the *ISS*. *Discovery* returned to Earth on December 22 after 202 orbits. It was the 117th space shuttle flight, the thirty-third by *Discovery* and the twentieth shuttle mission to visit the *ISS*.

Advanced transportation systems activities. In 2006, NASA's Constellation program entailed the Crew Launch Vehicle (CLV) named Ares 1 with the Crew Exploration Vehicle (CEV) spaceship *Orion*, capable of delivering crew and supplies to the *ISS*, carrying four astronauts to the Moon, and supporting up to six crewmembers on future missions to Mars; and a crewless series of heavy cargo lifters (CaLV) named Ares V. These systems will take the place of the space shuttle some time after it is decommissioned by end-2010.

Space sciences and astronomy. In 2006, the U.S. launched seven civil science spacecraft (five more than in the previous year): *New Horizons*, *ST5*, *CloudSat*, *CALIPSO*, *GOES-N*, *STEREO-A*, and *STEREO-B*.

New Horizons. Humankind's first mission to dwarf planet Pluto was launched by NASA on January 19 aboard an Atlas V rocket. Named *New Horizons*, the 478-kg (1054-lb), piano-sized spacecraft was the fastest ever launched, speeding away from Earth at approximately 16 km/s (36,000 miles per hour) on a trajectory that will take it on an unprecedented journey of exploration to the distant member of the solar system, traveling more than 5×10^9 km (3×10^9 mi) toward its primary science target. *New Horizons*, which flew past Jupiter for a gravity assist and science studies in February 2007, will conduct the first close-up, in-depth study of Pluto and its moons in summer 2015.

ST5. NASA's three *Space Technology 5* (*ST5*) microsatellites, part of the New Millennium Program, focused on the design, development, integration, and operation of three full-service 25-kg-class spacecraft that implemented multiple new technologies. They were launched on March 22 on an Orbital Sciences Pegasus XL rocket from a L-1011 aircraft from Vandenberg Air Force Base, California, each one fully fueled and weighing approximately 25 kg (55 lb) and about the size of a 13-in television set. A major mission milestone was reached on May 24 when the spacecraft assumed a constellation formation, lining up in nearly identical orbits, like three pearls on a necklace, approximately 350 km (220 mi) apart. The mission demonstrated the benefits of using a constellation of spacecraft to perform scientific studies of the auroral displays occurring near Earth's polar regions. The spacecraft simultaneously traversed electric current sheets and measured the magnetic field using miniature magnetometers. The mission was completed on June 20, 2006.

STEREO-A and STEREO-B. *STEREO* (Solar TErrestrial RElations Observatory) is the third mission in NASA's Solar Terrestrial Probes program. The twin *STEREO* spacecraft were launched on October 25 on a Delta 2 7925-10L rocket from Cape Canaveral Air Force Station in Florida. The 2-yr mission of the two nearly identical space-based observatories, one ahead of Earth in its orbit around the Sun (*STEREO-A*, for "ahead"), the other trailing behind (*STEREO-B*, "behind"), will provide the first-ever stereoscopic measurements to study the Sun and the nature of its coronal mass ejections (CMEs).

Gravity Probe-B. *Gravity Probe-B* (*GP-B*) is a NASA mission to test two predictions of Albert Einstein's general theory of relativity (GTR) using four spherical gyroscopes and a telescope, housed in a satellite orbiting 642 km (400 mi) above the Earth, measuring in a new way and with unprecedented accuracy two extraordinary effects predicted by the GTR (the second having never before been directly measured): (1) the geodetic effect—the amount by which the Earth warps the local spacetime in which it resides, and (2) the frame-dragging effect—the amount by which the rotating Earth drags its local spacetime around with it. Throughout 2006, the *GP-B* science team worked through a three-phase analysis of the data, releasing their preliminary results in April 2007. Final results were planned for release in December 2007 after further data analysis and refinement and a critical review by the *GP-B* external Science Advisory Committee (SAC), as well as by other international experts.

MESSENGER. NASA's *MESSENGER* (*Me*rcury *S*urface, *S*pace *En*vironment, *Ge*ochemistry, and *R*anging), launched in 2004, is scheduled to become the first spacecraft to orbit the planet Mercury, beginning in 2011. On October 24, 2006 *MESSENGER* flew past Venus, coming within 2990 km (1860 mi) of the surface, and will do so again in June 2007, using the tug of Venus' gravity to resize and rotate its trajectory closer to Mercury's orbit.

Swift. NASA's *Swift* satellite, launched in 2004, was designed to solve the 35-year-old mystery of the origin of gamma-ray bursts (GRBs). These flashes are brighter than a billion suns, yet can last as little as a few milliseconds. With contributions of data from *Swift* and other instruments, scientists now believe these distant yet fleeting explosions are related to the formation of black holes throughout the universe: the "birth cries" of black holes. Through coordination of observations from several ground-based telescopes, *Swift*, and other satellites, scientists solved the mystery of the origin of the powerful split-second flashes of light called short GRBs.

On July 5, 2006, the *Swift* observatory began observing supernova 2006dm a few days after its explosion. The supernova is the result of the thermonuclear explosion of a white dwarf in the galaxy MCG-01-60-21, which is located some 300 million light-years from Earth. Also in 2006, *Swift* carried out exploding "star within a star" x-ray observations of the outburst from RS Ophiuchi, a binary system containing a white dwarf orbiting around a red giant star. An unusual burst detected by *Swift*, GRB 050801 in the distant universe, showed a bright afterglow that displayed steady emission at both x-ray and optical wavelengths, and lasted for 250 s after the end of the prompt gamma-ray emission before beginning its typical decline in brightness. Scientists have suggested that the end product of this GRB could have been a magnetar, rather than a black hole. Another discovery by *Swift* in 2006 was the discovery of x-rays from a comet that is now passing the Earth and rapidly disintegrating on what could be its final orbit around the Sun. *See* SWIFT GAMMA-RAY BURST MISSION.

GALEX. *GALEX* (Galaxy Evolution Explorer), launched by NASA on April 28, 2003, on a Pegasus XL rocket from a L-1011 aircraft into a nearly circular Earth orbit, is an orbiting space telescope for observing tens of millions of star-forming galaxies in ultraviolet light across 10 billion years of cosmic history. Its telescope has a basic design similar to the *Hubble Space Telescope* (*HST*), but while *HST* views the sky in exquisite detail in a narrow field of view—like a grain of sand held at arm's length—*GALEX* is tailored to view hundreds of galaxies in each observation. Thus, it requires a large field of view, rather than high resolution, in order to efficiently perform the mission's surveys. During 2006 *GALEX* caught a giant black hole red-handed dipping into a cosmic cookie jar of stars, the first time astronomers have seen the whole process of a black hole eating a star, from its first to nearly final bites. Supermassive black holes in some giant galaxies create such a hostile environment that they shut down the formation of new stars, according to *GALEX* findings.

Spitzer Space Telescope. Formerly known as SIRTF (Space Infrared Telescope Facility) and launched in 2003, the *Spitzer Space Telescope* (*SST*) carries an 85-cm (33-in.) cryogenic telescope and three cryogenically cooled science instruments capable of performing imaging and spectroscopy in the 3.6–160-micrometer range.

In 2006, *SST*, with its powerful infrared vision, was able to unearth nearly 2300 planet-forming disks in the Orion Cloud complex, a collection of turbulent star-forming clouds that includes the well-known Orion Nebula. The disks—made of gas and dust that whirl around young sunlike stars—are too small and distant to be seen by visible-light telescopes; however, the infrared glow of their warm dust is easily spotted by *Spitzer*'s infrared detectors.

Other *SST* observations in 2006 strongly suggested that infrared light detected in a prior study originated from clumps of the very first objects of the universe. The data indicate this patchy light is splattered across the entire sky and comes from clusters of bright, monstrous objects more than 13 billion light-years away. Astronomers believe the objects are either the first stars—huge stars more than 1000 times the mass of our Sun—or voracious black holes that are consuming gas and spilling out enormous amounts of energy.

Another *Spitzer* "first" in 2006 were the first measurements of the day and night temperatures of a planet outside our solar system. The infrared observatory revealed that a Jupiter-like gas giant planet circling very close to its parent star, Upsilon Andromedae, once every 4.6 days is always as hot as fire on one side and potentially as cold as ice on the other, a temperature difference of 1400°C (2550°F). Upsilon Andromedae, about 40 light-years away and visible to the naked eye in the constellation Andromeda, is circled by two other known planets.

RHESSI. *RHESSI* (Reuven Ramaty High Energy Solar Spectroscopic Imager), launched in 2002, continued its operation in Earth orbit, providing advanced images and spectra to explore the basic physics of particle acceleration and explosive energy release in solar flares.

Hubble Space Telescope. Sixteen years after it was placed in orbit, the *Hubble Space Telescope* (*HST*) continued to probe far beyond the solar system. In 2006 *HST* captured an unprecedented look at the Orion Nebula. This turbulent star formation region is one of astronomy's most dramatic and photogenic celestial objects. In a mosaic containing a billion pixels, *Hubble*'s Advanced Camera for Surveys (ACS) uncovered thousands of stars never seen before in visible light. Some are merely one-hundredth the brightness of previously viewed Orion stars.

Scientists using *Hubble* discovered that dark energy is not a new constituent of space, but rather has been present for most of the Universe's history. Dark energy is a mysterious repulsive force that causes the Universe to expand at an increasing rate.

Also in 2006, NASA approved the fifth servicing mission to the *Hubble* by a space shuttle, not only to ensure that *HST* can function for perhaps as much as another 10 years, but also to increase its capabilities significantly in key areas. As part of the servicing and upgrade, two new scientific instruments will be installed: the Cosmic Origins Spectrograph (COS) and the Wide Field Camera 3 (WFC3). The shuttle mission, with a crew of six, is planned for 2008.

In 2006, design activities continued on the *HST*'s successor, the *James Webb Space Telescope* (*JWST*). The contracting team responsible plans to launch the giant new cosmic telescope around 2017 on a European Ariane 5 toward the second Lagrangian point (L2), 1.5×10^6 kilometers (930,000 mi) beyond Earth's orbit on the Sun-Earth line, where effects of their light on its optics are minimized and gravitational pull is relatively well balanced.

Chandra Observatory. Launched on shuttle mission STS-93 on July 23, 1999, the massive *Chandra X-ray Observatory* uses a high-resolution camera, high-resolution mirrors, and a charge-coupled detector (CCD) imaging spectrometer to observe x-rays of some of the most violent phenomena in the universe that cannot be seen by the *Hubble*'s visual-range telescope.

In 2006 *Chandra* astronomers found direct proof of the existence of dark matter in the universe through a tremendous collision of two large clusters of galaxies that has wrenched apart dark matter and normal matter, the most energetic cosmic event known, besides the big bang. In galaxy clusters, the normal matter is bound by the gravity of an even greater mass of dark matter, which is invisible and can be detected only through its gravity, without which the fast-moving galaxies and the hot gas would quickly fly apart.

Also in 2006, scientists announced the x-ray detection of a proto supermassive binary black hole by *Chandra*. The two black holes have already been seen in radio images, and the new x-ray images provided unique evidence that these two black holes, located in the nearby galaxy cluster Abell 400, are in the process of forming a binary system; that is, they are gravitationally bound and orbit each other. Each black hole is located at the center of its respective host galaxy and the host galaxies appear to be merging. It is not, however, just the two host galaxies that are colliding—the whole cluster in which they live is merging into another neighboring galaxy cluster.

Cassini/Huygens. NASA's 5.4-metric-ton (6-ton) spacecraft *Cassini* continued its epic 6.7-yr, 3.2×10^9-km journey to and inside the planetary system of Saturn. During 2006, the spacecraft remained in excellent health, returning stunning imagery from its continuing journey through the Saturnian system.

Cassini discovered a new ring around the planet during a one-of-a-kind observation made possible by the longest solar occultation of the spacecraft's 4-yr mission that allowed *Cassini* to map the presence of microscopic particles that are not normally visible across the ring system. The newly discovered ring is a tenuous feature, visible outside the brighter main rings of Saturn and inside the G and E rings, and coincides with the orbits of Saturn's moons Janus and Epimetheus. The 12-h marathon enabled scientists to see the entire structure in one view. The moon Enceladus is seen sweeping through the E ring, extending wispy, fingerlike projections into the ring. These very likely consist of tiny ice particles being ejected from Enceladus' south polar geysers, and entering the E-ring.

In 2006 *Cassini* also saw something never before seen on another planet: a hurricanelike storm at Saturn's south pole with a well-developed eye, ringed by towering clouds. *See* TITAN.

WMAP. NASA's *Wilkinson Microwave Anisotropy Probe*, launched in 2001 on a Delta-2, is now located in orbit around the second Lagrange libration point L2. Its differential radiometers measure the temperature fluctuations of the cosmic microwave background radiation (CMBR) left over from the big bang.

Scientists peering back to the oldest light in the universe have new evidence for what happened within the first trillionth of a second of the universe at the outset of the big bang, when it grew from submicroscopic to astronomical size. In 2006, data from *WMAP*, based on 3 years of continuous observations, supported this scenario, called "inflation." The new *WMAP* observations give not only a more detailed map of temperature variations in the universe than did its previous data (which had provided an accurate age of the universe), but also the first full-sky map of the polarization of the CMBR. The *WMAP* polarization data allow scientists to discriminate between competing models of inflation for the first time.

ACE. During 2006 the *Advanced Composition Explorer* (*ACE*) continued to observe, determine, and compare the isotopic and elemental composition of several distinct samples of matter, including the solar corona, the interplanetary medium, the local interstellar medium, and galactic matter. Its elongated elliptical orbit affords *ACE* a prime view of the Sun and the galactic regions beyond. As of October 2006, 438 peer-reviewed papers have been published by *ACE* science team members.

Stardust. In January 2006, NASA's comet probe *Stardust* returned safely to Earth. Launched in 1999, the probe passed by Comet P/Wild 2 in 2004. Having weathered a strong "sandblasting" by cometary particles, it collected particles and began its 2-yr, 1.14×10^9-km (7.08×10^8-mi) trek back to Earth. On January 15, 2006, *Stardust*'s sample return capsule deployed its drogue and main parachutes in the Earth's atmosphere, and at 2:10 am Pacific time the capsule carrying cometary and interstellar particles successfully touched down in the desert salt flats of Utah.

Ulysses. In 2006, the joint European/NASA solar polar mission *Ulysses* celebrated its sixteenth launch anniversary. On November 17, 2006, the spacecraft began its third passage over the Sun's south pole. *Ulysses* is engaged in the exploration of the heliosphere, the bubble in space blown out by the solar wind. As *Ulysses* approached the polar regions for the third time, the Sun had settled down once again and was close to its minimum. *Ulysses* orbits the Sun once every 6.2 yr, making it perfect for studying the 11-yr solar activity cycle. Even though the Sun will be close to its activity minimum just as it was in 1994–1995, there is one fundamental difference: the Sun's magnetic field has reversed its polarity.

Voyager. The Voyager missions, now in their thirtieth year, continue their quest to push the bounds of space exploration. On August 15, 2006, the distance from the Sun of *Voyager 1*, already the most distant human-made object in the cosmos, surpassed 100 astronomical units.

Both Voyagers are still working, no longer able to use solar power but operating on their nuclear power sources. *Voyager 1* is now at the outer edge of our solar system, in an area called the heliosheath, the zone where the Sun's influence wanes, and could cross into interstellar space within the next 10 years, while *Voyager 2* continues the current mission to study the region in space where the Sun's influence ends and the dark recesses of interstellar space begin. Both spacecraft are expected to continue to operate and send back valuable data until at least the year 2020.

Mars exploration. The main event in 2006 for NASA's Mars program was the arrival of yet another crewless exploration probe, *MRO*, at the Red Planet, joining five other spacecraft currently studying Mars: *Mars Express*, *Mars Odyssey*, *Mars Global Surveyor*, and two Mars Exploration Rovers. This is the largest number of active spacecraft to study another planet in the history of space exploration.

Mars Reconnaissance Orbiter. *Mars Reconnaissance Orbiter* (*MRO*) is a multipurpose spacecraft designed to conduct reconnaissance and exploration of Mars from orbit. It arrived at Mars on March 10, 2006, after a 5×10^{-8}-km (3×10^8-mi) trip, taking more than 35 h to circle the planet in its initial very elongated (elliptical) orbit for subsequent aerobraking maneuvers to achieve a lower circular orbit. During the final weeks of aerobraking, it was flying more than 10 orbits each day, after dipping into the upper fringes of the atmosphere 426 times to exert drag with its large solar arrays to lower the orbit. The main science investigations began in November. First results provided provocative new evidence of watery habitats on Mars eons ago. *MRO* contains a host of scientific instruments such as the High Resolution Imaging Science Experiment (HiRISE) camera, the Compact Reconnaissance Imaging Spectrometer for Mars (CRISM), and the Shallow Subsurface Radar (SHARAD), which will be used to analyze the landforms, stratigraphy, minerals, and ice of Mars. It will study potential landing sites and test a new telecommunications system.

Spirit (MER-A) and Opportunity (MER-B). At end-2006, NASA's twin Mars rovers, nearing the third anniversary of their landings, were getting smarter as they got older. The unexpected longevity of *Spirit* and *Opportunity* is giving NASA a chance to field-test on Mars some new capabilities useful both to these missions and future rovers. *Spirit* began its fourth year on Mars on January 3, 2007, *Opportunity* on January 24. In addition to their continuing scientific observations, they are now testing four new skills included in revised flight software uploaded to their onboard computers. One of the new capabilities enables spacecraft to examine images and recognize certain types of features. It is based on software developed for NASA's Space Technology 6 "thinking spacecraft." *Spirit* has photographed dozens of dusty whirlwinds in action, and both rovers have photographed clouds. The rovers can recognize dust devils or clouds and select only the relevant parts of those images to send back to Earth. Another new feature, called visual target tracking, enables a rover to keep recognizing a designated landscape feature as the rover moves. Visual target tracking can be combined with a third new feature—autonomy in calculating where it is safe to reach out with the contact tools on the rover's robotic arm. The combination gives *Spirit* and *Opportunity* a capability called "go and touch," which was yet to be tested on Mars at end of 2006.

By end-2006, *Spirit* and *Opportunity* had worked on Mars for nearly 12 times as long as their originally planned prime missions of 90 Martian days. *Spirit* had driven about 6.9 km (4.3 mi), *Opportunity* had driven about 9.8 km (6.1 mi). *Spirit* had returned more than 88,500 images, *Opportunity* more than 80,700.

Opportunity's key discovery since landing has been mineral and rock-texture evidence that water drenched and flowed over the surface in at least one region of Mars long ago. *Spirit* has found evidence that water in some form has altered mineral composition of some soils and rocks in older hills above the plain where the rover landed. After *Spirit* and *Opportunity*, NASA's next-generation Mars mission, launched on August 4, 2007, is the *Phoenix* Mars Lander, followed by the next rover, the *Mars Science Laboratory*, currently in development for launch in 2009.

Mars Odyssey. NASA's *Mars Odyssey* probe, launched in 2001, mapped surface textures and minerals all over Mars, among other feats. *Odyssey*'s camera system obtained the most detailed complete global maps of Mars ever, with daytime and nighttime infrared images at a resolution of 100 m (328 ft). The spacecraft, which has been examining Mars in detail since February 2002 (more than two full Mars years of about 23 Earth months each), continued in its second extended mission in to 2007, despite losing the power-processing component of the backup, or "B-side," systems in March 2007. In 2006, imaging experts at the Jet Propulsion Laboratory combined hundreds of images taken via the orbiter to create a high-resolution, simulated video ride through the "Grand Canyon" of the solar system, *Valles Marineris on Mars*. Stretching for 5000 km (3000 mi), this canyon system, also known as the Mariner Valley, slices across the Red Planet near its equator, 10 times longer and deeper than Arizona's Grand Canyon, and 20 times wider, featuring steep walls nearly as high as Mount Everest that give way to numerous side canyons, possibly carved by water.

Mars Global Surveyor. *Mars Global Surveyor* (*MGS*) had been transmitting a steady stream of high-resolution images, which showed that the Red Planet is a world constantly being reshaped by forces of nature including shifting sand dunes, monster dust devils, wind storms, frosts, and polar ice caps that grow and retreat with the seasons. On November 2, 2006, after studying Mars four times as long as originally planned, *Mars Global Surveyor* succumbed to battery failure caused by a complex sequence of events involving the onboard computer memory and ground commands. On that day, after the orbiter was ordered to perform a routine adjustment of its solar panels, *MGS* reported a series of alarms, but indicated that it had stabilized. That was its final transmission. Subsequently, the spacecraft reoriented to an angle that exposed one of two batteries carried on the spacecraft to direct sunlight. This caused the battery to overheat and ultimately led to the depletion of both batteries. Within 11 h, empty batteries likely left the spacecraft unable to control its orientation.

Earth science. In 2006, NASA launched three Earth science satellites, two more than in 2005 (*CloudSat*, *CALIPSO*, and *GOES-N*), as the NASA-centered international Earth Observing System (EOS) continued to operate, with *Aqua* as the first member of a group of satellites termed the Afternoon Constellation (or sometimes the A-Train). The fourth and fifth members were *CloudSat* and *CALIPSO*, launched in 2006.

CloudSat. *CloudSat* is an experimental mission designed to study the effects of clouds on climate and weather with capabilities 1000 times more sensitive than typical weather radar. This information is providing the first global measurements of cloud properties that will help scientists compile a database of cloud measurements, aiding in global climate and weather prediction models. *CloudSat* was launched with the *CALIPSO* satellite on April 28 into a polar orbit.

CALIPSO. Highly complementary with *CloudSat*, the *CALIPSO* (Cloud-Aerosol Lidar and Infrared Pathfinder Satellite Observation) satellite is providing new insight into the role that clouds and atmospheric aerosols (airborne particles) play in regulating Earth's weather, climate, and air quality. *CALIPSO* provides the next generation of climate observations, including an advanced study of clouds and aerosols, drastically improving our ability to predict climate change. Its payload includes three coaligned nadir-viewing instruments. These instruments are designed to operate autonomously and continuously, although one of them, the Wide Field Camera, acquires data only under daylight conditions.

GOES-N. The National Oceanic and Atmospheric Administration (NOAA)/NASA joint mission *GOES-N* lifted off in May. After reaching its geosynchronous orbit of approximately 35,800 km (22,300 mi) and a successful postlaunch checkout, the satellite was placed in "on-orbit storage" mode where it will be able to more rapidly replace a failure of any existing operational GOES (Geostationary Operational Environmental Satellite).

NOAA-18. After its launch in 2005, on a Boeing Delta 2 expendable rocket, the *NOAA-18* environmental satellite for NOAA in 2006 continued to operate in excellent condition. To improve weather forecasting and monitor environmental events around the world, *NOAA-18* is collecting data about the Earth's surface and atmosphere. NOAA polar-orbiting satellites also detect emergency beacon distress signals and relay their locations to ground stations. In 2006, SARSAT (Search and Rescue Satellite-Aided Tracking System) was credited with rescuing 272 people in 105 incidents in the United States.

Aura. *Aura* (Latin for "breeze"), launched from Vandenberg Air Force Base in 2004 on a Delta 2 rocket, is NASA's third major Earth Observing System (EOS) platform to provide global data on the state of the atmosphere, land, and oceans, as well as their interactions with solar radiation and each other. During 2006, scientists used *Aura* and other satellites for tracking different chemicals present in Earth's atmosphere. These data are giving researchers a more complete picture of the causes and effects of atmospheric pollution and leading to a surprising result: the amount of biomass burning in subequatorial Africa and the Indonesia/Australia region was underestimated by a factor of between 2 and 3. In addition, data from the Ozone Monitoring Instrument on *Aura*, which measures the total amount of ozone from the ground to the upper atmosphere over the entire Antarctic continent, showed that in 2006 the ozone hole in the polar region of the Southern Hemisphere has broken records for area and depth.

ICESat. *ICESat* (Ice, Cloud, and land Elevation Satellite), also an Earth Observing System (EOS) spacecraft, is the benchmark mission for measuring ice sheet mass balance and cloud and aerosol heights, as well as land topography and vegetation characteristics. Launched in 2003, the spacecraft carries only one instrument—the Geoscience Laser Altimeter System (GLAS). Scientists trying to understand the dynamics of the Earth are using the lasers of *ICESat* to measure the height of ice sheets, glaciers, forests, rivers, clouds, and atmospheric pollutants from space. In 2006, NASA continued collecting measurements from *ICESat* for developing a comprehensive ice sheet survey of the polar ice caps. It confirmed that climate warming is changing how much water remains locked in Earth's largest storehouse of ice and snow. It documented for the first time extensive thinning of the West Antarctic ice shelves and an increase in snowfall in the interior of Greenland, as well as thinning at the edges. All are signs of a warming climate predicted by computer models.

Aqua. Launched in 2002, the NASA satellite *Aqua*, has been circling Earth in a polar, sun-synchronous orbit. During its 6-yr mission, *Aqua* is observing changes in ocean circulation and studies how clouds and surface water processes affect our climate. Incorporating the instrument's data into numerical weather prediction models improves the accuracy range of experimental 6-day Northern Hemisphere weather forecasts by up to 6 h.

GRACE. In 2006 the twin *GRACE* (Gravity Recovery and Climate Experiment) satellites, named "Tom" and "Jerry," continued to map the Earth's gravity fields using the Global Positioning System (GPS) and a microwave ranging system. This allows making detailed measurements of Earth's gravity field. In its almost 5 years of operation, *GRACE* may have found a crater deep under the Antarctic ice, measured the seafloor displacement that triggered the tsunami of 2004, and quantified changes in subsurface water in the Amazon and Congo river basins. In 2006, data from the *GRACE* satellites in the first-ever gravity survey of the entire Antarctic ice sheet showed that the sheet's mass decreased significantly from 2002 to 2005. The project is a joint partnership between NASA and the German DLR (Deutsches Zentrum für Luft- und Raumfahrt).

TOPEX/Poseidon. In early 2006, the joint U.S. (NASA)/France (CNES) *TOPEX/Poseidon* oceanography satellite ceased operations after nearly 62,000 orbits of Earth. The spacecraft, launched in 1992, lost its ability to maneuver, bringing to a close a successful 13-year mission. Its data have helped in hurricane and El Niño/La Niña forecasting, ocean and climate research, ship routing, offshore industries, fisheries management, marine mammals' research, modernizing global tide models, and ocean debris tracking.

Department of Defense space activities. United States military space organizations continued their efforts to make space a routine part of military operations across all service lines. One focus concerns plans for shifting the advanced technology base toward space in order to continue building a new foundation for more integrated air and space operations in the twenty-first century. The increased use of satellites for communications, observations, and—through the GPS—navigation and high-precision weapons targeting was and is of decisive importance for the military command structure.

In 2006, the U.S. Congress called for the establishment of an Office of Operationally Responsive Space (ORS) under the Department of Defense Executive Agent for Space. According to the DOD, this program aims to ensure that space-based efforts that support military users and operations are developed, acquired, and employed quickly and economically. AFRL (Air Force Research Laboratory) Space Vehicles Directorate extended the demonstration phase of the *Experimental Satellite System-11* (*XSS-11*) mission an additional 6 months to continue on-orbit demonstration activities. The *XSS-11* belongs to a new class of lightweight, low-cost "microsatellites," this one meant to approach and photograph inactive U.S. satellites or old rocket bodies. Highlights of military space in 2006 included the launch of the first Delta 4 from Vandenberg Air Force Base in California.

Commercial space activities. In 2006, commercial space activities in the United States rose over the downturn of the prior years which came in the wake of the slump in the communications space market caused by failures of satellite constellations for mobile telephony in 2001–2002 and a slight recovery in 2003. Some difficulties remained due to the export restrictions imposed to the U.S. industry on sensitive technologies, but in general, commercial ventures gained a significantly larger role in U.S. space activities.

Russian Space Activities

Russia in 2006 showed increased activity in space operations from 2005, launching 12 different carrier rockets (2005: seven). Out of 66 launch attempts worldwide in 2006, 25 space launches were attempted by Russia, again placing it in the lead of spacefaring countries including the United States. Russian cosmonauts continued to hold the worldwide lead in spaceflight endurance, headed by Sergei Krikalev with 803 days on six *Mir*/Shuttle/*ISS* missions.

The Russian space program's major push to enter into the world's commercial arena by promoting its space products on the external market, driven by the need to survive in an era of severe reductions of public financing continued in 2006. The Proton heavy lifter, originally intended as a ballistic missile (UR500), by end-2006 had flown 244 times since 1980, with 15 failures (reliability: 0.938). Of the six Protons launched in 2006 (2005: 7; 2004: 8), five were for commercial customers. The new Soyuz-2 rocket had three successful launches in 2006 (of three attempts).

European Space Activities

Europe's efforts to reinvigorate its faltering space activities remain at a low level. After the decline of

Europe's commercial activities in space to a low in 2004, five flights of the Ariane 5 heavy-lift vehicle were added in 2006, carrying a total of 11 satellite payloads. Altogether, 174 flights had been successfully completed by Ariane rockets by end-2006, including 30 by the Ariane 5.

The *Galileo* (GNSS) navigation and global positioning system has run into delays and disputes. Originally planned to enable Europe to be independent of the U.S. GPS system starting in 2008, the newly formed consortium has yet to overcome basic difficulties although the industry has made some moves in the right direction. *Galileo* (not to be confused with NASA's Jupiter probe) will consist of a constellation of 30 small satellites placed in medium orbit, 24,000 km (15,000 mi) above Earth.

A major event for ESA human space flight efforts in 2006 was the launch of German astronaut Reiter, on contract with the Russian Space Agency, on STS-121/*Discovery* to the *ISS* as Europe's first long-duration crewmember on the space station (171 days). Also spending some time on the *ISS* was the Swedish-ESA astronaut Christer Fuglesang, flying as a crewmember on the shuttle STS-116/*Discovery* and participating in three spacewalks, before returning to Earth with Reiter on December 22.

In a project led by France/CNES, *COROT* (Convection Rotation and Planetary Transits) was launched on a Soyuz-2 on December 27 into a polar orbit at 896 km (557 mi) altitude to conduct highly precise star photometry in search of planets outside the solar system. In Germany, the low governmental interest (unlike Italy's and France's) in this field continued in 2006. A German satellite launched in 2006 on a Russian Kosmos-3M rocket was *SAR-Lupe 1*, Germany's first satellite-based radar reconnaissance system, followed by *SAR-Lupe 2* in 2007. In the space science area, in 2006, besides France's *COROT*, there was the launch of *MetOp-A*, and the arrival of the planetary probe *Venus Express* at its target planet.

MetOp-A. *MetOp-A*, Europe's first polar-orbiting meteorological satellite, was launched on October 19 on a Russian Soyuz-2 (1A), an upgrade of the venerable Soyuz launcher equipped with digital electronics and a new inertial navigation system. When *MetOp-A* becomes operational in 2007, its instruments will sound the atmosphere throughout its depth, gathering essential global information, day and night, about the atmosphere, land, and ocean surfaces.

Venus Express. The 1240-kg (2734-lb) *Venus Express* spacecraft, launched in 2005, arrived at the planet Venus on April 11, 2006, after a 153-day cruise. At its closest, *Venus Express* reaches an altitude of 250 km (155 mi) and at its farthest, it is 66,000 km (41,000 mi) away from the planet. Its science objectives are to study the atmosphere, plasma environment, and surface of Venus in great detail. *Venus Express* is the first mission to visit Venus since NASA's 1989–1994 *Magellan* mission, but unlike the *Magellan* probe, *Venus Express* is not capable of imaging the surface of Venus with high spatial resolution through cloud-penetrating imaging radar. Instead, it is equipped with several instruments designed to study Venus in new ways. *See* VENUS EXPRESS.

Rosetta. ESA's comet intercept mission *Rosetta* was launched in 2004 to Comet 67P/Churyumov-Gerasimenko. In 2006 the spacecraft headed for Mars for a fly-by of the Red Planet in February 2007, and then was scheduled to return to Earth twice, in November 2007 and November 2009, for its second and third fly-bys of our planet. On arrival at 67P in 2014, *Rosetta* will enter orbit around the comet and stay with it on its journey in toward the Sun, to study the origin of comets, the relationship between cometary and interstellar material, and its implications with regard to the origin of the solar system.

Envisat. In 2006 ESA's operational environmental satellite *Envisat*, the largest Earth Observation spacecraft ever built, continued its observations after its launch in 2002. The 8200-kg (18,100-lb) satellite circles Earth in a polar orbit at 800 km (497 mi) altitude, completing a revolution of Earth every 100 min. Because of its polar sun-synchronous orbit, it flies over and examines the same region of the Earth every 35 days under identical conditions of lighting. The 25-m-long (82-ft) and 10-m-wide (33-ft) satellite is scanning the Earth similar to the way that vertical slices are peeled off an orange as it is turned in one's hand. This enables *Envisat* to continuously scrutinize the Earth's surface (land, oceans, ice caps) and atmosphere, gathering a huge volume of invaluable information for scientists and operational users for global monitoring and forecasting to protect the planet.

SPOT 5. Launched on May 4, 2002, by the 112th Ariane 4 from Kourou (French Guiana), the fifth imaging satellite of the commercial Spot Image Company in 2006 continued operations in its polar sun-synchronous orbit of 813 km (505 mi) altitude.

INTEGRAL. ESA's *INTEGRAL* (International Gamma-Ray Astrophysics Laboratory), a cooperative project with Russia and United States, continued successful operations in 2006. Launched on October 17, 2002, on a Russian Proton rocket into a 72-h orbit with 51.6° inclination, a perigee height of 9000 km (5600 mi) and an apogee height of 155,000 km (96,000 mi), the sensitive gamma-ray observatory provides new insights into the most violent and exotic objects of the universe, such as black holes, neutron stars, active galactic nuclei, and supernovae. In 2006, *INTEGRAL* operations continued smoothly, with the spacecraft, instruments and ground segment performing nominally. Its instruments have produced the first all-sky map of the 511-keV line emission produced when electrons and their antimatter equivalents, positrons, collide and annihilate. The nature of the sources responsible for the antimatter is clearly one of the key areas for further investigation by *INTEGRAL* over the coming years. One intriguing possibility is that the antimatter is produced by the annihilation or decay of an exotic form of dark matter. In 2006, *INTEGRAL* detected surprisingly powerful x-ray and gamma-ray emission from a special class of neutron star. This discovery makes them the magnetically most active bodies in the universe. The *INTEGRAL* imager (IBIS) has also been used to detect a new persistent soft gamma-ray source, IGR J18135-1751, which is coincident with one of several sources of extreme energy in the inner part of the Galaxy. Such gamma rays are emitted by short-lived radioactive isotopes that occur in supernovae and their remnants.

XMM-Newton. Europe's *XMM* (X-ray Multi Mirror)-*Newton* observatory, launched in 1999, is the largest European science research satellite ever built. The telescope has a length of nearly 11 m (36 ft), with a mass of almost 4 metric tons (8800 lb). In 2006, *XMM-Newton* continued to probe into the unknown. After 6 years of operations, its research has resulted in more than 1000 scientific papers in top-class scientific journals, corresponding to an equivalent number of results. Scientific results based on *XMM-Newton* data are now being published at a steady rate of almost 300 papers per year, comparable to the famous *Hubble Telescope*. Among its discoveries, it characterized for the first time x-ray spectra and light curves of some classes of protostars (stars being born) and provided an unprecedented insight into the x-ray variability of the corona of stars similar to our Sun. In 2006, *XMM* and *Chandra* observations uncovered evidence that helps to confirm the identity of the remains of one of the earliest stellar explosions recorded by humans—the remnant of the supernova RCW86 that appears to have been observed by Chinese (and possibly Roman) astronomers in 185 AD. During their XMM Cluster Survey (XCS) in 2006, astronomers also discovered the most distant galaxy cluster ever found, at a distance of 10^{10} light-years from Earth. Since there appear to be old galaxies in the cluster, scientists are seeing their understanding challenged as to how such a well developed massive cluster can exist at a relatively early era in the universe.

Smart-1. On September 3, 2006, after final maneuvers on June 19 and September 1, *Smart-1* (Small Missions for Advanced Research in Technology 1), Europe's first lunar spacecraft, ended its highly successful mission after almost 2 years in orbit around the Moon by crashing itself on the lunar surface in an area called the "Lake of Excellence." The 370-kg (816-lb) spacecraft was intended to demonstrate new technologies for future missions, in particular, the use of solar-electric propulsion as the primary power source for its ion engine, fueled by xenon gas. Thanks to *Smart-1*, scientists now have the best-resolution surface images ever from lunar orbit, as well as a better knowledge of the Moon's minerals. For the first time from orbit, they have detected calcium and magnesium using an x-ray instrument. In addition, the spacecraft found an area near the north pole where the Sun always shines, even in winter.

Mars Express. *Mars Express* was Europe's entry into the ongoing and slowly expanding robotic exploration of the Red Planet from Earth as precursors to later missions by human explorers. Launched in 2003, it continued to produce stunning close-up imagery of the Mars surface through 2006. Its instruments have shown that many of the upper layers of Mars contain water ice. They detected claylike minerals that form during long-term exposure to water, but only in the oldest regions of Mars. These observations suggested that water flowed only during the first few hundred million years of the planet's history. When these bodies of water were lost, water then occasionally burst from inside the planet but quickly evaporated, and such bursts still may occur. Residual water ice, in the form of a frozen lake, was discovered in the open Vastitas Borealis Crater. *Mars Express* continues to remotely explore the Red Planet with a sophisticated instrument package.

Asian Space Activities

China, India, and Japan have space programs capable of launch and satellite development and operations. Kazakhstan is in the early stages of joining the space-faring community.

China. With a total of six launches in 2006 (2005: five; 2004: eight; 2003: six; 2002: 4; 2001: 1), China currently shares the third place of space-faring nations with Japan, after Russia and United States, having made worldwide headlines in 2003 with its successful orbital launch of the first Chinese "Taikonaut" in the 4760-kg (10,500-lb) spacecraft *Shenzhou 5* ("Divine Vessel 5") on a 21-h mission, and in 2005 with the launch of the 7700-kg (17,000-lb) two-seater *Shenzhou 6* carrying the two Taikonauts Fei Junlong and Nie Haisheng on a 75-orbit, 115 h 32 min flight, designed to further China's human spaceflight experience as it works toward developing a crewed space station and to serve as a symbol of national pride, demonstrating China's technological prowess.

The launch vehicle of the *Shenzhou* spaceships was the new human-rated Long March 2F rocket. China's Long March (Chang Zheng, CZ) series of launch vehicles consists of 12 differing versions, which by the end of 2006 had made 94 flights, sending 107 payloads (satellites and spacecraft) into space, with 91% success rate. Its six major launches in 2006 demonstrated China's growing space maturity.

Japan. In 2006, Japan's space program came back strong after the failure of its H-2A launch vehicle in 2003 and its return to flight in 2005. Of the total of six launches by Japan's space agency JAXA in 2006, four involved the H-2A and two the M-V rocket. A major JAXA launch on an M-V, on February 21, was the 925-kg (2035-lb) Astro-F ("*Akari*") infrared observatory in a 744-km-high (462-mi) polar orbit. Its 68.5-cm (27-in) telescope, cooled to 6 K, in 2006 provided images of a galaxy and nebula thousands of light-years from Earth that established the satellite as an important new infrared observatory.

The x-ray astronomy satellite *Suzaku* ("red bird of the South", or *Astro-E2*) began losing the cryogenic helium required for cooling the x-ray spectrometer. However, two other instruments continue to provide new data.

Also in the headlines in 2006 was Japan's Space Engineering Spacecraft *Hayabusa* ("peregrine falcon," or *MUSES-C*). *Hayabusa* encountered serious technical difficulties in November 2005 but in March 2006 JAXA announced that communication with *Hayabusa* had been recovered. In June 2006, two out of four ion engines were reportedly working normally, which will be sufficient for the return journey to Earth, in June 2010.

In March 2006, the Japanese National Institute of Information and Communication Technology (NICT) conducted a successful laser transmission test between the JAXA satellite *Oicets* (Optical Inter-orbit Communications Engineering Test Satellite) and a ground station in Japan in an ongoing research program aimed at developing a global broadband communications network that is "wireless, seamless, and ubiquitous," using ground, aviation, and satellite systems.

India. India's emerging space program, with only one launch attempt in 2006, suffered a severe setback with the failure of its first Geosynchronous Satellite Launch Vehicle (GSLV-F02) carrying the *INSAT-4C* communication satellite, when the rocket veered off its trajectory when its first stage failed to separate after its launch on July 10 from the Satish Dhawan Space Center in Sriharikota and fell into the Bay of Bengal. The 2168-kg (4780-lb) *INSAT-4C* satellite, designed to boost direct-to-home television service and digital news gathering, was the heaviest in its class, and *INSAT 4C* was to be the first of 15 INSAT satellites, all to be launched by the Indian Space Research Organization (ISRO), each with a mission life of 10 years.

India plans to explore the Moon, with the launch of a crewless probe scheduled for 2008. ISRO calls the Moon flight project *Chandrayaan Pratham*, which has been translated as "First Journey to the Moon" or "Moonshot One." The 525-kg (1157-lb) *Chandrayaan-1* would be launched on a PSLV (Polar Space Launch Vechicle) rocket. After first circling Earth in a geosynchronous transfer orbit (GTO), the spacecraft would fly on out into a polar orbit of the Moon some 100 km (60 mi) above the surface, carrying x-ray and gamma-ray spectrometers and sending data back to Earth for producing a high-resolution digital map of the lunar surface. The project's main objectives are high-resolution photography of the lunar surface using remote-sensing instruments sensitive to visible light, near-infrared light, and low-energy and high-energy x-rays.

Other Countries' Space Activities

The former Soviet Republic of Kazakhstan is aspiring to join the select group of space-faring nations. Home of Baikonur, the world's largest rocket launching site, Kazakhstan has been in lengthy political negotiations with Russia, the real owner and operator of Baikonur, who has leased the 6717 km^2 (2593 mi^2) of the Cosmodrome from Kazakhstan. Kazakhstan is making plans to use these facilities for two satellite launcher projects of its own. Kazakhstan's first satellite, the comsat *KazSat 1*, was launched on June 18, 2006.

In 2006, Canada's *Radarsat-1* "eye-in-the-sky" celebrated its eleventh anniversary in orbit. The sophisticated radar platform is still operating and surpassing the standards. *Radarsat-1* was built to catalog the vast expanse of Canada's Arctic. *Radarsat-2*, equipped with a Synthetic Aperture Radar (SAR) with multiple polarization modes, was scheduled for launch in December 2007.

For background information *see* CHANDRA X-RAY OBSERVATORY; COMET; COMMUNICATIONS SATELLITE; COSMIC BACKGROUND RADIATION; EARTH, GRAVITY FIELD OF; EXTRASOLAR PLANETS; GALAXY, EXTERNAL; GAMMA-RAY ASTRONOMY; GAMMA-RAY BURSTS; HUBBLE SPACE TELESCOPE; INFRARED ASTRONOMY; MARS; MERCURY (PLANET); METEOROLOGICAL SATELLITES; MILITARY SATELLITES; MOON; PLUTO; RELATIVITY; REMOTE SENSING; SATELLITE NAVIGATION SYSTEMS; SATURN; SCIENTIFIC AND APPLICATIONS SATELLITES; SOLAR WIND; SPACE FLIGHT; SPACE PROBE; SPACE SHUTTLE; SPACE STATION; SPACE TECHNOLOGY; SPITZER SPACE TELESCOPE; SUN; ULTRAVIOLET ASTRONOMY; VENUS; WILKINSON MICROWAVE ANISOTROPY PROBE; X-RAY ASTRONOMY in the McGraw-Hill Encyclopedia of Science & Technology. Jesco von Puttkamer

Bibliography. *Aerospace Daily*, various 2006 issues; *AIAA Aerospace America*, December 2006 issue; *Aviation Week & Space Technology (AW&ST)*, various 2006 issues; ESA *Press Releases*, 2006; NASA Public Affairs Office *News Releases*, 2006; *SPACE NEWS*, various 2006 issues; various Internet sites.

Spintronics

The term "spintronics" usually refers to the branch of physics concerned with the manipulation, storage, and transfer of information by means of electron spins in addition to or in place of the electron charge as in conventional electronics. Introduced in 1996, spintronics (the word coined by S. Wolf) was originally the name for a Defense Advanced Research Projects Agency (DARPA) program managed by Wolf. In conventional electronics, only the charge of the electrons is of consequence for device operation, but using the electron's other fundamental property, its spin, has opened up the new field of spintronics. Major advances in electron spin transport started in 1979–1980 with the discovery of large low-temperature magnetoresistance in metallic superlattices. Later demonstrations of the "giant" effect at room temperature evolved toward application in practical devices.

Spintronics promises the possibility of integrating memory and logic into a single device. In certain cases, switching times approaching a picosecond are possible, which can greatly increase the efficiency of optical devices such as light-emitting diodes (LEDs) and lasers. The control of spin is central as well to efforts to create entirely new ways of computing, such as quantum computing, or analog computing that uses the phases of signals for computations.

Spin is a fundamental quantum-mechanical property. It is the intrinsic angular momentum of an elementary particle, such as the electron. Of course,

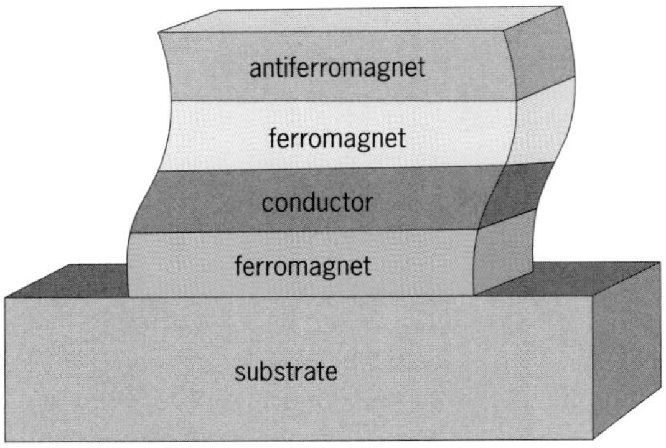

Fig. 1. Simple trilayer GMR structure.

any charged object possessing spin also possesses an intrinsic magnetic moment. It has been known for decades that in ferromagnetism the spins of electrons are preferentially aligned in one direction. Then, in 1988, it was demonstrated that currents flowing from a ferromagnet into an ordinary metal retain their spin alignment for distances longer than interatomic spaces, so that spin and its associated magnetic moment can be transported just as charge. This means that magnetization as well can be transferred from one place to another.

Giant magnetoresistive effect. The first practical application of this phenomenon is in the giant magnetoresistive effect (GMR). The GMR is observed in thin-film materials composed of alternate ferromagnetic and nonmagnetic layers (**Fig. 1**). The resistance of the material is lowest when the magnetic moments in ferromagnetic layers are aligned in the same direction, and highest when they are antialigned. This is because the spin-aligned currents from one layer are scattered strongly when they encounter a layer that is magnetically aligned in the opposite direction, creating additional resistance. But when the magnetic fields are oriented in the same direction, the spin-aligned currents pass through easily.

Current GMR materials operate at room temperature and exhibit significant changes in resistivity when subjected to relatively small external magnetic fields. Thus they can be used as magnetic field sensors. The imposed magnetic field changes the magnetic orientation of one of the two layers, disrupting their relative orientation and thus changing the resistivity. The first GMR-based magnetic field sensor was created in 1994, and high-performance disk drives utilizing GMR-based read heads to detect magnetic fields were realized in 1997 and now are ubiquitous. These read heads are responsible for the very rapid growth in magnetic storage densities that has occurred in the last decade.

Spin-dependent tunneling device. A spin-dependent tunneling device (SDT) is similar to a GMR cell but replaces the metal between the two ferromagnetic layers with a very thin insulator through which a current can tunnel preferentially when the two magnetic orientations are aligned. The difference in resistance between the spin-aligned and nonaligned cases is much greater than for GMR devices and large enough that the low-resistance state can encode, say, a "1" and the high-resistance state, a "0." Recently, an SDT device was used in the first commercial magnetoresistive random access memory (MRAM), a fast RAM that is nonvolatile, meaning it does not require power to retain information.

Spin momentum transfer effect. Significant developments ensure that MRAM will be able to scale down to 60 nm and below. The most notable of these was the discovery of the spin momentum transfer effect (SMT), predicted theoretically in 1996, in which the angular momentum carried by a spin-polarized current can exert a torque on the magnetization of a magnetic film that is magnetized in any nonparallel direction. This effect, also known as spin torque, was experimentally observed in 2000.

SMT-MRAM. Conventional MRAM utilizes current-generated magnetic fields to rotate the magnetization in the free layer. In spite of advances in the switching methodology to make the switching robust to disturbances, increase the yield, and lower the switching current by magnetic cladding of the word and bit lines, SMT potentially offers orders-of-magnitude lower switching currents and concomitantly much lower energy per bit to write. Apparently SMT switching can significantly improve the performance of MRAM and make it a truly universal memory. Key areas of research to be addressed in the near future include using MRAM in embedded memory and for multibit memory (stacked memory).

A summary of the projected performance of MRAM and SMT-MRAM is presented in the **table**, in

Projected performance of MRAM, SMT-MRAM, and more conventional semiconductor memories

	Standard MRAM, 90 nm*	DRAM, 90 nm	SRAM, 90 nm	SMT MRAM, 90 nm*	FLASH, 90 nm†	FLASH, 32 nm†	SMT MRAM, 32 nm*
Cell size, μm^2 (cell density)	0.25 (256 Mbit/cm^2)	0.25 (256 Mbit/cm^2)	1–1.3‡ (64 Mbit/cm^2)‡	0.12 (512 Mbit/cm^2)	0.1 (512 Mbit/cm^2)	0.02 (2.5 Gbit/cm^2)	0.01 (5 Gbit/cm^2)
Read time	10 ns	10 ns	1.1 ns	10 ns	10–50 ns	10–50 ns	1 ns
Program time	5–20 ns	10 ns	1.1 ns	10 ns	0.1–100 ms‡	0.1–100 ms‡	1 ns
Program energy/bit	120 pJ	5 pJ, needs refresh‡	5 pJ	0.4 pJ	30–120 nJ‡	10 nJ‡	0.02 pJ
Endurance	$>10^{15}$	$>10^{15}$	$>10^{15}$	$>10^{15}$	$>10^{15}$ read, $>10^6$ write‡	$>10^{15}$ read, $>10^6$ write‡	$>10^{15}$
Nonvolatility	Yes	No‡	No‡	Yes	Yes	Yes	Yes

*MRAM values are projected.
†These values are from the ITRS roadmap.
‡Technological shortfall for this memory device.

which the performance of the more conventional semiconductor memories is included. SMT-MRAM has the potential to dominate this aspect of memory technology, particularly because of its nonvolatility and very low power. Even at the 90-nm node, SMT-MRAM competes favorably with flash memory in density and has advantages over flash in speed, energy, and endurance.

Domain-wall "racetrack" memory. Spin momentum transfer can also provide a pathway to other novel memories as solid-state replacements for magnetic hard drives. This type of memory involves storing information by the presence or absence of a domain wall or boundary between oppositely magnetized regions of a ferromagnetic film. The domain walls form a linear array in a U-shaped magnetic film confined to a trench or channel in a silicon chip that is similar to the trench used to form the storage capacitor in a dynamic random access memory (DRAM; **Fig. 2**).

Oscillators. Another important property of spin momentum transfer is the generation of radio-frequency and microwave radiation by the conversion of the momentum of the spin-polarized current into coherent (that is, having a well-defined phase) spin waves in a magnetic host subject to a magnetic field. These spin waves radiate significant power in the frequency range from a few to tens of gigahertz. Theoretical predictions indicate a much larger bandwidth than has been experimentally observed to date. A frequency-agile nanoscale source of electromagnetic radiation in the frequency range from tens to potentially hundreds of gigahertz is attractive for a host of applications. For example, these oscillators can provide tunable sources for phased-array transceivers, sources for chip-to-chip and on-chip clocks, and local oscillators for handheld wideband radios.

Sensing devices. GMR-based sensing devices have also been developed, including analog magnetic field sensors, differential magnetic field sensors (gradiometers), digital magnetic field sensors, digital signal isolators, and isolated bus transceivers. The cost and power are extremely low, making these devices highly competitive. The performance of the isolators, in particular, can be much better than their optical counterparts at lower cost. These devices are sold in large numbers.

Spin-polarized field-effect transistors. While fast nonvolatile memories could be very important to increasing computer capabilities, a key bottleneck is moving information between memories and logic circuits. Ideally, if individual devices could both process and store information, transfer delays could be eliminated, at least for data in immediate use. A spin-based device that could accomplish this dual task is a spin-polarized field-effect transistor (spin FET). In a conventional FET, when a bias voltage is applied, a conducting channel is created between the source and the drain regions, allowing the transistor to act as a switch. If source and drain contacts are made from ferromagnetic materials, the electrons emitted from each contact have a preferential spin. Thus the current can be controlled either by applying a bias voltage as in a conventional FET, or by changing the

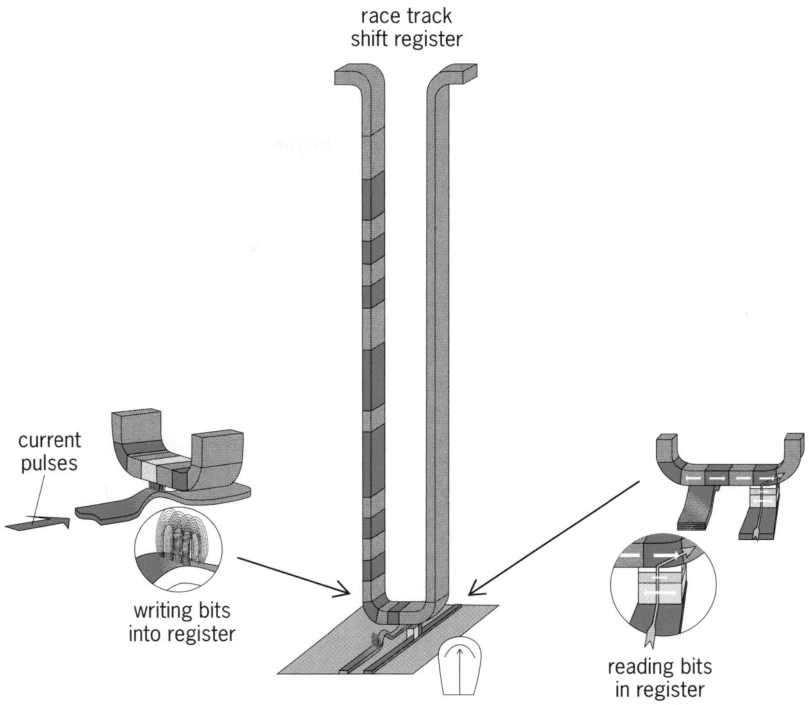

Fig. 2. Domain-wall "racetrack" memory device concept. To write bits into the racetrack shift register, current pulses move domain walls in nearby wire, and fringing fields from domain walls write bits. Magnetic tunneling junction (MTJ) sensor reads bit pattern in register.

orientation of the spins as they move from the source to the drain by rotation or by electric-field-controlled scattering.

There is, however, a serious difficulty that has so far prevented the development of practical spin FETs. The conductivity of ferromagnetic materials, generally metals, is much higher than that of the semiconductors that make up the rest of the FET. This means that there are far more mobile electrons in the ferromagnet than in the semiconductors, so only a few of the spin-aligned electrons are able to enter the semiconductor. For a large transfer of spin-aligned electrons, the conductivity of the ferromagnets and the semiconductors must be closely matched, or there must be a tunneling contact between the ferromagnet and the semiconductor to match the conductivities. One way to achieve this match is to utilize ferromagnetic semiconductors as the source and the drain.

The first ferromagnetic semiconductors with Curie temperatures (T_C) above 50 K (-223°C or -370°F), developed in 1996, were diluted magnetic semiconductors—alloys in which some atoms are randomly replaced by magnetic atoms, such as manganese. However, these early materials still had to be cooled to cryogenic temperatures to exhibit ferromagnetism. Subsequent research has shown that other types of semiconductors can exhibit ferromagnetism at much higher temperatures. In 1998 ferromagnetic behavior of GaMnAs was reported with a Curie temperature of about 110 K (-163°C or -262°F), which was subsequently raised to nearly 200 K (-73°C or -100°F). In 2000 room-temperature ferromagnetism in $TiCoO_2$ was discovered in Japan. There have also been reports, not widely reproduced, of ferromagnetism in many other semiconductors near or above room temperature. One of the key features of some of these materials is that they exhibit carrier-mediated ferromagnetism, in which the ferromagnetism is caused by the interaction of the magnetic ions with the carriers—electrons or holes. The Curie temperature and other magnetic properties can be modified by changing the carrier concentration with electric fields (gates) or with optical excitation. This ability to gate the magnetism by changing carrier concentration presents a new paradigm for novel devices in which carrier concentration and spin polarization are controlled concurrently. These discoveries appear to bring practical spin FETs within reach.

Spin resonant tunneling diodes. The addition of spin sensitivity can potentially produce devices that switch faster than any transistor. One such extremely fast switch is a spin resonant tunneling diode (RTD). This consists of a quantum well sandwiched between two insulating barriers. Current can flow only when the applied voltage reaches a precise value that allows a quantum-mechanically resonant state to exist within the quantum well. Such switches can turn on and off within less than 1 ps. However, conventional RTDs cannot substitute for transistors because they lack a third terminal that allows an input signal to alter the switch's functioning.

A spin RTD, in contrast, can act like a transistor. In such a device, the effective barrier height is different for spin-up and spin-down electrons in a modest magnetic field because of Zeeman splitting, generating two resonant voltages, one for each spin state. This allows the spin of the electron to be uniquely determined by measuring the tunneling probabilities. By using ferromagnetic contacts and thus varying the spin states of electrons in the current, the RTD can be switched between two states with different resonant voltage, allowing the third input to affect the current flowing across the RTD, and thus potentially creating an ultrafast logic device. Estimates show that, when scaled, the device could be 100 times better than CMOS in power, with the same speed (or be 10 times better in power and 10 times better in speed).

Quantum computing. Another avenue for using the spins of elementary particles comes from the rapidly developing field of quantum computing. The states of spin of electrons or other spin-$^1/_2$ particles can be used as an implementation of a qubit (quantum bit, the unit of quantum information). Information can be encoded using the polarization of the spin, manipulation (computation) can be done using external magnetic fields or laser pulses, and readout can be done by measuring spin-dependent transport. Quantum computers execute a series of simple unitary operations (gates) on one or two qubits at a time. The computation on a quantum computer is a sequence of unitary transformations of an initial state of a set of qubits. After the computation is performed, the qubits can be measured, and the outcome of the measurement is the result of the quantum computation. Quantum effects such as interference and entanglement are used as computational resources and make quick solutions to hard problems possible. For some very special problems, such as factorization of large prime numbers or exhaustive database searches, quantum-computing algorithms have been developed that show a very significant speed-up in computation time and a reduction in complexity. For certain calculations that find global properties of functions such as factoring and discrete logarithms, the speed-up for a quantum processor is dramatic. For these operations, a 30-logical-qubit quantum processor can perform the same calculation in the same time as a 10^9-bit classical computer.

Scientists are searching for quantum-mechanical two-state systems with long dephasing times, which would provide the ability to carry out computations before stored information is lost. It must be possible to readily farbricate and scale these quantum systems if they are to perform quantum algorithms. One very viable candidate for quantum information is electron spins in coupled quantum dots. However, other two-level systems have been proposed for implementing qubits, and include nuclear magnetic resonance (NMR), which involves nuclear spins in special molecules; excited states of ions in traps; cavity quantum electrodynamic systems; Josephson junctions; and SQUIDs (superconducting quantum interference devices). The potential uses of quantum qubit systems range from quantum key distribution, quantum encryption, and quantum dense coding to quantum teleportation and ultraprecise clock synchronization.

For background information *see* COMPUTER STORAGE TECHNOLOGY; DOMAIN (ELECTRICITY AND MAGNETISM); FERROMAGNETISM; MAGNETIC RECORDING; MAGNETISM; MAGNETIZATION; MAGNETORESISTANCE; QUANTIZED ELECTRONIC STRUCTURE (QUEST); QUANTUM COMPUTATION; SEMICONDUCTOR MEMORIES; SPIN (QUANTUM MECHANICS); TRANSISTOR; ZEEMAN EFFECT in the McGraw-Hill Encyclopedia of Science & Technology.

Daryl Treger; Stuart A. Wolf

Bibliography. Special issue on spintronics, *Proc. IEEE*, vol. 91, no. 5, May 2003; S. A. Wolf et al., Spintronics: A spin-based electronics vision for the future, *Science*, 294:1488–1495, 2001; S. A. Wolf and D. Treger, Spintronics: A new paradigm for electronics for the new millennium, *IEEE Trans. Magnet.*, 36:2748–2751, 2000.

Suicidal behavior

Suicide, from the Latin phrase *sui caedere* (to kill one's self), is defined as the intentional termination of one's own life, and is a significant public health problem that encompasses a continuum of thoughts and behaviors ranging from suicide ideation to suicide attempts to death by suicide. Suicide is the 5th leading cause of years of potential life lost before age 65, and is the 11th overall leading cause of death in the United States. Approximately 32,000 individuals in the United States die by suicide annually, and the number of suicide attempts, although believed to be an underestimate, is undoubtedly much higher. In fact, whereas the number of suicide deaths in the United States has slightly decreased since the 1990s, the number of suicide attempts remains the same. This is an important consideration because suicide attempts are one of the strongest predictors of future death by suicide.

The methodology of suicidal death varies greatly by age and gender; however, sociocultural characteristics, geographical locale, and governmental policies can also affect the means that a person uses to take his or her own life. In the United States, where guns are easily accessible, firearm deaths are the leading method used in suicide, particularly for males. In 2004, suicide by firearm accounted for 51% of all suicide deaths, followed by suffocation and hanging (22%), poisoning (18%), cutting, and drowning. In countries that have stricter gun-control laws, such as Japan, firearm deaths are much rarer; however, Japan continues to have a very high rate of suicide due to other lethal methodology. In many countries, such as China, Malaysia, Sri Lanka, and Trinidad, particularly in rural areas where farming chemicals are accessible, pesticide ingestion is the most common means of suicide.

International and sociocultural variation in suicide rates is not limited to methodology and is also expressed in a number of other factors, including the perception of and reaction to suicide by family and the community, stigmatization of suicide, beliefs about help-seeking behaviors, and one's own understanding of suicidal thoughts and behaviors. For instance, social norms and religious perspectives may influence an individual's decision on whether or not to consider suicide, but may also influence important community responses to suicidal activity, such as ignoring suicide as a problem, underreporting of suicidal behavior, or classification of suicide as an accident. This latter phenomenon, whether conscious or unconscious, often results in the misclassification of drug overdoses, falls and other mishaps, and automobile accidents as nonintentional rather than purposeful suicide deaths. The marked international and cultural difference in suicide rates and patterns around the world also speaks to the rich diversity of psychological belief systems that exist, as well as to the current political and socioeconomic status of a particular country or group of people. The United States ranks approximately 46th in the world for deaths by suicide (11 suicides per 100,000 people), whereas Russia (38 per 100,000) and the countries of the former Soviet bloc and other Eastern European countries have some of the highest rates of suicide in the world. In contrast, many Middle Eastern and South American and Latin countries have comparatively low rates of suicide. Depending on the cultural perspective, suicide may be prohibited or tolerated, viewed as a sign of bravery or as an indication of despair or weakness, demonized or even romanticized, as is sometimes the case with artists or musicians who die by suicide. In many cases, suicidal behavior is viewed with some sense of taboo, and has been socially, religiously, and legally penalized throughout history. Such beliefs may promote a stigma toward suicidal behavior and may inhibit help-seeking behaviors.

Etiology. Suicide is often viewed as the result of a mental illness or as the aftermath of some difficult life experience or stressor, and there is empirical support for these beliefs. There is a strong association between psychopathology and suicide. Mood disorders, including depression and bipolar disorder, alcohol and substance abuse, schizophrenia, and (for adolescents) conduct disorder are the mental illnesses most frequently related to suicide. Of individuals who die by suicide, 80–90% have some form of psychiatric disorder. The risk of suicide is increased by more than 50% in individuals with a mood disorder, and research suggests that 60–90% of those who committed suicide were depressed at the time of death. Recurrent and untreated episodes of depression also magnify risk for suicide. From a neurobiological perspective, altered function of the serotonin neurotransmitter system in the central nervous system is also implicated in suicide. Serotonin, also known as 5-hydroxytryptamine (5-HT), is a naturally occurring compound derived from tryptophan that is generated in the brainstem and appears to play a role in regulating mood and behavior. Reduced levels of 5-HT and its markers, serotonin platelet transporters (5-HTT) and cerebrospinal fluid serotonin [CSF 5-HIAA (5-hydroxyindoleacetic acid)], are associated with disordered behavioral inhibition, including aggression and impulsivity, as well as the more common, endogenous symptoms of clinical depression.

There is also a significant relationship between the experience and perception of stress and suicide, as well as between internal emotional states and traits and suicide. The experience of negative and potentially traumatic life events, such as neglect, abuse and violence, relationship or financial difficulties, the loss of a spouse or loved one, medical illness, uncontrollable pain, or functional impairment, is often significantly associated with the development of suicidal behavior. The emotional distress of negative or traumatic life events that seem uncontrollable and irresolvable may lead to the development of feelings of hopelessness and helplessness that, in turn, are strongly associated with suicidal behavior. Poor evaluations of self-efficacy and self-esteem, as well as a thwarted ability to achieve life goals or ambitions, often lead to a sense of failure or devalued sense of self-worth. Such beliefs about the self may create a predisposed vulnerability in an individual that weakens his or her capacity to buffer against negative outcomes, such as depression and suicide.

Risk and protective factors. Risk factors may be thought of as leading to or being associated with suicide; that is, people "possessing" the risk factor are at greater potential for suicidal behavior. Protective factors, on the other hand, enhance resilience and reduce the likelihood of suicide. Risk and

TABLE 1. Warning signs and risk factors for suicide

Warning signs (acute danger)	Risk factors (sustained vulnerability)
Abrupt changes in personality	Previous suicide attempt
Acting reckless	Family history of suicide; exposure to suicide
Giving away possessions	Major depression, bipolar disorder, schizophrenia
Change in eating or sleeping patterns	Legal or academic problems
Unwillingness or inability to communicate	Stressful or abusive family or interpersonal relationships
Unusual sadness, discouragement, and loneliness	Acute or chronic stress
Withdrawal from people or favorite activities	Serious injury or illness
Talk of wanting to die	Drug and alcohol abuse
Restlessness, confusion, inability to concentrate	Loss of home, spouse, job, roles, or functionality
Neglect of academic work and/or personal appearance	Male gender, older adulthood
Rebelliousness, running away from home or truancy from school	Personality: low extraversion, low openness to experience, lack of optimism and future orientation, hopelessness

protective factors may be biopsychosocial, environmental, or sociocultural in nature (see **Table 1**). Some risk factors are remediable and can be reduced via interventions such as providing lithium for manic-depressive illness or strengthening social support in a community. Risk factors that cannot be changed (such as a previous suicide attempt) can alert others to the heightened threat of suicide during periods of distress. Protective factors might include an individual's attitude and behaviors, as well as attributes of the environment and culture.

There are many risk factors associated with suicide; however, the impact of a risk factor may vary with age, gender, and other sociodemographic characteristics. Broadly, the presence of negative life events, medical illness and functional impairment, and mood-related dysfunction, such as depression, hopelessness, and pessimism, are well-established risk factors for suicide. Other markers of vulnerability, such as substance abuse and past psychiatric treatment, contribute significantly to risk for suicide. A previous suicide attempt is one of the strongest predictors of a future attempt and should be taken seriously in the evaluation of a suicidal individual. Past experiences may also place an individual at risk. Negative and stressful life experiences in childhood and adolescence are associated with increased suicide ideation and suicide attempts in adulthood. Sociodemographic characteristics are useful in understanding the profile of suicidal individuals, who are often teenagers, young adults or older adults, more often male than female, more often Native American or Caucasian than other ethnicities, and more often divorced, separated, or widowed. Cultural and geographic elements may also be associated with increased risk for suicide. In the United States, rates of suicide are highest in the intermountain states [that is, the states between the Rocky Mountains (to the east) and the Sierras (to the west in the south) and the Cascades (to the west in the north), including Utah, Idaho, and Nevada] and, over the last two decades, rates of suicide in rural areas (17.9 per 100,000) have often surpassed those of urban areas (14.9 per 100,000) [see **Table 2**]. Rural areas may be potentially isolating and struggle with workforce migration, an aging population, and economic decline that contribute to suicide risk. Urban areas have their own unique set of risk factors, which might include overcrowding, poverty, violent crime, and inadequate familial and social support.

A more comprehensive understanding of suicide is provided by considering the role of individual and environmental protective factors in suicidal behavior. Protective characteristics may buffer against suicidal thoughts and behaviors by directly acting against risk factors or moderating their effect, thereby providing some defense against self-harm. Personal protective factors might include sufficient coping skills, adaptive personality characteristics, the presence of positive future-oriented thoughts, setting and achieving goals, and the existence of successful, interpersonal relationships. Protective factors may also exist at the environmental level, such as access to and receipt of effective clinical care for mental illness, restricted access to lethal means of suicide, and adequate interpersonal and community support networks.

Age, gender, ethnicity, and suicide. Sociodemographic characteristics, such as age, gender, and ethnicity, play a complex role in epidemiological patterns of suicidal thoughts and behaviors (**Fig. 1**). Young and old adults, males and females, and members of different ethnic groups manifest distinct differences in suicide. In the United States, Caucasians (12.3 per 100,000) have higher rates of suicide deaths than African-Americans (5.2 per 100,000). Other ethnic and cultural groups have also exhibited increased risk for suicide, including Native Americans and other indigenous groups, and recent immigrants, such as foreign university students.

Gender differences in suicide rates also exist. For all racial groups, women attempt suicide more than males, whereas males have a higher rate of death by suicide than females. This discrepancy may be due to gender differences in suicide methodology; males are more likely to use more lethal means, such as

TABLE 2. Suicide rates by region in the United States (2004)

Region	Rate (per 100,000 population)	Number
West	12.4	8373
South	11.8	12,557
Midwest	10.8	7114
Northeast	8.1	4395
Nation (total)	11.1	32,439

a firearm, whereas females are more likely to use other methods, such as poison or hanging. In the United States, males die by suicide at a rate four times that of females; however, females attempt suicide three times more often than males. For African-Americans, males (8.98 per 100,000) have a much higher rate of suicide than females (1.8 per 100,000), who have the lowest suicide rate of all racial/gender groups. For African-American youth, the rate of male suicide (12.19 per 100,000) is approximately six times higher than that of females (2.16 per 100,000), and older African-American adult males account for 90.5% of deaths by suicide in this ethnic group. For older adults in general, approximately 84% of elderly suicides are male, and the rate of male suicides in late life is approximately 7.7 times greater than for female suicides. In fact, the suicide rate for women typically declines after age 60.

Different age groups typically manifest unique rates and patterns of suicide and risk factors for suicide (**Fig. 2**). Suicide is the third leading cause of death during late adolescence and early adulthood, accounting for 12.9% of all deaths among those between 15 and 24 years of age. For older adults, over the age of 50, there are more than 12,000 deaths by suicide annually in the United States; in fact, although the elderly comprise just 12.4% of the U.S. population, they account for almost 16% of suicide deaths. In the past 60 years, the suicide rate has quadrupled for males who are 15 to 24 years old, and has doubled for females of the same age range; however, there has been some decline (28.5%) since the mid-1990s. Elderly white males over the age of 65 have a rate of suicide greater than most other groups (~31 per 100,000), and white males over the age of 85 are at the greatest risk of all age-gender-race groups (48.4 per 100,000).

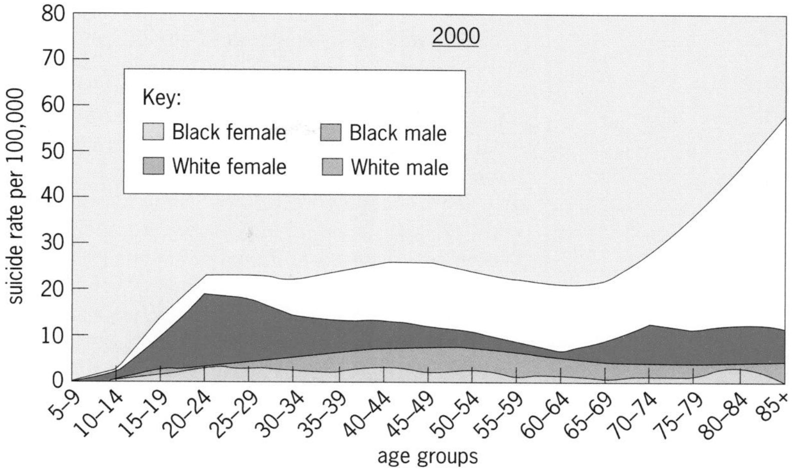

Fig. 1. Relationship of age, gender, and ethnicity with suicide. (*Source: National Institute of Mental Health; Data: Centers for Disease Control and Prevention, National Center For Health Statistics*)

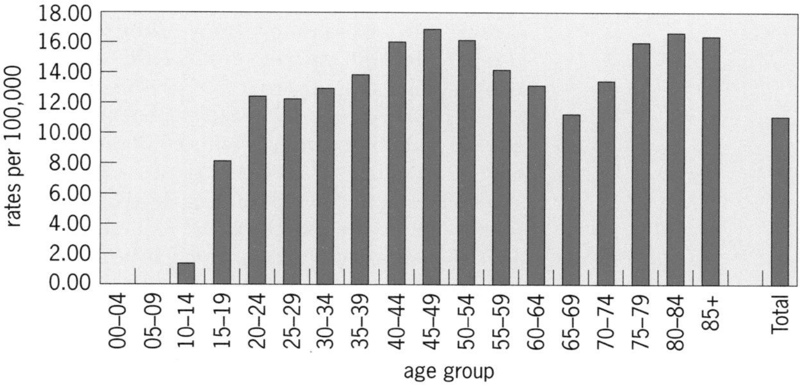

Fig. 2. Suicide rates for all ages. (*Source: Centers for Disease Control and Prevention, 2006*)

Prevention. In order to successfully prevent suicide, there must be ample understanding of the potential origins, mechanisms, and manifestations of suicidal thoughts and behaviors. Knowledge about warning signs for suicide and how to recognize them and assist a suicidal person are important elements of many suicide prevention programs. The most frequently identified warning signs include thoughts of self-harm; obsession with death; writing about death; sudden changes in personality, behavior, eating, or sleeping patterns; feelings of guilt; and decreased work or academic performance. One important approach to suicide prevention is gatekeeper training, which instructs participants to be vigilant in the recognition and reporting of suicidal symptoms. Gatekeepers may be anyone that could come into contact with a suicidal individual, such as teachers and bus drivers, and doctors and nurses. Training the public to recognize suicidal symptoms and increasing public knowledge about available mental health services are simple, yet helpful, prevention strategies.

Studies examining the effectiveness of suicide prevention programs have had mixed results. There are, however, many types of interventions that have been successful in reducing suicide, including community-based and individually focused treatment approaches. At the community level, screening programs and school-based efforts for youth and adolescents, home and primary care–based interventions for older adults, and workplace and community-based programs have been successful in reducing the occurrence of suicidal behavior. Most of these programs have in common the promotion of awareness of risk factors related to suicide, education of the public regarding available mental health services, and a reduction of stigma related to mental illness and help-seeking behavior. Involving local community members and organizations as gatekeepers is also a vital strategy toward preventing suicide, as are programs designed to target at-risk individuals at a known "point of access"—for instance, screening adolescents at school or older adults at a primary care office. The reduction of lethal means of committing suicide, or the restriction of access to those means, is also an important community-based type of prevention strategy. Gun safety and control, limiting the amount of medications that a patient is given, safety barriers on bridges or other potential jump sites, and elimination of coal gas use are just a few examples of means-reduction strategies that have successfully reduced suicide attempts or deaths by suicide.

Regulation of media coverage of suicide deaths is also an important community-based suicide prevention strategy. Suicide deaths are often portrayed dramatically or heroically, or are romanticized, and the detailed story and graphic images of the suicide death are typically repeated at length, with headline status. These representations of a death by suicide may, for vulnerable individuals, evoke identification with the suicide victim and may promote a phenomenon called suicide contagion. Suicide contagion, or copycat suicide, occurs when individuals attempt to mimic the method of the publicized suicide or attempt to commit suicide at the same location as the publicized suicide, often in the time period immediately following media reports of a suicide death. Media guidelines have been suggested that could help to reduce suicide contagion by educating the public about the nature of mental illness and suicide, instead of sensationalizing the death by suicide.

Suicide prevention and intervention strategies also often occur at the level of the individual and may include the provision of therapy or the administration of psychopharmacological agents, such as antidepressants. Cognitive and interpersonal therapies are effective treatments for suicidal thoughts and behaviors and, for individuals with borderline personality disorder, dialectical behavior therapy has been shown to be effective in suicide prevention. Therapeutic efforts often focus on the reduction of negative thoughts and emotions, such as hopelessness, but may also emphasize the promotion of positive thoughts and emotions, such as future orientation or optimism. Treating depressive symptoms with antidepressants, such as selective serotonin reuptake inhibitors (SSRIs), or schizophrenia with an antipsychotic medication, such as clozapine, typically has the effect of reducing suicidal thoughts and behaviors. Although the treatment of depression and other mental disorders through the use of counseling or medications often has the desirable secondary effect of reducing suicide, a direct focus on ameliorating suicidal thoughts and behaviors is also effective.

Although suicide is a difficult public health problem, it is generally preventable. Protective characteristics, in addition to appropriate treatment, can help a person who is feeling suicidal. Increased social, political, and economic support for the study and prevention of suicide over the past decade has resulted in many advances in the understanding of suicidal behavior; however, there is a need for continued clinical and research-based investigation of the complexities of suicide.

For background information *see* AFFECTIVE DISORDERS; ALCOHOLISM; PSYCHOLOGY; PSYCHOPHARMACOLOGY; PSYCHOSIS; PSYCHOTHERAPY; PUBLIC HEALTH; SCHIZOPHRENIA; SEROTONIN; SUICIDE in the McGraw-Hill Encyclopedia of Science & Technology.

Jameson K. Hirsch

Bibliography. D. Blazer, Selected annotated bibliography on depression and suicide, *South. Med. J.*, 7:761–763, 2007; K. R. Conner et al., Psychological vulnerability to completed suicide: A review of empirical studies, *Suicide Life-Threat. Behav.*, 31:367–385, 2001; Y. Conwell, P. R. Duberstein, and E. D. Caine, Risk factors for suicide in later life, *Biol. Psychiat.*, 52:193–204, 2002; T. E. Joiner and M. D. Rudd (eds.), Adolescent suicide: Risk, assessment, and treatment [special issue], *J. Adolescence*, 21(4), 2002; National Center for Health Statistics (NCHS), NCHS data on injuries, Centers for Disease Control and Prevention, Washington, D.C., 2006.

Supernova 2006gy

A supernova is a brilliant cosmic explosion marking the violent death of a star. These explosions are extremely bright, and in its final blaze of glory, the death of a single star can outshine the combined light from the billions of other stars in its home galaxy. One recent supernova, named SN 2006gy, surprised astronomers because it was intrinsically much brighter than any previous explosion, and it lasted much longer as well. The tremendous amount of total energy released in the visual light of the explosion exceeded other supernovae by a factor of 100 or more, and traditional explanations of supernova power sources seem inadequate. (Astronomers use a simple naming scheme for supernovae, with SN and the year of explosion. The year is followed by a letter or two in order of discovery, starting with A through Z, then aa, ab, ac, . . . , ba, bb, bc, and so on.)

Different types of explosions. Over time, astronomers have pieced together the physical mechanisms that power supernova explosions. Most supernovae can be broadly categorized into two basic types.

The first type is the thermonuclear explosion of a white dwarf star (called a Type Ia), which in turn is the remnant core of a star initially eight times the mass of the Sun or less. If the white dwarf can gain additional mass from a nearby companion star in a binary star system, it may exceed the Chandrasekhar mass limit of 1.4 times the mass of the Sun. At that point, a thermonuclear detonation destroys the white dwarf, leaving nothing behind. The light we see is powered mainly by the radioactive decay of the isotope ^{56}Ni, which heats the ejected debris, with an amount of ^{56}Ni created in the explosion corresponding to about one-tenth the mass of the Sun or less.

The second type of supernova is caused by the gravitational collapse of the iron core of a massive star, initially more than eight times the mass of the Sun. What astronomers see from a core-collapse supernova can vary greatly, depending on how much mass has been lost from the star in its lifetime. If a supernova shows a large quantity of hydrogen in its spectrum (called Type II), it is thought to be the explosion of a relatively cool red supergiant that retained much of its initial hydrogen envelope. Other core-collapse supernovae show no hydrogen, and instead have spectra dominated by helium or even heavier elements (Types Ib and Ic). These are thought to be the deaths of more massive stars that have shed their outer hydrogen envelopes through strong stellar winds. Still others have an intermediate fate, dying as blue supergiants with some of their initial hydrogen envelopes intact (also called Type II; SN 1987A is a famous example). The set of physical parameters that determine the end fate of a massive star are not yet fully understood.

Discovery and observation. In September 2006, R. Quimby discovered a new object that earned the designation SN 2006gy. Correcting for the known distance of 240 million light-years to its home galaxy, called NGC 1260, it became clear early on that SN 2006gy was extremely luminous compared to most supernova explosions.

At first it was thought that SN 2006gy might not be a supernova at all, that it might instead be a bright active galactic nucleus. Although SN 2006gy was located close to the nucleus of its home galaxy, subsequent high-resolution images obtained with ground-based telescopes as well as x-ray images from the National Aeronautics and Space Administration's (NASA's) *Chandra X-ray Observatory* showed that the source of SN 2006gy was in fact separate from the nucleus of the galaxy NGC 1260 and was therefore a genuine supernova. Meanwhile, SN 2006gy continued to brighten, and its spectrum confirmed that it was a Type II supernova.

Most supernovae reach their peak brightness within a couple of weeks and then start fading, but SN 2006gy started as bright or brighter than most ever get, and just kept getting brighter (as can be seen from the light curve of SN 2006gy compared to some other representative supernovae in **Fig. 1**). After about 70 days, SN 2006gy reached its peak of roughly 5×10^{10} times the Sun's luminosity and faded very slowly thereafter. That peak luminosity was about 10 times brighter than a Type Ia (normally the brightest type of supernova) and about 100 times brighter than most Type IIs. In short, SN 2006gy was more luminous than any other supernova, and it remained bright longer than any other. Integrated over time, the total energy released in visual light was more than 1.5×10^{44} joules, more than 100 times that of a normal supernova. That is roughly the amount of energy that will have been radiated by the Sun in its entire 10^{10}-year lifetime.

Extreme energy budget. The tremendous radiated power output of SN 2006gy in visual light exceeds all other known examples of supernovae so far. As a result, well-established theoretical explanations that are successful in explaining the engines of most supernovae come up short. There are basically two remaining ideas for power sources that might potentially explain SN 2006gy, one more exotic than the other.

Conversion of mechanical shock energy into light. In most supernovae, the radiated visual light represents a small fraction (a few percent) of the total energy budget. Most of the total energy (ignoring neutrinos) goes into the kinetic energy of the mass of expanding stellar debris, which almost always seems to be about 10^{44} J. If this reservoir of mechanical energy could be tapped, it might be able to power SN 2006gy. For this to happen, the blast wave of the explosion must run into dense material in the immediate surroundings of the star, giving up its mechanical energy into heat and radiation.

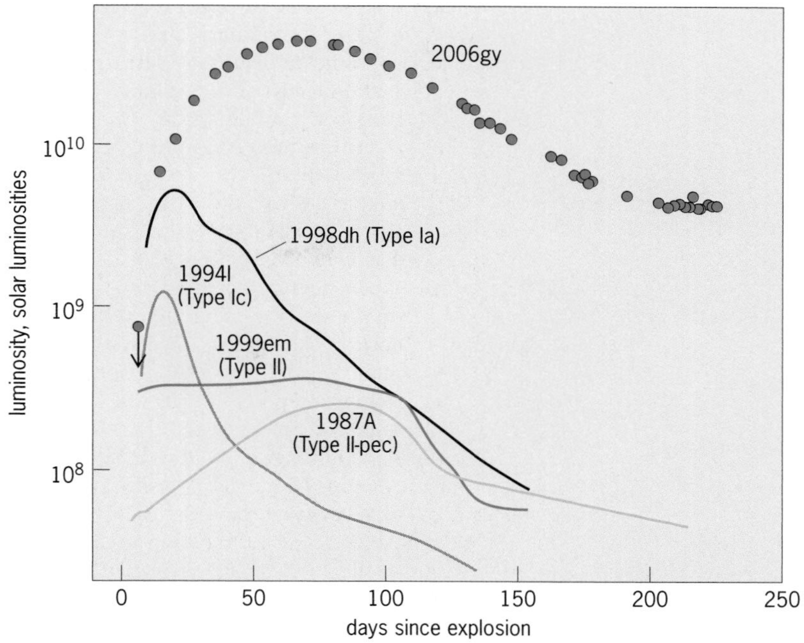

Fig. 1. The light curve of SN 2006gy compared to a few other representative supernovae. SN 1998dh is a Type Ia (the explosion of a white dwarf), SN 1994I is a type Ic (explosion of a stripped-envelope massive star), and SN 1999em is a Type II (explosion of a red supergiant). SN 1987A is the nearest supernova in recent years and the best studied; it was a somewhat peculiar explosion of a blue supergiant. (*N. Smith, University of California, Berkeley*)

Where would this dense environment come from? Some very massive stars can have less violent presupernova surface explosions that do not destroy the star. One example is a famous star in our own Milky Way Galaxy called Eta Carinae, which has a mass of about 100 times the Sun's mass and is probably the most massive star in the Milky Way. It had a violent explosion in the nineteenth century, but the star survived and today we see it surrounded by a spectacular nebula (**Fig. 2**). If the star shed a large amount of surface mass right before the supernova in this way, then the powerful blast wave from the final supernova would plow into that material and give a brilliant fireworks display. By converting the shock energy into radiated light, this mechanism can make the core-collapse supernova much brighter visually than it would have been otherwise. These supernovae show bright and narrow lines of hydrogen in their spectra, arising when the blast wave is decelerated by the matter that it plows into.

Can this mechanism explain SN 2006gy? In order to be able to convert enough shock energy into light, the blast wave needs to plow into more than 10 times the mass of the Sun situated conveniently outside the star. That is similar to the amount of mass ejected in the nineteenth-century eruption of Eta Carinae. However, this huge mass reservoir needs to have been ejected (coincidentally) just a decade or so immediately before the final supernova explosion. SN 2006gy does show tell-tale signs of this kind of blast-wave interaction: it has narrow hydrogen lines in its spectrum indicating expansion speeds of about 4000 km/s (2500 mi/s), and the supernova was also

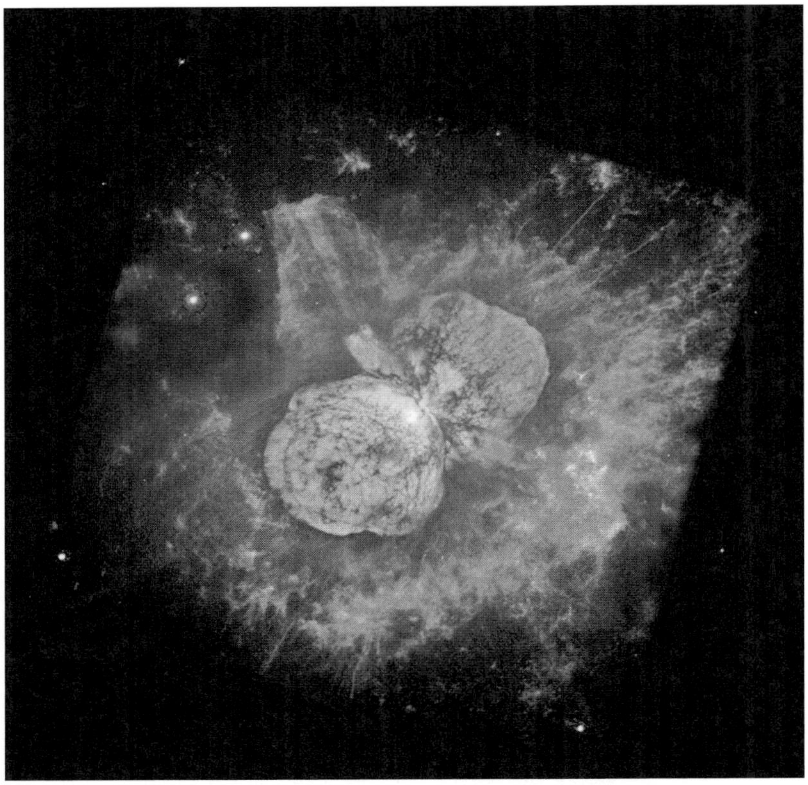

Fig. 2. The nebula around the massive star Eta Carinae, probably the most massive star in our own Milky Way galaxy. It is surrounded by clouds containing about 10 times the Sun's mass, caught in this image from the *Hubble Space Telescope*. (*N. Smith, University of California, Berkeley; NASA*)

detected in x-rays that come from hot shocked gas. However, the strength of that interaction indicated by the hydrogen lines and x-rays is hundreds of times too weak compared to what is needed to power the visual luminosity of the supernova.

There are also more fundamental physical limitations to consider. In general, the total available energy reservoir of a core-collapse supernova (or the explosion of a white dwarf) is the 10^{44} J of kinetic energy. Even with an overly generous conversion efficiency of 100%, this could barely account for the total radiated energy of more than 1.5×10^{44} J observed from SN 2006gy. The phrase "overly generous" is probably a gross understatement here, because observations indicate that during the first 200 days of SN 2006gy, the observed expansion speed did not drop. That would introduce a paradox that the blast wave did not slow down as it was drained of all its energy, which would mean that energy had to be continually added to avoid any contradiction with the laws of physics. The only solution would be if the blast wave initially had much more than the canonical 10^{44} J of kinetic energy, requiring that SN 2006gy was a superenergetic event.

Pair-instability supernova hypothesis. A more exotic explanation for the power source of SN 2006gy could be that it was not a core-collapse supernova at all, but an entirely different type of explosion mechanism called a pair-instability supernova. This mechanism is based on a theoretical idea proposed in the 1960s, which has not been confirmed by observations yet. In the interior of a sufficiently massive star, the pressure and temperature can be so high that high-energy gamma rays spontaneously split into pairs of electrons and positrons. This pair production removes pressure support inside the star, initially causing its core to collapse. This collapse triggers explosive thermonuclear burning that obliterates the star, sending all the fragments of the star into space and leaving behind no remnant like a black hole or neutron star. In this case, the power source for the radiated light we see is the radioactive decay of ^{56}Ni, which heats the surrounding stellar debris, much as in the explosion of a white dwarf. However, in this case the mass of ^{56}Ni is much higher; it can be several times the Sun's mass just in ^{56}Ni.

While exotic, this mechanism has some advantages over the conversion of shock energy into light for explaining the power source of SN 2006gy. First, the high luminosity, the slow rise to maximum, the long duration, the presence of hydrogen in the spectrum, and the slow expansion speed observed in SN 2006gy are all properties that are predicted by theoretical models of pair-instability supernovae. Also, since the visual light is powered by radioactive decay of ^{56}Ni instead of shock energy, it is no longer problematic that the tell-tale signatures of shock-wave interaction are hundreds of times too weak. Similarly, it is also no longer a paradox that the blast wave did not decelerate in the first 200 days, because the blast wave is not being drained of its energy. While far from proven, SN 2006gy does present the best candidate so far for a pair-instability supernova explosion. This idea has broad importance, because theoretically, pair-instability supernovae are predicted to occur only in the very early universe, when stars are thought to have been more massive on average, and when they are expected to make it to the ends of their lives with much of that mass intact. Such stars and their supernovae are too faint to be seen directly with current capabilities, but if SN 2006gy is representative of pair-instability supernovae, it might be possible to see them with future observatories like NASA's planned *James Webb Space Telescope*.

Then again, in general it is important to be cautious in appealing to physical mechanisms that have no empirical verification, and for which this one object would be the only example. Fortunately, this is a case where continued observations may be able to verify whether SN 2006gy continues to show the characteristic rate of decline in luminosity that is expected for radioactive decay in the pair-instability hypothesis. If not, some more thinking will need to be done about how to extract more radiated energy from conventional mechanisms. In either case, one thing is clear: the star that exploded to produce SN 2006gy was an extremely massive star, perhaps even the most massive star ever seen to explode. That is correct, in principle, because while such extremely massive stars are rare, SN 2006gy seems to stand alone among supernovae. As astronomers continue to search for supernovae, it will be interesting to see whether more objects like SN 2006gy are discovered.

Indeed, in October 2007, after the last sentence was written, Quimby announced that a supernova that he had discovered in 2005, named SN 2005ap, had an extremely high peak luminosity, about twice that of SN 2006gy. Like SN 2006gy, SN 2005ap was a Type II supernova, but its spectrum was not unusual. Although SN 2005ap was more luminous at its peak than SN 2006gy, it faded very quickly (in 20–30 days, like a normal supernova), so its total energy budget is not stupendous like SN 2006gy, and it is not problematic for more conventional explanations of supernovae.

For background information *see* CHANDRA X-RAY OBSERVATORY; ELECTRON-POSITRON PAIR PRODUCTION; GALAXY, EXTERNAL; STELLAR EVOLUTION; STELLAR POPULATION; SUPERNOVA in the McGraw-Hill Encyclopedia of Science & Technology.

Nathan Smith

Bibliography. A. Heger and S.E. Woosley, The nucleosynthetic signature of population III, *Astrophys. J.*, 567:532–543, 2002; E. O. Ofek et al., SN 2006gy: An extremely luminous supernova in the galaxy NGC1260, *Astrophys. J.*, 659:L13–L16, 2007; N. Smith et al., SN 2006gy: Discovery of the most luminous supernova ever recorded, powered by the death of an extremely massive star like Eta Carinae, *Astrophys. J.*, 666:1116–1128, 2007.

Sustainable development in mining

Mining of nonrenewable mineral resources is not a sustainable activity in the long term because resources become depleted and mines eventually close. On the other hand, mining can contribute to sustainable development through conversion of mineral capital into other forms of more durable social or human capital. This conversion only partially involves material utilization of the mineral products themselves, for example in infrastructure development, because much of the product is effectively consumed and sent to waste. High-value ornamental minerals such as gold and gem diamonds are anomalous in being of little practical value and are mostly hoarded. Their physical contributions as minerals to sustainable development are therefore minimal, although they are clearly of societal value. In contrast, the key contribution that the minerals industry makes to longer-term sustainable development is through conversion of its products to wealth. A portion of that wealth ideally returns to and benefits wider society through taxes, royalties, employment, and other payments and services that reflect the value and quantity of the commodity mined. A critical link in this conversion is the role of governments, which will decide how royalty and tax revenues are spent or invested. Stable and efficient governments, committed to sustainable development principles, coupled with fair and effective enforcement of laws, are essential to this process. In the absence of good governance or the capacity for governance, industry (of any kind) cannot effectively contribute to sustainable development, and may in fact help produce the opposite result by bankrolling and sustaining inefficient or corrupt administrations.

Minerals and society. Modern society is totally dependent on mined materials, ranging from sand and gravel for concrete and roads, to steel and copper for buildings, vehicles, and electrical wiring, and gold and other precious metals for currency and ornamentation. Mining also supplies energy materials, such as uranium, coal, and bitumen (from oil sands). Much of this material is effectively consumed, and as populations and standards of living continue to grow, so does the demand for mined products. Despite this demand and dependency, mining is commonly viewed by society as a dirty business, causing both environmental damage and social disruption, and is increasingly being marginalized from the developed world (for example, by restrictive environmental legislation). Consequently, much mineral exploration and new mine construction is now focused in the developing world or in frontier geographical regions (such as northern Canada), regions populated by indigenous peoples and characterized by fragile ecosystems. This situation poses both challenges and opportunities for mining companies: challenges to meet environmental standards in often harsh conditions and to deal with societal values and expectations that may differ greatly from "Western" sensibilities; and opportunities for long-term sustainable development of societies at both the local and national scale through the wealth that mining generates. Mining projects also present challenges (risks) and opportunities for societies at both local and national levels, and communities are increasingly saying "No!" to mining where the risks are perceived to be too high or the rewards too low. In addition, societies have a role in ensuring that the rewards are not simply consumed but are invested in a sustainable future, both through personal lifestyle choices and (in functioning democracies) through the people elected to government.

The globalization of knowledge and information, and the effective campaigning of special-interest nongovernmental organizations, in recent decades has brought a critical focus on industries that pollute or commit human rights abuses, whether they are operating in the developed or developing world. In the 1980s, industry learned to address (although not necessarily completely resolve) environmental issues, but only recently has begun to understand the importance of social and socioeconomic issues, to its cost where mining operations have been blocked or shut down by civil opposition and unrest. This understanding is encapsulated in the concept of corporate social responsibility.

Corporate social responsibility. At root is a basic expectation of fair treatment and compensation for people who have been displaced or otherwise economically disadvantaged by mining, or any other operations. Such expectations are largely taken for granted in the developed world, where they are protected by law, and the magnitude of compensation is of a scale that the disturbance of large populations rarely if ever now takes place. But in the developing world (or in marginalized parts of developed countries), it is often assumed that the cost of compensation is much lower. In the past (and sometimes still today) at best there were token one-time cash payments and at worst people were driven forcibly from the land. Forced closure of the giant Panguna copper mine on Bougainville Island, Papua New Guinea, in 1989 by Bougainville Islanders disgruntled with low compensation payments (among other issues) dramatically illustrated the inadequacy of simple monetary compensation for the loss of land and livelihoods. It is now accepted that compensation must be of a form and quantity that will enable the recipients to participate in the new economy generated by the mine, and that the benefits should endure beyond the life of the mine.

The opportunity that this form of compensation provides is for long-term and potentially sustainable economic advancement of local and regional societies, which is clearly of particular value in developing countries.

Governance. Although the complexities of developing effective compensation strategies are challenging, they are theoretically manageable, assuming a functioning and responsible governance structure with the capacity to act. However, this assumption is commonly unfounded, especially in relation to capacity for governance. Such capacity issues arise even in the developed world, where laws may exist but are not enforced. (An example is found today in Alberta, Canada, where the rate of new applications for tar-sands mining projects challenges the government's capacity to fully evaluate their impacts.) Under these circumstances, industry may find itself obliged to assume the role of de facto government, which is a risky proposition because such a role has no democratic foundation, and because industry is unqualified to conduct it. An even greater risk is that industry's financial contributions may, in fact, support and sustain corrupt or inefficient governments, making development of any kind difficult.

Typically, a balance is sought between the separate obligations on industry and governments. On one hand, industry should provide not only direct compensation for loss of land and livelihood, but also investment into local and national society through infrastructure development, royalty payments, and taxes. On the other hand, governments at various levels should reinvest mineral revenues to the long-term benefit of the societies they represent.

While industry can be expected, and can be compelled by law, to deliver on its side of this bargain, there are rarely any unqualified guarantees that governments will deliver on their side. Failure by governments to invest in local communities invariably reflects negatively on the company concerned, which may be the only visible authority or source of economic power present at the local level. This can lead to opposition to mining or even mine closure if the situation is not resolved. Such concerns are not limited to the developing world; for example, debates about the reinvestment of government revenues from oil sands mining are raging today in Alberta.

Industry's role. What can industry do to ensure that sustainable development follows from its activities? Squarely within its powers and responsibilities are the minimization of environmental and social impacts on directly affected communities, and maximization of benefits in both the short and long terms to those communities and the environment (for example, through continuous and effective site remediation and containment). A major contribution that can be made is the fostering of secondary local industries that can survive beyond the life of the mine. Other possible contributions include direct funding or sponsorship of education and health programs in the broader project-affected areas (preferably in perpetuity through trust funds). Whether an individual company can influence the investment decisions of regional and national governments is likely to depend on the specific circumstances, including the jurisdiction and governments in question, the size of the mine (that is, its economic importance to the region or country), and the size and standing of the company. In extreme cases, where corruption and human rights abuses are endemic or where there is a complete absence of law and order (such as civil war, which occurred in the Democratic Republic of the Congo, Sierra Leone, and Angola), socially responsible companies have little or no justification for initiating operations. Indeed, few companies or investors would be willing to accept the political and security risks of operating directly in such situations. However, indirect operations (such as product buying) are more common, and were the focus of the "Blood Diamond" campaign in 1998 to halt the sale of diamonds from Angola (and later other countries) that were being used to fund civil wars, and more recently the "No Dirty Gold" campaign, similarly aimed at jewelry retailers.

On a more general note, there may be more that industry can do to achieve sustainability goals at a higher, collective level. Mining companies are not natural collaborators, but commonly join forces to lobby governments through national and international mining organizations. The International Council on Mining and Metals (http://www.icmm.com) was established in 2001 to provide industry leadership toward sustainable development goals, and might do more in terms of influencing governments to adopt sustainable development objectives, based on the revenues generated from mining. A place to start might be to lobby for international agreements on standard royalty rates, which would remove the "race to the bottom" that minimizes returns to host countries from mining in the hope of attracting competitive foreign investment. The reality is that it is not so much the absolute taxation or royalty rates that concern investors, as uncertainty about changes in those rates. Current royalty rates on minerals are typically 3% or less, and mostly charged on profits, not on revenues or net smelter return, which would more accurately reflect the real value of the extracted commodity. They are too low to contribute significantly to sustainable development. Uniformly higher royalty rates would see increased prices passed on to consumers, but the cost of mineral components in most consumer products is a small percentage of the overall cost, so this effect would not be large. The overall benefit to supplier countries, which are usually developing countries or jurisdictions, could be huge if those increased revenues were wisely invested in sustainable development programs.

For background information *see* CONSERVATION OF RESOURCES; ENVIRONMENTAL MANAGEMENT; MINERAL RESOURCES; MINING in the McGraw-Hill Encyclopedia of Science & Technology.

Jeremy P. Richards

Bibliography. J. Connell, Compensation and conflict: The Bougainville copper mine, Papua New Guinea, in J. Connell and R. Howitt (eds.), *Mining and Indigenous Peoples in Australasia*, pp. 55–75, Sydney University Press, Sydney, Australia, 1991; C. Filer (ed.), *Dilemmas of Development: The Social and Economic Impact of the Porgera Gold Mine, 1989–1994*, Asia Pacific Press, Canberra, Australia, 1999; M. Hopkins, *Corporate Social Responsibility*, Earthscan, London, 2006, and Stylus, New York, 2007; H. M. Peskin, Sustainable resource accounting, in *Assigning Economic Value to Natural Resources*, pp. 59–66, National Academy Press, Washington, DC, 1994; J. P. Richards, "Precious" metals: The case for treating metals as irreplaceable, *J. Cleaner Prod.*, 14:324–333, 2006; M. E. Sandbu, Natural wealth accounts: A proposal for alleviating the natural resource curse, *World Develop.*, 34:1153–1170, 2006.

Swift gamma-ray burst mission

The *Swift* gamma-ray burst (GRB) satellite was launched by the National Aeronautics and Space Administration (NASA) into low-Earth orbit on November 20, 2004. *Swift*'s primary scientific objectives are to study gamma-ray bursts, brief cosmic blasts of very high energy electromagnetic radiation, and their afterglows, longer-lasting emissions in x-ray, ultraviolet, and visible light, which are produced by the cooling embers from the initial explosion. *Swift*'s unique design features three different telescopes on board: the Burst Alert Telescope (BAT), the X-ray Telescope (XRT), and the Ultraviolet-Optical Telescope (UVOT) [**Fig. 1**].

Instruments. The BAT is the main instrument on board *Swift*, continuously viewing about one-sixth of the sky to provide the initial location when a gamma-ray burst is detected. It is a coded-aperture telescope (that is, it has holes in a plate through which the gamma rays must pass, with spacing among holes determined in ways that allow directions to be found), with solid-state detectors made from a combination of cadmium, zinc, and tellurium, and it is sensitive to energies from 15 to 150 keV. Within about 20 s following a burst trigger, the BAT provides a celestial location that is accurate to within a few arcminutes. Using six on-board momentum wheels, the *Swift* spacecraft then "swiftly" and autonomously turns to center the burst in the narrower fields-of-view of the XRT and the UVOT, obtaining images and spectra of the afterglow emission. The BAT location for the burst is also sent down to Earth for the use of ground-based observers at many wavelengths. Occasionally, these locations are also sent to other space-based facilities such as the *Hubble Space Telescope* or the *Chandra X-ray Observatory*.

The XRT is an imaging x-ray telescope, using optics in which the x-rays strike the mirror at a low angle (so-called grazing incidence, with an optical design known as Wolter I), with charge-coupled-device (CCD) detectors sensitive to energies from 0.2 to 10 keV and positional accuracy of 3–5 arcsec. The UVOT is a 30-cm-diameter (12-in.) modified Ritchey-Chrétien telescope, with intensified CCD detectors augmented with six different filters spanning wavelengths from 170 to 650 nm. The UVOT can locate the burst's position to within less than 1 arcsec, and can also provide redshift information that is used to determine the distance to the burst.

Coordination with ground-based telescopes. Further detailed studies of the afterglows are conducted by a worldwide network of over 30 teams of scientists using larger ground-based infrared and visible-light telescopes, as well as radio observatories. Scientists using these facilities are rapidly alerted by the Gamma-Ray Burst Coordinates Network (GCN), which relays the initial locations determined by the BAT and, shortly thereafter, the more accurate XRT and UVOT positions. The GCN also provides a repository for the ground- and satellite-based observers to post their scientific results. In many cases the burst alerts are directly transmitted into the control systems for robotic telescopes linked through the Internet. The telescopes then autonomously repoint to observe the sky locations of the bursts.

GRB characteristics. *Swift* detects about 100 gamma-ray bursts per year. At typical distances of billions of light-years, the energy emitted in a gamma-ray burst is more than a billion billion (10^{18}) times the

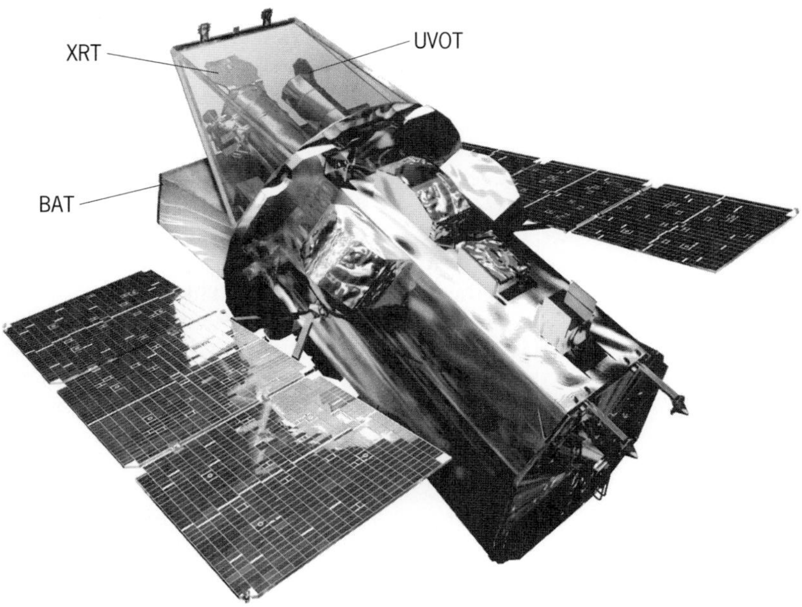

Fig. 1. Artist's rendering of *Swift* with solar panels extended. The sunshade has been rendered transparent so that the X-ray Telescope (XRT) and the Ultraviolet-Optical Telescope (UVOT) can be seen. A portion of the Burst Alert Telescope (BAT) is visible below the sunshade. (*Courtesy of NASA E/PO, Sonoma State University, Aurore Simonnet*)

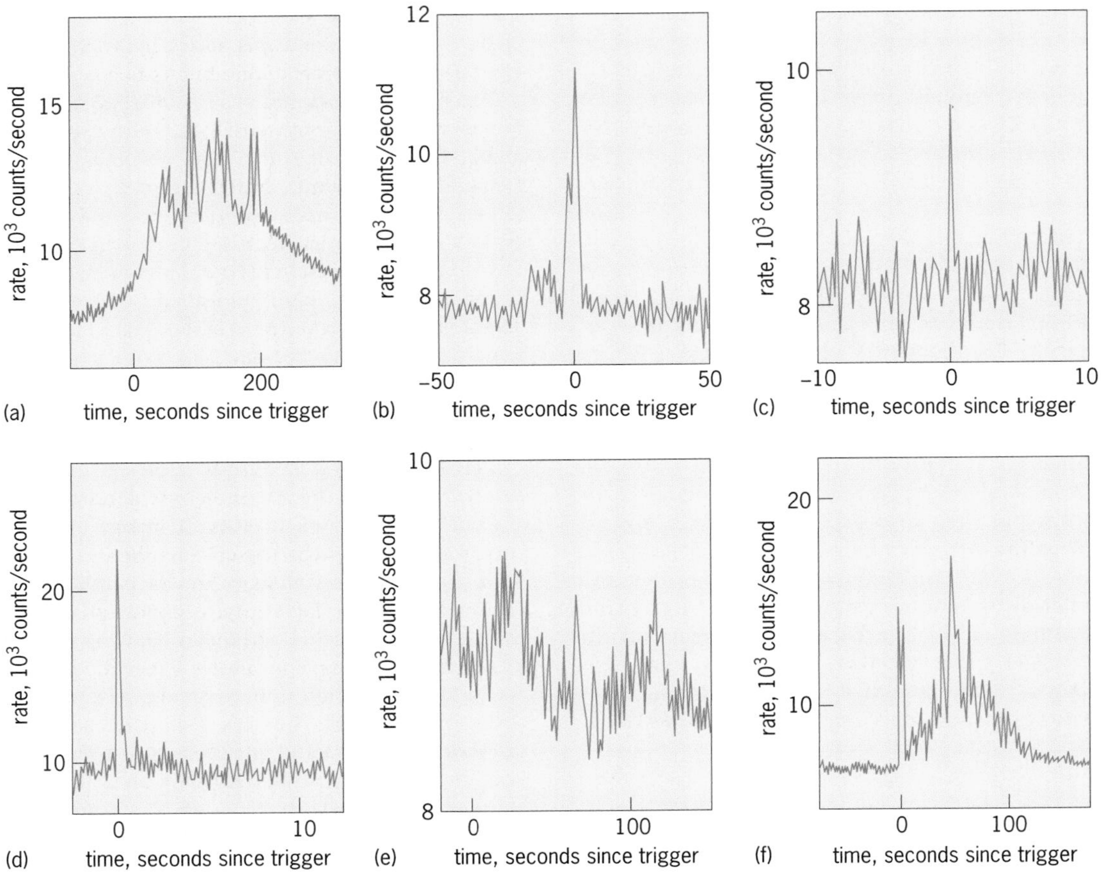

Fig. 2. Assortment of bursts detected by *Swift*. (*a*) GRB050117a, a very long burst that allowed *Swift* to detect x-ray emission while the burst was still flaring in gamma rays, the first truly simultaneous multiwavelength observation of a GRB. (*b*) GRB050502b, a relatively simple GRB that was followed by intense x-ray flaring (not shown). (*c*) GRB050509b, the first short GRB to have a detected x-ray afterglow. (*d*) GRB050724, a short GRB followed by long, flaring afterglow emission, presumably due to pieces of a neutron star getting swallowed up by the newly formed black hole. (*e*) GRB050904, the most distant stellar explosion ever recorded, at a distance of 12.8×10^9 light-years. (*f*) GRB060614, one of the new class of hybrid bursts: The initial short spike is followed by a longer period of emission, and had spectral characteristics similar to other short, hard bursts. However, despite its relatively nearby location, no supernova was detected.

energy emitted each second by the Sun. Most of the bursts that have been well-studied to date are consistent with the following interpretation: Those lasting more than 2 s appear to be associated with the supernova explosions of massive stars in distant galaxies, while shorter gamma-ray bursts can be understood as the result of merging neutron stars or black holes. In both scenarios, it has been thought that the gamma-ray bursts lead to the birth of black holes. However, *Swift* observations have recently provided evidence that the precursor stars that account for a small fraction of the "long" bursts may collapse into highly magnetized neutron stars known as magnetars. The bursts have a variety of time profiles, ranging from tens of milliseconds to more than several minutes (**Fig. 2**).

GRB discoveries in the *Swift* era. In 1993, it was suggested (based primarily on observations made with the BATSE detectors on board NASA's now-defunct *Compton Gamma-Ray Observatory*) that two classes of gamma-ray bursts exist: short bursts, with durations less than 2 s and more energetic (harder) spectra, and bursts longer than 2 s with less energetic (softer) spectra. X-ray afterglows for long bursts were discovered in 1997 by scientists using the *Beppo-SAX* satellite, and long-burst optical afterglows were discovered shortly thereafter by a team led by Jan van Paradijs using positional information provided by *Beppo-SAX*.

However, prior to the launch of *Swift*, no similar information existed about the shorter, hard bursts. On May 9, 2005, *Swift* detected the first x-ray afterglow from a short burst, and on July 7, 2005, NASA's *HETE-2* satellite detected a short burst that led to the first optical afterglow; this afterglow was seen in data obtained with ground-based telescopes. The galaxies (and lack of detectable supernovae) seen at the locations of the short bursts confirmed the general idea that short bursts arise from the mergers of compact objects: two neutron stars, two black holes, or one of each. In any of these scenarios, a resulting black hole is formed, with little detectable afterglow emission.

Swift has seen many bursts with repeated and intense x-ray flares following the initial gamma-ray blast. These flares were observable due to *Swift*'s rapid response to the initial burst trigger, which allowed the XRT to get on target quickly enough to

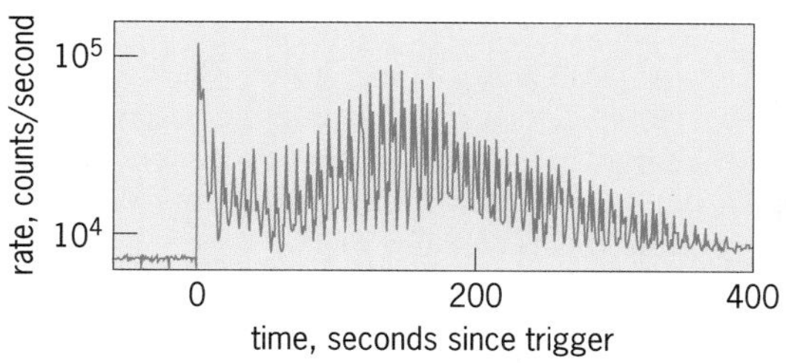

Fig. 3. ***Swift*** **data from the magnetic explosion by magnetar SGR1806-20, which affected the Earth's ionosphere. Vertical axis is logarithmic scale of intensity in counts per second. The intensity of the initial spike was too strong for the detectors to accurately measure, but it is estimated to have produced about 10^6 gamma-ray photons/cm^2/s at Earth, a distance of 50,000 light-years. Characteristic 8-s pulsations produced by the rotation of a hot spot on the neutron star surface following the outburst are evident.**

detect the lower-energy flaring behavior. Two examples are given in Fig. 2 (GRB050502b and GRB050724).

On September 4, 2005, *Swift* detected gamma rays from the most distant cosmic explosion seen so far, occurring 12.772×10^9 light-years away (a redshift of 6.29). This distance is directly comparable to that of the most distant object known to date, a quasar located at 12.797×10^9 light-years (or a redshift of 6.43). This record-breaking long gamma-ray burst, with intense and long-lasting x-ray activity following the burst, appears to be the result of a massive star collapsing into a black hole.

On June 14, 2006, *Swift* detected a new type of burst, called a hybrid burst. Lasting over 100 s, it was relatively nearby, yet showed no signs of an underlying supernova. This became the prototype of a new class of bursts: It does not fit into the long or short classification scheme that has been used for more than a decade. Scientists remain divided on whether this hybrid burst was a long-short burst from a merger or a long burst from a star explosion with no supernova. Most conclude, however, that some new process must be at play—either the model of mergers creating second-long bursts needs a major overhaul, or the progenitor star from an explosion is intrinsically different from the kind that make supernovae.

Non-GRB science with *Swift*. When not studying gamma-ray bursts, *Swift* observes a wide variety of other time-varying and exotic high-energy phenomena as it conducts a survey of the hard-x-ray sky. *Swift* has discovered a wealth of information on objects as diverse as highly magnetized neutron stars (magnetars), supernovae, and comets.

In December 2004, one of the brightest-ever cosmic explosions was observed by over a dozen spacecraft, including *Swift*. This flare from the magnetar and soft-gamma repeater SGR 1806-20 saturated the detectors of almost every satellite with gamma-ray or particle detectors, as well as affecting the ionization profile of the Earth's day-lit ionosphere, a phenomenon that had never been previously observed. **Figure 3** shows the light curve of this flare event as detected by *Swift*. Although the initial flare was too strong to be measured accurately, the characteristic "ringing" of the rotating neutron star can be seen in the approximately 8-s pulsations that follow the tremendous explosion. Unlike classical gamma-ray bursts, which do not repeat due to the annihilation of the underlying source of the explosion, soft-gamma repeaters flare every few years, with the release of a tremendous amount of magnetically powered energy due to the reconnection of magnetic field lines, similar to (but far more extreme) than solar flares.

Swift observed the probe of NASA's *Deep Impact* spacecraft crash into Comet Tempel 1 on July 4, 2005, using both the XRT and the UVOT. Comets can emit x-rays as the result of a charge-exchange mechanism between cometary atoms and energetic particles in the solar wind—the positive ions in the

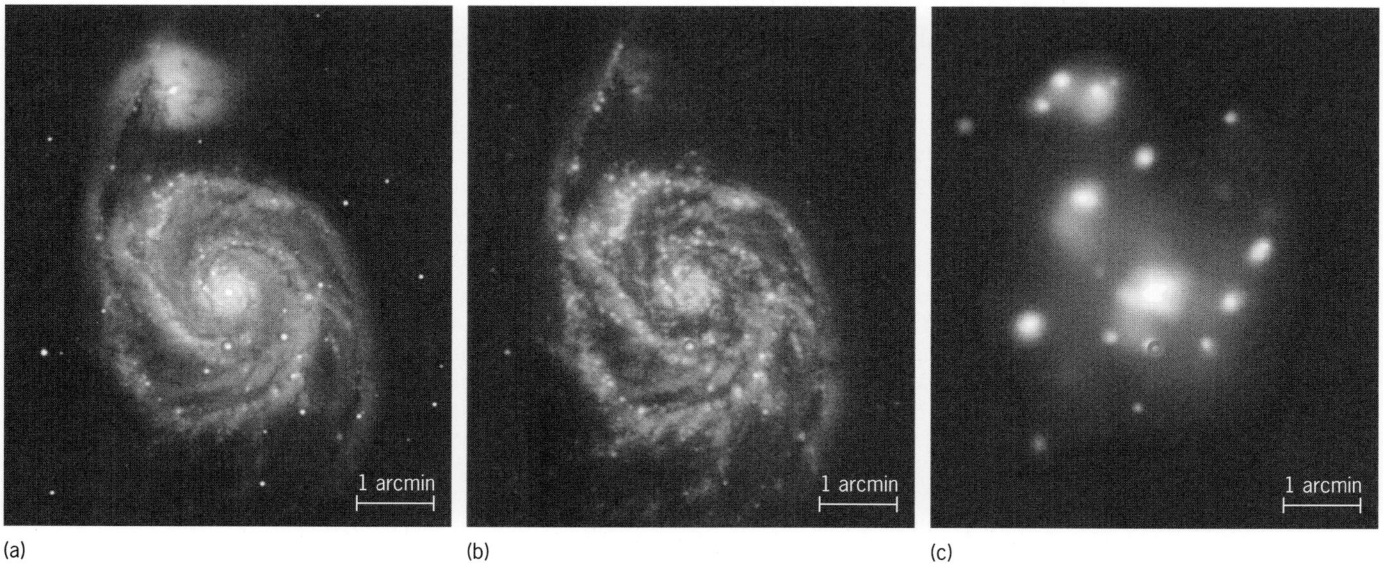

Fig. 4. Multiwavelength images of the supernova 2005cs in the galaxy M51 observed with *Swift* at (*a*) visible, (*b*) ultraviolet, and (*c*) x-ray wavelengths. The supernova is the circled bright spot immediately below the core of the galaxy in each of the images.

solar wind strip electrons off the comet, and as the electrons lose energy they emit x-rays. For about three days after the impact, the XRT did not detect much emission from Comet Tempel 1; however, the comet then brightened by almost an order of magnitude as a result of a solar storm that enveloped the comet in high-velocity energetic particles. Meanwhile, the UVOT took time-resolved images in an ultraviolet band that is sensitive to emission from dissociated cometary molecules, including water, and saw the brightness of the comet start to rise immediately after the collision with the *Deep Impact* probe, increasing by a factor of 4 in the first hours after impact as material thrown up by the impact filled out the comet's coma. The comet's ultraviolet brightness then gradually faded over the course of the next day. *Swift* also observed x-rays in May 2006 from Comet 73P/Schwassmann-Wachmann 3 during what may be its last orbit around the Sun before entirely disintegrating. *See* DEEP IMPACT.

In February 2006, scientists using *Swift* and a combination of orbiting and ground-based observatories for the first time caught a supernova in the act of exploding. Careful analysis revealed that the event, originally believed to be an unusual type of gamma-ray burst, occurred about 25 times closer and lasted about 100 times longer than a typical burst. The initial burst, in the form of a jet of high-energy x-rays, pierced through a doomed star from its core and sent out a warning within minutes that a supernova was imminent. The massive star then blew apart, as an international team of observers collected data throughout the event, which appeared to leave behind a magnetar. *Swift* has also accumulated detailed observations of many supernovae, primarily using the UVOT (**Fig. 4**).

In October 2006, NASA scientists using *Swift* announced that they completed the first census of galaxies with active, central black holes, a project that scanned the entire sky several times over a 9-month period. They found over 200 black holes, using data from the BAT. Many of these were not previously known due to obscuration at lower energies.

Administration. The design and development of *Swift* was led by Principal Investigator Neil Gehrels of NASA's Goddard Space Flight Center. An international team of scientists from the United States, Italy, and the United Kingdom helped to develop and build the instruments on board the satellite, which is now controlled from the Mission Operations Center (MOC) located at Pennsylvania State University. Data are returned to the MOC via a link from the Italian Space Agency's Malindi, Kenya ground station.

For background information *see* BLACK HOLE; COMET; GAMMA-RAY BURSTS; NEUTRON STAR; SUPERNOVA; TELESCOPE; X-RAY TELESCOPE in the McGraw-Hill Encyclopedia of Science & Technology.

Lynn Cominsky

Bibliography. R. Cowen, Farthest bang: a burst that goes the distance, *Sci. News*, 168(12):179, September 17, 2005; R. Cowen, Fleeting flash: Pinpointing a short gamma-ray burst, *Sci. News*, 167(20):308, May 14, 2005; R. Cowen, Unique explosion: gamma-ray burst leads astronomers to supernova, *Sci. News*, 169(9):133, March 4, 2006; R. Naeye, Fireworks from the far side of the Galaxy, *Sky Telesc.*, 109(5):28–32, May 2005; G. Schilling, Rare flare, *Sci. Amer.*, 292(6):24–25, June 2005.

System interoperability

Although the important concept of interoperability has been in existence for three decades, notions of interoperability continue to be refined. There are a number of definitions of the term "interoperability." The meaning of this term is very much dependent upon the context associated with its usage, and some authors have identified numerous different but related definitions, depending upon the context. A definition appropriate to the context of systems engineering might be: Interoperability is the process of ensuring that all of the systems, processes, procedures, and human efforts in an organization are managed such as to maximize the opportunities for internal and external exchange and reuse of information and knowledge within the organization.

This definition emphasizes that much more is needed to ensure interoperability than simply to use compatible software and hardware, although this compatibility will be much needed. It associates interoperability with the notion of integration. It emphasizes the fact that, to be effective, interoperability will often require radical change and transformation of the way in which organizations and enterprises function, including the way in which they deal with information and knowledge.

Operational and technical interoperability. The U.S. Department of Defense accepts the notion of two levels of interoperability: operational and technical.

Operational interoperability is the ability of systems, units, or forces to provide services to and accept services from other systems, units, or forces and to use the services so exchanged to enable them to operate effectively together.

Technical interoperability is the condition achieved among communications-electronics systems or items of communications-electronics equipment when information or services can be exchanged directly and satisfactorily between them or their users.

Operational interoperability addresses support to military, or more generally, enterprise operations and, as such, goes beyond systems to include people and procedures, interacting on an end-to-end basis. Implementation of operational interoperability therefore implies both the traditional approach of defining standards as well as enabling and assuring activities such as testing and certification, configuration, version management, and training. Interoperability at the technical level is essential to achieving operational interoperability. Technical interoperability occurs between technical systems.

LISI Model. These two types of interoperability are the basis for a Levels of Information System Interoperability (LISI) Model that was initiated in 1993.

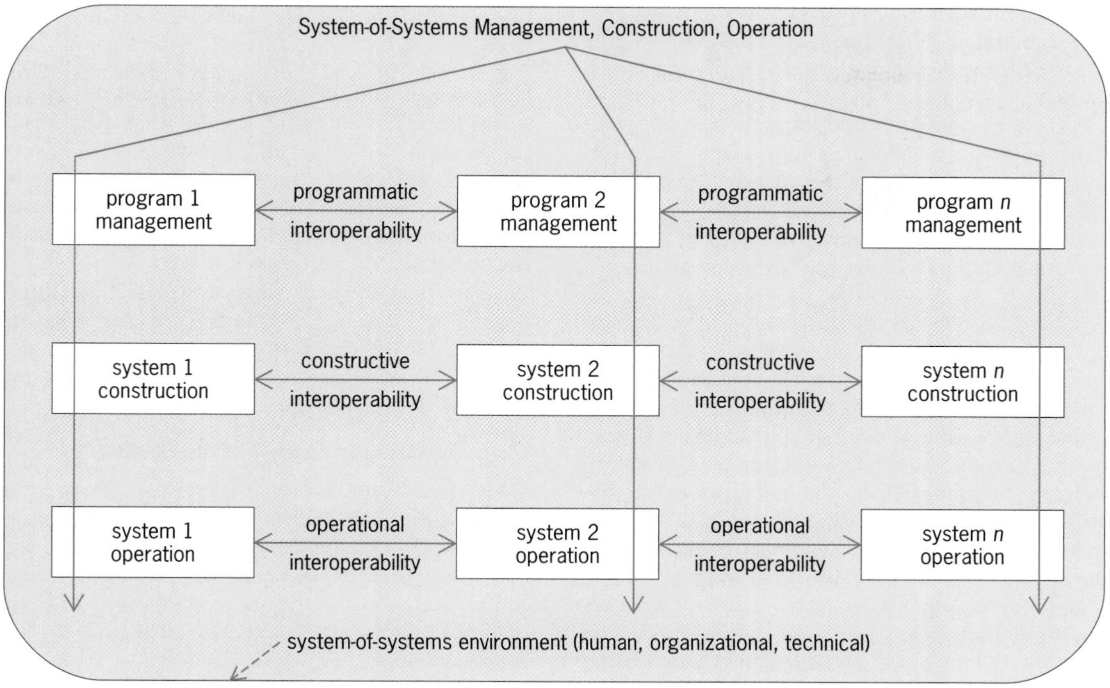

SOSI Model of interoperability in a system of systems.

LISI is a reference model to assess information system interoperability. It is a process-based approach for defining, measuring, assessing, and certifying degrees of interoperability required or achieved between systems. LISI uses a common frame of reference and common measures of performance. It is intended to be applied throughout an information systems engineering life cycle, from requirements definition to system development and subsequent deployment, including both fielding and subsequent modifications and reengineering. Thus, LISI is intended to facilitate a common understanding of interoperability, to provide an interoperability maturity model as the basis for making comparisons between systems, and to provide a process for guiding assessment of and improving interoperability throughout the process of engineering the system.

The LISI Maturity Model identifies stages through which a system should progress in order to improve its interoperability capabilities. There are five increasing levels of interoperability: level 0, isolated or nonconnected operation; level 1, systems connected through electronic connections such as email and other similar exchanges; level 2, functional interoperability with separate data so as to allow group collaboration; level 3, domain interoperability through such mechanisms as shared databases to allow relatively sophisticated collaboration; and level 4, enterprise interoperability allowing advanced collaboration, interactive manipulation capability, and shared data.

Each capability level must cover what are designated as four enabling attributes of interoperability, the so-called PAID attributes: (1) procedures, which represent the policies that guide a system's development through the use of established standards and procedures to influence system integration; (2) applications, which represent the functions a system is engineered to perform; (3) infrastructure, which represents the connections required to support system operations and which contains four subcomponents that are also defined in increasing levels of sophistication; and (4) data, which represents both the data and the information structures that support both functional applications and system infrastructure. Thus, each of the five LISI levels is associated with each of the four interoperability (PAID) attributes for a 20-element scoring matrix to use in evaluating interoperability.

In using the LISI approach, interoperability measures are typically identified using a scorecard approach. If precise measures are unavailable, human judgment may be used to estimate how well a system is doing relative to technical compliance, system-to-system interactions, and operational mission effectiveness. Each of these three elements needs to be measured.

SOSI Model. Much recent attention has been devoted to a System-of-Systems Interoperability (SOSI) Model. This model addresses programmatic interoperability, constructive or system interoperability, and operational interoperability across the organizations and enterprises engineering and utilizing interoperable systems of systems. This model is based on the realization that all three types of interoperability are important, especially when dealing with a system of systems:

1. Program management defines activities that should be accomplished well to manage the acquisition of an individual system as well as a system of systems.

2. System construction defines the activities that are needed to engineer a system of systems, such as the use of process standards, and architectural frameworks.

3. System operation defines activities that must exist within an individual system and between this individual system and its environment, including interoperation with other systems. In this description of a system activities model, end users of a system are considered part of the operational system.

When we consider the interaction between several programs that comprise a system of systems, the result may be depicted as shown in the **illustration**. This immediately leads to the notion of several different types of interoperability, including programmatic interoperability, or interoperability between different program offices, and constructive interoperability, or interoperability between organizations that are responsible for the implementation (detailed construction and maintenance) of a system.

In order to demonstrate interoperability in an operational system of systems, it is necessary to consider the full scope of interoperability between organizations that contribute to the acquisition of a system of systems. It is primarily this that leads to the need for programmatic interoperability and constructive interoperability. We need to consider interoperability issues between all relevant organizations responsible for any segment of a system of systems. This SOSI model extends the existing earlier models specifically by suggesting the need for a programmatic interoperability model of activities performed to manage the acquisition of a system of systems. The model must also include environmental factors and concerns affecting the system of systems. In the SOSI model, programmatic, constructive, and operational interoperability issues must be managed within this environmental sphere.

For background information *see* SYSTEM FAMILIES; SYSTEMS ENGINGEERING; SYSTEMS INTEGRATION in the McGraw-Hill Encyclopedia of Science & Technology.

Andrew P. Sage

Bibliography. T. C. Ford et al., A survey of interoperability measurement, in *Proceedings of the 12th International Command and Control Research and Technology Symposium*, 2007; T. C. Ford et al., The interoperability score, in *Proceedings of the 5th Annual Conference on Systems Engineering Research*, March 14–16, 2007, Hoboken, New Jersey, 2007; M. Kasunic and W. Anderson, *Measuring Systems Interoperability: Challenges and Opportunities*, Tech. note CMU/SEI-2004-TN-003, 2004; K. Kosanke, ISO standards for interoperability: A comparison, in D. Konstantas et al. (eds.), *Interoperability of Enterprise Software and Applications*, pp. 55–64, Springer, London, 2006; E. Morris et al., *System of Systems Interoperability (SOSI): Final Report*, Tech. rep. CMU/SEI-2004-TR-004, 2004; J. T. Pollock and R. Hodgson, *Adaptive Information: Improving Business Through Semantic Interoperability, Grid Computing, and Enterprise Integration*, Wiley, Hoboken, New Jersey, 2004.

Titan

Titan, 5150 km (3200 mi) in diameter, is Saturn's largest moon and the second largest satellite in the solar system. In the last few years there have been significant advances in our knowledge of Titan's atmosphere and surface, aided by the *Cassini-Huygens* mission of the National Aeronautics and Space Administration (NASA) and the European Space Agency (ESA). NASA's *Cassini* Saturn orbiter has flown by Titan every few weeks since October 2004 (three dozen times by mid-2007). In January 2005, *Cassini* released the ESA-built *Huygens* probe, which descended by parachute through Titan's atmosphere and landed on its surface.

Atmosphere and Methane Weather

Titan stands out in the solar system as the sole moon with a substantial atmosphere. This atmosphere hosts a methane-based meteorology that we have just begun to understand in the past decade. While Titan was discovered in 1655 by Christiaan Huygens, it was not until 1943–1944 that Gerard Kuiper made the definitive discovery of Titan's atmosphere with spectral observations that revealed the signature of gaseous methane. (Earlier observational evidence for the presence of an atmosphere, dating from 1907, is controversial.)

The *Voyager 1* spacecraft flyby of Titan in 1980 returned a wealth of information on Titan's atmosphere, including basic information about atmospheric composition and structure. Because of its design, *Voyager 1* could observe only at shorter wavelengths where Titan's surface and lower atmosphere are obscured by a thick stratospheric haze layer. This circumstance contributed to the idea, held by most astronomers until just a few years ago, that Titan's lower atmosphere was quiescent, with little activity or weather. Observations by Caitlin Griffith in the late 1990s showed otherwise. Unlike *Voyager 1*, Griffith observed at near-infrared wavelengths that are long enough to penetrate through the stratospheric haze. Her observations revealed reflectivity variations in the lower atmosphere on time scales of hours to days. The most likely, and now accepted, explanation was that clouds of methane were forming and dissipating.

Since the time of Galileo the resolution of telescopic observations has been limited by the distortions induced by Earth's atmospheric turbulence. Even at the best sites in the world the best resolution ever achieved is about 0.25 arcsecond, and 1 arcsecond is much more typical at major observatories. Unfortunately, as viewed from Earth, Titan is never more than 0.9 arcseconds across. While Titan can be easily seen with even a modest pair of binoculars (it is only about 20 times fainter than what can be seen with the unaided eye), it is nearly impossible to see any spatial detail of Titan with a traditional ground-based telescope. The launch of the *Hubble Space Telescope*, which sits in orbit above Earth's atmosphere and therefore receives an undistorted view of Titan, and the development of the new

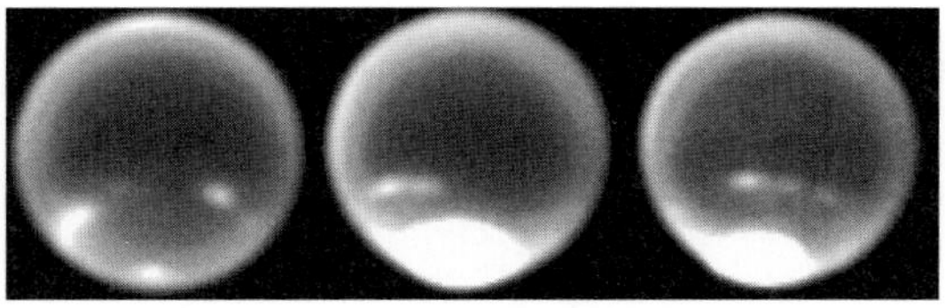

Fig. 1. Images of Titan's clouds taken with the Keck 10-m (400-in.) and Gemini 8-m (320-in.) telescopes using adaptive optics on (*a*) September 2, (*b*) October 2, and (*c*) October 7, 2004. They were taken using a carefully chosen near-infrared filter that blocks light reflected from Titan's surface and allows observations of atmospheric features. The background halo is reflection off of the stratospheric haze. These images show three classes of Titan's clouds: small, long-lived south polar clouds (part *a*); rare, enormous south polar storms (parts *b* and *c*); and sporadic linear clouds at 40° south latitude (all three images). (*Courtesy of H. Roe*)

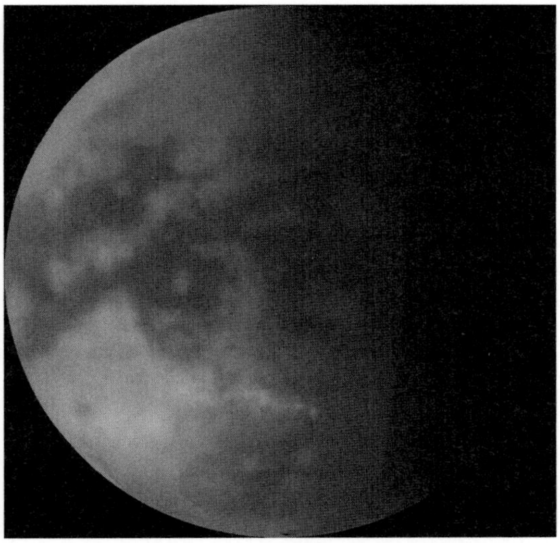

Fig. 2. Image of Titan as seen by *Cassini* during the July 22, 2006 flyby. Most of the visible details are surface features, but the features at lower center are clouds of the midlatitude linear type. (*Courtesy of NASA/JPL/University of Arizona*)

technology of adaptive optics, which allows for the real-time correction of the distorting effects of Earth's atmosphere on large Earth-bound telescopes, finally gave astronomers the high spatial resolution necessary to study Titan's surface and lower atmosphere. The combination of this high spatial resolution with the ability to observe in the near-infrared first revealed hints of Titan's clouds in data taken in the mid-to-late 1990s and has been used regularly since 2001 to study in detail how Titan's clouds and weather form, dissipate, and evolve with the changing seasons (**Fig. 1**).

Since its arrival in orbit about Saturn in 2004, the *Cassini* spacecraft has sent back extremely high-resolution images of Titan's clouds during its fly-bys of the satellite (**Fig. 2**). The *Huygens* probe returned spectacular images of Titan's surface and the best-yet information about the composition and structure of Titan's atmosphere.

Structure and composition. The physical structure of Titan's atmosphere resembles that of Earth's (**Fig. 3**). At first glance the two atmospheres appear quite similar, except offset significantly in temperature. At Titan's surface the pressure is about 1.5 times greater than at Earth's, but the temperature is only 93 K (−180°C or −292°F), about 200 K (200°C or 360°F) colder than is typical on Earth. In both atmospheres the lowest layer is the troposphere, where most of the weather occurs. Because of the low temperature and Titan's low surface gravity (about 14% that on Earth), the density of Titan's atmosphere at the surface is almost five times greater than the atmospheric surface density on Earth. A further consequence of Titan's low surface gravity is that the atmosphere is greatly extended compared to Earth's. In spite of Titan's smaller diameter (5150 km or 3200 mi, compared to Earth's 12,800 km or 7950 mi), the mass of Titan's atmosphere is about twice that of Earth's.

Titan's atmosphere is composed primarily of molecular nitrogen (N_2), as is Earth's. The second most abundant species is methane (CH_4) at 1.4%. As on Earth, methane is a potent greenhouse gas in Titan's atmosphere, raising the surface temperature of Titan more than 10 K (10°C or 18°F) above equilibrium. Methane is broken apart high in Titan's atmosphere by ultraviolet photons, primarily from the Sun, which leads into a complicated network of chemical and photochemical reactions. The end products are numerous heavier hydrocarbons

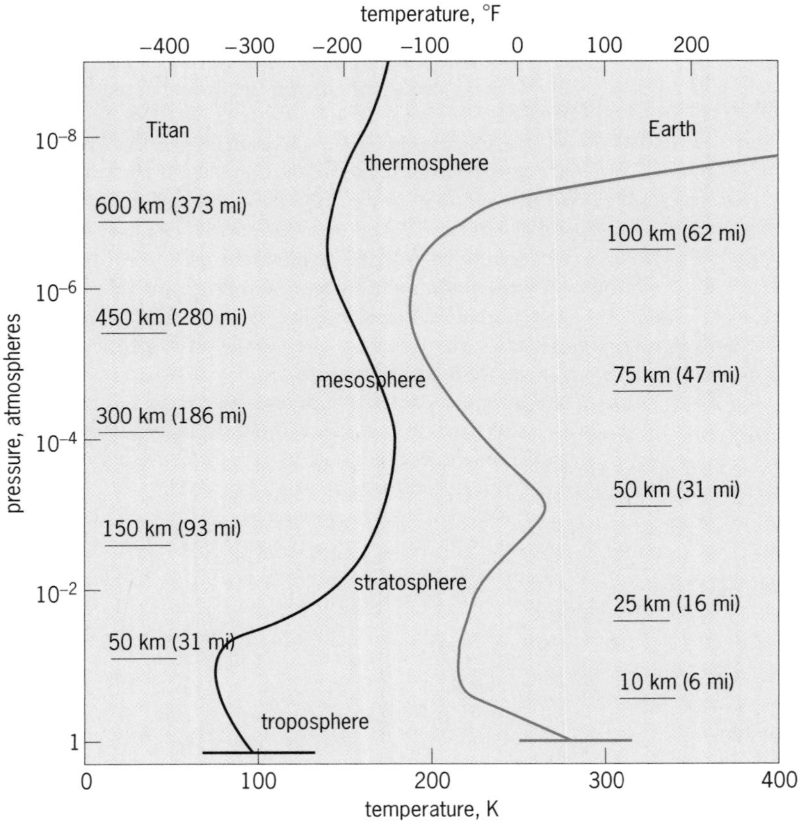

Fig. 3. Physical structures of the atmospheres of Titan and Earth. The vertical axis is logarithmic pressure. The two atmospheres appear similar, except offset in temperature. Also marked are the corresponding altitude levels in each atmosphere, showing Titan's atmosphere to be about six times more extended than Earth's. (*After H. Roe*)

such as ethane (C_2H_6), acetylene (C_2H_2), propane (C_3H_8), and benzene (C_6H_6); nitriles such as hydrogen cyanide (HCN) and dicyanoacetylene (C_2N_2); and a thick stratospheric organic haze layer.

In Earth's lower atmosphere, temperature and pressure conditions lie near the triple point of water, the temperature and pressure at which all three phases of water (solid, liquid, and gas) can coexist in equilibrium. The combination of the driving force of energy deposited by sunlight with the easy phase transitions between the different states of water are what lead to all of the interesting weather phenomena that are familiar on Earth. On Titan conditions in the lower atmosphere are far too cold for water to exist as anything but solid. However, conditions in Titan's atmosphere are near the triple point of methane. In direct analogy with the water-based weather seen on Earth, there is a methane-based meteorology in Titan's lower atmosphere.

Source of methane. The loss of methane to photochemistry at the top of Titan's atmosphere is significant. Without some replenishing source, the methane in Titan's atmosphere would be depleted within a few tens of millions of years (much less than the 4.5×10^9-year age of the solar system). Comet impacts are far too infrequent to supply enough methane to explain the present-day abundance. The source of Titan's atmospheric methane must be the interior of Titan. Images and radar scans from the *Cassini* spacecraft show surface features that appear to be the remains of cryovolcanoes, where the "lava" would be a water-ammonia mixture with dissolved methane gas. During eruptions these cryovolcanoes should release significant quantities of methane into Titan's atmosphere.

A further problem is to explain where the methane inside Titan came from originally. If the methane is primordial and simply condensed with the rest of Titan out of the proto-Saturnian nebula (which certainly had plenty of methane gas), then the noble gases xenon and krypton should also be present in Titan's atmosphere. The *Huygens* probe found no trace of these two gases in the atmosphere, suggesting that the methane must have formed inside Titan after the satellite condensed. Earlier in Titan's history, there was a subsurface liquid ocean of water and ammonia in contact with Titan's inner solid rocky core. [This ocean may still exist, but 100 km (60 mi) under Titan's surface or even deeper.] The most likely place for producing Titan's methane was at this ocean-rock interface where a reaction between water and rock, known as serpentinization, would have occurred. Serpentinization creates hydrogen gas, which in turn reacts with carbon to form methane.

Seasons and time scales. Titan is tidally locked to Saturn and orbits once every 16 days. Thus, Titan's rotation period, and its "day," are 16 days long as well. Saturn takes 29.5 years to orbit the Sun and thus Titan, which is itself in orbit around Saturn, has a 29.5-year-long annual seasonal cycle. Seasons on Earth arise because Earth's rotation axis (the line running through the North and South poles) is not perpendicular to the plane of Earth's orbit about the Sun. Titan's rotation axis is similarly tilted over (26.4°, compared with Earth's 23.5° tilt) with respect to its motion about the Sun. Thus, Titan's year is 29.5 times longer than Earth's. Titan has seasons similar to those on Earth, including long periods of total darkness and continuous daylight in the polar regions. The northern winter solstice occurred on Titan in October 2002 and the northern spring equinox will be in August 2009.

Because Earth's orbit around the Sun is nearly circular, its distance to the Sun varies only a small amount and Earth's seasons arise almost entirely due to the tilt of its rotation axis. Similarly, the dominant reason behind Titan's seasons is its rotation axis tilt. However, on Titan a second effect comes into play. Saturn's orbit about the Sun (and therefore Titan's) is somewhat eccentric. The distance between Titan and the Sun varies between 9.02 and 10.05 astronomical units (the mean distance between the Sun and Earth). An important implication of this is that the flux of sunlight hitting Titan varies with the season. When Titan is closest to the Sun (roughly in Northern winter) the Sun appears 1.23% as bright as what is seen from Earth, while when Titan is furthest from the Sun this drops to 0.99%. The combined effect of Titan's tilt and Saturn's eccentricity is that the South Pole at southern summer solstice receives 53% more sunlight per square meter than the annual average for a square meter on the equator. Because Titan's northern summer coincides with being further from the Sun, the North Pole at northern summer solstice receives only 35% more sunlight than the equatorial mean.

Clouds and circulation. Titan's tropospheric methane clouds that have been observed thus far divide into three general categories: smaller long-lived south polar clouds, giant south polar storms, and sporadic linear midlatitude clouds (Fig. 1). A fourth category of cloud composed of ethane particles has also been identified.

The smaller south polar clouds were present almost continuously from their discovery in 2001 until 2005 and have been sighted only occasionally since then. This corresponds with the period of late southern spring and early southern summer on Titan. The current understanding of these clouds is that they are a seasonal phenomenon. During this period the south polar region is in constant sunlight and experiences the maximum mean daily solar heating of any region on Titan's surface at any time of year. (That the pole receives more heating than the equator at the summer solstice may be counterintuitive, but is the result of the pole being in continuous sunlight while a point on the equator receives sunlight for only about half of the diurnal cycle.) Meanwhile the north polar region is in constant darkness. This north-south imbalance in heating is thought to drive a single pole-to-pole circulation cell, with an upwelling zone at the South Pole and downwelling at the North Pole. This upwelling zone leads to the formation of localized convective clouds in the south polar region, much as the equatorial upwelling zone on Earth of the Hadley cells leads to clouds.

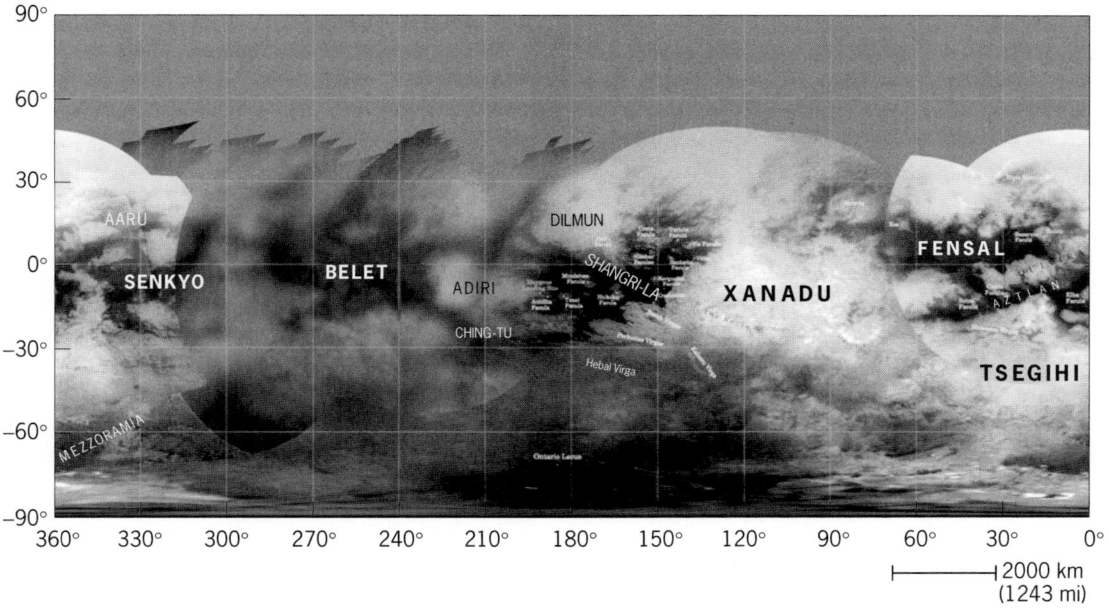

Fig. 4. Map of Titan, with names indicated for major regions, compiled in October 2006 from images of the *Cassini* Imaging Science System (ISS) at a wavelength of 940 nm. Some areas are mapped at rather higher resolution than others, and latitudes 50°N and higher are insufficiently illuminated in the present season to map. (*Courtesy of NASA/JPL/Ciclops*)

Occasionally, during this same time period, the small south polar clouds erupted into enormous storms covering much of the south polar region. The triggering mechanism for these storms is unknown. After several of these storms the smaller south polar clouds disappeared or were significantly reduced for several months. These storms probably result in significant rainfall to the surface. *Cassini* images show several possible methane lakes in the south polar region, which fits with this scenario.

In 2003 observations from the 10-m (400-in.) Keck and 8-m (320-in.) Gemini telescopes on the island of Hawaii revealed a third class of methane clouds, which always appear near 40°S latitude, are often long and streaky, and are most often seen over the region around 350°W longitude. (Titan is tidally locked to Saturn, as noted above, and 0° longitude is defined as the sub-Saturnian longitude.) The strong clustering of these clouds over a small region on Titan's surface strongly suggests some process related to the surface is driving their formation. One possible explanation is that these clouds are being formed by release of methane from a region of geologic activity, possibly active cryovolcanism. Curiously, *Cassini* recently discovered the largest, highest mountain range yet seen on Titan [32 km (20 mi) wide by 145 km (90 mi) long by 1.6 km (1 mi) high] directly under this region of cloud activity. An additional factor driving the formation of these clouds may be the breakdown of the seasonal pole-to-pole circulation cell.

As Titan moves further into northern winter, *Cassini* has discovered a thin cloud made of ethane particles in the upper tropopause covering much of the north polar region. The formation of this cloud is likely due to the seasonal downwelling in this region leading to supersaturated concentrations of ethane.

Henry Roe

Surface

Although Titan's surface remained largely hidden for 350 years after the satellite's discovery, it is now known to be the most Earth-like landscape in the solar system. Although *Voyager 1* flew close past Titan in 1980, its cameras barely managed to penetrate the thick haze in Titan's dense atmosphere, and scientists could only guess that it might be like other icy satellites underneath. However, the presence of large amounts of methane in Titan's atmosphere, and the surface temperature of 93 K (−180°C or −292°F, which is, as noted above, near the triple point of methane, much as Earth's temperatures are close to the triple point of water) led to speculation that Titan's surface might be covered in a deep ocean of that liquid. The first revelations about Titan's surface came in the early 1990s. First, radar echoes (using large radio telescopes on Earth) showed a somewhat reflective surface, more so than a liquid methane ocean would be. Soon thereafter, telescopic measurements showed selected "windows" of wavelengths that could penetrate the haze and detect the surface, which had both bright and dark areas of infrared reflectivity.

These were disk-integrated measurements; that is, they provided no surface resolution and effectively summed the radiation from the satellite's visible disk. They were followed in 1994 by images with the *Hubble Space Telescope* that allowed the first crude maps to be developed, notably showing a large bright region (named Xanadu) on Titan's leading face. (As noted above, Titan rotates synchronously, like Earth's Moon, always showing the same face to its parent planet.) Improving capabilities of large ground-based telescopes with adaptive optics systems allowed better maps to be made prior to *Cassini's* arrival, but even these telescopes show about as much detail on Titan as observers see on our Moon with the unaided eye, and so ideas about what the bright and dark regions might be remained speculative.

A planet's surface reflects the processes operating to shape it. Calculations of the possible rates of erosive processes, tectonics, cryovolcanism (that is, ice volcanism), and impact cratering suggested that unlike, for example, Io, which is dominated by volcanism, and smaller satellites dominated by impact craters, Titan should have a complex Earth-like balance of all of these processes. Measurements from the *Cassini* orbiter as it has flown past Titan, coupled with in situ measurements from the *Huygens* probe, have revealed most of what is known about Titan. This torrent of data has borne out the prediction: Titan's landscape is incredibly diverse and Titan appears to be the most Earth-like body in the solar system, despite being made of different material (ice and organics instead of rock) and shaped by a different fluid (liquid methane rather than water).

Cassini results: A young surface. *Cassini's* cameras, sensitive to near-infrared light and able to make maps with a resolution of a few kilometers (1 km = 0.6 mi), show Titan's surface to be varied at all scales (**Fig. 4**). The strongest bright-dark contrasts appear so far at the equator, with more muted features at midlatitudes. So far there are only distant views of the polar regions. (Because *Cassini's* arrival at Saturn in 2004 occurred during late southern summer, the high northern latitudes are in winter darkness until the equinox in 2009.) The first near-infrared images and the first *Cassini* radar observations (which are able to penetrate the haze and map the surface with resolutions down to 350 m or 1150 ft) showed a striking lack of impact features. Since impact features are continuously generated throughout the solar system, hundreds of craters should have been seen if the surface were old. Thus the lack of craters suggested that, like Earth, Titan's surface is young, being continuously shaped by processes such as erosion or cryovolcanism; analysis suggests an effective age for the surface of about 500 million years, rather similar to that for Venus or the Earth.

Spectral mapping by *Cassini's* Visual and Infrared Mapping Spectrometer (VIMS) has found several distinct terrain compositions. In addition to the bright terrain making up Xanadu, there seem to be two types of dark terrain, a 'brown' unit characteristic of dune-covered areas and a 'blue' one apparently associated with river channels. Work is under way to determine what chemical composition is responsible for these different units. A few patches on Titan are particularly bright at 5 μm wavelength (spectral identification of materials from orbit is complicated on Titan by the thick atmosphere), suggesting a distinct composition, perhaps related to cryovolcanism. The possible interaction, albeit short-lived, of cryolava (that is, water) with the organic haze material on Titan's surface is of particular interest in studies of the origin of life, as laboratory studies show that such reactions can produce prebiotic molecules such as amino acids and pyrimidines, which living things on Earth use in proteins and in DNA.

Huygens probe: Titan up close. The descent of the *Huygens* probe in January 2005 gave close-up images of an area about 30 km (19 mi) across just south of Titan's equator. A surprising and immediate result was

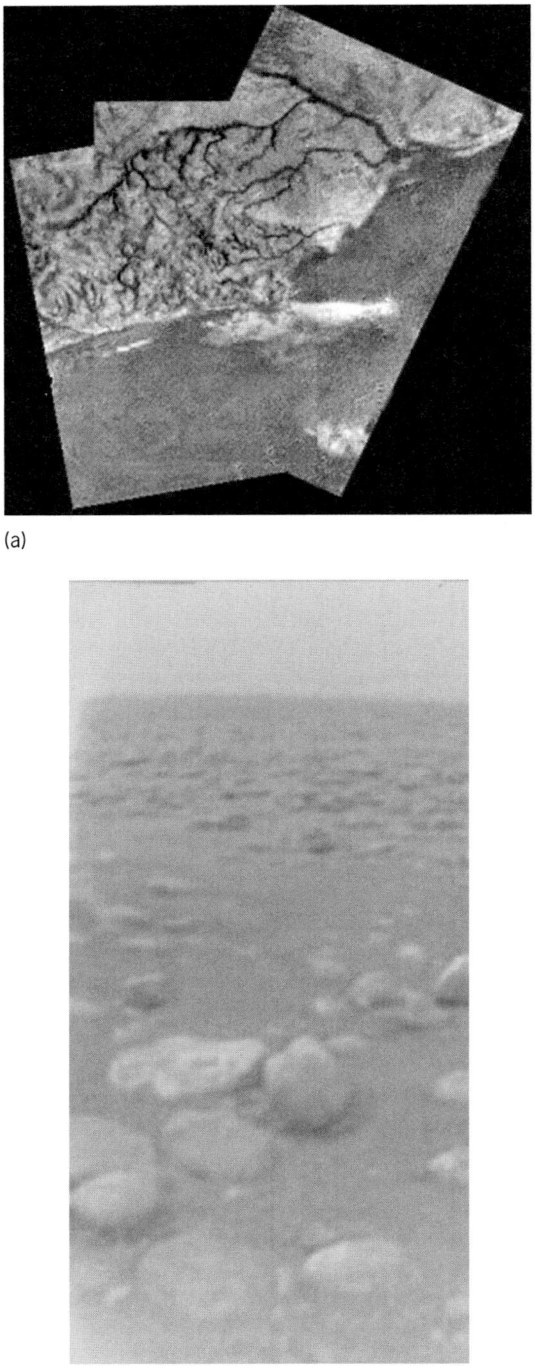

(a) (b)

Fig. 5. Images obtained by the *Huygens* probe. (*a*) Descent image, about 7 km (4.3 mi) across. (*b*) Side-looking image acquired after impact, showing rounded cobbles between about 5 and 15 cm (2 and 6 in.) across. (*NASA/JPL/ESA/University of Arizona*)

that this area had been shaped by fluvial processes—rain and rivers (**Fig. 5**). Not only were some bright highlands and dark plains dissected with networks of steep-sided dark channels, but the images from knee-height by the probe after its impact (although not designed explicitly to survive impact, the probe's landing was quite soft) showed a broadly flat plain with a fine-grained dark substrate on which sat rounded bright cobbles between 5 and 15 cm (2 and 6 in.) in diameter; in other words, *Huygens* had landed in a stream bed. The rounding of the cobbles and the absence of small pebbles suggested the area had been last shaped by a rapid flow of liquid, most probably the result of a methane thunderstorm.

The impact force on the probe suggested the surface texture was like wet sand (except here the "sand" means sand-sized particles of organic materials and ices). The impact also embedded the heated inlet of *Huygens's* gas analyzer into the ground, and this fortuitous surface sampling revealed methane and ethane, showing that the ground was damp with these liquids. Tracking of the probe just prior to impact showed that the near-surface winds were gentle, less than about 0.5 m/s (1 mi/h).

Dunes and lakes. A surprising finding in radar images in late 2005 was that large dark areas near Titan's equator were not (as was once thought) liquid, but rather are giant sand seas. These sand seas are filled with massive linear dunes, tens of kilometers long, a couple of kilometers apart, and up to 150 m (500 ft) high—exactly the size and style of linear dunes seen in the Namib or Arabian deserts on Earth. Linear dunes, which run along the vector mean wind direction rather than across it, form in winds that alternate between two general directions. On Earth, this fluctuation is usually seasonal, relating to monsoonal flows, but on Titan the fluctuation in wind direction may be due to Saturn's massive gravitational tide in Titan's atmosphere. These tidal currents of air exist on Earth, but are tiny compared with the solar-driven winds on our rapidly rotating planet that heats and cools on a daily time scale. On slowly rotating Titan with its massive atmosphere so far from the Sun, the response to changing sunlight is heavily damped, and the comparatively large tidal effects may dominate.

In summer 2006, the first radar images at high northern latitudes showed dozens of lakes (**Fig. 6**). Some of these were simply round, while others had the irregular appearance of flooded river valleys. The lakes were the darkest things seen to radar on Titan's surface, consistent with a liquid hydrocarbon composition. It remains to be seen whether the lakes are concentrated at high latitudes because of lower surface temperatures there, or because liquid ethane is preferentially deposited near the poles, or because the poles are lower in elevation.

Exploring a diverse, dynamic landscape. River channels, running for several hundred kilometers, have been observed in radar and near-infrared data. Some have the characteristics of rough desert washes, formed by rare but heavy downpours. [Models of methane thunderstorms on Titan show they can deposit several tens of centimeters (1 cm = 0.4 in.) of liquid in only a few hours, much like the heaviest downpours on Earth.] Some others are more heavily incised valleys, and yet others are shallow and meandering, suggesting they deposit fine-grained sediment.

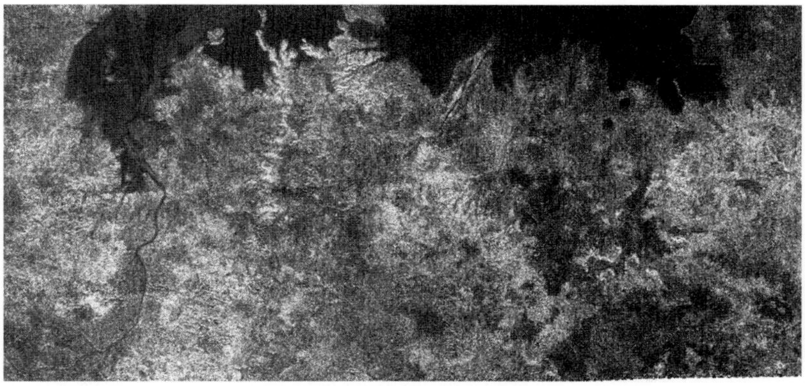

Fig. 6. *Cassini* radar image (300 × 140 km or 186 × 87 mi) showing a large dark lake of liquid methane and ethane at top. A meandering river drains into the left end of the lake. (*NASA/JPL*)

Although many areas of Titan appear quite flat (as indicated by radar altimetry) with elevation changes of only a few tens of meters (1 m = 3.3 ft) over hundreds of kilometers, there are steep mountains over a kilometer high in patches or chains all over Titan. The Xanadu region is particularly rugged. How the mountains form is not yet known.

The *Cassini* mission is still ongoing, and new data come in regularly; in mid-2007, only about 20% of the surface had been mapped by radar. Expected forthcoming results include more infrared and radar mapping, perhaps to show surface changes, as well as gravity measurements which will probe Titan's internal structure. (As noted above, Titan, like Europa, may have an internal water ocean beneath its ice crust.) Plans are already under way for a future mission to explore Titan, perhaps with a balloon and long-lived lander as well as an orbiter, able to make more complete and detailed maps, to measure the composition of Titan's surface, and to observe wind patterns and seismic activity.

For background information *see* ADAPTIVE OPTICS; ATMOSPHERE; ATMOSPHERIC GENERAL CIRCULATION; DUNE; GREENHOUSE EFFECT; HUBBLE SPACE TELESCOPE; METHANE; PREBIOTIC ORGANIC SYNTHESIS; RADAR ASTRONOMY; SATELLITE (ASTRONOMY); SATURN; SEASONS; SPACE PROBE; TRIPLE POINT in the McGraw-Hill Encyclopedia of Science & Technology.

Ralph D. Lorenz

Bibliography. S. K. Atreya, The mystery of methane on Mars and Titan, *Sci. Amer.*, 296(5):42–51, May 2007; J. K. Beatty, C. C. Petersen, and A. Chaikin (eds.), *The New Solar System*, 4th ed., Cambridge University Press, 1998; A. Coustenis and F. Taylor, *Titan: The Earth-Like Moon*, World Scientific, 1999; D. M. Harland, *Cassini at Saturn: Huygens Results*, Praxis-Springer, 2007; R. D. Lorenz and J. Mitton, *Lifting Titan's Veil: Exploring the Giant Moon of Saturn*, Cambridge University Press, 2002.

Traditional Chinese medicine authentication

For millennia, physicians' literature has addressed means of ensuring that purchased medicines are what they are claimed to be. Herbs are much more likely than cultivated vegetables to be accidentally confused: most botanicals are not highly modified in form and may closely resemble related or even unrelated species, and many are collected from the wild, where similar-looking species may also occur. Furthermore, inexpensive botanicals have sometimes been fraudulently substituted for or commingled with rare and costly botanicals. While most botanicals are correctly identified, substitution of the wrong species can cause serious harm. At worst, people have been poisoned after using a toxic species that was confused with an unrelated plant; in a relatively recent case, contamination of cultivated plantain with digitalis resulted in several reported illnesses. More commonly, the substituted plant may provide the consumer lesser benefits than the desired species, or none at all. In modern practice, every marketer of crude botanicals or manufacturer of botanical products should have explicit quality-control procedures to ensure that materials sold are correctly identified and of acceptable purity.

Traditional formulas. Traditional Chinese medicine was one of the first cultural medical systems to be codified in written form, and is probably the most complex. Its underlying philosophy relates health to a balance of vital and elemental energies and "pathogenic factors" (for example, "heat," "dampness") within bodily "organs" not corresponding precisely to the physical organs of the same name. Chinese healers usually use not single herbs but codified multiherb formulas that may contain a few to over two dozen ingredients, often including fungal, animal, or mineral products as well as botanicals. The primary herb in a formula is believed to act directly to correct an imbalance, whereas other herbs may increase the activity of the primary herb, reduce its side effects, or have a general tonic effect.

The official circumscription of a particular herb frequently includes more than one species: licorice root (*gan cao*) officially may be obtained from *Glycyrrhiza glabra, G. inflata*, or *G. uralensis*, while epimedium (*yin yang huo*) may include five species of *Epimedium*. Use of related species for similar purposes is common worldwide, and with reason as closely related species frequently share similar chemical constituents and activities; for example, in Western practice, several species of *Crataegus* (hawthorn) or *Plantago* (psyllium) may be used interchangeably. However, conflation of species from related genera occurs more frequently in traditional Chinese medicine, and in traditional practice unrelated herbs believed to have similar actions have occasionally been more or less freely substituted, depending upon local availability (see **table**). For example, in northern China the rhizome of *Menispermum dahuricum* (*bei dou gen* or "northern bean root") is sometimes substituted for the root of *Sophora tonkinensis* (*shan dou gen* or "mountain bean root"). Both of these botanicals have the primary activity of "clearing heat and eliminating toxins," but they are less likely to have similar chemical constituents as they are from very different plant families.

Health risks of Aristolochia substitution. *Aristolochia fangchi*, known as *guang fang ji* (family Aristolochiaceae), was believed by many traditional healers to be substitutable for the unrelated *Stephania tetrandra* (*fang ji* or *fen fang ji*, Menispermaceae); both have the actions of promoting diuresis and "dispelling wind and dampness." However, *A. fangchi* contains aristolochic acids, which can cause tubular interstitial fibrosis, a progressive kidney disease, and kidney cancer in both animals and humans. Worldwide case reports have documented instances of kidney failure following prolonged medicinal use of various *Aristolochia* species. Products sold as *Stephania* have repeatedly been found to contain aristolochic acids, proving that they were adulterated with *Aristolochia* species; consumers, believing the product to be the safer herb, might unknowingly consume hazardous amounts. (Contrarily, one survey has found that many "*Aristolochia*" products contain no detectable aristolochic acids.) Over 100 patients of a Belgian weight loss clinic developed interstitial fibrosis after using a regimen that combined apparently adulterated "*Stephania*" with other herbs and pharmaceutical drugs, including acetazolamide, which may have increased aristolochic acid toxicity. Western regulatory agencies have therefore banned both *Aristolochia* species and their harmless substitutes, unless those species are proven free of aristolochic acids.

Pharmaceutical drugs in herbal preparations. In recent decades, Chinese herbal medicines have also been adulterated with pharmaceutical drugs. One survey in Taiwan in the early 1990s found that 24% of sampled products were adulterated. Traditional philosophy may have condoned adding to a mixture any available substance that might improve its activity, but the unlabeled inclusion of potentially toxic drugs represents both fraud and a significant health hazard. Reported pharmaceutical adulterants with known health risks or contraindications include corticosteroids, appetite suppressants, acetaminophen, nonsteroidal anti-inflammatory drugs (NSAIDs), codeine, and diazepam; a number of consumer illnesses have been attributed to such adulteration. Because pharmaceuticals are added to an extract or finished product, they can be detected only by chemical analyses. Methods for detecting many potential adulterants by chromatographic methods, such as high-pressure liquid chromatography (HPLC), have been published.

The best-known case of pharmaceutical adulteration is PC-SPES (PC for prostate cancer; *spes* is Latin for "hope"), a Chinese herbal mixture that gained an impressive reputation for inhibiting prostate cancer. PC-SPES was noted to have estrogenic side effects, such as breast enlargement; further investigation revealed that early lots of PC-SPES were adulterated with synthetic estrogens, notably diethylstilbestrol

Several Chinese herbs for which unrelated or distantly related unofficial regional substitutes have been documented

Chinese name	Official or most common species	Substituted species
zi cao	*Lithospermum erythrorhizon* or *Arnebia euchroma* (family Boraginaceae)	*Onosma paniculatum* (family Boraginaceae)
guan zhong	*Dryopteris crassirhizoma* (Dryopteridaceae)	*Matteuccia struthiopteris* (Dryopteridaceae), *Osmunda japonica* (Osmundaceae), *Blechnum orientale* (Blechnaceae), etc. [class Filicopsida]
da qing ye	*Isatis indigotica* (Brassicaceae)	*Isatis tinctoria* (Brassicaceae), *Strobilanthes cusia* (Acanthaceae), *Polygonum tinctorium* (Polygonaceae), *Clerodendrum cyrtophyllum* (Lamiaceae)
shan dou gen	*Sophora tonkinensis* (Fabaceae)	*Menispermum dahuricum* (Menispermaceae)
jin qian cao	*Lysimachia christinae* (Primulaceae)	*Glechoma longituba* (Lamiaceae), *Desmodium styracifolium* (Fabaceae), *Hydrocotyle sibthorpioides* (Apiaceae), *Dichondra repens* (Convolvulaceae)
wang bu liu xing	*Vaccaria hispanica* (Caryophyllaceae)	*Ficus pumila* (Moraceae)

(DES), which is used to treat prostate cancer. Some later lots included detectable amounts of the anticoagulant warfarin—perhaps added in the hope of preventing the blood clots that DES might otherwise cause. Discovery of these adulterants led to a product recall and a premature end to one clinical trial. Ironically, it is likely that the herbal ingredients actually worked, at least to boost the activity of DES. Case reports and preliminary clinical trial results indicated that PC-SPES was superior to standard DES treatment, even though the DES present in the PC-SPES had been a small fraction of the normal dose, and in-vitro studies also demonstrated anticancer activity in the herbal mixture alone. By hindering the study and use of promising herbal adjuvants, this manufacturer's lack of quality standards may have led ultimately to more deaths and illnesses than those directly caused by any contaminated botanical.

Good manufacturing practices. Good manufacturing practices for a modern herbal product must ensure, among other things, that the identity, purity, and quality of each raw material included in the product are confirmed. For most botanicals, authentication is most easily accomplished by observing morphological and organoleptic (taste, odor, fracture) characteristics before the material is processed. Preservation of voucher samples from each batch then permits identity to be reconfirmed later in case a question about product quality arises. Reference works are available in Chinese and English that provide descriptions and photographs of popular Chinese herbal drugs. If preprocessing identification is neglected, or if the intact material is not unambiguously identifiable (say, if the part harvested and supplied is a root without any unique features that distinguish it from similar-looking species), more elaborate methods can be used to confirm identity of extracted material. Chemical fingerprinting methods such as HPLC and thin-layer chromatography (TLC) are the most commonly used. These methods confirm the presence of expected marker compounds and thus simultaneously serve as a measure of quality.

However, it is virtually impossible to fully assess the quality of a finished multiherb formula by affordable chemical means. The number of compounds present becomes so great that peaks inevitably overlap and obscure one another; nor can any one HPLC protocol be expected to visualize appropriate marker compounds from a dozen different materials. Therefore it is particularly essential that the raw materials for such complex formulas be properly authenticated and documented, and that manufacturers scrupulously abide by label claims regarding content. Reputable Western dietary supplement manufacturers follow such practices as a matter of course. The Chinese government has taken laudable steps to promote quality control, such as issuing pharmacopoeial standards for commonly used botanicals, but these steps will not be fully effective so long as they coexist with traditional practices that tolerate contamination at levels from the wildcrafter to the factory floor. Sadly, at the moment, traditional Chinese medicine formulas compounded by foreign manufacturers may sometimes be of higher quality than Chinese-made products. To a significant extent, the message for the traditional Chinese medicine consumer remains *caveat emptor*.

For background information *see* AGRICULTURAL SCIENCE (PLANT); ARISTOLOCHIALES; BOTANY; INSPECTION AND TESTING; LIQUID CHROMATOGRAPHY; PHARMACOLOGY; PHARMACY; QUALITY CONTROL TOXICOLOGY in the McGraw-Hill Encyclopedia of Science & Technology.

Wendy L. Applequist

Bibliography. D. Bensky, A. Gamble, and T. Kaptchuk, *Chinese Herbal Medicine: Materia Medica*, Eastland Press, Seattle, 1986; J. K. Chen and T. T. Chen, *Chinese Medical Herbology and Pharmacology*, Art of Medicine Press, City of Industry, CA, 2004; J. P. Cosyns, Aristolochic acids and "Chinese herbs nephropathy": A review of the evidence to date, *Drug Safety*, 26:33–48, 2003; E. Ernst, Adulteration of Chinese herbal medicines with synthetic drugs: A systematic review, *J. Intern. Med.*, 252:107–113, 2002; W. K. Oh et al., Prospective, multicenter, randomized phase II trial of the herbal supplement, PC-SPES, and diethylstilbestrol in patients with androgen-independent prostate cancer, *J. Clin. Oncol.*, 22:3705–3712, 2004.

Trans fatty acids

Trans fatty acids (TFA), so named because they contain a trans rather than a cis configuration at the unsaturation sites within the fatty acid structure, have become an increasing health concern to the public and health professionals alike as their levels in typical diets in North America and elsewhere have increased markedly during the past decade. Unlike naturally occurring monounsaturated (monounsaturates) and polyunsaturated fatty acids (polyunsaturates) as found in many liquid nonhydrogenated vegetable oils, which have cis double bonds at their unsaturation sites, TFA have trans double bonds at the unsaturation sites within their structure. [Unsaturated acids can be converted to saturated acids by the addition of hydrogen. Hydrogenation is the chemical binding of hydrogen to the double bonds of unsaturated fatty acids, carried out in the presence of hydrogen, pressure, and typically nickel as a metal catalyst. In cis bonds, adjacent carbons are on the same side of the molecule; in trans bonds, they are on opposite sides (see **illustration**).] Whereas "natural" monounsaturates and polyunsaturates have curvilinear structures and are liquid at room temperatures, the "industrial" TFA are typically linear structures that are solid fats at room temperatures due to their much higher melting points. Thus, TFA approach saturated fats in many of their physical-chemical properties.

The industrial process known as partial hydrogenation (partial since only enough hydrogen atoms are added until the desired consistency is reached), as performed on highly unsaturated liquid oils (such as soybean oil or canola oil) containing natural or cis double bonds free of TFA, results in the chemical transformation of these liquid oils into solid partially hydrogenated oils and vegetable shortenings. These latter commercially processed oil products are used extensively in assorted processed and fast foods because they impart the desired solidity and a greater resilience to oxidation and rancidity, including longer shelf lives, while allowing for labeling or marketing terms including "cholesterol-free," "low in saturated fat," and "free of animal fat," etc., which have receptive appeal to consumers. Consequently, a considerable portion of the so-called monounsaturates in a typical diet in North America and elsewhere is now represented by TFA (trans monounsaturates) in addition to the naturally occurring cis monounsaturates. It is also noteworthy that partial hydrogenation of vegetable oils gives rise to a diverse mixture of several TFA isomers (types) that are included in the collective TFA designation. In contrast to partial hydrogenation, edible oil refining (including deodorization), heating, and conventional frying with oils generate very small amounts of TFA (usually $<1\%$ of total fat).

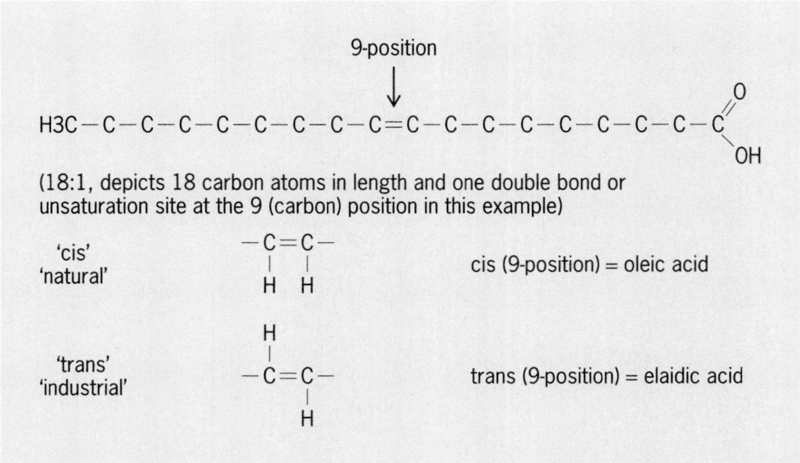

Generalized structure of double bond in natural and common industrial monounsaturated fatty acids (18:1): configuration of the unsaturation site in monounsaturated fatty acids as found as a cis structure in natural fats/oils or as a common trans structure found along with other trans isomers (forms) following the industrial process of partial hydrogenation.

Origins, food sources, and intake levels. Approximately 90% of the total TFA consumed per person daily in North America and many countries is derived from processed and fast-food products, with TFA representing ~2–4% of the total fat. Animal sources and fats contribute the remaining 10% to the dietary intake of TFA in the form of milk, butter, and beef. The natural biohydrogenation process as performed by microorganisms in the stomach of ruminant animals (dairy cows, beef cattle) produces some TFA from the unsaturated dietary fat consumed. The pattern of TFA isomers (rich in vaccenic acid) in such natural TFA is rather different from the types of TFA (rich in elaidic acid plus others) that predominate in partially hydrogenated vegetable oils (industrial TFA). (Whereas both elaidic acid and vaccenic acids are monounsaturated TFA having 18 carbons with one unsaturation site or double bond between adjacent carbon atoms, the former has the double bond adjacent to the 9th carbon from the carboxyl end while the latter has it adjacent to the 11th carbon.) Margarines represent ~20% of the total TFA intake in the North American diet, with fried plus processed foods combined being the major contributors (~70%). Various snack foods, including crackers, croissants, cookies, and potato chips, often contain up to 25–45% of their total fat as TFA, although such levels are now decreasing in many commercial products due to current and pending labeling and regulatory requirements on TFA in such foods. Other TFA-rich sources from many but not all brand name products include, include cake and pancake mixes, frozen breakfast waffles, doughnuts, french fries, breaded meats, fish, or chicken produce, and even baby biscuits. Some processed foods have up to 5–6 g of TFA per serving.

The **table** gives the estimated intakes of the different types of fats and fatty acids and sources in a typical North American diet. Estimates during the past decade on the per-capita (adult) intakes of total TFA in the North American diet have ranged from an average of 5 up to 10 g/person per day. Estimates of average TFA intakes across numerous countries (g/person/day) have indicated much higher intakes in Canada (at 8) and the United States (at 6) as compared to Japan and China (at 1). With an aggressive regulatory approach on industrial TFA content and

Estimated intakes and sources of lipids (fats) in a typical North American diet

Fat component	Common food source	Average daily consumption	
		g/day	% of energy (avg.)
Total fat (fatty acids)	Mixed (animal and plant)	70–105	30–35
Saturated fatty acids	Dairy products, fatty meats, palm oil, coconut oil	30–40	14.0
Monounsaturated fatty acids (natural)	Canola oil, olive oil, animal fats	21–30	10.0
"Trans" fatty acids (monounsaturated)	Hydrogenated vegetable oils, shortenings, processed and fast foods, ruminant fats (minor)	5–10	3–4
Polyunsaturated fatty acids (mostly omega-6 as linoleic acid)	Corn, safflower and sunflower oils, others	10–17	6.0
Polyunsaturated fatty acids (mostly omega-3 as alpha-linolenic acid)	Canola oil, soybean oil, flaxseed, others	1.4–2.0	0.6
Polyunsaturated fatty acids [omega-3 as eicosapentaenoic acid (EPA) and docosahexaenoic acid (DHA)]	Fish, fish oil	0.15	0.06
Cholesterol	Animal foods, animal fats	0.4	0

restrictions in Denmark, the average intake of TFA there has now dropped to ~1 g/person per day. Almost all of this very low intake is natural TFA mostly from dairy and beef produce. The higher intakes of TFA in the North American population appear to be in the younger sections (for example, up to ages 18–34 years). Unfortunately, one of the highest dietary sources of industrial TFA in the North American food supply is mothers' breast milk; these high levels relate directly to the high dietary intake of TFA during pregnancy and lactation. A Canadian report revealed that the average total TFA content was found to be 7.2% of the total milk fat, with the TFA ranging up to 17.2% (corresponding to average intakes in lactating women of 10.6 g/person/day, with intakes as high as 20.3 g/person per day in some women). Elevations of TFA levels in breast milk are very closely related to the intake of industrial TFA in the diet since the body itself generates no significant amount of TFA. Restriction or elimination of industrial TFA in the diet results in the corresponding depletion of TFA from the body. TFA can undergo metabolic oxidation (beta-oxidation) in the body and can be used as an energy source (as is done for natural monounsaturates).

Effects on cardiovascular/diabetes disease risk. Epidemiological studies have indicated that dietary TFA, as consumed mostly in processed and fast foods, represent major dietary risk factors for coronary heart disease (CHD) in the population. A 14-year perspective study on over 80,000 women from the United States indicated that the relative risk for developing CHD was almost doubled for every 2% increase in energy intake from TFA (that is, ~4–5 g daily). These findings are rather disturbing when considered in the context of the aforementioned high intakes of TFA in the North American population. Furthermore, dietary TFA on an equal intake basis (by weight) was found to be a significantly greater dietary risk factor for CHD when compared with saturated fats (up to five- to tenfold).

Controlled intervention trials in humans have indicated that both saturated fatty acids and TFA increase total and low-density lipoprotein (LDL, or commonly termed "bad") cholesterol levels in the circulation, thereby increasing the risk of cardiovascular disease (CVD). However, TFA also lowered protective high-density lipoprotein (HDL, or "good") cholesterol levels, whereas saturated fats did not; lowering of HDL-cholesterol levels has been associated with a substantially increased risk of CVD. In addition, TFA (but not saturates) tend to increase the levels of a highly atherogenic (artery-hardening) blood lipoprotein known as lipoprotein(a) which further increases the risk for CVD. The more profound increase in the ratio of circulating LDL-cholesterol:HDL-cholesterol with TFA intakes as compared to equivalent amounts of saturated fats further supports population data indicating TFA to be a much greater risk factor for CVD than saturated fats.

The impact, if any, of natural monounsaturated TFA as found in ruminant fats on the risk of CVD is still under investigation. The very small amount of a unique polyunsaturated trans fat known as CLA (conjugated linoleic acid) in the diet (including ruminant fats) is also under investigation with respect to any health implications, although such is unlikely based on the very low levels in our food supply.

High TFA intakes have also been implicated in other health effects such as the risk of type 2 diabetes. A major population study suggested that increased intakes of saturated fats were not significantly associated with the risk of type 2 diabetes in women, whereas a 2% increase in energy consumed as TFA (~4–5 g/day) appeared to increase the risk by 39%. There is also some evidence that high TFA intakes during pregnancy may interfere with the metabolism and deposition of important omega-3 fatty acids that are needed for the growth and development of infants during their neonatal period.

Food labeling and regulatory policies. Traditional approaches with respect to food regulation and labeling requirements and protection against premature CVD have focused upon labeling for cholesterol (in mg units) and saturated fats (in gram units). Consequently, labeling or marketing terms such as "cholesterol-free" and "low in saturated fats," when allowed by regulatory agencies, have implied to many in the public sector that such products have been deemed to be of potential benefit with respect to the prevention and/or management of CVD. Unfortunately, many of these products contained substantial levels of industrial TFA that might potentially promote rather than prevent the development of CVD via their aforementioned deleterious effects on risk factors such as LDL-cholesterol, HDL-cholesterol, and lipoprotein(a). Currently, numerous countries have instituted or are implementing mandatory labeling for TFA on various food products and/or restrictions on the amounts allowed per food product serving. In order to qualify for a potential trans-free claim on processed foods, Canadian labeling laws require <0.2 g of TFA per serving (plus restrictions on the amount of saturates present), while the corresponding cutoff is <0.5 g per serving in the United States. However, the actual intakes of certain foods (particularly snack foods) often surpass the official serving sizes as listed on food packages. Mandatory labeling for the amount of TFA is often included next to labeling for the amount of saturated fat on food labels, giving the amounts of each of these two fat types in gram units. While such labeling for TFA is an improvement over traditional labeling wherein TFA contents were not made available, this might be confusing if different quantitative units (mg versus g) are used. Many in the public sector cannot readily differentiate that 1 g of TFA is five times greater than 200 mg of cholesterol.

Mandatory labeling for TFA on processed foods does not protect the consumer from prepared foods as purchased in restaurants and other outlets. Consequently, many jurisdictions (including New York City) are instituting or considering legal bans on industrial TFA in processed and restaurant/fast foods. The argument for such a ban often includes restricting the availability of industrial TFA at the source (supply) prior to entry into a wide range of processed and fast foods, which would also reduce the costs of inspections and analytical monitoring of numerous food products. Consumer protection through a legislative ban on industrially produced TFA in foods in Denmark has been found to be accomplished without noticeable effects on the availability, price, or quality of foods previously containing high amounts of industrial TFA in that country.

Alternatives to trans fats. Numerous alternatives to the use of industrial TFA in processed and fast foods are available. Some of these will require a return to some increased usage of palm, palm kernel, or other tropical oils (containing saturated fatty acids, but free of TFA from partial hydrogenation) alone or mixed with unsaturated vegetable oils to produce the necessary properties for food applications. While a partial replacement of saturated fats for TFA may be of concern, it should be emphasized that saturated fats are generally considered to be a much lower dietary risk factor for CVD as compared to equivalent amounts of industrial TFA. Other alternatives to TFA include high-monounsaturated vegetable oils (some via genetic modification) with lowered susceptibility to oxidative deterioration as compared to highly polyunsaturated oils and blends (interesterification) with more saturated fats for appropriate textures. While butter fats typically contain ~62% of their total fat found as saturates, a 50:50 to 67:33 blend of butter fat with highly unsaturated liquid vegetable oils (low in saturates) can give rise to mixtures that often have desirable physical properties (including spreadability, taste, etc.) and lower levels of saturated fats (<42% of total fat without industrial TFA) and that may not increase risk for CVD. In some alternatives, unsaturated vegetable oils (for example, soybean oil and canola oil) can be subjected to complete hydrogenation as compared to partial hydrogenation, thereby giving rise to a predominant fatty acid product (18 carbons in chain length) known as stearic acid (fully saturated) which, unlike other saturates, has been found not to increase blood cholesterol levels in controlled human trials. Thus, mixing the high stearic acid product from complete hydrogenation with nonhydrogenated vegetable oils by blending or interesterification techniques can result in a usable fat product that is TFA-free and that does not increase total blood cholesterol and LDL-cholesterol levels. Numerous other options becoming available in the marketplace should help to accelerate the marked reduction or elimination of industrial TFA from the food supply if supported by required legislation, consumer education, and reasonable pricing. It is anticipated that the established health concerns with industrial TFA will continue to drive the impetus towards an "industrial TFA-free" global society.

For background information *see* CHOLESTEROL; DIABETES; FAT AND OIL; FAT AND OIL (FOOD); FOOD; FOOD ENGINEERING; FOOD MANUFACTURING; HEART DISORDERS; LIPOPROTEIN; NUTRITION in the McGraw-Hill Encyclopedia of Science & Technology.

Bruce J. Holub

Bibliography. A. Ascherio, Trans fatty acids and blood lipids, *Atheroscler. Suppl.*, 7:25–27, 2006; B. Holub, Hydrogenated fats and serum cholesterol levels, *N. Engl. J. Med.*, 341:1396–1397, 1999; J. Moss, Labeling of trans fatty acid content in food, regulations and limits—the FDA view, *Atheroscler. Suppl.*, 7:57–59, 2006; D. Mozaffarian et al., Trans fatty acids and cardiovascular disease, *N. Engl. J. Med.*, 354:1601–1613, 2006; S. Stender et al., High levels of industrially produced trans fat in popular fast foods, *N. Engl. J. Med.*, 354:1650–1652, 2006.

Tropics: latitudinal biodiversity gradient

The latitudinal diversity gradient (LDG) is the most pervasive pattern exhibited by the living biota on a global scale. Tropical latitudes are rich in species and in groups of organisms on higher levels of the tree of life, a richness that declines nearly monotonically toward polar latitudes, where the biota is most species-poor. This gradient of diversity is found in all major groups or organisms, such as fungi, land plants, vertebrates, and invertebrate groups, and characterizes both the terrestrial and marine realms. In the oceans, the gradient is recorded for organisms living on the sea floor of the continental shelves, in the deep sea, in the water column over the shelves, and in the open ocean. The LDG was well known in Darwin's day, and in 1878 Alfred Wallace, the coauthor of natural selection, suggested that the tropics were diverse because they were climatically more stable than higher latitudes, permitting evolution to accumulate species there. The diversity gradient has spawned many studies—some supporting Wallace's notion, others offering alternative explanations. A number of explanations center around temperature, which of course varies with latitude. The higher tropical temperatures have been held to support more rapid evolutionary activity, which is then progressively damped in colder climates. However, mechanisms that might underlie this relationship have never been satisfactorily demonstrated.

Cradle or museum of diversity? In 1974, Ledyard Stebbins famously posed a question summarizing many aspects of the problem: Are the tropics a cradle or a museum of diversity? That is, are the tropics diverse because the speciation rate is high there (the "cradle" hypothesis), or because they have accumulated species over long stretches of time (the "museum" hypothesis)? This deceptively simple question proved difficult to answer for it requires knowledge of the ages of origin of a representative sample of species across the LDG, which cannot reliably be estimated from the diversity pattern alone. However, paleoecologists and biogeographers, by drawing on the fossil record, have begun to piece together the history of originations of lineages, especially for the marine environment (where the fossil record is best for this purpose), and a new, dynamic picture of the LDG is coming into focus.

Fossil record. A particularly comprehensive analysis has been made of the LDG of the marine Bivalvia (a class of Mollusca containing clams, scallops, oysters, etc.), represented today by several thousand species (some are as yet undescribed, especially in the tropics). Because the fossil record is far better for genera than species and is best for the relatively shallow waters of the continental shelves and platforms [<200 m (<660 ft)], the record of shelf genera was studied; there are about 1300 living bivalve genera described that seem valid, most of which have species that live at shelf depths. The ages of origin of 773 of these genera were inferred from their first appearances in the fossil record, and their geographic distributions over the last 11 million years were traced. It turns out that most genera have originated within the tropics, although many of them have migrated toward the poles, usually by producing new species in higher latitudes, while retaining their tropical presence. At all extratropical latitudes in both northern and southern hemispheres combined, genera of tropical origin outnumber those with extratropical originations, as inferred from their first fossil occurrences, by about three to one. This is a most impressive figure, for the fossil record of the tropics is far less well known than that of the extratropics, so the number of genera with tropical origins should be significantly underestimated. Moreover, it is not clear that any of these bivalve genera have originated extratropically and then migrated into the tropics; at most, only a handful have done so.

The average age of genera in the tropics is lower than that of genera in extratropical faunas, and the average age of genera found in latitudinal bins increases with increasing latitude. Yet the tropics also contain older as well as younger lineages, and indeed the younger average generic age of the tropics is due to the large numbers of very young genera found there, not to any lack of older genera. Proceeding toward the poles, progressively fewer young genera are found, both absolutely and proportionately, and therefore the bivalves have an increasingly ancient generic aspect. Thus, the fauna becomes increasingly museumlike, even as it shrinks in size. The LDG is created by a dynamic process, one incorporating not only lineage originations, but migrations; the process can be epitomized as primarily "out of the tropics." The dichotomy implicit in Stebbins' question is thus misleading; the tropics are a cradle and a museum combined.

Ecologic factors. Although the dynamic outlined above answers Stebbins' question, it does not explain why diversity declines so significantly from the tropics toward the poles. There is a concept in ecology termed the "carrying capacity," which when applied to lineages states that, in a given environment, the number of different lineages that can be accommodated is limited by environmental features, so that there is a ceiling on lineage diversity. Limiting factors in this case would be those resources that could be used up by the incumbent lineages so that they would not be available to invaders; these are diversity-dependent factors. Diversity-dependent factors are usually applied to species. However, as the average number of bivalve species per genus is not less at the poles than in the tropics, data on generic diversity can be used as a conservative proxy in this case. A diversity-independent factor is something that is not used up by organisms, such as temperature. As noted, the gradient of warm to cold temperatures from the equator to the poles in the shallow sea has been widely hypothesized as responsible for the LDG, inferring that it is more difficult to adapt to lower temperatures. However, deep-sea temperatures are also cold, but diversity there is generally high, indicating that evolution is capable of overcoming the physiological problems involved with the low end of the marine temperature range. There would seem to be no reason why the presence of any number of species would prevent others from adapting to the same temperature.

The chief factors that can be used up in marine environments are food supplies and habitats. Habitat heterogeneity—roughly, the number of distinctive habitat types available—does not appear to change significantly with latitude along shelves that lack tropical reefs, which show strong LDGs. It certainly does not change by a factor of 10 or so as is needed to account for the LDG, so the food supply—basically, primary productivity—becomes the prime suspect. An important latitudinal change in productivity is due to a gradient in seasonality, with only modest variation in solar radiation and therefore primary productivity in low latitudes, grading to a high degree of seasonality with no direct solar radiation during the long nights of winter in high latitudes. Annual productivity in high latitudes tends to peak once in a major bloom that varies in timing following thawing and other local features, while in low latitudes, with year-round sunshine, variations in productivity are much smaller, although in some regions local conditions create exceptions.

One notion that relates the seasonality of primary productivity to diversity is that the adaptations required to cope with high seasonality involve generalist strategies. It is hypothesized that species in regions with unstable productivities tend to share a number of features, namely to live in a relatively wide range of habitats, feed on a wide range of items, feed low on the food pyramid, and live over broad areas. Such species thus maintain relatively large populations in many settings and may survive by persisting in some more favorable localities during inclement periods. Each species therefore requires a larger fraction of the available food resources than in more stable environments, limiting the number of species in the seasonal regimes and, when such contrasting regions have similar species-genus ratios, limiting the number of genera as well. To the extent that this hypothesis holds, the marine LDG results from a gradient in adaptive strategies.

Whatever the mechanism that limits marine diversity at higher latitudes, the dynamic associated with the recruitment of lineages primarily from the tropics is well documented and has important implications for conservation sciences. Not only genera have the preponderance of their first occurrences in tropical waters; so too do higher animal groups as well, such as families and orders. In the sea, many tropical communities are being degraded as a consequence of human activity. If tropical marine ecosystems are not well conserved, the main source of evolutionary novelties at all latitudes will be correspondingly reduced. This will be a problem not only for the future, but should be clearly understood when considering the legacy of our era.

For background information *see* BIODIVERSITY; BIVALVIA; CONSERVATION OF RESOURCES; ECOSYSTEM; FOSSIL; MACROEVOLUTION; ORGANIC EVOLUTION; PALEOECOLOGY; POPULATION ECOLOGY; SPECIATION; TROPICAL METEOROLOGY in the McGraw-Hill Encyclopedia of Science & Technology.

James W. Valentine

Bibliography. T. M. Blackburn and K. J. Gaston (eds.), *Macroecology: Concepts and Consequences*, Blackwell Science, Oxford, 2003; J. A. Crame, Evolution of taxonomic diversity gradients in the marine realm: A comparison of Late Jurassic and Recent bivalve faunas, *Paleobiology*, 28:184–207, 2002; D. Jablonski, The tropics as a source of evolutionary novelty: The post-Paleozoic fossil record of marine invertebrates, *Nature*, 364:142–144, 1993; D. Jablonski, K. Roy, and J. W. Valentine, Out of the tropics: Evolutionary dynamics of the latitudinal diversity gradient, *Science*, 314:102–106, 2006; K. Roy and E. E. Goldberg, Origination, extinction, and dispersal: Integrative models for understanding present-day diversity gradients, *Am. Naturalist*, 170:S71–S85, 2007; D. Schluter and J. Weir, Letters: Explaining latitudinal diversity gradients, *Science*, 317:451, 2007; G. L. Stebbins, *Flowering Plants: Evolution above the Species Level*, Belknap Press, Cambridge, Massachusetts, 1974; J. W. Valentine, *Evolutionary Paleoecology of the Marine Biosphere*, Prentice-Hall, Englewood Cliffs, New Jersey, 1973; G. J. Vermeij, *Biogeography and Adaptation*, Harvard University Press, Cambridge, Massachusetts, 1978; J. T. Weir and D. Schluter, The latitudinal gradient in recent speciation and extinction rates of birds and mammals, *Science*, 315:1574–1576, 2007.

2005 Kashmir earthquake

The October 8, 2005 Kashmir earthquake was the second major event with moment magnitude (Mw) ~7.6 within 5 years to have rocked the west Indian continent and the third since 1935. Each of these events exacted a heavy toll on life and property with 35,000 dead in 1935, 20,000 in 2001, and 75,000 in 2005, leaving many more homeless and destitute. These earthquakes occurred in a specific tectonic context of the Indian plate, although powered by the same process—its approximately 2 m per century north-eastward penetration into the Tibetan plate whose elevation and bulk is sustained by the dynamics of the collision process. To the north, the collision results in massive thrust earthquakes along the Himalayas as India slides beneath Tibet. The indentation of the Indian plate into Asia results in earthquakes along its eastern and western flanks that separate it from the Asian plate by broad zones of horizontal and compressional tectonics similar to fault systems that prevail in California. To the west-southwest, the submarine plate boundary consists of a simple oceanic fracture zone between the Indian plate and the Arabian plate, and to the southeast an oceanic apron of heavier rocks dives obliquely beneath southeast Asia and the modest Andaman plate (**Fig. 1**). In December 2004, almost the entire southeast plate boundary ruptured in a 1600-km-long magnitude 9.2 earthquake, creating giant tsunamis that struck the Indian ocean shores and killed more than 300,000 people, raising the fatality count of the three twenty-first century Indian earthquakes far above the total of all historical events.

The 2005 Kashmir earthquake was caused by approximately 5 m of southwest-directed slip of the western Himalayan ranges along a 30° northeast dipping fault that had been mapped by geologists, but had not slipped in historical times. The resultant slip raised the mountains locally by 2.5 m. Long-term slip on the fault is clearly responsible for the presence of substantial step in the mean topography in this part of the range. The surface rupture was mapped for at least 90 km along the strike, but aftershocks suggest that the rupture in the subsurface extended 20 km farther to the northwest. The rupture zone partly filled a previously identified seismic gap at the western extremity of the Himalayan collision zone, leaving a much larger gap to its east remaining to be relieved (**Fig. 2**). The 2005 rupture was outside the region considered typical of earthquakes in the central Himalayas, being close to the transition from pure radially directed thrusting to strike-slip faulting along the Chaman fault system of Afghanistan.

Its epicenter at 34.402°N, 73.560°E, as determined by the U.S. Geological Survey (USGS), lay about 19 km northeast of Muzaffarabad, the capital town of the Pakistan-administered Kashmir, which also suffered the heaviest casualties. The time of its occurrence, just before 9 o'clock in the morning [03:50:38 UTC (Coordinated Universal Time)] when children were at school and older people in bed after the predawn Ramadan meal, resulted in very large numbers of people being buried under the debris of collapsed buildings. As a result, 7300 schools were destroyed. Massive landslides reamed many hill slopes, wiping out entire villages and roads and aggravating rescue missions in difficult terrain. The severity of the rupture process resulted in damage over a large area, in one case bringing down a 10-story residential building housing 60 families in Islamabad, Pakistan's capital, about 100 km south-southwest of the epicenter. This was the earthquake's farthest reach southward, targeting a structure that was very poorly built in this modern city of more than a million people. Aftershocks continued for several months, including one of magnitude 6.2. Early aftershocks were sufficiently severe to further damage already weakened structures.

Tectonics of earthquakes surrounding the Indian plate and within the Indian subcontinent. As is now well known, earthquakes are the result of a mechanical instability in the Earth's upper, colder brittle crust, where elastic strains steadily created by relative plate movements can accumulate up to about 1 part in 10,000 before fracturing rock along faults. Earthquakes are thus understood to constitute a periodic phenomenon with a cyclicity determined by the rate of strain accumulation in the region and failure strength of the potential rupture surface(s). In principle, knowledge of this strain rate and the precise strain required to promote failure would permit earthquakes to be forecast. Global Positioning System (GPS) geodesy, when compared with geological data on plate motions, indicates that the velocity of tectonic plates have changed little over the past 3 million years. We therefore have a good present-day measure of surface strain accumulation rates. The absence of definitive knowledge about rock failure conditions or the absolute level of strain, however, makes it currently impossible to identify the location and timing of future earthquakes.

Earthquakes along the northern compression boundary generally occur on a northward gently dipping interface between the underthrusting Indian plate and the overlying Himalayan stack of thrust sheets formed by slivers shaved off its relentlessly penetrating front (upper right inset in Fig. 2). The largest of these earthquakes occur on a ≈100-km-wide gently dipping surface (6–9°) lying within 20 km of the ground surface, but the Kashmir earthquake rupture was steeper and narrower and penetrated to greater depth.

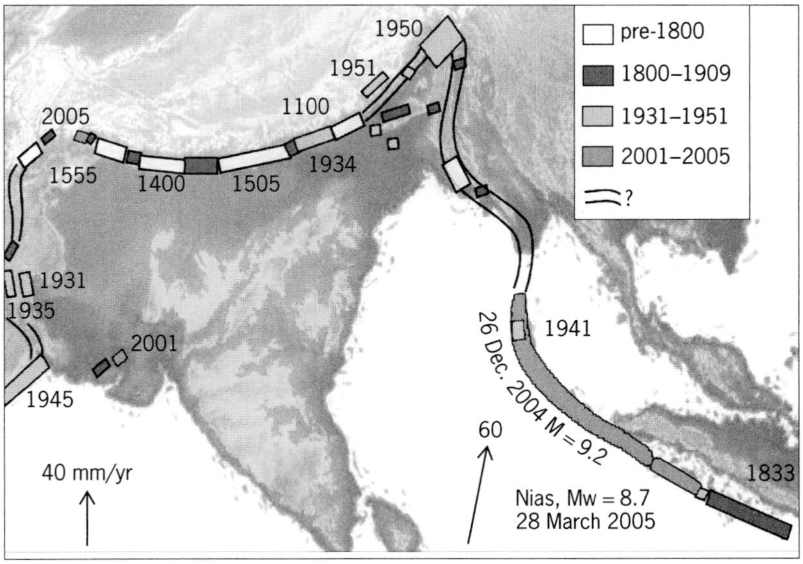

Fig. 1. Historically damaging earthquakes surround the Indian plate. One-fifth of the plate boundary shown here has ruptured in the past 5 years. With the exception of the Mw 8.2 Bihar Nepal earthquake in 1934, none of the earthquakes shown are a repeat of a previously known historical earthquake, although theoretical studies suggest a ≈500 year recurrence interval. Unshaded parts of the plate boundary will no doubt slip in the next few decades and centuries, as will regions of the central Himalaya that have not slipped for more than 500 years. Note that the Kashmir earthquake occurred at the extreme western end of the Himalayan arc near the transition to strike-slip faulting along its boundary with Afghanistan. (*Adapted from Bilham, 2006*)

Earthquakes along India's eastern boundary, such as the 2004 Andaman-Sumatra event, occur on an obliquely compressed interface separating the denser descending oceanic Indian plate from regions of lighter Asian continental plate or those of recently formed buoyant oceanic crust that form the Andaman plate. Steady descent of the old and therefore colder Indian plate beneath south Asia creates a stick-slip cycle of strain accumulation and release in the upper brittle crust. No similar megaquake is known to have occurred along this boundary in the past 200 years, but there are numerous examples of M > 7 earthquakes, which in 1881 and 1941 resulted in tsunamis along the Indian coast. Oblique motion was indeed observed on the main plate boundary during the 2004 earthquake, but moderate seismicity (Mw 7.5) in the previous century is characterized by thrust earthquakes on the subduction zone and strike-slip earthquakes on the Sumatra fault and its offshore continuation northwards. The process of separation of these components of slip is known as strain partitioning that, once established, carves out grooves on the fault zone that preferentially facilitate slip along them.

Earthquakes along India's western boundary permit northward slip of the plate relative to a promontory of Asia in Afghanistan and Baluchistan at a rate of 2–4 cm/yr. Though the plate boundary here is marked by a well-defined strike-slip fault (the

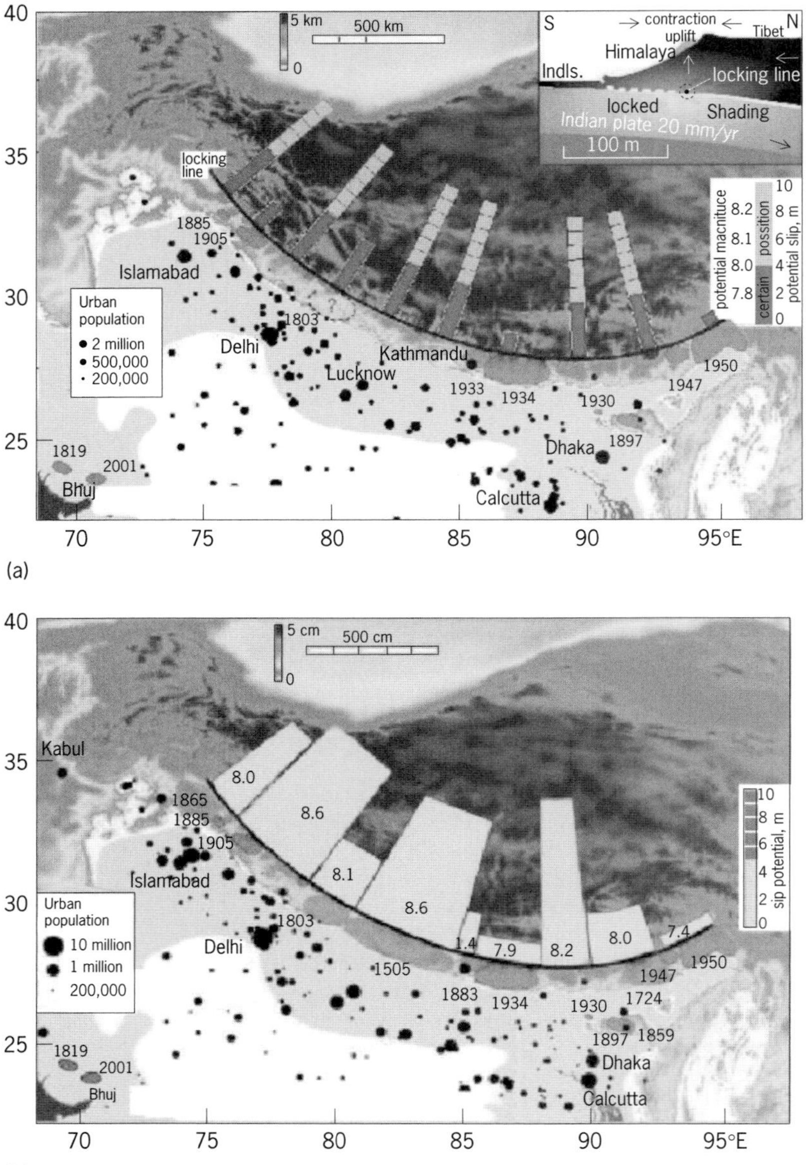

Fig. 2. Two views of the Himalayan earthquake hazard. (*a*) Conservative estimate of slip potential since 1800 using no information about past rupture dimensions, and (*b*) nonconservative estimate of maximum credible earthquake inferred from earthquakes since 1400 (dark gray areas are inferred rupture zones). Neither represents earthquake forecasts, since no indication of earthquake occurrence time is associated with these figures. The Mw 7.6 Kashmir earthquake occurred at the western end of the Himalaya, beyond the light gray region labeled Mw 8 in the right-hand panel. Black dots indicate population centers. Right upper inset in (a) shows in cross section the underthrusting Indian plate beneath Himalaya and Tibet.

Chaman fault) to the west, at places it is more than 150 km wide, and earthquakes in the Siestan ranges accommodate the compressional stress of indentation that result in a fold/thrust belt with a contraction rate of probably less than 1 cm/yr. The convergent and strike-slip components of this complex motion, as in the Nicobar/Andaman segment of India's eastern boundary, are separated into pure thrust on steep east- or west-dipping faults, or sinistral sliding motions on vertical faults. In 1931, a sequence of Mw > 7 earthquakes occurred that demonstrate the importance of strain partitioning. The 1931 Mach/Sharigh earthquakes were reverse slipping events, allowing roughly 1 m of extension of the Siestan range front. The transient decompression of the range resulted 3.5 years later in a reduction of fault-normal stress on a strike-slip fault near Quetta that apparently triggered slip in a Mw 7.7 earthquake. Although the reason for the delay is not yet understood, the association of these events is too unusual to be a coincidence and holds promise for forecasts of similar large strike-slip events to the north and to the south of the Quetta sequence.

Far from the plate boundaries, earthquakes such as the 1819 and 2001 Rann of Kachchh Mw > 7.6 earthquakes are also caused by the release of accumulated compressive strain. But since they are less frequent, their causal mechanisms are less clear. A proposed physical mechanism (**Fig. 3**) for earthquakes within the continent interior invokes the role of flexural stresses associated with the collision of India with the Tibetan plateau. A 450-m-high bulge has long been known to exist in central India, and was originally termed "the hidden range" by India's early geodesists because of the anomalous gravity field it produces where it crosses India's central plateau. The bulge is partly caused by the buckling forces of collision and partly by the weight of the plateau pushing down on India's northern edge, which has resulted in a 4-km depression beneath the frontal thrusts of the Himalayas, now filled by the sediments of the Punjab, Ganges, and Brahmaputra river systems. Tensile stresses occur near India's surface north of the bulge, resulting in normal faulting earthquakes with magnitudes less than Mw = 6.5. At depth below a neutral axis, stresses at around 20 km are compressive, giving rise to thrust or strike-slip earthquakes such as the 1988 Udaypur event on the Nepal-India border. A secondary depression occurs south of the bulge, and this depression results in reverse faulting near the surface and tensile normal faulting at depth. It is in this region that the devastating Latur earthquake occurred in 1993, causing 7500 deaths. Compressional stresses in a concave depression of the Indian plate are largest at the surface. It is therefore no accident that the Latur earthquake was the only earthquake on the Indian continent to produce a surface rupture.

New understanding and implications. First results from the epicenter of the Kashmir earthquake were in the form of studies of teleseismic waveforms that provided estimates of moment release, magnitude, and mechanism. These were followed shortly afterward by images of the epicentral region, using SAR (synthetic aperture radar) and ASTER (advanced spaceborne thermal emission and reflection radiometer), that confirmed the strike, the presence of near-surface rupture, and the extent of uplift. Some weeks later a sparse set of preseismic GPS observations were repeated, providing coseismic displacements south, west, and north of the epicenter. These GPS measurements confirmed the SAR data and suggested secondary faulting of unusual complexity.

In recent years, geophysicists have become accustomed to examining surface deformation through interferograms formed from pairs of temporally separated radar images, known as interferometric synthetic aperture radar (InSAR). The steep terrain and unstable slopes as well as vegetative changes of the Kashmir Himalayas meant that normal InSAR procedures were not amenable and special techniques had to be applied to recover the deformation field from ENVISAT (European Space Agency satellite) and ASTER imageries. However, the large vertical and horizontal displacements in the epicentral region facilitated these approaches. A useful guide to the methodology may be found at http://comet.nerc.ac.uk/news_kashmir.html.

While the 90-km-long northwest-southeast strip of coseismic deformation from Balakot to Muzaffrabad, gleaned from SAR, data coincide with the trace of a mapped active plate boundary fault, its rupture geometry differs markedly from those of Himalayan arc earthquakes to the east. In particular, its steeply dipping rupture plane at 37° is much larger than the ~9° gently dipping rupture planes that are known to fit the observed data for Himalayan earthquakes elsewhere along the arc. Another significant feature of this earthquake is a prominent cluster of shallow aftershocks along a more westerly trend within the earlier-identified Indus-Kohistan seismic zone, which is offset from the main rupture by about 30 km to the southwest (**Fig. 4**). This implies that the Kashmir earthquake was caused by a complex rupture process involving subsurface accommodation of northeast/southwest convergence.

Analysis of GPS data around the region gathered from a half-dozen sites in June 2001 and repeated immediately after the earthquake in October-November 2005 using a composite model of the main rupture as well as that constrained by the distribution and focal mechanism of aftershocks (Fig. 4) suggest that the Kashmir rupture may have triggered a further slip of 1.8 m on a blind (subsurface) wedge thrust to its northwest. Although this result is controversial, the interpretation is consistent with the spatial distribution of aftershocks as well as the trend of the historically active Indus-Kohistan seismic zone and reveals a pattern of slip propagation across weak sedimentary strata unsuspected in Himalayan ruptures.

Significantly, all models of this rupture predict large consequent increases in the Coulomb stress to the southeast and therefore enhanced hazard potential of the entire region west of the Kangra rupture. This is best appreciated by a reference to Fig. 1 and **Fig. 5** which show that the 2005 Kashmir rupture flanked by the smaller 1974 Pattan and 1885 Kashmir earthquakes filled only a small part of the large seismic gap west of the 1905 Kangra rupture that itself had been only partially relieved by the 1555 Kashmir Mw 7.6 or larger earthquake. The history of Himalayan earthquakes, which would allow one to set the date from which the strain budget along various segments of the Himalayas may be reckoned, is fairly well known since 1800 from writings in Persian, Arabic, and Tibetan texts, and travelers' accounts but less so before that time. The latter are rarely quantitative enough to estimate magnitudes and rupture areas. By piecing together the various threads of consistent accounts, a chronology of Indian plate-boundary earthquake ruptures has been attempted.

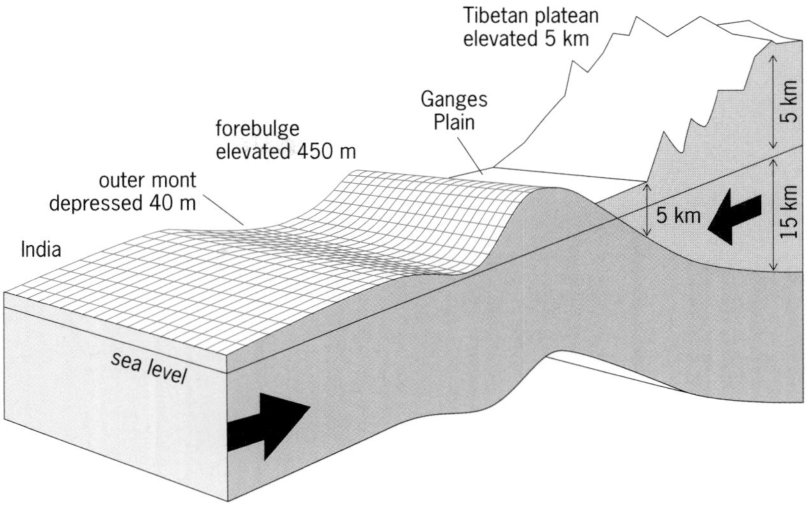

Fig. 3. The flexure of India caused by its collision with the Tibetan Plateau. India streams through this flexural geometry at a rate of 2 cm/year, resulting in slow changes in stress far from the Himalayan plate boundary. The Latur earthquake occurred near the outer moat where compressional stresses are a maximum near the surface. Shallow earthquakes near Delhi are caused by the tensile stress near the top and north of the flexural bulge that forms the central Indian plateau.

These estimates, plotted in Fig. 2, clearly draw attention to seismic hazard both in western and Central Himalayas, each bordering populous cities. A conservative estimate of the spatial extent and slip of earthquakes in the Kashmir region in 1501 and 1555 implies that the recent Kashmir earthquake released only a small part, <25%, of the accumulated slip along the segment flanked by the 1974 Pattan and the 1905 Kangra earthquake ruptures, and in so doing significantly augmented the prevailing stress at its unruptured eastern end. While we are unable to forecast when this eastern Kashmir segment will rupture and with what magnitude, there is little doubt that compressional stresses in the Panjal ranges are sufficiently mature to drive a great earthquake. Estimates of its magnitude range from a low of Mw 7.8 to a high greater than Mw 8.0. Studies of slip in trenches in the Himalayan foothills suggest that a megaquake had occurred near and west of Dehra Dun around 1400 that may have exceeded Mw 8.4 and may have extended west of the Kangra earthquake rupture into the present seismic gap.

Outlook. Earthquakes in ancient times used to be considered "acts of God" through the trauma they exert on innocent populations. Today, we are no longer ignorant of the causes and probable locations of future earthquakes. And while successive earthquakes come as no surprise to scientists, each one delivers a punishing and unacceptable blow to local communities. There is thus an ironic gulf between what the scientific community knows and what society does with this knowledge.

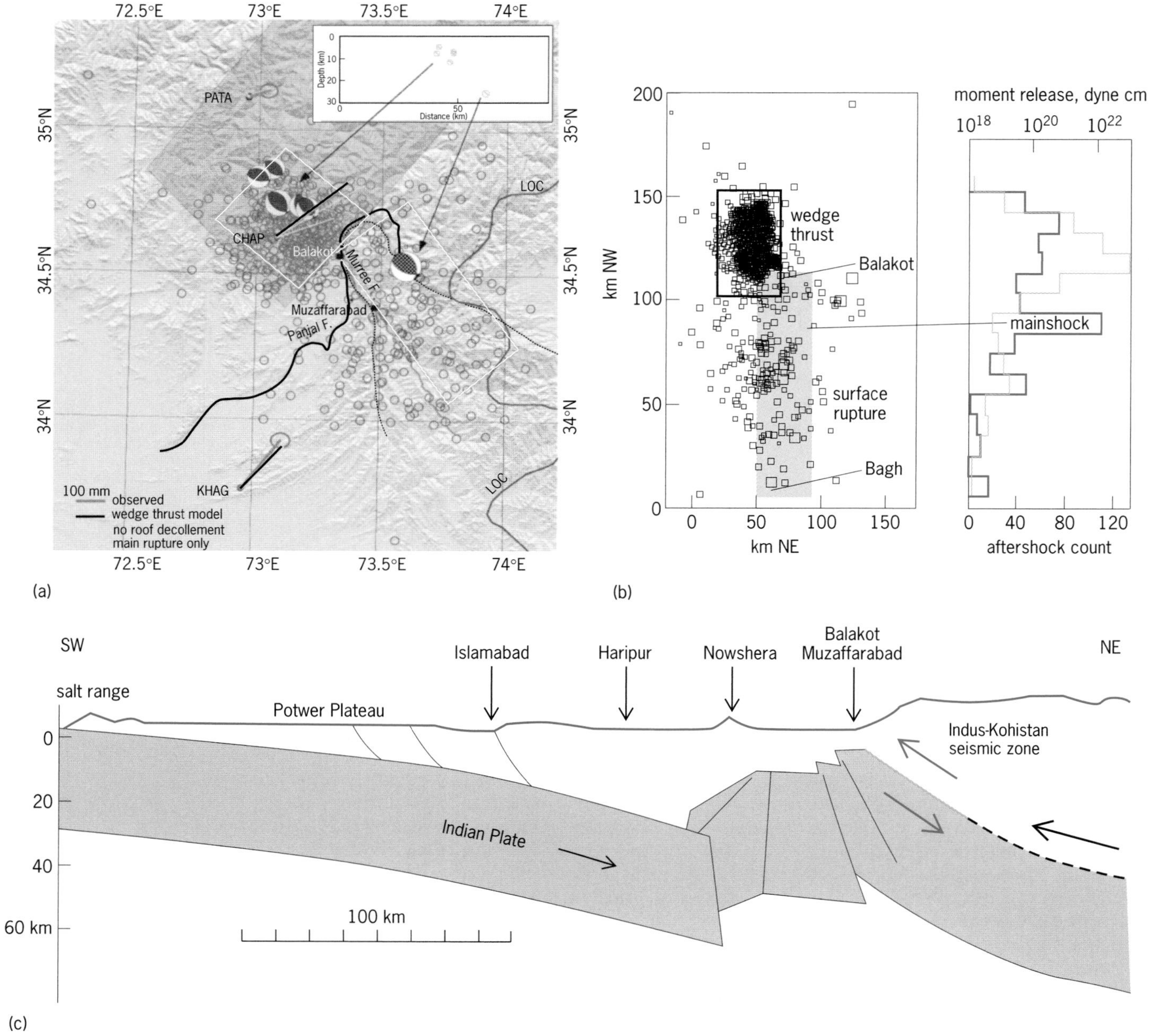

Fig. 4. (*a*) Observed and modeled displacements for the geodetic sites. Dipping dislocations are outlined in dashes. Recorded aftershocks are open, gray circles; those with focal mechanisms are gray beach balls. The focal mechanism of the main rupture is the hatched beach ball. The mapped trace of the MBT (locally the Muree fault) is framed by dotted line; the MCT (locally the Panjal fault) is shown by heavy black line; and the nodal plane from the radar image is shown with white line. Inset: focal mechanisms for the main shock and some aftershocks projected into a cross section normal to the strike of the main dislocation. (*b*) Aftershocks reported by National Earthquake Information Center (NEIC) plotted as function of transverse and longitudinal distance along the rupture to illustrate the offset between the surface rupture and the knot of aftershocks at its northwest end. The region corresponds to Indus-Kohistan seismic zone and 75% of all aftershocks occur there releasing ~3% of the total main shock moment release. Aftershock moment (black line), and numbers of events (grey line) are summed in 10 km segments along the strike. (*c*) Section of the 2005 Kashmir earthquake rupture geometry (*modified from Seeber and Armbruster, 1979*). (*After Bendick et al., 2006*)

Part of the reason for this regrettable gap is that scientists have yet to forecast earthquakes with any useful degree of credibility. For example, Fig. 2 contains no information about the timing of future earthquakes, but only about how big they could be were they to occur today. An extensive history of earthquakes extending back several thousands of years is a basic requirement for making a probabilistic assessment of hazard, so that effective measures are taken to minimize vulnerabilities of the communities exposed to earthquake risk. An authentic historical record of repeat earthquakes in India has yet to be created. This is an unfortunate situation that offers little hope of early remedy. The discovery of palaeoearthquakes in the Himalayas in the past 10,000 years will no doubt correct some of this deficiency, albeit the uncertainties may still be large and poorly constrained.

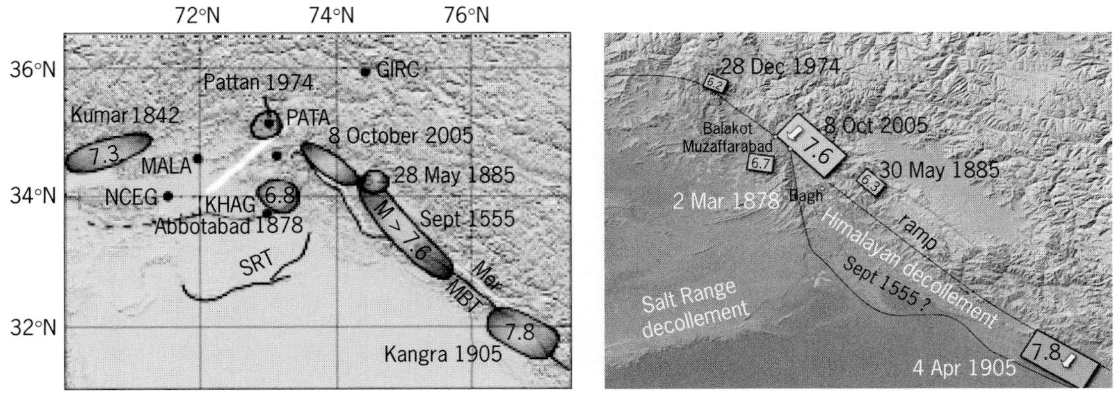

(a) (b)

Fig. 5. Historical earthquake ruptures (*a*) represent by the Medvedev-Sponheuer-Karnik (MSK) scale of estimated intensity MSK > VII (very strong) over the past 500 years. The inferred rupture zone for the 1555 earthquake is shifted SE of the reports to account for a possible bias in reporting weighted by populations in the valley. Black circles with four-letter codes indicate points with pre- and postseismic GPS observations. MCT is the Main Central Thrust, MBT the Main Boundary Thrust, and SRT is the Salt Range Thrust. (*Bilham et al., 2004*) (*b*) Some of these same rupture zones shown on an imagery of the area.

Notwithstanding these shortcomings, much can be done to mitigate the possible adverse impact of future earthquakes on communities that are arguably at grave risk, notably those in the populous towns situated on the soft alluvial plains bordering the Himalayan plate boundary. Earthquake risk to local populations is primarily posed by buildings liable to collapse when subjected to ground shaking. Regrettably, although building codes exist in many parts of India and Pakistan, they are rarely applied stringently enough to prevent shoddy construction. Corruption can be overt in the form of contractors bribing officials not to report code infringements, or covert in the form of poor materials being incorporated into ongoing construction and hidden from inspection. More often, unsound building practices are incorporated from lack of knowledge about building assembly among residents anxious to save money. The scarcity of inexpensive wood for building houses and its replacement by concrete is much to blame for many recent disasters.

Thus, while natural hazards cannot be prevented, their disaster potential and much of the attendant risk to populations can be greatly minimized. It is essential that new construction incorporate earthquake resistance, usually incurring no more than a mere 10% increase in construction costs. If this policy were universally adopted, the replacement of building stock through old age would likely eliminate half of the world's most vulnerable buildings in the next 50 years. The retrofit of existing houses would be a more costly undertaking, especially since many structures would require rebuilding. But responsible government has an important role in retrofitting schools, hospitals, and critical support systems, whose collapse in earthquakes is inexcusable but currently widespread.

If the warning offered by the three disastrous earthquakes in India in recent years is unheeded by city planners, politicians, and architects, the disastrous impact of a future earthquake could be far more grievous than any we have yet seen. Three inexorably growing trends of our social dynamic conspire to force this conclusion: increased populations (10 to 100 times larger than that exposed to previous Indian earthquakes), increased vulnerability (the building stock is taller and far less well-constructed than ever before), and the demographic pattern of high population density in regions expected to experience abnormally high ground-shaking intensity (thick sediments near river valleys like the Ganges and Brahmaputra systems, and ancient lakes like those in the Katmandu and Kashmir valleys). Given that several Mw > 8 earthquakes are anticipated along the Himalaya, the population at risk exceeds 50 million people. An expedidited program to ensure and wherever desirable enforce the incorporation of earthquake-resistant design and construction practices, using all possible means and retrofitting of all community buildings and support systems, has the promise of safeguarding a large proportion of those potentially at risk and preventing trillions of dollars of economic damage that a future earthquake in the region may otherwise exact.

[*Acknowledgements*: This paper has greatly benefited from numerous discussions with Rebecca Bendick, Peter Molnar, and Roger Bilham.]

For background information *see* ASIA; EARTHQUAKE; EARTHQUAKE ENGINEERING; FAULT AND FAULT STRUCTURES; PLATE TECTONICS; SEISMIC RISK; SEISMOLOGY in the McGraw-Hill Encyclopedia of Science & Technology.

Vinod K. Gaur

Bibliography. N. Ambraseys and R. Bilham, Earthquakes and crustal deformation in northern Baluchistan, *Bull. Seism Soc. Am.*, 93(4):1573–160, 2003; J.-P. Avouac et al., The 2005, Mw 7.6 Kashmir earthquake: Sub-pixel correlation of ASTER images and seismic waveforms analysis, *Earth Planet. Sci. Lett.*, 249(3–4):514–528, 2006; R. Bendick et al., Slip on an active wedge thrust from geodetic observations of the 8 October 2005 Kashmir earthquake, *Geology*, 2007; R. Bilham, Dangerous tectonics, fragile buildings, and tough decisions, *Science*, 311(5769):1873–1875, 2006; R. Bilham, Earthquakes in India and the Himalaya: Tectonics, geodesy and history, *Ann. Geophys.*, 47(2):839–858, 2004;

R. Bilham and K. Wallace, Future Mw > 8 earthquakes in the Himalaya: Implications from the 26 Dec 2004 Mw = 9.0 earthquake on India's eastern plate margin, *Geol. Surv. India Spec. Pub.*, 85:1–14, 2005; R. Bilham, R. Bendick, and K. Wallace, Flexure of the Indian Plate and intraplate earthquakes, *Proc. Indian Acad. Sci. (Earth Planet Sci.)*, 112(3):1–14, 2003; N. Feldl and R. Bilham, Great Himalayan earthquakes and the Tibetan Plateau, *Nature*, 444:165–170, 2006; S. Fujiwara et al., Satellite data gives snapshot of the 2005 Pakistan earthquake, *Eos, Transactions American Geophysical Union*, 87(7):73, 2006; T. Nakata et al., *Active Faults of Pakistan, Spec. Publ. 21*, Research Center for Regional Geography, Hiroshima University, Hiroshima, Japan, 1991; T. Parsons et al., Static stress change from the 8 October, 2005 M = 7.6 Kashmir Earthquake, *Geophys. Res. Lett.*, 33:L06304, 2006; E. Pathier et al., Displacement field and slip distribution of the 2005 Kashmir earthquake from SAR imagery, *Geophys. Res. Lett.*, 33(20):L20310, 2006; L. Seeber and J. G. Armbruster, Seismicity of the Hazara arc in northern Pakistan: Decollement vs. basement faulting, in A. Farah and K. A. DeJong (eds.), *Geodynamics of Pakistan [C]*, Geologic Survey of Pakistan, Quetta, pp. 131–142, 1979; V. C. Thakur et al., 8 October, 2005 Muzaffarabad earthquake and seismic hazard assessment of Kashmir gap in Northwestern Himalaya, *J. Geol. Soc. India*, 68(2):187–200, 2006; H. Wang, G. Linlin, and C. Xu, 3-D coseismic displacement field of 8 October 2005 Kashmir earthquake inferred from satellite radar imagery, *EOS Trans.*, 87(7):73–77, 2005.

Underground research laboratories

Underground research laboratories include a wide range of facilities for scientific and engineering research at depths ranging from 15 m (50 ft) to more than 2 km (1.25 miles) below the surface. Scientists and engineers labor at these depths to gain access to an environment that both shelters their experiments from the constant shower of cosmic radiation encountered at the surface and provides a window into the geological and biological processes that operate there. The reasons for underground laboratories fall into three general categories. (1) The depths of the Earth are explored to produce the minerals that lie within, to excavate relatively shallow tunnels for road and underground construction, or to dispose of hazardous substances. (2) Modern particle accelerators necessary for investigating subatomic structure require long paths to accelerate charged particles, and in the process produce ionizing radiation that can effectively be shielded by relatively shallow burial. (3) A new type of physics experiment requires a location that has reduced levels of cosmic radiation to detect interactions that are very infrequent, very weak, or both. Some of the more recent research projects involve the search for neutrinos, one of the more elusive of the subatomic particles. Therefore, the underground provides shielding from cosmogenic backgrounds due to the overlying rock or soil.

Mines and tunnels. The first underground laboratories were operating mines and tunnels. Because the life of a mine can be quite long, measured in tens or even hundreds of years, considerable experience can be gained through the process of mining the ore and observing the effects of removing portions of the supporting rock. As mining progressed from the Middle Ages onward, economics and safety demanded a better understanding of the behavior of the rock because of the changes in stresses associated with mining. The development of the Comstock Lode in Nevada in 1859 sped the progress of stabilizing large excavations in relatively weak rock as mines became deeper and more extensive.

Geologic disposal. A major step forward was the development of underground laboratories for the safe and responsible disposal of radioactive products from nuclear power and defense activities. The potentially damaging effects on the environment were recognized and a wide range of solutions put forward. The usefulness of underground nuclear waste repositories with favorable geological conditions spurred investigations into the construction problems, physical and chemical containment of the wastes, and potential pathways for contaminants to escape into the environment. The early history of this research was dominated by facilities such as the Canadian Underground Research Laboratory which began construction in 1982, and the U.S. Waste Isolation Pilot Plant Project which was authorized in 1980. Although many geologic disposal facilities have been examined around the world, two of the most active examples of this technology within the United States are the Waste Isolation Pilot Plant (WIPP) and the Yucca Mountain Nuclear Waste Repository. The WIPP in southeastern New Mexico is used for disposing of defense-generated radioactive wastes and the Yucca Mountain Waste Repository in southern Nevada is proposed as a site for storing high-level radioactive waste generated in the nuclear fuel cycle. In the case of WIPP, the location of the actual repository is 655 m (2150 ft) below the surface. The rocks of the repository include rock salt, anhydrite, and various other potassium and sodium salts associated with evaporite deposits. Rock salt's unique properties include low hydraulic permeability and high ductility under elevated confining pressures. Research on these properties has contributed greatly to the methods of containing hazardous material and the understanding of rock mechanics engineering (the study of how rocks deform under the influence of varying stresses) as well as how fluids move in the subsurface, particularly in materials with extremely low permeabilities.

Particle accelerators. Whereas the previous laboratories were established to promote industrial objectives and to protect the environment from contaminants, particle-accelerator laboratories use the underground as a convenient construction medium and take advantage of the shielding characteristics of the overlying soil and rock. These laboratories accelerate charged particles to speeds approaching that of light, which then are directed into targets to produce showers of elementary particles. Because the acceleration paths are kilometers in length, it is convenient to place them underground in circular rings, thus freeing the surface for other uses. The overlying geologic material also shields the environment at the surface through absorption and attenuation of any radiation generated as part of the experimentation. One example of this type of facility is operated by CERN, the European Organization for Nuclear Research, located northwest of Geneva, Switzerland. Although CERN has been in existence since 1954, the newest and largest project for the consortium is the Large Hadron Collider (LHC), which should be ready for physics experiments by 2008. The LHC is located in a circular tunnel with a circumference of 27 km (17 mi) located 100 m (330 ft) below the surface. Although this is a relatively shallow depth, the entire complex has over 40 km (25 miles) of mined tunnels, plus six large cavities with spans greater than 20 m (65 ft). Physics experiments using this facility are expected to produce the Higgs boson particle and to confirm predictions made within the Standard Model of particle physics.

Particle detectors. Another type of underground laboratory involves rare interactions between fundamental particles and subsurface detectors. The objective of these detectors is to measure the extremely small number of interactions between the detector medium and a fundamental particle, such as a neutrino that has a mass many orders of magnitude less than that of having an electron and no electrical charge. Although these types of particles interact only rarely with other particles and can pass easily through the entire Earth, they are sufficiently numerous that a few interactions per year per detector can be predicted. The likelihood of having an interaction and identifying the resulting detection is increased by having a larger detector with more mass with which to interact and reducing the interactions due to background radiation. Cosmic radiation is the largest source of background radiation. Interaction of cosmic radiation with the detector mass can result in occurrences that must be investigated to determine whether they are legitimate interactions. Because the cosmic ray flux decreases with depth due to absorption, the process of identifying neutrinos is simplified greatly with deeper emplacement of the detectors. **Figure 1** shows the expected decrease in cosmic radiation as a function of depth.

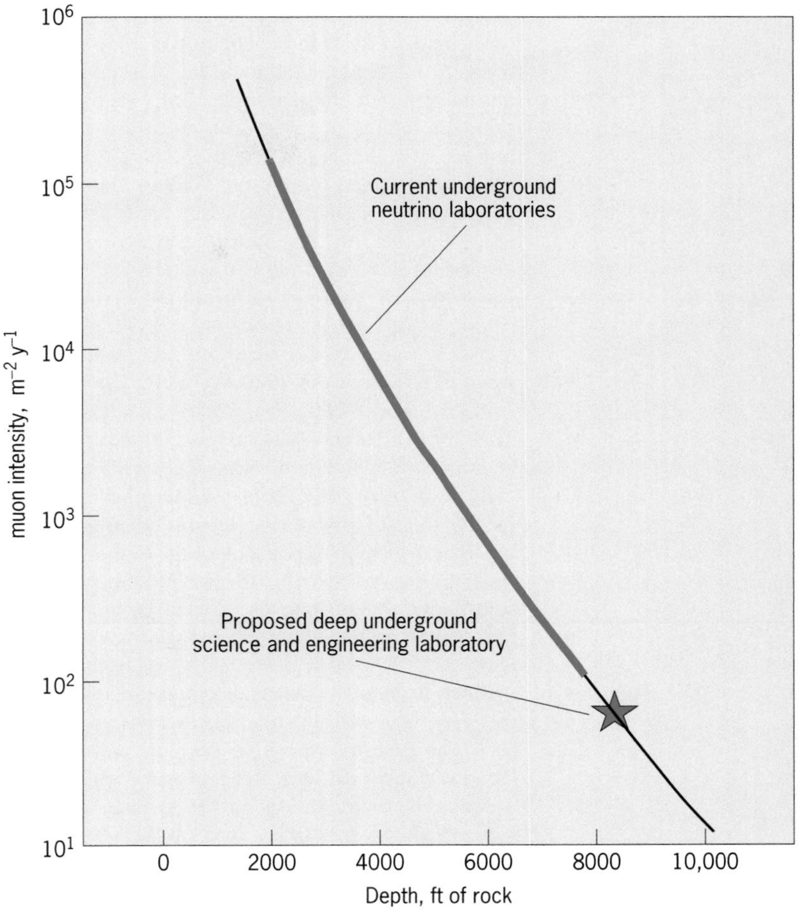

Fig. 1. Relationship between depth and the intensity of cosmic radiation (1 ft = 0.3 m). The rapid decrease of the cosmic radiation with depth results in many fewer nondesirable interactions. (*Data from D.-M. Mei and A. Hime, Muon-induced background study for underground laboratories, Phys. Rev. D, 73, 053004, 2006*)

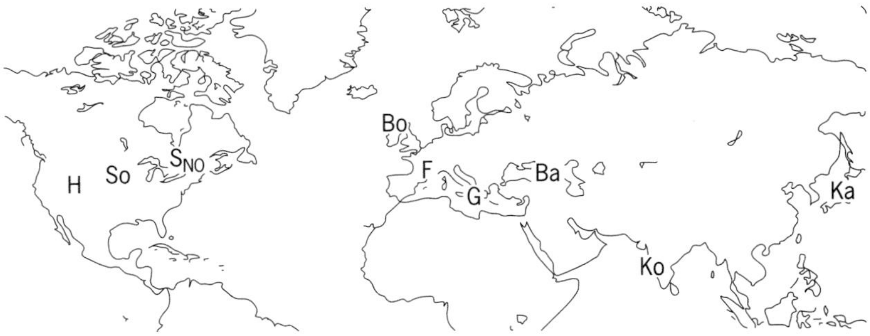

Fig. 2. Locations of neutrino experiments and laboratories. (H) Brookhaven Solar Neutrino Laboratory, Homestake Mine, (So) Soudan Underground Laboratory, (SNO) Sudbury Neutrino Laboratory, (Bo) Boulby Underground Facility, UK, (F) Frejus Detector, (G) Gran Sasso Laboratory, (Ba) Baksan Neutrino Observatory, (Ko) Kolar Gold Fields Neutrino Experiment, (Ka) Kamiokande Neutrino Detector.

Numerous neutrino laboratories are located around the world (**Fig. 2**), and the descriptions shown below for three of these illustrates the scope of their work.

Gran Sasso. This laboratory is devoted to particle physics, particle and nuclear astrophysics, and neutrino research. The subsurface facilities are located beneath the Gran Sasso Mountain in the Apennines Mountains about 120 km (75 mi) northeast of Rome, Italy. The laboratory extends beneath the mountain, which results in an average rock cover of 1400 m (4600 ft). The three experimental halls can be reached through horizontal accesses that are each about 20 m (65 ft) wide and 100 m (330 ft) long. Current projects include the detection of neutrinos produced by nuclear processes that provide energy from the Sun and supernovas, the search for dark matter in the universe, and the identification of products of nuclear reactions of astrophysical interest.

Kamiokande. This detector is located in the Kamioka mine in Japan at a depth of about 1 km (0.6 mi) and spans an excavation of 40 m (130 ft). The detector is made of two parts. The outer detector, which consists of a water shell, identifies incoming cosmic muons through the production of Cerenkov radiation; the effects of the cosmic radiation can be removed. The inner detector consists of a container holding a mineral oil that emits light when charged particles interact with it. Approximately 2100 photomultiplier tubes are distributed around the container of mineral oil to measure the emitted light.

SNOLab. The Sudbury Neutrino Observatory is located in the active Creighton nickel mine near Sudbury, Ontario, Canada. The laboratory is constructed at a depth of 2070 m (6800 ft) beneath the surface. It uses 1000 tons of heavy water (D_2O) in a 12-m (39-ft) diameter container. Cerenkov radiation is produced when neutrinos react with the heavy water. An array of 9600 photomultiplier tubes surrounds the heavy-water vessel and detects the emitted light resulting from interactions. The results from this laboratory have already provided important insights into the properties of neutrinos and fusion reactions that are occurring within the core of the Sun.

Other research activities. Depending on the primary mission of the underground research laboratory, the research tends to support that mission. However, most underground laboratories welcome other work that requires access to the subsurface. Many of these activities center on the composition and movement of water and gases in the subsurface, the distribution of rock types, and the geochemistry of fluids in the subsurface. A new and exciting avenue of investigation is that associated with the presence of life in even the deepest mines and laboratories. Investigations of life in these extreme environments require special attention to ensure that the samples are collected without contaminating them.

Future neutrino laboratories. New approaches for detecting neutrinos are being planned at laboratories deep within the Earth, including the newly designated laboratory at the site of the former Homestake gold mine in Lead, South Dakota (Figs. 1 and 2). A series of facilities are envisioned at this laboratory that use detectors based on liquefied noble gases and ultrapure water. Laboratory installations are planned for initially at a depth of 1480 m (4850 ft) below the surface and later at 2260 m (7400 ft) or deeper. Chambers of 50–60 m (160–200 ft) in diameter may be required. Ensuring the stability of these large cavities at the pressures found at these great depths will be challenging. The anticipated laboratory will be a multidisciplinary project and will encompass experiments for engineering, geosciences, and geomicrobiology. Experiments in the geosciences and geomicrobiology will examine the effects of pressure on the long-term stability of the underground structures, the flow of fluids through the rock, and how microbial life maintains itself at high temperatures and in a nutrient-limited environment. Laboratories of this type require innovation in construction design, bringing together the experience and long history of underground construction and the requirements of the physicists who probe the fundamental nature of matter.

For background information *see* CERENKOV RADIATION; COSMIC RAYS; ELEMENTARY PARTICLE; HIGGS BOSON; MINING; NEUTRINO; PARTICLE ACCELERATOR; SALINE EVAPORITES; STANDARD MODEL in the McGraw-Hill Encyclopedia of Science & Technology.

William Roggenthen

Bibliography. Board on Physics and Astronomy, *Connecting Quarks with the Cosmos: Eleven Science Questions for the New Century*, The National Academies Press, 2003; H. C. Hoover (1912 translator of the first Latin edition of 1556), *Georgius Agricola De Re Metallica*, Kessinger Publishing, 2003; B. J. McPherson and the EarthLab Steering Committee, *EarthLab: A Subterranean Laboratory and Observatory to Study Microbial Life, Fluid Flow, and Rock Deformation*, Geosciences Professional Services, Inc., 2003; Neutrino Facilities Assessment Committee, *Neutrinos and Beyond: New Windows on Nature*, The National Academies Press, 2003; W. D. Weart, N. T. Rempe, and D. W. Powers, The Waste Isolation Pilot Plant, *Geotimes*, vol. 43, no. 10, pp. 14–19, 1998.

Unfolded protein response

The secretory pathway of eukaryotic cells is responsible for delivery of plasma membrane receptors, blood coagulation factors, secreted signaling proteins (for example, insulin), and many other proteins to the cell surface. The entry point for all these proteins into the secretory pathway is the endoplasmic reticulum (ER). Proteins destined for the secretory pathway enter the ER through an aqueous pore, the SEC61 translocation channel, in the ER membrane as an extended, completely unfolded polypeptide chain (**Fig. 1**). In the ER, the polypeptide chains fold into their native three-dimensional conformations and are posttranslationally modified by trimming of oligosaccharides that were added during polypeptide elongation and translocation through the pore and by disulfide bond formation. Protein folding is assisted by two classes of protein folding helper proteins: the protein foldases and the molecular chaperones (Fig. 1). Protein foldases such as protein disulfide isomerases (PDIs) and *cis-trans* peptidyl-prolyl isomerases (PPIs) catalyze disulfide bond formation and *cis-trans* isomerization of peptidyl-prolyl bonds, respectively. Oxidative protein folding by PDIs is responsible for ~25% of all reactive oxygen species in a eukaryotic cell. Three chaperone systems operate in the higher eukaryotic ER: (1) BiP, originally identified as a binding partner of unassembled immunoglobulin heavy chains in the ER lumen of antibody-producing cells; (2) GRP94 (glucose-regulated protein 94); and (3) the oligosaccharide-binding lectin chaperones, calnexin (CNX) and calreticulin (CRT). BiP and the lectin chaperones cycle through energy [adenosine triphosphate (ATP)]-consuming rounds of substrate binding and release until the unfolded substrate is folded. BiP preferentially works on

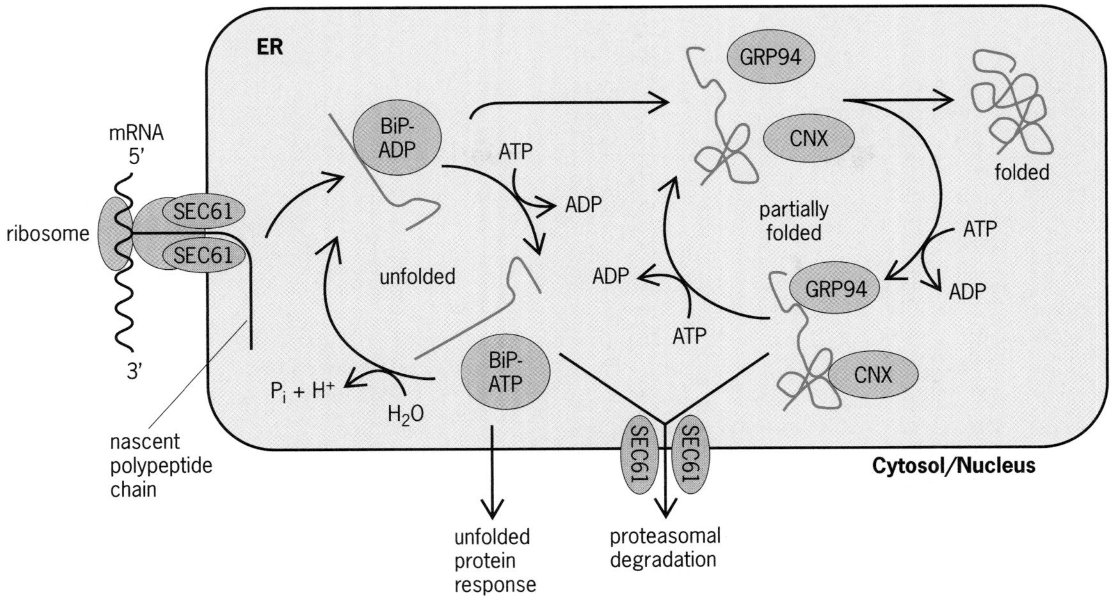

Fig. 1. Protein folding and quality control in the ER. (*Left*) A polypeptide synthesized on a ribosome bound to the ER membrane is translocated during the course of chain elongation (that is, cotranslationally) into the lumen of the ER. A nonglycosylated released unfolded polypeptide or nonglycosylated regions of a nascent glycoprotein bearing exposed hydrophobic polypeptide regions, which in a fully folded protein are buried in the interior of the protein, bind to BiP carrying ADP (adenosine diphosphate), preventing it from aggregating with other unfolded proteins. An enzyme, called an exchange factor, exchanges the ADP for ATP (adenosine triphosphate) on the bound BiP, which leads to release of BiP from the unfolded protein. While the protein is not bound to BiP, it has a chance to continue on its folding pathway. At the same time, BiP hydrolyzes the bound ATP to yield BiP-ADP and inorganic phosphate [HPO_4^{2-}; abbreviated P_i]. Cycles of BiP binding, ATP for ADP exchange, BiP release, and ATP hydrolysis repeat until the protein is folded and no longer binds to BiP. During ER stress, the amount of unfolded proteins in the ER lumen exceed the amount of BiP that can handle them, a condition resulting in activation of the UPR. (*Right*) After release of a partially folded protein from BiP, the protein is bound by the chaperone GRP94 and the lectin chaperones CNX and CRT. CNX and CRT bind to oligosaccharides attached to the protein. They are released from glycoproteins upon enzymatic removal of a glucose moiety. Readdition of this glucose moiety to unfolded glycoproteins triggers a second round of interaction with the lectin chaperones. These cycles of glucose removal and addition continue until folding of the glycoprotein is complete. Glycoproteins that take a long time to fold and are intermittently bound to the lectin chaperones over long time periods are delivered to a channel in the ER membrane, possibly the SEC61 channel, where they are translocated to the cytosol for degradation in the proteasome.

completely unfolded substrates, whereas GRP94, calnexin, and calreticulin preferentially interact with partially folded substrates. In these folding cycles, the substrate conformation is continuously probed until the substrate has folded into its native conformation, with the chaperone serving to prevent the aggregation of unfolded or partially folded proteins. This quality control mechanism retains unfolded and partially folded proteins in the ER until they are completely folded, and targets slowly folding proteins for export to the cytosol and proteasomal degradation in a process called ER-associated protein degradation (ERAD) [Fig. 1]. This chaperone machinery also relays a signal for perturbation of ER homeostasis due to excessive levels of unfolded or incompletely folded proteins, as manifested by an imbalance between the folding demand on the ER and its folding capacity, to a signal transduction network called the unfolded protein response (UPR) [Fig. 1]. The UPR coordinates adaptive responses to ER stress that serve to return the ER to homeostasis. If this adaptive response fails to restore homeostasis, the UPR activates apoptosis (cell death) to eliminate unhealthy cells from an organism.

Signal transduction by the UPR. The ER membrane harbors at least three classes of transmembrane proteins functioning as proximal signal transducers in the UPR: basic leucine zipper (bZIP) transcription factors (ATF6α, ATF6β, CREB-H, and others) that form homo- or heterodimers through their leucine zipper domains; the protein kinase PERK; and the protein kinase endoribonucleases IRE1α and IRE1β (**Fig. 2**). These proximal signaling molecules share a common transmembrane topology and activation mechanism. They are composed of an extended ER-luminal domain that, in unstressed cells where there is an adequate amount of free BiP, associates with that molecular chaperone, a transmembrane segment, and a cytosolic effector domain, which encodes a transcription factor, a protein kinase, or a protein kinase that also has endoribonuclease activity. Upon ER stress, BiP is released from the ER-luminal domains as it binds to the excess unfolded or misfolded proteins, resulting in activation of the proximal signal transducers of the UPR. Transmembrane bZIP transcription factors then translocate to the Golgi complex, where their cytosolic domains encoding the transcription factor are proteolytically released from the Golgi membrane. The cytosolic segments containing the bZIP domains then translocate to the nucleus to regulate transcription (Fig. 2). Transcriptional activation of genes encoding ER-resident chaperones and foldases by ATF6α and ATF6β is an immediate response to ER stress. An ATF6·CREB-H heterodimer

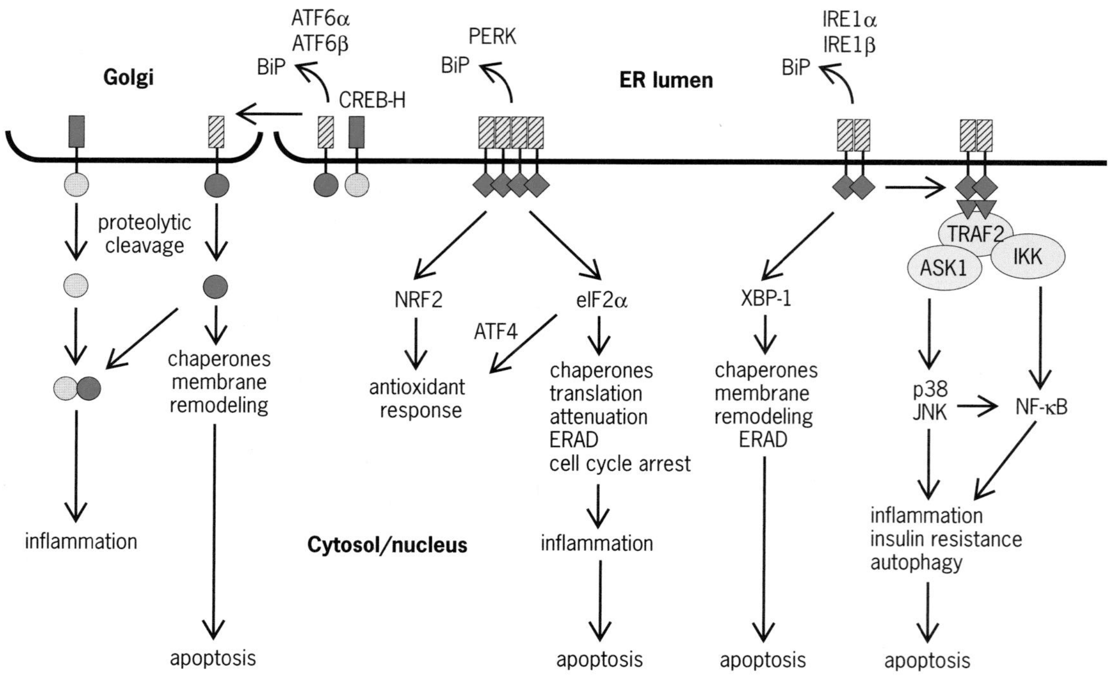

Fig. 2. Principal signaling pathways in the human UPR. (*Reprinted from Current Molecular Medicine, copyright 2006, with permission from Bentham Science Publishers*)

activates expression of acute phase response genes (whose products are rapidly produced to combat tissue damage and infection) and an inflammatory response.

PERK and IRE1 dimerize or oligomerize to activate their protein kinase domains. Activated PERK phosphorylates the bZIP transcription factor NRF2, resulting in translocation of NRF2 to the nucleus and activation of genes involved in an antioxidant response, which counteracts increased formation of reactive oxygen species originating from increased disulfide bond isomerization reactions on accumulated unfolded or partially folded proteins in ER stressed cells (Fig. 2). Through phosphorylation of the eukaryotic translation initiation factor 2α (eIF2α), PERK attenuates general translation to decrease the influx of newly synthesized, unfolded polypeptide chains into the ER. Attenuation of general translation clears short-lived proteins, which require continued synthesis to maintain normal steady-state levels, from the cell, resulting in cell cycle arrest in G_1 because of loss of cyclin D1 (a key regulator of cell cycle progression), and activation of the proinflammatory transcription factor NF-κB. Paradoxically, eIF2α phosphorylation also results in stimulation of translation of messenger ribonucleic acids (mRNAs) harboring several short open reading frames in their 5′ untranslated leader sequences—in particular, the mRNA encoding the bZIP transcription factor ATF4. ATF4, as a heterodimer with NRF2, activates an antioxidant response, and also activates transcription of the pro-apoptotic bZIP transcription factor CHOP. CHOP represses expression of the anti-apoptotic protein BCL-2 and induces transcription of the pro-apoptotic protein TRB3. eIF2α phosphorylation by PERK is countered by several phosphatases that remove the added, inhibitory phosphate. Late in the UPR, the regulatory subunit GADD34 of protein phosphatase 1 is induced by ATF4 to allow for recovery from prolonged ER stress.

IRE1 is a protein kinase endoribonuclease. Dimerization or oligomerization of IRE1 following its release of BiP activates its protein kinase domain, which in turn activates its endoribonuclease domain. The substrate for the IRE1 endoribonuclease is the mRNA encoding the bZIP transcription factor XBP-1. After cleavage of both exon-intron junctions in *XBP-1* mRNA by activated IRE1, an unknown mammalian RNA ligase joins the *XBP-1* exons. In yeast, this RNA ligase is transfer RNA (tRNA) ligase. This unconventional cytosolic splicing reaction circumvents the conventional mRNA splicing machinery of eukaryotic cells, the spliceosome, and by introducing a frameshift in *XBP-1* mRNA introduces a more potent transcriptional activation domain into XBP-1. XBP-1 induces transcription of the ER-resident molecular chaperone BiP, phospholipid biosynthetic genes, and several genes encoding proteins involved in ERAD. Activation of XBP-1 follows activation of ATF6 and PERK, thus providing for an initial phase in which increased chaperone synthesis promotes folding of unfolded proteins, followed by a second phase, in which folding and degradation of unfolded proteins are increased (Fig. 2). In addition to its role in restoring the ER to homeostasis via induction of BiP and of ERAD components that are a consequence of its activation of *XBP-1* mRNA, phosphorylated IRE1 contributes to proinflammatory and pro-apoptotic components of the UPR as well as the induction of autophagy (a specialized form of protein degradation, in which bulk cytosol including

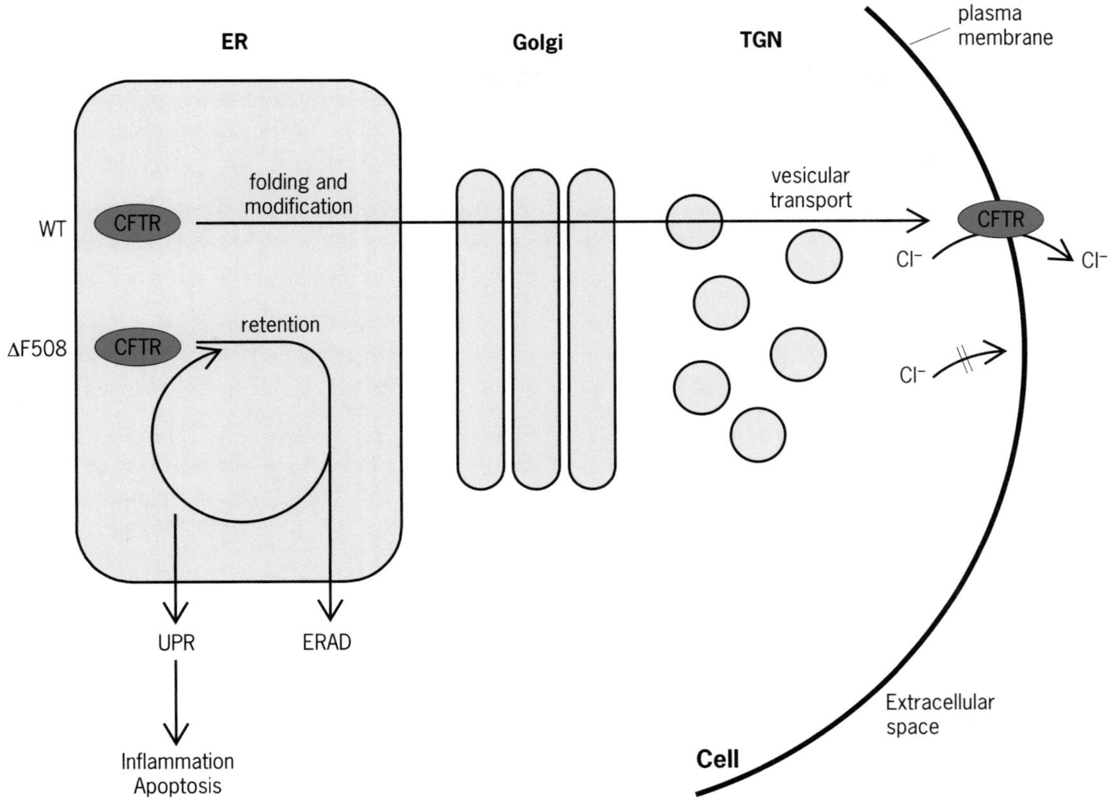

Fig. 3. Conceptualization of ER storage diseases. In the case of CFTR, the wild-type (WT) protein functions as a chloride channel at the plasma membrane. Mutant ΔF508 CFTR, in which the phenylalanine at position 508 has been deleted, is recognized in the ER as unfolded, resulting in its retention in the ER, its targeting for ERAD, and activation of the UPR. TGN, *trans*-Golgi network.

organelles is engulfed by a membranous structure called an autophagosome and proteolytically degraded). IRE1 assembles a signaling complex consisting of tumor necrosis factor receptor associated factor 2 (TRAF2), apoptosis signal-regulating kinase 1 (ASK1), and the inhibitor of NF-κB kinase (IKK) [Fig. 2]. ASK1 activates two protein kinase signaling pathways leading to the protein kinases p38 and JNK, which have pro-apoptotic and proinflammatory functions in the UPR. IKK, by phosphorylating and inducing proteasomal degradation of the inhibitor of NF-κB (IκB), activates the proinflammatory transcription factor NF-κB and an inflammatory response. JNK protein kinases are also required for induction of autophagy in the UPR (Fig. 2).

Physiology and pathophysiology of the UPR. The UPR is important for the normal function of secretory cell types. Knockout of *XBP-1* or *IRE1* in mice is lethal to the embryo and associated with defects in liver development, a major secretory organ. Differentiation of B cells (B lymphocytes) into antibody-secreting plasma cells requires an intact UPR. The UPR is also important for normal function of pancreatic β-cells that secrete large amounts of insulin to control blood glucose levels. Missense mutations in PERK cause an early-onset juvenile form of diabetes called Wolcott-Rallison syndrome, and mouse knockout models for PERK and knock-in models for a nonphosphorylatable version of eIF2α also develop diabetes.

Further evidence for a critical role of the UPR in secretory cell types comes from numerous studies reporting activation of the UPR in human diseases caused by mutations that impair protein folding in the ER, resulting in retention of the unfolded protein in the ER by its quality control machinery. An example is cystic fibrosis where deletion of phenylalanine 508 in the cystic fibrosis transmembrane conductance regulator (CFTR) allows for synthesis of a largely functional protein that is nevertheless retained in the ER because it is recognized as being unfolded by the ER-resident quality control machinery. As a consequence, mutant CFTR does not reach its place of function, the plasma membrane (**Fig. 3**). Other examples are blood coagulation disorders, neurodegenerative diseases, and viral infections. Many viral infections induce the synthesis of large amounts of highly glycosylated viral proteins, which, like cellular plasma membrane proteins, are initially inserted into the ER membrane, resulting in activation of the UPR. Mutant viruses that encode envelope proteins impaired in their folding activate a stronger UPR and are usually associated with a higher degree of cell death than their wild-type counterparts. These observations show that the host cell's ability to process viral glycoproteins in the ER is an important determinant of the cytotoxicity of some viral infections. Activation of inflammatory and acute phase responses by the UPR may help to combat viral and bacterial infections.

The UPR is also involved in nutrient sensing, regulation of cellular responses to nutrient starvation, regulation of lipid metabolism, biogenesis, and regulation of membrane composition of the ER via activation of XBP-1. The function of the UPR in nutrient sensing is conserved from unicellular eukaryotes such as yeast to humans. For example, obesity activates the UPR in adipocytes and hepatocytes, and via activation of JNK by IRE1 inhibits insulin signaling, resulting in insulin resistance (Fig. 2). Activation of the proinflammatory transcription factors NF-κB and CREB-H by the UPR contributes to inflammatory diseases such as obesity, atherosclerosis, arthritis, and cystic fibrosis. The importance of the UPR in this multitude of disabling diseases justifies future research into the UPR and its role in human diseases.

For background information *see* CELL (BIOLOGY); CELL MEMBRANES; ENDOPLASMIC RETICULUM; GOLGI APPARATUS; MOLECULAR CHAPERONE; PEPTIDE; PROTEIN; PROTEIN FOLDING; RIBOSOMES; SECRETION in the McGraw-Hill Encyclopedia of Science & Technology. Martin Schröder

Bibliography. S. Bernales, F. R. Papa, and P. Walter, Intracellular signalling by the unfolded protein response, *Annu. Rev. Cell. Dev. Biol.*, 22:487–508, 2006; M. Schröder and R. J. Kaufman, The mammalian unfolded protein response, *Annu. Rev. Biochem.*, 74:739–789, 2005; M. Schröder and R. J. Kaufman, Divergent roles of IRE1α and PERK in the unfolded protein response, *Curr. Mol. Med.*, 6:5–36, 2006; N. Strudwick and M. Schröder, The unfolded protein response, in M. Al-Rubeai and M. Fussenegger (eds.), *Cell Engineering 5: Systems Biology*, Springer-Verlag, Dordrecht, 2007.

Use of optics by Renaissance artists

An extensive visual investigation by the artist David Hockney, supported by optical evidence detailed in subsequent technical papers, shows that important artists began using optical devices as aids for creating their work early in the Renaissance, approximately 175 years before the time of Galileo. These discoveries show there has been a continuous use of optics for artistic purposes continuing until today, that started about 1425 with Jan van Eyck and Robert Campin in Flanders, followed by such well-known artists as Bartholomé Bermejo in Spain about 1474, Hans Holbein in England about 1530, and Caravaggio in Italy about 1600. Before the optical evidence in representative Renaissance paintings is described, the state of optical knowledge at the time will be discussed to establish the context for these discoveries.

Medieval knowledge of optics. The optical principles of the camera obscura were described by the Arab scientist, philosopher, and mathematician Abu Ali al-Hasan ibn al-Haytham (965–1039), known in the West as Alhazen or Alhacen. A camera obscura (Latin for "dark room") is simply a small opening in a dark curtain, through which an image of the outside scene is projected, upside down and reversed, onto the opposite wall in a darkened room or enclosure (**Fig. 1**). Alhazen explained the image of the crescent Sun in his treatise *The Shape of the Eclipse*, based on the optics of the camera obscura. By the early thirteenth century, Alhazen's writings had been translated into Latin and incorporated into the optical manuscripts of Roger Bacon (about 1265), Witelo (about 1275), and John Peckham (about 1280). This was a period of rapid development of optical knowledge, with 61 books on optics written between the year 1000 and the time of van Eyck. As a result, by the end of the fourteenth century, knowledge of the camera obscura was widely distributed through Europe.

Concurrent with the growing theoretical understanding of optics were practical developments such as spectacles. These were invented in Italy around 1276, and by the end of the fourteenth century were being exported in large numbers to all parts of Europe. Evidence within paintings shows that at some point toward the end of this period someone realized that replacing the small opening in a camera obscura with a lens resulted in a projected image that was both brighter and sharper. One lens from a pair of spectacles is all that is required to turn a simple camera obscura into a device that can create images of the size, brightness, and sharpness necessary to be useful to artists.

Concave mirrors also can project images, and there is strong circumstantial evidence that, at least initially, artists used projections made by them. An image projected by a concave mirror (**Fig. 2**) maintains the parity of the scene (that is, it does not flip the image left for right as is the case for the image viewed in a flat mirror), which is an advantage for an artist. All that is required is for an artist to project that image onto a canvas, trace enough of the features to provide the necessary detail and perspective, and then turn the canvas over and paint in the overall scene. Witelo's *Perspectiva* (about 1275) is closely modeled on Alhazen's writings, bringing into European science the study of the imaging properties of mirrors. By 1430, the writings of Alhazen and Witelo were a required part of the curriculum at Oxford University, further spreading knowledge of the optical properties of mirrors. Alhazen was so well known at the time that he and his optical works are described in *The Romance of the Rose* by Guillaume de Lorris and Jean de Meun (about 1275), one of the most widely read works of the French language. He is also referred to by name in Geoffrey Chaucer's *The Canturbury Tales* (about 1400), the first widely read work in the vernacular English language.

The first visual depiction of lenses and concave mirrors is in Tomaso da Modena's 1352 paintings of *Hugh of Provence* and *Cardinal Nicholas of Rouen*, which show, respectively, spectacles and a magnifying glass, either of which would have been capable of projecting an image useful for an artist. His *St. Jerome* and *Isnardo of Vicenza* both show concave mirrors as well. Pilgrims at the time carried small convex mirrors into cathedrals where they used them as wide-angle lenses to enable a much larger area of the scene to be visualized, a practice that shows how widespread and common the uses of optics had become.

Effects of refocusing. Before turning to the evidence of the use of optics in the production of art, it is important to realize that while a simple lens can be focused at only one specific distance at a time, the brain causes the muscles of a human eye to contract or expand as it refocuses and traverses the various depths of a given scene. Because of this, we do not simultaneously see part of a scene in focus and part of it out of focus. In contrast with the eye, no matter what the distance of focus of a simple lens, only a certain depth of field of the scene on either side of that distance remains acceptably sharp, depending on the diameter of the lens and the magnification of the projected image. To change the focus distance requires physically altering the position of the lens with respect to the subject and the image plane (or, in an artist's case, the canvas). This results in a small but measurable change in the magnification of the projected scene, as well as a slight change in the vanishing points (that is, the distance into an image where parallel lines converge to a point). Both effects are quite small for low magnifications, such as in typical photographs, so they usually go unnoticed. However, such effects are fundamental characteristics of images projected by lenses, but are extremely unlikely to occur in paintings that are either done by eye alone or created using the geometrical rules first articulated in the fifteenth century.

Optical evidence in paintings. Considerable evidence for the use of optics is found in the painting *Husband and Wife* by Lorenzo Lotto from about 1524, roughly 75 years before Galileo. Calculations from geometrical optics show that, if Lotto used a lens to project the octagonal pattern of the central feature of the carpet in this painting, there would have to be three regions in the painting corresponding to the depths into the scene where he was forced to refocus his lens (**Fig. 3**). Each time the depth of field of his lens forced him to refocus, it would result in a change in magnification and in the sets of vanishing points. All of these complex features are found in the octagonal pattern of the carpet in the painting, and all are in agreement to within 0.5% with calculations of geometrical optics, providing extremely strong evidence indeed that a lens was used as an aid for creating this portion of the painting.

Another example is the complex chandelier in van Eyck's painting *Arnolfini Marriage* of 1434. The optical evidence that this portion of the painting is based on an optical projection includes the following:

1. The perspective-corrected length, width, and shape of the main arc of all six arms of the chandelier are identical to within about 2%.

2. The perspective-corrected radii of all six candle holders are within ±1.5% of the radius of a perfect circle centered on the axis of the chandelier.

3. The perspective-corrected angular positions of all six candle holders are within ±4° of the points of a perfect hexagon.

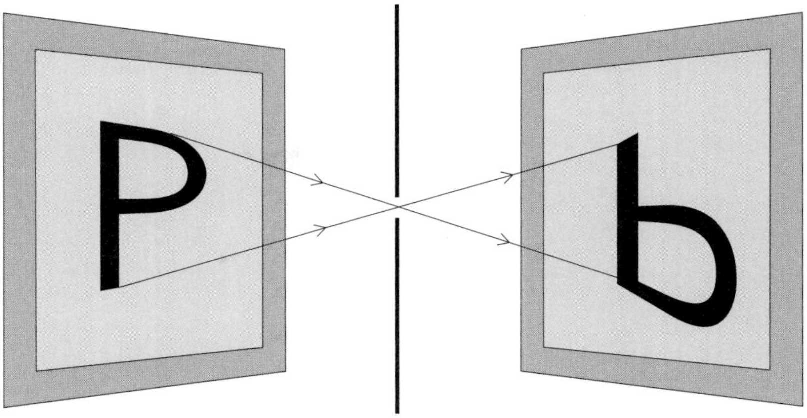

Fig. 1. Projection of an image by a camera obscura.

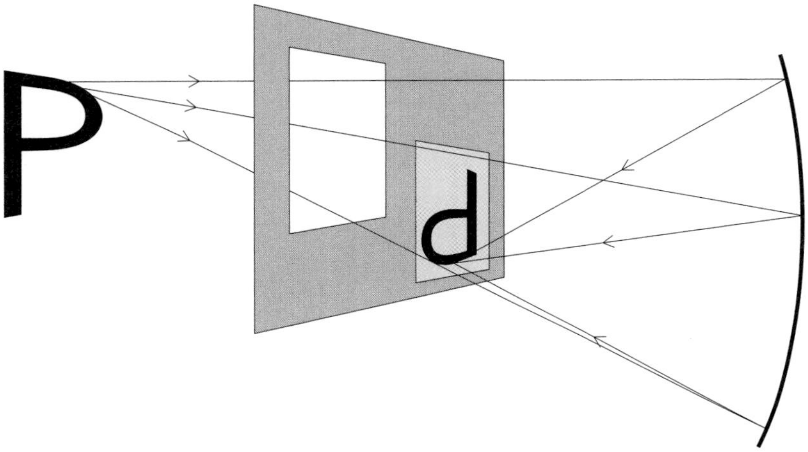

Fig. 2. Projection of an image by a concave lens.

4. To within ±1 mm (0.04 in.) on the painting, the positions of the lowest points on the arcs of five of the six arms have the identical perspective-corrected hexagonal symmetry as the candle holders, with both sets of data rotated by the same 6°.

5. The vanishing points defined by the candle holders as well as by the lowest positions of the arcs converge to the same horizon to within the accuracy expected for the imperfections in a real chandelier.

6. The overall diameter of the chandelier of approximately 1 m (40 in.) that can be estimated by assuming that the candle flame is approximately 3 cm (1.2 in.) high is physically reasonable for a real object of this type, as are the minor variations in the lengths and angles for such an ornate, handmade object. Also, analysis of photographs of modern six-arm lighting fixtures taken from approximately the same viewpoint as the chandelier finds angular and radial deviations of the perspective-corrected positions of their light-bulb holders of relative size comparable to that of van Eyck's chandelier.

Going back even further in time, the earliest evidence that optics were used to project a portion of a painting is found in the center and right panels of Campin's *Merode Altarpiece* of about 1425, 175 years before Galileo. Both of these panels exhibit small but measurable changes in perspective

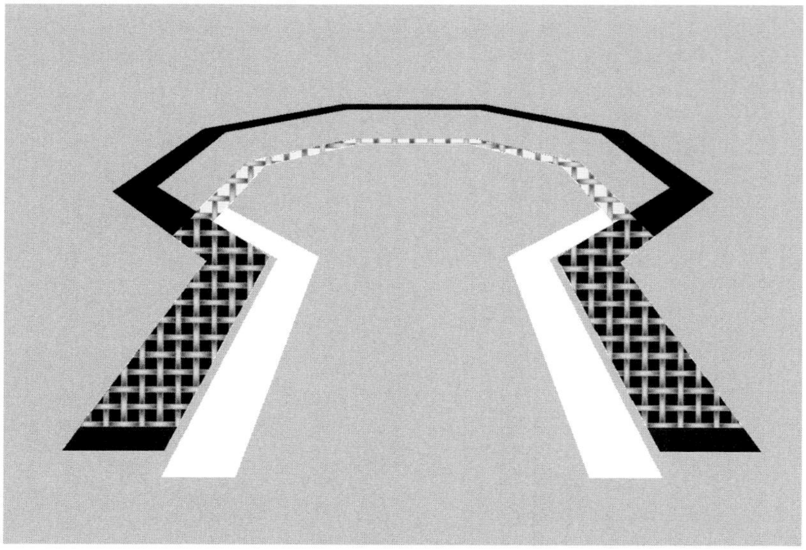

Fig. 3. Simplified diagram showing the distortion of a simple octagonal pattern that occurs as a result of having to refocus due to exceeding the depth of field of a lens, resulting in a change of magnification of the projected image. The artist sketches the lattice in the front first (guided by the black area), then refocuses and matches the pattern as best as possible to sketch the back (guided by the white area). The resulting overall octagonal pattern is severely distorted, yet still appears to be a reasonable octagon.

resulting from Campin having been forced to refocus twice because of the depth-of-field of the lens that this evidence shows he used. The complex perspective exhibited by the latticework of the bench depicted in the right-hand panel of the painting is a direct and inevitable outcome from the depth of field of a lens, and it would be extremely unlikely to have resulted from any geometrical construction.

These discoveries show that optical devices (lenses and concave mirrors) were in use for projecting images nearly 200 years earlier than previously thought possible, and account for the remarkable transformation of realism in portraiture that occurred early in the fifteenth century. However, useful as it is as a tool, a lens does not arrange a composition, fill in the colors or shadings, or make any of the various other artistic decisions that are essential to create a painting. It aids, but in no way replaces, artistic talent.

For background information *see* GEOMETRICAL OPTICS; LENS (OPTICS); MIRROR OPTICS in the McGraw-Hill Encyclopedia of Science & Technology.

Charles M. Falco

Bibliography. C. M. Falco, Frequently Asked Questions, 2007; D. Hockney, *Secret Knowledge: Rediscovering the Lost Techniques of the Old Masters*, expanded ed., Viking Studio, 2006; D. Hockney and C. M. Falco, Optical insights into Renaissance art, *Optics Photonics News*, 11(7):52–59, July 2000 (http://www.optics.arizona.edu/ssd/OPN.edu); V. Ilardi, *Renaissance Vision, from Spectacles to Telescopes*, American Philosophical Society, 2007; D. Lindberg, *A Catalogue of Medieval and Renaissance Optical Manuscripts*, Pontifical Institute of Medieval Studies, 1975; D. C. Lindberg, *Theories of Vision from Al-Kindi to Kepler*, 1976; B. Steffens, *Ibn al-Haytham: First Scientist*, Morgan Reynolds, 2006.

Venus Express

Venus Express is the European Space Agency's (ESA's) first mission to Venus and only its second to another planet in the solar system (after *Mars Express*). The spacecraft was launched in November 2005, entering Venus orbit in April 2006, and has been operating successfully ever since (as of October 2007). Its primary mission is to observe the thick atmosphere of Venus and study its evolving dynamics and chemistry over the course of 2 Venus sidereal days (approximately 500 Earth days).

Venus has been visited by spacecraft since the dawn of the space age. Since the first mission to Venus, the National Aeronautics and Space Administration's (NASA's) *Mariner 2* in 1962, more than 20 successful spacecraft have visited our "sister planet." The last mission to enter Venus orbit was NASA's *Magellan* in 1990; by using radar to probe below the thick atmosphere, it was able to map the surface terrain in unprecedented detail. However, a great number of scientific questions remained unanswered, especially concerning the atmospheric circulation and its dynamics and chemistry. *Venus Express* was envisaged to try to fill this void in knowledge with modern, state-of-the-art cameras and spectrometers.

Objective. The most puzzling questions to answer concerning Venus are related to its atmosphere. The atmosphere is made up of 96% carbon dioxide (CO_2), with trace amounts of carbon monoxide (CO), water vapor (H_2O), sulfur dioxide (SO_2), and other minor species. The surface pressure is a staggering 90 bars (90 times the Earth's surface pressure), and the temperature on the surface is approximately 740 K (870°F or 460°C). This high surface temperature is caused by the greenhouse effect, whereby infrared solar radiation is trapped by the high concentrations of CO_2. One of the key questions is how this greenhouse effect has evolved and altered the atmosphere of Venus to the state we see it today.

Venus is covered by a global layer of sulfuric acid clouds, which extends from approximately 55 to 65 km (34 to 40 mi) above the surface (**Fig. 1**). These clouds follow the prevailing winds, which rotate around the planet at speeds in excess of 320 km/h (200 mi/h) at the cloud tops at the equator. This phenomenon is termed superrotation, which means that clouds can rotate around Venus in less than 5 days. This rate is in stark contrast to the rotation of the solid body below, which rotates some 60 times slower. *Venus Express* will study what causes this superrotation phenomenon, as well as the cyclonic vortices at the poles of the planet that are created by the superrotation. The structure, dynamics, and chemistry of the clouds will also be studied in great detail.

Related to all these topics are the following questions: What are the escape processes of the minor gases and their associated chemical cycles in the atmosphere? Is there volcanic or seismic activity on the surface, and what caused the apparent global volcanic resurfacing some 500 million years ago? Is

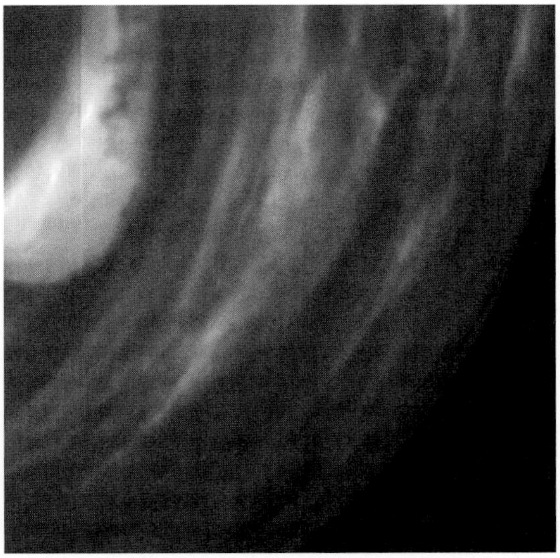

Fig. 1. Cloud features of Venus viewed at wavelength of 1.73 μm in the infrared by VIRTIS on *Venus Express*. (*ESA/VIRTIS/INAF–ASF/Paris Observatory–LESIA*)

there lightning activity in the atmosphere? What is the nature of the Hadley cell? [On Earth, three meridional circulation cells are present (the Hadley, Ferrel, and polar cells), but on Venus, the current consensus is for the existence of a single, large cell (the Hadley cell) in each hemisphere. If so, what is the structure of this cell?] Answering these questions is at the heart of the *Venus Express* mission.

Design. In 2001, ESA invited scientists from Europe to propose a mission. In an effort to reduce costs and speed up its development, it was decided to reuse components and parts from ESA's past missions. The result, *Venus Express*, is a virtual twin of the *Mars Express* spacecraft, which entered Mars orbit in December 2003. The spacecraft's aluminium body is a 1.5-m (5-ft) square box, with solar arrays made from gallium arsenide, which extend 3 m (10 ft) from the main body. Additional modifications to the *Mars Express* spacecraft needed to successfully operate in Venus orbit include a redesign of the thermal insulation and the addition of more radiators for effective heat rejection. (Since Venus is much closer to the Sun than Mars, the heating load is four times higher than on *Mars Express*.) Other modifications include 20% increased fuel load to enter the gravitationally stronger Venus orbit and the addition of a second high-gain antenna for communications with Earth during inferior conjunctions.

There are seven science instruments onboard, composed of three that were flown on *Mars Express*; three from another ESA mission called *Rosetta*, which is visiting Comet 67 P/Churyumov-Gerasimenko in 2014; and one new instrument. The instruments include a plasma and energetic atoms instrument (ASPERA), a magnetometer (MAG), a high-resolution infrared spectrometer (PFS), an ultraviolet/infrared stellar and solar occultation spectrometer (SPICAV), a radio-science experiment (VeRa), a visible and infrared imaging spectrometer (VIRTIS), and a camera (VMC). These instruments complement each other in both spectral range and resolution, covering wavelengths from the ultraviolet through the visible and thermal infrared. The use of multiple wavelengths will make it possible to "see" the atmosphere at different heights. In addition, many of these instruments are designed to probe, for the first time, the near-infrared "windows." These are regions in the night-side infrared spectrum, from 1.0 to 2.5 μm, where radiation is able to permeate into space. Radiation at these wavelengths is emitted from the lower atmosphere, as well as from the surface. Discovered in the 1980s, these "spectroscopic window" regions have never been viewed from the orbit of Venus. The study of absorption features resulting from trace gases and the tracking of winds at these wavelengths will yield unprecedented information concerning the chemistry and dynamics of the lower atmosphere.

Implementation. *Venus Express* was launched on a Soyuz-Fregat launcher from Baikonur Cosmodrome in Kazakhstan on November 9, 2005. Once in space, the Fregat upper stage fired to place *Venus Express* in a parking orbit around Earth, before firing a second time to propel the spacecraft on an interplanetary trajectory to Venus. The interplanetary flight took 153 days. On arrival in the vicinity of Venus on April 11, 2006, the orbit insertion burn took 50 min, after which *Venus Express* entered a highly elliptical orbit (**Fig. 2**). This orbit was then lowered during the following week with the burning of the main engines.

The nominal orbit is a highly elliptical polar one, with an apoapsis (furthest point from the planet) height of 66,000 km (41,000 mi), located below the South Pole, and a periapsis (closest point from the planet) altitude of approximately 250 km (150 mi). An orbit of this nature is essential to viewing the polar regions, which is one of the primary goals of *Venus Express*. The orbital period is 24 (terrestrial) h. During this period, 16 h are dedicated to observing Venus, while the remaining 8 h are used to communicate with Earth. Communication with *Venus Express* is conducted through the Venus Express Missions Operations Center (VMOC) at the European Space Operations Center (ESOC) in Darmstadt, Germany,

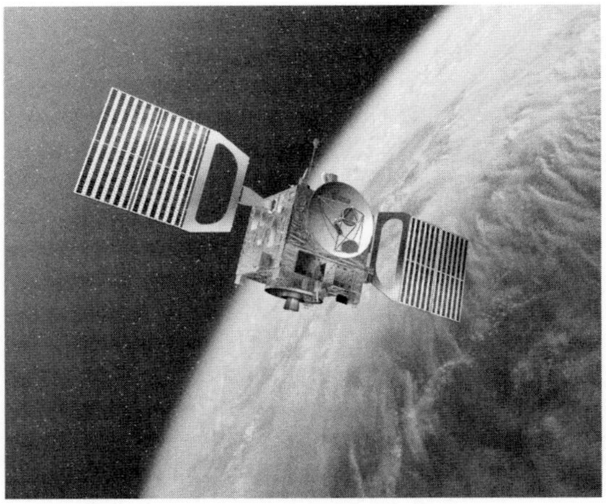

Fig. 2. Artist's concept of *Venus Express* in orbit around Venus. (*ESA*)

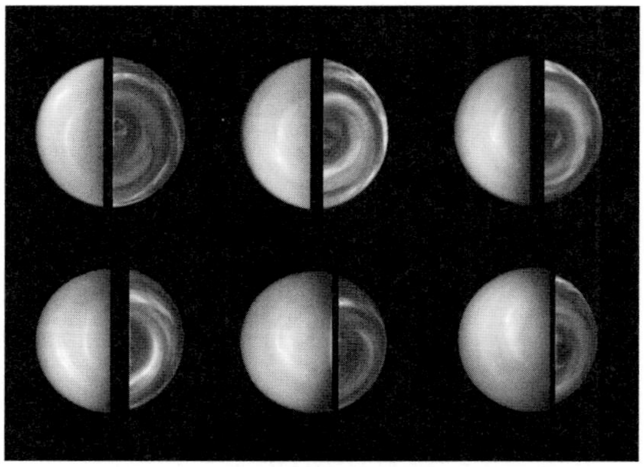

Fig. 3. Orbit insertion observations of Venus from VIRTIS on *Venus Express*. Left hemispheres show day-side reflections from the cloud tops (at wavelength of 360 nm in the ultraviolet), while right hemispheres show thermal emission (at wavelength of 1.73 μm in the infrared) from below the clouds. (*ESA/VIRTIS/INAF–IASF/Paris Observatory–LESIA*)

in association with the deep space ground station at Cebreros, Spain.

The science observations are divided into sequences depending on the orientation of the spacecraft, its payload configuration, and its thermal constraints. During periods near the apoapsis, the cameras and imaging spectrometers take multiple images to cover the entire field of view of the planet, to create mosaics and movies of the moving cloud features. Near the periapsis, the spacecraft is moving so quickly over the planet that imaging is impossible. Instead, spectrometer measurements are preferred. These are either nadir (downward) or limb (horizon) viewing, and also include stellar occultations. In situ measurements of the magnetic field and plasma environment are taken continuously throughout the observational part of the orbit.

Results. *Venus Express* has taken an enormous amount of data. The near-infrared windows have provided the first glimpse of thermal emission from deep in the atmosphere and of the surface. Surface topography and features can clearly be seen and readily identified against radar topography. It is hoped that continued observation of the surface will identify localized "hot spots," which are indicative of active volcanism. Measurements of the emissivity of the surface might also be possible. Night-side thermal emissions observed at wavelengths of 1.73 and 2.30 μm have revealed in detail the movement of cloud features caused by superrotation (**Fig. 3**). Some of the images have also shown cloud wave structures in the atmosphere. Measurements of zonal (east-to-west) and meridional (equator-to-pole) winds at different altitudes have been made, indicating vertical wind shears. An important atmospheric phenomenon, located at the poles of the planet, is the polar dipole, which visually manifests itself as two eyes of a vortex, swirling around the poles. The observations of the Southern Hemisphere polar dipole have been remarkable. Analysis of the cloud-top temperatures, rotation rates, and morphology of this dipole are continuing, giving insights into the chaotic circulation as air moves poleward and downward. The detection of oxygen airglow at 1.27 μm in the high stratosphere is also fascinating, and hints at the complex relationship between the dynamics and chemistry of the atmosphere. Unprecedented global Southern Hemisphere mapping and excellent vertical resolution with limb-viewing geometries are making it possible to study this dramatic oxygen airglow phenomenon. The continuation of the mission until well into 2008 promises even more exciting science results.

For background information *see* AIRGLOW; ASTRONOMICAL SPECTROSCOPY; ATMOSPHERIC GENERAL CIRCULATION; GREENHOUSE EFFECT; PLANET; SOLAR SYSTEM; SPACE PROBE; VENUS in the McGraw-Hill Encyclopedia of Science & Technology.

Constantine Tsang

Bibliography. K. Baines et al., To the depths of Venus: Exploring the deep atmosphere and surface of our sister world with *Venus Express*, *Planet. Space Sci.*, 54:1263–1278, 2006; D. H. Grinspoon, *Venus Revealed*, Helix Books (Basic Books), 1997; M. Ya. Marov and D. H. Grinspoon, *The Planet Venus*, Yale Planetary Exploration Series, Yale University Press, New Haven and London, 1998; F. Taylor, D. Crisp, and B. Bézard, Near-infrared sounding of the lower atmosphere of Venus, in S. W. Bougher, D. M. Hunten, and R. J. Phillips (eds.), *Venus II: Geology, Geophysics, Atmosphere, and Solar Wind Environment*, University of Arizona Press, Tucson, pp. 325–351, 1997; D. V. Titov et al., Venus Express science planning, *Planet. Space Sci.*, 54:1279–1297, 2006.

Voltage stability

In recent years, voltage instability of large electric power systems has caused costly blackouts. Worldwide, voltage instability and collapse have caused major blackouts in France (1978), Belgium (1982), Sweden (1983 and 2003), western France (1987), Tokyo (1987), São Paulo (1997), and Greece (2004). Voltage control and stability problems were prominent in the August 14, 2003 blackout in northern Ohio, Michigan, New York, and Ontario. There have been many lesser incidents in the United States and elsewhere. In order to avoid instability, power companies are forced to limit power imports and to use more expensive generating plants.

Power system voltage stability, sometimes called load stability, is associated with the load dynamics in large power systems. This contrasts with the better-known electromechanical or rotor-angle stability associated with maintaining synchronism of interconnected generators, sometimes called generator stability. Active power balance governs synchronous stability, while reactive power balance governs voltage stability. Many blackouts involve aspects of both synchronous and voltage instability.

Power companies try to keep voltages within about ±5% of rated voltage. Voltage instability may result in either high or low voltage, but the major concern is low voltage occurring during heavy load conditions such as during very hot or very cold weather. Voltage collapse is a condition of either abnormally low voltage (brownout) or zero voltage (blackout).

Reactive power. Voltage and "reactive" power are closely related. Active power, measured in watts, kilowatts, or megawatts, is associated with current that is in phase with the voltage 60- or 50-Hz waveform. Reactive power is associated with the electric and magnetic fields in transmission lines, transformers, motors, capacitors, and other equipment. For example, the alternating-current sinusoidal waveform of current drawn by an induction motor lags the applied voltage waveform due to magnetizing current requirements. At no load, energized motors and transformers require magnetizing current that lags the applied voltage by 90°; multiplying this magnetizing current by the voltage gives reactive power. For a loaded motor, the current waveform lags the voltage waveform by about 35°.

On the other hand, capacitors draw current that leads voltage by 90°, resulting in reactive power with the opposite sign. The lagging or leading angle is termed the power factor angle and the power factor is the cosine of this angle. Rather than dealing with voltage and current waveforms, engineers use effective or root-mean-square (RMS) values of voltage and current. Related to the power-factor angle, complex numbers and phasors are used (**Fig. 1**). A phasor is a rotating vector that describes the voltage or current root-mean-square magnitude and phase relationship. Although reactive power is mathematically imaginary, it cannot be ignored.

Capacitor banks produce or generate reactive power to support voltages, while inductive equipment consumes or absorbs reactive power. For high active power loading, transmission lines are net consumers of reactive power. Generators with high field current produce reactive power, while generators with low field current consume reactive power.

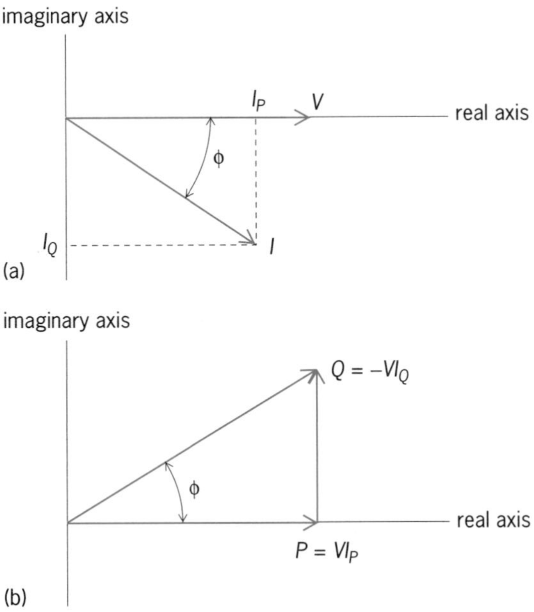

Fig. 1. Voltage and current phasors, and active and reactive power. (*a*) Current phasor *I* lagging the reference voltage phasor *V* by the power factor angle φ, as for an induction motor. The current phasor has component I_P in phase with the voltage phasor, and component I_Q in quadrature (90° out of phase) with the voltage phasor. (*b*) Corresponding active power *P* and reactive power *Q*, using the convention that the motor is consuming positive active and reactive power.

Transmission. A key aspect of voltage stability is the need to provide reactive power to support voltages. It is possible to transmit active power over long distances, but it is not possible to transmit reactive power over long distances, especially during heavy load conditions. This is because of large reactive power losses along the transmission path caused by the series inductance of transmission lines and transformers. These losses are proportional to the square of the line or transformer current. Also, reactive power transmission requires the voltage at the sending (generation) end to be significantly higher than the voltage at the receiving or load end.

Compensation. To avoid transmitting reactive power, utilities and industrial customers apply reactive power compensation near the points of need. Industrial customers install shunt capacitor banks to provide for motor magnetizing currents. (Shunt means connections between the three phases, or between the phases and ground.) This improves the overall power factor and avoids utility charges for low power factor. Similarly, power companies use capacitor banks on the transmission and distribution networks during heavy load conditions.

Capacitor banks are inexpensive, have very low losses, and are widely used to support voltage. They can be rapidly switched in or out of service. They are, however, destabilizing since the reactive power output is proportional to the square of the voltage. During voltage decay, output goes down when it is needed most.

The other important source of reactive power and voltage control is automatic excitation control of synchronous generators. Generator voltages are continuously controlled by adjusting the field circuit voltage. In addition to active power production, generators provide reactive power to support the lagging power factor loads and the reactive power losses of the transmission network. *See* REACTIVE POWER COMPENSATION TECHNOLOGIES.

Voltage instability mechanisms. There are two types of voltage instability that occur in different time frames—a few seconds, and tens of seconds or minutes. In both types, outages weaken the power system causing voltage sags. Immediately after a disturbance, load reduces because of voltage sag (which is stabilizing), but then load is at least partially restored (which is destabilizing by increasing the burden on the transmission and generation).

Short-term voltage instability. Short-term voltage instability occurs within a few seconds. For example, a transmission-line short circuit occurs that is rapidly isolated by the opening of circuit breakers, weakening the power system. Motors initially are an impedance (resistive) load, with a drop in power proportional to the square of the voltage. Because of the short circuit, inertial dynamics (Newton's second law applied to rotation) rapidly reduces the motor speeds, with electrical torques and powers reestablished to meet the load mechanical torque demand regardless of the voltage. Motors draw much higher current at reduced speed, burdening the weakened power system as they try to reaccelerate after the short circuit is removed. This slows voltage recovery, and motors may stall. Stalled motors that are not disconnected draw about six times more current than during normal operation, causing further voltage sags and stalling of other motors. The load area may then suffer a voltage collapse, or suffer abnormally long voltage recovery until motors disconnect.

Based on many incidents and on the increasing use of residential air conditioning, short-term voltage stability is now of greater interest. Residential air conditioners are particularly onerous loads since they require near-constant mechanical torque, and because stalled air conditioners usually do not trip until they overheat after some seconds.

As an example, a near blackout of the Phoenix, Arizona metropolitan area occurred on July 29, 1995 during very hot weather (44°C, 112°F). Much of the Saturday afternoon load was residential air conditioning. A short circuit with delayed clearing at a major substation resulted in loss of five 230-kV lines and two 230/69-kV transformers. Normally voltage would recover almost instantaneously after short-circuit isolation, but voltage recovery actually took around 20 s (**Fig. 2**). Many residential air conditioners stalled, and then tripped off after some seconds to allow eventual recovery of the remaining power system (about 2100 MW of load was lost). Recordings show high reactive power output of area generators during the recovery period. High reactive power output from generators at the nearby Palo Verde nuclear plant was essential for the recovery.

Short-term instability phenomena may occur in the final stages of a slower collapse. Large wind power plants employing induction generators are of concern since induction generators, like induction motors, also draw high current during abnormal speed operation during short circuits and outages.

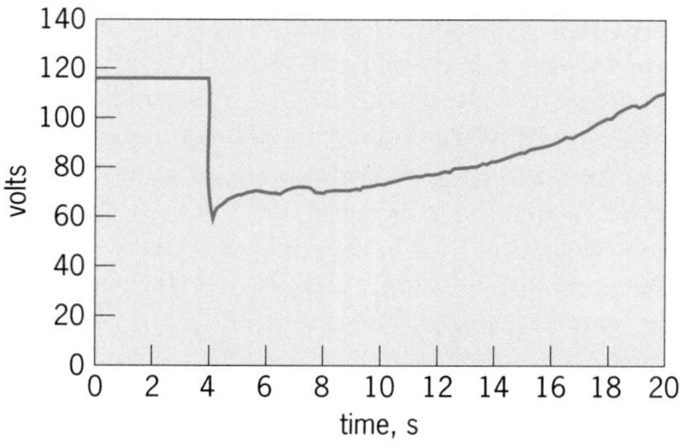

Fig. 2. Very slow residential voltage recovery for Phoenix area incident on July 29, 1995. (*Copyright © 2000 IEEE; from J. A. Diaz de Leon II and C. W. Taylor, Understanding and solving short-term voltage stability problems, Proceedings of IEEE/PES 2002 Summer Meeting, vol. 2, pp. 745–752, 2002*)

Long-term voltage instability. Long-term voltage instability occurs over 0.5–30 min. Following outage of transmission lines or generators, voltages in load areas sag. Load is reduced during voltage sags, and this load reduction or load relief stabilizes voltage. Most bulk power delivery transformers, however, incorporate automatic under-load tap changing. A typical transformer converts 115-kV transmission voltage to 12.5-kV distribution voltage, and delivers around 20 megawatts of power. Following a drop in voltage, tap-changing transformers have an initial delay time of 0.5–2 min, and then 5–10 s mechanism time between steps. A ±10% voltage tap range is common. The tapping regulates the load-side voltage to restore voltage to the desired value, and thus restores the voltage-sensitive load. But the load restoration further stresses the transmission and generation, possibly causing instability.

An even slower load restoration mechanism is the automatic or manual control of loads that must deliver a certain amount of energy. Electric heating loads are constant-energy loads—the power for resistive heating is proportional to the square of the voltage, and the energy is proportional to the time heaters are on. Electric space heating, electric water heating, ovens, and cooktops are either thermostatically or manually controlled. Following a drop in voltage, heaters must stay on longer to meet the energy demand. This means that more heaters will be on at a given time, reducing the load "diversity" of an area. Thermostatically controlled electric space heating loads are important in wintertime voltage stability situations.

Another important mechanism for long-term voltage instability occurs at the generating plants. Generators have many seconds or a few minutes of time-overload (overheating) capability in the field and armature windings. This time-overload capability, along with the initially voltage-sensitive loads, allows stable operation for perhaps several minutes following severe outages. Field winding time-overload limits are usually reached first, and are typically controlled automatically by reducing synchronous generator field excitation voltage and current. This reduces the generator reactive power output, increasing the reactive power demand on other generators. With continued voltage decay, armature windings rapidly reach their time-overload limit. Armature current limiting (usually performed manually by power plant operators) requires a much more drastic reduction in generator reactive power leading to, together with current limiting at other generators, a cascading voltage collapse.

Long-term voltage instability is the traditional or classical type of voltage instability, and was responsible for most of the blackouts mentioned earlier. As an example, a collapse of the southern Sweden and Denmark (Zealand island including Copenhagen) power grid occurred on September 23, 2003. An unusual and very severe short circuit at a major substation caused loss of a large power plant and transmission lines. After the outages, voltages continued to decay, driven by load restoration via tap-changing

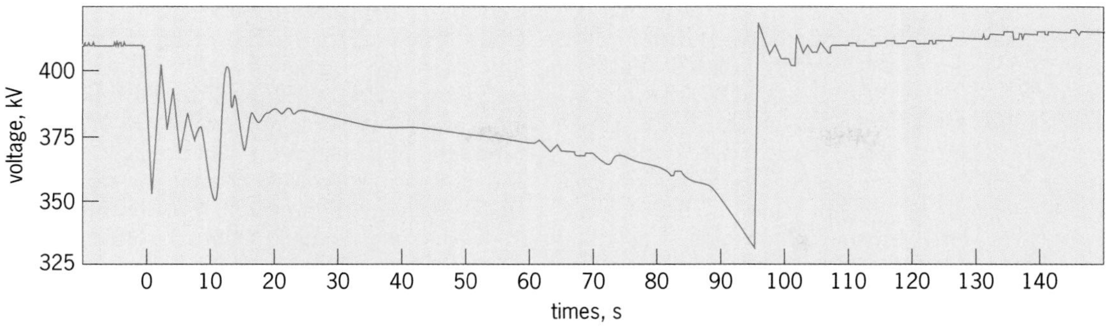

Fig. 3. Voltage recording at a 400-kV substation on the northern side of the system separation during incident in southern Sweden and Denmark on September 23, 2003. (*Courtesy of Svenska Kraftnät*)

transformers (**Fig. 3**). About 97 s after the short circuit, the northern and southern Swedish grid separated, with the northern side recovering as shown in Fig. 3 and the southern side collapsing.

Voltage instability countermeasures. Electric power engineers must find cost-effective solutions to prevent voltage instability and to facilitate electrical commerce. For example, power companies install capacitor banks to allow generators to operate near unity power factor with a large reserve of automatically controlled reactive power capability for emergencies. This reduces the likelihood of field-current limiting.

For infrequently occurring extreme disturbances, many power companies install automatic undervoltage load shedding. This is analogous to underfrequency load shedding that companies installed following the 1965 blackout in northeastern North America.

In the future, power companies may economically ensure voltage stability by direct control of customer load over information technology (IT) infrastructure. During emergencies, air conditioners, water heaters, space heaters, and other noncritical loads are rapidly turned off until other stabilizing measures can be taken. *See* WIDE-AREA POWER SYSTEM PROTECTION AND CONTROL.

For background information *see* ALTERNATING CURRENT; ALTERNATING-CURRENT CIRCUIT THEORY; ALTERNATING-CURRENT GENERATOR; ALTERNATING-CURRENT MOTOR; ELECTRIC POWER SYSTEMS; REACTIVE POWER; STATIC VAR COMPENSATOR; VOLT-AMPERE in the McGraw-Hill Encyclopedia of Science & Technology. Carson W. Taylor

Bibliography. J. A. Diaz de Leon II and C. W. Taylor, Understanding and solving short-term voltage stability problems, in *Proceedings of IEEE/PES 2002 Summer Meeting*, vol. 2, pp. 745–752, 2002; IEEE/CIGRE Joint Task Force on Stability Terms and Definitions, Definition and classification of power system stability, *IEEE Trans. Power Syst.*, 19:1387–1401, 2004; IEEE Power Engineering Society, *Blackout Experiences and Lessons, Best Practices for System Dynamic Performance, and Role of New Technologies*, 2007; C. W. Taylor, *Power System Voltage Stability*, McGraw-Hill, 1994; T. Van Cutsem and C. Vournas, *Voltage Stability of Electric Power Systems*, Kluwer Academic Publishers, 1998.

Warship design trends

Western navies have spent nearly two decades absorbing the effects of the end of the Cold War. The main post–Cold War naval missions seem to be expeditionary, which means primarily either operating against land targets or imposing maritime interdiction far from home. Antisubmarine warfare now means largely dealing with limited numbers of diesel-electric submarines operating in relatively shallow warm water, dramatically unlike the North Atlantic for which NATO long trained. Mine countermeasures has changed from home defense to mine clearance or evasion-conducted on short notice in very unfamiliar places. The change shows in naval building programs and in the shape of fleets. Several navies have discarded large numbers of frigates designed for the Cold War escort role, such as the U.S. *Perry* class, the British Type 23, and the Dutch M-frigate. Although quite successful, they were no longer very relevant. Instead there came smaller numbers of larger ships with a greater ability to deal with the air threats that may be expected near foreign coasts. Examples include the British *Daring* class destroyers (Type 45) and the Dutch *De Zeven Provincien*.

The fundamental change was from a defensive stance, in which the Western Alliance had to maintain its free use of the sea in the face of an expected Soviet challenge, to a more offensive one in which the Western navies use the sea to transport power to distant places. The war in Afghanistan is a case in point: The attacks of September 2001 demonstrated that events in that remote place could directly affect United States security. There was no question of any naval power preventing the U.S. force from appearing in waters within range of Afghanistan. However, local powers were often reluctant to offer base rights, so the fleet in the Arabian Sea was essential in the war. Much the same could be said of the attack on Iraq in 2003. It seems that operations in remote areas will continue to be the major role of Western navies, because these events remain vitally important, and because only navies can ensure access.

Expeditionary warfare often involves either attacks on land targets or landing and supporting troops. Strikes can be carried out either by aircraft or by missiles fired from ships. The advent of, and continued development of, short- or vertical-takeoff aircraft, such as the British Sea Harrier and the evolving U.S. Joint Strike Fighter (JSF), makes it possible to operate useful tactical aircraft from ships much smaller than full aircraft carriers, albeit in smaller numbers and with considerably reduced capabilities. At the same time, it increasingly seems that troops should normally be brought ashore by helicopters rather than by the beaching craft of the past. Several navies have, therefore, built ships that can function both as short-takeoff and vertical-landing (STOVL) carriers and as amphibious ships. Examples include the HMS *Ocean*, the Italian *Conte di Cavour*, and the new Spanish "force projection" ship. U.S. large-deck amphibious ships also operate in this way. At the same time, many more Western navies are investing heavily in more traditional amphibious ships, capable of loading and launching beaching craft.

Land attack missiles. These deserve separate mention. Until a few years ago the surface-to-surface missiles on virtually all warships were designed to attack only other ships. Now such missiles are offered with, among other things, guidance based on the Global Positioning System (GPS), which gives the missiles an ability to attack designated points ashore. Because their numbers are limited (typically eight or fewer per ship), they cannot provide sustained fire support, but they do symbolize increased interest in supporting troops ashore. The U.S. Navy, which has been interested in this role for much longer, has land-attack Tomahawk missiles, up to 100 on board a destroyer. Guided projectiles are being developed for shipboard guns. Although they cannot deliver the punch of a Tomahawk at anything like its range, they have a crucial advantage. Shells are relatively easy to transfer at sea, whereas missiles are difficult or impossible to transfer.

Communications. Land attack in support of troops ashore requires communication between the troops and the fleet offshore, ideally beyond the horizon so that the ships can stand off and also cover the widest possible area. The amount of data involved and the over-the-horizon requirement make satellite links the most suitable for the purpose. Such links can also provide a fleet with essential information for attacking land targets, such as those targets gathered from satellite reconnaissance. Satellite radomes thus sprout from ships, but placing them is a problem, because shipboard space is limited. Each of the multiple satellite systems requires its own radomes. As a ship maneuvers, parts of its superstructure get in the way of the spaceward path needed by each radome. Published drawings of the new U.S. *Zumwalt* design show a projected solution. The top of the ship's superstructure carries electronically scanned satellite antennas, which always have a clear spaceward path. The number of such antennas is reduced by using broadband elements, which cover multiple satellite systems.

Many ships now enjoy what amounts to Internet access. For example, from about 2000 on, the U.S. Navy found that Internet-style chat rooms were an excellent way of disseminating the commander's tactical requirements around a task force or battle group. NATO surface ships have long exchanged tactical data using computer-to-computer links. However, the new kind of computer data are sent on a far more massive scale (comparable to that of a broadband Internet connection), and in many cases are also more highly classified because the data can include intelligence information. Security often seems to demand that the more highly classified information be handled in a separate command space, filtered before it can be passed into the tactical computer used by the ship. Again, there are physical consequences, in the forms of extra (or, in some ships, greatly enlarged) command spaces and satellite dishes with their radomes.

Stealth. This is another important trend. The goal is to make it difficult for a radar aboard a missile or attack airplane to detect a ship, which is normally readily visible to radar because the side of its hull makes a right angle with the water, forming a corner reflector (the brightest kind of radar reflector), and because of the clutter on its decks. Many ships now show sloped hull sides, which turn to the vertical only well above the waterline. Much more of the ship's equipment is placed inside the carefully shaped hull. Items that cannot be placed inboard are shielded by bulkheads or screens from radars looking at the ship at a shallow angle (as from a missile or airplane). The first prominent example of this kind of design was the French *Lafayette* class frigate of the late 1990s.

The U.S. *Zumwalt* design is an extreme example. Instead of slanting the sides of the ship outward, as is common, the designers slanted them inward. Even so, as the ship rolled, at times its sides would be nearly vertical, so that there would be moments when the ship would be visible to radar. A computer-equipped airplane or missile might connect up the positions thus revealed to indicate the position of the ship. To avoid this problem, the designers stabilized the hull and they also added internal ballast tanks that would keep the ship at a chosen waterline. Critics have cited fears that the tumble-home design is inherently unseaworthy, safe only because the ship is computer-stabilized. Moreover, some radars would not be affected by the special shaping. The new U.S. Littoral Combat Ship, which is intended to operate much closer to enemy shores than a *Zumwalt*, seems to show a much less extreme version of stealth shaping, presumably still well worthwhile but much less expensive.

Modularity. Another trend is an attempt to reduce ship and force cost by modularity. One application of this idea is to make it possible to use the same hull-machinery combination for various purposes, so that more hulls are built and each is less expensive. The most prominent example is the very successful German MEKO export frigate, in which much the same hull and machinery are adapted to any of a wide variety of containerized weapons and sensors. Thus the ship need not be redesigned for each customer.

The Royal Danish Navy found that the same ship could be used for multiple functions by changing containers on board. In the 1980s the Danes had to replace their numerous fleet of small ships, all commissioned during the early Cold War buildup of the 1950s. The key was that war, if it came, would be fought in phases. For example, coastal surveillance would precede a phase of minelaying and one of missile and torpedo combat. Why were different hulls needed for each distinct phase? The combat direction systems for different roles were considered as distinct entities. By this time, however, computer power was growing to the point where a single computer system could incorporate the software required for many different missions. Once that was done, it was no great problem to develop containers corresponding to the different missions, conversion taking less than 24 h. The result was the *Flyvefisken* class of 300- ton "Stanflex" corvettes. Larger ships followed, including a current class in which the containers of the past are replaced by trailers in a "hangar" below decks.

The U.S. Littoral Combat Ship (LCS) is a different version of this idea. It has a "plug and play" combat direction system that can accommodate alternative modular payloads. However, the U.S. modules are very different. Typically the LCS operates to support crewless vehicles—air, surface, and underwater. In some applications they strew sensors that the LCS uses to form a tactical picture. On the basis of that picture, it or other ships in the group offshore deal with threats. One such sensor is an upward-looking acoustic array laid on the bottom. A submarine passing over such an array announces its presence. If it passes over multiple arrays, they register its course and speed; the resulting track can be the basis for attacking the submarine. The hope is that separating the modules from the basic hull will make it possible to design an inexpensive hull that can be built in large numbers.

All-electric ship. In this type of ship, the prime mover (diesel, gas turbine, or nuclear) generates electricity used for all ship functions, from turning a propeller (using a lightweight electric motor) to driving a catapult (on the next-generation carrier), driving pumps, and opening and closing watertight doors. It is far easier for a computer to control electricity than to control the hydraulic power typically used to drive auxiliary machinery such as pumps. Western navies are under enormous pressure to reduce the number of sailors per ship, because personnel accounts for an enormous and growing fraction of their budgets. One way to do so is to fill a ship with sensors connected to computers, which can monitor the internal condition of the ship and react accordingly. Over a decade ago the U.S. Navy's David Taylor Research Center estimated that a cruiser in which computers had been embedded could be operated with a crew of only about 115, compared to over 300 for a conventional ship. Computers could operate a ship using hydraulic power, but only with considerable difficulty.

An all-electric ship can be far tougher than a conventional ship. In a conventional ship, propeller shafts connecting the prime mover and the propellers (at the stern) fill about a third or more of the total length of the ship. Severe damage in that rear third of the ship will distort or break a shaft and immobilize the ship. By way of contrast, the ship can have several alternative paths (buses) to carry electric power from engines (which need not all be concentrated amidships) to propellers, and the propellers need not all be at the stern. This potential has not been realized in the U.S. Navy's new all-electric *Zumwalt* design, but it is certainly present.

An electric design is also adapted to a future change in the powerplant. For example, for years there has been intense interest in high-powered fuel cells. They can operate at low temperatures, and thus eliminate the infrared signature that ships currently create. They may also operate on hydrogen and oxygen, and hence may be viable even if oil has to be abandoned as a fuel. It is easy to imagine replacing a gas turbine and generator with a fuel cell. To replace a conventional powerplant linked mechanically to a propeller would be far more difficult, and would require a complete redesign of a ship. Conventional ships would be almost impossible to modify.

Moreover, the electric ship is adapted to a potential future generation of electric weapons: rail guns and electric lasers. This vision was probably responsible for the decision to make the new *Zumwalt* class destroyer an all-electric ship, the hope being that a rail gun will ultimately replace the currently planned 155-mm Advanced Gun System (AGS). In a rail gun, electric current propels a projectile up a pair of charged rails. This is a future weapon; it cannot become reality without a source of vast electric power, like a ship's powerplant. The weapon would be energized by briefly shunting the ship's power into it. Because its shells would need no propellant, a ship armed with an electric gun could accommodate far more projectiles than a conventional one. Like conventional shells, they could easily be transferred at sea. Yet the impulse provided by the electric gun would far exceed that of a conventional one, so the shells would fly much farther. Electricity seems to be the only viable way to create high-energy shipboard lasers, which are an attractive means of defense against fast antiship missiles; the chemicals used in the alternative chemical type are far too toxic.

For background information *see* ANTENNA (ELECTROMAGNETISM); ANTISUBMARINE WARFARE; FUEL CELL; GUIDED MISSILE; MARINE MACHINERY; NAVAL ARMAMENT; RADAR; SHORT TAKEOFF AND LANDING (STOL); VERTICAL TAKEOFF AND LANDING (VTOL) in the McGraw-Hill Encyclopedia of Science & Technology.

Norman Friedman

Bibliography. M. H. H. Evans, *Amphibious Operations: The Projection Of Sea Power Ashore*, Brassey's, London, 1990; N. Friedman, *The Naval Institute Guide to World Naval Weapon Systems*, 5th ed., Naval Institute Press, Annapolis, 2006; N. Friedman, *Seapower and Space*, Chatham, London, 2000; W. J. R. Gardner, *Anti-Submarine Warfare*, Brassey's, Washington, 1996; D. G. Kiely, *Naval Electronic Warfare*, Brassey's, London, 1988; D. H. Wagner,

W. C. Mylander, and T. J. Sanders, *Naval Operations Analysis*, 3d ed., Naval Institute Press, Annapolis, 1999.

Web spam identification and blocking

Web spam is defined as behavior with the effect of manipulating search engines' ranking algorithms. A frequent user of search engines may search for pages about a subject such as a pop star, but find that some top-ranked results do not contain anything relevant at all. Web spam is typically responsible for this disappointing situation.

Why do people bother to generate spam? Most people use search engines to find information on the web. However, most users examine only the top results. This will undoubtedly bring more traffic to the Web sites that have high-ranking positions. More traffic generally leads to more profit for Web site owners. Thus, it is hardly surprising that many Web site owners want their sites to be ranked as high as possible. However, instead of working hard to improve the quality of their Web sites, some content providers try to find a shortcut and utilize deceptive methods to manipulate search engine ranking algorithms, with the purpose of generating better ranking positions for their own sites. This is how Web spam originates. In general, the persons or organizations who generate Web spam are called spammers.

Factors used by search engines. Before introducing spam, it is worth discussing how search engines rank Web pages. From the search engine's perspective, it is not trivial to decide which page should rank higher than other pages from a repository containing billions of Web pages. Search engines use statistical methods based on hundreds of factors to rank Web pages for given queries. Two category factors stand out: One is term-based and the other is link-based.

Term-based factors will judge a page's relevance to a given query based purely on the textual content of the page. Briefly speaking, a page containing more instances of query terms or more unusual terms will be given a higher term factor score. In contrast, link-based factors usually help to calculate an importance score for each page. Generally speaking, a page pointed by more pages or more important pages will have a better link factor score. For a given query, all the factor scores are ultimately combined into one number, which is used to decide the order of the final response list returned to web surfers.

Known spamming techniques. Spammers usually design their spamming techniques by attacking known factors used by search engines in their ranking algorithms. For example, to attack the term factor described above, spammers will repeat a popular query term many times in their Web pages. The **table** lists several of the most common spamming techniques that have been discovered by experts from both industry and academia.

Popular spamming techniques

Spam technique	Symptom
Keyword stuffing	Spammers repeat the same keyword many times or add many unrelated keywords on their pages, so their pages will match more queries.
Link farms	Spammers register many Web sites and make them all point to each other, so all of these sites will have a higher link score.
Link bombing	Many spammers join together to use misleading anchor texts when pointing to a certain target page, so the target page will be ranked very high when these misleading anchor texts are used as queries.
Cloaking	Spammers send different content to search engines than to Web surfers' browsers.
Redirection	Spammers use tricky redirection to some target spam pages, so users will see only these target pages.

Detecting spam. Search engines have been fighting Web spam for many years, but it continues to grow. Since the approaches used by search engines are considered sensitive, experts within industry are typically reluctant to give details of how they detect and block spam. Obviously, it might be quite easy for spammers to bypass the antispam approaches if too much detail is given.

On the other hand, researchers from universities and research centers have recently published many ideas to detect search engine spam. Generally speaking, spam pages have some abnormal features not usually found in normal web pages. Experts expend considerable effot to find these features, and then different algorithms or tools are proposed or applied to detect spam based on the abnormal features.

Generating features. Some features can be generalized by comparing the artificial example of a spam page shown in the **illustration** (with many repeated keywords and links) with a normal web page. Possible features that may appear on spam pages include:

Features related to text on the page:

1. The spam page contains more words than a normal web page.

2. The spam page contains more unique "hot terms" (such as the terms "online gambling" and "popular" in the illustration) than a normal web page.

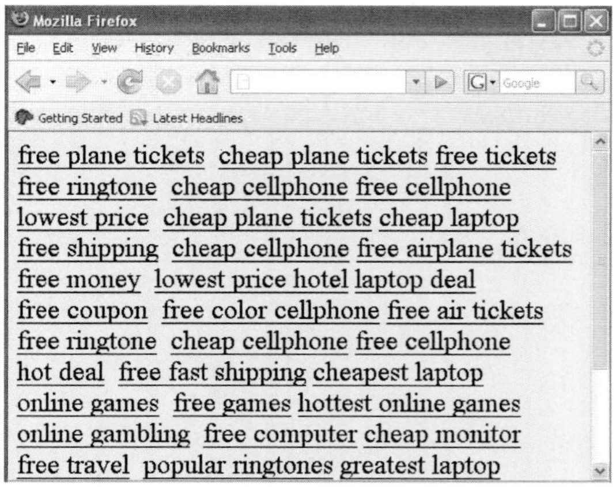

An example of a spam page.

3. The spam page may repeat hot terms (such as the term "free" in the illustration) more times than a normal web page.

Features related to links:

1. The spam page contains more links than a normal web page.
2. The spam page has more common pages in a parent page set and a children page set than a normal web page does.
3. The spam page may point to more pages using the same IP address.

For another example, possible features of a page that uses the cloaking technique include:

1. The copy from the search engine's copy contains more terms than the one from the user's browser.
2. The two copies use different languages.
3. The copy from the search engine's copy contains more hot terms than the one from the user's browser.

Machine-learning algorithms. When enough features have been discovered and a page set containing enough examples of both spam pages and normal Web pages has been collected, experts will apply machine-learning algorithms to sample spam and nonspam pages to build a classifier, which will be used later to determine the value of new pages from the Web. Some well-known classifiers, such as support vector machines and decision trees, have proved quite useful for detecting spam pages.

The use of statistical metrics in building models to detect spam has also been explored. For example, if one takes all the pages on the Web, then a curve showing the relationship of pages and the number of terms within them should follow some statistical distribution (such as a power-law distribution). Pages deviating from the standard distribution curve are very likely to be spam pages.

Other approaches. Researchers have proposed other approaches besides machine-learning algorithms to combat spam. For example, the whole Web can be considered as a graph, with pages as nodes and links as edges among them. Z. Gyöngyi and his colleagues have proposed using the concept of trust from sociology to combat spam. The basic idea is that trust will be propagated iteratively along the links from preselected good pages (such as yahoo.com or mit.edu) to the rest of web. Finally, pages with high trust scores are good but pages with poor scores are usually spam pages.

In a patent filing, A. Acharya and his colleagues first publicly proposed using historical data to identify link spam pages. Here historical data are the collection of pages over a timeline. For example, a search engine may download all the pages on the Web once per month; then, all the data within a year can be used for historical signal analysis. Some possible historical signals include a sudden jump for pages in the ranking for a query or a spiky rate of growth in the number of incoming links from other pages.

In addition, methods such as natural language processing, distrust propagation, revised link analysis algorithms, and random walk models have been explored to detect different types of spam.

Subjective judgments and detection policies. There are still no definitive, final definitions of Web spam. In most cases a subjective judgment, based on an appreciation of the intention of the author of a page, is involved when experts are evaluating spam versus nonspam. Thus, it is not surprising that different search engines will apply different policies when detecting spam.

Blocking spam. Upon the detection of spam, different mechanisms can be applied to punish spam pages, depending on the specific policy of the search engine. The most common and severe punishment is to block spam pages and never show them to Web surfers. This can be accomplished at two points in the operation of a search engine: One option is for the search engines to decide not to visit such pages any more; hence the spam pages will not exist in the search engine's index. Another option is to filter spam (based on a pregenerated blacklist) just before sending the result set to users.

Search engines can also apply less severe policies. For example, some pages have useful content but utilize spamming techniques. A search engine may keep such pages in the response list but demote them to lower ranking positions as punishment. Since the goal of a search engine is to provide high-quality results to users, this punishment has a short-term/long-term tradeoff for spammers with good content. Showing the results is good in the short term, but generates the wrong incentive in the long term.

Prospects. The war between spammers and experts fighting them is like an arms race. New techniques constantly appear from each side, requiring the other to respond and adapt. As long as generating spam is profitable for spammers, this war will continue. If one day it costs more for a spammer to generate a spam page than a high-quality page, spammers may disappear and the war will end.

For background information *see* ALGORITHM; ARTIFICIAL INTELLIGENCE; GRAPH THEORY; INTELLIGENT MACHINE; NATURAL LANGUAGE PROCESSING; WORLD WIDE WEB in the McGraw-Hill Encyclopedia of Science & Technology. Baoning Wu; Brian D. Davison

Bibliography. A. Acharya et al., Information retrieval based on historical data, U.S. Patent Application no. 20050071741, 2005; Z. Gyöngyi and H. Garcia-Molina, Spam: It's not just for inboxes anymore, *IEEE Comp. Mag.*, 38(10):28–34, October 2005; Z. Gyöngyi, H. Garcia-Molina, and J. Pederson, Combating Web spam with TrustRank, in M. A. Nascimento et al. (eds.), *Proceedings of the Thirtieth International Conference on Very Large Data Bases, Toronto, Canada*, Morgan Kaufmann, pp. 576–587, 2004; A. Ntoulas et al., Detecting spam Web pages through content analysis, in *Proceedings of the 15th International World Wide Web Conference,*

Edinburgh, Scotland, ACM Press, pp. 83–92, 2006; B. Wu and B. D. Davison, Detecting semantic cloaking on the web, in *Proceedings of the 15th International World Wide Web Conference, Edinburgh, Scotland*, ACM Press, pp. 819–828, 2006.

Wide-area power system protection and control

Electric power system breakdown is a rather rare event. However, when it occurs the influence on society and the costs are tremendous. The electric power system is the most complex, widespread, and powerful interconnected process ever designed and constructed by humankind. Power systems normally cover many countries in synchronous operation, which means that changes of the operational conditions in one part of the system affect those in the rest of the system. In normal stable operation these changes are small, coordinated, and well planned for, and do not affect the stability or the integrity of the system. Power systems will always be exposed to faults, such as earth faults and short circuits, caused by lightning strokes or insulation breakdown. Ever since the beginning of the electrification era, protection has been used to rapidly disconnect a power system element, such as a line, transformer, or generator, if it is faulted or overloaded, in order to save the element from total destruction and to continue the power supply to the healthy part of the system. However, power systems may, slowly or more rapidly, slide towards instability. Such a transition, from stable operational conditions, can be triggered by a very severe disturbance (such as the loss of a number of lines or of a whole substation) in a stressed operational situation (high load level and large power transfers). It could also be a result of a very rapid load growth, where the corresponding increase in generation cannot be achieved. To preserve the integrity of the power system and avoid a widespread system breakdown, wide-area protection systems are now being introduced. Other names of systems aimed at taking actions to preserve the power system operation, even though no specific element is overloaded or faulted, are remedial action schemes (RAS); system protection schemes (SPS); and wide-area monitoring, protection, and control systems (WAMPAC).

Protection system principles. Protection aiming at disconnection of faulted power system elements has to be both quick and reliable. A short circuit in a transmission system is normally cleared within 60–100 ms. Equipment protection is therefore always decentralized and installed in the substations, with a minimum of communication requirements. However, the performance and availability of communication has increased considerably. Based on communication, more sophisticated protection systems can be designed for both power system elements and power system integrity. There are two different philosophies on how to design wide-area protection systems: (1) centralized schemes, where information from all over the power system is sent to a central place for execution of the protection algorithms, and action orders are then sent to appropriate control devices; and (2) decentralized schemes, where the decision for each action is taken locally, based on information from all over the power system. In practice the two solutions can have the same performance with respect to speed and reliability. Very often the design includes a "bottom-level safety-net" based on local criteria only.

In the past, protection engineers were very focused on tripping faulted elements, and to increase the probability of tripping a faulted element the protections were often doubled (main and backup). This action improved the dependability of the protection system, that is, the capability of the protection system "to trip when you want it to trip." However, with increasing dependability, the risk for "overtripping" also increases, and focus among relay engineers has now shifted toward the security of the protection system, that is, the capability of the protection system "to not trip when you do not want it to trip." As power systems become more complex and society becomes ever more dependent on a reliable electric power supply, it becomes more important not to trip a healthy element when there is a fault in the power system.

Power system phenomena to counteract. Utility needs and problems are often formulated in very loose terms, such as "intelligent load shedding," "protection system against major disturbances," and "counteract cascaded line tripping." These needs have to be broken down to physical phenomena, such as protection against transient angle instability (first swing), small-signal angle instability (damping), frequency instability, short-term voltage instability, long-term voltage instability, and cascading outages.

Violation of transient angle stability is often preceded by a short circuit in the transmission system, causing the machines in the generation end of the system to accelerate, and the machines in the load end to decelerate. When the fault is cleared, the tension between the two ends of the system has increased, and the question is whether they will come back to stable synchronous operation (the generator end to retard and the load end to accelerate), or

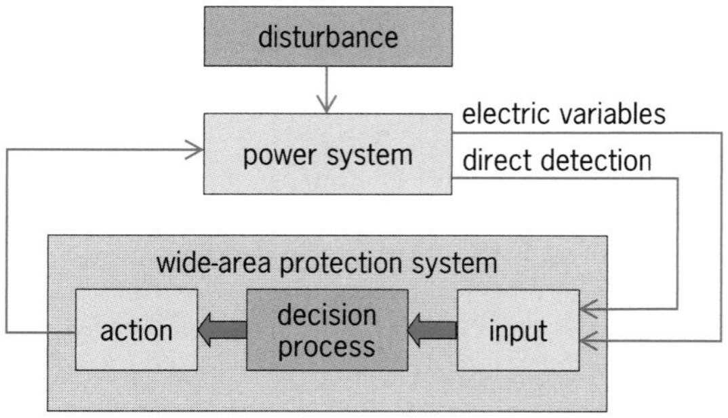

Fig. 1. General design of a wide-area protection system, composed of inputs, a decision-making process, and actions. The power system is exposed to different kinds of disturbances. The power system state is continuously measured both via electric variables, such as voltages and currents, and via direct detection, such as circuit-breaker statuses (on or off). The inputs are processed by the protection system, and appropriate protective actions are taken when called upon.

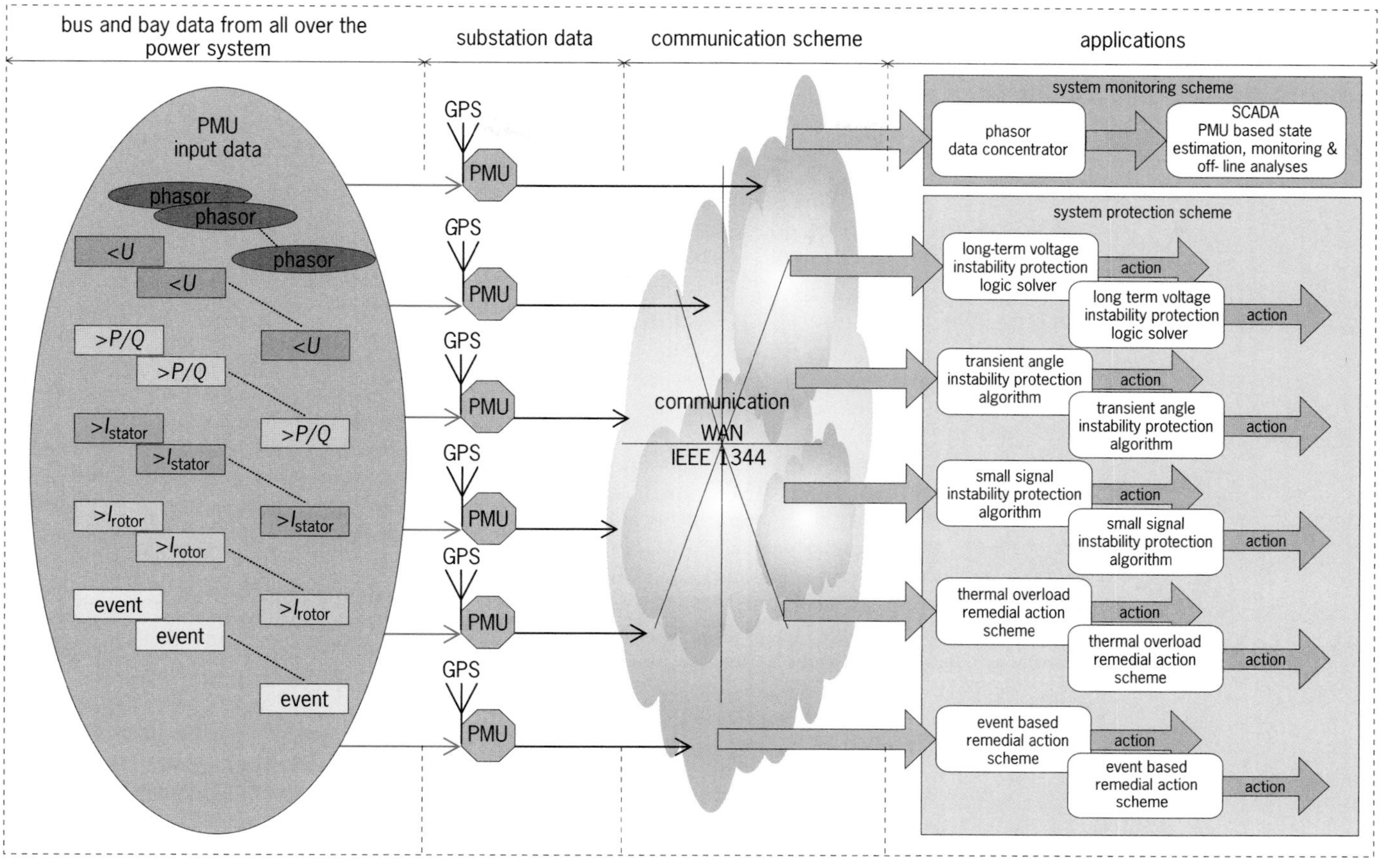

Fig. 2. Multilayered wide-area protection design in more detail. Bus and bay data from all over the power system are picked as inputs to the phasor measurement units (PMUs). Data are then communicated via a wide-area network (WAN), in a standardized format, to the application algorithms. These algorithms can be implemented either at a central location, from where the action orders are remotely sent to the appropriate substations where the actions are to be taken, or at different substations, from where the actions are taken locally. $<U$ = undervoltage function; $>P/Q$ = high power function; I_{rotor} = high rotor current function; I_{stator} = high stator current function; GPS = Global Positioning System; SCADA = supervisory control and data acquisition.

synchronism will be lost, with a system breakdown as a consequence.

Small-signal angle stability refers to the ability of the system to damp out minor oscillations caused by any credible triggering event, such as a switching or fault clearance.

The frequency stability is related to the balance between generation and load. If the generation increases without a matching increase in load, the frequency will increase, and vice versa. Small and slow variations are taken care of by the frequency controllers of the generators and the natural control of the load; if the frequency increases, rotating loads will go faster, thereby increasing their power consumption.

Short-term voltage instability is normally associated with an extremely severe reduction of the network capacity, for example, caused by the tripping of several parallel lines due to a bush fire. A characteristic of the short-term voltage instability is that there is no stable equilibrium point for the power flow immediately after the clearance of the initial fault or faults. Remedial actions to save the system in such a situation therefore have to be fast (a few seconds or fractions of a second) and powerful (for example, a large amount of load shedding). *See* VOLTAGE STABILITY.

When a power system is in transition toward a long-term voltage instability, it has "survived" the initial disturbance: that is, there was a stable equilibrium point immediately after the clearance of the disturbance. However, load recovery and tap-changer operations cause the transmission system voltage to decrease, and the collapse normally occurs on a time scale of 10 s to about 30 min. Without any initial disturbance, a long-term voltage instability might occur due to a very large and rapid load increase.

Cascading outages of lines or generators might have different origins, but are mainly associated with some kind of overload, followed by the trip of one line or generator unit, which cause an increased overload on the remaining units, and so on. In such situations, load shedding or generation rejection might be required to preserve the integrity of the power system.

For each phenomenon, a reliable (based on redundancy and robustness) protection system has to be designed with respect to input variables, decision criteria, and output actions. Parallel systems counteracting different phenomena and different layers of safety nets can be designed and coordinated.

Wide-area protection system design. The protection system has to be designed according to the phenomena it is intended to counteract. In general the

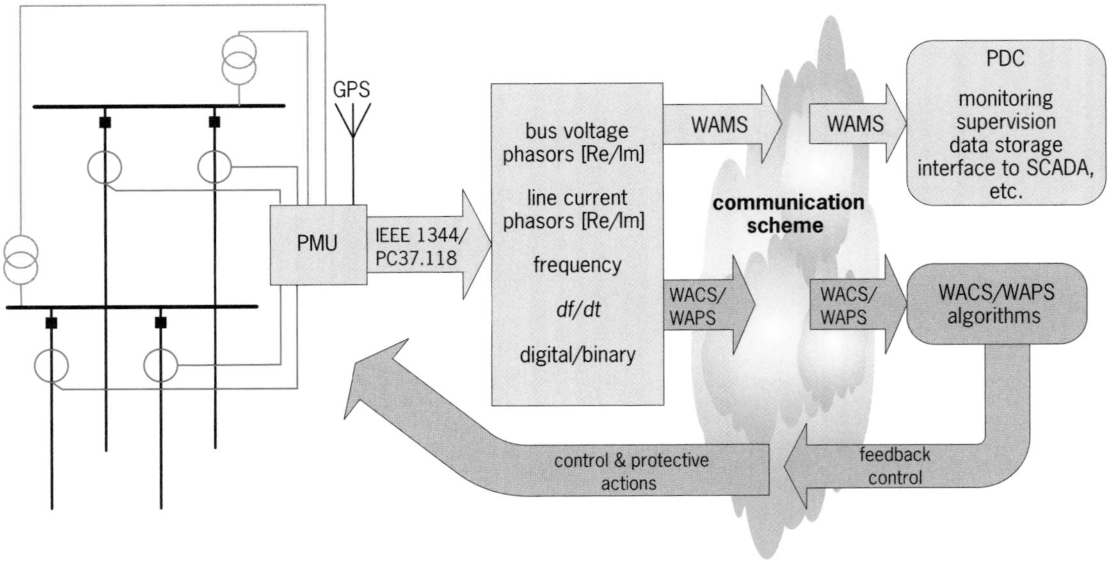

Fig. 3. A wide-area monitoring, control, and protection scheme design. The phasor measurement unit (PMU) is directly connected to the substation current and voltage transformers. The phasors are then sent via the communication network to the wide-area measurement system (WAMS), wide-area control system (WACS), and wide-area protection system (WAPS), wherever they might be implemented. From the control and protection functions, a feedback loop back to the power system for control and protective measures is used. (Re and Im are the real and imaginary parts of the phasors; *df/dt* is the rate of change of frequency with respect to time; PDC = phasor data concentration; SCADA = supervisory control and data acquisition.)

protection system comprises inputs (measurements from the power system), a decision-making process (the protection algorithms), and outputs (the actions to be performed in the power system, such as load shedding) [**Figs. 1–3**].

Underfrequency-controlled load shedding, to reestablish the frequency balance after an extremely large loss of generation, has been widely applied since the 1960s. The frequency is by its nature a universal variable, available throughout the system. Therefore, such an underfrequency-controlled load shedding scheme does not need any communication.

To damp power oscillations, frequency or phase-angle measurements are necessary, as well as devices for active damping, most likely damping resistors that are switched in and out according to the oscillations. A damping resistor capacity of about 5% of the oscillation power magnitude is normally sufficient for smooth and reliable damping.

The large disturbances in August and September 2003 (in the United States, Italy, and Sweden/Denmark) were all of the voltage instability kind, where an initial disturbance weakened the transmission system. Voltages then dropped in the load end, a

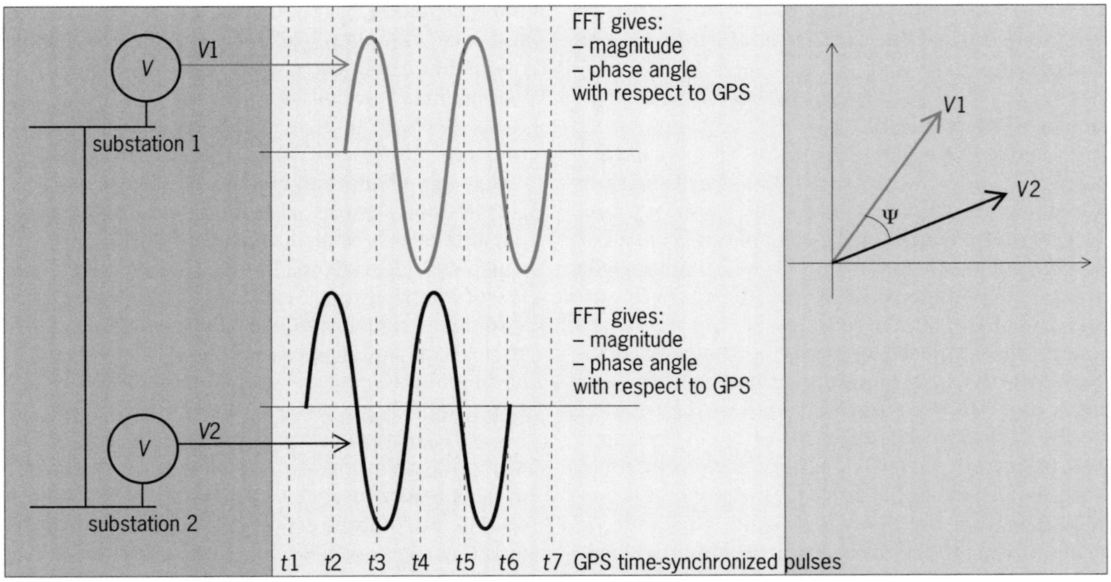

Fig. 4. Synchronization of phase measurements. The GPS satellites provide a very accurate time synchronization signal, available, via an antenna input, throughout the power system. This means that that voltage and current recordings from different substations can be directly displayed on the same time axis and in the same phasor diagram. FFT = fast Fourier transform.

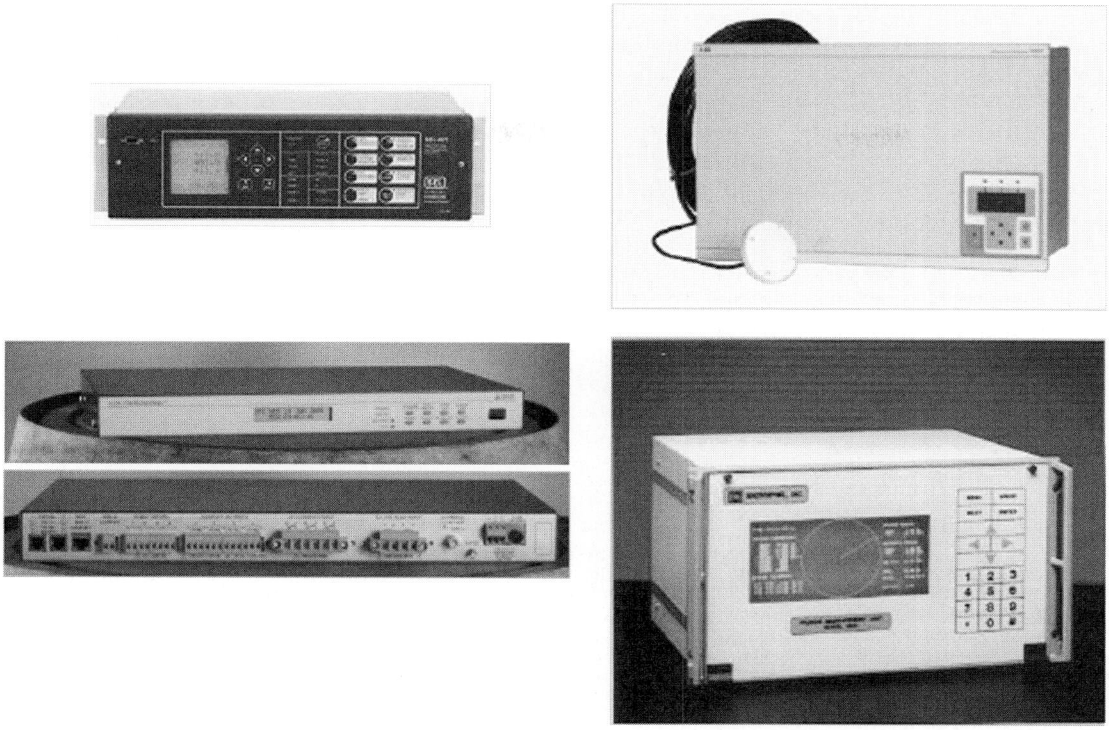

Fig. 5. Phasor measurement units (PMUs) from different manufacturers with slightly different properties and performances. The PMU is a central component in a wide-area protection system. A GPS antenna is also shown in the unit in the upper right.

load recovery followed the initial load relief, and the transmission system was further stressed, until a protection device tripped one of the most heavily loaded lines, resulting in an even higher stress on the remaining lines, followed by quick trips of these lines, and finally a system breakdown. An undervoltage-controlled shedding of about 10–20% of the load would probably have saved the integrity of the power systems in Italy and Sweden.

Wide-area synchronized measurements. The potential for wide-area protection is highly dependent on precise and timely measurements. The ability to make measurements synchronized by the Global Positioning System (GPS) satellites, with an accuracy better than 1 microsecond, has made it possible to directly measure phase-angle differences between voltages and currents at different locations throughout an interconnected power system (**Fig. 4**). The measuring devices are called phasor measurement units (PMUs), and are now commercially available, with angle measurement accuracy better than 0.1°. Such PMUs have been installed in power systems all over the world (**Fig. 5**). So far the recordings have been used mainly for monitoring and post-fault analyses, but power system engineers are just about to take the step toward automatic protective actions to preserve the power system integrity in situations approaching instability, based on PMU measurements.

Future trends and challenges. The technology, with respect to measurements, synchronization, communication, wide-area protection devices, and actuators, is already available for large-scale implementation of power-system protection schemes. The protection algorithm design and tuning might, however, be a challenge, since the wide-area protection system is supposed to operate very seldom, perhaps once every 20 years. Therefore, the operational experience will be limited. The reliability requirements on the system are also extremely high with respect to both security and dependability. Since the protection system is supposed to save the integrity of the power system on the edge of a breakdown, to save a lot of money and effort, it is extremely important that it operates as intended when called upon. On the other hand, since the actions taken by the wide-area protection system are very powerful, often including shedding of high-priority load, unnecessary operation is not acceptable.

Daniel Karlsson

Wide-area control of power systems. The typical electric power grid is a geographically large interconnected network consisting of components such as generators, transmission lines, real and reactive power compensators, and loads. The high-voltage transmission system links the generators to substations, which supply power to the user through the distribution systems. In a modern power system, power sources and loads typically are widely dispersed. The number of bulk power exchanges over long distances has increased as a consequence of the deregulation of the electric power industry, the growing demand for electricity, and the dependence of critical infrastructures on the supply of electricity. The stability and security of the power system are critical. Usually, distributed control agents are employed to provide reactive control at several places on the power network through power system stabilizers (PSSs), automatic voltage regulators (AVRs), flexible alternating-current transmission systems (FACTS) devices, and so forth. The inherent nonlinearity in the system becomes a major source of model uncertainty. The model uncertainty includes the inaccuracies in modeling the power system devices such as the transformers, the transmission lines, and the loads. The loads are dynamic and continuously changing. Final control settings are made using field tests at a couple of operating points of the power system on the distributed control devices. Although local optimization is realized by these control agents (PSSs, AVRs), the lack of coordination among the local agents may cause serious problems, such as system oscillations (interarea) in the power network.

In order to minimize the problems encountered in a large distributed power network control, a centralized wide-area control system (WACS) can be deployed. The WACS coordinates the actions of the distributed agents/local controllers by, for example, using supervisory control and data acquisition (SCADA), phasor measurement units (PMUs), or other available information. Employing PMUs makes it possible to deliver remote signals at a speed as high as the 30-Hz sampling rate. The WACS receives information on different areas in the power system and, based on some predefined objective functions, sends appropriate control (feedback) signals to the distributed agents in the power network (**Fig. 6**). A WACS can also be viewed as a supervisory control system.

Owing to the complexity, high nonlinearity, and fast-changing nature of the power system, a fast, accurate online monitoring system, a wide-area monitor (WAM), for effective control of power networks

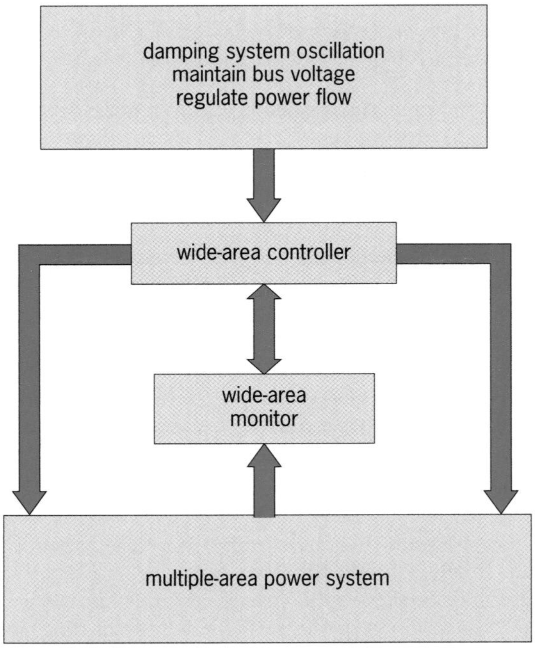

Fig. 6. Schematic of wide-area control system (WACS) with a wide-area monitor and a wide-area controller.

with an adaptive/optimal wide-area controller (WAC) is required and would offer the following benefits:

1. Transmission capacity enhancement can be achieved by on-line monitoring of the system stability limits and capabilities.
2. Compensates for uncertainties in simulation results used in setting operating criteria.
3. Better understanding of the dynamic and complex behavior of the system.
4. Reinforcement of the power system based on accurate feedback signals obtained during analysis of system dynamics.
5. Minimization of conditions not primarily controllable by local agents/controllers.
6. Possibility of a coordinated approach for the execution of fast stabilizing actions in case of severe network disturbances.
7. Triggering of additional functions by a WAC.

Wide-Area Monitor. The WAC is highly dependent on the accuracy and timing of the data received, which emphasizes the need for a reliable wide-area monitor capable of estimating the states of the system in real-time. Any system or technique designed for such a purpose should address the following issues.

1. There is a signal transmission delay associated with the communication channels used in power networks. The average transmission delay could be as long as seconds for the Internet or as short as tens of milliseconds in a fiber-optic link. The retrieval of real-time information from the received delayed values should be possible.
2. The signal transmission delay is not necessarily static. In the worst case, it can be due to a missing sensor or a failed communication channel. The methodology should be robust in case of partial loss of information and should be able to restore the required data using the available information.
3. Any changes to the power system configuration can largely diminish the effectiveness of the state estimator. Some mechanism should be in place such that the estimator adaptively adjusts itself to the ever changing nature of the power network.

Analytical methods are traditionally used to solve the state estimation problem in power systems. However, as the size of the power system increases, the problem becomes complex and obtaining a reasonable solution gets difficult. In particular, this complexity can be a problem when a large number of redundant measurements are used in a power network. In addition, incorporating change in the system topology and conditions into the mathematical formulation of the state estimation problem can be a tedious process. Dependence on the model of the network is the major disadvantage of this technique, which can weaken its robustness and practicality in real-world problems. In addition, the analytical state estimation method converges slowly, can become trapped in local minima, or produce ill-conditioned and unreliable solutions. Ideally, state estimation should run at the scanning rate; however, due to computational limitations most practical

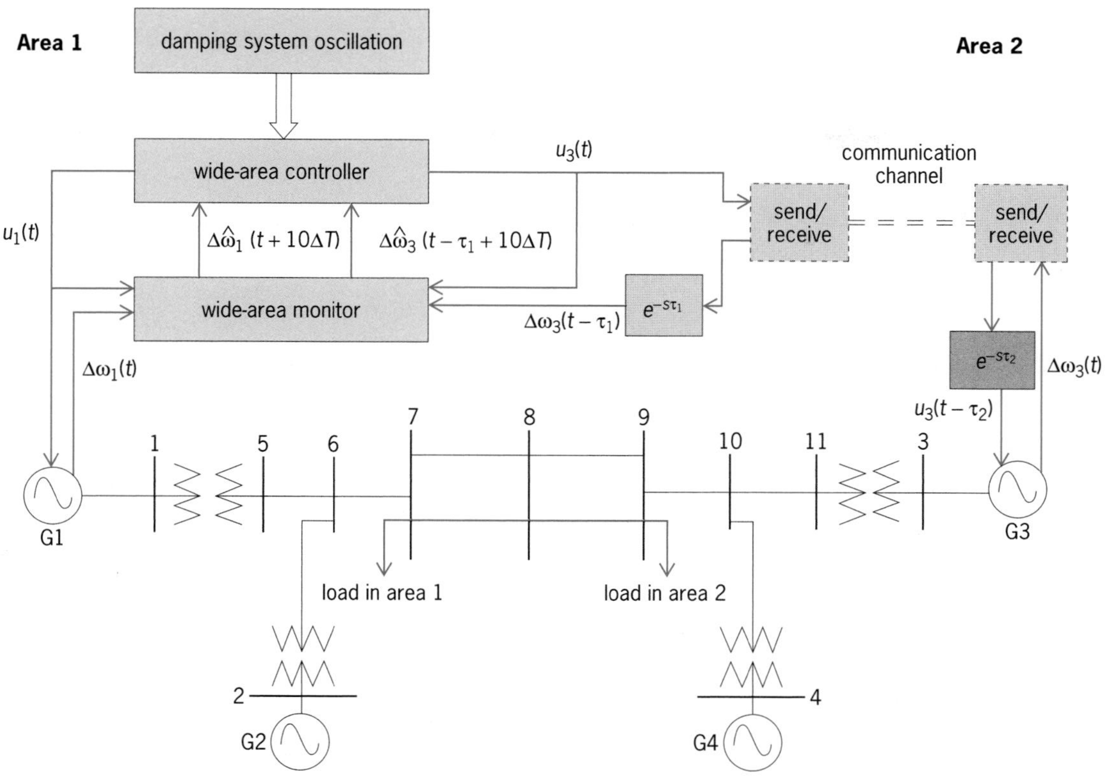

Fig. 7. Two-area power system with a WAM and a WAC.

estimators run every few minutes or whenever a major change occurs.

Neural networks can serve as a universal dynamical system representation, and they have a great potential for input-output modeling especially for highly nonlinear systems. They are very versatile in the tasks they can solve, and without doubt neural networks represent a solution concept that is inspired by biology and extremely successful. Neural networks such as the multilayer perceptrons, radial basis functions, recurrent neural networks, and echo state networks are able to identify or model multiple-input–multiple-output time-varying systems. With continuous online training these models can track the dynamics of these systems, thus yielding adaptive identification for changes in operating points and conditions. The adaptive online identification provides an up-to-date system model to adapt controller parameters, thus providing the desired control signals for any given disturbance.

Wide-area controller. Several classical control techniques have been proposed for the design of a WACS. Multiple-model-based adaptive and hierarchical wide-area control and H-infinity-based robust control have been reported for damping postdisturbance oscillations. A few of the robust control techniques have also shown promise to compensate for signal transmission delays. But, all these designs require an accurate high-order mathematical model of the system that is difficult to obtain for a practical large-scale power system network. Classical linear control techniques used for WACSs are limited over a small operating region, because a linear model of the system is valid only for a particular operating point. As the system goes through different operating conditions and disturbance scenarios, linear control degrades. Hence, a nonlinear WACS is necessary to model, predict, and control an increasingly complex nonlinear system such as the North American power system.

Several paradigms of computational intelligence have been investigated for control of nonlinear complex systems such as the electric power system. Different intelligent agents based on adaptive critic designs, neural networks, and fuzzy logic, along with various optimization algorithms such as particle swarm optimization and evolutionary algorithms, have shown a lot of promise for various applications including control of nonlinear systems. Among the paradigms of computational intelligence, neural networks have the closest resemblance to the functioning of the human brain. Neural networks are capable of approximating nonlinear functions that have the potential to be widely utilized for prediction and control of complex systems.

Figure 7 shows a two-area four-machine power system with a WACS. The objective of the WACS here is to damp out system oscillations, mainly the interarea oscillation. The WACS actions are based on remote signal measurements from areas 1 and 2 of the power system. These remote signals are exchanged over communication channels that have transmission delays resulting in late arrival of receiving signals by τ_1 s at the wide-area controller location. Thus, the

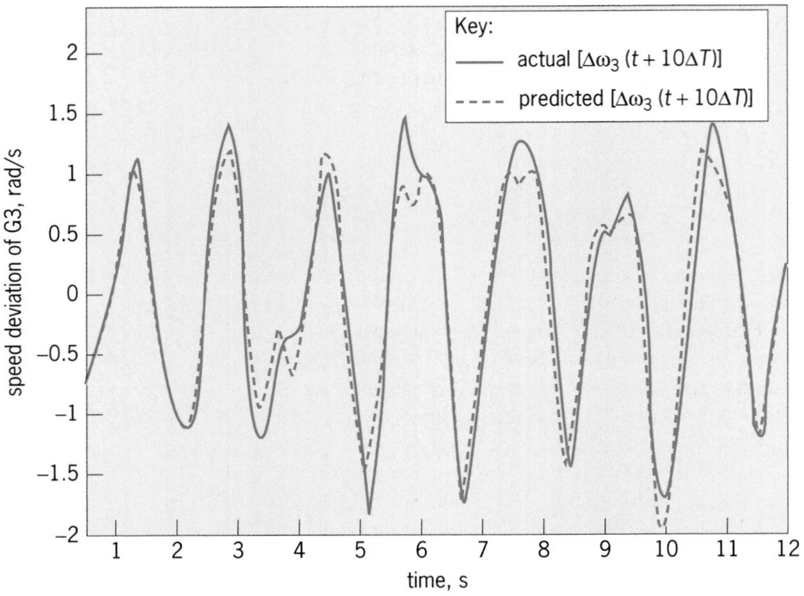

Fig. 8. Actual and a neural network predicted speed deviation of generator G3 in Fig. 7 for small perturbations applied to the excitation systems of generators G1 and G3.

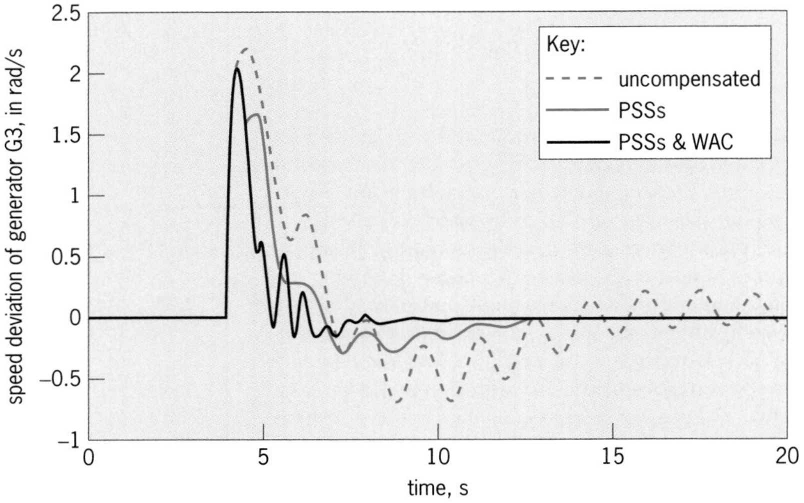

Fig. 9. Speed deviation of generator G3 in Fig. 7 without power system stabilizers (PSSs), with PSSs, and with PSSs and WAC for a 10-cycle three-phase short circuit.

actions taken by the WACS based on these signals are no longer valid for the current state of the system. Similarly, the control signals of the WACS communicated to the respective remote areas/generators also have late arrivals by τ_2 s. Hence, the stabilizing control signal generated by WACS is out of phase and it adversely affects the stability of the system. If possible, it is assumed that the WACS can be located in one of the areas so that the delay from the signals communicated in that area (local signals) is zero, as in Fig. 7. **Figure 8** shows the multistep predictions of the speed deviation of generator G3 in Fig. 7 by a recurrent-neural-network-based WAM when small perturbations are applied to the excitation systems of generators G1 and G3. The WACS is designed based on adaptive critic designs, based on combined concepts of approximate dynamic programming and reinforcement learning. **Figure 9** shows the speed deviation of generator G3 for a 10-cycle three-phase short circuit applied at bus 8 in Fig. 7. The responses are shown for cases without PSSs, with PSSs, and with a combination of PSSs and a WAC on the power system. It is clear that the neural-network-based WAM and WAC are able to accurately identify the dynamics of the power system and provide additional damping to the power system oscillations.

Summary. The occurrence of large blackouts is a general concern with the increasing complexity and reduced investment in the electric power infrastructure. The use of wide-area monitors overcomes the difficulties and limitations of having up-to-date analytic equations of the power network every time the operating points and the configuration change. WAMs help wide-area controllers to generate appropriate auxiliary control signals to generators and transmission devices to damp oscillations and maintain bus voltages at their respective levels, thus providing an additional layer of defense against outages and blackouts.

For background information *see* ADAPTIVE CONTROL; ALTERNATING CURRENT; ALTERNATING-CURRENT CIRCUIT THEORY; COMPUTATIONAL INTELLIGENCE; ELECTRIC POWER SYSTEMS; ELECTRIC PROTECTIVE DEVICES; ESTIMATION THEORY; EVOLUTIONARY COMPUTATION; FUZZY SETS AND SYSTEMS; NEURAL NETWORK; NONLINEAR CONTROL THEORY; OPTIMIZATION; SATELLITE NAVIGATION SYSTEMS in the McGraw-Hill Encyclopedia of Science & Technology.

Ganesh K. Venayagamoorthy

Bibliography. M. Begovic et al., Wide-area protection and emergency control, *Proc. IEEE*, 93:876–891, 2005; D. Karlsson, M. Hemmingsson, and S. Lindahl, Wide area system monitoring and control—Terminology, phenomena, and solution implementation strategies, *IEEE Power Energy Mag.*, 2(5):68–76, September-October, 2004; P. Kundur, *Power System Stability and Control*, McGraw-Hill, 1994; S. Mohagheghi, G. K. Venayagamoorthy, and R. G. Harley, Optimal wide area controller and state predictor for a power system, *IEEE Trans. Power Systems*, vol. 22, issue 2, pp. 693–705, May 2007; P. Pourbeik, P. S. Kundur, and C. W. Taylor, The anatomy of a power grid blackout—Root causes and dynamics of recent major blackouts, *IEEE Power Energy Mag.*, 4(5):68–76, September-October, 2006; S. Ray and G. K. Venayagamoorthy, Real-Time Implementation of a Measurement based Adaptive Wide Area Control System Considering Communication Delays, in *IET Proceedings on Generation, Transmission and Distribution*, in press; C. W. Taylor, *Power System Voltage Stability*, McGraw-Hill, 1993; C. W. Taylor et al., WACS-wide-area stability and voltage control system: R&D and online demonstration, *Proc. IEEE*, vol. 93, pp. 892–906, May 2005; T. Van Cutsem and C. Vournas, *Voltage Stability of Electric Power Systems*, Springer, 1998; G. K. Venayagamoorthy and S. Ray, A neural network based optimal wide area control for a power system, in *IEEE IAS Annual Meeting*, Hong Kong, October 2–6, 2005; G. K. Venayagamoorthy, Online design of an echo state network based wide area monitor for a multimachine power system, in *Neural Networks*, vol. 20, issue 3, pp. 404–413, April 2007; M. Zima et al., Design aspects for wide-area monitoring and Control System, in *Proceedings of the IEEE*, vol. 93, pp. 980–996, May 2005.

Wireless sensor networks

The rapid progress of wireless communication, embedded micro-electro-mechanical systems (MEMS), and microsensing technologies has made wireless sensor networks (WSNs) possible. A WSN consists of many inexpensive wireless sensors, which are capable of collecting, storing, processing environmental information, and communicating with neighboring nodes. In the past, sensors were connected by wired networks. With the development of ad hoc networking technologies, tiny sensors can communicate through wireless links in a more convenient manner. Much research has been dedicated to this area, including energy-efficient media access control (MAC) protocols, routing and transport protocols, self-organizing schemes, sensor deployment and coverage issues, and localization algorithms.

Many WSN platforms have been developed, such as MICA2, MICAz, Telos mote, and Dust Network. To allow heterogeneous devices and different systems to work together, standards such as ZigBee™ have been developed. ZigBee is a global hardware and software standard designed for WSN requiring high reliability, high scalability, low power, low data rate, and low cost. ZigBee adopts the physical and data link layer protocols of the IEEE 802.15.4 standard and handles interoperability issues from the physical layer to the application layer. **Figure 1** shows the ZigBee protocol stack. In the physical layer, there are three frequency bands with 27 channels. The data link layer is divided into two sublayers: the logical link control (LLC) sublayer and the medium access control (MAC) sublayer. The LLC sublayer in IEEE 802.15.4 follows the IEEE 802.2 standard. The MAC sublayer manages superframes, controls channel access, validates frames, and sends acknowledgements. The IEEE 802.15.4 MAC sublayer also supports low-power operations and security mechanisms. The network layer provides reliable and secure transmissions among devices. In order to support different kinds of applications, three network topologies, namely star, tree, and mesh, are supported. A star network has a coordinator with all other devices directly connected to the coordinator. In tree and mesh networks, devices can communicate with each other in a multihop fashion. Nodes in a ZigBee tree network can operate in low-power mode. In the application layer, ZigBee defines application profiles, which are the collections of device descriptors. For example, a dimmer node can communicate with a lighting device node to form a remote light control application.

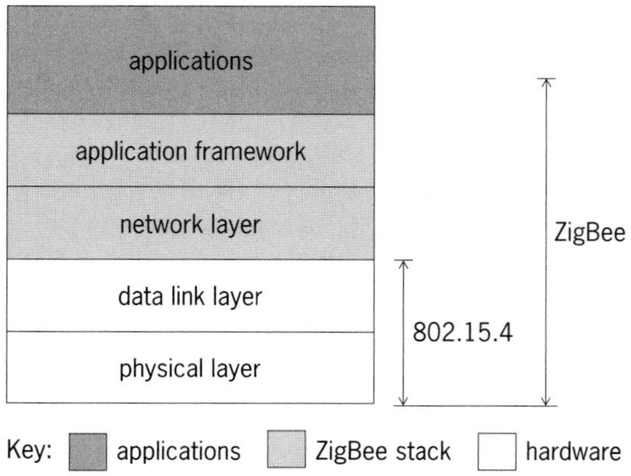

Fig. 1. ZigBee protocol stack.

Wireless sensor networks can support various kinds of applications. For outdoor applications, wireless sensors can be used to monitor habitat or wildfires, track mobile objects, navigate helicopters, and so on. For home networking, wireless sensors can be used for light control, heating, ventilation, air conditioning (HVAC), and security monitoring. For health care, wireless sensors can be integrated with sphygmomanometers or electronic thermometers to monitor patient status. For industrial control, wireless sensors can be used to improve manufacturing control systems, detect unstable situations, control production pipelines, and so on.

System architecture of emergency navigation applications. One potential application of wireless sensor networks is indoor emergency navigation and monitoring service. At normal time, the network can be responsible for monitoring the environment. When emergency events are detected, the network can adaptively modify its topology to ensure reliability of data transportation, quickly identify hazardous regions that should be avoided, and find safe navigation paths that can lead people to exits. In the following, we will introduce a possible system architecture of this service. The system is divided into server side and node side. The server side (**Fig. 2*a***) is responsible for maintaining, configuring, and monitoring the system. The server is equipped with two communication interfaces, which allow the server to communication with the wireless sensors and outside world. Above the communication interfaces, there are two main components. The deployment tool is for the system manager to plan sensor locations in a sensing field. It establishes a mapping between sensor IDs and their locations and initializes the network. This tool helps the system manager to plan the network to ensure both coverage and connectivity. If there are coverage holes or disconnected components in the network, the tool should be able to recommend how to add more sensors. The management tool is used to control and monitor the network. A graphical user interface (GUI) should be provided for the system manager to send query commands to sensor nodes. The reported readings from sensors can be shown

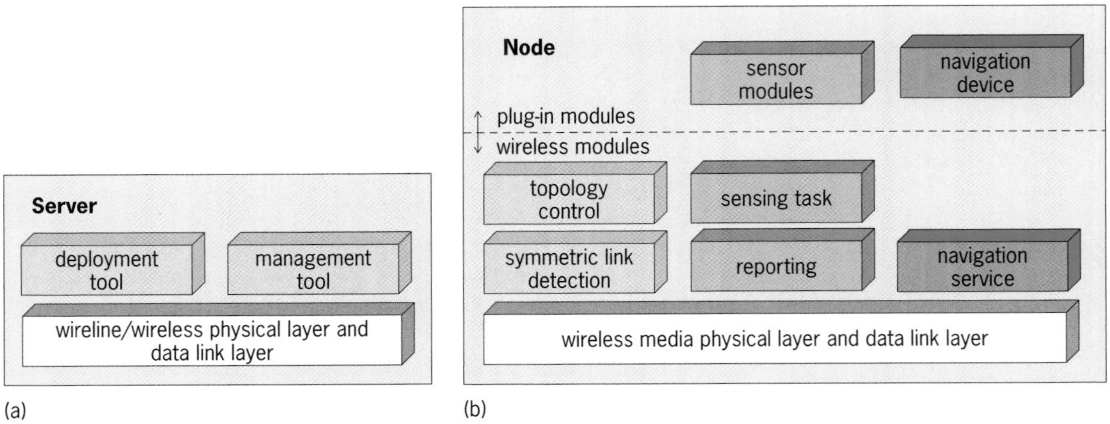

Fig. 2. *(a)* **The components in the server side.** *(b)* **The components in the node side.**

on the GUI. When emergency events are detected, such as fires, the management tool can broadcast alarm signals to people in the sensing field and notify fire departments. In addition, the management tool should be able to configure the network and show communication statuses.

The node side (Fig. 2*b*) senses the environment, reports sensory data to the server, and shows navigation directions. The components in a sensor node can be divided into three categories. The leftmost two components are communication-related components to guarantee the reliability of communication links. The middle three components are sensing-related components to deal with user queries and ensure the correctness of sensory data. The rightmost two components are navigation-related components, which are triggered as emergency events happen.

The topology control component is to guarantee network reliability. Since the link quality of wireless sensors may not be stable and emergencies that usually accompany damage to communication links, the topology control component is triggered when a node loses connectivity. Each sensor maintains a neighbor table, and together sensors form a data-reporting tree that is rooted at the server node as the network backbone. If a sensor's parent is removed from its neighbor table, this component triggers a tree reconstruction algorithm to find a new parent for the sensor. The symmetric link detection component ensures the communication links between sensors are bidirectional. This is achieved by sensors periodically exchanging packets with their neighbors. If a sensor finds a neighbor's signal quality is below a threshold for some time, this neighbor will be removed from the sensor's neighbor table. When selecting a routing path, only symmetric links are considered.

At normal time, sensors periodically collect environmental data and report to the server. The sensing task component is responsible for fulfilling the query rate required by the system manager. More importantly, the sensing task guarantees the accuracy of the sensory readings. Because of random effects in the environment, the collected data usually contain noises. Therefore, it is very important to filter out noisy data before reporting to the server. Assuming the sensory readings to be trustworthy, sensors can identify potential emergency events. When a sensor detects an emergency event, the sensing task component will trigger the navigation service components.

The main object of the navigation service component is to find navigation paths leading people to safe exits. The navigation algorithm should be distributed. When a sensor detects an emergency event, this sensor and its surroundings should be marked as hazardous regions. After locating hazardous regions, each sensor will compute a safe guidance direction to one of the exits. Sensors should avoid leading people through hazardous regions. If passing hazardous regions is inevitable, sensors can also guide people as far away from emergency locations as possible. Navigation devices are attached to wireless sensors. When a sensor determines its guidance direction, it will send a control signal to the attached navigation device. The navigation device can show a safe escape path to an exit. When the system is deployed in a multifloor building, the navigation service and devices need additional designs to support navigation in a three-dimensional (3D) environment. **Figure** 3*a* shows two navigation examples. In the left part of Fig. 3*a*, a fire emergency occurs on the ground floor. Sensors on the second floor should avoid guiding people to go through stair A, even if it is a shorter escape way to the exit. People escaping through stair B would be safer. In the right part of Fig. 3*a*, the stair is connected to the roof. Again, when an emergency occurs near the stair on the ground floor, sensors on the second floor should navigate people to the roof. In order to support 3D navigation, navigation devices also need to support directing people to go upstairs or downstairs. Figure 3*b* is a prototype that can show up to six directions.

Extensions. The navigation service can also be applied to guide rescuers to particular locations to relieve the emergency or to help victims. When rescuers go inside the building, wireless sensors can provide real-time emergency-related information to

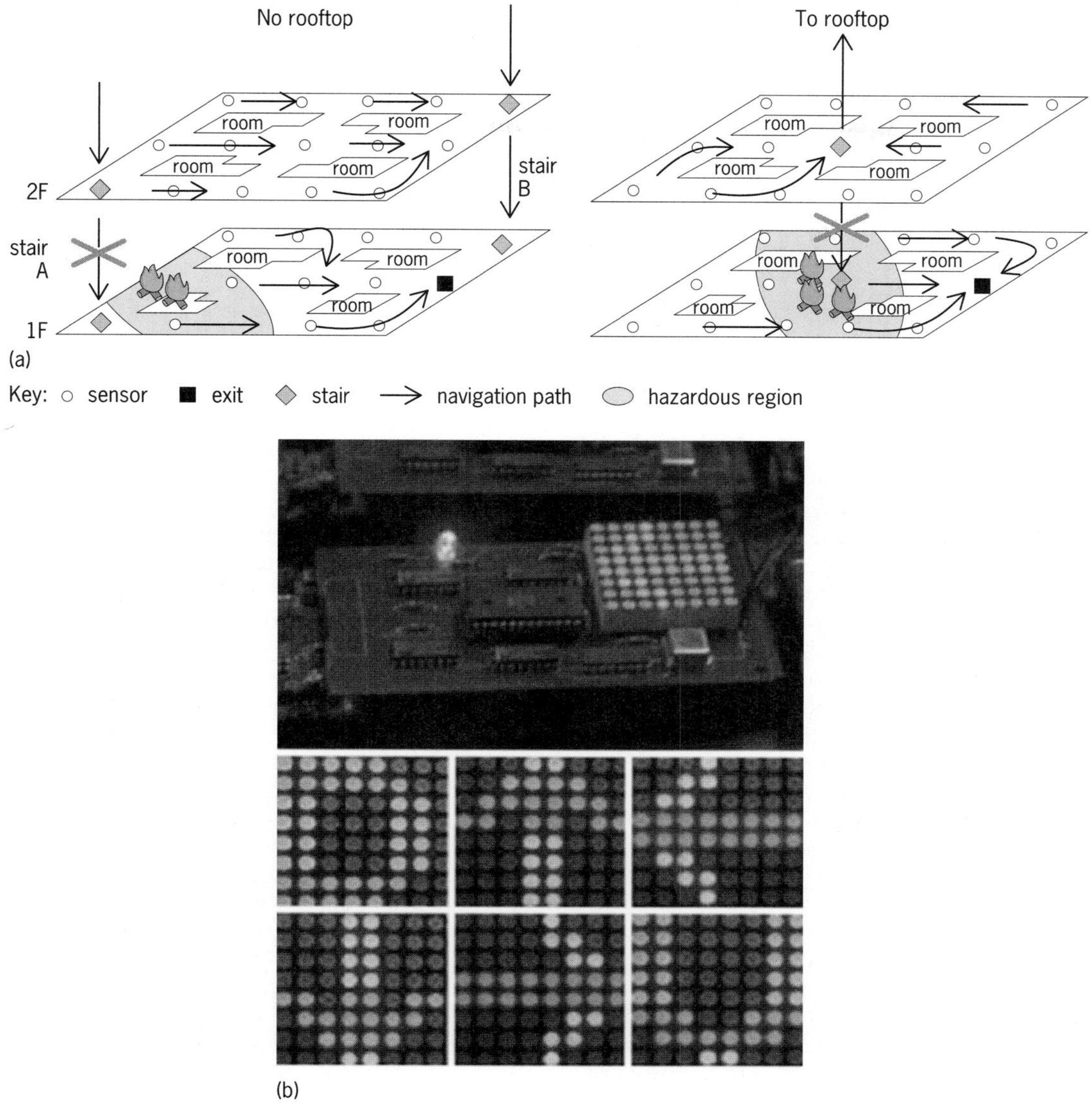

Fig. 3. ***(a)*** **Two navigation scenarios that in 3D buildings.** ***(b)*** **A navigation device for 3D navigation (where "D" means downstairs and "U" means upstairs).**

be shown on the personal digital assistants (PDAs) or the navigation devices held by the rescuers. In some situations, rescuers may not be able to go into the emergency scenes. The navigation service can also be integrated with mobile robots. For example, when an emergency is detected, a robot can use the information reported from sensors to find the shortest path to the emergency site. With more sophisticate sensors, a robot can collect data that are more detailed and send them back through the wireless network.

The system architecture stated above also is suited for health-care applications. The main difference is that users (patients or elders) can wear wireless sensors. Sensors deployed in the building can support location-tracking services. In a hospital, when sensors detect abnormal events of patients, the system automatically alerts nearby doctors by PDA of the exact positions of the patients.

For background information *see* DATA COMMUNICATIONS; ELECTRONIC NAVIGATION SYSTEMS; MICRO-ELECTRO-MECHANICAL SYSTEMS (MEMS); MICROSENSOR; TELEMETERING; WIRELESS FIDELITY (WI-FI) in the McGraw-Hill Encyclopedia of Science & Technology.

Yu-Chee Tseng; Meng-Shiuan Pan; Chih-Wei Yi

Bibliography. I. F. Akyildiz et al., A survey on sensor networks, *IEEE Commun. Mag.*, 40(8):102–114, August 2002; A. Mainwaring et al., Wireless sensor networks for habitat monitoring, *Proceedings of the 1st ACM International Workshop on Wireless Sensor Networks and Applications (WSNA), ACM Press, Atlanta*, 2002; Y.-C. Tseng, M.-S. Pan, and Y.-Y. Tsai, Wireless Sensor Networks for Emergency Navigation, *IEEE Comp.*, 39(7):55–62, July 2006.

Contributors

Contributors

The affiliation of each Yearbook contributor is given, followed by the title of his or her article. An article title with the notation "coauthored" indicates that two or more authors jointly prepared an article or section.

A

Akiyama, Prof. Takahiko. *Department of Chemistry, Gakushuin University, Tokyo, Japan.* CHIRAL BRØNSTED ACID CATALYSIS.

Alario-Franco, Prof. Miguel Á. *Facultad de Química, Universidad Complutense de Madrid, Ciudad Universitaria, Spain.* HIGH-TEMPERATURE SUPERCONDUCTIVITY.

Applequist, Dr. Wendy L. *Missouri Botanical Garden, William L. Brown Center, St. Louis.* TRADITIONAL CHINESE MEDICINE AUTHENTICATION.

Armitage, Lynne. *Department of Biology, University of York, United Kingdom.* AUXIN AND THE AUXIN RESPONSE—coauthored.

Ayora, Prof. Carlos. *Institut de Ciències de la Terra Jaume Almera, Consejo Superior de Investigaciones Científicas, Barcelona, Spain.* GYPSUM MEGACRYSTALS—coauthored.

B

Balaban, Dr. Evan. *Behavioral Neuroscience Program, McGill University, Montreal, Quebec, Canada.* AVIAN BRAIN CHIMERAS.

Balaguru, Prof. Perumalsamy N. *Department of Civil and Environmental Engineering, Rutgers University, Piscataway, New Jersey.* GREEN ENGINEERING—coauthored.

Barton, Dr. Hazel A. *Department of Biological Sciences, Northern Kentucky University, Highland Heights.* MICROBIAL DIVERSITY IN CAVES.

Beard, Michael. *RTI International, Research Triangle Park, North Carolina.* ASBESTOS MEASUREMENT—coauthored.

Beck, Dr. Pierre. *Geophysical Laboratory, Carnegie Institution of Washington, District of Columbia.* OPTICAL PROPERTIES OF DEEP-EARTH MINERALS—coauthored.

Behl, Dr. Marc. *GKSS Research Centre Geesthacht, Institute of Polymer Research, Teltow, Germany.* SHAPE-MEMORY POLYMERS—coauthored.

Bellegarde, Jennifer D. *Department of Psychiatry, Yale University School of Medicine, New Haven, Connecticut.* PATHOLOGICAL GAMBLING—coauthored.

Beroza, Dr. Gregory C. *Department of Geophysics, Stanford University, California.* SLOW EARTHQUAKES.

Birchler, Dr. James A. *Department of Biological Sciences, University of Missouri, Columbia.* B CHROMOSOMES IN PLANTS.

Blanton, Dr. Michael. *Department of Physics, New York University, New York.* SLOAN DIGITAL SKY SURVEY.

Boehm, Prof. Barry. *Center for Systems and Software Engineering, University of Southern California, Los Angeles.* LEAD SYSTEMS INTEGRATOR—coauthored.

Bruynseraede, Prof. Yvan. *Department of Physics and Astronomy, Katholieke Universiteit Leuven, Belgium.* INTEGRATED NANOSENSORS—coauthored.

Buchholz, Dr. Florian. *Department of Computer Science, James Madison University, Harrisonburg, Virginia.* CYBER FORENSICS.

C

Canals, Dr. Angel. *Departament de Cristal.lografia, Mineralogia i Dipòsits Minerals, Universitat de Barcelona, Barcelona, Spain.* GYPSUM MEGACRYSTALS—coauthored.

Caron, Dr. Jean-Bernard. *Department of Natural History, Royal Ontario Museum, Toronto, Canada.* ODONTOGRIPHUS: EARLIEST MOLLUSK—coauthored.

Chappell, Dr. Alan R. *National Security Directorate, Pacific Northwest National Laboratory, Richland, Washington.* CONCEPT-BASED USER INTERFACES—coauthored.

Chase, Dr. Mark W. *Molecular Systematics Section, Jodrell Laboratory, Royal Botanic Gardens, Kew, Richmond, Surrey, United Kingdom.* BARCODING LIFE.

Che, Dr. Ye. *Laboratory of Computational Biology, National Heart, Lung and Blood Institute, National Institutes of Health, Bethesda, Maryland.* MOLECULAR MODELING FOR DRUG DESIGN—coauthored.

Chew, Dr. Teng-Leong. *Director, Cell Imaging Facility, Northwestern University Feinberg School of Medicine, Chicago, Illinois.* CELLULAR IMAGING.

Chimenti, Prof. Dale E. *Department of Aerospace Engineering, Iowa State University, Ames.* AIR-COUPLED ULTRASONIC TESTING.

Chin, Dr. George. *Fundamental and Computational Sciences Directorate, Pacific Northwest National Laboratory, Richland, Washington.* CONCEPT-BASED USER INTERFACES—coauthored.

Chong, Dr. Ken P. *Director of Mechanics and Structures of Materials, National Science Foundation, Arlington, Virginia.* GREEN ENGINEERING—coauthored.

Christie, Dr. John M. *Plant Science Group, Division of Biochemistry and Molecular Biology, Institute of Biomedical and Life Sciences, University of Glasgow, Scotland, United Kingdom.* PHOTOTROPIN.

Citro, Dr. Francesco. *The Clean Fuels Institute, The City College of New York.* CONCENTRATING SOLAR POWER—coauthored.

Clack, Dr. Jennifer A. *Department of Zoology, University of Cambridge, United Kingdom.* DEVONIAN MISSING LINK.

Clouse, Dr. Steven D. *Department of Horticultural Science, North Carolina State University, Raleigh.* BRASSINOSTEROIDS.

Cogni, Rodrigo. *Department of Ecology and Evolution, State University of New York, Stony Brook.* BIOLOGICAL DIVERSITY—coauthored.

Cominsky, Prof. Lynn. *Department of Physics and Astronomy, Sonoma State University, Rohnert Park, California.* SWIFT GAMMA-RAY BURST MISSION.

Coulouvrat, Prof. François. *Institut Jean Le Rond d'Alembert, Centre National de la Recherche Scientifique, Université Pierre et Marie Curie, Paris, France.* FOCUSED SONIC BOOMS—coauthored.

Crameri, Prof. Reto. *Department of Molecular Allergology, Swiss Institute of Allergy and Asthma Research, Davos, Switzerland.* FUNGAL ALLERGIES—coauthored.

D

Dal Cin, Dr. Valeriano. *Department of Plant Molecular and Cellular Biology, University of Florida, Gainesville.* ETHYLENE (PLANT PHYSIOLOGY)—coauthored.

Das, J. C. *AMEC, Inc., Tucker, Georgia.* ARC FLASH HAZARD.

Davis, Prof. Paul K. *RAND and Pardee RAND Graduate School, Santa Monica, California.* CAPABILITIES-BASED PLANNING.

Davison, Dr. Brian D. *Department of Computer Science and Engineering, Lehigh University, Bethlehem, Pennsylvania.* WEB SPAM IDENTIFICATION AND BLOCKING—coauthored.

Delclòs, Dr. Xavier. *Departament d'Estratigrafia, Paleontologia i Geociències Marines, Facultat de Geologia, Universitat de Barcelona, Spain.* EARLY SPIDER WEB—coauthored.

Devreese, Prof. Jozef T. *Departement Fysica, Universiteit Antwerpen, Belgium.* INTEGRATED NANOSENSORS—coauthored.

Diez, Dr. F. Javier. *Department of Mechanical and Aerospace Engineering, Rutgers University, Piscataway, New Jersey.* BUOYANT PLUMES IN CROSSFLOWS.

Dixon, Prof. Juan W. *Department of Electrical Engineering, Pontificia Universidad Católica de Chile, Santiago.* REACTIVE POWER COMPENSATION TECHNOLOGIES—coauthored.

E

Erickson, Dr. David. *Sibley School of Mechanical and Aerospace Engineering, Cornell University, Ithaca, New York.* AUTONOMOUS MICROSYSTEMS.

F

Falco, Prof. Charles M. *College of Optical Sciences, University of Arizona, Tucson.* USE OF OPTICS BY RENAISSANCE ARTISTS.

Farrington , Dr. Robert. *Center for Transportation Technologies and Systems, National Renewable Energy Laboratory, Golden, Colorado.* HYBRID AUTOMOTIVE POWER SYSTEMS—coauthored.

Fastovsky, Dr. David E. *Department of Geosciences, University of Rhode Island, Kingston.* DINOSAUR BEHAVIOR.

Finlayson, Dr. Clive. *Director, The Gibraltar Museum, Gibraltar.* NEANDERTHAL EXTINCTION.

Fisher, Dr. John P. *Fischell Department of Bioengineering, University of Maryland, College Park.* BIODEGRADABLE MATERIALS FOR TISSUE ENGINEERING—coauthored.

Flagg, Dr. Thomas P. *Department of Cell Biology and Physiology, Washington University School of Medicine, St. Louis, Missouri.* ION CHANNELS (GENETICS)—coauthored.

Fletcher, Dr. Jacqueline. *Department of Entomology and Plant Pathology, Oklahoma State University, and National Institute for Microbial Forensics & Food and Agricultural Biosecurity, Stillwater, Oklahoma.* PLANT PATHOGEN FORENSICS.

Friedman, Dr. Norman. *New York, New York.* WARSHIP DESIGN TRENDS.

Frisch, Harold P. *Emeritus, Systems Engineering Division, NASA/Goddard Space Flight Center, Greenbelt, Maryland.* MODEL-BASED SYSTEMS.

G

García-Ruiz, Prof. Juan Manuel. *Consejo Superior de Investigaciones Cientificas: Laboratorio de Estudios Cristalográficos, Granada, Spain.* GYPSUM MEGACRYSTALS—coauthored.

Gaur, Prof. Vinod K. *Distinguished Professor, Indian Institute of Astrophysics, Bangalore, India.* 2005 KASHMIR EARTHQUAKE.

Geeta, Dr. R. *Department of Ecology and Evolution, State University of New York, Stony Brook.* BIOLOGICAL DIVERSITY—coauthored.

Goncharov, Dr. Alexander. *Geophysical Laboratory, Carnegie Institution of Washington, District of Columbia.* OPTICAL PROPERTIES OF DEEP-EARTH MINERALS—coauthored.

Gonder, Jeffrey. *Advanced Vehicle Systems, National Renewable Energy Laboratory, Golden, Colorado.* HYBRID AUTOMOTIVE POWER SYSTEMS—coauthored.

Goswami, Dr. Anjali. *Department of Earth Sciences, University of Cambridge, United Kingdom.* CARNIVORAN EVOLUTION.

Gracio, Deborah. *Fundamental and Computational Sciences Directorate, Pacific Northwest National Laboratory, Richland, Washington.* CONCEPT-BASED USER INTERFACES—coauthored.

Gräsbeck, Dr. Ralph. *Minerva Foundation Institute for Medical Research, Biomedicum Helsinki, Finland.* REFERENCE VALUES AND REFERENCE INTERVALS.

Grimaldi, Dr. David A. *Department of Entomology, American Museum of Natural History, New York, New York.* EARLY SPIDER WEB—coauthored.

Guo, Dr. Jing. *Department of Electrical and Computer Engineering, The University of Florida, Gainesville.* CARBON-BASED ELECTRONICS.

Gusse, Adam C. *Department of Entomology, University of Wisconsin-La Crosse.* MYCORESTORATION—coauthored.

H

Hage, Dr. David S. *Department of Chemistry, University of Nebraska, Lincoln.* AFFINITY MONOLITH CHROMATOGRAPHY—coauthored.

Ham, Dr. Donhee. *School of Engineering and Applied Sciences, Harvard University, Cambridge, Massachusetts.* SOLITONS IN ELECTRICAL NETWORKS—coauthored.

Hardie, Dr. D. Grahame. *Division of Molecular Physiology, College of Life Sciences, University of Dundee, Scotland, United Kingdom.* AMP-ACTIVATED PROTEIN KINASE (AMPK).

Harley, Prof. John P. *Department of Biological Sciences, Eastern Kentucky University, Richmond.* ESCHERICHIA COLI OUTBREAKS.

Harmon, Dr. Elizabeth. *Department of Anthropology, Hunter College, New York, New York.* INFANT AUSTRALOPITHECUS FROM DIKIKA.

Hedden, Dr. Peter. *Rothamsted Research, Harpenden, Herts, United Kingdom.* GIBBERELLIN BIOSYNTHESIS AND SIGNAL TRANSDUCTION.

Hegarty, Dr. Christopher J. *The MITRE Corporation, Bedford, Massachusetts.* INTEGRITY MONITORING (SATELLITE NAVIGATION).

Herbst, Dr. Jan F. *Materials and Processes Laboratory, General Motors Research and Development Center, Warren, Michigan.* HYDROGEN STORAGE MATERIALS—coauthored.

Hirsch, Dr. Jameson K. *Department of Psychology, East Tennessee State University, Johnson City, Tennessee.* SUICIDAL BEHAVIOR.

Holt, Dr. R. Glynn. *Department of Aerospace and Mechanical Engineering, Boston University, Massachusetts.* LASER-ULTRASONIC CAVITATION—coauthored.

Holub, Dr. Bruce J. *Department of Human Health and Nutritional Sciences, University of Guelph, Ontario, Canada.* TRANS FATTY ACIDS.

Homola, Timothy J. *Color Change Corporation, Streamwood, Illinois.* COLOR-CHANGING INKS.

Hornsby, Dean. *Integrated Solutions, Matthews International, Pittsburgh, Pennsylvania.* LASER MARKING.

Huber, Dr. Alan. *National Oceanic & Atmospheric Administration, Research Triangle Park, North Carolina.* DISPERSION MODELING IN COMPLEX URBAN SYSTEMS—coauthored.

J

Jacobsen, Dr. Steven D. *Department of Earth and Planetary Sciences, Northwestern University, Evanston, Illinois.* OPTICAL PROPERTIES OF DEEP-EARTH MINERALS—coauthored.

K

Kamel, Dr. Stephanie J. *Department of Zoology, University of Toronto, Ontario, Canada.* CLIMATE CHANGE AND SEX DETERMINATION.

Kanamori, Prof. Hiroo. *Seismological Laboratory, California Institute of Technology, Pasadena..* EARTHQUAKE EARLY WARNING—coauthored.

Karlsson, Dr. Daniel. *Power Systems, Gothia Power AB, Göteborg, Sweden.* WIDE-AREA POWER SYSTEM PROTECTION AND CONTROL—IN PART.

Katz, Prof. Joseph. *Department of Aerospace Engineering, San Diego State University, California.* RACE-CAR AERODYNAMICS.

Käufl, Dr. Hans-Ulrich. *Infrared Instrumentation Department, European Southern Observatory, Garching bei München, Germany.* DEEP IMPACT.

Kevany, Dr. Brian M. *Department of Plant Molecular and Cellular Biology, University of Florida, Gainesville.* ETHYLENE (PLANT PHYSIOLOGY)—coauthored.

Khuri-Yakub, Prof. Butrus T. *E. L. Ginzton Laboratory, Department of Electrical Engineering, Stanford University, Stanford, California.* MICROMACHINED ULTRASONIC TRANSDUCERS—coauthored.

Klee, Dr. Harry J. *Department of Plant Molecular and Cellular Biology, University of Florida, Gainesville.* ETHYLENE (PLANT PHYSIOLOGY)—coauthored.

Kominsky, John R. *Environmental Quality Management, Inc., Cincinnati, Ohio.* ASBESTOS MEASUREMENT—coauthored.

Kuliev, Dr. Anver. *Reproductive Genetics Institute, Chicago, Illinois.* PREIMPLANTATION GENETIC DIAGNOSIS AND THERAPY—coauthored.

Kupnik, Dr. Mario. *E. L. Ginzton Laboratory, Department of Electrical Engineering, Stanford University, Stanford, California.* MICROMACHINED ULTRASONIC TRANSDUCERS—coauthored.

L

Lane, Jo Ann. *Center for Systems and Software Engineering, University of Southern California, Los Angeles.* LEAD SYSTEMS INTEGRATOR—coauthored.

Larson, Dr. Greger. *Department of Archaeology, University of Durham, United Kingdom.* DNA AND ANIMAL DOMESTICATION.

Lendlein, Prof. Dr. Andreas. *GKSS Research Centre Geesthacht, Institute of Polymer Research, Teltow, Germany.* SHAPE-MEMORY POLYMERS—coauthored.

Levinson, Dr. David. *Department of Civil Engineering, University of Minnesota, Minneapolis.* ELECTRONIC TOLL COLLECTION.

Leyser, Prof. Ottoline. *Department of Biology, University of York, United Kingdom.* AUXIN AND THE AUXIN RESPONSE—coauthored.

Li, Xiaofeng. *School of Engineering and Applied Sciences, Harvard University, Cambridge, Massachusetts.* SOLITONS IN ELECTRICAL NETWORKS—coauthored.

Lidgard, Dr. Scott. *Department of Geology, Field Museum of Natural History, Chicago, Illinois.* POST-PALEOZOIC ECOLOGICAL COMPLEXITY.

Lindström, Dr. U. Marcus. *Department of Chemistry, McGill University, Montreal, Quebec, Canada.* ORGANIC REACTIONS IN WATER.

Lioy, Dr. Paul J. *Environmental and Occupational Health Sciences Institute, Piscataway, New Jersey.* DISPERSION MODELING IN COMPLEX URBAN SYSTEMS—coauthored.

Litchinitser, Dr. Natalia M. *Department of Electrical Engineering and Computer Science, University of Michigan, Ann Arbor.* NEGATIVE REFRACTION—coauthored.

Lorenz, Dr. Ralph D. *Space Department, Johns Hopkins University Applied Physics Laboratory, Laurel, Maryland.* TITAN—IN PART.

Loucks, Dr. Colby J. *World Wildlife Fund, Conservation Science Program, Washington, DC.* GIANT PANDA.

Lowenstein, Prof. Tim. *Department of Geological Sciences and Environmental Studies, Binghamton University, New York.* ANCIENT MICROORGANISMS IN SALT.

M

Manilova, Dr. Elena P. *Polzunov Central Boiler and Turbine Institute, St. Petersburg, Russia.* METALLOGRAPHIC PREPARATION FOR ELECTRON BACKSCATTERED DIFFRACTION—coauthored.

Manson, Dr. Jamie L. *Department of Chemistry and Biochemistry, Eastern Washington University, Cheney.* POLYMERIC MOLECULAR MAGNETS—coauthored.

Marchiano, Dr. Régis. *Institut Jean Le Rond d'Alembert, Centre National de la Recherche Scientifique, Université Pierre et Marie Curie, Paris, France.* FOCUSED SONIC BOOMS—coauthored.

Masten, Dr. David A. *Fuel Cell Activities Center, General Motors Corporation, Honeoye Falls, New York.* FUEL CELLS FOR AUTOMOBILES.

Materna, Stefan C. *Department of Biology, California Institute of Technology, Pasadena.* SEA URCHIN: GENOMIC INSIGHTS.

McSpadden Gardener, Dr. Brian B. *Department of Plant Pathology, Ohio State University, Wooster.* BIOCONTROL OF PLANT DISEASES WITH BIOFUNGICIDES.

Moon, Dr. Robert J. *USDA Forest Service, Forest Products Laboratory, Madison, Wisconsin.* NANOMATERIALS IN THE FOREST PRODUCTS INDUSTRY.

Moore, Dr. Brian R. *Department of Ecology and Evolutionary Biology, Environmental Science Center, Yale University, New Haven, Connecticut.* INFERRING PATTERNS OF DIVERSIFICATION.

Morán, Prof. Luis A. *Department of Electrical Engineering, Universidad de Concepción, Chile.* REACTIVE POWER COMPENSATION TECHNOLOGIES—coauthored.

Murray, Dr. Todd W. *Department of Aerospace and Mechanical Engineering, Boston University, Massachusetts.* LASER-ULTRASONIC CAVITATION—coauthored.

N

Naiche, Dr. L. A. *Department of Genetics and Development, Columbia University, New York, New York.* LIMB IDENTITY DURING DEVELOPMENT.

Nichols, Dr. Colin G. *Department of Cell Biology and Physiology, Washington University School of Medicine, St. Louis, Missouri.* ION CHANNELS (GENETICS)—coauthored.

Nickrent, Dr. Daniel L. *Department of Plant Biology, Southern Illinois University, Carbondale.* PARASITIC PLANTS.

O

O'Keefe, Dr. F. Robin. *Department of Biology, Marshall University, Huntington, West Virginia.* MESOZOIC MARINE REPTILES.

Oliphant, Dr. Andrew J. *Department of Geography & Human Environmental Studies, San Francisco State University, California.* HYDROCLIMATOLOGY.

Oralkan, Dr. Ömer. *E. L. Ginzton Laboratory, Department of Electrical Engineering, Stanford University, Stanford, California.* MICROMACHINED ULTRASONIC TRANSDUCERS—coauthored.

Organ, Dr. Chris. *Department of Organismic and Evolutionary Biology, Harvard University, Cambridge, Massachusetts.* PALEOGENOMICS.

Orsini, Dr. James A. *Department of Clinical Studies, New Bolton Center, University of Pennsylvania, Kennett Square.* EQUINE LAMINITIS.

P

Pan, Meng-Shiuan. *Department of Computer Science, National Chiao-Tung University, Taiwan, Republic of China.* WIRELESS SENSOR NETWORKS—coauthored.

Patel, Dr. Minal. *Fischell Department of Bioengineering, University of Maryland, College Park.* BIODEGRADABLE MATERIALS FOR TISSUE ENGINEERING—coauthored.

Peñalver, Dr. Enrique. *Museo Geominero, Instituto Geológico y Minero de España, Madrid, Spain.* EARLY SPIDER WEB—coauthored.

Pierce, Dr. Marcia M. *Department of Biological Sciences, Eastern Kentucky University, Richmond.* ANTHRAX BACILLUS AND THE IMMUNE RESPONSE; BRONCHIOLITIS OBLITERANS ORGANIZING PNEUMONIA (BOOP); CLOSTRIDIUM DIFFICILE OUTBREAKS.

Pinkerton, Dr. Frederick E. *Materials and Processes Laboratory, General Motors Research and Development Center, Warren, Michigan.* HYDROGEN STORAGE MATERIALS—coauthored.

Piomelli, Dr. Daniele. *Department of Pharmacology, University of California, Irvine.* ENDOCANNABINOIDS.

Potenza, Dr. Marc N. *Department of Psychiatry, Yale University School of Medicine, New Haven, Connecticut.* PATHOLOGICAL GAMBLING—coauthored.

Pratt, Edward. *Titan Wood Limited, Kensington Centre, London, United Kingdom.* Accoya wood.

Prevedouros, Prof. Panos D. *Department of Civil and Environmental Engineering, University of Hawaii at Manoa, Honolulu, Hawaii.* Reversible express lanes—coauthored.

Pugno, Dr. Nicola. *Department of Structural Engineering, Politecnico di Torino, Torino, Italy.* Space elevator.

Purdy, Dan C. *FDES, Ottawa, Ontario, Canada.* Forensic document analysis.

R

Rauchwerk, Michael D. *LGS, Whippany, New Jersey.* Software-defined radio.

Ree, Dr. Richard H. *Department of Botany, Field Museum of Natural History, Chicago, Illinois.* Ancestral ranges and historical biogeography.

Rhyner, Dr. Claudio. *Department of Molecular Allergology, Swiss Institute of Allergy and Asthma Research, Davos, Switzerland.* Fungal allergies—coauthored.

Richards, Prof. Jeremy Peter. *Earth & Atmospheric Sciences, University of Alberta, Edmonton, Canada.* Sustainable development in mining.

Ricketts, Dr. David S. *Department of Electrical & Computer Engineering, Carnegie Mellon University, Pittsburgh, Pennsylvania.* Solitons in electrical networks—coauthored.

Robinson, Dr. Gene. *Department of Entomology and Institute for Genomic Biology, University of Illinois at Urbana-Champaign.* Honeybee genome.

Roe, Dr. Henry. *Lowell Observatory, Flagstaff, Arizona.* Titan—in part.

Roggenthen, Prof. William. *Department of Geology and Geological Engineering, South Dakota School of Mines and Technology, Rapid City.* Underground research laboratories.

Rouse, Prof. William B. *Tennenbaum Institute, Georgia Institute of Technology, Atlanta.* Complex systems.

Roy, Dr. Ronald A. *Department of Aerospace and Mechanical Engineering, Boston University, Massachusetts.* Laser-ultrasonic cavitation—coauthored.

Rudkin, David. *Department of Natural History, Royal Ontario Museum, Toronto, Canada.* Odontogriphus: earliest mollusk—coauthored.

S

Saarela, Dr. Jeffery M. *Research Division, Canadian Museum of Nature, Ottawa, Ontario, Canada.* Hydatellaceae.

Sage, Prof. Andrew P. *Systems Engineering and Operations Research Department, George Mason University, Fairfax, Virginia.* System interoperability.

Scarborough, Dr. Victoria. *Sherwin Williams, New Technology, Cleveland, Ohio.* New coatings for wood.

Scawthorn, Prof. Charles. *Department of Urban Management, Kyoto University, Kyoto, Japan.* Earthquake early warning—coauthored.

Schander, Dr. Christoffer. *Department of Biology, University of Bergen, Norway.* Odontogriphus: earliest mollusk—coauthored.

Schauer, Dr. Nicolas. *De Ruiter Seeds, Research and Development, Bergschenhoek, the Netherlands.* Plant metabolomics.

Scheltema, Dr. Amélie. *Woods Hole Oceanographic Institution, Massachusetts.* Odontogriphus: earliest mollusk—coauthored.

Schiel, John E. *Department of Chemistry, University of Nebraska, Lincoln.* Affinity monolith chromatography—coauthored.

Schlueter, Dr. John A. *Materials Science, Argonne National Laboratory, Argonne, Illinois.* Polymeric molecular magnets—coauthored.

Schröder, Dr. Martin. *School of Biological and Biomedical Sciences, Durham University, United Kingdom.* Unfolded protein response.

Setvák, Dr. Martin. *Satellite Department, Czech Hydrometeorological Institute, Prague, Czech Republic.* Multispectral satellite observations of severe storms.

Shalaev, Prof. Vladimir M. *School of Electrical and Computer Engineering and Birck Nanotechnology Center, Purdue University, West Lafayette, Indiana.* Negative refraction—coauthored.

Sharma, Dr. Abhay. *School of Graphic Communications Management, Ryerson University, Toronto, Ontario, Canada.* Color management.

Shinnar, Prof. Reuel. *Department of Chemical Engineering, City College of New York.* Concentrating solar power—coauthored.

Smith, Dr. Nathan. *Department of Astronomy, University of California, Berkeley.* Supernova 2006gy.

Solomon, Dr. Stanley C. *High Altitude Observatory, NCAR, Boulder, Colorado.* Effects of carbon dioxide on the upper atmosphere.

Stone, Dr. Martin. *Tampa-Hillsborough County Expressway Authority, Tampa, Florida.* Reversible express lanes—coauthored.

Struzhkin, Dr. Viktor. *Geophysical Laboratory, Carnegie Institution of Washington, District of Columbia.* Optical properties of deep-earth minerals—coauthored.

T

Taylor, Dr. Carson W. *Portland, Oregon.* Voltage stability.

Thomas, Dr. Jean-Louis. *Institut des NanoSciences de Paris, Centre National de la Recherche Scientifique, Université Pierre et Marie Curie, Paris, France.* Focused sonic booms—coauthored.

Treger, Dr. Daryl. *Strategic Analysis, Inc., Arlington, Virginia.* Spintronics—coauthored.

Tsai, Dr. Mindy. *Department of Pathology, Stanford University, Stanford, California.* Mast cells.

Tsang, Dr. Constantine. *Atmospheric, Oceanic and Planetary Physics, Department of Physics, Clarendon Laboratory, University of Oxford, United Kingdom.* Venus Express.

Tseng, Prof. Yu-Chee. *Department of Computer Science, National Chiao-Tung University, Taiwan, Republic of China.* Wireless sensor networks—coauthored.

Turgeon, Dr. Robert. *Department of Plant Biology, Cornell University, Ithaca, New York.* Phloem loading.

V

Valentine, Dr. James W. *Department of Integrative Biology, University of California, Berkeley.* Tropics: latitudinal biodiversity gradient.

Vallero, Dr. Daniel A. *U.S. Environmental Protection Agency, National Exposure Research Laboratory, Research Triangle Park, North Carolina.* Asbestos measurement—coauthored; Dispersion modeling in complex urban systems—coauthored.

Vander Voort, George F. *Buehler Ltd., Lake Bluff, Illinois.* Metallographic preparation for electron backscattered diffraction—coauthored.

Veis, Dr. Arthur. *Department of Cell and Molecular Biology, Northwestern University, Chicago, Illinois.* Gene expression during tooth development.

Venayagamoorthy, Dr. Ganesh K. *Department of Electrical and Computer Engineering, University of Missouri-Rolla.* Wide-area power system protection and control—in part.

Verlinsky, Dr. Yury. *Reproductive Genetics Institute, Chicago, Illinois.* Preimplantation genetic diagnosis and therapy—coauthored.

Volk, Dr. Thomas J. *Department of Biology, University of Wisconsin-La Crosse.* Mycorestoration—coauthored.

von Puttkamer, Dr. Jesco. *NASA Headquarters, Office of Space Flight, Washington, DC.* Space flight.

W

Whitcomb, Dr. Clifford. *Department of Systems Engineering, Naval Postgraduate School, Monterey, California.* Multihull ships.

Wiseman, Dr. Justin M. *Prosolia, Inc. Indianapolis, Indiana.* Desorption electrospray ionization mass spectrometry.

Wolf, Prof. Stuart A. *Department of Materials Science and Engineering, and Department of Physics, University of Virginia, Charlottesville.* Spintronics—coauthored.

Wu, Dr. Baoning. *Snap.com, Idealab, Pasadena, California.* Web spam identification and blocking—coauthored.

Wulffraat, Stephan. *Department of Ecology, World Wide Fund for Nature-Indonesia, Jakarta.* New species of Borneo.

X

Xiang, Dr. Zhexin. *Center for Molecular Modeling, Center for Information Technology, National Institutes of Health, Bethesda, Maryland.* Molecular modeling for drug design—coauthored.

Y

Yannas, Prof. Ioannis. *Department of Biological Engineering, Massachusetts Institute of Technology, Cambridge.* Organ regeneration.

Yao, Dr. David D. *Department of Industrial Engineering & Operations Research, Columbia University, New York.* Financial engineering—coauthored.

Yi, Prof. Chih-Wei. *Department of Computer Science, National Chiao-Tung University, Taiwan, Republic of China.* Wireless sensor networks—coauthored.

Z

Zerbe, Dr. John I. *USDA Forest Service, Forest Products Laboratory, Madison, Wisconsin.* Ethanol from wood.

Zhou, Prof. Xun Yu. *Mathematical Finance, Department of Systems Engineering and Engineering Management, Oxford University, United Kingdom.* Financial engineering—coauthored.

Index

Index

Asterisks indicate page references to article titles.

A

B

C

F

N

O

T

U

V

W

X

Y

Z